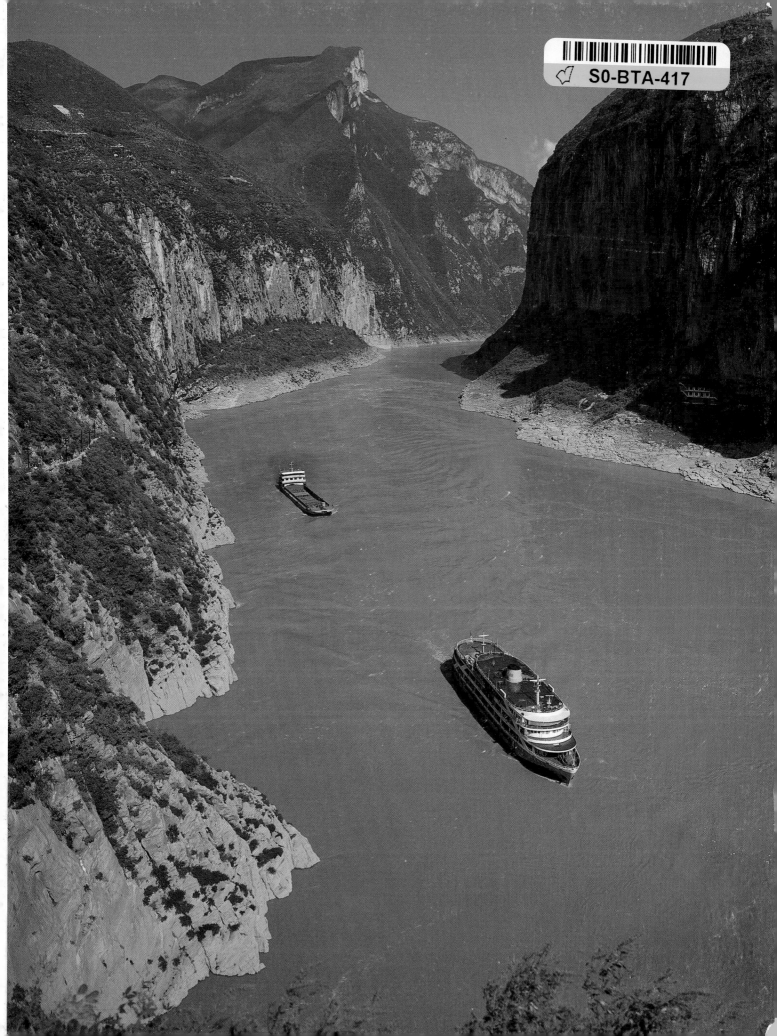

A LEGACY OF SERVICE TO GEOGRAPHY

WILEY
Publishers Since 1807

John Wiley & Sons is the worldwide leader in geography publishing. We began our partnership with geography education in 1911 and continue to publish college texts, professional books, journals, and technology products that help teachers teach and students learn. We are committed to making it easier for students to visualize spatial relationships, think critically about their interactions with the environment, and appreciate the earth's dynamic landscapes and diverse cultures.

To serve our customers we have partnered with the AAG, the AGS, and the NCGE while building extraordinary relationships with Microsoft and Rand McNally that allow us to bundle discounted copies of Encarta and the Goode's Atlas with all of our textbooks.

Wiley Geography continues this legacy of service during each academic year with with outstanding first editions and revisions.

Regional Geography

▶ de Blij/Muller — *Geography: Realms, Region, and Concepts 11e* (0-471-15224-2)
▶ de Blij/Muller — *Concepts and Regions in Geography* (0-471-09303-3)
▶ Blouet/Blouet — *Latin America and the Caribbean 4e, Update* (0-471-48052-5)
▶ Weightman — *Dragons and Tigers: Geography of South, East and Southeast Asia 1e, Update* (0-471-48476-8)

Physical Geography

▶ Strahler/Strahler — *Physical Geography: Science and Systems of the Human Environment 3e* (0-471-48053-3)
▶ Strahler/Strahler — *Introducing Physical Geography 3e, Update* (0-471-66969-5)
▶ Marsh/Grossa — *Biogeography: Introduction to Space, Time and Life* (0-471-24193-8)
▶ MacDonald — *Biogeography: Introduction to Space, Time and Life* (0-471-24193-8)
▶ Cutter/Renwick — *Exploration, Conservation, Preservation: A Geographic Perspective on Natural Resource Use* (0-471-15225-0)

Human Geography

▶ de Blij/Murphy — *Human Geography 7e* (0-471-44107-4)
▶ Kuby/Harner/Gober — *Human Geography in Action 3e* (0-471-43055-2)

GIS and Remote Sensing

▶ Chrisman — *Exploring GIS 2e* (0-471-31425-0)
▶ DeMers — *Fundamentals of GIS 3e* (0-471-20491-9)
▶ Lillesand/Kiefer/ Chipman — *Remote Sensing and Image Interpretation 5e* (0-471-45152-5)

Political Geography

▶ Glassner/Fahrer — *Political Geography 3e* (0-471-35266-7)

Natural Resources

▶ Cutter/Renwick — *Exploitation, Conservation, Preservation 4e* (0-471-15225-0)
▶ Cech — *Principles of Water Resources 2e* (0-471-48475-X)

Third Edition

INTRODUCING PHYSICAL GEOGRAPHY

MEDIA VERSION

Third Edition

INTRODUCING PHYSICAL GEOGRAPHY

MEDIA VERSION

Alan Strahler
Boston University

Arthur Strahler

John Wiley & Sons, Inc.

Acquisitions Editor	Jerry Correa
Associate Editor	Denise Powell
Production Manager	Jeanine Furino
Production Editor	Sandra Dumas
Marketing Manager	Clay Stone
Senior Designer	Karin Kincheloe
Photo Editor	Jennifer MacMillan
Illustration Editor	Sigmund Malinowski
Production Management Services	Hermitage Publishing Services
Text Design	Jerry Wilke
Cover Design	David Levy
Cover Photo	Christopher Liu/China Stock Photo Library
Cover Inset Photo	Courtesy NASA/GSFC/MITI/ERSDAC/JAROS, and U.S./Japan

This book was typeset in 10/12 Times New Roman by Hermitage Publishing Services and printed and bound by Von Hoffmann Corporation. The cover was printed by Von Hoffmann Corporation.

The paper in this book was manufactured by a mill whose forest management programs include sustained yield harvesting of its timberlands. Sustained yield harvesting principles ensure that the number of trees cut each year does not exceed the amount of new growth.

This book is printed on acid-free paper. ⊗

Strahler, Alan, N., Strahler, Arthur/Introducing Physical Geography, Third Edition—Media Version
ISBN 0-471-66969-5

Printed in the United States of America.

10 9 8 7 6 5 4 3 2 1

PREFACE

It is with great pleasure that we present the Third Edition Update "Media Version" of Introducing Physical Geography. Our new updated edition includes 2 powerful media ensembles, eGrade Plus and GeoDiscoveries on the web.

For instructors, eGrade Plus is a powerful online tool that provides an integrated suite of teaching and learning resources in one easy-to-use website.

Organized around the essential activities instructors perform in class, eGrade Plus provides the tools you need to:

- **Prepare & Present:** Create class presentations using a wealth of Wiley-provided resources—such as an online version of the textbook, PowerPoint slides, and animations—making your preparation time more efficient. You may easily adapt, customize, and add to this content to meet the needs of your course.
- **Create Assignments:** Automate the assigning and grading of homework or quizzes by using Wiley-provided question banks, or by writing your own. Student results will be automatically graded and recorded in your gradebook. eGrade Plus can link homework problems to the relevant section of the online text, providing students with context-sensitive help.
- **Track Student Progress:** Keep track of your students' progress via an instructor's gradebook, which allows you to analyze individual and overall class results to determine their progress and level of understanding.
- **Administer Your Course:** eGrade Plus can easily be integrated with another course management system, gradebook, or other resources you are using in your class, providing you with the flexibility to build your course, your way.

For more information about the features and benefits of eGrade Plus, please view our online demonstration at www.wiley.com/college/egradeplus.

Keep All of Your Class Materials in One Location

Course Administration tools allow you to manage your class and integrate your eGrade Plus resources with most Course Management Systems, allowing you to keep all of your class materials in one location.

Enhance the Power of Your Class Preparation and Presentations

A **Prepare and Present tool** contains all of the Wiley-provided resources, such as **PowerPoint slides** and **student activities,** making your preparation time more efficient. You may easily adapt, customize, and add to Wiley content to meet the needs of your course.

Create Your Own Assignments or Use Ours, All with Automatic Grading

An **Assignment** area allows you to create **student homework** and **quizzes** by using **Wiley-provided question banks,** or by writing your own. You may also assign readings, activities and other work you want your students to complete. One of the most powerful features of eGrade Plus is that student assignments will be automatically graded and recorded in your gradebook. This will not only save you time but will provide your students with immediate feedback on their work.

Assess Student Understanding More Closely

An **Instructor's Gradebook** will keep track of your students' progress and allow you to analyze individual and overall class results to determine their progress and level of understanding

Students,
eGrade Plus Allows You to:

Study More Effectively

Get Immediate Feedback When You Practice on Your Own

eGrade Plus problems link directly to relevant sections of the **electronic book content,** so that you can review the text while you study and complete homework online. Additional resources include **interactive simulations, study guide** and other problem-solving resources.

Complete Assignments / Get Help with Problem Solving

An **Assignment** area keeps all your assigned work in one location, making it easy for you to stay "on task." In addition, many homework problems contain a **link** to the relevant section of the **multimedia book,** providing you with a text explanation to help you conquer problem-solving obstacles as they arise. You will have access to a variety of **interactive problem-solving tools,** as well as other resources for building your confidence and understanding.

Keep Track of How You're Doing

A **Personal Gradebook** allows you to view your results from past assignments at any time.

Significant changes in the third edition include:

- *Introducing Physical Geography*, a new chapter that introduces geography as a discipline and identifies the realm of physical geography within geography as a whole. After briefly introducing key geographer's tools, the chapter continues with the conceptual themes of spheres, scales, systems, and cycles as they will appear in the remainder of the book. It concludes by motivating the study of physical geography with the global change topics of climate change, carbon cycle, biodiversity, pollution, and extreme events.

- *Biogeographic Processes,* a new chapter that adds key concepts from historical biogeography as well as additional depth to our previous coverage of ecological biogeography. Positioned between our surveys of global climate (Chapter 7) and global vegetation (Chapter 9), the new chapter better ties the concepts that explain the geographical distribution of plants and animals to the global biogeographic patterns we see today.

The remaining parts of our text have also seen some significant revisions. Especially noteworthy are major changes in:

- *The Earth's Global Energy Balance (Chapter 2),* revised for flow and content and updated with new values for the solar constant and new flux rates for the global energy balance.

- *Winds and Global Circulation (Chapter 5),* extensively revised to provide better flow and more detail in relating surface winds to upper atmosphere circulation. The chapter includes new figures and text on global air pressure gradients, the geostrophic wind, and upper air wind motion and pressure.

- *Global Climates (Chapter 7),* revised to provide a better introduction to global temperature and rainfall regimes with accompanying new figures.

- *Earth Materials and Plate Tectonics (Chapters 12 and 13),* revised for a clearer description of the rock cycle and better flow of topics across the two chapters.

- *Weathering and Mass Wasting (Chapter 14),* restructured for better topical flow, including a new separate section on processes and landforms of alpine tundra.

Also updated are many references to **environmental events and global change**, including ozone layer coverage, earthquakes, hurricanes, and landslides. New data appear in our air pollution feature (Chapter 4). We also provide new information on greenhouse gases and their impact on global warming, as well as updated coverage of environmental treaties (Buenos Aires, Kyoto) and new climate change projections from the International Panel on Climate Change (Chapter 3).

Another new development is **much more extensive use of remotely sensed images**. Our senior author, Alan Strahler, has used his expertise in this field to select many new images from the latest sources, including abundant coverage

from NASA's Earth Observing System. The new images are used extensively as text figures and also appear in special boxed features entitled *Focus on Remote Sensing*. To support these new uses of remotely sensed images, we have expanded our appendix coverage of the basic principles of remote sensing and placed it as a special boxed feature within the text, *Geographer's Tools 2.1 • Remote Sensing for Physical Geography*, which appears in Chapter 2. Our previous appendix on Geographic Information Systems has also been moved to the text as *Geographer's Tools 1.3 • Geographic Information Systems* in Chapter 1, where it is accompanied by *Geographer's Tools 1.2 • Focus on Maps*.

THE LEARNING ENVIRONMENT

In developing *Introducing Physical Geography*, we paid special attention to the learning environment for students. This includes an emphasis on clearly communicating the structure of the text. The first few paragraphs of each chapter are especially written to provide an orientation to the material to come, and transitional paragraphs continue within the text whenever an important new subject begins. To enhance the forward linkages, we provide a brief final section, *A Look Ahead*, that puts the present chapter in perspective and leads into the next.

Another important part of the learning environment is our series of *Putting You in the Picture* features that open each chapter. The first such feature, in our *Introducing Physical Geography* chapter, puts the student in the place of an astronaut stationed on the Moon who embarks on a journey home to Earth. By following this journey, the student sees the planet at ever-increasing scales, first as a whole, then as a realm of the four great spheres, then as regions differentiated by climate, vegetation, and soils, and finally as a collection of individual landscape features as the return spacecraft proceeds to a touchdown. The journey is both a metaphor and model for the learning to come.

Other *Putting You in the Picture* features continue the involvement of the student, personalizing the chapter with an experience or topic relevant to the chapter's main themes. For example, in Chapter 6 the student reads an account of experiencing Hurricane Andrew as it devastated the South Florida coast in 1992. In Chapter 16, the student is introduced to the erosive power of rivers by riding a raft down the Colorado River through the Grand Canyon. Most of these essays are written in the second person to draw the student in more directly.

Students enjoy learning that their new knowledge is important and useful. Our *Eye on the Environment* and *Eye on Global Change* features help reinforce text material and provide relevance. These essays focus the concepts of the chapter on specific topics in environmental and global change, ranging from the ozone layer to ice sheets and global warming.

The book concludes with an *Epilogue* that raises the environmental implications of the physical geography that the student has learned. In an extended script for a news broadcast from the mid-twenty-first century, students picture how life might change in the future as human impacts continue to reshape our planet. The news broadcast leads into a discussion of the most serious environmental issues facing the human species and the prognosis for the Earth's immediate future.

Learning Tools

To help students master text material, we provide a number of learning tools. To help build the necessary vocabulary, we have identified the dozen or so key terms that are most important in each chapter with the use of **boldface**. Less important terms are identified in *italics*. All terms are defined in our *Glossary*, which contains over 1000 individual entries and includes many terms that are part of a basic science vocabulary.

As another aid to learning, we have embedded brief *key statements* in the text that summarize important concepts in one or two sentences. They are set as type blocks between text columns.

Our suite of end-of-chapter materials is also designed to reinforce key concepts and ideas of the chapter. A summary, *In Review*, covers all the major concepts of the chapter. The list of *Key Terms* gathers together the critical vocabulary of the chapter. *Review Questions* are designed as oral or written exercises that require description or explanation of important ideas and concepts. They are provided both for text and *Eye on the Environment/Global Change* boxes. *Visualizing Exercises* are review questions that utilize sketching or graphing as a way of focusing attention on key illustrations. Also provided are *Essay Questions* that require more synthesis or the reorganization of knowledge in a new context.

Users of our third edition update will recognize the key elements of our visualization strategy to learning, which focuses on helping students form lasting visual images of the processes of physical geography. The strategy involves three components: illustrations, word pictures, and the Geo-Discoveries website. As noted earlier in this preface, each line art illustration in our third edition has been restyled with a larger, more readable typeface for labels and enhanced colors for better clarity and more visual appeal. Building word pictures that stay in students' minds is another aspect of visualization, and our *Putting You in the Picture* features and the *Epilogue* are key parts of that effort.

GEODISCOVERIES

The GeoDiscoveries media program offers a rich electronic medium for exploring the core concepts of physical geography. This robust assembly of media resources allows students and professors to move beyond the textbook with interactive exercises, animations, videos, self-assessments and an interactive globe.

Building off the success of our present/interact/assess framework, our GeoDiscoveries program has been integrated into our on-line book companion web sites. Web sites tailored to the needs of both students and instructors include the following resources:

Resources specific to the study of physical geography have been created and developed by S. Mary P. Benbow of the University of Manitoba.

For Students:

Animated Globe: This 3D interactive globe allows you to overlay different layers of geographic information, including plate tectonics, topography, political boundaries, and population to help visualize the spatial relationships between these features and elements.

Animations: Key diagrams and drawings from our rich signature art program have been animated to provide a virtual experience of difficult concepts. These animations have proven crucial to the understanding of this content for visual learners.

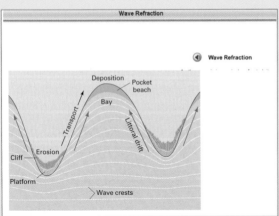

Videos: Brief video clips provide real-world examples of geographic features, and put these examples into context with the concepts covered in the text.

Simulations: Computer-based models of geographic processes allow you to manipulate data and variables to explore and interact with virtual environments.

Interactive Exercises: Learning activities and games build off our presentation material. They give students an opportunity to test their understanding of key concepts and explore additional visual resources.

For Instructors:

Our animations, videos, simulations, and interactive exercises provide ideal in-class presentation tools to help engage students in the classroom. All these resources can be easily integrated into instructors' customized PowerPoint presentations.

In addition to our multi-media content, we offer an on-line image gallery, containing both line art and comprehensive photographic materials. These resources can be easily uploaded into PowerPoint presentations, course management systems, and on-line tutorials and web sites.

We also offer prepared PowerPoint presentations centered on key concepts and hot topics in the field of physical geography. These lecture materials can be customized to suit the instructor's individual needs, or can be used as is to add context and relevancy to any lecture.

GeoDiscoveries also offers a wealth of assessment and assignment tools that can be used for tutorials as well as quizzes purposes. These question banks are closely tied to our multi-media materials to help gauge understanding of the resources. Easy integration of this content into course management systems and homework assignments gives instructors the opportunity to integrate multimedia with their syllabuses and with more traditional reading and writing assignments.

BOOK COMPANION SITE

In addition to our GeoDiscoveries multimedia content, our student companion site offers a wealth of study and practice materials, including:

Student On-line Resources:
- *Self-Quizzes*—chapter based multiple choice and fill-in-the-blank questions
- *Annotated Weblinks*—useful weblinks selected to enhance chapter topics and content
- *Chapter Summaries*—brief overviews of chapter content
- *Flashcards*—drill and practice application to help study key terms
- *Learning Objectives*—list of key concepts covered in each chapter
- *Learning Styles Survey*—a survey developed to help determine your own particular way of learning, and study skill suggestions to help each kind of learner
- *Geographers at Work*—weblinks to geographers' web sites to give you an opportunity to explore research being done in the field today
- *Virtual Field Trips*—web sites devoted to the exploration and virtual experience of landscapes and environments around the world.

Instructor Resources:
- *PowerPoint Lecture Slides*—chapter-oriented slides including lecture notes and text art
- *Computerized Test Bank*—including multiple choice and fill-in-the-blank test items
- *Teaching Tips*—for enhancing the classroom experience
- *On-line Essay and In-class Activities*—Materials for in-class projects, with corresponding essay-based homework assignments
- *Instructor's Manual*—including lecture notes, chapter objectives, and guides to additional resources

ACKNOWLEDGMENTS

The preparation of *Introducing Physical Geography, Third Edition*, was greatly aided by the following reviewers who read and evaluated the various parts of the manuscript:

Peter Blanken, University of Colorado, Boulder
Robert E. Lee, Seattle Central Community College
Denyse Lemaire, Rowen University
Eugene J. Palka, U.S. Military Academy, West Point

We would also like to thank the reviewers who read and commented on the second editon:

Tanya Allison, Montgomery College
Joseph M. Ashley, Montana State University
Roger Balm, Rutgers State University
David Bixler, Chaffey Community College
Kenneth L. Bowden, Northern Illinois University
David Butler, University of North Carolina, Chapel Hill
Les Dean, Riverside Communicty College
Christopher H. Exline, University of Nevada
Jerry Green, Miami University
Clarence Head, University of Central Florida
Peter W. Knightes, Central Texas College
Christopher S. Larsen, State University of New York, Buffalo
James F. Petersen, Southwest Texas State University
Clinton Rowe, University of Nebraska-Lincoln
Carol Shears, Ball State University
Carol I. Zinser, State University of New York, Plattsburgh

It is with particular pleasure that we thank the staff at Wiley for their long hours of careful work in the preparation and production of our *Third Edition*. They include our editor, Ryan Flahive, our developmental editor Suzanne Thibodeau, photo researchers Alexandra Truitt and Jerry Marshall, photo research associate Jennifer MacMillan, designer Karin Kincheloe, illustration editor Sigmund Malinowski, the artists of Precision Graphics illustration studio, marketing manager Clay Stone, new media product manager Tom Kulesa, Larry Meyer and the staff of Hermitage Publishing Services for production services and page layout, and our production editor Sandra Russell, who made it all come together.

Alan Strahler
Arthur Strahler
Cambridge, Massachusetts
February 1, 2002

ABOUT THE AUTHORS

Alan Strahler received his B.A. degree in 1964 and his Ph.D degree in 1969 from The Johns Hopkins University, Department of Geography and Environmental Engineering. He has held academic positions at the University of Virginia, the University of California at Santa Barbara, and Hunter College of the City University of New York, and is now Professor of Geography at Boston University. With Arthur Strahler, he is a coauthor of seven textbook titles with nine revised editions on physical geography and environmental science. He has published over 250 articles in the refereed scientific literature, largely on the theory of remote sensing of vegetation, and has also contributed to the fields of plant geography, forest ecology, and quantitative methods. His work has been supported by over $6 million in grant and contract funds, primarily from NASA. In 1993, he was awarded the Association of American Geographers/Remote Sensing Specialty Group Medal for Outstanding Contributions to Remote Sensing. In 2000, he received the honorary degree Doctorem Scientiarum Honoris Causa (D.S.H.C.) from the Université Catholique de Louvain, Belgium, for his academic accomplishments in teaching and research.

Arthur Strahler received his B.A. degree in 1938 from the College of Wooster, Ohio, and his Ph.D. degree in geology from Columbia University in 1944. He was appointed to the Columbia University faculty in 1941, serving as Professor of Geomorphology from 1958 to 1967 and as Chairman of the Department of Geology from 1959 to 1962. He was elected as a Fellow of both the Geological Society of America and the Association of American Geographers for his pioneering contributions to quantitative and dynamic geomorphology, contained in over 30 major papers in leading scientific journals. He is the author or coauthor with Alan Strahler of 16 textbook titles with 15 revised editions in physical geography, environmental science, the Earth sciences, and geology. His most recent new title, *Plate Tectonics* (GeoBooks Publishing, Cambridge, Massachusetts), has received critical praise as an upper-level college text on plate tectonics as the grand paradigm in Earth science.

Brief Contents

CONTENTS

An astronaut on the Moon would see the Earth suspended in space in a similar perspective to this Meteosat photo, which records clouds, water vapor patterns, and land masses as seen from a geostationary orbit.

INTRODUCING PHYSICAL GEOGRAPHY

PUTTING YOU IN THE PICTURE

Physical geography is the study of the surface of our home planet, Earth. It is not just a catalog of mountains and rivers, but rather a branch of science that investigates how and why the surface changes from day to day, from year to year, and over millions of years.

Perhaps the easiest way to introduce physical geography is to examine the features of our planet—first as a whole and then by zooming in on continents, regions, and, finally, local areas. But to make it more interesting, imagine that the year is 2050 and that you are an astronaut ending a tour of duty on a lunar base station. Relax and follow along as you move from the lunar base to a space station orbiting the Earth and finally to the descent and landing of a space shuttle carrying you safely home.

It's eight hundred hours, on December 17, Houston time, and your duty watch is just beginning. Taking your place at the lab's environmental control desk, you check the oxygen and carbon dioxide balances and review the pressure and water vapor concentration log from the previous watch. All is well. You rise from your desk and turn. A small push on your right foot produces a giant but graceful step. You are standing in front of the viewport, looking out across the moonscape. Not much scenery here at Lunar Base Alpha, at least not when you're in the middle of the long lunar night. But in the black sky above, there is a truly majestic sight—a huge blue, green, and brown disk veiled by swirls and films of white. It is Earth, your home.

From your lunar viewpoint, the Earth's environmental realms are easily identified by their colors. The blue of the world oceans—the hydrosphere—dominates the area of the disk. Rich browns characterize the lithosphere—the solid rocky portion of the planet—where it rises above the seas. Vast white cloud expanses form and fade in the gaseous portion of the planet—the atmosphere. The lush greens of plants are the visible portion of the biosphere, the realm of life on

Earth, which dominates the life layer—the interface between air, land, and ocean.

The right-hand edge of the disk is in shadow, and you see the dramatic contrast between the day and night portions of the globe. The land and water masses of the illuminated side bask in the sunlight, and here the temperature of the life layer is increasing. On the shadowed side of the planet, however, the heat of the oceans, atmosphere, and continents is continuously being lost—radiated into outer space. Without sunlight, the Earth's surface cools.

Over the next few hours, you notice the planet's slow rotation. The continent of South America glides across the center of the disk to its edge, then disappears. Australia emerges from behind the curtain of night and makes its way out onto the illuminated portion of the disk. As you watch, each land and water mass slowly takes its turn in the Sun, absorbing and reflecting its quota of solar energy.

Africa and South America receive the lion's share of the sunlight striking the Earth. These continents lie near the equator, where they meet the strong, near-vertical rays of the Sun. North America and Europe, on the other hand, are short-changed because they intercept the Sun's rays at a low angle. At the lower rim of the Earth's disk, the stark, white continent of Antarctica rotates slowly but remains sunlit, experiencing 24-hour daylight.

As the Earth days go by, you often gaze at your planet, marking the changes. Away from the equator, huge swirls of clouds form, dissolve, and reform in its atmosphere, making their way eastward across the disk. First one portion of a continent, and then another, is obscured. These swirls mark the passage of weather fronts and storms that are moved eastward by persistent westerly winds above the surface.

Near the Earth's equator in Africa you notice a band of patchy, persistent cloudiness bracketed by reddish-brown areas of earth that are normally clear. The clouds result when warm, moist air is heated by excess solar energy and rises. As the air rises, it cools, and the moisture it contains condenses, forming clouds and producing rain. The lush, green landscape that is occasionally visible underneath the cloud belt seems to thrive on the warm temperature and abundant rainfall it receives. The reddish-brown areas are vast deserts. These receive the air that rises over the equator and becomes depleted of moisture. As the air descends toward the desert, it warms. Showing the colors of rock and soil, the hot, dry deserts are barren of plants.

During the remainder of your six-month tour of duty at the lunar base, you watch the slow changes of the planet with the season. By June, the Earth has changed its position in the Sun. Antarctica has moved below the southern rim. The sunlit northern rim of the disk now includes northern Canada, Siberia, and the Arctic Ocean. In previous months, you watched the green wave of vegetation sweep northward, up and across North America, Europe, and eastern Asia, following the warming temperatures of spring. The band of tropical cloudiness has also moved northward, bringing a green wave in Africa along with it as well. Clearly, each region of the Earth has its unique climate, responding to the rhythms of the season in different ways. You remember fondly the days of June on Earth—the smell of warm soil mixed with the scent of flowers, the hot sidewalks and streets at noon, the long evenings and warm nights alive with quiet rustlings. As the end of your tour draws near, you count the days eagerly, thinking of your return to your true home, planet Earth.

It's sixteen hundred hours, June 18, Houston time, aboard the Space Station *Enterprise,* in low earth orbit. Your duty watch is just over. Being an environmental specialist in orbit around the Earth is not the hardest job on the space station, but it isn't the easiest either. Today's shift required changing the filter on the nutrient pump for the algae, and in the weightlessness of orbit, it was a ticklish, delicate job.

It's been quite a trip so far. You recall the events of the last few weeks—the arrival of the new crew on the lunar shuttle at Moon Base Alpha's spaceport, the preparations for the departure of your crew, the relaunch of the lunar shuttle, the slow days in transit to Earth orbit, and, at long last, the docking of the lunar craft at the space station. But the most memorable part of the return trip was approaching the Earth ever

Deforestation from space *A Landsat satellite photo of deforestation in the Rondonia region, Brazil. In this false-color image, vegetation appears red and forest clearings appear blue.*

The San Francisco Bay area *A Landsat satellite view from Earth orbit.*

green islands are visible, but you know that from this distance they are actually ranges of hills or diked cities. Where the Yangtze meets the sea, a plume of sediment-laden river water curls lazily across the ocean waters. In this view, the work of rivers is especially evident—moving sediment from highlands to lower lands and the ocean, under the power of rains generated ultimately by unequal solar heating of the globe.

The following day, your view is of the San Francisco Bay area. The San Andreas fault is clearly visible on the peninsula as a line marked by narrow valleys and linear lakes at the eastern edge of the Santa Cruz mountains. Crossing seaward of the Golden Gate, it heads northwest, producing the long arm of Tomales Bay. As you have

closer. With each day and hour, it grew larger and larger. Continents became regions of forests and deserts, mountain ranges, and inland seas. At last the planet became a sphere, its edges falling away and out of view as the lunar shuttle neared the space station. Now, in low Earth orbit, finer detail is visible—systems of rivers, agricultural patterns, large cities.

Still drawn to the majesty and serenity of the Earth turning below you, you float gently to the viewport for yet another few minutes of observation. Beneath you stretches Amazonia, its lush green rainforests visible through gaps in the clouds. The verdant pattern is interrupted by patches and streaks of recently cleared land, some with hazy smoke plumes at their edges. The rainforest is but one of the many types of vegetation covers you have noted from space, ranging from spruce forests to tropical grasslands. The vegetation patterns have fit the cloud patterns well and have also marked the seasonal changes of solar illumination you noted from your lunar base.

Some hours later, your view is of China. Finally, the clouds have cleared, and a disaster is in view—the Yangtze River is in flood. The river is now a wide, shining ribbon, filling its broad valley from edge to edge. A few tiny

Touchdown view *The Space Shuttle runway at Edwards Air Force Base, California, as photographed from the Space Shuttle Challenger.*

surveyed the world's landforms from your orbital vantage point, you have noted how many seem to be related to the great structural features of the Earth's surface—its great upwarps and downwarps creating mountain zones and inland seas, its folded and faulted crust producing mountains, valleys, and plains.

It's nine hundred hours, June 26, Houston time, aboard the *Space Shuttle Discovery*. The tension is building strongly now. Gravity is starting to return, and the Earth shuttle is beginning to feel the slowing of friction with the atmosphere. The communications blackout period is about to begin. Strapped firmly into your seat, you contemplate the bumpy ride ahead with

a racing heart. Fortunately, you are just a passenger on this segment and can leave the tricky duties of navigating the shuttle to a safe landing to the flight crew and their computers. You'll be landing at Edwards Spaceport, in the California desert, since the two other spaceports are experiencing winds and rain. For now, there is nothing much to see out of the nearby viewport—it is night over the Pacific. The minutes pass slowly—too slowly. The attitude-control jets fire from time to time as the shuttle moves through its slowing maneuvers.

Soon the line of the horizon begins to redden. The Sun rises quickly, spectacularly, moving rapidly skyward. Below, the California coast approaches, the white stripe of its beaches gleaming in the Sun. The light brown scars of recent debris avalanches on the steep

marine cliffs dissect the gray-green of the chaparral. Off to the right is Los Angeles. The pink bubble of polluted air over the city is plainly visible at this early hour.

Still descending, you cross the Tehachapi Mountains, which are now very close, and drop toward the Mojave Desert. A patch of stripes flashes by—lines of sand dunes silhouetted in the early light. The landing gear descends, shaking your craft with a thud as it locks in place. Your grip on the handrest tightens as your body tenses in anticipation. The shuttle flares out over the runway. Bang! The wheels hit the ground hard. Touchdown! The brakes and drogue chute press you against the seat. You're home at last! ∎

INTRODUCING PHYSICAL GEOGRAPHY

You're invited! To a learning journey devoted to physical geography—the science of life environments from local to global scale. Your learning experience will take you from the top layers of the Earth's atmosphere to the rocks underlying the ocean basins to the forests of the farthest continents. It will focus primarily on the land surface, helping you understand and explain many common natural phenomena that you experience every day—ranging from the formation of thunderstorms to the building of sand dunes on a beach. When your journey is complete, you will come away with both a new appreciation of the landscape around you and a better understanding of the global environmental problems faced by the human race as its population expands into the twenty-first century.

This introductory chapter will provide you with some key concepts and ideas of physical geography that you will use along your journey. But before we begin that discussion, you may be curious about what geography is and what geographers study.

> Geography is the study of the evolving character and organization of the Earth's surface. Regional geography is concerned with the differentiation of the surface into places. Systematic geography studies the human and physical processes that make this differentiation.

about how, why, and where human and natural activities occur and how these activities are interconnected.

To get a better understanding of geography, think of it as having two sides. One side, which we can term **regional geography,** is concerned with how the Earth's surface is differentiated into unique **places.** What makes Vancouver (Figure I.1) unique? Is it its spectacular setting where the Pacific Ranges meet the Pacific Ocean? the marine west-coast climate that provides its mild and rainy winters and blue summer skies? its position as a seaport gateway to Asia? its English roots combined with dashes of French, Asian, and even Russian culture? In fact, all of these attributes contribute to making Vancouver the unique place that it is.

Although places are unique, the physical, economic, and social processes that form them are not. Thus, geographers are concerned with discovering, understanding, and modeling the processes that differentiate the Earth's surface into places. This is the other side of geography, which we can term **systematic geography.** Why are pineapples cheap in Hawaii and expensive in Georgia? oranges

INTRODUCING GEOGRAPHY

What is geography?[1] **Geography** is the study of the evolving character and organization of the Earth's surface. It is

[1] For descriptions and perspectives of geography, see *Rediscovering Geography: New Relevance for Science and Society* (Washington, DC: National Academy Press, 1997), 234 pp., and *Geography in America,* G. L. Gaile and Cort J. Wilmott, eds. (Columbia, OH: Merrill, 1989), 840 pp., which were used in preparing this introduction.

I.1 Eye on the Landscape *Vancouver, British Columbia This cosmopolitan city enjoys a spectacular setting on the Strait of Georgia, flanked by the Pacific and Vancouver Island Ranges.* **What else would the geographer see?...Answers at the end of the chapter.**

cheap in Florida and expensive in North Dakota? steak cheap in Kansas City and expensive in Boston? These are examples of a simple principle of economic geography—that prices include transportation costs and that when goods travel a longer distance, they are usually more expensive. Discovering such principles and extending them to model and predict spatial phenomena is the domain of systematic geography. To summarize, geographers study both the "vertical" integration of characteristics that define a place and the "horizontal" connections between places.

Realms of Geography

As a field of study, systematic geography is often divided into two broad realms—**human geography,** which deals with social, economic, and behavioral processes that differentiate places, and **physical geography,** which examines the natural processes occurring at the Earth's surface that pro-

vide the physical setting for human activities. Figure I.2 is a diagram showing the principal fields of physical and human geography. Reading downward from the left, we see five fields of physical geography, from climatology to biogeography, which are illustrated in Figure I.3. These are the main focus of this text.

Climatology is the science that describes and explains the variability in space and time of the heat and moisture states of the Earth's surface, especially its land surfaces. Since heat and moisture states are part of what we call weather, we can think of climate as a description of average weather and its variation at places around the world. Chapters 1–7 will familiarize you with the essentials of climatology, including the processes that control the weather we experience daily. Climatology is also concerned with climate change, both past and future. One of the most rapidly expanding and challenging areas of climatology is global climate modeling, which we touch on in several chapters.

I.2 Fields of systematic geography

This field attempts to predict how human activities, such as converting land from forest to agriculture or releasing CO_2 from fossil fuel burning, will change global climate.

Geomorphology is the science of Earth surface processes and landforms. The Earth's surface is constantly being altered under the combined influence of human and natural factors. The work of gravity in the collapse and movement of Earth materials, as well as the work of flowing water, blowing wind, breaking waves, and moving ice, acts to remove and transport soil and rock and to sculpt a surface that is constantly being renewed though volcanic and tectonic activity. The closing chapters of our book (Chapters 14–19) describe these geomorphic processes, while the basic geologic processes that provide the raw material are covered in Chapters 11–13. Modern geomorphology also focuses on modeling landform-shaping processes to predict both short-term, rapid changes, such as landslides, floods, or coastal storm erosion, and long-term, slower changes, such as soil erosion in agricultural areas or as a result of strip mining.

Coastal and marine geography combines the study of geomorphic processes that shape shores and coastlines with their application to coastal development and marine resource utilization. Chapter 18 describes these processes and provides some perspectives on problems of human occupation of the coastal zone.

Geography of soils includes the study of the distribution of soil types and properties and the processes of soil formation. It is related to both geomorphic processes of rock breakup and weathering, and to biological processes of growth, activity, and decay of organisms living in the soil (Chapter 10). Since both geomorphic and biologic processes are influenced by the surface temperature and availability of

---

Five major fields of physical geography include climatology (study of climate), geomorphology (study of landforms), coastal and marine geography, geography of soils, and biogeography (study of the distribution patterns of plants and animals). Water resources and hazards assessment are important applied fields of physical geography.

---

moisture, broad-scale soil patterns are often related to climate.

Biogeography, covered in Chapters 8 and 9, is the study of the distributions of organisms at varying spatial and temporal scales, as well as the processes that produce these distribution patterns. Local distributions of plants and animals typically depend on the suitability of the habitat that supports them. In this application, biogeography is closely aligned with *ecology,* which is the study of the relationship between organisms and environment. Over broader scales and time periods, the migration, evolution, and extinction of plants and animals are key processes that determine their spatial distribution patterns. Thus, biogeographers often seek to reconstruct past patterns of plant and animal communities from fossil evidence of various kinds. *Biodiversity*—the assessment of biological diversity from the perspective of maintaining the diversity of life and life-forms on Earth—is a biogeographic topic of increasing importance as human impact on the environment continues. The present global-scale distribution of life-forms as the great biomes of the Earth provides a basic context for biodiversity.

In addition to these five main fields of physical geography, two others are strongly involved with applications of physical geography—water resources and hazards assessment. **Water resources** is a broad field that couples both basic study of the location, distribution, and movement of water, for example, in river systems or as ground water, with the utilization and quality of water for human use. This field involves many aspects of human geography, including regional development and planning, political geography, and agriculture and land use. We touch on water resources briefly in this book by discussing water wells, dams, and water quality in Chapters 15 and 16.

Hazards assessment is another field that blends physical and human geography. What are the risks of living next to a river, and how do inhabitants perceive those risks? What is the role of government in protecting citizens from floods or assisting them in recovery from flood damages? Answering questions such as these requires not only knowledge of how physical systems work, but also how humans perceive and interact with their physical environment as both individuals and as societies. In this text, we develop an understanding of the physical processes of floods, earthquakes, landslides, and other disaster-causing natural events as a background for appreciating hazards to humans and their activities.

The many remaining fields of human geography are also shown in Figure I.2. Although they are not covered in this book, most have linkages with physical geography. For example, climatic and biogeographic factors may determine the spread of disease-carrying mosquitoes (medical geography). Or mountain barriers may isolate populations and increase the cost of transporting goods from one place to another (cultural geography, transportation geography). Or unique landforms and landscapes may be the objects of tourism (geography of recreation, tourism, and sport). Nearly all human activities take place in a physical environment that varies in space and time, so the physical processes that we examine in this text provide a background useful for further learning in any of geography's fields.

Tools in Geography

Because geographers deal with spatial phenomena, they use unique tools to represent spatial information (Figure I.4). The map is a common example. A **map** is a paper representation of space showing point, line, or area data—that is, locations, connections, and regions. It typically displays some characteristics or features of the Earth's surface that are positioned on the map in much the same way that they occur on the surface. The map's scale links the true distance between places with the distance on the map. The art and science of map-making is called **cartography.** Chapter 1 provides some basic information about maps as well as a special supplement, *Geographer's Tools 1.2 • Focus on Maps,* that adds depth on this important subject.

Maps, like books, are very useful methods for storing information, but they have limitations. In the past two decades, striking advances in data collection, storage, analysis, and display have led to the development of **geographic information systems (GISs).** These spatial databases rely on computers for analysis, manipulation, and display of spatial data. We will turn to an examination of GISs in Chapter 1 with our special supplement, *Geographer's Tools 1.3 • Geographic Information Systems.*

Another important geographic technique for acquiring spatial information is **remote sensing,** in which aircraft or spacecraft provide images of the Earth's surface. Depend-ing on the scale of the remotely sensed image, the information obtained can range from fine local detail—such as the arrangement of cars in a parking lot—to a global-scale picture—for example, the "greenness" of vegetation for an entire continent. As you read this textbook, you will see many examples of remote sensing, especially images from orbiting satellites, as well as boxed features highlighting applications of remote sensing in physical geography. We will return to remote sensing in our special supplement, *Geographer's Tools 2.1 • Remote Sensing for Physical Geography,* which is included in Chapter 2.

Tools in geography also include **mathematical modeling and statistics.** Using math and computers to model geographic processes is a powerful approach to understanding both natural and human phenomena. Statistics provides methods that can be used to manipulate geographic data so that we can ask and answer questions about differences, trends, and patterns. Because these tools rely heavily on specialized knowledge, we only touch on them lightly in this book.

Maps, geographic information systems (GISs), and remote sensing are important geographic tools to acquire, display, and manipulate spatial data. Mathematical modeling and statistics are also helpful tools for the geographer.

Understanding Physical Geography

From the start of your study of physical geography, you will find that it differs from other sciences in that it focuses on the world around you, from the changes in the weather to the landforms you travel over every day. When you finish your study, you should be able to see the landscape in a new way. To help you preview what you will learn about our environment, we provide a new feature—Eye on the Landscape—which takes many of our photos and identifies prominent features that a physical geographer would recognize and understand.

As another aid to comprehension and understanding of our text material, we have placed **summary statements** on many pages. These short statements restate the most important concepts of physical geography as they are being explained and described in the text. They should help you keep focused on the key points as you read along.

We have also included a series of *Eye on the Environment* and *Eye on Global Change* features that highlight particular topics of environmental interest. These range from human impact on the ozone layer to the effect of global warming on ice sheets Greenland and the Antarctic.

Our *GeoDiscoveries* CD-ROM, along with its associated worldwide web site and web links, is yet another important resource that will help you to understand physical geography. It includes animations, movies, photos, diagrams, and interactive text in modules keyed to each chapter. Throughout this book you will find icons and pathways pointing to topics in *GeoDiscoveries* that are particularly appropriate to the text material. We've worked very hard on *GeoDiscoveries* and hope that it can make a real difference in your mastery of physical geography.

I.3 Fields of Physical Geography

a...Climatology Climate is largely dependent on atmosphere-surface interaction.

b...Geomorphology
Geomorphology is the study of landform-making processes.

c...*Eye on the Landscape*
Coastal and marine geography
Coastal and marine geography examines coastal processes, marine resources, and their human interface. **What else would the geographer see?...Answers at the end of the chapter.**

d...Eye on the Landscape
Biogeography Biogeography
examines the distribution patterns of
plants and animals and relates them
to environment, migration, evolution,
and extinction. **What else would the
geographer see?...Answers at the end of
the chapter.**

e...Geography of soils Soils are
influenced by their parent material,
climate, biota, and time.

a...Cartography *A portion of the U.S. Geological Survey 1:24,000 topographic map of Green Bay, Wisconsin. Using symbols, the map shows creeks and rivers, a bay, swampy regions, urban developed land, streets, roads, and highways.*

b...Geographic information systems (GIS) *Computer programs that store and manipulate geographic data are essential to modern applications of geography. This screen from the ARCInfo GIS program package shows earthquake centers in eastern Asia superimposed on a political map underlain by a shaded relief map of undersea topography.*

c...Remote sensing *Remote sensing includes observing the Earth from the perspective of an aircraft or spacecraft. Wildfires on the Greek island of Peloponnesos, seen in a Landsat image from July 2000, are an example.*

d... ***Mathematical modeling***
By describing a phenomenon using a
mathematical model, a geographer can
predict outcomes and examine "what-if"
scenarios. These equations demonstrate the
calculation of an exponential growth factor.

$$M = e^{(R \times T)}$$
$$= e^{(0.04 \times 20)}$$
$$= 2.718^{0.80}$$
$$= 2.26$$

d

e... ***Statistics*** Statistical tools, such as this
graph, allow the exploration of geographic
data to determine trends and develop
mathematical models. The plot shows the
value of the Southern oscillation Index, an
indicator of El Niño conditions (see Eye on
the Environment 5.2 • El Niño).

Southern Oscillation Index

e

SPHERES, SCALES, SYSTEMS, AND CYCLES

As a part of your introduction to physical geography, it will be useful to take a look at the big picture and examine some ideas that arch over all of physical geography—that is, spheres, scales, systems, and cycles. The first of these ideas is that of the four great physical realms, or *spheres* of Earth—atmosphere, lithosphere, hydrosphere, and biosphere. These realms are distinctive parts of our planet with unique components and properties. Another important idea is that of *scales* in physical geography, which range from global to local. When we look at the Earth at different scales, we often see different processes operating.

Another big idea is that of *systems*—viewing the processes that shape our landscape as a set of interrelated components that comprise a system. The systems viewpoint stresses linkages and interactions and helps to understand complex problems, such as global climate change or loss of biodiversity. The last big idea is that of *cycles*—regular changes in systems that recur in a fixed period of time.

The Spheres—Four Great Earth Realms

The natural systems that we will encounter in the study of physical geography operate within the four great realms, or **spheres,** of the Earth. These are the atmosphere, the lithosphere, the hydrosphere, and the biosphere (Figure I.5).

The **atmosphere** is a gaseous layer that surrounds the Earth. It receives heat and moisture from the surface and redistributes them, returning some heat

> The four great realms of natural systems are the atmosphere, hydrosphere, lithosphere, and biosphere. The life layer is the shallow surface zone that contains most of the biosphere.

I.6 The life layer *As this sketch shows, the life layer is the layer of the Earth's surface that supports nearly all of the Earth's life. This includes the land and ocean surfaces and the atmosphere in contact with them.*

and all the moisture to the surface. The atmosphere also supplies vital elements—carbon, hydrogen, oxygen, and nitrogen—that are needed to sustain life-forms.

The outermost solid layer of the Earth, or **lithosphere,** provides the platform for most Earthly life-forms. The solid rock of the lithosphere bears a shallow layer of soil in which nutrient elements become available to organisms. The surface of the lithosphere is sculpted into landforms. These features—such as mountains, hills, and plains—provide varied habitats for plants, animals, and humans.

The liquid realm of the Earth is the **hydrosphere,** which is principally the mass of water in the world's oceans. It also includes solid ice in mountain and continental glaciers, which, like liquid ocean and fresh water, is subject to flow under the influence of gravity. Within the atmosphere, water occurs as gaseous vapor, liquid droplets, and solid ice crystals. In the lithosphere, water is found in the uppermost layers in soils and in ground water reservoirs.

The **biosphere** encompasses all living organisms of the Earth. Life-forms on Earth utilize the gases of the atmosphere, the water of the hydrosphere, and the nutrients of the lithosphere, and so the biosphere is dependent on all three of the other great realms. Figure I.5 diagrams this relationship.

Most of the biosphere is contained in the shallow surface zone called the **life layer.** It includes the surface of the lands and the upper 100 meters or so (about 300 ft) of the ocean (Figure I.6). On land, the life layer is the zone of interactions among the biosphere, lithosphere, and atmosphere. The hydrosphere is represented on land by rain, snow, still water in ponds and lakes, and running water in rivers. In the ocean, the life layer is the zone of interactions among the hydrosphere, biosphere, and atmosphere. The lithosphere is represented by nutrients dissolved in the upper layer of sea water. Throughout our exploration of physical geography, we will often refer to the life layer and the four realms that interact within it.

I.5 The Earth realms *The Earth realms, shown as intersecting circles.*

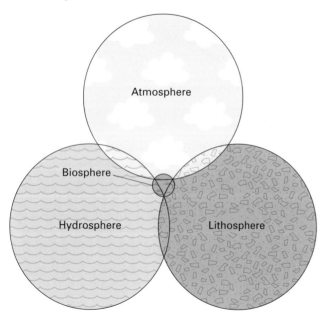

Scales in Physical Geography

The processes of the four great realms and the life layer operate at various **scales** (Figure I.7). On a *global scale,* our planet is a nearly spherical body that turns endlessly within the radiance of our Sun. Because the Sun is the energy source that powers most of the phenomena that occur within the life layer, Earth–Sun relationships are very important. In studying these relationships, we need to consider the planet and its global energy balance as a whole and view the Sun and Earth from a vantage point far from the Earth itself.

> Scales in physical geography vary from global to continental, regional, local, and individual. A systems approach to physical geography looks for linkages and interactions among physical processes. Time cycles are regular rhythms in natural systems that range from hours to millions of years.

The Sun's energy is not absorbed evenly by the Earth's land and water surface. Unequal solar heating produces currents of air and water, just as a candle flame creates a current of hot, rising air. These currents constitute the global atmospheric and oceanic circulation system. To study this system, we need to move to the *continental scale,* where we can distinguish continents and oceans and track winds and ocean currents.

Moving still closer to the Earth to the *regional scale,* we distinguish patterns of mountain ranges and valleys, of lakes, coasts, and bays. We also see the cloud patterns of weather systems and their regular movements over time. These movements, along with solar control of surface temperature, form the basis for the climates of the world.

Regional climate strongly influences the nature of the Earth's cover of plants and the soil layers beneath them. But factors at the *local scale* are important in determining the exact patterns of vegetation and soils. Is the location on a mountain slope or in a flat valley? Is it being farmed or used as grazing land? This information is too specific to use at the global, continental, or regional scales, but it is important when we consider the nature of the vegetation cover and soils within a specific region.

At the very finest scale, we see *individual-scale* landscape features such as a grassy sand dune on a beach or a moss-covered bank on the side of a river. These individual landforms and their associated plant and animal communities are produced by unique activities of wind or water, and develop distinctive biological communities and soil properties.

Time scales vary widely, too. It may take millions of years for Earth forces to produce a chain of mountains like the Himalayas. But a fracturing of the Earth's crust and the resulting earthquake may last only a minute or two.

Systems in Physical Geography

The processes that interact within the four realms to shape the life layer and differentiate global environments are varied and complex. A helpful way to understand the relationships among these processes is to study them as **systems.** "System" is a common English word that we use in everyday speech. It typically means a set or collection of things that are somehow related or organized. An example is the solar system—a collection of planets that revolve around the Sun. In the text, we will use the word *system* in this way quite often. Sometimes it refers to a scheme for naming things. For example, we will introduce a climate system in Chapter 7 and a soil classification system in Chapter 10. However, we will also use *system* to mean a group of interrelated processes that operate simultaneously in the physical landscape.

When we study physical geography using a *systems approach* we look for linkages and interactions among processes. For example, global warming should enhance the process of evaporation of water from oceans and moist land surfaces, generating more clouds. But an increase in clouds also affects the process of solar reflection, in which white, fleecy clouds reflect solar radiation back out to space. This leaves less radiation to be absorbed by the atmosphere and surface and so should tend to cool our planet, reducing global warming. This is actually an example of negative feedback, in which one process counteracts another process to reduce its impact. (We'll present more information about this topic in Chapter 6.) Throughout the text there will be more examples of this systems viewpoint in physical geography.[2]

Time Cycles

Many natural systems show **time cycles**—rhythms in which processes change in a regular and repeatable fashion. For example, the annual revolution of the Earth around the Sun generates a time cycle of incoming solar energy flow. We speak of this cycle as the rhythm of the seasons. The rotation of the Earth on its axis sets up the night-and-day cycle of darkness and light. The Moon, in its monthly orbit around the Earth sets up its own time cycle. We see the lunar cycle in the range of tides, with higher high tides and lower low tides ("spring tides") occurring both at full moon and at new moon.

The astronomical time cycles of Earth rotation and solar revolution will appear at several places in our early chapters. Other time cycles with durations of tens to hundreds of thousands of years describe the alternate growth and shrinkage of the great ice sheets. Still others, with durations of millions of years, describe cycles of the solid Earth in which supercontinents form, break apart, and reform anew.

---

[2] For a more careful and rigorous treatment of systems as flow systems of energy and matter, see our *Focus on Systems* features in *Physical Geography: Science and Systems of the Human Environment,* 2nd ed., Wiley, 2001.

I.7 Scales in Physical Geography

a...Global scale *A model of the Earth's topography as it would be seen from a viewpoint in space.*

b...Continental scale *A topographic model of North America, enhanced with clouds.*

c...Regional scale *Topography of the western United States and southern Canada.*

d...Local scale *A Landsat image of the Willamette River Valley, Oregon and Washington.*

e...Individual-scale landscape *The Willamette River, near Newburg, Oregon, seen from the air.*

PHYSICAL GEOGRAPHY, ENVIRONMENT, AND GLOBAL CHANGE

Physical geography is concerned with the natural world around us—in short, with the human environment. Because natural processes are constantly active, the Earth's environments are constantly changing (Figure I.8). Sometimes the changes are slow and subtle, as when crustal plates move over geologic time to create continents and ocean basins. At other times, the changes are rapid, as when hurricane winds flatten vast areas of forests or even tracts of houses and homes.

Environmental change is now produced not only by the natural processes that have acted on our planet for millions of years but also by human activity. The human race has populated our planet so thoroughly that few places remain free of some form of human impact. Global change, then, involves not only natural processes, but also human processes that interact with them. Physical geography is the key to understanding this interaction.

Environment and global change are sufficiently important that we have set off these topics from ordinary text in this book by placing them in special sections identified with **Eye on the Environment.** As you read, watch for these sections as your key to the application of physical geography to environmental and global change topics.

What are some of the important topics in global change that lie within physical geography? Let's examine a few.

> **Environmental change is produced by both natural and human processes. Human activities are currently changing both the Earth's climate and the global flows of carbon from Earth to ocean to atmosphere.**

Global Climate Change

Are human activities changing global climate? It seems that almost every year we hear that it has been the hottest year, or one of the hottest years, on record. But climate is notoriously variable. Could such a string of hot years be part of the normal variation? This is the key question facing scientists studying global climate change. Over the past decade, most scientists have come to the opinion that human activity has, indeed, begun to change our climate. How has this happened?

The answer lies in the greenhouse effect. As human activities continue to release gases that block heat radiation from leaving the Earth, the greenhouse effect intensifies. The most prominent of these gases is CO_2, which is released by fossil fuel burning. Others include methane (CH_4), nitrous oxide (NO), and the chlorofluorocarbons that until recently served as coolants in refrigeration and air conditioning systems and as aerosol spray propellants. Taken together, these gases are acting to raise the Earth's surface temperature, with consequences including dislocation of agricultural areas, rise in sea level, and increased frequency of extreme weather events, such as severe storms or record droughts.

Climate change is a recurring theme throughout this book, ranging from the urban heat island effect that tends to raise city temperatures (Chapter 3) to the El Niño phenomenon that alters global atmospheric and ocean circulation (Chapter 5), to the effect of clouds on global warming (Chapter 6), and to rising sea level due to the expansion of sea water with increasing temperature (Chapter 18).

The Carbon Cycle

One way to reduce human impact on the greenhouse effect is to slow the release of CO_2 from fossil fuel burning. In fact, the Kyoto Protocol of 1997, an international agreement first proposed at the 1992 Earth Summit in Rio de Janeiro, commits the world's industrialized nations to limit current CO_2 emissions to levels at or below those of 1990. But since modern civilization depends on the energy of fossil fuels to carry out almost every task, reducing fossil fuel consumption to stabilize the increasing concentration of CO_2 in the atmosphere is not easy. However, some natural processes reduce atmospheric CO_2. Plants withdraw CO_2 from the atmosphere by taking it up in photosynthesis to construct plant tissues, such as cell walls and wood. In addition, CO_2 is soluble in sea water. These two important pathways, by which carbon flows from the atmosphere to lands and oceans, are part of the carbon cycle. So, biogeographers and ecologists are now focusing in detail on the global carbon cycle in order to better understand the pathways and magnitudes of carbon flow. They hope that this understanding will suggest alternative actions that can reduce the rate of CO_2 buildup without penalizing economic growth.

As nations work to determine how much CO_2 they release and to develop plans to reduce that amount, scientists are focusing on the role of CO_2 uptake by natural vegetation. An important question now under study concerns the balance between forest growth and organic matter decay in the northern hemisphere forest zone. At present, there are large areas of young forests that are actively growing and taking up significant amounts of CO_2. In an opposite effect, this uptake is offset by the decay of organic matter in soils, which releases CO_2. Because the decay process depends on temperature, global warming is thought to be increasing the rate of decay and thus CO_2 release. Scientists have not yet determined the balance between these two pathways but are intensively studying these flows. The processes of carbon fixation and release are covered in Chapter 8. This chapter also describes how changes in temperature and CO_2 concentration affect plant growth and decay, and documents the process of succession by which plant communities change through time.

a...Global climate change
Is the Earth's climate changing? Most global change scientists think that human activities have resulted in climate warming and that weather patterns, shown here in this satellite image of clouds and weather systems over the Pacific ocean, are changing.

b...Carbon cycle Clearcutting of timber, shown here near the Grand Tetons, Wyoming, removes carbon from the landscape, while regrowth returns carbon through photosynthesis.

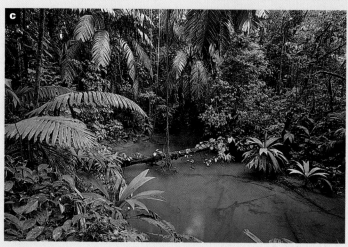

c...Biodiversity Reduction in the area and degradation of the quality of natural habitats is reducing biodiversity. The banks of this stream in the rainforest of Costa Rica are lined with several species of palms.

d...Pollution Human activity can create pollution of air and water, causing change in natural habitats as well as impacts on human health. The discharge from this pulp mill near Port Alice, British Columbia, is largely water vapor, but pulp mill pollutants often include harmful sulfur oxides.

e...Extreme events Hurricanes, severe storms, droughts, and floods may be becoming more frequent as global climate warms. A tornado flattened this neighborhood in Dell City, Oklahoma, in May of 1999.

Biodiversity

Among scientists, environmentalists, and the public, there is a growing awareness that the diversity in the plant and animal forms harbored by our planet—the Earth's biodiversity—is an immensely valuable resource that will be cherished by future generations. One important reason for preserving as many natural species as possible is that, over time, species have evolved natural biochemical defense mechanisms against diseases and predators. These defense mechanisms involve bioactive compounds that can sometimes be very useful, ranging from natural pesticides that increase crop yields to medicines that fight human cancer.

Another important reason for maintaining biodiversity is that complex ecosystems with many species tend to be more stable and to respond better to environmental change. If human activities inadvertently reduce biodiversity significantly, there is a greater risk of unexpected and unintended human effects on natural environments.

Biogeographers focus on both the processes that create and maintain biodiversity and the existing biodiversity of the Earth's many natural habitats. These topics are treated in Chapters 8 and 9.

Pollution

As we all know, unchecked human activity can degrade environmental quality. In addition to CO_2 release, fuel burning can release gases that are hazardous to health, especially when they react to form such toxic compounds as ozone and nitric acid in photochemical smog. Water pollution from fertilizer runoff, toxic wastes of industrial production, and acid mine drainage can severely degrade water quality. Such degradation impacts not only the ecosystems of streams and rivers, but also the human populations that depend on rivers and streams as sources of water supply. Ground water reservoirs can also be polluted or turn salty in coastal zones when drawn down excessively.

Environmental pollution, its causes, its effects, and the technologies used to reduce pollution, form a subject that is broad in its own right. As a text in physical geography that emphasizes the natural processes of the Earth's land surface, we touch on air and water pollution in several chapters—Chapter 4 for air pollution and Chapter 15 for surface water pollution, irrigation effects, and ground water contamination.

Extreme Events

Catastrophic events—floods, fires, hurricanes, earthquakes, and the like—can have great and long-lasting impacts on both human and natural systems. Are human activities increasing the frequency of these extreme events? As our planet warms in response to changes in the greenhouse effect, global climate modelers predict that weather extremes will become more severe and more frequent. Droughts and consequent wildfires and crop failures will happen more often, as will spells of rain and flood runoff. In the last decade, we have seen numerous examples of extreme weather events, from Hurricane Andrew in 1992—the most costly storm in U.S. history—to the southeast drought of 2000, which devastated crops in large parts of the southeastern United States. Is human activity responsible for the increased occurrence of these extreme events? Significant evidence now points in that direction. (See *Eye on the Environment 3.2 • The IPCC Report of 2001.*)

Other extreme events, such as earthquakes, volcanic eruptions, and seismic sea waves (wrongly called tidal waves), are produced by forces deep within the Earth that are not affected by human activity. But as the human population continues to expand and comes to rely increasingly on a technological infrastructure ranging from skyscrapers to the Internet, we are becoming more sensitive to damage and disruption of these systems by extreme events.

This text describes many types of extreme events and their causes. In Chapters 4 and 6, we discuss thunderstorms, tornadoes, cyclonic storms, and hurricanes. Droughts in the African Sahel are presented in Chapter 7. Earthquakes, volcanic eruptions, and seismic sea waves are covered in Chapter 13. Floods are described in Chapter 15, including the great Mississippi flood of 1993.

Human activity is reducing the biodiversity of many of the Earth's natural habitats. Environmental pollution degrades habitat quality for humans as well as other species. Extreme weather events, which are thought to be more frequent with human-induced climate change, as well as other rare natural events, are increasingly destructive to our expanding human population.

A LOOK AHEAD The past few pages have presented an introduction to geography, to physical geography, to some of the tools and approaches that physical geographers use in studying the landscape, and to some of the big ideas that overarch physical geography. We have also introduced some of the key environmental and global change topics that will appear in our text.

With this introduction now complete, we can turn to our topical coverage of physical geography. We begin in Chapter 1 with some fundamental principles about Earth–Sun relationships that drive global climate, which in turn is a key factor in many fields of physical geography, ranging from biogeography to geomorphology.

Humans are now the dominant species on the planet. Nearly every part of the Earth has felt human impact in some way. As the human population continues to grow and rely more heavily on natural resources, our impact on natural systems will continue to increase. Each of us is charged with the responsibility to treat the Earth well and respect its finite nature. Understanding the processes that shape our habitat as they are described by physical geography will help you carry out this mission.

In Review | Introducing Physical Geography

- **Geography** is the study of the evolving character and organization of the Earth's surface. **Regional geography** is concerned with how the Earth's surface is differentiated into unique **places,** while **systematic geography** is concerned with the processes that differentiate places in time and space.

- Within systematic geography, **human geography** deals with social, economic, and behavioral processes that differentiate places, and **physical geography** examines the natural processes occurring at the Earth's surface that provide the physical setting for human activities. **Climatology** is the science that describes and explains the variability in space and time of the heat and moisture states of the Earth's surface, especially its land surfaces. **Geomorphology** is the science of Earth surface processes and landforms. **Coastal and marine geography** combines the study of geomorphic processes that shape shores and coastlines with their application to coastal development and marine resource utilization. **Geography of soils** includes the study of the distribution of soil types and properties and the processes of soil formation. **Biogeography** is the study of the distributions of organisms at varying spatial and temporal scales, as well as the processes that produce these distribution patterns. **Water resources** and **hazards assessment** are applied fields that blend both physical and human geography.

- Important tools for studying the fields of physical geography include **maps, geographic information systems, remote sensing, mathematical modeling,** and **statistics.**

- **Spheres, scales, systems,** and **cycles** are four overarching themes that appear in physical geography. The four great Earth realms are **atmosphere, hydrosphere, lithosphere,** and **biosphere.** The **life layer** is the shallow surface layer where lands and oceans meet the atmosphere and where most forms of life are found. The processes of the four great realms operate at multiple scales, including *global, continental, regional, local,* and *individual.*

- Physical processes often act together in an organized way that we can view as a **system.** A *systems approach* to physical geography looks for linkages and interactions between processes.

- Natural systems may undergo periodic, repeating changes that constitute **time cycles.** Important time cycles in physical geography range in length from hours to millions of years.

- Physical geography is concerned with the natural world around us—the human environment. Natural and human processes are constantly changing that environment. Global climate is changing in response to human impacts on the greenhouse effect. Global pathways of carbon flow can influence the greenhouse effect and are the subject of intense research interest. Maintaining global biodiversity is important both for maintaining the stability of ecosystems and guarding a potential resource of bioactive compounds for human benefit. Unchecked human activity can degrade environmental quality and create pollution. Extreme events take ever-higher tolls on life and property as populations expand. Extreme weather—storms and droughts, for example—may be more frequent with global warming caused by human activity.

Key Terms

geography, p. 6
regional geography, p. 6
places, p. 6
systematic geography, p. 6
human geography, p. 7
physical geography, p. 7
climatology, p. 7

geomorphology, p. 8
coastal and marine geography, p. 8
geography of soils, p. 8
biogeography, p. 8
water resources, p. 8
hazards assessment, p. 9
map, p. 9

cartography, p. 9
geographic information system (GIS), p. 9
remote sensing, p. 9
mathematical modeling and statistics, p. 9
spheres, p. 14
atmosphere, p. 14

lithosphere, p. 14
hydrosphere, p. 14
biosphere, p. 14
life layer, p. 14
scales, p. 15
systems, , p. 15
time cycles, p. 15

Review Questions

1. What is geography? regional geography? systematic geography?
2. How does human geography differ from physical geography?
3. Identify and define five important fields of science within physical geography.

4. Identify and describe three tools that geographers use to acquire, display, and manipulate spatial data.
5. Name and describe each of the four great physical realms of Earth. What is the life layer?
6. Provide two examples of processes or systems that operate at each of the following scales: global, continental, regional, local, and individual.

7. How is the word *system* used in physical geography? What is a systems approach?

8. What is a time cycle as applied to a system? Give an example of a time cycle evident in natural systems.

9. Identify and describe two interacting components of global change.

10. How is global climate change influenced by human activity?

11. Why are current research efforts focused on the carbon cycle?

12. Why is loss of biodiversity a concern of biogeographers and ecologists?

13. How does human activity degrade environmental quality? Provide a few examples.

14. How do extreme events affect human activity? Is human activity influencing the size or reoccurrence rate of extreme events?

Essay Question

1. Go to the extended table of contents for this book. With a pencil, check off every topic listed that you know something about. Now think about where you learned about these topics. Which of the following sources has provided you with the most information about physical geography? television, books and magazines, schoolwork, courses in college or university, learning by experience, other? How do you think the information sources by which you have learned about physical geography have affected what you have learned?

Eye on the Landscape

I.1 Vancouver, British Columbia With its deep embayments and steep topography, this is a coastline of submergence **(A)**. Much of the terrain was carved into steep slopes and wide valleys by glacial ice descending from nearby peaks during the Ice Age, when sea level was as much as 100 m (about 300 ft) lower. With the melting of ice sheets, sea level rose, drowning the landscape. Also note the snowcapped peaks in the far distance **(B)**. Rugged landscapes like this often indicate recent plate tectonic activity in which immense crustal plates collide to push up mountain chains like the Cascade Range. Chapters 18 and 19 describe coastlines and glacial erosion features. Plate tectonics is covered in Chapter 12.

I.3c Coastal and marine geography (A) This rock platform, shown at low tide, is a feature formed by wave action and is found on many rocky shorelines. At times of high water and surf, waves attack the cliffs, releasing rock fragments that are carried back and forth by the waves, thus cutting the flat bench. Chapter 18 provides more information on wave action. At **(B)**, horizontal lines revealed in the rock faces of these cliffed promontories mark individual layers (beds) of sedimentary rock. These beds were probably formed when silt and clay settled from turbid waters in a shallow arm of the ocean or in an inland sea. After millions of years of burial and consolidation into rock, the beds were uplifted and are now being eroded by waves. Sedimentary rock formation is covered in Chapter 11, and uplift (tectonic processes) is covered in Chapters 12 and 13.

I-3d Biogeography These patches of fog at **(A)** illustrate that fog is simply cloud touching the ground. They are probably patches of morning fog, formed when the moist air of the rainforest is chilled below the dew point by radiation cooling on a clear night. As the air warms, they will evaporate. Chapter 4 covers clouds, fog, and radiation cooling.

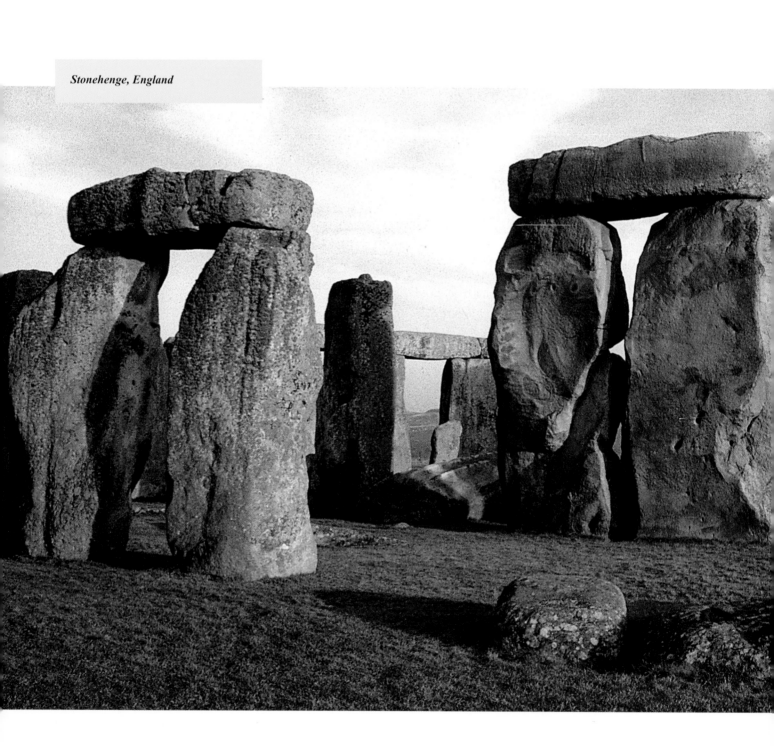

Stonehenge, England

1 | THE EARTH AS A ROTATING PLANET

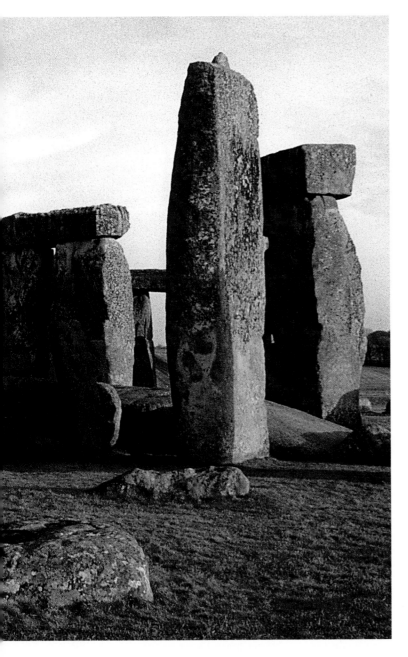

. . . .
PUTTING YOU IN THE PICTURE

This chapter is about the two great rhythms of solar illumination that affect every location on Earth—the daily rhythm of daylight and darkness, and the annual rhythm of lengthening and shortening of the daylight period with the seasons. They are vital to life on Earth, and nearly every form of life is adapted in some way to these solar rhythms. As we will see, these two rhythms arise from the Earth's rotation on its axis and its revolution around the Sun. To an observer at a fixed point on the Earth, however, the motions appear to be those of the Sun, not the Earth. As the Sun moves through the heavens, it rises and sets each day, marking the daily cycle of rotation. But its position of rising and setting on the horizon changes slowly day by day on an annual cycle, as does its elevation above the horizon at noon and the length of time that it remains in the sky. This annual rhythm marks the revolution of the Earth around the Sun during a year.

The motions of the Sun in the sky have been the subject of observation and study by human civilizations for thousands of years. An example is Stonehenge, a collection of huge rock slabs put in place thousands of years ago by a long-lost culture in a remote part of southwest England. Imagine yourself now at Stonehenge in the early morning of the first day of summer, waiting for first light…

At last the sky is brightening. Perhaps it won't be much longer now, you think, until the first rays of the Sun become visible on the horizon next to the Slaughter Stone. Exhaling slowly, you watch the condensation from your breath float away on the still air. A shiver overwhelms you and you grasp your jacket more tightly, pulling it closer to your body.

Although the morning is cold, the sky is mostly clear. June weather in the southwest corner of England is notoriously changeable, but it looks like the viewing conditions will be good. According to the information you noted yesterday on your visit to the Greenwich Observatory, the summer solstice will occur at 3:28 P.M. this afternoon. So, today will be the longest day of the year—almost sixteen and a half hours at this latitude of more than 50 degrees north. It will also be the day of the year on which the Sun rises at its most northerly position from due east. You check your watch. Sunrise will be at 4:59. Still about 10 minutes to go.

You pause for a moment to survey your surroundings. Not far away is a trilithon—two immense vertical slabs of brown sandstone crowned by a massive horizontal sandstone called a lintel. This trilithon is one of five, arranged in a horseshoe. Three are still standing, and two are represented only by a single, still-erect stone. Surrounding the trilithons by four or five yards are the remaining vertical slabs of the sarsen circle, arranged in a ring of about 100 feet in diameter. Stonehenge was quite an accomplishment, considering that it was built three to four thousand years ago using only the most primitive methods.

The low voices of the crowd are suddenly quiet in anticipation as a ruby gleam brightens the sky. The first solar rays are soon visible. Slowly, the red solar disk emerges. Beyond the Sun, the sky is washed with red and pink, fading to deep blue. Small, puffy clouds above glow a spectacular rosy pink. What a sight!

From your location near the center of the ruin, the red solar ball is visible just to the left of the immense Slaughter Stone that marks the entrance to the ancient structure. You visualize the missing twin slab to the Slaughter Stone, nothing that the Sun would have been centered perfectly between the two huge vertical monoliths. The long shadow of the Slaughter Stone to your right would have been joined by a second shadow on your left, marking a broad sunlit path extending from the entrance to the center of the structure. Quite amazing, you think, that the ancient peoples that built Stonehenge were so fascinated with the paths of the Sun and the Moon in the sky that they marked them by hauling huge stones hundreds of miles to this hallowed place. Suddenly, a wave of reverence for Earth and nature hits you, and you realize that the scene has touched something primitive deep inside you that springs from the common heritage of humanity. ■

THE EARTH AS A ROTATING PLANET

Since the dawn of civilization, humans have observed the movements of the Sun in the sky. However, only since the time of Copernicus—the sixteenth century—have we known that the daily and seasonal motions of the Sun are a consequence of the motion of the Earth rather than the Sun. From day to day, the Earth's rotation on its axis produces the alternation of the light and darkness that we experience in 24 hours. From month to month, the revolution of the Earth around the Sun produces slow changes in the length of daylight that create the rhythm of the seasons. These two great cycles continue, day after day, year after year, regulating the processes of Earthly life. We begin our systematic study of physical geography by devoting a chapter to the motions of the Earth and Sun and the implications of these motions for Earth location, timekeeping, and the seasons.

GEODISCOVERIES

> The Earth rotates on its axis and revolves around the Sun. The rotation produces the day–night cycle, while the revolution produces the seasonal cycle.

by astronauts and by orbiting satellites also show us that the Earth is a round body. We learn this fact so early in life that it seems quite unremarkable.

Many of our ancestors, however, were not aware of our planet's spherical shape. To sailors of the Mediterranean Sea in ancient times, the shape and breadth of the Earth's oceans and lands were unknown. On their ships and out of sight of land, the sea surface looked perfectly flat and bounded by a circular horizon. Given this view, many sailors concluded that the Earth had the form of a flat disk and that their ships would fall off if they traveled to its edge. Of course, had the telescope been invented, they could have seen a distant ship with its decks seemingly below the water line—a demonstration of the curvature of the Earth's surface.

The curvature of the Earth is also evident as you view a sunset with clouds in the sky (Figure 1.1*d*). Although the Sun is below the horizon from your viewpoint and no longer illuminates the land around you, it is above the horizon at the altitude of the clouds and still bathes them in its red and pinkish rays. As the Sun progressively washes the clouds overhead in red light, the red band of illumination slowly moves farther and farther toward your horizon. Soon only

THE SHAPE OF THE EARTH

The Earth's shape is very close to that of a sphere, as we all learn early in school (Figure 1.1). Pictures taken from space

1.1 Our Spherical Earth

a...*Photo of Earth curvature* This astronaut photo shows the Earth's curved horizon from low-Earth orbit.

b...*Eye on the Landscape* Space view of Earth From a geostationary weather satellite, the Earth is a disk obscured by swirls of moving clouds. What else would the geographer see?...Answers at the end of the chapter.

d...*Sunset photo*
In this dramatic sunset photo, the far distant clouds are still directly illuminated by the Sun's last red rays. For the clouds directly overhead, however, the Sun has left the sky.

c...*Distant Ship* Seen through a telescope, the decks of a distant ship seem to be under water.

the far distant clouds can still "see" the Sun. This movement of solar illumination across the clouds is easily explained by a rotating spherical Earth.

Actually, the Earth is not perfectly spherical. Because the outward force of the Earth's rotation causes the Earth to bulge slightly at the equator and flatten at the poles, the Earth assumes the shape of an *oblate ellipsoid.* ("Oblate" means flattened.) Thus, the Earth's equatorial diameter, about 12,756 km (7926 mi), is very slightly larger than the polar diameter, about 12,714 km (7900 mi). The difference is very small—about three-tenths of 1 percent.

Scientists still study the Earth's shape and attempt to measure it as precisely as possible. Satellite navigation systems need this important information for aircraft, ocean vessels, and ground vehicles seeking to determine their exact location (see *Focus on Remote Sensing 1.1 • The Global Positioning System,* later in this chapter).

EARTH ROTATION

Another fact about our planet that we learn early in life is that it spins slowly, making a full turn with respect to the Sun every day. We use the term **rotation** to describe this motion. One complete rotation with respect to the Sun defines the *solar day.* By convention, the solar day is divided into exactly 24 hours.

The Earth rotates on its *axis,* an imaginary straight line through its center. The two points where the axis of rotation intersects the Earth's surface are defined as the poles. To distinguish between the two poles, one is called the *north pole.* and the other, the *south pole.*

To determine the direction of Earth rotation, you can use one of the following guidelines (Figure 1.2):

- Imagine yourself looking down on the north pole of the Earth. From this position, the Earth is turning in a counterclockwise direction (Figure 1.2*a*).

> The direction of the Earth's rotation is counterclockwise when viewed from above the north pole, or from west to east if viewed with the north pole up.

- Imagine yourself off in space, viewing the planet much as you would view a globe in a library, with the north pole on top. The Earth is rotating from left to right or in an eastward direction (Figure 1.2*b*).

The Earth's rotation is important for three reasons. First, the axis of rotation serves as a reference in setting up the geographic grid of latitude and longitude, which we will discuss later in the chapter. Second, it provides the day as a convenient measure of the passage of time, with the day in turn divided into hours, minutes, and seconds. Third, it has important effects on the physical and life processes on Earth.

Environmental Effects of Earth Rotation

The first—and perhaps most obvious—effect of the Earth's rotation is that it imposes a daily, or *diurnal,* rhythm in daylight, air temperature, air humidity, and air motion.

All surface life responds to this diurnal rhythm. Green plants receive and store solar energy during the day and consume some of it at night. Among animals, some are active during the day, others at night. The daily cycle of incoming solar energy and the corresponding cycle of fluctuating air temperature are topics for analysis in Chapters 2 and 3.

A second environmental effect is that the flow paths of both air and water are turned consistently in a sideward direction because of the Earth's rotation. Flows in the northern hemisphere are turned toward the right and in the southern hemisphere toward the left. This phenomenon is called the *Coriolis effect.* It is of great importance in studying the Earth's systems of winds and ocean currents and is discussed in Chapter 5.

A third physical effect of the Earth's rotation is the movement of the tides. The Moon exerts a gravitational attraction on the Earth, while at the same time the Earth turns with respect to the Moon. These forces induce a rhythmic rise and fall of the ocean surface known as the *tide.* The tide in turn causes water currents to flow in alternating directions in the shallow salty waters of the coastal zone. This ebb and flow of tidal currents is a life-giving pulse for many plants and animals that live in coastal saltwater environments. The tidal cycle is a clock regulating many daily human activities in the coastal zone as well. When we examine the tide and its currents further in Chapter 18, we will see that the Sun also has an influence on the tides.

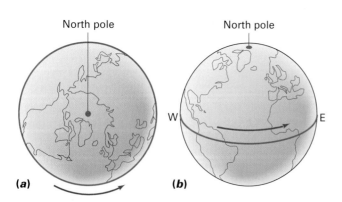

1.2   Direction of Earth rotation *The direction of rotation of the Earth can be thought of as (a) counterclockwise at the north pole, or (b) from left to right (eastward) at the equator.*

THE GEOGRAPHIC GRID

The geographic grid provides a system for locating places on the Earth's surface. Because the Earth's surface is curved, and not flat, we cannot divide it into a rectangular grid, like a sheet of graph paper. Instead, we divide it using imaginary circles set on the surface that are perpendicular to the axis of

Focus on Remote Sensing | 1.1

The Global Positioning System

The latitude and longitude coordinates of a point on the Earth's surface describe its position exactly. But how are those coordinates determined? For the last few hundred years, we have known how to use the position of the stars in the sky coupled with an accurate clock to determine the latitude and longitude of any point. Linked with advances in mapping and surveying, these techniques became highly accurate, but they were impractical for precisely determining desired locations in a short period of time.

Thanks to new technology originally developed by the U.S. Naval Observatory for military applications, there is now in place a *global positioning system (GPS)* that can provide location information to an accuracy of about 20 meters within a minute or two. The system uses 24 satellites that orbit the Earth every 12 hours, continuously broadcasting their position and a highly accurate time signal.

GEODISCOVERIES

To determine location, a receiver listens simultaneously to signals from four or more satellites. The receiver compares the time readings transmitted by each satellite with the receiver's own clock to determine how long it took for each signal to reach the receiver. Since the radio signal travels at a known rate of speed, the receiver can convert the travel time into the distance between the receiver and the satellite. Coupling the distance to each satellite with the position of the satellite in its orbit at the time of the transmission, the receiver calculates its position on the ground to within about 20 m (66 ft) horizontally and 30 m (98 ft) vertically.

The accuracy of the location is affected by several types of errors. First are small perturbations in the orbits of the satellites, caused by such unpredictable events as solar particle showers. These cause errors in the information about satellite position. Another source of error is small variations in the atomic clock that each satellite carries. A larger source of error, however, is the effect of the atmosphere on the radio waves of the satellite signal as they pass from the satellite to the receiver. The layer of charged particles at the outer edge of the atmosphere (ionosphere) and water vapor in the lowest atmospheric layer (troposphere) act to slow the radio waves. Since the conditions in these layers can change within a matter of minutes, the speed of the radio waves varies in an unpredictable way. Another transmission problem is that the radio waves may bounce off local obstructions and then reach the receiver, causing two slightly different signals to arrive at the receiver at the same time. This "multipath error" creates noise that confuses the receiver.

There is a way, however, to determine location within about 1 m (3.3 ft) horizontally and 2 m (6.6 ft) vertically. The method uses two GPS units, one at a base station and one that is mobile and used to determine the desired locations. The base station unit is placed at a position that is known with very high accuracy. By comparing its exact position with that calculated from each satellite signal, it determines the small deviations from orbit of each satellite, any small variations in each satellite's clock, and the exact speed of that satellite's radio signal through the atmosphere at that moment. It then broadcasts that information to the GPS field unit, where it is used to calculate the position more accurately.

Because this method compares two sets of signals, it is known as *differential GPS*.

Differential GPS is now in wide use for coastal navigation, where a few meters in position can make the difference between a shipping channel and a shoal. It is also required for the new generation of aircraft landing systems that will allow much safer instrument landings with equipment that is much lower in cost than existing systems.

As GPS technology has developed, costs have fallen exponentially. It is now possible to buy a small, hand-held GPS receiver for less than $100. Worldwide sales of GPS products and services reached $8 billion in 2000 and are expected to double by 2003. Besides plotting your progress on a computer-generated map as you drive your car or sail your boat, GPS technology can even keep track of your children at a theme park. With the coupling of wireless telephones and GPS, you may never have the chance to claim you're at a client meeting when the readout shows your employer that you're on the golf links!

A GPS satellite as it might look in orbit high above the Earth.

rotation in one direction and parallel to the axis of rotation in the other direction.

Parallels and Meridians

Imagine a point on the Earth's surface. As the Earth rotates, the point traces out a path in space, following an *arc*—that is, a curved line that forms a portion of a circle. With the completion of one rotation, the arc forms a full circle. This circle is known as a parallel of latitude, or a **parallel** (Figure 1.3*a*). Imagine cutting the globe much as you might slice an onion to produce onion rings—that is, perpendicular to the onion's main axis. Each cut creates a circle or parallel crosswise through the globe. The Earth's longest parallel of latitude lies midway between the two poles and is designated the **equator.** The equator is a fundamental reference line for measuring the position of points on the globe.

Imagine now slicing the Earth with a plane that passes through the axis of rotation instead of across it. This is the way you might cut up a lemon to produce wedges. Each cut outlines a circle on the globe passing through both poles. Half of this circular outline, connecting one pole to the other, is known as a meridian of longitude, or, more simply, a **meridian** (Figure 1.3*b*).

Meridians and parallels define geographic directions. Meridians are north-south lines, so if you walk north or south you are following a meridian. Parallels are east-west lines, and so if you walk east or west you are following a parallel. Any number of parallels and meridians are possible. Every point on the globe is associated with a unique combination of one parallel and one meridian. The position of the point is defined by their intersection. The total system of parallels and meridians forms a network of intersecting circles called the **geographic grid.**

Looking more closely at meridians and parallels, we see that they are made up of two types of circles—great and small (Figure 1.4). A **great circle** is created when a plane passing through the center of the Earth intersects the Earth's

> The geographic grid consists of an orderly system of circles—meridians and parallels—that are used to locate position on the globe. Parellels are east-west lines, while meridians are north-south lines.

surface. A great circle always has the Earth's center as its own center, and it bisects the globe into two equal halves. A **small circle** is created when a plane passing through the Earth, but not through the Earth's center, intersects the Earth's surface. By looking at Figure 1.3, you can easily see that meridians are actually halves of great circles, while all parallels, except the equator, are small circles. Great circles can be aligned in any direction on the globe. This means that if we choose two points on the globe, we can always find a great circle that passes through both points. As we will see shortly in our discussion of map projections, the portion of the great circle between the two points is the shortest distance between them.

Latitude and Longitude

To label parallels and meridians, we use a special system—latitude and longitude. Parallels are identified by latitude and meridians by longitude. **Latitude** is an indicator of how far north or south of the equator a parallel is situated. The latitude of a parallel is measured by the angle between a point on the parallel, the center of the Earth, and a point on the equator intersected by a meridian passing through the point of interest on the parallel (Figure 1.5*a*). Note that

1.4 *Great and small circles* *A great circle is created when a plane passes through the Earth, intersecting the Earth's center. Small circles are created when a plane passes through the Earth but does not intersect the center point.*

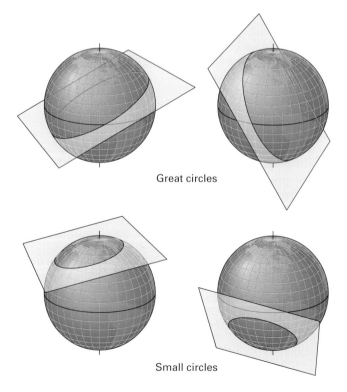

Great circles

Small circles

1.3 *Parallels and meridians* *(a) Parallels of latitude divide the globe crosswise into rings. (b) Meridians of longitude divide the globe from pole to pole.*

(a)   *(b)*

(a)

(b)

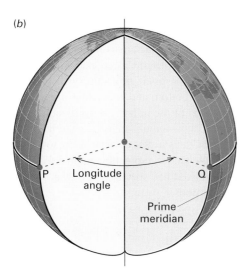

1.5 Latitude and longitude angles *(a) The latitude of a parallel is the angle between a point on the parallel (P) and a point on the equator at the same meridian (Q) as measured from the Earth's center. (b) The longitude of a meridian is the angle between a point on that meridian at the equator (P) and a point on the prime meridian at the equator (Q) as measured at the Earth's center.*

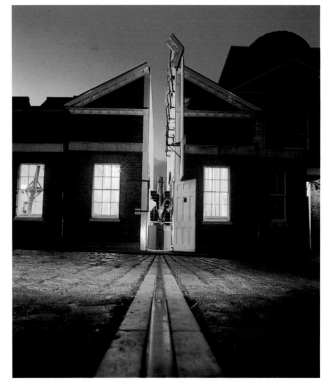

1.6 The prime meridian *This photograph, taken at dusk at the old Royal Observatory at Greenwich, England, shows the prime meridian, which has been marked as a stripe on the forecourt paving.*

the equator divides the globe into two equal portions, or *hemispheres.* All parallels north of the equator—that is, in the *northern hemisphere*—are designated as having north latitude, and all points south of the equator—in the *southern hemisphere*—are designated as having south latitude (N or S).

Longitude is a measure of the position of a meridian eastward or westward from a reference meridian, called the *prime meridian.* As Figure 1.5*b* shows, longitude is the angle, measured in degrees, between a plane passing through the meridian and a plane passing through the prime meridian. The prime meridian passes through the

> **Latitude and longitude uniquely determine the position of a point on the globe. Latitude records the parallel, and longitude the meridian, associated with the point.**

location of the old Royal Observatory at Greenwich, near London, England (Figure 1.6). For this reason it is also referred to as the Greenwich meridian. It has the value long. 0°. The longitude of a meridian on the globe is measured eastward or westward from the prime meridian, depending on which direction gives the smaller angle. Longitude thus ranges from 0° to 180°, east or west (E or W).

Figure 1.7 shows how the location of a point on the Earth's surface is determined by its latitude and longitude. The point *P* lies on the parallel of latitude at 50 degrees north, which we can abbreviate as lat. 50° N. It also lies on the meridian of longitude at 60 degrees west, which we can abbreviate as long. 60° W. When both the latitude and longitude of a point are known, it can be accurately and precisely located on the geographic grid.

When latitude or longitude angles are measured other than in full-degree increments, *minutes* and *seconds* can be used. A minute is 1/60 of a degree, and a second is 1/60 of a minute, or 1/3600 of a degree. Thus, the latitude 41°, 27 minutes ('), and 41 seconds (") north (lat. 41° 27' 41" N) means 41° north plus 27/60 of a degree plus 41/3600 of a degree. This cumbersome system has now largely been replaced by decimal notation. In this example, the latitude 41° 27' 41" N translates to 41.4614° N.

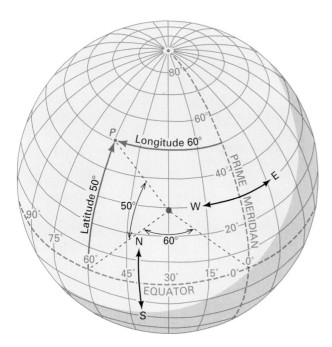

1.7   *Latitude and longitude of a point* *The point P lies on the parallel of latitude at 50° north (50° from the equator) and on the meridian at 60° west (60° from the prime meridian). Its location is therefore lat. 50° N, long. 60° W.*

Degrees of latitude and longitude can also be used as distance measures. A degree of latitude, which measures distance in a north-south direction, is equal to about 111 km (69 mi). The distance associated with a degree of longitude, however, will be progressively reduced with latitude because meridians converge toward the poles. For example, at 60° latitude, a degree of longitude has a length exactly half of that at the equator, or 55.5 km (34.5 mi).

The latitude and longitude of any point can now be determined quickly and accurately with the help of the *global positioning system (GPS)*—a system of satellites that constantly sends radio signals to Earth with information that allows a GPS receiver to calculate its position on the Earth's surface. *Focus on Remote Sensing 1.1* ● *The Global Positioning System* provides more information about GPS and how it works.

MAP PROJECTIONS

With a working understanding of the geographic grid, we can consider how to display the locations of continents, rivers, cities, islands, and other geographic features on maps. This will take us briefly into the realm of *cartography,* the art and science of making maps. The discussion in the main text will focus on a few simple types of maps that are used widely and found in our chapters. At the end of this chapter we provide a special supplement, *Geographer's Tools 1.2* ● *Focus on Maps,* with more information about how maps are made and how they display information.

As we observed earlier, the Earth's shape is nearly spherical, and so the Earth's surface is curved. However, maps are flat. It is impossible to copy a curved surface onto a flat surface without cutting, stretching, or otherwise distorting the curved surface in some way. So, making a map means devising an orderly way of changing the globe's geographic grid of curved parallels and meridians into a grid that lies flat. We refer to a system for changing the geographic grid to a flat grid as a **map projection.** *GEODISCOVERIES*

Associated with every map is a *scale fraction*—a ratio that relates distance on the map to distance on the Earth's surface. For example, a scale fraction of 1:50,000 means that one unit of map distance equals 50,000 units of distance on the Earth.

Because a curved surface cannot be projected onto a flat surface without some distortion, the scale fraction of a map holds only for one point or a single line on the map. Away from that point or line, the scale fraction will be different. However, this variation in scale is a problem only for maps that show large regions, such as continents or hemispheres.

We will concentrate on the three most useful map projections. The first is the polar projection, which pictures the globe from top or bottom and is essential for scientific uses like weather maps of the polar regions. Second is the Mercator projection, a navigator's map invented in 1569 by Gerhardus Mercator. It is a classic that has never gone out of style. Third is the Goode projection, named for its designer, Dr. J. Paul Goode. It has special qualities not found in the other two projections.

Polar Projection

The *polar projection* (Figure 1.8) can be centered on either the north or the south pole. Meridians are straight lines radiating outward from the pole, and parallels are nested circles centered on the pole. Spacing of the parallels increases outward from the center. The map is usually cut off to show only one hemisphere so that the equator forms the outer edge of the map. Because the intersections of the parallels with the meridians always form true right angles, this projection shows the true shapes of all small areas. That is, the shape of a small island would always be shown correctly, no matter where it appeared on the map. However, because the scale fraction increases in an outward direction, the island would look larger toward the edge of the map than near the center.

Mercator Projection

The *Mercator projection* (Figure 1.9) is a rectangular grid of meridians as straight vertical lines and parallels as straight horizontal lines. Meridians are evenly spaced, but the spacing between parallels increases at higher latitude so that the spacing at 60° is double that at the equator. Closer to the poles, the spacing increases even more, and the map must be cut off at some arbitrary parallel, such as 80° N. This change of scale enlarges features when they near the pole, as can easily be seen in Figure 1.9. There

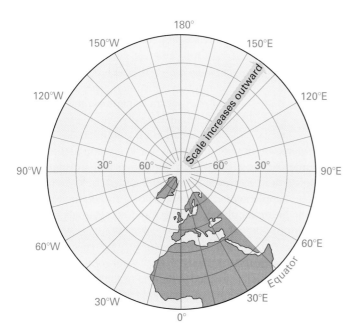

1.8 A polar projection *The map is centered on the north pole. All meridians are straight lines radiating from the center point, and all parallels are concentric circles. The scale fraction increases in an outward direction, making shapes toward the edges of the map appear larger.*

Greenland appears larger than Australia and is nearly the size of Africa! In fact, Greenland is very much smaller, as you can see on a globe.

The Mercator projection has several special properties. One is that a straight line drawn anywhere on the map is a line of constant compass direction. A navigator can therefore simply draw a line between any two points on the map and measure the *bearing,* or direction angle of the line, with respect to a nearby meridian on the map. Since the meridian is a true north-south line, the angle will give the compass bearing to be followed. Once aimed in that compass direction, a ship or an airplane can be held to the same compass bearing to reach the final point or destination.

This line will not necessarily follow the shortest actual distance between two points. As we noted earlier, the shortest path between two points on the globe is always a portion of a great circle. On a Mercator projection, a great circle line curves (except on the equator) and can falsely seem to represent a much longer distance than a compass line.

Because the Mercator projection shows the true compass direction of any straight line on the map, it is used to show many types of straight-line features. Among these features are flow lines of winds and ocean currents, directions of crustal features (such as chains of volcanoes), and lines of equal values, such as lines of equal air temperature or equal

1.9 The Mercator projection *The compass line connecting two locations, such as Portland and Cairo, shows the compass bearing of a course directly connecting them. However, the shortest distance between them lies on a great circle, which is a longer, curving line on this map projection. The diagram at the right shows how rapidly the map scale increases at higher latitudes. At lat. 60°, the scale is double the equatorial scale. At lat. 80°, the scale is six times greater than at the equator.*

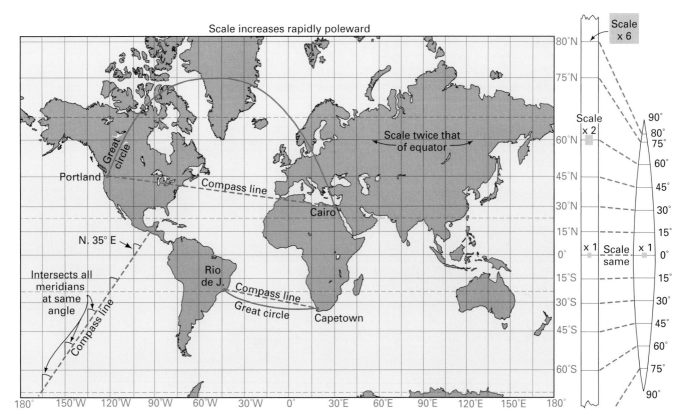

air pressure. This explains why the Mercator projection is chosen for maps of temperatures, winds, and pressures.

> The Mercator projection shows a line of constant compass bearing as a straight line and so is used to display directional features such as wind direction.

Goode Projection

The *Goode projection* (Figure 1.10) uses two sets of mathematical curves (sine curves and ellipses) to form its meridians. Between the 40th parallels, sine curves are used, and beyond the 40th parallel, toward the poles, ellipses are used. Since the ellipses converge to meet at the pole, the entire globe can be shown. The straight, horizontal parallels make it easy to scan across the map at any given level to compare regions most likely to be similar in climate.

The Goode projection has one very important property—it indicates the true sizes of areas of the Earth's surface. That is, if we drew a small circle on a sheet of clear plastic and moved it over all parts of the Goode world map, all the areas enclosed by the circle would have the same value in square kilometers or square miles. Because of this property, we use the Goode map to show geographical features that occupy surface areas. Examples of useful Goode projections include maps of the world's climates, soils, and vegetation.

The Goode map suffers from a serious defect, however. It distorts the shapes of areas, particularly in high latitudes and at the far right and left edges. To minimize this defect, Dr. Goode split his map apart into separate, smaller sectors, each centered on a

> The Goode projection displays the relative areas of land masses correctly but distorts their shape— especially near the poles.

different vertical meridian. These were then assembled at the equator. This type of split map is called an *interrupted projection*. Although the interrupted projection greatly reduces shape distortion, it separates parts of the Earth's surface that actually lie close together, particularly in the high latitudes.

Maps are in wide use today for many applications as a simple and efficient way of compiling and storing spatial information. However, in the past two decades, maps are being supplemented by more powerful computer-based methods for acquiring, storing, processing, analyzing, and outputting spatial data. These are contained within *geographic information systems (GISs)*. Our special supplement, *Geographer's Tools 1.3 • Geographic Information Systems*, which is located at the end of the chapter, presents some basic concepts of geographic information systems and how they work. GEODISCOVERIES

GLOBAL TIME

Maps and map projections are a practical application of the Earth's geographic grid. Another practical application, which involves both the grid and the Earth's rotation, is global time.

Our planet requires 24 hours for a full rotation with respect to the Sun. Put another way, humans long ago decided to divide the solar day into 24

1.10 **The Goode projection** *The meridians in this projection follow sine curves between lat. 40° N and lat. 40° S, and ellipses between lat. 40° and the poles. Although the shapes of continents are distorted, their relative surface areas are preserved. (Copyright © by the University of Chicago. Used by permission of the Committee on Geographical Studies, University of Chicago.)*

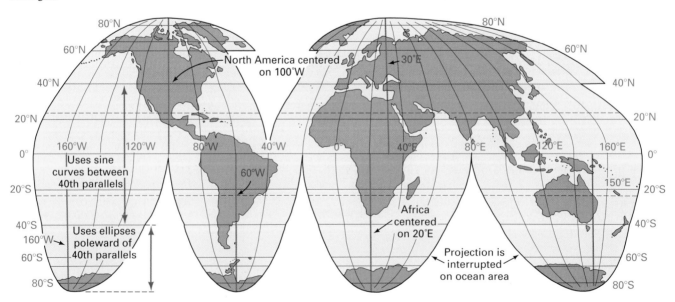

units, called hours, and devised clocks to keep track of hours in groups of 12. Yet, different regions set their clocks differently. For example, when it is 10:03 A.M. in New York, it is 9:03 A.M. in Chicago, 8:03 A.M. in Denver, and 7:03 A.M. in Los Angeles. Note that these times differ by exactly one hour. How did this system come about? How does it work?

Our global time system is oriented to the Sun. Think for a moment about how the Sun appears to move across the sky. In the morning, the Sun is low on the eastern horizon, and as the day progresses, it rises higher until at *solar noon* it reaches its highest point in the sky. If you check your watch at that moment, it will read a time somewhere near 12 o'clock (12:00 noon). After solar noon, the Sun's elevation in the sky decreases. By late afternoon, the Sun appears low in the sky, and at sunset it rests on the western horizon.

Imagine for a moment that you are in Chicago, the time is noon, and the Sun is at or near its highest point in the sky. Imagine further that you call a friend in New York and ask about the time there and the position of the Sun. You will receive a report that the time is 1:00 P.M. and that the Sun is already past solar noon, its highest point. Calling a friend in Vancouver, you hear that it is 10:00 A.M., there and that the Sun is still working its way up to its highest point. However, a friend in Mobile, Alabama, will tell you that the time in Mobile is the same as in Chicago and that the Sun is at about solar noon. How can we explain these different observations?

The difference in time between Chicago, New York, and Vancouver makes sense because solar noon can occur simultaneously only at locations with the same longitude. In other words, only one meridian can be directly under the Sun and experience solar noon at a given moment. Locations on meridians to the east of Chicago, like New York, already will have passed solar noon, and locations to the west of Chicago, like Vancouver, will not yet have reached solar noon. Since Mobile and Chicago have nearly the same longitude, they experience solar noon at approximately the same time.

Figure 1.11 indicates how time varies with longitude. In this figure, the inner disk shows a polar projection of the world, centered on the north pole. Meridians are straight lines (radii) ranging out from the pole. The outer ring indicates the time in hours. The figure shows the moment in time when the prime meridian is directly under the Sun—that is, the 0° meridian is directly on the 12:00 noon mark. This means that, at this instant, the Sun is at the highest point of its path in the sky in Greenwich, England. The alignment of meridians with hour numbers tells us the time in other locations around the globe. For example, the time in New York, which lies roughly on the 75° W meridian, is about 7:00 A.M. In Los Angeles, which lies roughly on the 120° W meridian, the time is about 4:00 A.M.

1.11 The relation of longitude to time *The outer ring shows the time at locations identified by longitude meridians on the central map. The diagram is set to show noon conditions in Greenwich, England—that is, on the prime meridian. Clock time is earlier to the west of Greenwich and later to the east.*

Notice that 15° of longitude equates to an hour of time. Since the Earth turns 360° in a 24-hour day, the rotation rate is 360° ÷ 24 = 15° per hour.

Standard Time

We've just seen that locations with different longitudes experience solar noon at different times. But consider what would happen if each town or city set its clocks to read 12:00 at its own local solar noon. All cities and towns on different meridians would have different local time systems. In these days of instantaneous global communication, chaos would soon result.

> In the standard time system, we keep global time according to nearby standard meridians that normally differ by one hour from each other.

The use of standard time simplifies the global timekeeping problem. In the **standard time system,** the globe is divided into 24 **time zones.** All inhabitants within a zone keep time according to a *standard meridian* that passes through their zone. Since the standard meridians are usually 15 degrees apart, the difference in time between adjacent zones is normally one hour. In some geographic regions, however, the difference is only one-half hour.

Seven time zones cover the United States and its Caribbean possessions. Six zones cover Canada. Their names and standard meridians of longitude are as follows:

U.S. Zones	Meridian	Canadian Zones
	52¹/₂°	Newfoundland
Atlantic	60°	Atlantic
Eastern	75°	Eastern
Central	90°	Central
Mountain	105°	Mountain
Pacific	120°	Pacific-Yukon
Alaska-Bering	135°	
Hawaii	150°	

If carried out strictly, the standard time system would consist of belts exactly 15°, extending to meridians 7¹/₂° east and west of each standard meridian. However, this system could be inconvenient since the boundary meridians could divide a state, county, or city into two different time zones. As a result, time zone boundaries are often routed to follow agreed-upon natural or political boundaries.

Figure 1.12 presents a map of time zones for the contiguous United States and southern Canada. From this map, you can see that most time zone boundaries are conveniently located along an already existing and widely recognized line. For example, the eastern time–central time boundary line follows Lake Michigan down its center, and the mountain time–Pacific time boundary follows a ridge-crest line also used by the Idaho–Montana state boundary.

World Time Zones

Figure 1.13 shows the 24 principal standard time zones of the world. In the figure, 15° meridians are dashed lines, while the 7¹/₂° meridians, which form many of the boundaries between zones, are bold lines. The figure also shows the time of day in each zone when it is noon at the Greenwich meridian. The country spanning the greatest number of time zones from east to west is Russia, with 11 zones, but these are grouped into eight standard time zones. China spans five time zones but runs on a single national time using the standard meridian of Beijing.

A few countries keep time by a meridian that is midway between standard meridians, so that their clocks depart from those of their neighbors by 30 or 90 minutes. India and Iran are examples. The Canadian province of Newfoundland and the interior Australian states of South Australia and Northern Territory are examples of regions within countries that keep time by 7¹/₂° meridians.

World time zones are numbered to indicate the number of hours difference between time in a zone and time in Greenwich. A number of +7, for example, indicates that adding seven hours to local time will give Greenwich time, while a –3 indicates that subtracting three hours from local time will give Greenwich time.

1.12 Time zones of the contiguous United States and southern Canada *The name, standard meridian, and number code are shown for each time zone. Note that time zone boundaries often follow preexisting natural or political boundaries.*

1.13 Time zones of the world *Dashed lines represent 15° meridians, and bold lines represent 7¹/₂° meridians. Alternate zones appear in color. (U.S. Navy Oceanographic Office.)*

International Date Line

When we take a world map or globe with 15° meridians and count them in an eastward direction, starting with the Greenwich meridian as 0, we find that the 180th meridian is number 12 and that the time at this meridian is therefore 12 hours later than Greenwich time. Counting in a similar manner westward from the Greenwich meridian, we find that the 180th meridian is again number 12 but that the time is 12 hours earlier than Greenwich time. How can the same meridian be both 12 hours ahead of Greenwich time and 12 hours behind? This paradox is explained by the fact that different days are observed on either side of this meridian.

Imagine that you are on the 180th meridian on June 26th and it is exactly midnight. Let's stop the world for a moment and examine the situation. On the 180th meridian at the exact instant of midnight, the same 24-hour calendar day covers the entire globe. Stepping east will place you in the very early morning of June 26, while stepping west will place you very late in the evening of June 26. You are in the same calendar day on both sides of the meridian but 24 hours apart in time.

Doing the same experiment an hour later, at 1:00 A.M., stepping east you will find that you are in the early morn-

ing of June 26. But if you step west you will find that midnight of June 26 has passed, and it is now the early morning of June 27. So on the west side of the 180th meridian, it is also 1:00 A.M. but one day later than on the east side. For this reason, the 180th meridian serves as the *International Date Line.* This means that if you travel westward across the date line, you must advance your calendar by one day. If traveling eastward, you set your calendar back by a day.

Air travelers between North America and Asia cross the date line. For example, flying westward from Los Angeles to Sydney, Australia, you may depart on a Tuesday evening and arrive on a Thursday morning after a flight that lasts only 14 hours. On an eastward flight from Tokyo to San Francisco, you may actually arrive the day before you take off, taking the date change into account!

Actually, the International Date Line does not follow the 180th meridian exactly. Like many time zone boundaries, it deviates from the meridian for practical reasons. As shown in Figure 1.13, it has a zigzag offset between Asia and North America, as well as an eastward offset in the South Pacific to keep clear of New Zealand and several island groups.

Daylight Saving Time

Especially in urban areas, many human activities being well after sunrise and continue long after sunset. Therefore, we adjust our clocks during the part of the year that has a longer daylight period to correspond more closely with the modern pace of society. This adjusted time system, called *daylight saving time,* is obtained by setting all clocks ahead by one hour. The effect of the time change is to transfer the early morning daylight period, theoretically wasted while schools, offices, and factories are closed, to the early evening, when most people are awake and busy. Daylight saving time also yields a considerable savings in power used for electric lights. In the United States, daylight saving time comes into effect on the first Sunday in April and is discontinued on the last Sunday of October.

Precise Timekeeping

Many scientific and technological applications require precise timekeeping. Today, a worldwide system of master atomic clocks measures time to better than one part in 1,000,000,000,000. However, our Earth is a much less precise timekeeper, demonstrating small changes in the angular velocity of its rotation on its axis and variations in the time it takes to complete one circuit around the Sun. As a result, constant adjustments to the timekeeping system are necessary. The legal time standard recognized by all nations is Coordinated Universal Time, which is administered by the Bureau International de l'Heure, located near Paris.

THE EARTH'S REVOLUTION AROUND THE SUN

So far, we have discussed the importance of the Earth's rotation on its axis. Another important motion of the Earth is its **revolution,** or its movement in orbit around the Sun. GEODISCOVERIES

The Earth completes a revolution around the Sun in 365.242 days—about one-fourth day more than the calendar year of 365 days. Every four years, the extra one-fourth days add up to about one whole day. By inserting a 29th day in February in leap years, we largely correct the calendar for this effect. Further minor corrections are necessary to perfect the system.

The Earth's orbit around the Sun is shaped like an ellipse, or oval. This means that the distance between the Earth and Sun varies somewhat through the year. The Earth is nearest to the Sun at *perihelion,* which occurs on or about January 3. It is farthest away from the Sun at *aphelion,* on or about July 4. However, the distance between Sun and Earth varies only by about 3 percent during one revolution because the elliptical orbit is shaped very much like a circle. For most purposes we can regard the orbit as circular.

> The axis of the Earth's rotation is tilted by 23^1/$_2$° away from the plane of the ecliptic. This tilt causes the seasons.

1.14 Revolution of the Earth and Moon *Viewed from a point over the Earth's north pole, the Earth both rotates and revolves in a counterclockwise direction. From this viewpoint, the Moon also rotates counterclockwise.*

In which direction does the Earth revolve? Imagine yourself in space, looking down on the north pole of the Earth. From this viewpoint, the Earth travels counterclockwise around the Sun (Figure 1.14). This is the same direction as the Earth's rotation.

The Moon rotates on its axis and revolves about the Sun in the same direction as the Earth. However, the Moon's rate of rotation is such that one side of the Moon is always directed toward the Earth. Thus, astronomers had never seen the far side of the Moon until a Soviet spacecraft passing the Moon transmitted photos back to Earth in 1959. The phases of the Moon are determined by the position of the Moon in its orbit around the Earth, which in turn determines how much of the sunlit Moon is seen from the Earth. In Figure 1.14, the Moon is behind the Earth and we see the Moon as full. In Figure 1.15, the Moon is nearly full. From the way that the Sun illuminates the Moon as a sphere, it is easy to see that the Sun is to the right.

Tilt of the Earth's Axis

The seasons we experience on Earth are related to the orientation of the Earth's axis of rotation as it revolves around the Sun. We usually describe this situation by stating that the Earth's axis is tilted with respect to the *plane of the ecliptic*—the plane containing the Earth's orbit around the Sun. Figure 1.16 shows the plane of the ecliptic as it intersects the Earth, and Figure 1.17 shows the full Earth orbit traced on the plane of the ecliptic.

Now consider the angle of the axis of the Earth's rotation as shown in Figure 1.16. Note that the axis is tilted at an angle of 23^1/$_2$° away from a right angle to the plane of the ecliptic. That is, the angle between the axis and the plane of the ecliptic is 66^1/$_2$°, not 90°. In addition, the direction toward which the axis points is fixed in space—it aims toward Polaris, the north star. The direction of the axis does not change as the Earth revolves around the Sun. Let's investigate this phenomenon in more detail.

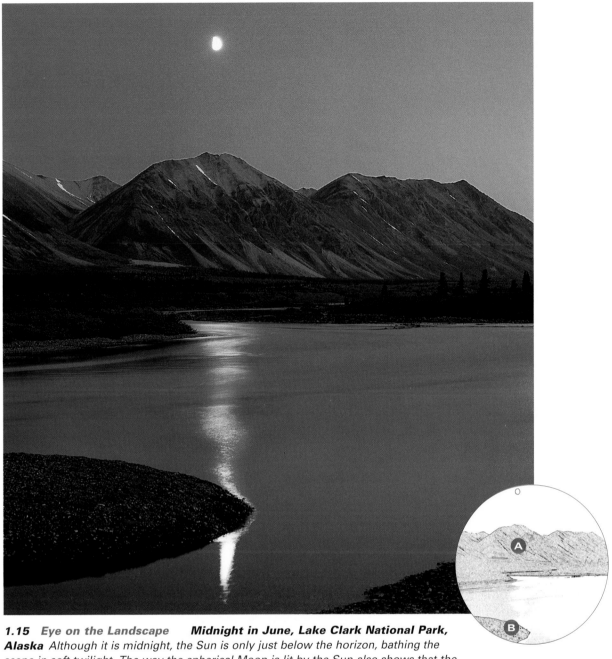

1.15 Eye on the Landscape **Midnight in June, Lake Clark National Park, Alaska** *Although it is midnight, the Sun is only just below the horizon, bathing the scene in soft twilight. The way the spherical Moon is lit by the Sun also shows that the Sun is located below the horizon and to the right.* **What else would the geographer see? ... Answers at the end of the chapter.**

Solstice and Equinox

Figure 1.17 diagrams the Earth as it revolves in its orbit through the four seasons. Because the direction of the Earth's axis of rotation is fixed, the north pole is tilted away from the Sun during one part of the year and is tilted toward the Sun during the other part. Consider first the

---

At the June solstice, the north pole is tilted toward the Sun. At the December solstice, it is tilted away from the Sun.

---

event on December 22, which is pictured on the far right. On this day, the Earth is positioned so that the north polar end of its axis leans at the maximum angle away from the Sun, $23\frac{1}{2}°$. This event is called the **winter solstice** in the northern hemisphere. (While it is winter in the northern hemisphere, it is summer in the southern hemisphere, so

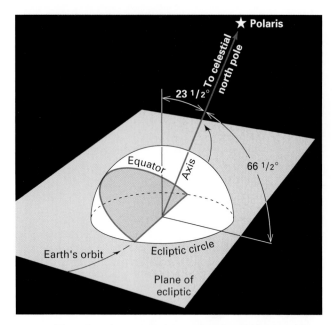

1.16 **The tilt of the Earth's axis of rotation with respect to its orbital plane** *As the Earth moves in its orbit on the plane of the ecliptic around the Sun, its rotational axis remains pointed toward Polaris, the north star, and makes an angle of 66¹/₂° with the ecliptic plane.*

you can use the term *December solstice* to avoid any confusion.) At this time, the southern hemisphere is tilted toward the Sun and enjoys strong solar heating. **GEODISCOVERIES**

Six months later, on June 21, the Earth is on the opposite side of its orbit in an equivalent position. At this event, known as the **summer solstice** *(June solstice)* in the northern hemisphere, the north polar end of the axis is tilted at 23¹/₂° toward the Sun. Thus, the north pole and northern hemisphere are tilted toward the Sun, while the south pole and southern hemisphere are tilted away.

Midway between the solstice dates, the equinoxes occur. At an **equinox,** the Earth's axial tilt is neither toward nor away from the Sun. The axis makes a right angle with a line drawn to the Sun, and neither the north nor south pole is tilted toward the Sun. The *vernal equinox* occurs on March 21, and the *autumnal equinox* occurs on September 23. Conditions are identical on the two equinoxes as far as Earth–Sun relationships are concerned. We should also note that the date of any solstice or equinox in a particular year may vary by a day or so, since the revolution period is not exactly 365 days. Let's look at equinoxes and solstices in more detail.

1.17 **The four seasons** *The four seasons occur because the Earth's tilted axis keeps a constant orientation in space as the Earth revolves about the Sun. This tips the northern hemisphere toward the Sun for the summer solstice and away from the Sun for the winter solstice. Both hemispheres are illuminated equally at the spring equinox and the fall equinox.*

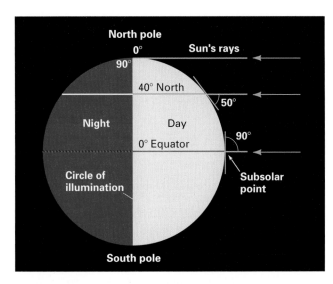

1.18 Equinox conditions *At this time, the Earth's axis of rotation is exactly at right angles to the direction of solar illumination. The subsolar point lies on the equator. At both poles, the Sun is seen at the horizon. Note that the viewpoint for this diagram is away from the plane of the ecliptic, as shown in Figure 1.17, so that both poles may be seen.*

Equinox Conditions

The conditions at an equinox, shown in Figure 1.18, form the simplest case. The figure illustrates two important concepts of global illumination that we use for describing equinoxes and solstices. The first concept is the *circle of illumination*. Note that the Earth is always divided into two hemispheres with respect to the Sun's rays. One hemisphere (day) is lit by the Sun, and the other (night) lies in the darkness of the Earth's shadow. The circle of illumination is the circle that separates the day hemisphere from the night hemisphere. The second concept is the *subsolar point*,

the single point on the Earth's surface where the Sun is directly overhead at a particular moment.

At equinox, the circle of illumination passes through the north and south poles, as we see in Figure 1.18. The Sun's rays graze the surface at either pole, and the surface there receives little or no solar energy. The subsolar point falls on the equator. Here, the angle between the Sun's rays and the Earth's surface is 90°, and the solar illumination is received in full force. At an intermediate latitude, such as 40° N, the rays of the Sun at noon strike the surface at a lesser angle. This *noon angle*—the elevation of the Sun above the horizon at noon—can be easily determined. Some simple geometry shows that the noon angle is equal to 90° minus the latitude, or 50° in this example, for equinox conditions.

Imagine yourself at a point on the Earth, say, at a latitude of 40° N. Visualize the Earth rotating from left to right, so that you turn with the globe, completing a full circuit in 24 hours. At the equinox, you spend 12 hours in darkness and 12 hours in sunlight. This is because the circle of illumination passes through the poles, dividing every parallel exactly in two. Thus, one important feature of the equinox is that day and night are of equal length everywhere on the globe.

Solstice Conditions

Now examine the solstice conditions shown in Figure 1.19. Summer solstice is on the left. Consider yourself back at a point on the lat. 40° N parallel. The circle of illumination does not divide your parallel in equal halves because of the tilt of the northern hemisphere toward the Sun. Instead, the larger part is in daylight. For you, the day is now considerably longer (about 15 hours) than the night (about 9 hours).

The farther north you go, the more the effect increases. In fact, the entire area of the globe north of lat. 66½° is on the daylight side of the circle of illumination. This parallel is known as the **arctic circle.** Even though the Earth

> **The Sun's declination describes the latitude of the subsolar point as it ranges from 23½° S (December solstice) to 23½° N (June solstice) throughout the year.**

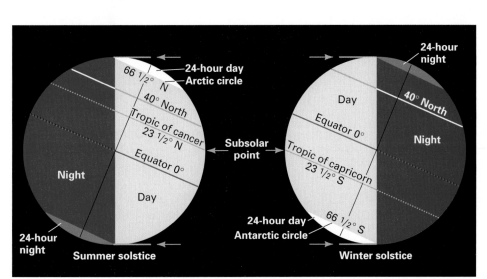

1.19 Solstice conditions *At the solstice, the north end of the Earth's axis of rotation is fully tilted either toward or away from the Sun. Because of the tilt, polar regions experience either 24-hour day or 24-hour night. The subsolar point lies on one of the tropics, at lat. 23½° N or S.*

rotates through a full cycle during a 24-hour period, the area north of the arctic circle remains in continuous daylight. We also see that the subsolar point is at a latitude of $23\frac{1}{2}°$ N. This parallel is known as the **tropic of cancer.** Because the Sun is directly over the tropic of cancer at this solstice, solar energy is most intense here.

At the winter solstice, conditions are exactly reversed from those of the summer solstice. If you imagine yourself back at lat. 40° N, you find that the night is now about 15 hours long while daylight lasts about 9 hours. All the area south of lat. $66\frac{1}{2}°$ S lies under the Sun's rays, inundated with 24 hours of daylight. This parallel is known as the **antarctic circle.** The subsolar point has shifted to a point on the parallel at lat. $23\frac{1}{2}°$ S, known as the **tropic of capricorn.**

The solstices and equinoxes represent conditions at only four times of the year. Between these times, the latitude of the subsolar point travels northward and southward in an annual cycle between the tropics of cancer and capricorn. We refer to the latitude of the subsolar point as the Sun's *declination.* Figure 1.20 plots the Sun's declination through the year. As the year progresses, the declination varies between $23\frac{1}{2}°$ S lat. at the December solstice and $23\frac{1}{2}°$ N lat. at the June solstice.

As the seasonal cycle progresses in polar regions, areas of 24-hour daylight or 24-hour night shrink and then grow. At other latitudes, the length of daylight changes slightly from one day to the next, except at the equator. In this way, the Earth experiences the rhythm of the seasons as it continues its revolution around the Sun.

A LOOK AHEAD

The flow of solar energy to the Earth powers most of the natural processes that we experience every day, from changes in the weather to the work of streams in carving the landscape. Our next chapter examines in detail solar energy and its interaction with the Earth's atmosphere and surface. The solar radiation intercepted by the Earth constitutes the biggest and most important energy flow system for us, the land dwellers of the Earth. What have we learned about the Sun–Earth energy

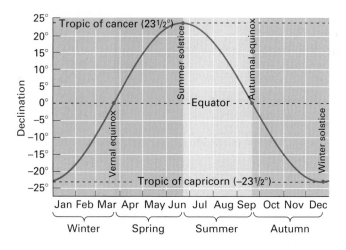

1.20 The Sun's declination through the year *The latitude of the subsolar point marks the Sun's declination, which changes slowly through the year from $-23\frac{1}{2}°$ to $+23\frac{1}{2}°$ to $-23\frac{1}{2}°$. Solstices and seasons are labeled for the northern hemisphere.*

flow system thus far? Here are two simple, important facts we can glean from this chapter:

- *Half of the Earth is always receiving solar energy.* So, there is a constant flow of light and heat from Sun to Earth.
- *However, the solar energy flow is not received uniformly over the surface.* This is because of the Earth's spherical shape, its constant rotation on its axis, the tilt of its axis, and its constant revolution around the Sun.

As you will see in Chapter 2, the amount of incoming solar energy at a location depends on the length of the daylight portion of the day and on the height of the Sun's path in the sky during the daylight period. Both of these factors depend on the latitude and season of the location. So, your understanding of the Earth's rotation and revolution will come in quite handy in the chapter to come.

Geographer's Tools | 1.2

Focus on Maps

Maps play an essential role in the study of physical geography because much of the information content of geography is stored and displayed on maps. Map literacy—the ability to read and understand what a map shows—is a basic requirement for day-to-day functioning in our society. Maps appear in almost every issue of a newspaper and in nearly every TV newscast. Most people routinely use highway maps and street maps. The purpose of this special supplement is to provide additional information on the subject of cartography, the science of maps and their construction.

More about Map Projections

Recall from the chapter text that a *map projection* is an orderly system of parallels and meridians used as a base on which to draw a map on a flat surface. As noted earlier, a projection is needed because the Earth's surface is not flat but, rather, curved in a shape that is very close to the surface of a sphere. All map projections misstate the shape of the Earth in some way. It's simply impossible to transform a spherical surface to a flat (planar) surface without violating the true surface as a result of cutting, stretching, or otherwise distorting the information that lies on the sphere. GEODISCOVERIES

Perhaps the simplest of all map projections is a grid of perfect squares. In this simple map, horizontal lines are parallels, and vertical lines are meridians. They are equally spaced in degrees. This grid is often used in modern computer-generated world maps displaying data that consist of a single number for each square, representing, for example, 1, 5, or 10 degrees of latitude and longitude. A grid of this kind can show the true spacing (approximately) of the parallels, but it fails to show how the meridians converge toward the two poles. This grid fails dismally in high latitudes, and the map usually has to be terminated at about 70° to 80° north and south.

Early attempts to find satisfactory map projections made use of a simple concept. Imagine the spherical Earth grid as a cage of wires (a kind of bird cage). A tiny light source is placed at the center of the sphere, and the image of the wire grid is cast upon a surface outside the sphere. This situation is like a reading lamp with a lamp-

S1.1 Simple ways to generate map projections *Rays from a central light source cast shadows of the spherical geographic grid on target screens. The conical and cylindrical screens can be unrolled to become flat maps. (A. N. Strahler.)*

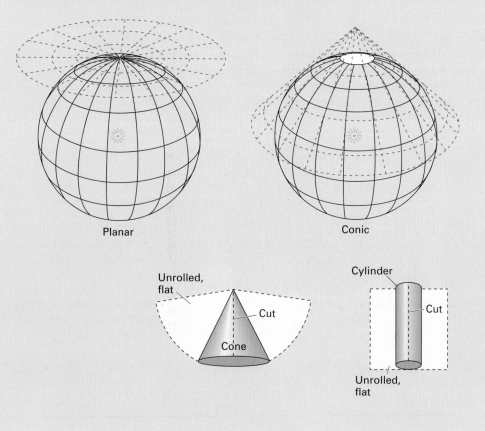

Planar

Conic

Unrolled, flat — Cut — Cone

Cylinder — Cut — Unrolled, flat

Cylindric

shade. Basically, three kinds of "lampshades" can be used, as shown in Figure S1.1.

First is a flat paper disk balanced on the north pole. The shadow of the wire grid on this plane surface will appear as a combination of concentric circles (parallels) and radial straight lines (meridians). Here we have a polar-centered projection. Second is a cone of paper resting point-up on the wire grid. The cone can be slit down the side, unrolled, and laid flat to produce a map that is some part of a full circle. This is called a conic projection. Parallels are arcs of circles, and meridians are radiating straight lines. Third, a cylinder of paper can be wrapped around the wire sphere so as to be touching all around the equator. When slit down the side along a meridian, the cylinder can be unrolled to produce a cylindrical projection, which is a true rectangular grid.

Take note that none of these three projection methods can show the entire Earth grid, no matter how large a sheet of paper is used to receive the image. Obviously, if the entire Earth grid is to be shown, some quite different system must be devised. Many such alternative solutions have been proposed. In our main chapter text, we described three types of projections used throughout the book— the the *polar projection;* the *Mercator projection,* which is a cylindrical projection; and the *Goode projection,* which uses a special mathematical principle.

Scales of Globes and Maps

All globes and maps depict the Earth's features in much smaller size than the true features they represent. Globes are intended in principle to be perfect scale models of the Earth itself, differing from the Earth only in size. The *scale* of a globe is the ratio between the size of the globe and the size of the Earth, where "size" is some measure of length or distance (but not of area or volume).

Take, for example, a globe 20 cm (about 8 in.) in diameter, representing the Earth, with a diameter of about 13,000 km. The scale of the globe is the ratio between 20 cm and 13,000 km. Dividing 13,000 by 20, we see that one centimeter on the globe represents 650 kilometers on the Earth. This relationship holds true for distances between any two points on the globe.

Scale is more usefully stated as a simple fraction, termed the *scale fraction.* It can be obtained by reducing both Earth and globe distances to the same unit of measure, which in this case is centimeters. (There are 100,000 centimeters in one kilometer.) The advantage of the scale fraction is that it is entirely free of any specified units of measure, such as the foot, mile, meter, or kilometer. It is usually written as a fraction with a numerator of one using either a colon or with the numerator above the denominator. For the example shown above, the scale fraction is obtained by reducing 20/1300000000 to 1/65000000 or 1:65,000,000.

Being a true-scale model of the Earth, a globe has a constant scale everywhere on its surface, but this is not true of a map projection drawn on a flat surface. In flattening the curved surface of the sphere to conform to a plane surface, all map projections stretch the Earth's surface in a nonuniform manner, so that the map scale changes from place to place. So we can't say about any world map: "Everywhere on this map the scale is 1:65,000,000." It is, however, possible to select a meridian or parallel—the equator, for example—for which a scale fraction can be given, relating the map to the globe it represents. In Figure 1.9, the scale of the Mercator projection along its equator is about 1:325,000,000.

Small-Scale and Large-Scale Maps

When geographers refer to small-scale and large-scale maps, they refer to the value of the scale fraction. For example, a global map at a scale of 1:65,000,000 has a scale fraction value of 0.00000001534, which is obtained by dividing 1 by 65,000,000. A hiker's topographic map might have a scale of 1:25,000, for a scale value of 0.000040. Since the global scale value is smaller, it is a small-scale map, while the hiker's map is a large-scale map. Note that this contrasts with many persons' use of the terms *large-scale* and *small-scale.* When we refer in conversation to a large-scale phenomenon or effect, we typically refer to something that takes place over a large area and that is usually best presented on a small-scale map.

Maps of large scale show only small sections of the Earth's surface. Because they "zoom in," they are capable of carrying the enormous amount of geographic information that is available and that must be shown in a convenient and effective manner. Most large-scale maps carry a *graphic scale,* which is a line marked off into units representing kilometers or miles. Figure S1.2 shows a portion of a large-scale map on which sample graphic scales in miles, feet, and kilometers are superimposed. Graphic scales make it easy to measure ground distances.

For practical reasons, maps are printed on sheets of paper usually less than a meter (3 ft) wide, as in the case of the ordinary highway map or navigation chart. Bound books of maps—atlases, that is— consist of pages usually no larger than 30 by 40 cm (about 12 by 16 in.), whereas maps found in textbooks and scientific journals are even smaller.

Informational Content of Maps

The information conveyed by a map projection grid system is limited to one category only: absolute location of points on the Earth's surface. To be more useful, maps also carry other types of information. Figure S1.2 is a portion of a large-scale *multipurpose map.* Map sheets published by national governments, such as this one, are usually multipurpose maps. Using a great variety of symbols, patterns, and colors, these maps carry a high information content. A

S1.2 Graphic scales on a topographic map *A portion of a modern, large-scale topographic map for which three graphic scales have been provided. (U.S. Geological Survey.)*

larger example of a multipurpose map is reproduced on the last pages of this book. It is a portion of a U.S. Geological Survey topographic quadrangle map for San Rafael, California.

In contrast to the multipurpose map is the *thematic map,* which shows only one type of information, or *theme.* We use many thematic maps in this text. Some examples include Figure 4.23, mapping the frequency of severe hailstorms in the United States; Figure 5.17, atmospheric surface pressures; Figure 7.2, mean annual precipitation of the world; and Figure 7.6 world climates.

Map Symbols

Symbols on maps associate information with points, lines, and areas. To express points as symbols, they can be renamed "dots." Broadly defined, a dot can be any small device to show point location. It might be a closed circle, open circle, letter, numeral, or a little picture of the object it represents (see "church with tower" in Figure S1.3). A line can vary in width and can be single or double. The line can also consist of a string of dots or dashes. A specific area of surface can be referred to as a "patch." The patch can be shown simply by a line

marking its edge, or it can be depicted by a distinctive pattern or a solid color. Patterns are highly varied. Some consist of tiny dots and others of parallel, intersecting, or wavy lines.

A map consisting of dots, lines, and patches can carry a great deal of information, as Figure S1.3 shows. In this case, the map uses two kinds of dot symbols (both symbolic of churches), three kinds

S1.3 Multipurpose map *A multipurpose map of an imaginary area with 10 villages illustrating the use of dots, lines, and patches. (After J. P. Cole and C.A.M. King, Quantitative Geography, copyright © John Wiley & Sons, London. Used by permission.)*

- - - - Parish	▨ Woodland	⛪ Church
—— Contour	▧ Water Meadow	⛪ Church w/ tower
══ Main Road	▩ Village	

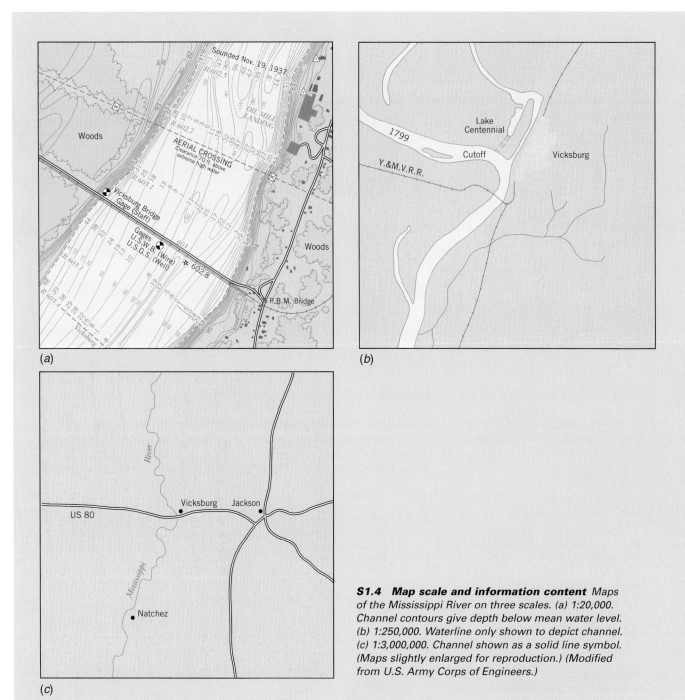

(a)

(b)

(c)

S1.4 Map scale and information content *Maps of the Mississippi River on three scales. (a) 1:20,000. Channel contours give depth below mean water level. (b) 1:250,000. Waterline only shown to depict channel. (c) 1:3,000,000. Channel shown as a solid line symbol. (Maps slightly enlarged for reproduction.) (Modified from U.S. Army Corps of Engineers.)*

of line symbols, and three kinds of patch symbols. Altogether, eight types of information are offered. Line symbols freely cross patches, and dots can appear within patches. Two different kinds of patches can overlap. For more examples of symbols, consult the display of topographic map symbols facing the map of San Rafael on the last pages of this book.

The relation of map symbols to map scale is of prime importance in cartography. Maps of very large scale, along with architectural and engineering plans, can show objects to their true outline form. As map scale is decreased, representation becomes more and more generalized. In physical geography, an excellent example is the depiction of a river, such as the lower

Mississippi. Figure S1.4 shows the river channel at three scales, starting with a detailed plan, progressing to a double-line symbol that generalizes the channel form, and ending with a single-line symbol. As generalization develops, the details of the river banks and channel bends are simplified as well. The level of depiction of fine details is described by the term *resolution.* Maps of large scale have much greater resolving power than maps of small scale.

Presenting Numerical Data on Thematic Maps

How are numerical data shown on maps? In physical geography, much of the information collected about particular areas is in the form of numbers. The numbers might represent readings taken from a scientific instrument at various places throughout the study area. A simple example is the collection of weather data, such as air temperature, air pressure, wind speed, and amount of rainfall. Another example is the set of measurements of the elevation of a land surface, given in meters above sea level. Another category of information consists merely of the presence or absence of a quantity or attribute. In such cases, we can simply place a dot to mean "present," so that when entries are completed, we have before us a field of scattered dots (Figure S1.5).

In some scientific programs, measurements are taken uniformly, for example, at the centers of grid squares laid over a map. For many classes of data, however, the locations of the observation points are predetermined by a fixed and nonuniform set of observing stations. For example, data of weather and climate are collected at stations typically located at airports. Whatever the sampling method used, we end up with an array of numbers and dots indicating their location on the field of the base map.

Although the numbers and locations may be accurate, it may be difficult to see the spatial pattern present in the data being displayed. For this reason, cartographers often simplify arrays of point values into isopleth maps. An *isopleth* is a line of equal value (from the Greek *isos,* "equal," and *plethos,* "fullness" or "quantity"). Figure 3.19 shows how an isopleth map is constructed for temperature data. In this case, the isopleth is an

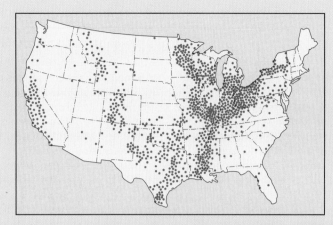

S1.5   Dot map *A dot map showing the distribution of soils of the order Alfisols in the United States. (From P. Gersmehl, Annals of the Assoc. of Amer. Geographers, vol. 67. Copyright © Association of American Geographers. Used by permission.)*

isotherm, or line of constant temperature. In drawing an isopleth, the line is routed among the points in a way that best indicates a uniform value, given the observations at hand.

Isopleth maps are important in various branches of physical geography. Table S1.1 gives a partial list of isopleths of various kinds used in the Earth sciences, together with their special names and the kinds of information they display. Examples are cited from our text. A special kind of isopleth, the *topographic contour* (or *isohypse*), is shown on the maps in Figures S1.2, S1.4*(a),* and in the portion of the San Rafael topographic map at

the close of the book. Topographic contours show the configuration of land surface features, such as hills, valleys, and basins.

In contrast to the isopleth map is the *choropleth* map, which identifies information in categories. Our global maps of vegetation (Figure 9.3) and soils (Figure 10.13) are examples of thematic choropleth maps.

Cartography is a rich and varied field of geography with a long history of conveying geographic information accurately and efficiently. If you are interested in maps and map-making, you might want to investigate cartography further.

Table S1.1	Examples of Isopleths		
Name of Isopleth	Greek Root	Property Described	Examples in Figures
Isobar	*barros,* weight	Barometric pressure	5.17
Isotherm	*therme,* heat	Temperature of air, water, or soil	3.20
Isotach	*tachos,* swift	Fluid velocity	5.26
Isohyet	*hyetos,* rain	Precipitation	4.17
Isohypse (topographic contour)	*hypso,* height	Elevation	19.9, 19.10

Geographic Information Systems

A primary activity of geographers is the acquisition, processing, and display of *spatial data*—pieces of information that are in some way associated with a specific location or area of the Earth's surface. Recent advances in computing capability have enabled geographers to develop a powerful new tool to acquire, analyze, and display large volumes of spatial data—the *geographic information system (GIS)*. Geographic information systems have allowed geographers, geologists, geophysicists, ecologists, planners, landscape architects, and others to develop applications of spatial data processing ranging from planning land subdivisions on the fringes of suburbia to monitoring the deforestation of the Amazon Basin. This special supplement provides a brief introduction to some of the key concepts of geographic information systems. *GEODISCOVERIES*

What Is a GIS?

In essence, a geographic information system is a system for acquiring, processing, storing, querying, creating, analyzing, and displaying spatial data. A simple example of a GIS is a map overlay system. Imagine that a planner is deciding how to divide a tract of land into building lots. Appropriate inputs would include a topographic map, giving contour lines of equal elevation; a vegetation map, showing the type of existing vegetation cover; a map of existing roads and trails; a map of streams and watercourses; a map of wetlands areas; a map of utility corridors crossing the area (power transmission lines and gas pipelines); and so forth. From these, the planner would locate the positions of roads and bridges; identify steep slopes, utility corridors, wetlands,

and regions near streams as conservation areas; and then lay out individual lots having a minimum area of land suitable for building, access to roads and utilities, and other necessary attributes.

To do this, the planner needs to be able to overlay the various maps. Normally this would be done by tracing them onto clear layers, then stacking the layers as needed while drawing the subdivision plans on the top layer. However, some problems may be encountered in this process. One common problem is that the maps may not be drawn to the same scale or compiled on the same map projection. Another is that sometimes some maps need further work, such as, for example, to identify areas of steep slopes given a map of topographic contours. Modern, computer-based geographic information systems are designed to solve these problems and allow ready manipulation of spatial data for diverse applications.

Spatial Objects in Geographic Information Systems

Geographic information systems are designed to manipulate spatial objects. A *spatial object* is a geo-

graphic area to which some information is attached. This information may be as simple as a place name or as complicated as a large data table with many types of information. A spatial object will normally have a boundary, which is described by a *polygon* that outlines the object. Some spatial objects are illustrated in Figure S1.6.

Spatial objects also include points and lines. A *point* can be thought of as a special type of spatial object with no area. A *line* is also a spatial object with no area, but it has two points associated with it, one for each end of the line. These special points are often referred to as *nodes*. Normally a line is straight, but it can also be defined as a smooth curve having a certain shape. If the two nodes marking the ends of the line are differentiated as starting and ending, then the line has a direction. If the line has a direction, then its two sides can be distinguished. This allows information to be attached to each side—for example, labels for land on one side and water on the other. Lines connect to other lines when they share a common node. A series of con-

S1.6 Spatial objects Spatial objects in a GIS can include points, lines of various types, intersecting lines, and polygons.

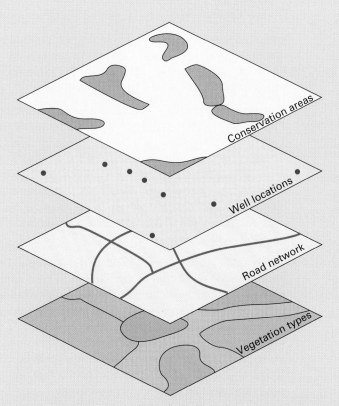

S1.7 **Data layers in a GIS** *A GIS allows easy overlay of spatial data layers for such queries as "Identify all wells on conservation land."*

nected lines that form a closed chain is a polygon.

By defining spatial objects in this way, computer-based geographic information systems allow easy manipulation of the objects and permit many different types of operations to compare objects and generate new objects. As an example, suppose we have a GIS data layer composed of conservation land in a region represented as polygons and another layer containing the location of preexisting water wells as points within the region (Figure S1.7). It is very simple to use the GIS to identify the wells that are on conservation land. Or the conservation polygons containing wells may also be identified and even output as a new data layer. By comparing the conservation layer with a road network layer portrayed as a series of lines,

we could identify the conservation polygons containing roads. We could also compare the conservation layer to a layer of polygons showing vegetation type, and tabulate the amount of conservation land in forest, grassland, brush, and so forth. We could even calculate distance zones around a spatial object, for example, to create a map of buffer zones that are located within, say, 100 meters of conservation land. Many other possible manipulations exist.

Key Elements of a GIS

A geographic information system consists of five elements: data acquisition, preprocessing, data management, data manipulation and analysis, and product generation. Each is a component or process needed to ensure the functioning of the system as a whole.

In the *data acquisition* process, data are gathered together for the particular application. These may include maps, air photos, tabular data, and other forms as well. In *preprocessing,* the assembled spatial data are converted to forms that can be ingested by the GIS to produce data layers of spatial objects and their associated information. The *data management* component creates, stores, retrieves, and modifies data layers and spatial objects. It is essential to proper functioning of all parts of the GIS. The *manipulation and analysis* component is the real workhorse of the GIS. Utilizing this component, the user asks and answers questions about spatial data and creates new data layers of derived information. The last component of the GIS, *product generation,* produces output products in the form of maps, graphics, tabulations, statistical reports, and the like that are the end products desired by the users. Taken together, these components provide a system that can serve many geographic applications at many scales.

Many new and exciting areas of geographic research are associated with geographic information systems, ranging from development of new ways to manipulate spatial data to the modeling of spatial processes using a GIS. An especially interesting area is understanding how outputs are affected by errors and uncertainty in spatial data inputs, and how to communicate this information effectively to users.

Geographic information systems is a new, emerging field of geographic research and application that relies heavily on computers and computer technology. Given the rate at which computers become ever more powerful as technology improves, we can expect great strides in this field in future years.

In Review | The Earth as a Rotating Planet

- The **rotation** of the Earth on its axis and the revolution of the Earth in its orbit about the Sun are fundamental topics in physical geography. The Earth is nearly spherical—it is slightly flattened at the poles into the shape of an *oblate ellipsoid.* It rotates on its *axis* once in 24 hours. The intersection of the axis of rotation with the Earth's surface marks the *north* and *south poles.* The direction of rotation is counterclockwise when viewed from above the north pole.

- The Earth's rotation provides the first great rhythm of our planet—the daily alternation of sunlight and darkness. The tides, and a sideward turning of ocean and air currents, are further effects of the Earth's rotation.

- The Earth's axis of rotation provides a reference for the system of location on the Earth's surface—the **geographic grid,** which consists of **meridians** and **parallels.** This system is indexed by our system of **latitude** and **longitude,** which uses the **equator** and the *prime meridian* as references to locate any point on Earth.

- We require a **map projection** to display the Earth's curved surface on a flat map. The *polar projection* is centered on either pole and pictures the globe as we might view it from the top or bottom. The *Mercator projection* converts the curved geographic grid into a flat, rectangular grid and best displays directional features. The *Goode projection* distorts the shapes of continents and coastlines but preserves the areas of land masses in their correct proportion.

- We monitor the Earth's rotation by daily timekeeping. Each hour the Earth rotates by 15°. In the **standard time system** we keep time according to a nearby *standard meridian.* Since standard meridians are normally 15° apart, clocks around the globe usually differ by whole hours. At the *International Date Line,* the calendar day changes—advancing for westward travel, dropping back a day for eastward travel. *Daylight saving time* advances the clock by one hour.

- The seasons are the second great Earthly rhythm. They arise from the **revolution** of the Earth in its orbit around the Sun, combined with the fact that the Earth's rotational axis is tilted with respect to its orbital plane. The **solstices** and **equinoxes** mark the cycle of this revolution. At the **summer** *(June)* **solstice,** the northern hemisphere is tilted toward the Sun. At the **winter** *(December)* **solstice,** the southern hemisphere is tilted toward the Sun. At the equinoxes, day and night are of equal length.

Key Terms

rotation, p. 26	great circle, p. 28	standard time system, p. 33	equinox, p. 38
parallel, p. 28	small circle, p. 28	time zones, p. 33	arctic circle, p. 39
equator, p. 28	latitude, p. 28	revolution, p. 36	tropic of cancer, p. 40
meridian, p. 28	longitude, p. 29	winter solstice, p. 37	antarctic circle, p. 40
geographic grid, p. 28	map projection, p. 30	summer solstice, p. 38	tropic of capricorn, p. 40

Review Questions

1. What is the approximate shape of the Earth? How do you know? What is the Earth's true shape?
2. What is meant by Earth rotation? Describe three environmental effects of the Earth's rotation.
3. Describe the geographic grid, including parallels and meridians.
4. How do latitude and longitude determine position on the globe? In what units are they measured?
5. Name three types of map projections and describe each briefly. Give reasons why you might choose different map projections to display different types of geographical information.
6. Explain the global timekeeping system. Define and use the terms *standard time, standard meridian,* and *time zone* in your answer.
7. What is meant by the "tilt of the Earth's axis"? How is the tilt responsible for the seasons?

Geographer's Tools 1.2 • Focus on Maps

1. Explain three types of map "projections" as they might occur by projecting a wire globe onto a flat sheet of paper.
2. What is the scale fraction of a map or globe? Can the scale of a flat map be uniform everywhere on the map? Do large-scale maps show large areas or small areas?
3. What types of symbols are found on maps, and what types of information do they carry?
4. How are numerical data represented on maps? Identify three types of isopleths. What is a choropleth map?

Geographer's Tools 1.3 • Geographic Information Systems

1. What is a geographic information system?
2. Identify and describe three types of spatial objects.
3. What are the key elements of a GIS?

Visualizing Exercises

1. Sketch a diagram of the Earth at an equinox. Show the north and south poles, the equator, and the circle of illumination. Indicate the direction of the Sun's incoming rays and shade the night portion of the globe.

2. Sketch a diagram of the Earth at the summer (June) solstice, showing the same features. Also include the tropics of cancer and capricorn, and the arctic and antarctic circles.

Essay Question

1. Suppose that the Earth's axis were tilted at 40° to the plane of the ecliptic instead of $23\frac{1}{2}°$. What would be the global effects of this change? How would the seasons change at your location?

Eye on the Landscape

1.1b Space view of Earth Unusually clear summer weather provides a good look at Europe **(A)** in this image acquired by the Meteosat geostationary meteorological satellite. It's easy to recognize Great Britain and Ireland, as well as the Italian Peninsula, by their distinctive shapes. Further east, Greece, Turkey, and the Black Sea are visible. To the north of Germany lie Denmark and Sweden, enclosing the western end of the Baltic Sea. The Pyrenees, at the top of the Iberian Peninsula, and the Alps, above Italy, are white with either clouds or snow. They delineate the edges of crustal Earth plates that are slowly colliding to push up mountain chains (see plate tectonics, covered in Chapter 12). The comma-shaped cloud patterns **(B)** mark a progression of cyclones with trailing frontal boundaries as they move eastward in the midlatitudes. We'll see more cloud images in Chapters 5 and 6.

1.15 Lake Clark National Park, Alaska These mountain landforms **(A)** show the effects of glacial ice and frost action during the most recent Ice Age. The patches of snow mark the sites where small glaciers once formed, carving shallow basins in the bedrock of the peaks (see mountain glaciers, Chapter 19). At the base of the mountains lie sloping fields of angular rock fragments, called talus, that are broken loose by frost action and carried downslope by gravity (covered in Chapter 14, Weathering and Mass Wasting). The gravel in the foreground **(B)** has a particularly fresh look and was probably deposited recently by running water, perhaps during the spring flood season. Chapter 16 provides more information on how water in streams and rivers moves sediment such as this.

EYE ON THE LANDSCAPE

A beach on St. Croix, U.S. Virgin Islands.
What else would the geographer see? … Answers at the end of the chapter.

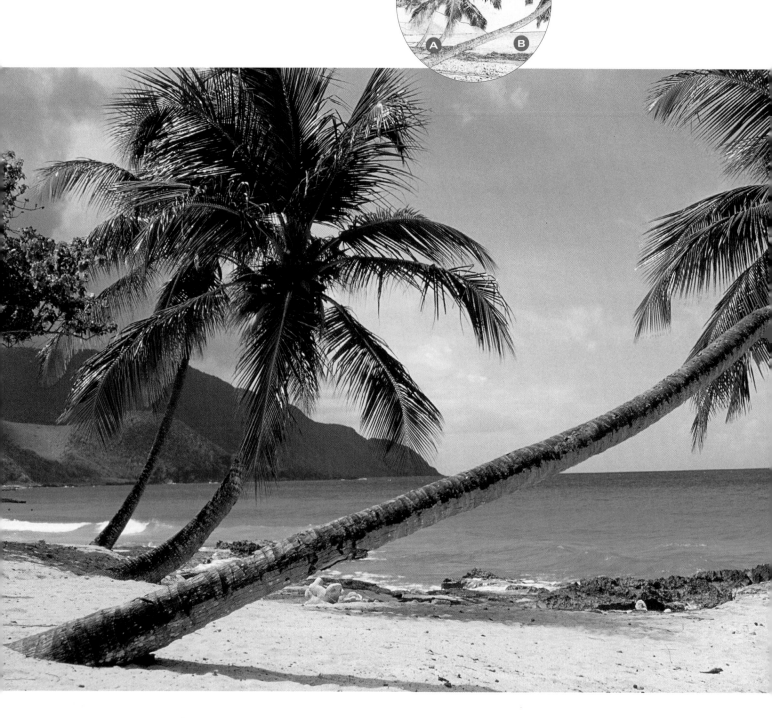

2 | THE EARTH'S GLOBAL ENERGY BALANCE

• • • • • • • • • • • • • • • • • •

PUTTING YOU IN THE PICTURE

The Sun is deliciously warm on your back. The sound of the surf is subdued as you lie stretched full-length on your towel. The beach here at Cancun is everything you thought it would be—soft, fine sand and clear blue water, with hazy tropical skies overhead. What a place to spend spring break!

It's not long before the Sun on your back seems a little too warm, and so you roll over, pulling the brim of your hat down over your eyes to screen out the bright sunlight. Soon you are hot all over, and the sound of the gently breaking waves seems very inviting. Shaking off a few clinging grains of sand, you head for the surf. A wave breaks over your feet, splashing cool drops of foam on your legs. A few quick steps, a shallow dive, and your skin is tingling with the shock of the Gulf waters. What a marvelous feeling!

Thoroughly refreshed, you emerge from the waves and walk back up to your place on the beach. As you towel the water from your skin, the tropical breeze takes on a cool edge. You lie down again, keeping low, and exposing your body to the direct rays of the sun overhead. Soon your skin dries, and you are warm once more.

Later, in the evening, you return to the beach. The stars are out now, and the white crescent is framed by palm trees behind you and the dark ocean in front. The gentle breakers splash phosphorescent foam up the beach, which quickly sinks into the fine sand. After a few minutes, you notice that your shoulders and arms are a bit cold. The sand, however, is still warm to your feet. You lie down, burrowing slightly into the soft surface to create a smooth support for your back and legs. Comfortable now, you relax and watch the sky overhead. Slowly, the stars grow brighter and the night behind them deepens to a velvet black. A shooting star flashes across the sky. This will be a spring break you'll always remember.

This description of a beach holiday probably seems far from the subject of this chapter—the Earth's global energy balance. But, in fact, many of the principles involved in the global radiation balance are the same as those that warm or cool your skin on a Caribbean beach. Like your skin, the Earth's surface absorbs sunlight and is warmed. At night, lacking sunlight, it loses heat and is cooled. The surface, as well as your skin, also loses heat when cooled by evaporation. And the surface can be cooled or warmed by direct contact with a cool or warm atmosphere—just as your skin can be cooled or warmed by direct contact with beach sand or ocean water.

No matter whether the surface is your skin or a layer of grass-covered soil, it is constantly emitting radiation. For example, our bodies emit radiation in the form of heat. Very hot objects, such as the Sun or the filament of a light bulb, give off radiation that is nearly all in the form of light. The surface and atmosphere of our planet constantly emit heat energy. Of course, the Earth also receives a flow of energy—radiation from the Sun, largely in the form of light. Over the long run, the Earth emits exactly as much energy as it absorbs, forming a global energy balance.

In this chapter, we examine the energy balance of the Earth, including the energy balances of land and ocean surfaces and the atmosphere. These balances control the seasonal and daily changes in the Earth's surface temperature. They also produce differences in energy flow rates from place to place that drive currents of air and ocean water. These, in turn, produce the changing weather and rich diversity of climates we experience on the Earth's surface. ■

THE EARTH'S GLOBAL ENERGY BALANCE

Lounging on an exotic tropical beach, you might be particularly aware of the Sun and impressed by its ability to light your world and warm your body across millions of miles of black space. Actually, the Sun's power is even more impressive, considering that most natural phenomena that take place at the Earth's surface are directly or indirectly solar-powered. From the downhill flow of a river to the movement of a sand dune to the growth of a forest, solar radiation drives nearly all of the natural processes that shape the world around us. As we will see here and in many other chapters of this book, it is the power source for wind, waves, weather, rivers, and ocean currents. This chapter explains how solar radiation is intercepted by our planet, flows through the Earth's atmosphere, and interacts with the Earth's land and ocean surfaces.

A primary topic of this chapter is the *energy balance* of the Earth. The energy balance refers to the balance between the flows of energy reaching the Earth, which includes land and ocean surfaces and the atmosphere, and flows of energy leaving it. The Earth's energy balance controls the seasonal and daily changes in the Earth's surface temperature. Differences in energy flow rates from place to place also drive currents of air and ocean water. All these, in turn, produce the changing weather and rich diversity of climates we experience on the Earth's surface.

Human activities now dominate many regions of the Earth, and we have irreversibly modified our planet by changing much of its surface cover and adding carbon dioxide to its atmosphere. Have we shifted the balance of energy flows? Is our Earth absorbing more solar energy and becoming warmer? Is it absorbing less and becoming cooler? Before we can understand human impact on the Earth–atmosphere system, we must examine the global energy balance in detail.

ELECTROMAGNETIC RADIATION

Our study of the global energy system begins with the subject of radiation—that is, **electromagnetic radiation.** This form of energy is emitted by all objects. Light and radiant heat are two familiar examples of electromagnetic radiation. Light is radiation that is visible to our eyes. Heat radiation, though not visible, is easily felt when you hold your hand near a warm object, such as an oven or electric stove burner. In this chapter, we are concerned primarily with electromagnetic energy in the forms of light and heat.

Think of electromagnetic radiation as a collection, or spectrum, of waves of a wide range of wavelengths traveling quickly away from the surface of an object. *Wavelength* describes the distance separating one wave crest from the next wave crest (Figure 2.1). The unit we will use to measure wavelength is the *micrometer* (formerly called the micron). A micrometer is one millionth of a meter, or 10^6 m. This is such a small unit that the tip of your little finger is about 15,000 micrometers wide. In this text, we use the abbreviation μm for the micrometer. The first letter is the Greek letter μ, or mu. It is used in metric units to denote micro-, meaning one-millionth (10^{-6}). Radiant energy can exist at any wavelength. Heat and light are identical forms of electromagnetic radiation except for their wavelengths.

Figure 2.2 shows how electromagnetic waves differ in wavelength throughout their entire range, or *spectrum.* At

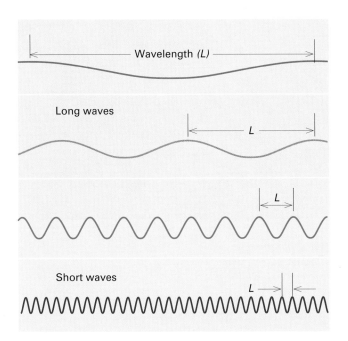

2.1 Wavelength of electromagnetic radiation
Electromagnetic radiation can be described as a collection of energy waves with different wavelengths. Wavelength L is the crest-to-crest distance between successive wave crests.

the short wavelength end of the spectrum are *gamma rays* and *X rays*. Their wavelengths are normally expressed in *nanometers*. A nanometer is one one-thousandth of a micrometer, or 10^{-9} m, and is abbreviated nm. Gamma and X rays have high energies and can be hazardous to health.

Gamma and X-radiation grade into *ultraviolet* radiation, which begins at about 10 nm and extends to 400 nm or 0.4 μm. Like gamma and X rays, ultraviolet radiation can be damaging to living tissues.

The **visible light** portion of the spectrum starts at about 0.4 μm with violet. Colors then grade through blue, green, yellow, orange, and red, reaching the end of the visible spectrum at about 0.7 μm. Next is *near-infrared* radiation, with wavelengths from 0.7 to 1.2 μm. This radiation is very similar to visible light in that most of it comes from the Sun. However, human eyes are not sensitive to radiation beyond about 0.7 μm, so we do not see near-infrared light. Healthy plant leaves and green tissues reflect near-infrared light strongly, a fact that is often used in remote sensing with near-infrared light.

Shortwave infrared radiation lies in the range of 1.2 to 3.0 μm. Like near-infrared light, it comes mostly from the Sun. In remote sensing, this radiation can distinguish clouds from snow. It is also useful for differentiating rock types, which may have different "colors" in this portion of the spectrum. From 3 μm to about 6 μm is *middle-infrared* radiation. It comes from the Sun as well but is also emitted by fires burning at the Earth's surface, such as forest fires or gas well flames.

At wavelengths between 6 and about 300 nm, we have *thermal infrared* radiation. This radiation includes the heat radiation given off by bodies at temperatures normally found at the Earth's surface. As we'll see shortly, most heat radiation lies in the range of 8 to 12 μm. Beyond infrared wavelengths lie the domains of microwaves, radar, and communications transmissions, such as radio and television.

2.2 The electromagnetic spectrum *Electromagnetic radiation can exist at any wavelength. By convention, names are assigned to specific wavelength regions as shown in the figure.*

2.3 A thermal infrared image *This suburban scene was imaged at night in the thermal infrared spectral region. Black and blue tones show lower temperatures, while yellow and red tones show higher temperatures. Ground and sky are coldest, while the windows of the heated homes are warmest.*

Thermal infrared radiation is no different in principle from visible light radiation. Although photographic film is not sensitive to thermal radiation, it is possible to acquire an image of thermal radiation using a special sensor. Figure 2.3 shows a thermal infrared image of a suburban scene obtained at night. Here the image brightness is displayed using color, with red tones indicating the warmest temperatures and black tones the coldest. Windows appear red because they are warm and radiate more intensely. House walls are intermediate in temperature and appear blue. Roads and driveways are cool, as are the trees, shown in purple tones. Ground and sky are coldest (black). GEODISCOVERIES

> **Visible light includes colors ranging from violet to red and spans the wavelength range of about 0.4 to 0.7 μm.**

Radiation and Temperature

There are two important physical principles that concern the emission of electromagnetic radiation. The first is that an inverse relationship exists between the range of wavelengths of the radiation that an object emits and the temperature of the object. For example, the Sun, a very hot object, emits radiation with short wavelengths. In contrast, the Earth, a much cooler object, emits radiation with longer wavelengths.

The second principle is that hot objects radiate more energy than cooler objects—much more. In fact, the flow of radiant energy from the surface of an object is directly related to the absolute temperature of the surface raised to the fourth power. (Absolute temperature is temperature measured on the Kelvin scale, with zero being the absence of all heat.) Thus, if the absolute temperature of an object is doubled, the flow of radiant energy from its surface will be 16 times larger. Because of this relationship, a small increase in

> **Hotter objects radiate energy at shorter wavelengths than cooler objects. Hotter objects also radiate substantially more energy than cooler objects.**

temperature can mean a large increase in the rate at which radiation is given off by an object or surface. For example, water at room temperature emits about one-third more energy than when it is at the freezing point.

Solar Radiation

Our Sun is a ball of constantly churning gases that are heated by continuous nuclear reactions. It is about average in size compared to other stars, and it has a surface temperature of about 6000°C (about 11,000°F). Like all objects, it emits energy in the form of electromagnetic radiation. The energy travels outward in straight lines, or rays, from the Sun at a speed of about 300,000 km (about 186,000 mi) per second. At that rate, it takes the energy about $8\frac{1}{3}$ minutes to travel the 150 million km (93 million mi) from the Sun to the Earth.

As solar radiation travels through space, none of it is lost. However, the rays spread apart as they move away from the Sun. This means that a planet farther from the Sun, like Mars, receives less radiation than one located nearer to the Sun, like Venus. The Earth intercepts only about one-half of one-billionth of the Sun's total energy output.

The Sun's interior is the source generating solar energy. Here, hydrogen is converted to helium at very high temperatures and pressures. In this process of nuclear fusion, a vast quantity of energy is generated and finds its way to the Sun's surface. Because the rate of production of energy is nearly constant, the output of solar radiation is also nearly constant. So, given the average distance of the Earth from the Sun, the amount of solar energy received on a small, fixed area of surface held at right angles to the Sun's rays is almost constant. This rate of incoming energy, known as the *solar constant,* is measured beyond the outer limits of the Earth's atmosphere, before energy has been lost in passing through the atmosphere. The solar constant has a value of about 1370 watts per square meter (W/m^2).

The *watt* (W), which describes a rate of energy flow, is a familiar measure of power. You've probably seen it applied to light bulbs or to home appliances, such as stereo amplifiers or microwave ovens. When we use the watt to measure the intensity of a flow of radiant energy, we must specify the unit of surface area that receives or emits the energy. This cross section is assigned the area of 1 square meter (1 m^2). Thus, the measure of intensity of received (or emitted) radiation is given as watts per square meter (W/m^2). Because there are no common equivalents for this energy flow rate in the English system, we will use only metric units.

The intensity of solar radiation is greatest in the visible portion of the spectrum. In this range, most of the solar radiation penetrates the Earth's atmosphere to reach the surface.

Characteristics of Solar Energy

Let's look in more detail at the Sun's output as it is received by the Earth, which is shown in Figure 2.4. Energy intensity is shown on the graph on the vertical scale. Note that it is a logarithmic scale—that is, each whole unit marks an intensity 10 times greater than the one below. Wavelength is shown on the horizontal axis, also on a logarithmic scale.

The left side of Figure 2.4 shows how the Sun's incoming electromagnetic radiation varies with wavelength. The uppermost line shows how a "perfect" Sun would supply solar energy at the top of the atmosphere. By "perfect," we mean a Sun radiating as a *blackbody,* which follows physical theory exactly. The solid line shows the actual output of the Sun as measured at the top of the atmosphere. It is quite close to the "perfect" Sun, except for ultraviolet wavelengths, where the real Sun emits less energy. Note that the Sun's output peaks

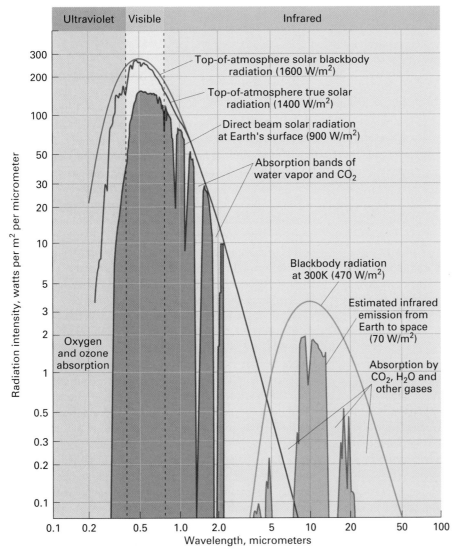

2.4 Spectra of solar and Earth radiation This figure plots both shortwave radiation, which comes from the Sun (left side), and longwave radiation, which is emitted by the Earth's surface and atmosphere (right side). (After W. D. Sellers, Physical Climatology, University of Chicago Press. Used by permission.)

in the visible part of the spectrum. Thus, our human vision system is adjusted to the wavelengths where solar light energy is most abundant.

The line showing solar radiation reaching the Earth's surface is quite different from the others, however. As solar radiation passes through the atmosphere, it is affected by **absorption** and scattering. Absorption occurs when molecules and particles in the atmosphere intercept and absorb radiation at particular wavelengths. This absorption is shown by the steep "valleys" in the graph—for example, at about 1.3 μm and 1.9 μm. At these wavelengths, molecules of water vapor and carbon dioxide absorb solar radiation very strongly and keep nearly all of it from reaching the Earth's surface. Note also that oxygen and ozone absorb almost all of the ultraviolet radiation at wavelengths shorter than about 0.3 μm. Atmospheric absorption is important because it warms the atmosphere directly and constitutes one of the flows of energy in the global energy balance that we will discuss toward the end of this chapter.

Solar radiation passing through the atmosphere can also be scattered. By **scattering,** we refer to a turning aside of radiation by a molecule or particle so that the direction of a scattered ray is changed. Scattered rays may go upward toward space or downward toward the surface. The line on Figure 2.4 is for direct beam energy only, so it does not include any scattered radiation that makes its way to the surface.

From the wavelengths shown on the horizontal axis, we see that solar energy received at the surface ranges from about 0.3 μm to 3 μm. This is known as **shortwave radiation.** These wavelengths are "short" as compared to the longer wavelengths of energy that are emitted by the Earth and atmosphere, which we discuss next.

Shortwave radiation refers to wavelengths emitted by the Sun, which are in the range of about 0.3 to 3 μm. Longwave radiation refers to wavelengths emitted by cooler objects, such as Earth surfaces, and ranges from about 3 to 30 μm.

Longwave Radiation from the Earth

Recall that both the range of wavelengths and the intensity of radiation emitted by an object depend on the object's temperature. Since the Earth's surface and atmosphere are much colder than the Sun's surface, we can deduce that our planet radiates less energy than the Sun and that the energy emitted has longer wavelengths.

The right side of Figure 2.4, which shows outgoing radiation, confirms these deductions. The upper line shows the radiation of a blackbody at a temperature of about 300 K (23°C, 73°F), which is not a bad approximation for the Earth as a whole. At this temperature, radiation ranges from about 3 to 30 μm and peaks at about 10 μm in the thermal infrared region. We refer to the thermal infrared radiation emitted by the Earth as **longwave radiation.**

Beneath the blackbody curve is an irregular line, shaded underneath, that shows upwelling energy emitted by the Earth and atmosphere as measured at the top of the atmosphere.

Comparing this line with the blackbody curve shows that much of the Earth's surface radiation is absorbed by the atmosphere, especially between about 6 and 8 μm and between 14 and 17 μm, where absorption is nearly complete. Above 21 μm, absorption is also essentially complete. Absorption of Earth radiation by the atmosphere is an important part of the greenhouse effect, which we will discuss shortly. Water and carbon dioxide are the primary absorbers.

The atmospheric absorption regions of the spectrum leave three regions in which outgoing energy flow is significant—4 to 6 μm, 8 to 14 μm, and 17 to 21 μm. We can call these *windows* through which longwave radiation leaves the Earth and flows to space.

An examination of Figure 2.4 shows that significant amounts of radiant energy, both shortwave by scattering and longwave by radiation, leave the Earth's surface and pass upward through the atmosphere. This energy, whose wavelength ranges from visible to thermal infrared, is therefore available for aircraft and satellite instruments to measure and compose into images of the Earth's surface through the process of remote sensing. Our *Geographer's Tools 2.1 • Remote Sensing for Physical Geography,* placed at the end of this chapter, provides some basic concepts of remote sensing that will prove useful in understanding both our *Focus on Remote Sensing* features and the uses of remotely sensed images in the main text.

The Global Radiation Balance

As we have seen, the Earth constantly absorbs solar shortwave radiation and emits longwave radiation. Figure 2.5 presents a diagram of this energy flow process, which we refer to as the Earth's **global radiation balance.**

The Sun provides a nearly constant flow of shortwave radiation toward Earth that is received at the top of the atmosphere. Part of this radiation is scattered away from the Earth and heads back into space without absorption. Clouds and dust particles in the atmosphere contribute to this scattering. Land and ocean surfaces also reflect some shortwave radiation back to space. The shortwave energy from the Sun that is not scattered or reflected is absorbed by either atmosphere, land, or ocean. Once absorbed, solar energy raises the temperature of the atmosphere as well as the surfaces of the oceans and lands.

The atmosphere, land, and ocean also emit energy in the form of longwave radiation. This radiation ultimately leaves the planet, headed for outer space. The longwave radiation outflow tends to lower the temperature of the atmosphere, ocean, and land, and thus cool the planet. In the long run, these flows balance—incoming energy absorbed and outgoing radiation emitted are equal. Since the temperature of a surface is determined by the amount of energy it absorbs and emits, the Earth's overall temperature tends to remain constant.

2.5 The global radiation balance *Shortwave radiation from the Sun is transmitted through space, where it is intercepted by the Earth. The absorbed radiation is then ultimately emitted as longwave radiation to outer space. (From A. N. Strahler, "The Life Layer," Journal of Geography, vol. 69, p. 72. Copyright 1970 © by The Journal of Geography. Used by permission.)*

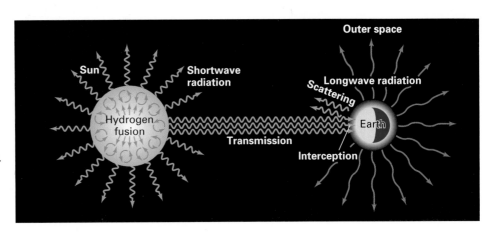

INSOLATION OVER THE GLOBE

The flow of solar radiation to the Earth as a whole remains constant, but the amount received varies from place to place and time to time. We turn now to this variation and its causes. In this discussion, we will refer to **insolation** (**in**coming **sol**ar radi**ation**)—the flow of solar energy intercepted by an exposed surface assuming a uniformly spherical Earth with no atmosphere. Insolation is a flow rate and has units of watts per square meter (W/m²). *Daily insolation* refers to the average of this flow rate over a 24-hour day. Annual insolation, discussed in a later section, is the average flow rate over the entire year. Insolation is important because it measures the amount of solar power available to heat the land surface and atmosphere.

Insolation refers to the flow rate of incoming solar radiation. It is high when the Sun is high in the sky.

Insolation depends on the angle of the Sun above the horizon. Figure 2.6 shows that insolation is greatest when the Sun is directly overhead and the Sun's rays are vertical (position *A* in the figure). When the Sun is lower in the sky, the same amount of solar energy spreads over a greater area of ground surface, so insolation is lower (position *B* in the figure).

Daily insolation at a location depends on two factors: (1) the angle at which the Sun's rays strike the Earth, and (2) the length of time of exposure to the rays. Recall from Chapter 1 that both of these factors are controlled by the latitude of the location and the time of year. For example, in midlatitude locations in summer, days are long and the Sun rises to a position high in the sky, thus heating the surface more intensely. Let's investigate these variations in more detail.

Refer back to Chapter 1 and Figure 1.18, showing equinox conditions. Only at the subsolar point does the Earth's spherical surface present itself at a right angle to the Sun's rays. To a viewer on the Earth at the subsolar point, the Sun is directly overhead. But as we move away from the subsolar point toward either pole, the Earth's curved surface becomes turned at an angle with respect to the Sun's rays. To the Earthbound viewer at a new latitude, the Sun appears to be nearer the horizon. When the circle of illumination is reached, the Sun's rays are parallel with the surface. That is, the Sun is viewed just at the horizon. Thus, the angle of the Sun in the sky at a particular location depends on the latitude of the location, the time of day, and the time of year. *GEODISCOVERIES*

2.6 Insolation and Sun angle *The angle of the Sun's rays determines the intensity of insolation on the ground. The energy of vertical rays A is concentrated in square a by c, but the same energy in the slanting rays B is spread over a larger rectangle, b by c.*

Insolation and the Path of the Sun in the Sky

How does the angle of the Sun vary during the day? The angle depends on the Sun's path in the sky. When the Sun is high above the horizon—near noon, for example—the Sun's angle is greater, and so insolation will be greater. Figure 2.7 shows the Sun's daily path in the sky at various latitudes and how this path changes from season to season.

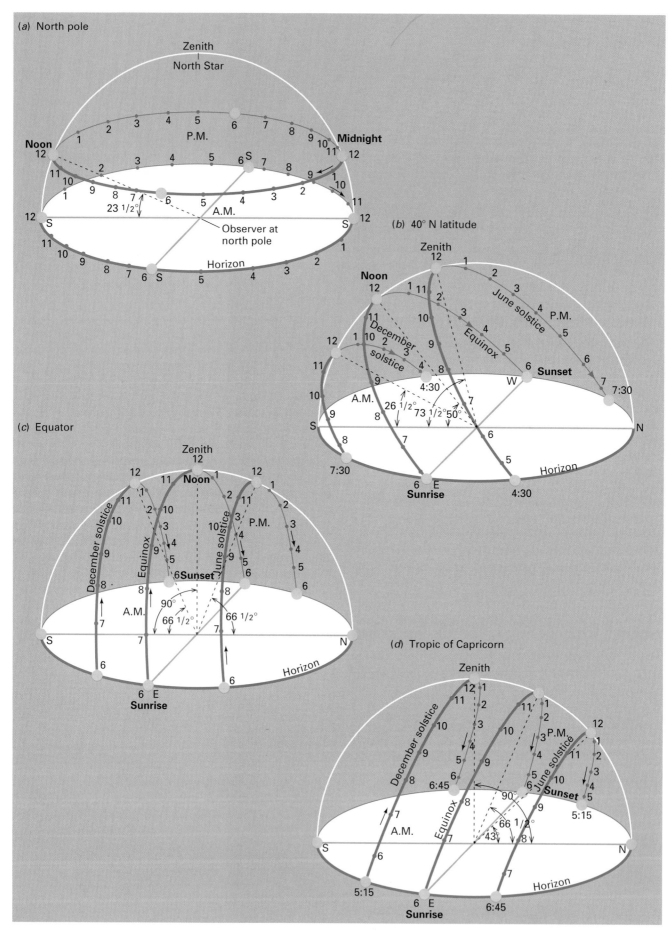

2.7   The path of the Sun in the sky at different latitudes and seasons *The Sun's path in the sky can change greatly in position and height above horizon through the seasons. (a) North pole. (b) 40° N latitude. (c) Equator. (d) Tropic of capricorn. (Parts a, c, d, Copyright © A. N. Strahler. Used by permission.)*

Let's look at part *b* first, since it is likely to be the most familiar. This diagram is drawn for latitude 40° N and is typical of conditions found in midlatitudes in the northern hemisphere—for example, at New York or Denver. The diagram shows a small area of the Earth's surface bounded by a circular horizon. This is the way things appear to an observer standing on a wide plain. The Earth's surface appears flat, and the Sun seems to travel inside a vast dome in the sky. GEODISCOVERIES

At equinox, the Sun rises directly to the east and sets directly to the west. The noon Sun is positioned at an angle of 50° above the horizon in the southern sky. The Sun is above the horizon for exactly 12 hours, as shown by the hour numbers on the Sun's path. At the June solstice, the Sun's path rises much higher in the sky—at noon it will be $73^1/_2$° above the horizon. The Sun is above the horizon for about 15 hours, and it rises and sets at points on the horizon that are well to the north of east and west. Clearly, daily insolation will be greater at the June solstice than at the equinox since the Sun is in the sky longer and reaches a higher angle at noon. At the December solstice, the Sun's path is low in the sky, reaching only $26^1/_2$° above the horizon, and the Sun is visible for only about 9 hours. Sunrise and sunset are to the south of east and west. Thus, daily insolation reaching the surface at the December solstice will be less than at the equinox and much less than at the June solstice.

Figure 2.8 shows daily insolation values for a selection of latitudes in the northern hemisphere. Following the curve for 40° N, we see that daily average insolation ranges from about 160 W/m^2 at the December solstice to about 460 W/m^2 at the June solstice. Equinox values are around 350 W/m^2. Thus, the actual values confirm our expectation, based on our analysis of how high the Sun rises during the day and how long it stays above the horizon.

Part *a* of Figure 2.7 depicts conditions at the north pole. At this latitude, the Sun seems to trace a daily circle in the heavens that is parallel to the horizon. This motion is actually part of a long, flat spiral, as shown in Figure 2.9. At the June solstice, the Sun is elevated at $23^1/_2$ degrees above the horizon. As time proceeds to the September equinox, the Sun's elevation slowly falls to the horizon. Between the September and March equinoxes, the Sun is below the horizon. At the March equinox, it emerges above the horizon to work its way slowly up to the June solstice position during the next three months.

From this analysis of the Sun's motion, we see that daily insolation is greatest at the June solstice, when the Sun is at its highest position in the sky (Figure 2.7). Insolation then falls to zero at the September equinox, remains at zero between September and March equinoxes, and then increases as the Sun rises above the horizon to reach its position at the June solstice again. Note from Figure 2.8 that summer solstice daily insolation, over 500 W/m^2, is higher at the pole than at any other location at any other time of the

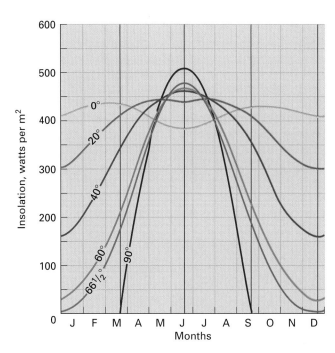

2.8 Daily insolation through the year at various latitudes (northern hemisphere) *Black lines mark the equinoxes and solstices. Latitudes between the equator (0°) and tropic of cancer ($23^1/_2$° N) show two maximum values; others show only one. Poleward of the arctic circle ($66^1/_2$° N), insolation is zero for at least some period of the year. (Copyright © A. N. Strahler. Used by permission.)*

year. This shows the importance of day length in determining daily insolation.

At the equator (part *c*), the Sun at the September equinox rises exactly east at 6:00 A.M., moves to the zenith, or position directly overhead, at noon, and then moves to set due west at 6 P.M., 12 hours later. At the December solstice, the Sun also rises at 6 A.M., but from a point south of east. At noon, it is not directly overhead but at an elevation of $66^1/_2$° above the horizon. It sets at 6:00 P.M. at a point south of west. Comparing the solstice and equinox, we see that the Sun spends the same amount of time above the horizon (12 hours) but is not as high above the horizon on the solstice. Daily insolation is therefore smaller at the solstice.

> The path of the Sun in the sky varies with latitude and season. In the summer, the Sun rises higher in the sky and remains in the sky for a longer part of the 24-hour day.

As December moves to March, the Sun returns to the equinox position and then passes on to the June solstice position. Here the situation is the same as the December solstice but reversed in that the Sun rises to the north of east and sets to the north of west. Again, daily insolation is less at solstice than at equinox.

Note that for the equator, there are two maximums of insolation—one at each equinox—and two minimums—one at each solstice. This means that for locations near the equator, two hot seasons may be experienced. Figure 2.8 clearly shows the two peaks in daily insolation, each centered around an equinox. (Note that the two peaks are not quite

2.9 Sun's path in the sky at the north pole
At the poles, the Sun's daily path traces part of a long, fine spiral beginning on the horizon at the equinox, proceeding to 23$\frac{1}{2}$° above the horizon at the solstice, and then descending to the horizon again at the following equinox. (For the south pole, the direction of the spiral is reversed.) (Copyright © A. N. Strahler. Used by permission.)

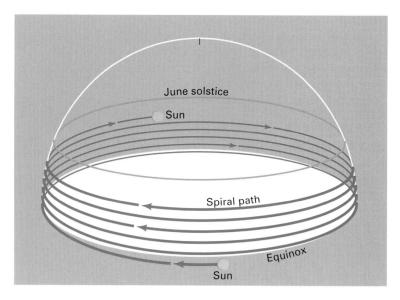

equal and are not exactly at the equinoxes. These small effects occur because the Earth's orbit is not exactly circular.) There are actually two maximum daily insolation values at all latitudes between the tropic of cancer (23$\frac{1}{2}$° N) and the tropic of capricorn (23$\frac{1}{2}$° S). However, as either tropic is approached in a poleward direction, the two maximum periods get closer and closer in time, and then merge into a single maximum. The latitude 20° line in Figure 2.8 shows this clearly, with two maximums about two months apart.

The tropic of capricorn (part *d*), at 23$\frac{1}{2}$°S. lat., provides an example from the southern hemisphere with one maximum in daily insolation. At the equinox, the Sun rises directly east, moves to a noon position 66$\frac{1}{2}$° above the horizon to the north, and then sets directly west. On the December solstice, the Sun rises to the south of east at about 5:15 A.M. and moves to the zenith point, where it is directly overhead at noon. During the afternoon, it moves toward its sunset point, at a position to the south of west at about 6:45 P.M. Day length is about 13$\frac{1}{2}$ hours. At the June solstice, the Sun only rises to 43° above the horizon, and day length is only about 10$\frac{1}{2}$ hours. So the summer (December) solstice is the time of maximum daily insolation at the tropic of capricorn.

Daily Insolation through the Year
Returning to Figure 2.8, what conclusions can we draw about daily insolation as it varies with latitude and season?

- Latitudes between the tropics of cancer and capricorn and arctic and antarctic circles show a wavelike pattern of greater daily insolation at the summer solstice and lower daily insolation at the winter solstice.
- Poleward of the arctic and antarctic circles, the Sun is below the horizon for at least part of the year, and daily insolation drops to zero during that period.
- Daily insolation is greatest of all latitudes at the pole at the summer solstice.

- There are two maximums and two minimums in daily insolation at the equator, occurring at the equinoxes and solstices, respectively.
- Poleward of the equator but between the two tropics, there are also two maximum and minimum daily insolation values, but as the tropic is approached, the two maximum periods get closer and closer in time, and then merge into a single maximum at the tropic.

These differences in daily insolation are important because daily insolation measures the flow of solar power

2.10 Annual insolation from equator to pole for the Earth
The red line shows the actual curve. The blue line shows what the annual insolation would be if the axis were perpendicular to the plane of the ecliptic.

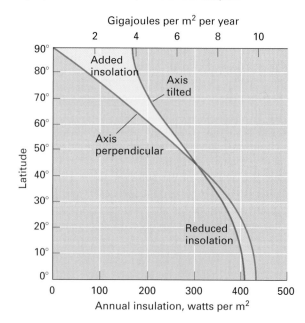

available to heat the Earth's surface. Thus, it is the most important factor in determining air temperatures, as we will see in later chapters. The change in daily insolation with the seasons at a location is therefore a major determinant of climate.

Annual Insolation by Latitude

What is the effect of latitude on annual insolation—the rate of insolation averaged over an entire year? Figure 2.10 shows two curves of annual insolation by latitude—one for the actual case of the Earth's axis tilted at $23^1/_2°$ and the other for an Earth with an untilted axis. Let's look first at the real case of a tilted axis. Annual insolation varies smoothly from the equator to the pole. As we might expect, annual insolation is greater at lower latitudes. But the high latitudes still receive a considerable flow of solar energy—the annual insolation value at the pole is about 40 percent of the value at the equator.

The case of the untilted axis—that is, for which the Earth's axis is perpendicular to the plane of the ecliptic—is quite different. In this case, there are no seasons. The Sun traces its equinox path in the sky every day of the year at every location. Under these circumstances, annual insolation changes from a high value at the equator, where the Sun is directly overhead at noon every day, to zero at the poles, where the Sun is always just on the horizon. The curve of annual insolation thus shows a progression from a value near 440 W/m² at the equator to zero at 90° latitude.

The comparison of the two cases shows that the tilting of the Earth's axis redistributes a very significant portion of the Earth's insolation from the equatorial regions toward the poles. Although the pole is in the dark for six months of the year, it still receives nearly half the amount of solar radiation received at the equator. Without a tilted axis, our planet would be quite a different place!

WORLD LATITUDE ZONES

The seasonal pattern of daily insolation can be used as a basis for dividing the globe into broad latitude zones (Figure 2.11). The zone limits shown in the figure and specified below should not be taken as absolute and binding, however. Rather, this system of names is a convenient way to identify general world geographic belts throughout this book.

The *equatorial zone* encompasses the equator and covers the latitude belt roughly 10° north to 10° south. Here the Sun provides intense insolation throughout most of the year, and days and nights are of roughly equal length. Spanning the tropics of cancer and capricorn are the *tropical zones,* ranging from latitudes 10° to 25° north and south. A marked seasonal cycle exists in these zones, combined with high annual insolation.

> The tilt of the Earth's axis of rotation causes insolation to be distributed more evenly from pole to pole. On an annual basis, the poles receive about 40 percent of the insolation received at the equator.

Moving toward the poles from each of the tropical zones are transitional regions called the *subtropical zones.* For convenience, we assign these zones the latitude belts 25° to 35° north and south. At times we may extend the label "subtropical" a few degrees farther poleward or equatorward of these parallels. These zones have a strong seasonal cycle combined with an annual insolation nearly as large as the tropical zones.

The *midlatitude zones* are next, lying between 35° and 55° north and south latitude. In these belts, the Sun's height in the sky shifts through a wide range annually. Differences in day length from winter to summer are also large. Thus seasonal contrasts in insolation are quite strong. In turn, these regions can experience a large range in annual surface temperature.

Bordering the midlatitude zones on the poleward side are the *subarctic* and *subantarctic zone,* 55° to 60° north and south latitudes. Astride the arctic and antarctic circles from latitudes 60° to 75° N and S lie the *arctic* and *antarctic zones.* These zones have an extremely large yearly variation in day lengths, yielding enormous contrasts in insolation from solstice to solstice.

The *polar zones,* north and south, are circular areas between about 75° latitude and the poles. Here the polar

2.11 *World latitude zones* *A geographer's system of latitude zones, based on the seasonal patterns of daily insolation observed over the globe.*

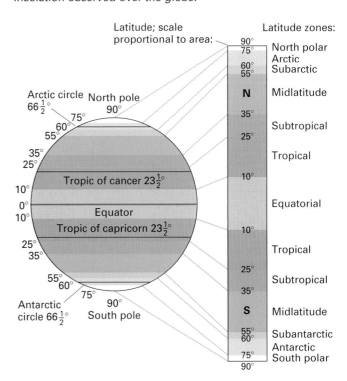

regime of a six-month day and six-month night is predominant. These zones experience the greatest seasonal contrasts of insolation of any area on Earth.

COMPOSITION OF THE ATMOSPHERE

A main goal of this chapter is to explain the flows of energy within the atmosphere and between the atmosphere and the Earth's surface. To understand this topic, you need to know some basic facts about the atmosphere.

The Earth's atmosphere consists of *air*—a mixture of various gases surrounding the Earth to a height of many kilometers. This envelope of air is held to the Earth by gravitational attraction. Almost all the atmosphere (97 percent) lies within 30 km (19 mi) of the Earth's surface. The upper limit of the atmosphere is at a height of approximately 10,000 km (about 6000 mi) above the Earth's surface, a distance approaching the diameter of the Earth itself.

From the Earth's surface upward to an altitude of about 80 km (50 mi), the chemical composition of air is highly uniform in terms of the proportions of its gases. Pure, dry air consists largely of nitrogen, about 78 percent by volume, and oxygen, about 21 percent (Figure 2.12). Other gases account for the remaining 1 percent.

Nitrogen in the atmosphere exists as a molecule consisting of two nitrogen atoms (N_2). Nitrogen gas does not enter easily into chemical union with other substances and can be thought of mainly as a neutral substance. Very small amounts of nitrogen are extracted by soil bacteria and are made available for use by plants. Otherwise, it is largely a "filler," adding inert bulk to the atmosphere.

In contrast to nitrogen, *oxygen gas* (O_2) is highly active chemically. It combines readily with other elements in the process of oxidation. Combustion of fuels represents a rapid

> Nitrogen is the most abundant of the atmosphere's gases. Oxygen is essential to living tissue. Water vapor and carbon dioxide absorb heat radiation and help warm the atmosphere.

form of oxidation, whereas certain forms of rock decay (weathering) represent very slow forms of oxidation. Living tissues require oxygen to convert foods into energy.

These two main component gases of the lower atmosphere are perfectly diffused so as to give pure, dry air a definite set of physical properties, as though it were a single gas. The remaining 1 percent of dry air is mostly argon, an inactive gas of little importance in natural processes. In addition, there is a very small amount of *carbon dioxide* (CO_2), amounting to about 0.035 percent. As we saw in Figure 2.4, CO_2 absorbs incoming shortwave radiation and outgoing longwave radiation in specific wavelength regions. Longwave absorption is particularly important for the greenhouse effect, in which the lower atmosphere is warmed by this absorption and reradiates a portion of its heat back to the surface, thus making surface temperatures warmer. (We'll return to the greenhouse effect in more detail later in this chapter.) Carbon dioxide is also used by green plants in the photosynthesis process. During photosynthesis, CO_2 is converted into chemical compounds that build up the plant's tissues, organs, and supporting structures.

Another important component of the atmosphere is *water vapor,* the gaseous form of water. Individual water molecules in the form of vapor are mixed freely throughout the atmosphere, just like the other atmospheric component gases. Unlike the other component gases, however, water vapor can vary highly in concentration. Usually, water vapor makes up less than 1 percent of the atmosphere. Under very warm, moist conditions, as much as 2 percent of the air can be water vapor. Since water vapor, like carbon dioxide, is a good absorber of heat radiation, it also plays a major role in warming the lower atmosphere.

The atmosphere also contains dust and tiny floating particles that absorb and scatter radiation. These are discussed more fully in Chapter 4.

Ozone in the Upper Atmosphere

Another small, but important, constituent of the atmosphere is **ozone,** a form of oxygen in which three oxygen atoms are bonded together (O_3). Ozone is found mostly in the upper part of the atmosphere, in a layer termed the *stratosphere.* This layer of the atmosphere lies about 14 to 50 km (9 to 31 mi) above the surface, and is described in more detail in the next chapter. Ozone in the stratosphere absorbs ultraviolet radiation from the Sun as this radiation passes through the atmosphere, thus shielding the life layer from its harmful effects.

Ozone is most concentrated in a layer that begins at an altitude of about 15 km (about 9 mi) and extends to about 55 km (about 34 mi) above the Earth. It is produced by gaseous chemical reactions in the stratosphere. The details of these chemical reactions are complicated, but in summary they are quite simple. An oxygen molecule (O_2) absorbs ultraviolet

2.12 Component gases of the lower atmosphere
Values show percentage by volume for dry air. Nitrogen and oxygen form 99 percent of our air, with other gases, principally argon and carbon dioxide, accounting for the final 1 percent.

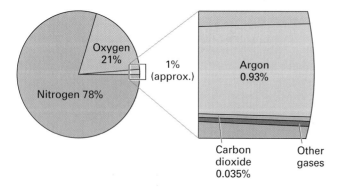

light energy and splits into two oxygen atoms (O + O). A free oxygen atom (O) then combines with an O_2 molecule to form ozone, O_3.

Once formed, ozone can also be destroyed. Like O_2, ozone absorbs ultraviolet radiation, and it splits to form O_2 + O. Furthermore, two oxygen atoms (O + O) can join to form O_2. The net effect is that ozone (O_3), molecular oxygen (O_2), and atomic oxygen (O) are constantly formed, destroyed, and reformed in the ozone layer, absorbing ultraviolet radiation with each transformation.

The absorption of ultraviolet radiation by the ozone layer protects the Earth's surface from this damaging form of radiation. If the concentration of ozone is reduced, fewer transformations among O, O_2, and O_3 occur, and so ultraviolet absorption is reduced. If solar ultraviolet radiation were to reach the Earth's surface at full intensity, bacteria exposed on the Earth's surface would be destroyed, and unprotected animal tissues would be severely damaged. The presence of the ozone layer is thus essential to maintaining a viable environment for life on this planet.

Certain forms of air pollution are currently reducing ozone concentrations substantially in some portions of the atmosphere at certain times of the year. This threat to the ozone layer is described in *Eye on Global Change 2.2 • The Ozone Layer—Shield to Life.*

Ozone is also found in the lowest layers of the atmosphere, where it is an air pollutant. It occurs in photochemical smog and is very damaging to lungs and other living tissues. *Eye on the Environment 4.2 • Air Pollution* provides more information about the ozone and smog.

SENSIBLE HEAT AND LATENT HEAT TRANSFER

Thus far, we have discussed energy in the form of shortwave and longwave radiation. However, energy can be transported in other ways. Two of them—sensible heat and latent heat transfer—are very important in discussing the global energy balance.

The most familiar form of heat storage and transport is known as **sensible heat**—the quantity of heat held by an object that can be sensed by touching or feeling. This kind of heat is measured by a thermometer. When the temperature of an object or a gas increases, its store of sensible heat increases. When two objects of unlike temperature contact each other, heat energy moves by *conduction* from the warmer to the cooler. This type of heat flow is referred to as **sensible heat transfer.** Sensible heat transfer can also occur by *convection*. In this process, a fluid is heated by a warm surface, expands, and rises, creating an upward flow. This flow moves heat away from the surface and into the fluid. We will return to atmospheric convection in Chapter 4.

Unlike sensible heat, **latent heat**—or hidden heat—cannot be measured by a thermometer. It is heat that is taken up and stored in the form of molecular motion when a substance changes state from a solid to a liquid, from a liquid to a gas, or from a solid directly to a gas. Water is an important example. For liquid water to make the transition to water vapor, energy is required. As the water evaporates from liquid to gas, it draws up heat from the surroundings. This is the latent heat. It is carried in the form of the fast random motion of the free water vapor molecules as they enter the gaseous state during the process of evaporation. For example, consider the cooling you feel in a breeze when your skin is wet. As the water evaporates from liquid to gas, it draws up the heat required to make the change of state and carries it away from your skin, keeping you cool. *GEODISCOVERIES*

Although the latent heat of water vapor in air cannot be measured by a thermometer, energy is stored there just the same. When water vapor reverts to a liquid or solid, the latent heat is released, making the surroundings warmer. In the Earth–atmosphere system, **latent heat transfer** occurs when water evaporates from a moist land surface or open water surface. This process transfers heat from the surface to the atmosphere. On a global scale, latent heat transfer by movement of moist air provides a very important mechanism for transporting large amounts of energy from one region of the Earth to another.

> **Sensible heat transfer refers to the flow of heat between the Earth's surface and the atmosphere by conduction or convection. Latent heat transfer refers to the flow of heat carried by changes of state of water.**

THE GLOBAL ENERGY SYSTEM

The flow of energy from the Sun to the Earth and then back out into space is a complex system involving not only radiant energy flow, but also energy storage and transport. Keep in mind that solar energy is the ultimate power source for the Earth's surface processes, ranging from wind systems to the cycle of evaporation and precipitation. So when we trace the energy flows between the Sun, surface, and atmosphere, we are really studying how these processes are powered.

Solar Energy Losses in the Atmosphere

Let's examine the flow of insolation through the atmosphere on its way to the surface. This is shown in Figure 2.13, which gives typical values for losses of incoming solar energy. As the shortwave radiation penetrates the atmosphere, its energy is absorbed or diverted in various ways. The thin outer layers of the atmosphere almost completely absorb gamma and X rays. Much of the ultraviolet radiation is absorbed as well, particularly by ozone.

As radiation moves through deeper and denser atmospheric layers, gas molecules cause some radiation to be scattered. As we noted earlier, scattered radiation is radiation that

Eye on Global Change | 2.2

The Ozone Layer—Shield to Life

The release of *chlorofluorocarbons,* or *CFCs,* from the Earth's surface into the atmosphere poses a serious threat to the ozone layer. CFCs are synthetic industrial chemical compounds containing chlorine, fluorine, and carbon atoms. Before 1976, when the U.S. Environmental Protection Agency issued a ban on aerosol spray cans charged with CFCs, many households used CFCs every day. CFCs are still in wide use as cooling fluids in existing refrigeration systems. When these appliances using CFCs are disposed of, their CFCs can be released to the air.

Molecules of CFCs in the atmosphere are very stable close to the Earth. They move upward by diffusion without chemical change until they eventually reach the ozone layer. There, they absorb ultraviolet radiation and are decomposed to chlorine oxide (ClO) molecules. The chlorine oxide molecules in turn attack molecules of ozone, converting them in large numbers into ordinary oxygen molecules by a chain reaction. In this way, the ozone concentration within the ozone layer is reduced, and so there are fewer ozone molecules to absorb ultraviolet energy in the stratosphere.

Other gaseous molecules can act to reduce the concentrations of ozone in the stratosphere as well. Nitrogen oxides, bromine oxides, and hydrogen oxides are examples. As with chlorine oxides, the concentrations of these molecules in the stratosphere are influenced by human activity.

Not all threats to the ozone layer can be attributed to human activity. Aerosols inserted into the stratosphere by volcanic activity also can act to reduce ozone concentrations. The June 1991 eruption of Mount Pinatubo, in the Philippines, reduced global ozone in the stratosphere by 4 percent during the following year, with reductions over midlatitudes of up to 9 percent.

Studies of the ozone layer based on satellite data during the 1980s showed a substantial decline in total global ozone. By late in the decade, scientists had agreed that the decrease in the ozone layer was escalating far faster than had been predicted.

Compounding the problem of global ozone decrease was the discovery in the mid-1980s of a "hole" in the ozone layer over the continent of Antarctica. Here, seasonal thinning of the ozone layer occurs during the early spring of

the southern hemisphere, and ozone reaches a minimum during the month of October.

The thinning occurs after the formation of a polar vortex in the stratosphere during the winter period. This vast whirlpool of wind traps the air it contains and keeps this air out of the Sun during the months of the long polar night. The air in the vortex eventually becomes very cold, and clouds of crystalline water (ice) or other water-containing compounds form within it. The crystals are important because they provide a surface on which chemical reactions can take place. One of these reactions allows chlorine to be converted from a stable form to chlorine oxide. As we noted with CFCs, ClO can rapidly destroy ozone in the presence of sunlight. As the southern hemisphere spring approaches, the polar vortex slowly becomes illuminated by the Sun, and the chlorine oxide reacts with ozone. Ozone concentrations are reduced, and the ozone hole is formed.

The figure on the facing page presents a satellite-derived map of stratospheric ozone over the south pole on September 17, 2001, clearly revealing the ozone hole. In this south polar projection, transparent

---

is turned aside in all directions. It is unchanged except for the direction of its motion. Dust and other particles in the air can also cause scattering. Scattered radiation moving in all directions through the atmosphere is known as *diffuse radiation.* Because scattering occurs in all directions, some energy is sent back into outer space, and some flows down to the Earth's surface. When the scattering turns radiation back to space, we call this process *diffuse reflection.* Under clear sky conditions, this diffuse reflection sends about 3 percent of the incoming solar radiation back to space.

What about absorption? As we saw earlier, molecules and particles can absorb radiation as it passes through the atmosphere. Carbon dioxide and water are the two primary absorbers, but because the water vapor content of air can vary greatly, absorption can vary from one global environ-

ment to another. Absorption results in a rise in air temperature. In this way, incoming solar radiation causes some direct heating of the lower atmosphere. Absorption accounts for about 15 percent of the incoming solar radiation. Thus, when skies are clear, diffuse reflection and absorption combined total about 17 percent, leaving as much as 80 percent of the solar radiation to reach the ground.

The presence of clouds can greatly increase the amount of incoming solar radiation reflected back to space. The bright, white surfaces of thick, low clouds are extremely good reflectors of shortwave radiation. Cloud reflection can account for a direct turning back to space of 30 to 60 percent of incoming radiation (Figure 2.13). Clouds also absorb radiation, perhaps as much as 5 to 20 percent. Under condi-

Map of ozone concentration for the south polar region on September 17, 2001 *Ozone concentration is measured in Dobson units. The lowest values over central Antarctica are about 90 units (blue). Prior to 1979, values were in the range of 250 to 300 units (green). Data were acquired by the Total Ozone Mapping Spectrometer (TOMS) on NASA's Earth Probe TOMS-EP Satellite. (Courtesy NASA.)*

color overlies lighter oceans and darker continents to depict ozone concentration. Blue tones are lowest, while red-brown tones are highest. The largest ozone hole through 2000 was recorded on September 19, 1998, and covered 26,200,000 km² (10,500,000 mi²). In typical years, the antarctic ozone hole forms in mid-September, grows to a maximum extent around the first of October, then slowly shrinks and ultimately disappears in early December.

In the northern hemisphere, conditions for the formation of an ozone hole are not as favorable. An arctic polar vortex forms as well but is much weaker than the antarctic vortex and is less stable. As a result, an early-spring ozone hole is not usually observed in the Arctic. However, in 1993, 1996, 1997, and 1999, notable arctic ozone holes occurred, and atmospheric computer models have projected more such events in the period 2010–2019.

Although the most dramatic reductions in stratospheric ozone levels have been associated with arctic and antarctic ozone holes, reductions have been noted elsewhere as well—including the northern midlatitudes. A number of studies during the 1990s showed reductions of 6 to 8 percent in stratospheric ozone concentrations occurring over North America.

As the global ozone layer thins, the rate of incoming ultraviolet solar radiation reaching the Earth's surface should increase. Recent studies of satellite data have documented this effect. Since 1978, the average annual exposure to ultraviolet radiation increased by 6.8 percent per decade along the 55° N parallel. At 55° S, the increase was 9.8 percent per decade. Over most of North America, the increase was about 4 percent per decade.

How might declining ozone concentrations affect the Earth? An increase in the incidence of skin cancer in humans from enhanced ultraviolet radiation is one of the predicted effects. Other possible effects include reduction of crop yields and damage to some forms of aquatic life.

Responding to the global threat of ozone depletion and its anticipated impact on the biosphere, 23 nations signed a United Nations-sponsored environmental treaty on protection of the ozone layer in Montreal in 1987, with developed nations cutting global CFC consumption by 50 percent by 1999. Halting the use of CFCs in developing nations, scheduled for 2010, is the next step. In 1999, developed nations pledged one billion dollars to help developing nations switch to safer alternatives.

In operation for more than 10 years, the treaty has had an effect. By late 1999, scientists confirmed that stratospheric chlorine concentrations had topped out in 1997 and were continuing to fall. Although the ozone layer is not expected to be completely restored until the middle of the twenty-first century, this observation bodes well for eliminating the threat of enhanced ultraviolet radiation to the Earth's life layer.

---

tions of a heavy cloud layer, as little as 10 percent of incoming solar radiation may actually reach the ground.

The amount of global cloud cover, including cloud type and altitude, is an important factor in determining our global climate. In Chapter 6, we will discuss this topic further.

Albedo

The proportion of shortwave radiant energy scattered upward by a surface is called its **albedo.** For example, a surface that reflects 40 percent of the shortwave radiation it receives has an albedo of 0.40. Albedo is an important property of a surface because it measures how much incident solar energy will be absorbed. A surface with a high albedo, such as snow or ice (0.45 to 0.85), reflects much or most of the solar radiation and absorbs only a smaller amount. A surface with a low albedo, such as black pavement (0.03), absorbs nearly all the incoming solar energy. Since the energy absorbed by a surface can warm the air immediately above it by conduction and convection, surface temperatures will be warmer over a low albedo surface (pavement) than over a high albedo surface (snow or ice).

The albedo of a water surface varies with the angle of incoming radiation. It is very low (0.02) for nearly vertical rays on calm water. However, when the Sun shines on a water surface at a low angle, much of the radiation is directly reflected as Sun glint, producing a higher albedo. For fields, forests, and bare ground, albedos are of intermediate value, ranging from as low as 0.03 to as high as 0.25 percent.

Certain orbiting satellites are equipped with instruments to measure the energy levels of shortwave and longwave

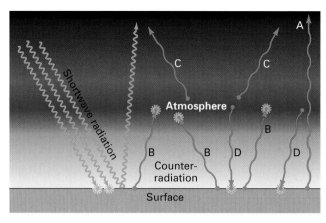

2.14 Counterradiation and the greenhouse effect
Shortwave radiation (left) passes through the atmosphere and is absorbed at the surface, warming the surface. The surface emits longwave radiation. Some of this flow passes directly to space (A), but most is absorbed by the atmosphere (B). In turn, the atmosphere radiates longwave energy back to the surface as counterradiation (D) and also to space (C). The return of outbound longwave radiation (B) by counterradiation (D) constitutes the greenhouse effect.

2.13 Fate of incoming solar radiation *Losses of incoming solar energy are much lower with clear skies (left) than with cloud cover (right). (Copyright © A. N. Strahler. Used by permission.)*

radiation emerging at the top of the atmosphere. Both incoming radiation from the Sun and outgoing radiation from the atmosphere and Earth's surfaces below are measured. Data from these satellites have been used to estimate the Earth's average albedo. This albedo includes reflection by the Earth's atmosphere as well as all its surfaces, so it includes diffuse reflection by clouds, dust, and atmospheric molecules. The albedo values obtained in this way vary between 0.29 and 0.34. This means that the Earth–atmosphere system directly returns to space slightly less than one-third of the solar radiation it receives. *Focus on Remote Sensing 2.3 ● CERES—Clouds and the Earth's Radiant Energy System* describes how satellite instruments map and monitor the reflection of solar shortwave radiation and the emission of longwave radiation by the Earth–atmosphere system.

Counterradiation and the Greenhouse Effect

The amount of shortwave energy absorbed by a surface is an important determinant of a surface's temperature. But the surface is also warmed significantly by longwave radiation emitted by the atmosphere and absorbed by the ground. Let's look at this in more detail.

Figure 2.14 is a diagram showing energy flows between the surface, atmosphere, and space. On the left is the flow of shortwave radiation from the Sun to the surface that we saw in more detail in Figure 2.13. Some of this radiation is reflected back to space (as measured by the albedo of the surface), but much is absorbed, warming the surface.

Looking now at longwave radiation, we see that the surface emits radiation upward as flows labeled A and B. Some of this radiation escapes directly to space (A), but the remainder is absorbed by the atmosphere (B). What about longwave radiation emitted by the atmosphere? Although the atmosphere is colder than the surface, it emits longwave radiation as well. This radiation is shown by flows C and D. It is emitted in all directions, and so some radiates upward to space (C) while the remainder radiates downward toward to surface (D). Since this downward flow is the opposite direction to longwave radiation leaving the surface, we call this flow **counterradiation.** It acts to replace some of the heat emitted by the surface, and so it makes the surface warmer.

The amount of counterradiation depends strongly on the presence of carbon dioxide and water vapor in the atmosphere. Recall from our discussion of longwave radiation earlier in this chapter that much of the longwave radiation emitted upward from the Earth's surface is absorbed by these two atmospheric constituents. This absorbed energy raises the temperature of the atmosphere, which causes it to emit more counterradiation. Thus, the lower atmosphere, with its longwave-absorbing gases, acts like a blanket that traps heat underneath it. Because liquid water

> **Counterradiation from the atmosphere to the Earth's surface helps warm the surface and creates the greenhouse effect. It is enhanced by carbon dioxide, water droplets, and water vapor in the atmosphere.**

is also a strong absorber of longwave radiation, cloud layers, which are composed of tiny water droplets, are even more important than carbon dioxide and water vapor in producing a blanketing effect.

This mechanism, in which the atmosphere traps longwave radiation and returns it to the surface, is termed the **greenhouse effect.** It is similar to the principle used in greenhouses and in homes with large windows to entrap solar heat. In the greenhouse, window glass permits entry of shortwave energy but absorbs and blocks the exit of longwave energy. This phenomenon is even better demonstrated by the intense heating of air in a closed car left parked in the Sun.

Global Energy Budgets of the Atmosphere and Surface

Thus far, we have described a number of important pathways by which incoming solar radiation is scattered, absorbed, and reradiated as longwave radiation. It is an important principle of physics that, except for nuclear reactions, energy is neither created nor destroyed. It is therefore possible to follow the initial stream of solar energy and account for its diversion into different system pathways and its conversion into different energy forms. A full accounting of all the important energy flows among the Sun, atmosphere, surface, and space forms the global energy budget of the Earth's atmosphere and surface. Like a household budget, it must balance, at least over the long term.

Incoming Shortwave Radiation

The global energy budget for the Earth's surface and atmosphere is diagrammed in Figure 2.15. We begin with a discussion of solar shortwave radiation, shown in the left part of the figure. Here we will trace only the downward flow of solar radiation and its fate. Reflection by molecules and dust, clouds, and the surface (including the oceans) totals 31 percentage units. Expressed as a proportion (0.31), this is the albedo of the Earth and atmosphere as a combined system. The right side of this diagram shows values for absorption in the atmosphere. The combined losses through absorption by molecules, dust, and clouds average 20 percentage units. With 31 percentage units of the incoming solar energy flow reflected and 20 percentage units absorbed in the atmosphere, 49 units are left. These units are absorbed by the Earth's land and water surfaces. We have now accounted for all of the incoming flow of insolation.

Surface Energy Flows

Let's now turn to energy flows to and from the surface. We've noted the flow of 49 units of shortwave radiation absorbed by the surface, but what about longwave radiation? The right part of Figure 2.15 shows the components of outgoing longwave radiation for the Earth's surface and the atmosphere. The large arrow on the left shows total longwave radiation leaving the Earth's land and ocean surface. The long, thin arrow indicates that 12 percentage units are lost to space, while 102 units are absorbed by the atmosphere. (These flows are equivalent to flows A and B in Figure 2.14.) The loss is therefore equivalent to 114 percentage units.

This means that the surface emits 114 percent of the total incoming solar radiation! How can this be correct? To answer this question, note that the surface also receives a flow of 95 units of longwave counterradiation from the atmosphere. This will be in addition to the 49 units of short-

2.15 *Diagram of the global energy balance* Values are percentage units based on total insolation as 100. The left figure shows the fate of incoming solar radiation. The right figure shows longwave energy flows occurring between the surface and atmosphere and space. Also shown are the transfers of latent heat, sensible heat, and direct solar absorption that balance the budget for Earth and atmosphere. Data of Kiehl and Trenberth, 1997.

CERES—Clouds and the Earth's Radiant Energy System

The Earth's global radiation balance is the primary determinant of long-term surface temperature, which is of great importance to life on Earth. Yet this balance can also be affected by human activities, such as converting forests to pasture lands or releasing greenhouse gases into the atmosphere. Thus, it is important to monitor the Earth's radiation budget over time as accurately as possible.

For nearly 20 years, the Earth's global radiation balance has been the subject of NASA missions that have launched radiometers—radiation-measuring devices—into orbit around the Earth. These devices scan the Earth and measure the amount of shortwave and longwave radiation leaving the Earth at the top of the atmosphere. An ongoing NASA experiment entitled CERES—Clouds and the Earth's Radiant Energy System— is placing a new generation of these instruments in space to monitor the Earth's global radia-

tion balance well into the twenty-first century.

The CERES instruments scan the Earth from horizon to horizon, measuring outgoing energy flows of reflected solar radiation (0.3 to 5.0 µm), outgoing longwave radiation (8 to 12 µm), and outgoing total radiation (0.3 to 100 µm). From these observations, a team of scientists at NASA's Langley Research Center, led by Dr. Bruce Barkstrom, produces daily global maps of the Earth's upwelling radiation fields. The two images presented in the figure show global reflected solar energy and emitted longwave energy averaged over the month of March 2000 as obtained from the CERES instrument aboard NASA's Terra satellite platform.

The top image shows average shortwave flux ("flux" means "flow"), ranging from 0 to 210 W/m². The largest flows occur over regions of thick clouds near the equator, where the bright, white

clouds reflect much of the solar radiation back to space. The Amazon Basin of South America is an example. In the midlatitudes, persistent cloudiness during this month also shows up as light tones. Tropical deserts, the Sahara for example, are also bright. Snow and ice surfaces in polar regions are quite reflective, but in March the amount of radiation received in polar regions is much less than at the equator or in the midlatitudes. As a result, they don't appear as bright in this image. Oceans, especially where skies are clear, absorb solar radiation and thus show low shortwave fluxes.

Longwave flux is shown in the bottom image on a scale from 100 to 320 W/m². Cloudy equatorial regions have low values, showing the blanketing effect of thick clouds that trap longwave radiation beneath them. Warm tropical oceans in regions of clear sky emit the most longwave flux. Poleward, surface and atmospheric tempera-

wave radiation that are absorbed (left figure). So the surface receives 95 (longwave) + 49 (shortwave) = 144 units in all. Thus, a loss of 114 units by longwave radiation out of a gain of 144 units of longwave and shortwave radiation is certainly possible.

On the far right of the figure are two smaller arrows. These show the flow of energy away from the surface as latent heat (23 units) and sensible heat (7 units). Evaporation of water from moist soil or ocean transfers latent heat from the surface to the atmosphere, as explained earlier. Sensible heat transfer occurs when heat is directly conducted from the surface to the adjacent air layer. When air is warmed by direct contact with a surface and rises, it takes its heat along with it by convection. In this process, heat flows from the surface to the atmosphere.

These last two heat flows are not part of the radiation balance, since they are not in the form of radiation. However, they are a very important part of the total energy budget of the surface, which includes all forms of energy. Taken together, these two flows account for 30 units leaving the surface. With their contribution, the surface energy balance

is complete. Total gains are 95 (longwave) + 49 (shortwave) = 144. Total losses are 114 (longwave) + 23 (latent heat) + 7 (sensible heat) = 144.

Energy Flows to and from the Atmosphere

What about the atmosphere? Let's first look at energy flowing into the atmosphere. Like the Earth's surface, the atmosphere gains energy by absorption of shortwave radiation, amounting to 20 units. The atmosphere also gains energy from latent heat transfer (23 units) and sensible heat transfer (7 units). In longwave radiation, the atmosphere absorbs 102 units emitted by the surface. (These are the flows labeled B in Figure 2.14.) These units total to 152.

Looking at energy leaving the atmosphere, we see a loss of 57 units of longwave radiation to space (flows labeled C in Figure 2.14) and a loss of 95 units in counterradiation to the surface (flows labeled D in Figure 2.14). Since these total 152 units, we see that the atmosphere's energy budget is balanced.

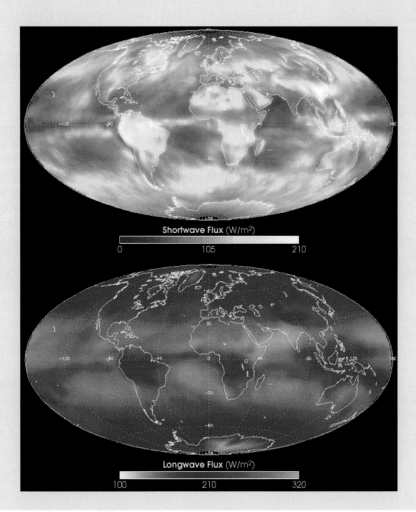

tures drop, so longwave energy emission also drops significantly.

As you can see from these images, clouds are very important determiners of the global radiation balance. A primary goal of the CERES experiment is to learn more about the Earth's cloud cover, which changes from minute to minute and hour to hour. This knowledge can be used to improve global climate models that predict the impact of human and natural change on the Earth's climate.

The most important contribution of CERES, however, is continuous and careful monitoring of the Earth's radiant energy flows. In this way, small, long-term changes, induced by human or natural change processes, can be detected in spite of large variations in energy flows from place to place and time to time caused by clouds.

This analysis helps us to understand the vital role of the atmosphere in trapping heat though the greenhouse effect. Without the 95 units of counterradiation of the atmosphere, the surface would have only 49 units of absorbed shortwave radiation to emit. The temperature needed to radiate so few units is well below freezing. So without the greenhouse effect, our planet would be a cold, forbidding place.

Our analysis also helps us to understand how global change might affect the Earth's climate. Suppose that clearing forests for agriculture and turning agricultural lands into urban and suburban areas decreases surface albedo. In that case, more energy would be absorbed by the ground, raising its temperature. That, in turn, would increase the flow of surface longwave radiation to the atmosphere, which would be absorbed and then boost counterradiation. So the effect would probably be to amplify the warming through the greenhouse effect. What if air pollution causes more low, thick clouds to form? Since low clouds increase shortwave reflection back to space, the effect will be to cool the surface and atmos-

phere. What if increasing condensation trails from jet aircraft cause more high, thin clouds? High, thin clouds absorb more longwave energy than they reflect shortwave energy, so they should make the atmosphere warmer. That should boost counterradiation, increasing the greenhouse effect. The energy flow linkages between the Sun, surface, atmosphere, and space are critical components of our climate system, and understanding them will help you better understand global climate change. GEODISCOVERIES

NET RADIATION, LATITUDE, AND THE ENERGY BALANCE

Solar energy is intercepted by our planet, and because some of it is absorbed, the heat level of our planet tends to rise. At the same time, our planet radiates energy into outer space, a process that tends to reduce its level of heat energy. Over time, as we have seen, these incoming and outgoing radiation flows must balance for the Earth as a whole. However, incoming and outgoing flows do not have to balance at any given surface location. At night, for example, there is no incoming radiation, yet the Earth's surface and atmosphere still emit outgoing radiation. In any one place and time, then, more radiant energy can be lost than gained, or vice versa.

Net radiation is the difference between all incoming radiation and all outgoing radiation. In places where radiant energy flows in faster than it flows out, net radiation is a positive quantity, providing an energy surplus. In other places, where radiant energy is flowing out faster than it is flowing in, net radiation is a negative quantity, yielding an energy deficit. Our analysis of the Earth's radiation balance has already shown that for the entire Earth and atmosphere as a unit, net radiation is zero on an annual basis.

In Figure 2.8, we saw that solar energy input varies strongly with latitude. What is the effect of this variation on net radiation? To answer this question, we will look at the net radiation profile spanning the entire latitude range from 90° N to 90° S. In this analysis we will use yearly averages for each latitude, so that the effect of seasons is concealed.

The lower part of Figure 2.16 presents a global profile of net radiation from pole to pole. Between about 40° N lat. and 40° S lat. there is a net radiant energy gain labeled "energy surplus." In other words, incoming solar radiation exceeds outgoing longwave radiation throughout the year. Poleward of 40° N and 40° S, the net radiation is negative

and is labeled "energy deficit"—meaning that outgoing longwave radiation exceeds incoming shortwave radiation.

If you carefully examine the lower part of Figure 2.16, you will find that the area on the graph labeled "surplus" is equal in size to the combined areas labeled "deficit." The matching of these areas confirms the fact that net radiation for the Earth as a whole is zero and that global incoming shortwave radiation exactly balances global outgoing longwave radiation.

The pattern of energy surplus at low latitudes and energy deficit at high latitudes creates a flow of energy from low latitudes to high (upper part of Figure 2.16). This energy flow is in the form of sensible and latent heat in poleward movements of warm ocean water and warm, moist air that are part of the global circulation patterns of the ocean and the atmosphere. Heat is exchanged when oceanic and atmospheric currents move warm water and warm, moist air poleward and cooler water and cooler, drier air equatorward. *GEODISCOVERIES*

We'll return to these flows in later chapters, but keep in mind that this **poleward heat transfer,** driven by the imbalance in net radiation between low and high latitudes, is thus the power source for ocean currents and broad-scale atmospheric circulation patterns. Without this circulation, low latitudes would heat up and high latitudes would cool down until a radiative balance was achieved, leaving our planet with much more extreme temperature contrasts. Our Earth would be a far different place than it is now. Figure 2.17 illustrates

2.16 Annual surface net radiation from pole to pole *Where net radiation is positive, incoming solar radiation exceeds outgoing longwave radiation. There is an energy surplus, and energy moves poleward as latent heat and sensible heat. Where net radiation is negative, there is an energy deficit. Latent and sensible heat are lost in the form of outgoing longwave radiation.*

a...Solar-powered call box
This emergency telephone is powered by the solar cell atop its pole.

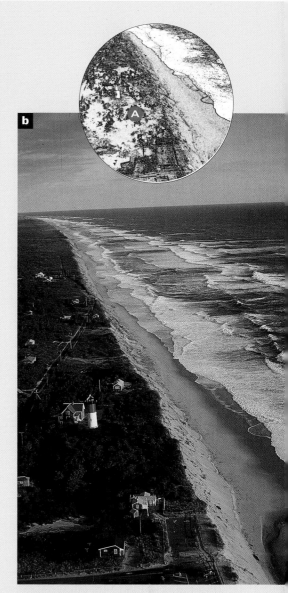

b...Eye on the Landscape Wave erosion Ocean waves, powered by the Sun through the Earth's wind system, have eroded the eastern bluffs of Cape Cod by hundreds of meters since the end of the Ice Age. **What else would the geographer see?...Answers at the end of the chapter.**

c...Eye on the Landscape Tropical cyclone Solar power also indirectly powers severe storms like Typhoon Odessa, shown here in a space photo. **What else would the geographer see?...Answers at the end of the chapter.**

d...Water power The hydrologic cycle, powered by solar evaporation of water over oceans, generates runoff from rainfall that erodes and deposits sediment.

Eye on the Environment | **2.4**

Solar Power

The Earth intercepts solar energy at the rate of 1.77×10^{14} kW, or about $1\frac{1}{2}$ quadrillion megawatt-hours per year. (A megawatt, mW, is one million—10^6—watts.) This quantity of energy is about 28,000 times as much as all the energy presently being consumed each year by human society. Thus, an enormous source of energy is near at hand waiting to be used. In addition to its abundance, solar energy has other advantages. Unlike fossil fuel consumption, solar power generation does not release carbon dioxide. As we will see in Chapter 3, CO_2 concentrations have been steadily increasing from fossil fuel combustion for the last century, and it is predicted that they will cause climatic warming through an enhancement of the greenhouse effect. Also, utilization of solar power does not involve the release of pollutant gases and airborne

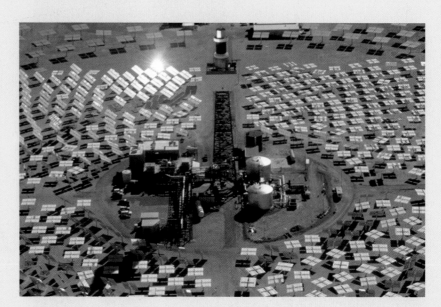

Solar II *This solar power plant, built as a demonstration project in 1994–1995, generated 100 megawatts of electric power when in operation. The mirrors, called heliostats, were moved by computer-controlled motors to focus the Sun's rays on a reservoir of molten nitrate salt located atop the central tower. Heat from the salt was exchanged to create steam that was used to generate the power.*

some of the ways that natural processes and human uses are driven by solar power.

Solar radiation to Earth is a truly huge energy flow, dwarfing human energy production by fossil fuel burning or other means of power generation. In the long run, it seems likely that we will have to tap more of the flow of the Sun's energy to reduce the buildup of CO_2 in the atmosphere that is caused by fossil fuel burning and threatens to warm global climate. *Eye on the Environment*

Poleward heat transfer moves heat from the low latitudes toward the poles. The heat transfer is in the form of warm ocean water and warm, moist air that flows poleward and is replaced by cooler water and cooler, drier air moving equatorward.

2.4 • Solar Power provides more information on how solar power is harvested.

A LOOK AHEAD Although our analysis of the Earth's energy balance is not yet complete, it is clear that the balance is a sensitive one involving a number of factors that determine how energy is transmitted and absorbed. Have our industrial activities already altered the components of the planetary radiation balance? An particles as does the burning of fossil fuel.

The simplest form of solar heating involves the use of large glass panes to admit sunlight into a room—the greenhouse principle. An overhanging roof admits solar rays during the winter, when the Sun's path is low in the sky, but blocks them in summer. Practical solar heating of interior building space and hot-water systems makes use of solar collectors. Panels of solar collectors are usually placed on the roof of the building The flat-plate collector consists of a network of aluminum or copper tubes carrying circulating water. The tubes, painted black, absorb solar energy and conduct it to the water. Water is heated to a temperature of about 65°C (about 150°F) and can be used to transfer heat to a hot-water supply, as well as to a conventional space-heating system.

A few solar energy power plants also operate by direct absorption of solar energy using reflecting mirror systems. Many movable mirrors, called heliostats, reflect solar rays to the top of a central tower where a steam boiler and an electric generator are located. The black metal of the boiler readily absorbs the heat. A heliostat power plant constructed in 1981 near Barstow, California, named Solar I, provided steam for five years, delivering 10 megawatts of electric power. In 1994 and 1995, the plant was modified by adding more heliostats and converting it to use molten nitrate salt to store the solar heat instead of steam. Its capacity was also increased to 100 megawatts. The power plant, shown in the photo above, is now known as Solar II.

Another way to produce electricity from solar radiation is to generate it directly with solar cells. These cells use special materials, such as crystalline silicon, that absorb solar radiation and convert it into electricity. Using present technology, arrays of solar cells would need to occupy large areas of ground surface to produce significant outputs of energy. Storage of the electricity is another problem, since the solar cells will not generate power at night. Storage as hydrogen fuel is an attractive possibility for large-scale systems, while batteries can be used to store power in small systems, such as those of homes, farms, and ranches. One small-system use is to drive electric water pumps needed for irrigation of fields in the sunny dry season of tropical countries, for example, on the wheat farms of Pakistan and northern India.

Note that the Sun also powers the natural motion of wind and water and that these moving fluids can be harnessed for power generation as well. We will touch on wind, water, and wave power in later chapters.

increase in carbon dioxide will increase the absorption of longwave radiation by the atmosphere, enhancing the greenhouse effect. Will this change cause a steady and permanent rise in average atmospheric temperature?

An increase in atmospheric particles at upper levels of the atmosphere will increase the scattering of incoming shortwave radiation and thus reduce the shortwave energy available to warm the surface. On the other hand, increased dust content at low levels will act to absorb more longwave radiation and so raise the surface temperature. Has either change occurred?

Human habitation, through cultivation and urbanization, has profoundly altered the Earth's land surfaces. Have these changes, which affect surface albedo and the transfer of latent and sensible heat to the atmosphere, modified the global energy balance? Answers to these questions require further study of the processes of heating and cooling of the Earth's atmosphere, lands, and oceans. Our next chapter concerns air temperature and how and why it varies daily and annually depending on the surface energy balance.

Remote Sensing for Physical Geography

Remote sensing refers to gathering information from great distances and over broad areas, usually through instruments mounted on aircraft or orbiting space vehicles. These instruments, or *remote sensors,* measure electromagnetic radiation coming from the Earth's surface and atmosphere as received at the aircraft or spacecraft platform. The data acquired by remote sensors are typically displayed as images—photographs or similar depictions on a computer screen or color printer—but are often processed further to provide other types of outputs, such as maps of albedo, vegetation condition or extent, or land-cover class.

All substances, whether naturally occurring or synthetic, are capable of reflecting, transmitting, absorbing, and emitting electromagnetic radiation. For remote sensing, however, we are only concerned with energy that is reflected or emitted by an object and that reaches the remote sensor. For remote sensing of reflected energy, the Sun is the source of radiation in many applications. As we saw earlier in this chapter, solar radiation reaching the Earth's surface is concentrated in wavelengths from about 0.3 to 2.1 μm in visible, near-infrared, and shortwave infrared wave bands (Figures 2.2, 2.4). Remote sensors are commonly constructed to measure radiation in all or part of this range. For remote sensing of emitted energy, the object or substance itself is the source of the radiation, which is related largely to its temperature.

Colors and Spectral Signatures

Most objects or substances at the Earth's surface possess color to the human eye. This means that they reflect radiation differently in different parts of the visible spectrum. Figure S2.1 shows typical reflectance spectra for water, vegetation, and soil in the solar wavelength range as they might be viewed through the atmosphere. (Note that water vapor in the atmosphere absorbs radiation strongly at wavelengths from about 1.2 to 1.4 μm and 1.75 to 1.9 μm, so it is not possible for a remote sensor to "see" the surface at those wavelengths.)

Let's examine the three spectral reflectance curves. Water surfaces are always dark but are slightly more reflective in the blue and green regions of the visible spectrum. Thus, water appears blue or blue-green to our eyes. Beyond the visible region, water absorbs nearly all shortwave radiation it receives and so looks black in images acquired in the near-infrared and shortwave infrared regions.

Vegetation appears dark green to the human eye, which means that it reflects more energy in the green portion of the visible spectrum while reflecting somewhat less in the blue and red portions. But vegetation also reflects very strongly in near-infrared wavelengths, which the human eye cannot see. Because of this property, vegetation will be bright in near-infrared images. This distinctive behavior of vegetation—appearing dark in visible bands and bright in the near-infrared—is the basis for much of vegetation remote sensing, as we will see in many examples of remotely sensed images throughout this book.

The soil spectrum shows a slow increase of reflectance across the visible and near-infrared spectral regions and a slow decrease

S2.1 Reflectance spectra of vegetation, soil, and water

through the shortwave infrared. Looking at the visible part of the spectrum, we see that soil is brighter overall than vegetation and is somewhat more reflective in the orange and red portions. Thus, it appears brown. (Note that this is just a "typical" spectrum—soil color can actually range from black to bright yellow or red.)

We refer to the pattern of relative brightness within spectral bands as the *spectral signature* of an object or substance. Spectral signatures can be used to recognize objects or surfaces in remotely sensed images in much the same way that we recognize objects by their colors. In computer processing of remotely sensed images, spectral signatures can be used to make classification maps, showing, for example, water, vegetation, and soil.

Aerial Photography

Aerial photography is the oldest form of remote sensing. Air photos have been in wide use since before World War II. Commonly, the field of one photograph overlaps the next along the plane's flight path, so that the photographs can be viewed stereoscopically for a three-dimensional effect. Because of its high resolution (degree of sharpness) and low cost, aerial photography is a widespread application of remote sensing.

Aerial photography often makes use of color infrared film. This special film is sensitive to near-infrared wavelengths in addition to visible wavelengths. Red color in the film is produced as a response to near-infrared light, green color is produced by red light, and blue color by green light. Because healthy, growing vegetation reflects much more strongly in the near-infrared than in the red or green regions of the spectrum, vegetation has a characteristic red appearance. Figure S2.2 shows a color infrared photo image of an agricultural region in California. The distinctive red color identifies crop fields. The image also shows areas of abnormally high ground water and high soil salinity, both of which are agricultural problems in this region. Other examples of color infrared photos and images will be found in later chapters of this book.

Photography has been extended to greater distances through the use of cameras on orbiting space vehicles. An example is the large-format camera, which was designed to produce very large, very detailed transparencies of the Earth's surface suitable for precise topographic mapping. An excellent example, acquired in a mission aboard the Space Shuttle, is shown in Figure S2.3. Astronauts have also made a number of striking photos of the Earth from space using hand-held cameras.

Thermal Infrared Sensing

Recall from Figure 2.2 that radiation leaving the Earth's surface is concentrated in the thermal infrared spectral region, ranging from about 8 to 12 μm. The amount of radiation emitted by the surface of an object or substance is proportional to the fourth power of its absolute temperature. This means that warm objects emit more thermal radiation than cold ones, so warmer objects appear brighter in thermal infrared images. We have already noted this principle in our discussion of Figure 2.3 in the main chapter text.

Besides absolute temperature, the intensity of infrared emission depends on the *emissivity* of an object or substance. Whereas the blackbody is an ideal perfect radiator of energy, all substances are imperfect radiators and emit less energy. Emissivity is the ratio of the actual energy emitted by an object or substance to that of a blackbody at the same temperature. For most natural Earth surfaces, emissivity is comparatively high—between 0.85 and 0.99. Differences in emissivity affect thermal images. For example, two different surfaces might be at the same temperature, but the one with the higher emissivity will look brighter because it emits more energy.

Some substances, such as crystalline minerals, show different emissivities at different thermal infrared wavelengths. In a way, this is like having a particular color, or spectral signature, in the thermal infrared spectral region. In Chapter 11 we will see examples of how some rock types can be distinguished and mapped using thermal infrared images from several wavelengths.

Radar

There are two classes of remote sensor systems: passive and active. *Passive systems* acquire images without providing a source of wave energy. The most familiar passive system is the camera, which uses film that is sensitive to solar energy reflected from the scene. *Active systems* use a beam of wave energy as a source, sending the beam toward an object or surface. Part of the energy is reflected back to the source, where it is recorded by a detector. A simple analogy would be the use of a spotlight on a dark night to illuminate a target, which then reflects light back to the eye.

Radar is an example of an active sensing system. Radar systems in remote sensing use the *microwave* portion of the electromagnetic spectrum, so named because the waves have a short wavelength compared to other types of radio waves (Figure 2.2). (However, these wavelengths are much, much longer than those of solar short-wave radiation.) Radar systems emit short pulses of microwave radiation and then "listen" for a returning microwave echo. By analyzing the time and strength of the return pulse, an image is created showing the surface as it is illuminated by the radar beam.

An advantage of radar systems is that they use wavelengths that are not significantly absorbed by liquid water. This means that radar systems can penetrate clouds to provide images of the Earth's surface in any weather. At some short wavelengths, however, microwaves are scattered by water droplets and can produce a return

S2.2 *High-altitude color infrared photograph* This color infrared photo of an area near Bakersfield, California, in the southern San Joaquin Valley, was taken by a NASA aircraft flying at approximately 18 km (60,000 ft). Photos such as this can be used to study problems associated with agriculture. Problems affecting crop yields arise in abnormally high ground water areas (A), which appear dark, and areas of high soil salinity (B), which appear light. The various red tones are associated with different types of crops. (Courtesy NASA, compiled and annotated by John E. Estes and Leslie W. Senger.)

S2.3 *Color infrared photograph from low Earth orbit* *A special large-format camera carried on a flight of the NASA Space Shuttle acquired this color infrared photograph. The area shown, about 125 km (77 mi) in width, includes the Sulaiman Range in western Pakistan (upper left). The Indus and Sutlej rivers, fed by snowmelt from the distant Hindu Kush and Himalaya Ranges, cross the scene flowing from northeast to southwest. A mottled pattern of green fields (red) interspersed with barren patches of saline soil (white) covers much of the lower fourth of the area. (Courtesy Geonex/Chicago Aerial Survey, Inc.)*

signal sensed by the radar apparatus. This effect is the basis for weather radars, which can detect rain and hail and are used in local weather forecasting.

Figure S2.4 shows a radar image of the folded Appalachian Mountains in south-central Pennsylvania. It is produced by an airborne radar instrument that sends pulses of radio waves downward and sideward as the airplane flies forward. Surfaces oriented most nearly at right angles to the slanting radar beam will return the strongest echo and therefore appear lightest in tone. In contrast, those surfaces facing away from the beam will appear darkest. The effect is to produce an image resembling a three-dimensional model of the landscape illuminated at a strong angle. The image shows long mountain ridges running from upper right to lower left and casting strong radar shadows to emphasize their three-dimensional form. The ridges curve and turn sharply, revealing the geologic structure of the region. Between the ridges are valleys of agricultural land, which are distinguished by their rougher texture in the image. In the upper left is a forested plateau that has a smoother appearance.

S2.4   Side-looking radar image from south-central Pennsylvania *The image shows a portion of the folded Appalachians with zigzag ridges and intervening valleys. The area shown is about 40 km (25 mi) wide. Compare this with Figure 17.15, which is a Landsat image of this region. (SAR image courtesy of Intera Technologies Corporation, Calgary, Alberta, Canada.)*

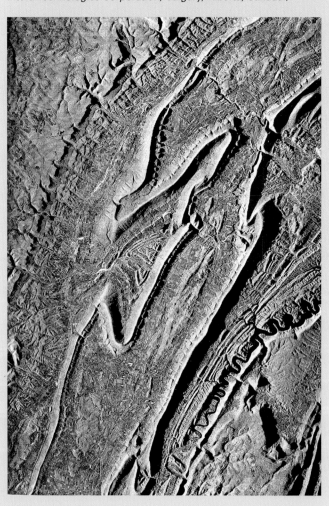

Digital Imaging

Modern remote sensing relies heavily on computer processing to extract and enhance information from remotely sensed data. This requires that the data be in the form of a *digital image* (Figure S2.5). In this format, the picture consists of a very large number of grid cells arranged in rows and columns to cover the scene. Each grid cell records a brightness value and is referred to as a *pixel,* a term that arises as a contraction of "picture element." Normally, low values code for dark (low reflectance), and high values code for light (high reflectance). To create an image that is viewable, as, for example, on a computer screen, the brightness values are fed to a special computer chip that generates a corresponding television signal fed to the monitor. Or the image is processed and sent to a computer printer to produce a hard copy print or transparency.

The great advantage of digital images over images on photographic film is that they can be processed by computer, for example, to increase contrast or sharpen edges (Figure S2.6). *Image processing* refers to the manipulation of digital images to extract, enhance, and display the information that they contain. In remote sensing, image processing is a very broad field that includes many methods and techniques for processing remotely sensed data.

Digital Imaging Systems

Many remotely sensed digital images are acquired by scanning systems, which may be mounted in aircraft or on orbiting space vehicles. *Scanning* is the process of receiving information instantaneously from only a very small portion of the area being imaged (Figure S2.7). The scanning instrument senses a very small field of view that runs rapidly across the ground scene. Light from the field of view is focused on a detector that responds very quickly to small changes in light intensity. Electronic circuits read out the detector at very short time intervals and

S2.5 Zooming in on a digital image *These four views of the San Francisco Bay Bridge and Yerba Buena Island acquired by Landsat in the near-infrared band show how a digital image is composed of individual pixels with different brightness values. (A–C) Progressively smaller subimages zooming on the east end of the island. (D) Actual brightness values for a small array of 25 pixels, scaled to range from 0 (darkest) to 255 (lightest). (A. H. Strahler.)*

record the intensities. Later, the computer reconstructs a digital image of the ground scene from the measurements acquired by the scanning system.

Most scanning systems in common use are *multispectral scanners*. These devices have multiple detectors and measure brightness in several wavelength regions simultaneously. An example is the Thematic Mapper instrument used aboard the Landsat series of Earth-observing satellites. This instrument simultaneously collects reflectance data in seven spectral bands. Six wavebands sample the visible, near-infrared and short-wave infrared regions, while a seventh records thermal infrared emissions. Figure S2.8 shows a color composite of the Boston region acquired by the Landsat Thematic Mapper on September 27, 1991. The image uses red, near-infrared, and shortwave infrared wavebands to show vegetation in green,

S2.6 *Image processing* (a) A red band image of Death Valley acquired by the Landsat Thematic Mapper. The semicircular feature in the upper left of the photo is a gravel fan, built at the base of the adjoining mountain front. The dark feature at the top of the fan is irrigated land with a green vegetation cover. In (b), the contrast has been enhanced. In (c), an edge detection procedure produces an image of white lines marking edges. Part (d) shows an edge-enhanced image made by adding images (b) and (c) together.

beaches and bare soils in pink, and urban surfaces in shades of blue.

Scanning provides images using a small number of detectors that are read out very quickly. An alternative to scanning is *direct digital imaging* using large numbers of detectors arranged in a two-dimen-sional array (Figure S2.9). This technology is in common use in digital cameras. Instead of a layer of light-sensitive film, a digital camera has an array of millions of tiny detectors arranged in rows and columns that individually measure the amount of light they receive during an exposure. Electronic circuitry reads out the measurement made by each detector, composing the entire image within a second or two. The image is stored on a memory chip for later downloading to a computer. Although photographic film in a

S2.7 Multispectral scanning from aircraft *As the aircraft flies forward, the scanner sweeps from side to side. The result is a digital image covering the overflight area. (A. H. Strahler.)*

S2.9 An area array of detectors *The center part of this computer chip is covered by an array of tiny light detectors arranged in rows and columns.*

conventional camera can record finer detail than a digital camera in a typical application, the digital camera measures brightness more accurately and outputs a digital image that can be read directly and processed by computer. Note that it is possible to scan a photograph or transparency to produce a digital image as well. This allows the photographic image to be processed in ways that could never be attempted in a photographic darkroom.

Orbiting Earth Satellites

With the development of orbiting Earth satellites carrying remote sensing systems, remote sensing has expanded into a major branch of geographic research. Because orbiting satellites can image and monitor large geographic areas or even the entire Earth, global and regional studies have become possible that could not have been carried out in any other way.

How does a satellite sensor image the entire Earth? One way is to use a polar orbit. As the satellite circles the Earth, passing over each pole, the Earth rotates underneath it, allowing all of the Earth to be imaged after repeated passes. However, a polar orbit remains fixed in space with respect to the stars, not to the Sun. So as the Earth revolves around the Sun in its yearly cycle, the satellite passes overhead about four minutes later each solar day. This means that illumination conditions slowly change from day to day and quite a lot from month to month or season to season.

S2.8 Landsat image of Boston *An image of Boston, acquired by Landsat Thematic Mapper on September 27, 1991. In this false-color composite, the red color is from the shortwave infrared band (1.55 to 1.75 μm), the green from the near-infrared (0.79 to 0.91 μm), and the blue from the red band (0.60 to 0.72 μm). (A. H. Strahler.)*

(a)

(b)

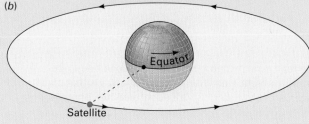

S2.10 Satellite orbits *(a) Earth track of a Sun-synchronous orbit. With the Earth track inclined at 80° to the equator, the orbit slowly swings eastward at about 30° longitude per month, maintaining its relative position with respect to the Sun. Between March 1 and May 1 (shown) the orbit moves about 60°. (© A. N. Strahler.) (b) Motion of a geostationary satellite. Because the satellite revolves around the Earth above the equator at the same rate as the Earth's rotation, it appears fixed in the sky above a single point on the equator.*

A solution to this problem is to use a *Sun-synchronous orbit* (Figure S2.10*a*). This type of near-polar orbit is oriented at a small angle to the polar axis and so crosses the equator at an angle less than 90 degrees. Recall from Chapter 1 that the Earth is not a perfect sphere but is slightly thicker at the equator than at the poles. Thus, the Earth's gravity is slightly greater at the equator. When the orbit crosses the equator at an angle, the difference in gravity acts to push the orbit very slightly eastward. This keeps the satellite orbit in time with the Sun instead of the stars, so that satellite overpasses continue to occur at the same time of day. Typical Sun-synchronous orbits take 90 to 100 minutes to circle the Earth and are located at heights of about 700 to 800 km (430 to 500 mi) above the Earth's surface.

Another orbit used in remote sensing is the *geostationary orbit* (Figure S2.10*b*). Instead of orbiting around the poles, a satellite in geostationary orbit constantly revolves above the equator. The orbit height, about 35,800 km (22,200 mi), is set so that the satellite makes one revolution in exactly 24 hours in the same direction that the Earth turns. Thus, the satellite always remains above the same point on the equator. From its high vantage point, the geostationary orbiter provides a view of nearly half of the Earth at any moment.

Geostationary orbits are ideal for observing weather, and the weather satellite images readily available on television and the Internet are obtained from geostationary remote sensors. Geostationary orbits are also ideal for communications satellites. Since a geostationary orbiter remains at a fixed position in the sky for an Earthbound observer, a high-gain antenna can be pointed at the satellite and fixed in place permanently, providing high-quality, continuous communications.

Remote sensing is an exciting, expanding field within physical geography and the geosciences in general. This special supplement has provided some basic information about remote sensing that will come in handy as you read our special boxed features, *Focus on Remote Sensing*, and as you view the various examples of remotely sensed images throughout the text.

In Review | The Global Energy Balance

- **Electromagnetic radiation** is a form of energy emitted by all objects. The wavelength of the radiation determines its characteristics. The hotter an object, the shorter the wavelengths of the radiation and the greater the amount of radiation that it emits.

- Radiation emitted by the Sun includes *ultraviolet, visible, near-infrared,* and *shortwave infrared radiation. Thermal infrared radiation,* which is emitted by Earth surfaces, is familiar as heat. The atmosphere **absorbs** and **scatters** radiation in certain wavelength regions. Radiation flows are measured in watts per square meter.

- The Earth continuously absorbs solar **shortwave radiation** and emits **longwave radiation.** In the long run, the gain and loss of radiant energy remain in balance, and the Earth's average temperature remains constant.

- **Insolation,** the rate of solar radiation flow available at a location at a given time, is greater when the Sun is higher in the sky. *Daily insolation* is also greater when the period of daylight is longer. Between the tropics and poles, the Sun rises higher in the sky and stays longer in the sky at the summer solstice than at the equinox and longer at the equinox than at the winter solstice. Near the equator, daily insolation is greater at the equinoxes than at the solstices. Annual insolation is greatest at the equator and least at the poles. However, the poles still receive 40 percent of the radiation received at the equator.

- The pattern of annual insolation with latitude leads to a natural naming convention for latitude zones: *equatorial, tropical, subtropical, midlatitude, subarctic (subantarctic), arctic (antarctic),* and *polar.*

- The Earth's atmosphere is dominated by *nitrogen* and *oxygen gases. Carbon dioxide* and *water vapor,* though lesser constituents by volume, are very important because they absorb longwave radiation and enhance the greenhouse effect.

- **Ozone** (O_3) is a tiny but important constituent of the atmosphere that is concentrated in a layer of the upper atmosphere. It is formed from oxygen (O_2) by chemical reactions that absorb ultraviolet radiation, thus sheltering the organisms of the Earth's surface from the damaging effects of ultraviolet rays.

- **Latent heat** and **sensible heat** are additional forms of energy. Latent heat is taken up or released when a change of state occurs. Sensible heat is contained within a substance. It can be transferred to another substance by *conduction* or *convection.*

- The global energy system includes a number of important pathways by which insolation is transferred and transformed. Part of the solar radiation passing through the atmosphere is absorbed or scattered by molecules, dust, and larger particles. Some of the scattered radiation returns to space as *diffuse reflection.* The land surfaces, ocean surfaces, and clouds also reflect some solar radiation back to space.

- The proportion of radiation that a surface absorbs is termed its **albedo.** The albedo of the Earth and atmosphere as a whole planet is about 30 percent.

- The atmosphere absorbs longwave energy emitted by the Earth's surface, causing the atmosphere to **counterradiate** some of that longwave radiation back to Earth, thereby creating the **greenhouse effect.** Because of this heat trapping, the Earth's surface temperature is considerably warmer than we might expect for an Earth without an atmosphere.

- **Net radiation** describes the balance between incoming and outgoing radiation. At latitudes lower than 40 degrees, annual net radiation is positive, while it is negative at higher latitudes. This imbalance creates **poleward heat transfer** of latent and sensible heat in the motions of warm water and warm, moist air, and thus provides the power that drives ocean currents and broad-scale atmospheric circulation patterns.

Key Terms |

electromagnetic
   radiation, p. 52
visible light, p. 53
absorption, p. 56
scattering, p. 56
shortwave radiation,
   p. 56

longwave radiation,
   p. 56
global radiation
   balance, p. 56
insolation, p. 57
ozone, p. 62
sensible heat, p. 63

sensible heat transfer,
   p. 63
latent heat, p. 63
latent heat transfer,
   p. 63
albedo, p. 65
counterradiation, p. 66

greenhouse effect, p. 67
net radiation, p. 78
poleward heat transfer,
   p. 70

Review Questions

1. What is electromagnetic radiation? How is it characterized? Identify the major regions of the electromagnetic spectrum.
2. How does the temperature of an object influence the nature and amount of electromagnetic radiation that it emits?
3. What is the solar constant? What is its value? What are the units with which it is measured?
4. How does solar radiation received at the top of the atmosphere differ from solar radiation received at the Earth's surface? What are the roles of absorption and scattering?
5. Compare the terms *shortwave radiation* and *longwave radiation.* What are their sources?
6. How does the atmosphere affect the flow of longwave energy from the Earth's surface to space?
7. What is the Earth's global energy balance, and how are shortwave and longwave radiation involved?
8. How does the Sun's path in the sky influence daily insolation at a location? Compare summer solstice and equinox paths of the Sun in the sky for 40° N lat. and the equator.
9. What influence does latitude have on the annual cycle of daily insolation? on annual insolation?
10. Identify the three largest components of the dry air. Why are carbon dioxide and water vapor important atmospheric constituents?
11. How does the ozone layer protect the life layer?
12. Describe latent heat transfer and sensible heat transfer.
13. What is the fate of incoming solar radiation? Discuss scattering and absorption, including the role of clouds.
14. Define albedo and give two examples.
15. Describe the counterradiation process and how it relates to the greenhouse effect.
16. Discuss the energy balance of the Earth's surface. Identify the types and sources of energy flows that the surface receives and do the same for energy flows that it loses.
17. Discuss the energy balance of the atmosphere. Identify the types and sources of energy flows that the atmosphere receives and do the same for energy flows that it loses.
18. What is net radiation? How does it vary with latitude?
19. What is the role of poleward heat transport in balancing the net radiation budget by latitude?

Geographer's Tools 2.1 ● Remote Sensing for Physical Geography

1. What is remote sensing? What is a remote sensor?
2. Compare the reflectance spectra of water, vegetation, and a typical soil. How do they differ in visible wavelengths? in infrared wavelengths?
3. What color is vegetation on color infrared film? Why?
4. What is emissivity, and how does it affect the amount of energy radiated by an object?
5. Is radar an example of an active or a passive remote sensing system? Why?
6. What is a digital image? What advantage does a digital image have over a photographic image?
7. Describe two ways of acquiring a digital image.
8. How does a Sun-synchronous orbit differ from a geostationary orbit? What are the advantages of each type?

Eye on the Environment 2.2 ● The Ozone Layer—Shield to Life

1. What are CFCs, and how do they impact the ozone layer?
2. When and where have ozone reductions been reported? Have corresponding reductions in ultraviolet radiation been noted?
3. What is the outlook for the future health of the ozone layer?

Eye on the Environment 2.4 ● Solar Power

1. Why is solar power in principle more desirable than power produced by burning fossil fuels?
2. Describe two different ways of harnessing solar power.

Visualizing Exercises

1. Place yourself in Figure 2.7*b*. Imagine that you are standing in the center of the figure where the N–S and E–W lines intersect. Turn so that you face south. Using your arm to point at the Sun's position, trace the path of the Sun in the sky at the equinox. It will rise exactly to your left, swing upward to about a 50° angle, and then descend to the horizon exactly at your right. Repeat for the summer and winter solstices, using the figure as a guide. Then try it for the north pole, equator, and tropic of capricorn.
2. Sketch the world latitude zones on a circle representing the globe and give their approximate latitude ranges.
3. Sketch a simple diagram of the Sun above a layer of atmosphere above the Earth's surface, somewhat like Figure 2.14. Using Figure 2.15 as a guide, draw arrows indicating flows of energy among Sun, atmosphere, and surface. Label each arrow using terms from Figure 2.15.

Essay Questions

1. Suppose the Earth's axis of rotation was perpendicular to the orbital plane instead of tilted at $23\frac{1}{2}°$ away from perpendicular. How would global insolation be affected? How would insolation vary with latitude? How would the path of the Sun in the sky change with the seasons?

2. Imagine that you are following a beam of either (a) shortwave solar radiation entering the Earth's atmosphere heading toward the surface, or (b) a beam of longwave radiation emitted from the surface heading toward space. How will the atmosphere influence the beam?

Eye on the Landscape

Chapter opener A beach on St. Croix, U.S. Virgin Islands These picturesque leaning palms **(A)** are probably the result of beach erosion in a storm that undermined their root systems, causing them to fall toward the surf. After the storm, high tides would have filled in the sand around them. The storm might have been a hurricane (covered in Chapter 6). For the action of waves on the shoreline, see Chapter 18. Also, note the intense blue color of the Caribbean Sea **(B)**. Clear water is blue because it scatters blue light while absorbing light of longer wavelengths, such as green and red. Since the green and red rays are absorbed, the water appears to have a blue color. Sediment and organic matter tend to absorb all colors more evenly, so that turbid water appears dark gray or brown. The colors and wavelengths of the electromagnetic spectrum are a subject of this chapter.

2.17b Wave erosion Note the low vegetation near these ocean bluffs **(A)**. In severe storms, heavy salt spray is driven across this area, killing young salt-intolerant plants and leaving only the salt-tolerant ones. The environment is also very dry, since the sandy soils of these bluffs don't hold much rainwater. As a result, a distinctive community of plants is found here, including such shrubby species as bayberry, beach plum, and wild rose. Environmental effects on plants are covered in Chapter 8.

2.17c Tropical cyclone This astronaut photo shows the structure of the hurricane, or typhoon, very nicely. You can easily see the central eye **(A)**, where air descends rapidly and ground wind speeds are light and variable. The radiating arms of the storm **(B)** are formed by bands of severe thunderstorms and rain spiraling inward toward the center of the storm. Hurricane structure and development are topics discussed in Chapter 6, Weather Systems.

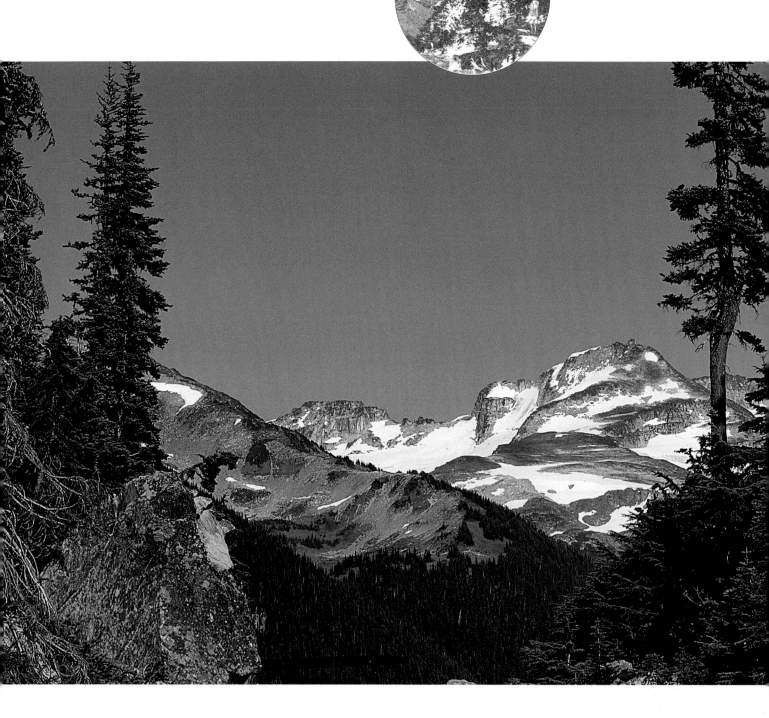

3 | AIR TEMPERATURE

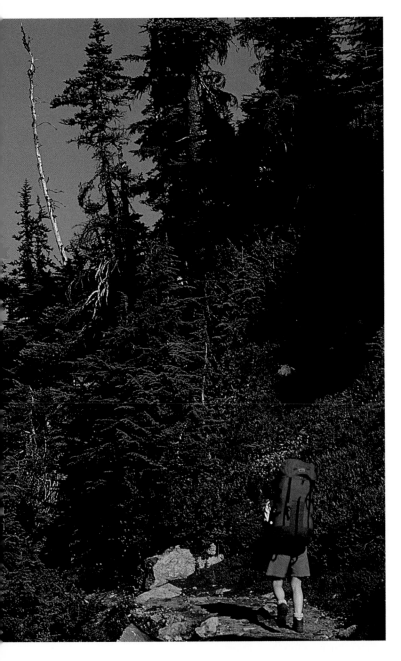

. . . .
PUTTING YOU IN THE PICTURE

The stars are bright and clear tonight. From your vantage point here in the grassy meadow, it's easy to trace the outline of the Big Dipper, with the North Star high above its rim. In fact, you can even see the white blur of the Milky Way, arching across the sky and disappearing behind the peaks of the tall firs that surround the meadow. Brrr! It's getting chilly now. With regret, you decide to abandon your stargazing and return to the campsite, just a few minutes away in the midst of the conifer forest.

Entering the forest, you feel the chill leaving your body as the sheltering conifers shield you from the cold night sky. As you approach the camp, the fire is burning brightly. You step close to it, enjoying its warm glow. A friend offers you a mug of hot cocoa. "I'll bet we'll have frost in the meadow tonight," you think as you sip from the steaming cup. Sure enough, the next morning the grass in the meadow sparkles with white crystals that slowly turn to glistening beads as they absorb the Sun's first rays.

A few hours later in the day, you are well along on your journey to the mountain summit. Here, the trail is steep and you are sweating. Although it was cool in the forest below, the low shrubs and scattered grasses nested in the cracks and crevices of the rocky, high terrain don't relieve the force of the Sun high overhead in the sky. A sudden breeze brings a pleasant cooling to your head and shoulders, as the moisture accumulating under your hat and shirt evaporates. You scramble over a low ledge, and the stark granite rocks are warm to your touch. Straightening up, you see the summit ahead, shimmering in the rising currents of hot air.

It is not long before you reach the summit, and the view is absolutely spectacular. In the thin, clear air, the surrounding peaks stand majestically against the deep blue sky, some with patches of white snow glistening in the intense sunlight. Below the peaks, the gray-brown tones of the barren summit slopes grade into the deep green mantle of conifer forests that line the valleys and lower slopes. It's been a long, hard climb, especially the last part up the crusted surface of the snowfield. The view, however, is worth every moment of exertion. You wouldn't have missed it for anything!

The subject of this chapter is air temperature, primarily the air temperature we experience every day within a few feet from the ground surface. What influences air temperature? Three factors are important. The first and most important one is the cyclic nature of solar radiation, produced by the two constant motions of the Earth—its daily rotation on its axis and its annual revolution in its solar orbit. These cycles produce cycles of air temperature that follow daily and annual rhythms.

The second is the nature of the ground surface. In our example above, the forest is a thick layer covering the surface that acts to insulate the ground from temperature changes. Thus, the grassy meadow loses heat more rapidly at night than does the forest, and frost forms first on the meadow grass.

The third is location on the continent. Locations near the ocean experience a narrower range of temperatures than locations in continental interiors. The reason is that water heats and cools more slowly than soil, so air temperatures near water bodies tend to change more slowly as well. Understanding air temperatures will be easy if you keep these three factors in mind. ■

AIR TEMPERATURE

This chapter focuses on **air temperature**—that is, the temperature of the air as observed at 1.2 m (4 ft) above the ground surface. Air temperature conditions many aspects of human life, from the clothing we wear to the fuel costs we pay. Air temperature and air temperature cycles also act to select the plants and animals that make up the biological landscape of a region. Even geological weathering and soil-forming processes are dependent on temperature over the long term. And air temperature, along with precipitation, is a key determiner of climate, which we will explore in more depth in Chapter 7.

Is air temperature changing? You may have heard that the Earth is becoming increasingly warmer and that, as a result, sea level is rising, climate boundaries are shifting, and severe weather is becoming more frequent. Global warming and its effects are a complicated story that we have already touched upon at several points and will return to again in future chapters. Toward the end of this chapter, we will examine the effect of increasing CO_2 and other gases on global warming and place the present record warmth in the perspective of the Earth's natural time cycles. But first you will need to understand how and why air temperature changes from day to day, month to month, and year to year.

What factors influence air temperatures? Five factors are important.

1. Insolation. The two motions of the Earth—its daily rotation on its axis and its annual revolution in its solar orbit—create the daily and annual cycles of insolation that we described in Chapter 2. In turn, these cycles produce the cycles of air temperature that distinguish day and night, winter and summer.

2. Latitude. As we noted in Chapter 2, daily and annual cycles of insolation vary systematically with latitude, causing air temperatures and air temperature cycles to vary as well. As shown in Figure 2.10, yearly insolation decreases toward the poles, so less energy is available to heat the air. Thus, temperatures generally fall as we move poleward. Temperatures also become more variable over the year as latitude increases. Recall from Figure 2.8 that as much, or more, solar energy may be received in a summer day at high latitudes than is received at the equator, while little or no solar energy is received at high latitudes during the winter. Because of this annual variation, high latitudes experience a much greater range in air temperatures through the year.

3. Surface type. Urban surfaces of asphalt, roofing shingles, and rubber are dry compared to the moist soil surfaces of rural areas and forests. They heat more rapidly because solar energy cannot be taken up in evaporation of water. They also absorb a greater portion of the Sun's energy than vegetation-covered surfaces. Because of these factors, urban air temperatures are generally higher than rural temperatures. The same is true for areas of barren or rocky soil surfaces, such as those of deserts.

Five key factors influence a station's air temperature, its daily variation, and its annual variation: insolation, latitude, surface type, coastal vs. interior location, and elevation.

---

4. Coastal vs. interior location. Locations near the ocean experience a narrower range of air temperatures than locations in continental interiors. The reason is that water heats and cools more slowly than land, so air temperatures over water bodies tend to be less extreme than temperatures over the land surface. And since winds can easily cause air to flow from water to land, a coastal location will more often feel the influence of the adjacent water.

5. Elevation. At high elevation, there is less atmosphere above the surface. This means that the greenhouse effect provides a less effective insulating blanket. Average temperatures are cooler because more surface heat is lost to space. This allows snow to accumulate and remain longer on high peaks, for example. The reduced greenhouse effect also results in greater daily temperature variation.

Understanding air temperatures will be easy if you keep these five factors in mind: insolation, latitude, surface type, location, and elevation.

SURFACE TEMPERATURE

Temperature is a familiar concept. It is a measure of the level of sensible heat of matter, whether it is gaseous (air), liquid (water), or solid (rock or dry soil). We know from experience that when a substance receives a flow of radiant energy, such as sunlight, its temperature rises. Similarly, if a substance loses energy, its temperature falls. This energy flow moves in and out of the substance at its **surface**—for example, the very thin surface layer of soil that actually absorbs solar shortwave radiation and radiates longwave radiation out to space.

The temperature of a surface is determined by the balance among the various energy flows that move across it. As we saw in Chapter 2, **net radiation**—the balance between incoming shortwave radiation and outgoing longwave radiation—produces a radiant energy flow that can heat or cool a surface. Recall that during the day, incoming solar radiation normally exceeds outgoing longwave radiation, so the net radiation balance is positive and the surface warms. Energy flows through the surface into the cooler soil below. At night, net radiation is negative, and the soil loses energy as the surface temperature falls and the surface radiates longwave energy to space.

Energy may also move to or from a surface in other ways. **Conduction** describes the flow of sensible heat from a warmer substance to a colder one through direct contact. When heat flows into the soil from its warm surface during the day, it flows by conduction. At night, heat is conducted back to the colder soil surface. **Latent heat transfer** is also important. When water evaporates at a surface, it removes the heat stored in the change of state from liquid to vapor, thus cooling the surface. When water condenses at a surface, latent heat is released, warming the surface. Another form of energy transfer is **convection,** in which heat is distributed in a fluid by mixing. If the surface is in contact with a fluid, such as a soil surface with air above, upward and downward flowing currents can act to warm or cool the surface. (We'll return to convection in more detail in Chapters 4 and 5.)

AIR TEMPERATURE

Thus far, we've discussed the temperature of surfaces. However, in this chapter we are concerned with air temperature, which is measured a short distance above the ground surface. As we will see shortly, air temperature can be different from surface temperature. Walking across a parking lot in sandals on a clear summer day, you will certainly notice that the pavement is quite a lot hotter than the air you feel on the upper part of your body. But because conduction and convection transfer sensible heat to and from the surface to the air, air temperatures will tend to follow surface temperatures. So we can safely conclude that the temperature changes and trends experienced at the ground surface will be reflected in the air temperatures that are observed and recorded a short distance above the surface.

Measurement of Air Temperature

Air temperature is a piece of weather information that we encounter daily. In the United States, temperature is still widely measured and reported using the Fahrenheit scale. The freezing point of water on the Fahrenheit scale is 32°F, and the boiling point is 212°F. This represents a range of 180°F. In this book, we use the Celsius temperature scale, which is the international standard. On the Celsius scale, the freezing point of water is 0°C and the boiling point is 100°C. Thus, 100 Celsius degrees are the equivalent of 180 Fahrenheit degrees (1°C = 1.8°F; 1°F = 0.56°C). Conversion formulas between these two scales are given in Figure 3.1.

You are probably familiar with the *thermometer* as an instrument for measuring temperature. In this device, a liquid inside a glass tube expands and contracts with temperature, and the position of the liquid surface indicates the temperature. Since air temperature can vary with height, it is measured at a standard level—1.2 m (4 ft) above the ground. A *thermometer shelter* (Figure 3.2a) is a louvered box that holds thermometers or other weather instruments at proper height while sheltering them from the direct rays of the Sun. Air circulates freely through the louvers, ensuring that temperatures inside the shelter are the same as the outside air.

Liquid-filled thermometers are being replaced by newer instruments for the routine measuring of temperatures. Perhaps you have used a digital fever thermometer, which reads out your body temperature directly as a number on a display. It uses a device called a *thermistor,* which changes its electrical resistance with temperature. By measuring this resistance, the temperature may be

---

Surface air temperature is measured under standard conditions at 1.2 m (4 ft) above the ground. Maximum and minimum temperatures are typically recorded. The mean daily temperature is taken as the average of the minimum and maximum temperatures.

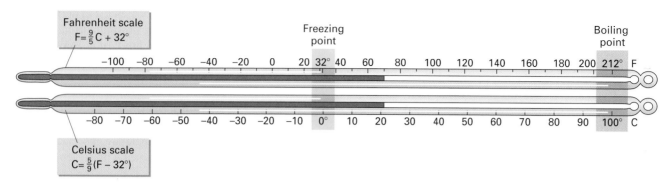

Fahrenheit scale
$F = \frac{9}{5}C + 32°$

Freezing point

Boiling point

Celsius scale
$C = \frac{5}{9}(F - 32°)$

3.1 Celsius and Fahrenheit temperature scales compared *At sea level, the freezing point of water is at Celsius temperature (C) 0°, while it is 32° on the Fahrenheit (F) scale. Boiling occurs at 100°C, or 212°F.*

obtained electronically. Many weather stations are now equipped with temperature measurement systems that use thermistors (Figure 3.2*b*).

Temperature measurements are reported to governmental agencies charged with weather forecasting, such as the U.S. Weather Service or the Meteorological Service of Canada. These agencies typically make available daily, monthly, and yearly temperature statistics for each station using the daily maximum, minimum, and mean temperature. Note that the mean daily temperature is defined as the average of the maximum and minimum daily values. These statistics, along with others such as daily precipitation, are used to describe the climate of the station and its surrounding area.

THE DAILY CYCLE OF AIR TEMPERATURE

Because the Earth rotates on its axis, incoming solar energy at a location can vary widely throughout the 24-hour period, while outgoing longwave energy remains more constant. During the day, net radiation is positive, and the surface gains heat. At night, net radiation is negative, and the surface loses heat by radiating it to the sky and space. Since the air next to the surface is warmed or cooled as well, air temperatures follow the same cycle. This results in the daily cycle of rising and falling air temperatures.

3.2 Weather recording instruments *In part (a), an open instrument shelter shows instruments inside. A maximum-minimum thermometer is mounted near the right edge of the door. In part (b), a thermistor-based maximum-minimum recording system (left) and a recording rain gauge (right) are shown.*

Daily Insolation and Net Radiation

Let's look in more detail at how insolation, net radiation, and air temperature are linked in this daily cycle. The three graphs in Figure 3.3 show average curves of daily insolation, net radiation, and air temperature that we might expect for a typical observing station at lat. 40° to 45° N in the interior of North America. The time scale is set so that 12:00 noon occurs when the Sun is at its highest elevation in the sky.

Graph *(a)* shows daily insolation. At the equinox (middle curve), insolation begins at about sunrise (6 A.M.), rises to a peak value at noon, and declines to zero at sunset (6 P.M.). At the June solstice, insolation begins about two hours earlier

3.3 Daily cycles of insolation, net radiation, and air temperature *These three graphs show idealized daily cycles for a midlatitude station at a continental interior location. Insolation (a) is a strong determiner of net radiation (b). Air temperatures (c) respond by generally increasing, while net radiation is positive and decreasing when it is negative.*

(4 A.M.) and ends about two hours later (8 P.M.). The June peak is much greater than at equinox, and the total insolation for the day is also much greater. At the December solstice, insolation begins about two hours later than the equinox curve (8 A.M.) and ends about two hours earlier (4 P.M.). Both the peak intensity and daily total insolation are greatly reduced in the winter solstice season.

Graph *(b)* shows net radiation for the surface. Recall that when net radiation is positive, the surface gains heat, and when negative, it loses heat. The curves for the solstices and equinox generally resemble those of insolation. Net radiation begins the 24-hour day at midnight as a negative value—a deficit. The deficit continues into the early morning hours. Net radiation shows a positive value—a surplus—shortly after sunrise and rises sharply to a peak at noon. In the afternoon, net radiation decreases as insolation decreases. A value of zero is reached shortly before sunset. With no incoming insolation, net radiation then becomes negative, showing a deficit, where it remains until morning.

Although the three net radiation curves show the same general daily pattern—negative to positive to negative—they differ greatly in magnitude. For the June solstice, the positive values are quite large, and the surplus period is much larger than the deficit period. This means that for the day as a whole, net radiation is positive. At the December solstice, the surplus period is short and the surplus is small. The total deficit, which extends nearly 18 hours, outweighs the surplus. This means that the net radiation for the entire day is negative. As we will see, this pattern of positive daily net radiation in the summer and negative daily net radiation in the winter drives the annual cycle of temperatures.

Daily Temperature

Graph *(c)* shows the typical, or average, air temperature cycle for a 24-hour day. The minimum daily temperature usually occurs about a half hour after sunrise. Since net radiation has been negative during the night, heat has flowed from the ground surface, and the ground has cooled the surface becomes positive, the surface warms quickly and transfers heat to the air above. Air temperature rises sharply in the morning hours and continues to rise long after the noon peak of net radiation.

We should expect the air temperature to rise as long as net radiation is positive. However, another process begins in the early afternoon on a sunny day. Large convection currents develop and mix the air within several hundred meters of the surface, carrying heated air aloft and bringing cooler air downward toward the surface. Therefore, the temperature peak usually occurs in the midafternoon. The peak is shown in the figure at about 3 P.M., but it typically occurs between 2 and 4 P.M., depending on local conditions. By sunset, air temperature is falling rapidly. It continues to fall, but at a decreasing rate, throughout the night.

Note also that the general level of the temperature curves varies with the seasons. In the summer, the daily curve is high, showing warm temperatures. In winter, the curve is low, showing cold temperatures. In between are the equinoxes, with their intermediate temperatures. Since the temperatures lag behind the seasonal changes in net radiation, the September equinox shows temperatures that are considerably warmer than the March equinox. Even though net radiation is the same for the two, the curves are not the same because each reflects earlier seasonal conditions.

The daily cycle of surface air temperature is controlled by the daily cycle of net radiation. During the day, positive net radiation causes temperatures to rise until afternoon highs are reached. At night, negative net radiation causes air temperature to fall.

Temperatures Close to the Ground

We noted at the beginning of this chapter that air temperatures, as measured at 1.2 m (4.0 ft) above the surface, generally reflect the same trends as those of the ground surface but that ground surface temperatures are likely to be more extreme. Let's examine how soil, surface, and air temperatures within a few meters of the ground might change through the day for a bare, dry soil surface.

Figure 3.4 presents a series of generalized temperature profiles from about 30 cm (12 in.) below the surface to 1.5 m (4.9 ft) above it at five times of day. At 8 A.M. (curve 1), the temperature of air and soil is uniform, producing a vertical line on the graph. By noon (curve 2), the surface is considerably warmer than the air at standard height, and the soil below the surface has been warmed as well. By 3 P.M. (curve 3), the soil surface is very much warmer than the air at standard height. By 8 P.M. (curve 4), the surface is cooler than the air, and by 5 A.M. (curve 5), it is much colder.

This analysis shows that daily temperature variation is greatest at the surface, while air temperature at standard height is less variable. Note also that within the soil, the daily cycle weakens with depth to a point where the temperature does not change during the 24-hour cycle.

Environmental Contrasts: Urban and Rural Temperatures

Human activity has altered much of the Earth's land surface. Cities are an obvious example. To build a city, vegetation is removed, and soils are covered with pavement or structures. Using the principles of the surface energy balance we have learned, we can understand the effects of these changes on the temperatures of urban areas as compared to rural regions.

Figure 3.5 illustrates some of the differences between urban and rural surfaces. In rural areas, the land surface is normally covered with a layer of vegetation. A natural activity of plants is **transpiration.** In this process, water is taken up by plant roots and moved to the leaves, where it evaporates. This cools leaf surfaces, which in turn tend to keep nearby air cooler. Water also evaporates directly from the soil, again cooling the surface and keeping air temperatures down.

The cooling effect of vegetation is even greater in a forest. Not only are transpiring leaves abundant, but also solar radiation is intercepted by a layer of leaves stretching from the tops of the trees to the forest floor. Thus, solar warming is not concentrated intensely at the ground surface, but rather serves to warm the whole forest layer.

Another feature of the rural environment is that the soil surfaces tend to be moist. During rainstorms, precipitation

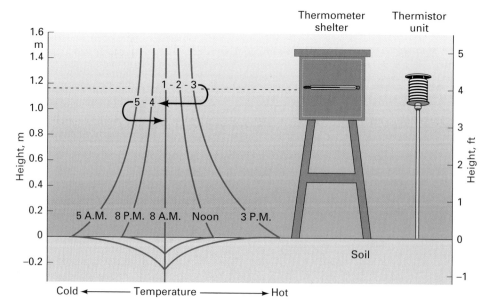

3.4 Daily temperature profiles close to the ground *This simplified diagram shows the temperature profile close to the ground surface at five times of day. The soil surface becomes very hot by midafternoon but cools greatly at night.*

b...*Eye on the Landscape* *Rural surfaces* Rural surfaces are composed of moist soil, covered largely by vegetation. Evapotranspiration keeps these surfaces cooler. **What else would the geographer see?...Answers at the end of the chapter.**

a...*Eye on the Landscape* **Urban surfaces** Urban surfaces are composed of asphalt, concrete, building stone, and similar materials. Sewers drain away rainwater, keeping urban surfaces dry. Because evapotranspiration is limited, surface and air temperatures are hotter. **What else would the geographer see?...Answers at the end of the chapter.**

seeps into the soil. When sunlight reaches the soil surface, the water evaporates, thus removing heat and keeping the soil cool. We refer to the combined effects of transpiration and evaporation as **evapotranspiration.**

In contrast are the typical surfaces of the city. Rain runs easily off the impervious surfaces of roofs, sidewalks, and streets. The rainwater is channeled into storm sewer systems, where it flows directly into rivers, lakes, or oceans, instead of soaking into the soil beneath the city surfaces. Because the impervious city surfaces are dry, the full energy of insolation warms the surfaces, with evaporation taking place only where street trees, lawns, and bare soil patches occur.

City surfaces are also darker and more absorbent than rural surfaces. In fact, asphalt paving and roofing absorb about twice as much solar energy as vegetation. Heat absorption is also enhanced by the many vertical surfaces in cities, which reflect radiation from one surface to another. Since some radiation is absorbed with each reflection, the network of vertical surfaces tends to trap solar energy as well as radiant heat more effectively than a single horizontal surface. Another factor is the ability of concrete, stone, and asphalt to conduct and hold heat better than soil, even when the soil is dry. Therefore, during the day, more heat can be stored by the city's building materials. At night, the heat is conducted back to the surface, keeping nighttime temperatures warmer.

Yet another important factor in warming the city is fuel consumption and waste heat. In summer, city temperatures are raised through the use of air conditioning. This equipment pumps heat out of buildings, releasing the heat to the air. The power used to run the air conditioning systems is also released as heat.

> Urban surfaces lack moisture and so are warmer than rural surfaces during the day. At night, urban materials conduct stored heat to the surface, also keeping temperatures warmer.

The Urban Heat Island

As a result of these effects, air temperatures in the central region of a city are typically several degrees warmer than those of the surrounding suburbs and countryside. Figure 3.6 shows a sketch of a temperature profile across an urban area in the late afternoon. Temperatures are hottest in the downtown area, followed by commercial, urban-residential, and suburban-residential areas. Parks and rural areas are coolest. The elevated temperatures identify an urban **heat island.** The heat island persists through the night because of the availability of a large quantity of heat stored in the ground during the daytime hours.

Figure 3.7 shows two remotely sensed images of the Atlanta central business district that demonstrate the heat island effect. The images were acquired by an aircraft with a thermal infrared scanning system that images longwave energy radiated by the surface. During the day, the main city area, in tones of red and yellow, is clearly warmer than the suburban area, which is to the right in tones of blue and green. At night, the contrast is even greater. The street pattern of asphalt pavement is shown very clearly as a red grid, with many of the downtown squares also filled with red.

The urban heat island effect has important economic consequences. Higher temperatures demand more air conditioning and more electric power in the summer. The fossil fuel

3.6 Urban heat island profile *This diagram shows how air temperatures might vary on a summer afternoon across an urban area. Downtown and commercial areas are warmest, while rural farmland is coolest. (After EPA.)*

°C	10.0	12.6	15.1	17.7	20.2	22.8	25.3	27.9	30.4	33.0	35.5	38.1	40.7	43.2	45.7	48.2
°F	50.0	54.7	59.2	63.9	68.4	73.0	77.5	82.2	86.7	91.4	95.9	100.6	105.3	109.8	114.3	118.8

Temperature

°C	2.0	3.5	5.0	6.4	7.9	9.4	10.8	12.3	13.8	15.2	16.7	18.2	19.6	21.1	22.6	24.0
°F	35.6	38.3	41.0	43.5	46.2	48.9	51.4	54.1	56.8	59.4	62.1	64.8	67.3	70.0	72.7	75.2

Temperature

3.7 Thermal infrared images of downtown Atlanta *These thermal infrared images, acquired in May 1997, over downtown Atlanta, Georgia, show the urban heat island. On the left is a daytime image, while a nighttime image is on the right. (Courtesy NASA/EPA. Provided by Dr. Dale Quattrochi, Marshall Space Flight Center.)*

burned to generate this power contributes CO_2 and air pollutants to the air. The increased temperatures are also more conducive to smog formation, which is unhealthy and damaging to materials. To reduce these effects and the magnitude of the heat island, cities are now taking steps to plant more vegetation and to encourage the use of reflective surfaces, such as concrete or bright roofing materials, where they can reflect solar energy back to space.

Note that the heat island effect does not necessarily apply to cities in desert climates. In the desert, the evapotranspiration of the irrigated vegetation of the city may actually keep the city cooler than the surrounding barren region.

TEMPERATURE STRUCTURE OF THE ATMOSPHERE

So far, we have discussed temperatures in the layer of the atmosphere that surrounds us—up to about 2 m (6 ft) above the ground surface. How different are air temperatures at increasing heights through the atmosphere? In general, temperatures are lower. Why is air cooler at higher altitudes? Recall from Chapter 2 that most of the incoming solar radiation penetrates the atmosphere and is absorbed at the surface. Thus, the atmosphere is largely warmed from below. In general, the farther the air is away from the Earth's surface, the cooler the air will be.

The decrease in measured air temperature with increasing altitude is called the **lapse rate.** This rate measures the drop in temperature in degrees C per 1000 m (or degrees F per 1000 ft). Figure 3.8 shows how temperature varies with altitude for a typical summer day in the midlatitudes. Altitude is plotted on the vertical axis and temperature on the horizontal axis. The curve shows average values—if we sent up a balloon that radioed the temperature of the air back to us every minute or two and sent up many balloons over a long period of time, we would obtain an average profile of temperature very much like that shown in the figure. Temperature drops with altitude at an average rate of 6.4°C/1000 m (3.5°F/1000 ft). This average value is known as the **environmental temperature lapse rate.** For example, when the air temperature near the surface is a pleasant 21°C (70°F), the air at an altitude of 12 km (40,000 ft) will be a bone-chilling –55°C (–67°F). Keep in mind that the environmental temperature lapse rate is an average value and that on any given day the observed lapse rate might be quite different.

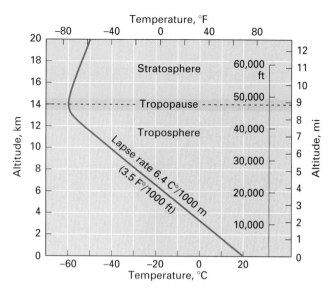

3.8 A typical atmospheric temperature curve for a summer day in the midlatitudes *Temperature decreases with altitude in the troposphere. The rate of temperature decrease with elevation, or lapse rate, is shown at the average value of 6.4°C/1000 m (3.5°F/1000 ft), which is known as the environmental temperature lapse rate. At the tropopause, the decreasing trend stops. In the stratosphere above, temperature is constant or increases slightly with altitude.*

Figure 3.8 shows another important feature of the atmosphere. For the first 12 km (7 mi) or so, temperature falls with increasing elevation. However, at 12 to 15 km (7 to 9 mi) in height, the temperature stops decreasing. In fact, above that height, temperature slowly increases with elevation. This feature has led atmospheric scientists to distinguish two different layers in the lower atmosphere—the troposphere and the stratosphere.

> In the lower atmosphere, air temperatures decrease with increasing altitude. The average rate of temperature decrease with height is termed the environmental temperature lapse rate and is 6°C/1000 m (3.5°F/1000 ft).

Troposphere

The **troposphere** is the lowest atmospheric layer, in which temperature decreases with increasing elevation. It is thickest in the equatorial and tropical regions, where it ranges from sea level to about 16 km (10 mi), and it thins toward the poles, where it is about 6 km (4 mi) thick. Since almost all human activity occurs in this layer, it is of primary importance to us. Everyday weather phenomena, such as clouds or storms, occur mainly in the troposphere.

One important feature of the troposphere is that it contains significant amounts of water vapor. When the water vapor content is high, vapor can condense into water droplets, forming low clouds and fog, or the vapor can be deposited as ice crystals, forming high clouds. When condensation or deposi-

> The troposphere and stratosphere are the two lowermost atmospheric layers. Clouds and weather phenomena occur in the troposphere. The stratosphere is the home of strong, persistent winds that flow from west to east.

tion is rapid, rain, snow, hail, or sleet—collectively termed *precipitation*—may be produced and fall to Earth. Regions where water vapor content is high throughout the year will therefore have moist climates. In desert regions, where water vapor is present only in small amounts, precipitation is infrequent. In Chapter 2, we also described the important role of water vapor in absorbing and reradiating heat emitted by the Earth's surface. In this way, water vapor helps to create the greenhouse effect, which warms the Earth to habitable temperatures.

The troposphere contains countless tiny particles that are so small and light that the slightest movements of the air keep them aloft. These are called *aerosols*. They are swept into the air from dry desert plains, lakebeds, and beaches, or they are released by exploding volcanoes. Oceans are also a source of aerosols. Strong winds blowing over the ocean lift droplets of spray into the air. These droplets of spray lose most of their moisture by evaporation, leaving tiny particles of watery salt as residues that are carried high into the air. Forest fires and brushfires are another important source of aerosols, contributing particles of soot as smoke. Meteors contribute dust particles as they vaporize from the heat of friction upon entering the upper layers of air. Industrial processes that incompletely burn coal or fuel oil release aerosols to the air as well.

The most important function of aerosols is that certain types serve as nuclei, or centers, around which water vapor condenses to form tiny droplets. When these droplets grow large and occur in high concentration, they are visible to the eye as clouds or fog. As we noted in Chapter 2, aerosol particles also scatter sunlight, thus brightening the whole sky while reducing slightly the intensity of the solar beam.

The height at which the troposphere gives way to the stratosphere above is known as the *tropopause*. Here, temperatures stop decreasing with altitude and start to increase. The altitude of the tropopause varies somewhat with season. This means that the troposphere is not uniformly thick at any location.

Stratosphere and Upper Layers

Above the tropopause lies the **stratosphere,** in which the air becomes slightly warmer as altitude increases. The stratosphere extends to a height of roughly 50 km (about 30 mi) above the Earth's surface. It is the home of strong, persistent winds that blow from west to east. There is little mixing of air between the troposphere and stratosphere, and so the stratosphere normally holds very little water vapor or dust.

One important feature of the stratosphere is that it contains the ozone layer. As we saw in *Eye on Global Change 2.2 • The Ozone Layer—Shield to Life*, the ozone layer absorbs solar ultraviolet radiation and thus shields earthly life

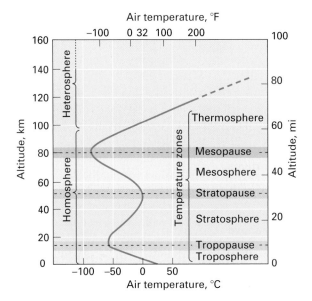

3.9 Temperature structure of the upper atmosphere
Above the troposphere and stratosphere are the mesosphere and thermosphere. The homosphere, in which air's chemical components are well mixed, ranges from the surface to nearly 100 km altitude.

from this intense, harmful form of energy. In fact, the warming of the stratosphere with altitude is caused largely by the absorption of solar energy by ozone molecules.

Above the stratosphere are two other layers—the mesosphere and thermospere. They are shown in Figure 3.9. In the *mesosphere,* temperature falls with elevation. This layer begins at the *stratopause,* the altitude at which the stratospheric temperature ceases to increase with altitude. It ends at the *mesopause,* the level at which temperature ceases to fall with altitude. Above the mesosphere is the *thermosphere,* a layer of increasing temperature. However, at the altitude of the thermosphere, the density of air is very thin and the air holds little heat.

From the viewpoint of gas composition, the atmosphere is uniform for about the first 100 km of altitude, which includes the troposphere, stratosphere, mesosphere, and lower portion of the thermosphere. This region is referred to as the *homosphere.* Above 100 km, gas molecules tend to become increasingly sorted into layers by molecular weight and electric charge. This region is referred to as the *heterosphere.*

High-Mountain Environments

Did you ever walk or drive to the top of a high mountain? Mount Whitney, in California, perhaps (Figure 3.10)? or New Hampshire's Mount Washington? or British Columbia's Mount Garibaldi? If so, you have probably made the simple observation that temperatures generally drop as you go up in elevation. We've already noted this fact in the discussion of

3.10 Eye on the Landscape Clear air at high elevation *Mount Whitney (elevation 4418 m, 14,495 ft) and surrounding peaks of the Southern Sierra Nevada provide a stunning panorama. The excellent visibility and dark blue sky are products of the thin air at this elevation.* **What else would the geographer see? ... Answers at the end of the chapter.**

the environmental temperature lapse rate. But what else might you notice as you climb? You might notice that you sunburn a bit faster. Perhaps you also find yourself getting out of breath more quickly. And if you camp out, you may notice that the nighttime temperature gets lower than you might expect, even given that temperatures are generally cooler.

These effects are related to the fact that at high elevations, there is significantly less air above you. With fewer air molecules and aerosol particles to scatter and absorb the Sun's light, the Sun's rays will be stronger. There is also less carbon dioxide and water vapor, and so the greenhouse effect is reduced. With the reduced warming effect, temperatures will tend to drop lower at night. At high elevations, the air pressure is lower because there is a smaller mass of air above you. This means that there are fewer gas molecules in a unit volume of air. You feel short of breath on a high mountain simply because the oxygen pressure in your lungs is lower than you are used to at lower elevations.

> At high elevations, air temperatures are generally cooler and show a greater day-to-night range. Both effects occur because the thickness and density of the air column above decrease with elevation, and therefore the greenhouse effect is weaker at high elevations.

What effect does elevation have on the daily temperature cycle? Figure 3.11 shows temperature graphs for five stations ascending the Andes Mountain range in Peru during the same 15 days in July. The 15-day mean for each station is shown as a horizontal line. Mean temperatures clearly decrease with elevation, from 16°C (61°F) at sea level to –1°C (30°F) at 4380 m (14,370 ft). The daily range also increases with eleva-

> A temperature inversion occurs when air temperature increases with altitude. Low-level inversions typically occur on clear, cold nights with still air and are often found over snow-covered surfaces in winter.

tion, except for Cuzco. This large city does not experience nighttime temperatures as low as you might expect because of its urban heat island.

Temperature Inversion

So far, air temperatures seem to decrease with height above the surface. Is this always true? Think about what happens on a clear, calm night. Under clear conditions, the ground surface radiates longwave energy to the sky, net radiation becomes negative, and the surface cools. This means that the air near the surface will also be cooled, as we saw in curves 4 and 5 on Figure 3.4 for the air within a few meters of the ground. However, if the surface continues to stay cold, a layer of cooler air can accumulate underneath a layer of warmer air.

This situation is illustrated in a graph of temperature with height (Figure 3.12). The straight, slanting line of the normal environmental lapse rate bends to the left in a "J" hook. In this example, the air temperature at the surface, point A, has dropped to –1°C (30°F). This value is the same as that at point B, some 750 m (about 2500 ft) aloft. As we move up from ground level, temperatures become warmer up to about 300 m (about 1000 ft). Here the curve reverses itself, and normal lapse rate takes over. The lower portion of the lapse rate curve is called a

3.11 The effect of elevation on air temperature cycles *Those graphs show daily maximum and minimum air temperatures for mountain stations in Peru, lat. 15° S. All data cover the same 15-day observation period in July. As elevation increases, the mean daily temperature decreases and the temperature range increases.*

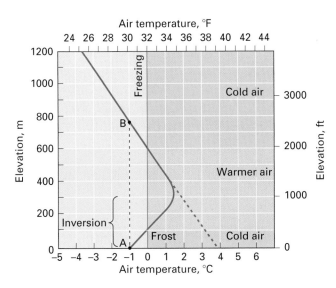

3.12 A low-level temperature inversion with frost
While air temperature normally decreases with altitude (dashed line), in an inversion temperature increases with altitude. In this example, the surface temperature is at –1°C (30°F), and temperature increases with altitude (solid line) for several hundred meters (1000 ft or so) above the ground. Temperature then resumes a normal, decreasing trend with altitude.

3.13 Frost-buster fan *This large fan is used to mix the air at low levels in a navel orange orchard to protect blossoms and fruit from freezing.*

low-level **temperature inversion.** Here, the normal cooling trend is reversed and temperatures increase with height.

In the case shown in Figure 3.12, the temperature of the lowermost air has fallen below the freezing point, 0°C (32°F). For sensitive plants during the growing season, this temperature condition is called a killing frost (even though actual frost may not form). Perhaps you have seen news reports about killing frosts damaging fruit trees or crops in Florida or California. Growers commonly use several methods to break up an inversion. Most commonly, large fans are used to mix the cool air at the surface with the warmer air above (Figure 3.13). Where air pollution is not a problem, oil-burning heaters are sometimes used to warm the surface air layer and create air circulation.

Low-level temperature inversions often occur over snow-covered surfaces in winter. Inversions of this type are very intense and can extend thousands of meters into the air. They build up over many long nights in arctic and polar regions, where the solar heat of the short winter day cannot completely compensate for nighttime cooling. Inversions can also result when a warm air layer overlies a colder one. This type of inversion is often found along the west coasts of major continents, and we will discuss it in more detail in Chapter 4. We will also see in that chapter that temperature inversions suppress vertical mixing of the air and so can act to seal air pollution within a thin surface layer.

> The annual cycle of net radiation, which results from the variation of insolation with the seasons, drives the annual cycle of air temperatures. Where net radiation varies strongly with the seasons, so does surface air temperature.

THE ANNUAL CYCLE OF AIR TEMPERATURE

As the Earth revolves around the Sun, the tilt of the Earth's axis causes an annual cycle of variation in insolation. This cycle produces an annual cycle of net radiation, which, in turn, causes an annual cycle to occur in mean monthly air temperatures. Although the annual cycle of net radiation is the most important factor in determining the annual temperature cycle, we will see that location—maritime or continental—also has an important influence. That is, places located well inland and far from oceans generally experience a stronger temperature contrast from winter to summer. Let's begin our study of the annual temperature cycle by looking at the relationship between net radiation and temperature at four stations, ranging from the equator almost to the arctic circle.

Net Radiation and Temperature

Graph *(a)* of Figure 3.14 shows the yearly cycle of the net radiation rate for the four stations. The average value of net radiation flow rate for the month is plotted, in units of W/m².

Graph *(b)* shows mean monthly air temperatures for these same stations. We will compare the net radiation graph with the air temperature graph for each station, beginning with Manaus, a city on the Amazon River in Brazil.

At Manaus, located nearly on the equator, the average net radiation rate is strongly positive in every month. However, there are two minor peaks. These coincide approximately with the equinoxes, when the Sun is nearly straight overhead. A look at the temperature graph of Manaus shows uniform air temperatures, averaging about 27°C (81°F) for the year. The annual temperature range, or difference between the highest and lowest mean monthly temperature is only 1.7°C (3°F). In other words, near the equator the temperature is similar each month. There are no temperature seasons.

We go next to Aswan, Egypt, a very dry desert location on the Nile River at lat. 24°N. The positive net radiation rate curve shows that a large radiation surplus exists for every month. Furthermore, the net radiation rate curve has a much stronger annual cycle, with values for June and July that are almost double those of December and January. The temperature graph shows a corresponding annual cycle, with an annual range of about 17°C (31°F). June, July, and August are terribly hot, averaging over 32°C (90°F).

Moving farther north, we come to Hamburg, Germany, lat. 54°N. The net radiation rate cycle here is also strongly developed. The rate is positive for nine months, providing a radiation surplus. When the rate becomes negative for three winter months, a deficit is produced. The temperature cycle reflects the reduced total insolation at this latitude. Summer months reach a maximum of just over 16°C (61°F), while winter months reach a minimum of just about freezing (0°C or 32°F). The annual range is about 17°C (31°F), the same as at Aswan.

Finally, we travel to Yakutsk, Siberia, lat. 62° N. During the long, dark winters, the net radiation rate is negative, and there is a radiation deficit that lasts about six months. During this time, air temperatures drop to extremely low levels. For three of the winter months, monthly mean temperatures are between –35 and –45°C (about –30 and –50°F). This is actually one of the coldest places on Earth. In summer, when daylight lasts most of a 24-hour day, the net radiation rate rises to a strong peak. In fact, this peak value is higher than those of the other three stations. As a result, air temperatures show a phenomenal rise beginning in the spring to summer-month values of over 13°C (55°F). Because of Yakutsk's high latitude and continental interior location, its annual temperature range is enormous—over 60°C (108°F).

Land and Water Contrasts

Have you ever visited San Francisco? If so, you probably noticed that this magnificent city has quite a unique climate. Fog is frequent, and cool, damp weather prevails for most of the year. The cool climate is due to its location—on the tip of a peninsula, with the Pacific Ocean on one side and San Francisco Bay on the other. Ocean and bay water temperatures are quite cool, since a southward-flowing current sweeps cold water from Alaska down along the northern California coast. Winds from the west move cool, moist ocean air, as well as clouds and fog, across the peninsula, keeping summer air temperatures low and winter temperatures above freezing. Figure 3.15 shows a typical record of temperatures for San Francisco for a week in the summer.

(a) **Net radiation**

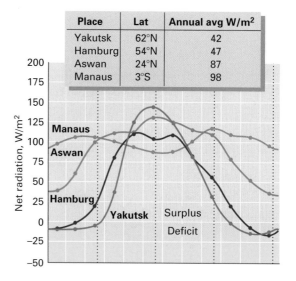

Place	Lat	Annual avg W/m²
Yakutsk	62°N	42
Hamburg	54°N	47
Aswan	24°N	87
Manaus	3°S	98

(b) **Monthly mean air temperature**

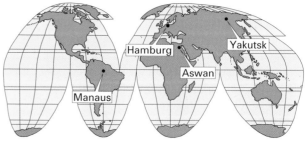

3.14 Net radiation and temperature cycles *Net radiation and temperature for Aswan (Egypt), Hamburg (Germany), and Yakutsk (Siberia) all show strong annual cycles with summer maximums and winter minimums. In contrast is Manaus (Brazil), very near the equator, which shows two net radiation peaks and a nearly uniform temperature. (Data courtesy of David H. Miller.)*

Temperatures hover around 13°C (55°F) and change only a little from day to night.

In contrast is a location far from the water, like Yuma, Arizona, also shown in Figure 3.15. Located in the Sonoran Desert, air temperatures here are much warmer on the average—about 28°C (82°F). Clearly, no ocean cooling is felt in Yuma! The daily range is also much greater, nearly 20°C (36°F). Hot desert days become cool desert nights, with the clear, dry air allowing the ground to lose heat rapidly.

Why do these differences occur? The important principle is this: the surface layer of any extensive, deep body of water heats more slowly and cools more slowly than the surface layer of a large body of land when both are subjected to the same intensity of insolation. Because of this principle, daily and annual air temperature cycles will be quite different at coastal locations than at interior locations.

Four important thermal differences between land and water surfaces account for the land–water contrast (see Figure 3.16). The most important difference is that much of the downwelling solar radiation penetrates water, distributing the absorbed heat throughout a substantial water layer. In contrast, solar radiation does not penetrate soil or rock, so its heating effect is concentrated at the surface. The radiation therefore warms a thick water surface layer only slightly, while a thin land surface layer is warmed more intensely.

A second thermal factor is that water is slower to heat than dry soil or rock. The *specific heat* of a substance describes how the temperature of a substance changes with a given input of heat. Consider an experiment using a small volume of water and the same volume of rock. Let's suppose we warm each volume by the same amount—per-

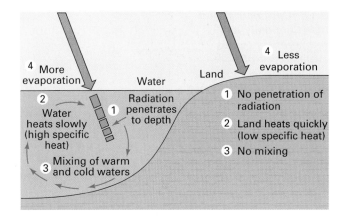

3.16 Land–water contrasts *These four differnces illustrate why a land surface heats more rapidly and more intensely than the surface of a deep water body. As a result, locations near the ocean have more uniform air temperatures—cooler in summer and warmer in winter.*

haps one degree. If we measure the amount of heat required, we will find that it takes about five times as much heat to raise the temperature of the water one degree as it does to raise the temperature of the rock one degree. That is, the specific heat of water is about five times greater than that of rock. The same will be true for cooling—after losing the same amount of heat, the temperature of water falls less than the temperature of rock.

A third difference between land and water surfaces is related to mixing. That is, a warm water surface layer can mix with cooler water below, producing a more uniform temperature throughout. For open water, the mixing is produced by wind-generated waves. Clearly, no such mixing occurs on land surfaces.

A fourth thermal difference is that an open water surface can be cooled easily by evaporation. Land surfaces can also be cooled by evaporation, but only if water is present at or near the soil surface. When the surface dries, evaporation stops. In contrast, a free water surface can always provide evaporation.

With these four important differences in mind, we can easily see that air temperatures above water will be less variable than those over land. But just what effect does the contrast in thermal characteristics have on air temperature?

DAILY TEMPERATURE CYCLE   Let's examine the daily cycle of air temperature first. Two sets of daily air temperature curves are shown in Figure 3.17—El Paso, Texas, and North Head, Washington. Four months are presented—January, April, July, and October.

The El Paso curves show the temperature environment of an interior desert in midlatitudes. Because soil moisture content is low and vegetation is sparse, evaporation and transpiration are not important cooling effects. Cloud cover is

> **Air temperatures at coastal locations tend to be more constant than at interior continental locations because water bodies heat and cool more slowly than the land surface.**

3.15 Maritime and continental temperatures *A recording thermometer made these continuous records of the rise and fall of air temperature for a week in summer at San Francisco, California, and at Yuma, Arizona. At San Francisco, on the Pacific Ocean, the daily air temperature cycle is very weak. At Yuma, a station in the desert, the daily cycle is strongly developed.*

3.17 Continental and maritime daily air temperature cycles *The average cycle of air temperature throughout the day for four different months shows the effect of continental and maritime location. Daily and seasonal ranges are great at El Paso, a station in the continental interior, but only weakly developed at North Head, Washington, a station on the Pacific coast. The seasonal effect on overall temperatures is stronger at El Paso.*

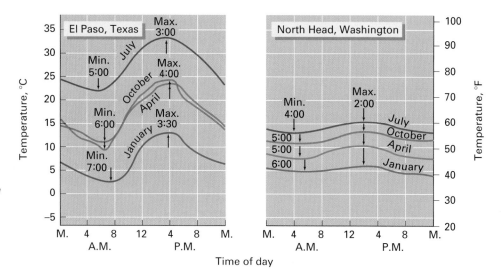

generally sparse. Under these circumstances, the ground surface heats intensely during the day and cools rapidly at night. Air temperatures show an average daily range of 11 to 14°C (20 to 25°F). This type of variation represents an interior temperature environment—typical of a station located in the interior of a continent, far from the ocean's influence.

North Head is located on the Washington coast. Here, prevailing westerly winds sweep cool, moist air off the adjacent Pacific Ocean. The average daily range at North Head is a mere 3°C (5°F) or less. Persistent fogs and cloud cover also contribute to the minimal daily range. Note further that the annual range is much restricted, especially when compared to El Paso. North Head exemplifies a coastal temperature environment, typical of a station located in the path of oceanic air.

ANNUAL TEMPERATURE CYCLE Let's now turn to the effects of land–water surface contrasts on the temperature cycle for the year. We have already noted for El Paso and North Head that the temperature cycle for the four months plotted shows a greater range for the interior station than for the coastal one. Let's look in more detail at the annual cycle for another pair of stations—Winnipeg, Manitoba, located in the interior of the North American continent, and the Scilly Islands, off the southwestern tip of England, which are surrounded by the waters of the Atlantic Ocean. This time, the two stations chosen are at the same latitude, 50°N. As a result, they have the same insolation cycle and receive the same potential amount of solar energy for surface warming. Figure 3.18 shows their annual cycles of temperature as well as the insolation curve common to both.

The temperature graphs for the Scilly Islands and Winnipeg confirm the effects we have already noted for North

> Because water bodies heat and cool more slowly than land surfaces, monthly temperature maximums and minimums occur later in the year at coastal stations than at interior stations.

Head and El Paso—that the annual range in temperature is much larger for the interior station (39°C, 70°F) than for the coastal station (8°C, 14°F). Note that the nearby ocean waters keep the air temperature at the Scilly Islands well above freezing in the winter, while January temperatures at Winnipeg fall to near –20°C (–4°F)!

Another important effect of land–water contrasts concerns the timing of maximum and minimum temperatures. Insolation reaches a maximum at summer solstice, but it is still strong for a long period afterward. This means that heat energy continues to flow into the ground well after the solstice. Therefore, the hottest month of the year for interior regions is July, the month following the solstice. Similarly, the coldest month of the year for large land areas is January, the month after the winter solstice. This is because the ground continues to lose heat even after insolation begins to increase.

Over the oceans and at coastal locations, maximum and minimum air temperatures are reached a month later than on land—in August and February, respectively. Because water bodies heat or cool more slowly than land areas, the air temperature changes more slowly. This effect is clearly shown in the Scilly Islands graph, where February is slightly colder than January.

WORLD PATTERNS OF AIR TEMPERATURE

We have learned some important principles about air temperatures in this chapter. Some of these principles are local, such as the effect of urban and rural surfaces on air temperatures. Others are on a continental scale, such as the effect of maritime or continental location. Still others are

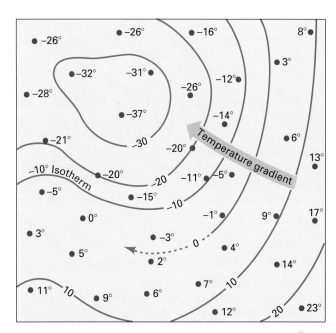

3.19 Isotherms *Isotherms are used to make temperature maps. Each line connects points having the same temperature. Where temperature changes along one direction, a temperature gradient exists. Where isotherms close in a tight circle, a center exists. This example shows a center of low temperature.*

3.18 Continental and maritime annual air temperature cycles *Annual cycles of insolation and monthly mean air temperature for two stations at lat. 50° N: Winnipeg, Canada, and Scilly Islands, England. Insolation (a) is identical for the two stations. In (b), Winnipeg temperatures clearly show the large annual range and earlier maximum and minimum that are characteristic of its continental location. Scilly Islands temperatures show its maritime location in the small annual range and delayed maximum and minimum.*

global, such as the effect of latitude or elevation on temperature. We now turn to world temperature patterns, tying our discussion closely to these principles. First, however, we will need a quick explanation of air temperature maps and their meaning.

The distribution of air temperatures is often shown on a map by **isotherms**—lines drawn to connect locations having the same temperature. Figure 3.19 shows a map on which the observed air temperatures have been recorded and placed at their proper location on the map. These may be single readings, such as a daily maximum or mini-

> **Isotherms are lines of equal temperature drawn on a map. Maps of isotherms show centers of high and low temperatures as well as temperature gradients.**

mum, or they may be averages of many years of records for a particular day or month of a year, depending on the purposes of the map. The isotherms are constructed by drawing smooth lines through and among the points in a way that best indicates a uniform temperature, given the observations at hand. Usually, isotherms representing 5- or 10-degree differences are chosen, but they can be drawn at any convenient temperature interval.

Isothermal maps are valuable because they show the important features of the temperature pattern clearly. Centers of high or low temperatures are visible, as are *temperature gradients*—directions along which temperature changes. Centers and gradients give the broad pattern of temperature, as shown in the following discussion of world patterns.

Factors Controlling Air Temperature Patterns

World patterns of isotherms are largely explained by three factors that we have already discussed. The first of these factors is latitude. As latitude increases, average annual insolation decreases, and so temperatures decrease as well, making the poles colder than the equator. The effect of latitude on seasonal variation is also important. For example, more solar energy is received at the poles at the summer solstice than at the equator. We must therefore consider the time of year as well as the latitude in understanding world temperature patterns.

The second factor is that of coastal-interior contrasts. As we've noted, coastal stations that receive marine air from prevailing winds have more uniform temperatures—cooler in summer and warmer in winter. Interior stations, on the other hand, show a much larger annual variation in temperature. Ocean currents, discussed further in Chapter 5, can also have an effect. By keeping coastal waters somewhat warmer or cooler than expected, temperatures at maritime stations will be influenced in a similar manner.

Elevation is the third important factor. At higher elevations, temperatures will be cooler. Therefore, we expect world temperature maps to show the presence of mountain ranges, which will be cooler than surrounding regions.

World Air Temperature Patterns for January and July

With these factors in mind, let's look at some world temperature maps in more detail. Figure 3.20 consists of maps of world temperatures for two months—January and July. The Mercator projections show temperature trends from the equator to the midlatitude zones, and the polar projections give the best picture for high latitudes. From the maps, we can make six important points about the temperature patterns and the factors that produce them. Be sure to follow along by examining the maps carefully.

1. Temperatures decrease from the equator to the poles. Annual insolation decreases from the equator to the poles, thus causing temperatures to decrease. This temperature gradient is most clearly seen in the polar maps for the southern hemisphere in January and July. On these maps, the isotherms are nearly circular, decreasing to a center of low temperature on Antarctica near the south pole. The center is much colder in July, when most of the polar region is in perpetual night. We can also see this same general trend in the north polar maps, but the continents complicate the pattern. The general temperature gradient from the equator poleward is also evident on the Mercator maps.

2. Large land masses located in the subarctic and arctic zones develop centers of extremely low temperatures in winter. The two large land masses we have in mind are North America and Eurasia. The January north polar map shows these low-temperature centers very well. The cold center in Siberia, reaching –50°C (–58°F), is strong and well defined. The cold center over northern Canada is also quite cold (–35°C, –32°F) but is not as well defined. Both features are visible on the January Mercator map. Greenland shows a low-temperature center as well, but it has a high dome of glacial ice, as discussed below in point 6. An important factor in keeping winter temperatures low in these regions is the high albedo of snow cover, which reflects much of the winter insolation back to space.

3. Temperatures in equatorial regions change little from January to July. Note the broad space between 25°C (77°F) isotherms on both January and July Mercator maps. In this region, the temperature is greater than 25°C (77°F) but less than 30°C (86°F). Although the two isotherms

move a bit from winter to summer, the equator always falls between them. This demonstrates the uniformity of equatorial temperatures. (An exception is the northern end of the Andes Mountains in South America, where high elevations, and thus cooler temperatures, exist at the equator.) Equatorial temperatures are uniform primarily because insolation at the equator does not change greatly with the seasons.

4. Isotherms make a large north-south shift from January to July over continents in the midlatitude and subarctic zones. Figure 3.21 demonstrates this principle. In the winter, isotherms dip equatorward, while in the summer, they arch poleward. This effect is shown in North America and Eurasia in the January and July Mercator maps. In January the isotherms drop down over these continents, and in June they curve upward. For example, the 15°C (59°F) isotherm lies over central Florida in January. But by July this same isotherm has moved far north, cutting the southern shore of Hudson Bay and then looping far up into northwestern Canada. In contrast are the isotherms over oceans, which shift much less. This striking difference is due to the contrast between oceanic and continental surface properties, which cause continents to heat and cool more rapidly than oceans.

5. Highlands are always colder than surrounding lowlands. You can see this by looking at the pattern of isotherms around the Rocky Mountain chain, in western North America, on the Mercator maps. In winter, the –5°C (23°F) and –10°C (14°F) isotherms dip down around the mountains, indicating that the center of the range is colder. In summer, the 20°C (68°F) and 25°C (77°F) isotherms also dip down, showing the same effect even though temperatures are much warmer. The Andes Mountains in South America show the effect even more strongly. The principle at work here is that temperatures decrease with an increase in elevation.

6. Areas of perpetual ice and snow are always intensely cold. Greenland and Antarctica contain our planet's two great ice sheets. Notice how they stand out on the polar maps as cold centers in both January and July. They are cold for two reasons. First, their surfaces are high in elevation, rising to over 3000 m (about 10,000 ft) in their centers. Second, the white snow surfaces reflect much of the insolation. Since little solar energy is absorbed, little is available to warm the snow surface and the air above it. The Arctic Ocean, bearing a cover of floating ice, also maintains its cold temperatures throughout the year. However, the cold is much less intense in January than on the Greenland Ice Sheet, since ocean water underneath the ice acts as a heat reservoir to keep the ice above from getting extremely cold.

The Annual Range of Air Temperatures

Figure 3.22 is a world Mercator map showing the annual range of air temperatures. The lines, resembling isotherms, show the difference between the January and July monthly means. We can explain the features of this map by using the same effects of latitude, interior-maritime location, and elevation.

JANUARY

JULY

3.20 Mean monthly air temperatures (°C) for January and July, Mercator and polar projections (Compiled by John E. Oliver.)

JANUARY

JULY

-70° -65° -60° -55° -50° -45° -40° -35° -30° -25° -20° -15° -10° -5° 0° 5° 10° 15° 20° 25° 30° 35°

-90° -80° -70° -60° -50° -40° -30° -20° -10° 0° 10° 20° 32° 40° 50° 60° 70° 80° 90°

3.20 (Continued)

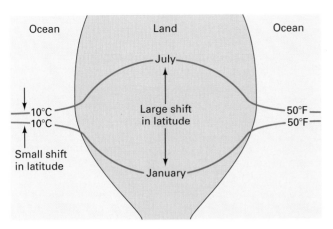

3.21 Seasonal migration of isotherms *Continental air temperature isotherms shift over a much wider latitude range from summer to winter than do oceanic air temperature isotherms. This difference occurs because oceans heat and cool much more slowly than continents.*

1. The annual range increases with latitude, especially over northern hemisphere continents. This trend is most clearly shown for North America and Asia. This is due to the contrast between summer and winter insolation, which increases with latitude.

The global pattern of annual temperature range shows that tropical oceans have the smallest range, while northern hemisphere continental interiors show the largest range.

---

2. The greatest ranges occur in the subarctic and arctic zones of Asia and North America. The map shows two very strong centers of large annual range—one in northeast Siberia and the other in northwest Canada–eastern Alaska. In these regions, summer insolation is nearly the same as at the equator, while winter insolation is very low.

3. The annual range is moderately large on land areas in the tropical zone, near the tropics of cancer and capricorn. These are regions of large deserts—North Africa (Sahara), southern Africa (Kalahari), and Australia (interior desert) are examples. Dry air and the absence of clouds and moisture allow these continental locations to cool strongly in winter and warm strongly in summer, even though insolation contrasts with the season are not as great as at higher latitudes.

4. The annual range over oceans is less than that over land at the same latitude. This can be seen by following a parallel of latitude—40° N, for example. Starting from the right, we see that the range is between 5 and 10°C (9 and 18°F) over the Atlantic but increases to about 30°C (54°F) in the interior of North America. In the Pacific, the range falls to 5°C (9°F) just off the California coast and increases to 15°C (27°F) near Japan. In Central Asia, the range is near 35°C (63°F). Again, these major differences

3.22 Annual range of air temperature in Celsius degrees *Data show differences between January and July means. The inset box shows the conversion of Celsius degrees to Fahrenheit for each isotherm value (Compiled by John E. Oliver.)*

are due to the contrast between land and water surfaces. Since water heats and cools much more slowly than land, a narrower range of temperatures is experienced.

5. The annual range is very small over oceans in the tropical zone. As shown on the map, the range is less than 3°C (5°F) since insolation varies little with the seasons near the equator and water heats and cools slowly.

GLOBAL WARMING AND THE GREENHOUSE EFFECT

Tune in to the evening news, and one of the recurring scientific subjects you will hear about is global warming. Is our climate changing? Many scientists have concluded that the temperature of our planet is warming significantly because of human activities. Other scientists argue that the warming of the last few years is part of a natural global cycle.

You may also hear that carbon dioxide (CO_2), produced by human activities, is a major cause of concern in climate warming. This gas is released to the atmosphere in large quantities by fossil fuel burning, and as human energy consumption increases, so does atmospheric CO_2 concentration. The increase in CO_2 with time follows an *exponential growth* principle. This is the growth principle behind the compound interest that causes a savings account to increase in value.

Why is the CO_2 buildup a problem? Recall from Chapter 2 that the greenhouse effect is caused by atmospheric absorption of outgoing longwave radiation, largely by carbon dioxide and water vapor, and that a portion of the absorbed energy is counterradiated back toward the Earth's surface. With more CO_2, more energy is absorbed and counterradiated, and the surface is warmed by an extra amount. For this reason, many scientists are concerned that the greenhouse effect is escalating and that global temperatures are rising—or will rise soon—in response. (See

---

Greenhouse gases act to warm the atmosphere and enhance the greenhouse effect. CO_2 is the most important greenhouse gas, but methane (CH_4), chlorofluorocarbons, tropospheric ozone, and nitrous oxide (N_2O), when taken together, are as important as CO_2.

Eye on Global Change 3.1 • Carbon Dioxide—On the Increase.)

Also of concern are other gases that are normally present in much smaller concentrations—methane (CH_4), the chlorofluorocarbons, tropospheric ozone (O_3), and nitrous oxide (N_2O). (See also *Eye on Global Change 2.2 • The Ozone Layer—Shield to Life,* Chapter 2.) Taken together with CO_2, they are known as **greenhouse gases.** Though less abundant, these gases are better absorbers of longwave radiation than CO_2, and so are more effective, molecule per molecule, at enhancing the greenhouse effect.

Factors Influencing Climatic Warming and Cooling

Figure 3.23 shows how a number of important factors have influenced global warming since about 1850. The five bars on the left show the greenhouse gases, ranging from CO_2 to N_2O. Although carbon dioxide is the largest, note that the other four together contribute about the same amount of warming. Taken together, the total enhanced energy flow to the surface produced by greenhouse gases is about 3 W/m^2, which is about 1.25 percent of the solar energy absorbed by the Earth and atmosphere.

Methane, CH_4, is naturally released by the decay of organic matter in wetlands. However, human activity generates about double that amount in rice cultivation, farm animal wastes, bacterial decay in sewage and landfills, fossil fuel extraction and transportation, and biomass burning. Chlorofluorocarbons (CFCs) are shown as having both a warming and cooling effect. The warming effect results because these compounds are very good absorbers of longwave energy. The cooling effect occurs because CFCs destroy ozone in the stratosphere, and ozone contributes to warming. (See *Eye on Global Change 2.2 • The Ozone Layer—Shield to Life.*) Ozone warming also occurs in the troposphere, where O_3 is created by air pollution. Nitrous oxide,

---

3.23 Factors affecting global warming and cooling *Greenhouse gases act primarily to enhance global warming, while aerosols, cloud changes, and land-cover alterations caused by human activity act to retard global warming. Natural factors may be either positive or negative. (After Hansen et al., 2000, Proc. Nat. Acad. Sci., used by permission.)*

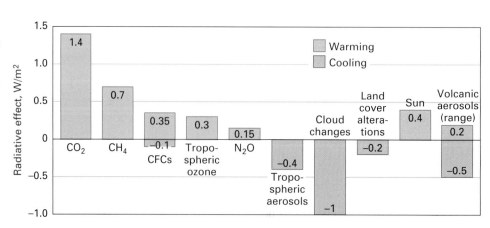

N₂O, is released by bacteria acting on nitrogen fertilizers in soils and runoff water. Motor vehicles also emit significant amounts of N₂O.

The next three factors are primarily a result of human activity, and all tend to cool the Earth–atmosphere system. Tropospheric aerosols are produced largely by fossil fuel burning and are considered to be a potent form of air pollution. They include sulfate particles, fine soot, and organic compounds. Aerosols act to scatter solar radiation back to space, thus reducing the flow of solar energy available to warm the surface. In addition, they enhance the formation of low, bright clouds that also reflect solar radiation back to space. These, and other changes in cloud cover caused directly or indirectly by human activity, lead to a significant cooling effect. Land-cover alteration, which has also induced cooling, includes conversion of forested lands to cropland and pastures, which are brighter and reflect more solar energy.

The last two factors are natural in origin. Solar output has increased slightly, causing warming. Volcanic aerosols, however, have at times caused warming and at other times cooling.

Taken together, the warming effect of the greenhouse gases has outweighed the cooling effects of other factors, and the result is a net warming effect of about 1.6 W/m², which is about 2/3 of 1 percent of the total solar energy flow absorbed by the Earth and atmosphere. Has this enhanced energy flow caused global temperatures to rise? If not, will it warm the Earth in the near future? Before addressing this important question, we need to review the record of the Earth's surface temperature over the last few centuries.

The Temperature Record

Figure 3.24 shows the Earth's mean annual surface temperature from 1866 to 1999 as obtained from surface air temperature measurements. The temperature is expressed as a difference from the average annual temperature for the period 1951–1980. Two curves are shown—yearly data and data smoothed over a five-year period. Although temperature has increased, and especially toward the end, there have been wide swings in the mean annual surface temperature.

Some of the variation is due to volcanic activity. We have already noted that volcanic activity can propel particles and gases—especially sulfur dioxide, SO₂—into the stratosphere, forming stratospheric aerosols. Strong winds spread the aerosols quickly throughout the entire layer. The aerosols have a cooling effect because they reflect incoming solar radiation. For example, the eruption of Mount Pinatubo in the Philippines in the spring of 1991 lofted 15 to 20 million tons of sulfuric acid aerosols into the stratosphere (Figure 3.25). The aerosol layer produced by the eruption reduced solar radiation reaching the Earth's surface between 2 and 3 percent for the year or so following the blast. In response, global temperatures fell about 0.3°C (0.5°F) in 1992 and 1993 (see Figure 3.24).

Although the period of direct air temperature measurement does not extend past the middle of the nineteenth century, the record can be extended further back by using

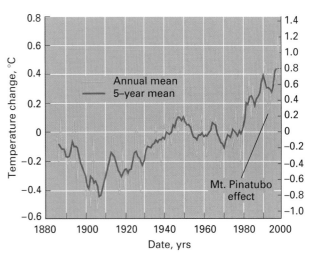

3.24 Mean annual surface temperature of the Earth, 1866–1999 *The vertical scale shows departures in degrees from a zero line of reference representing the average for the years 1951–1980. The yellow line shows the mean for each year. The red line shows a running five-year average. Note the effect of Mount Pinatubo in 1992–1993. (James Hansen/NASA Goddard Institute for Space Studies.)*

tree-ring analysis. The principle is simple—each year, trees grow in diameter. In climates where the seasons are distinct, this growth can appear as an annual ring. If growing conditions are good, the annual ring is wide. If poor, the annual ring is narrow. For trees along the timberline in North America, the width is related to temperature—the trees grow better when temperatures are warmer. Since only one ring is formed each year, the date of each ring is easy to determine by counting backward from the present. Because these trees live a long time, the temperature record may be extended backward several centuries.

Figure 3.26 shows a reconstruction of northern hemisphere temperatures from 1700 to about 1975 using tree-ring analysis. Like Figure 3.24, it is expressed as a difference from a recent average. From 1880 to present, the temperatures reconstructed from tree-ring analysis seem to fit the observed temperatures (Figure 3.24) quite well. Tree-ring analysis of an earlier period shows us another cycle of temperature increases and decline. The low point, around 1840, marks a historic cold event during which European alpine glaciers became more active and advanced. Other evidence indicates that the two cycles in Figure 3.26 are part of a natural global cycle of temperature warming and cooling lasting about 150 to 200 years. These cycles have occurred regularly over the last thousand years.

What causes these cycles? Again, a number of theories have been offered, but no consensus has yet been reached as to the actual cause or causes. Still longer cycles of temperature change are evident in the fossil record, inducing the advance and retreat of continental and mountain glaciers during the Ice Age. These cycles are discussed in more detail in Chapter 19. Thus, the temperature record shows that the

Carbon Dioxide—On the Increase

In the centuries before global industrialization, carbon dioxide concentration in the atmosphere was at a level slightly less than 300 parts per million (ppm), or about $3/100$ths of a percent by volume. During the last hundred years or so, that amount has been substantially increased. Why? The answer is fossil fuel burning. When fuels like coal, oil, or natural gas are burned, they yield water vapor and carbon dioxide. The release of water vapor does not present a problem because a large amount of water vapor is normally present in the global atmosphere. But because the normal amount of CO_2 is so small, fossil fuel burning has raised the level to about 360 ppm, According to studies of bubbles of atmospheric gases trapped in glacial ice, this level is the highest attained in the last 420,000 years and nearly double the amount of CO_2 present during glaciations of the most recent Ice Age.

Increase in atmospheric carbon dioxide, observed to 2000 and predicted into the twenty-first century

The figure above shows how CO_2 has increased with time since 1860. Until 1940 or so, the level remained nearly stable. Even with future concerted global action to reduce CO_2 emissions, scientists estimate that levels will stabilize at a value not lower than about 550 ppm by the late twenty-first century. This will nearly double preindustrial levels.

Predicting the buildup of CO_2 is difficult because not all the carbon dioxide emitted into the air by fossil fuel burning remains there. Instead, a complex cycle moves

Earth's climate is naturally quite variable, responding to many different influences on many different time scales.

Future Scenarios

The year 1998 was the warmest year on record since the middle of the nineteenth century, with an average temperature of 14.7°C (58.5°F) compiled by NASA scientists at New York's Goddard Institute of Space Science. It was also the warmest year of the millennium, according to reconstructions of past temperatures using tree rings and glacial ice cores by University of Massachusetts scientists. In fact, 1998, 1990, 1991, 1995, and 1988 ranked as the five warmest years since 1400. (1999 ranked eighth-warmest, with 1998 and 2000 tied for ninth-warmest.)

Is this recent warming an effect of human activity? In 1995 the Intergovernmental Panel on Climate Change, a United Nations-sponsored group of more than 2000 scientists, concluded for the first time that human activity has caused climatic warming. This judgment was based largely on computer simulations of global climate that account for the release of CO_2 and SO_2 from fossil fuel burning occurring since the turn of the century. The simulations agreed well with the patterns of warming observed over that period, leading to the conclusion that, in spite of natural variability, human influence has been felt in the climate record of the twentieth century. A new report, issued in 2001, cited increased confidence in this conclusion. *Eye on Global Change 3.2.* • *The IPCC Report of 2001* provides a summary of observed climate change and predictions of climate change for the remainder of the twenty-first century.

Given that greenhouse gases are warming the planet, what will be the effect? Using various projections of the continuing release of greenhouse gases coupled with computer climate models, this same group of scientists projected that global temperatures will warm between 1.4°C (2.5°F) and 5.8°C (10.4°F) by the year 2100.

CO_2 throughout the life layer. Photosynthesis is one part of this *carbon cycle.* Plants use CO_2 in photosynthesis, the process by which they use light energy to build their tissues. So, plants take in carbon dioxide, removing it from the atmosphere. When plants die, their remains are digested by decomposing organisms, which release CO_2. This process returns CO_2 to the atmosphere. Normally, these processes are in balance.

Human activity has tilted this balance by clearing land and burning the vegetation cover as new areas of forest are opened for development. This practice increases the amount of CO_2 in the air. When agricultural land is allowed to return to its natural forest state, CO_2 is removed from the air by growing trees. At present, scientists calculate that forests are growing more rapidly than they are being destroyed in the midlatitude regions of the northern hemisphere. This plant growth helps to counteract the buildup of CO_2 produced by fossil fuel burning. But it may be outweighed by the tropical deforestation that is taking place in South America, Africa, and Asia.

(See *Eye on Global Change 9.1* • *Exploitation of the Low-Latitude Rainforest Ecosystem.*)

Laboratory and field experiments have shown that plants exposed to increased concentrations of CO_2 will grow faster and better. The faster they grow, the more CO_2 they can take in, which helps to reduce the amount in the atmosphere. However, scientists are unsure whether increased CO_2 will stimulate plant growth under natural conditions. Early indications are that plants eventually adjust to the higher CO_2 levels and slow their growth rates.

Another part of the cycle involves the oceans. The ocean's surface layer contains microscopic plant life that takes in carbon dioxide. The CO_2 in the ocean water initially comes from the atmosphere and is mixed into the ocean by surface waves. When these microscopic floating plants die, their bodies sink to the ocean bottom. There they decompose and release CO_2, enriching waters near the ocean floor.

This CO_2 eventually returns to the surface through a system of global ocean current flows in

which CO_2-poor waters sink in the northernmost Atlantic and CO_2-rich bottom waters rise to the surface in the northernmost Pacific. In fact, the ocean acts like a slow conveyor belt, moving CO_2 from the surface to ocean depths and releasing it again in a cycle lasting about 1500 years. (We return to this long-term ocean circulation pattern in Chapter 5.)

At present, scientists estimate that ocean surface waters absorb more CO_2 than they release, owing to increased atmospheric levels of CO_2. Therefore, carbon dioxide may be accumulating in ocean depths. However, current studies of global climate computer simulations indicate that the oceans may not be as effective in removing excess CO_2 as scientists previously thought.

Although there is a great deal of uncertainty about the movements and buildup rate of excess CO_2 released to the atmosphere by fossil fuel burning, one thing is certain. Without conversion to solar, nuclear, and hydroelectric power, fuel consumption will continue to release carbon dioxide, and its effect on climate will continue to increase.

Why does this temperature rise cause concern? The problem is that many other changes may accompany a rise in temperature. One of these changes is a rise in sea level, as glaciers and sea ice melt in response to the warming. Current predictions call for a rise of 9 to 88 cm (3.5 to 34.6 in.) in sea level by the year 2100. This would place as many as 92 million people within the risk of annual flooding. Climate change could also promote the spread of insect-borne diseases such as malaria. Furthermore, climate boundaries may shift their positions, making some regions wetter while others become drier. Thus, agricultural patterns could shift, displacing large human populations as well as natural ecosystems.

A recently discovered effect of climatic warming is the enhancement of variability in climate. Events such as very

The 1990s brought a succession of very warm years, including the warmest year of the second millennium (1998). Most scientists agree that human activity, primarily through the enhanced release of greenhouse gases to the atmosphere, is responsible for the warming trend.

high 24-hour precipitation—extreme snowstorms, rainstorms, sleet and ice storms, for example—appear to be occurring more frequently since 1980. More frequent and more intense spells of hot and cold weather may also be related to climatic warming.

The world has become widely aware of the problem of the buildup in CO_2 and other greenhouse gases. At the Rio de Janeiro Earth Summit in 1992, nearly 150 nations signed a treaty limiting emissions of greenhouse gases. Many details were ironed out at a subsequent meeting in Kyoto, Japan, in 1997. The Kyoto Protocol calls on a group of 38 industrial nations to reduce their emissions of CO_2 and other greenhouse gases to about 5 percent below 1990 levels. Nations of the European Union, the United States, and Japan are to reduce average emissions during 2008 and 2012 to levels that

3.25 Eruption of Mount Pinatubo, Philippine Islands, April 1991 *Volcanic eruptions like this can inject particles and gases into the stratosphere, influencing climate for several years afterwards.*

3.26 Three centuries of northern hemisphere temperatures *A reconstruction of the departures of northern hemisphere temperatures from the 1950–1965 mean, based on analyses of tree rings sampled along the northern tree limit of North America. Original data are from 1700 to 1975; for completeness, data from Figure 5.24 are shown from 1975 to the present in a different color. (Courtesy of Gordon C. Jacoby of the Tree-Ring Laboratory of the Lamont-Doherty Geological Observatory of Columbia University.)*

Eye on Global Change | 3.2

The IPCC Report of 2001

In 2001, the United Nations Intergovernmental Panel on Climate Change (IPCC), a body of over 600 scientists nominated by countries from all over the globe, issued a series of reports on human-induced changes in global climate. Here are some of the panel's findings, taken from *Climate Change 2001: The Scientific Basis, A Report of the Working Group I of the Intergovernmental Panel on Climate Change,* Summary for Policymakers, IPCC, Geneva, Switzerland, 2001.

Recent Climate Change

- Global average surface temperature increased about 0.6°C (1.1°F) during the twentieth century. The 1990s was the warmest decade and 1998 the warmest year since 1861. Nighttime daily minimum air temperatures rose about twice as fast as daytime maximum temperatures. Since 1950, the frequency of extreme low temperatures has decreased. The frequency of extreme high temperatures has increased a smaller amount.
- The lowest 8 km (5 mi) of the atmosphere has warmed at a rate of about 0.05°C (0.09°F) per decade since at least 1979.
- Snow and ice cover has decreased by about 10 percent since the late 1960s. The duration of snow cover has been reduced by about two weeks in the mid- and high latitudes of the northern hemisphere. The extent of sea ice in spring and summer in the northern hemisphere has decreased by 10 to 15 percent since the 1950s. Summer and fall sea ice is thinner.
- Global average sea level rose between 10 and 20 cm (4 and 8 in.) during the twentieth century. Global ocean heat content has increased since at least the late 1950s.

- Precipitation increased by 0.5 to 1 percent per decade in the twentieth century over the mid- and high latitudes of the northern hemisphere. Rainfall decreased in subtropical regions. There was an increase of 2 to 4 percent in the frequency of heavy precipitation events in the latter half of the twentieth century. The frequency and intensity of droughts in some parts of Africa and Asia have also increased in recent decades.
- Cloud cover increased by about 2 percent in the mid- and high latitudes of the northern hemisphere during the twentieth century.
- Since 1970, El Niño episodes have been more frequent, intense, and persistent, compared to those in the previous 100 years.
- Some important aspects of climate have not changed. Some parts of the southern hemisphere oceans and parts of Antarctica have not warmed in recent decades. No significant changes have occurred in Antarctic sea ice coverage since 1978. Tropical and extratropical cyclone frequencies have not been shown to be changing.
- Concentrations of greenhouse gases in the atmosphere have increased as a result of human activities. CO_2 has increased by 31 percent since 1750. For the past two decades, CO_2 concentration has increased by about 1.5 parts per million (ppm) per year. Methane (CH_4) concentration has increased by 151 percent since 1750; nitrous oxide (N_2O) by about 17 percent; and ozone (O_3) by about 36 percent.
- There is newer and stronger evidence that human activities are responsible for most of the warming observed since 1950.

The IPCC reports also offer these predictions for the rest of the

twenty-first century. Although they may not all come true, they are considered likely given the state of present knowledge and current levels of confidence in global climate model predictions, as well as reasonable scenarios for economic growth, development, and emission control.

- Atmospheric composition will continue to change. CO_2 from fossil fuel burning will continue to increase. As this occurs, land and ocean will take up a decreasing fraction of CO_2 released, accentuating the increase. By 2100, CO_2 concentration will have increased by 90 to 250 percent above the 1750 value, depending on the scenario of economic growth, development, and CO_2 release.
- Other greenhouse gas concentrations are likely to change, increasing under most scenarios. Aerosols could contribute significantly to warming if their release is unabated.
- Global average temperature will rise between 1.4 and 5.8°C (2.5 and 10.4°F) from 1990 to 2100. The projected rate of warming is much larger than that of the late twentieth century and is likely to be greater than any warming episode in the last 10,000 years.
- Land areas will warm more rapidly than the global average, especially northern latitudes in the winter.
- Precipitation will increase in northern mid- to high latitudes and in Antarctica in winter. Larger year-to-year variations in precipitation are likely. At low latitudes, there will be decreases in precipitation in some areas and increases in others.
- Over nearly all land areas, there will be higher maximum temperatures and more hot days; higher minimum temperatures, fewer cold days and frost days;

reduced daily temperature range; more intense precipitation events; increased risk of summer drought; and an increase in peak wind and precipitation intensities of tropical cyclones.
• Snow and ice cover will decrease further, and ice caps and glaciers will continue their widespread retreat of the late twentieth century. The Antarctic Ice Sheet will gain mass from increased precipitation, while the Greenland Ice Sheet will thin.
• Sea level will rise from 9 to 88 cm (3.5 to 34.6 in.) depending on the scenario, due to thermal expansion and a gain in volume from melting ice.
• Human-generated climate change will persist for many centuries into the future. Even if concentrations of greenhouse gases are stabilized, oceans will continue to warm; sea level will continue to rise; and ice sheets and glaciers will continue to melt. Models suggest that a local warming of 3°C (5.4°F) over the Greenland ice cap, if sustained over 10 centuries, would completely melt the ice cap, raising sea level by about 7 m (9.8 ft).

are 6, 7, and 8 percent lower, respectively, than 1990 levels. Achieving such reductions could have economic consequences ranging from minor to severe, depending on the analysis.

Meeting again in Buenos Aires in 1998, the participating nations set timetables for the implementation of plans for greenhouse gas reduction in 1999–2000. Important breakthroughs occurred when the conference adopted a plan to allow countries to trade emission rights, as well as a "clean development" mechanism that would permit industrial nations to invest in emissions-reducing enterprises in developing countries. For the first time, some developing countries pledged specific greenhouse gas reductions of their own.

The international effort to reduce releases of greenhouse gases suffered a setback in the spring of 2001, when President George W. Bush rejected American participation in the Kyoto Protocol. However, American resistance was not sufficient to keep 178 nations from pledging to join in the effort to reduce greenhouse gas emissions.

Although these efforts will slow the buildup of greenhouse gases, the ultimate solution will certainly involve greater reliance on solar, wind, and geothermal energy sources, which produce power without releasing CO_2. Energy conservation and development of new methods of utilizing fossil fuels that produce reduced CO_2 emissions will also be important.

A LOOK AHEAD In this chapter we have developed an understanding of both air temperatures and temperature cycles, along with the factors that influence them—insolation, latitude, surface type, continental-maritime location, and elevation. Air temperatures are a very important part of climate, and looking ahead, we will study the climates of the Earth in Chapter 7.

The other key ingredient of climate is precipitation–the subject covered in the next chapter. Here temperature plays a very important role, since moist air must cool before condensation forms water droplets and, eventually, precipitation.

In Review | Air Temperature

- **Air temperature**—measured at 1.2 m (4 ft) above the surface—is influenced by insolation, latitude, surface type, location, and elevation. The energy balance of the ground surface is determined by **net radiation, conduction** to the soil, and **latent heat transfer** and **convection** to and from the atmosphere.

- Air temperature is measured using a *thermometer* or *thermistor*. Stations make daily minimum, maximum, and mean temperature measurements.

- The two major cycles of air temperature—daily and annual—are controlled by the cycles of insolation produced by the rotation and revolution of the Earth. These cycles induce cycles of net radiation at the surface. When net radiation in the daily cycle is positive, air temperatures increase; when negative, air temperatures decrease. This principle applies to both daily and annual temperature cycles.

- Temperatures of air and soil at or very close to the ground surface are more variable than air temperature as measured at standard height.

- Surface characteristics also affect temperatures. Rural surfaces are generally moist and slow to heat and cool, while urban surfaces are dry and absorb and give up heat readily. This difference creates an urban heat island effect.

- Air temperatures normally fall with altitude in the **troposphere.** The average value of decrease with altitude is the **environmental temperature lapse rate,** 6.4°C/1000 m (3.5°F/1000 ft). At the *tropopause,* this decrease stops. In the **stratosphere** above, temperatures increase slightly with altitude.

- Air temperatures observed at mountain locations are lower with higher elevation, and day-night temperature differences increase with elevation.

- When air temperature increases with altitude, a **temperature inversion** is present. This can develop on clear nights when the surface loses longwave radiation to space.

- Annual air temperature cycles are influenced by the annual pattern of net radiation, which depends largely on latitude.

- Maritime or continental location is another important factor. Ocean temperatures vary less than land temperatures because water heats more slowly, absorbs energy throughout a surface layer, and can mix and evaporate freely. Maritime locations that receive oceanic air therefore show smaller ranges of both daily and annual temperature.

- Global temperature patterns for January and July show the effects of latitude, maritime-continental location, and elevation. Equatorial temperatures vary little from season to season. Poleward, temperatures decrease with latitude, and continental surfaces at high latitudes can become very cold in winter. At higher elevations, temperatures are always colder.

- **Isotherms** over continents swing widely north and south with the seasons, while isotherms over oceans move through a much smaller range of latitude. The annual range in temperature increases with latitude and is greatest in northern hemisphere continental interiors.

- Our planet's global temperature changes from year to year. Within the last few decades, global temperatures have been increasing. CO_2 released by fossil fuel burning is important in causing warming, but so are the other **greenhouse gases,** CH_4, CFCs, O_3, and N_2O. Aerosols scatter sunlight back to space and induce more low clouds, so they tend to lower global temperatures. Solar output and volcanic activity also influence global temperatures.

- Most scientists agree that the human-induced buildup of greenhouse gases has begun to affect global climate. However, natural cycles, such as variations in the Sun's output, still provide strong influences. If we continue to release large quantities of greenhouse gases at increasing rates, we can expect a significant rise in global temperatures that will be accompanied by shifts of climate zones and increasing sea level.

Key Terms

air temperature, p. 88
surface, p. 89
net radiation, p. 89
conduction, p. 89
latent heat transfer, p. 89

convection, p. 89
transpiration, p. 92
evapotranspiration, p. 94
heat island, p. 94

lapse rate, p. 95
environmental temperature lapse rate, p. 95
troposphere, p. 96
stratosphere, p. 96

temperature inversion, p. 99
isotherms, p. 103
greenhouse gases, p. 108

Review Questions

1. Identify five important factors in determining air temperature and air temperature cycles.
2. What factors influence the temperature of a surface?
3. How are mean daily air temperature and mean monthly air temperature determined?
4. How does the daily temperature cycle measured within a few centimeters or inches of the surface differ from the cycle at normal air temperature measurement height?
5. Compare the characteristics of urban and rural surfaces and describe how the differences affect urban and rural air temperatures. Include a discussion of the urban heat island.
6. What are the two layers of the lower atmosphere? How are they distinguished? Name the two upper layers.
7. How and why are the temperature cycles of high mountain stations different from those of lower elevations?
8. Explain how latitude affects the annual cycle of air temperature through net radiation by comparing Manaus, Aswan, Hamburg, and Yakutsk.
9. Why do large water bodies heat and cool more slowly than land masses? What effect does this have on daily and annual temperature cycles for coastal and interior stations?
10. What three factors are most important in explaining the world pattern of isotherms? Explain how and why each factor is important, and what effect it has.
11. Turn to the January and July world temperature maps shown in Figure 3.20. Make six important observations about the patterns and explain why each occurs.
12. Turn to the world map of annual temperature range in Figure 3.22. What five important observations can you make about the annual temperature range patterns? Explain each.
13. Identify the important greenhouse gases and rank them in terms of their warming effect. What human-influenced factors act to cool global temperature? How?
14. Describe how global air temperatures have changed in the recent past. Identify some factors or processes that influence global air temperatures on this time scale.

Eye on Global Change 3.1 ● Carbon Dioxide—On the Increase

1. Why has the atmospheric concentration of CO_2 increased in recent years?
2. How does plant life affect the level of atmospheric CO_2?
3. What is the role of the ocean in influencing atmospheric levels of CO_2?

Eye on Global Change 3.2 ● The IPCC Report of 2001

1. What changes in global climate have been noted for the last half of the twentieth century in global temperature, snow and ice cover, precipitation, and greenhouse gas concentrations?
2. How is climate predicted to change by the end of the twenty-first century with respect to temperature, precipitation, snow and ice cover, and sea level?

Visualizing Exercises

1. Sketch graphs showing how insolation, net radiation, and temperature might vary from midnight to midnight during a 24-hour cycle at a midlatitude station such as Chicago.
2. Sketch a graph of air temperature with height showing a low-level temperature inversion. Where and when is such an inversion likely to occur?

Essay Questions

1. Portland, Oregon, on the north Pacific coast, and Minneapolis, Minnesota, in the interior of the North American continent, are at about the same latitude. Sketch the annual temperature cycle you would expect for each location. How do they differ and why? Select one season, summer or winter, and sketch a daily temperature cycle for each location. Again, describe how they differ and why.
2. Many scientists have concluded that human activities are acting to raise global temperatures. What human processes are involved? How do they relate to natural processes? Are global temperatures increasing now? What other effects could be influencing global temperatures? What are the consequences of global warming?

Eye on the Landscape

Chapter Opener Garibaldi Provincial Park, British Columbia High-mountain environments can be tough on trees **(A)**. This fir obviously lost its top some years ago but was still sufficiently alive to send a new sprout aloft from its split and twisted bole. The heavy weight of ice accumulating during an ice storm was most likely the cause of the damage. Other trees show broken branches or are simply dead snags—all victims of the harsh environment. In exposed locations, wind also shapes trees by stunting or killing windward branches (see Figure 9.24 for an example of this.)

3.5a Urban surfaces Beyond the urban surfaces in the foreground is a distant view of a semiarid landscape in the western United States **(A)**. The sparse and low vegetation cover reflects the semiarid climate (see Chapters 7 and 9 for semiarid climate and vegetation). Note also the smooth profile of the land surface from the mountain peak to the base of the slope **(B)**. This shape indicates long-continued erosion by running water (fluvial erosion), which we describe in Chapter 16.

3.5b Rural surfaces The human influence on this agricultural landscape **(A)** is strong. A dam has created a small fire pond, providing water for both livestock and fire-fighting. The fields are mowed for hay, providing a groomed look by cutting all plants. Grazing by livestock would be more selective and give a more tufted appearance to the pasture. The young trees dotting the landscape are probably all planted and have been pruned to pleasing shapes. The split-rail fencing is attractively rustic but too expensive for most farm applications. In short, this is the spread of a gentleman farmer.

3.10 Clear air at high elevation— Mount Whitney The jagged topography of these high peaks **(A)** indicates glacial, rather than fluvial, erosion. Imagine a dome of glacial ice covering this entire scene, leaving only the tip of Mount Whitney uncovered. Ice flowing in and around the rock masses below the summit gouges the rock, tearing huge chunks loose along cracks and joints in the rock. Then the climate warms and the ice melts, revealing rugged slopes and sheer rock faces stripped by the ice action. We'll cover glacial landscapes in Chapter 19.

Lightning heralds the approach of a summer thundershower.

4 | ATMOSPHERIC MOISTURE AND PRECIPITATION

● ● ● ● ● ● ● ● ● ● ● ● ● ● ● ●

PUTTING YOU IN THE PICTURE

Rain is a common event in most parts of the world. We don't think much about a rainy day, other than perhaps to grumble a bit at having to grab a raincoat or umbrella before leaving the house. On the other hand, precipitation is a vital part of life, for without rain and snow we'd have no fresh water to meet our daily needs.

Precipitation, along with temperature, is an essential ingredient of climate. The timing and duration of rainy periods condition many human activities, especially agriculture. They also influence the nature of plant and animal communities in a region, as well as the soil processes that provide the substrate needed for natural and cultivated plant growth.

Another important role for precipitation is the shaping of the landscape through the processes of erosion and deposition. As rain and melting snow run off the landscape under the influence of gravity, water carries and deposits soil and sediment, both carving the landscape by removal and building it up by accumulation.

What causes precipitation? As we'll see in this chapter, rain and snow form when warm, moist air is chilled and condensation or deposition takes place. The chilling is usually the result of uplift, for example, as air passes up and over a mountain barrier or is carried upward by convection.

When visualizing a rainfall, what usually comes to mind is a soggy day of gray skies, showers, or drizzle. While most rainstorms are pretty routine, a thunderstorm can be an exciting, even dangerous, event. You might experience it like this…

Suddenly you notice that the thick, humid air is perfectly still. The songbirds and insects are strangely silent. Not a single leaf trembles in the huge, ancient oak shading your lawn chair. You look up from the book you've been reading and notice that the puffy white clouds that had filled the sky seem to have merged into a darkening blanket. Across the grassy field in front of you, the rambling old farmhouse where you are spending the holiday weekend with friends looks pale and somehow vacant of life in the flat, gray light.

A low, distant rumble shatters the stillness. A thunderstorm is approaching! Tucking your book under your arm, you fold up your lawn chair and head for the house. As you walk through the field, you begin to hear a rushing sound. Looking over your

shoulder, you see the tops of the forest trees at the edge of the field shaking and swaying in a sudden wind. Then, the top of the old oak twists violently, its upper branches jerking chaotically as it is battered by strong gusts. A flood of cool air sweeps over you, lifting the lawn chair in your left hand. The sickly smell of ozone, now assaulting your nose, harkens of danger. Unconsciously, you quicken your steps toward the farmhouse.

The first large, heavy drops begin to fall as you reach the wide front porch. A rushing sound deepens as the first wave of heavy rainfall approaches through the distant trees and crosses the field. The sky explodes with light, and a loud crack of thunder startles you. The rain now drives intensely toward the ground, banging on the tin roof of the porch and swirling onto the porch in the strong gusts. The oak in the middle of the field is barely visible through the falling torrent.

A crackle pierces the din, and a blinding flash prints an image of the ancient oak onto your eyes. At the same moment, a deafening bang shatters the air. Lightning has struck the old oak! A huge branch crashes to the ground, the sound of its fall muffled by the wind and rain. That was too close for comfort! You retreat from the porch for the safety of the house to watch the storm from a window, hopeful that the lightning rods that project from the chimneys at either end of the house will save you from the fate of the ancient oak.

Later, you survey the damage. The top of the tree lies on the ground, its thick limbs splintered by the force of the lightning strike. The stout trunk is split almost to the base. Huge splinters are strewn on the surrounding ground. The smell of fresh-split green oak is pungent in the humid air. "Good thing I didn't decide to wait out the storm under the tree!" you think to yourself. ∎

ATMOSPHERIC MOISTURE AND PRECIPITATION

Water plays a key role in the energy flows that shape our planet's climate and weather. In Chapter 2 we noted how ocean currents act to carry the annual surplus of solar energy from equatorial regions toward the poles. Another water-assisted part of this global energy flow is the flow of latent heat in the form of water vapor conducted poleward by atmospheric circulation. When water evaporates over warm tropical oceans, it takes up latent heat. When it condenses, often in distant colder regions, that latent heat is released in a different location. Precipitation, then, is an important part of this global energy flow process. Precipitation over land also provides water (and ice) that can move under the influence of gravity, thus carving the landforms and landscapes that provide the Earth's varied human habitats. We will return to this role of water in Chapters 15–17.

In this chapter, we focus on water in the air, both as vapor and as liquid and solid water. **Precipitation** is the fall of liquid or solid water from the atmosphere to reach the Earth's land or ocean surface. It is formed when moist air is cooled, causing water vapor to form liquid droplets or solid ice particles. If cooling is sufficiently long and intense, liquid and solid water particles will grow to a size too large to be carried in the air by air motion. They can then fall to Earth.

What causes the cooling of moist air? The answer is upward motion. Whenever air moves upward in the atmosphere, it is cooled. This is not simply because surrounding air temperatures tend to get lower as you go up, as we saw

in Chapter 3. It is because of a simple physical principle stating that when air expands, it cools. And because atmospheric pressure decreases with altitude, air moving upward expands—and so at the same time it gets cooler.

What makes air move upward? There are three major causes. One occurs when winds move air up and over a mountain barrier. The second takes place when unequal heating at the ground surface creates a bubble of air that is warmer than the surrounding air. Since it is less dense, it is buoyed upward—like a cork under water bobbing to the surface. This chapter focuses on these two causes of uplift. The third occurs when a mass of cooler, denser air slides under a mass of warmer, lighter air, lifting the warmer air aloft. This type of upward movement occurs in weather systems, which we will cover in Chapter 6. But before we begin our study of atmospheric moisture and precipitation, we will briefly review the three states of water and the conversion of one state to another.

THREE STATES OF WATER

As shown in Figure 4.1, water can exist in three states—solid (ice), liquid (water), and gas (water vapor). A change of state from solid to liquid, liquid to gas, or solid to gas requires the input of heat energy. As we noted in Chapter 2, this energy is called *latent heat,* which is drawn in from the surroundings and stored within the water molecules. When the change goes the other way, from liquid to solid, gas to

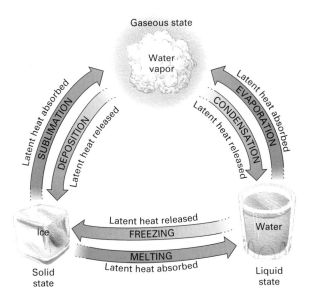

4.1 Three states of water *Arrows show the ways that any one state of water can change into either of the other two states. Heat energy is absorbed or released, depending on the direction of change.*

THE HYDROSPHERE AND THE HYDROLOGIC CYCLE

Let's now turn to the **hydrosphere,** the realm of water in all its forms, and the flows of water among ocean, land, and atmosphere. About 97.2 percent of the hydrosphere consists of ocean saltwater, as shown in Figure 4.2. The remaining 2.8 percent is fresh water. The next largest reservoir is fresh water stored as ice in the world's ice sheets and mountain glaciers. This water accounts for 2.15 percent of total global water.

Fresh liquid water is found both on top of and beneath the Earth's land surfaces. Water occupying openings in soil and rock is called *subsurface water.* Most of it is held in deep storage as *ground water,* at a level where plant roots cannot access it. Ground water makes up 0.63 percent of the hydrosphere, leaving 0.02 percent of the water remaining.

The right-hand portion of Figure 4.2 shows how this small remaining proportion of the Earth's water is distributed. This proportion is important to us because it includes the water available for plants, animals, and human use. *Soil water,* which is held in the soil within reach of plant roots, comprises 0.005 percent of the global total. Water held in streams, lakes, marshes, and swamps is called *surface water.* Most of this surface water is about evenly divided between freshwater lakes and saline (salty) lakes. An extremely small proportion is held in the streams and rivers that flow toward the sea or inland lakes.

Note that the quantity of water held as vapor and cloud water droplets in the atmosphere is also very small—0.001 percent of the hydrosphere. Though small, this reservoir of water is of enormous importance. It provides the supply of precipitation that replenishes all freshwater stocks on land. And, as we will see in the next chapter, the flow of water vapor from warm tropical oceans to cooler regions provides a global flow of heat, in latent form, from low to high latitudes.

The movements of water among the great global reservoirs constitute the **hydrologic cycle.** In this cycle, water moves from land and ocean to the atmosphere as water vapor and returns as precipitation. Because precipitation over land exceeds evaporation, water also runs off the land to the oceans. We will return to the hydrologic cycle in more detail in Chapter 15.

liquid, or gas to solid, this latent heat is released to the surroundings.

For each type of transition, there is a specific name, shown in the figure. *Melting, freezing, evaporation,* and *condensation* are all familiar terms. **Sublimation** is the direct transition from solid to vapor. Perhaps you have noticed that old ice cubes in your refrigerator's freezer seem to shrink away from the sides of the ice cube tray and get smaller with time. They shrink through sublimation, which is induced by the constant circulation of cold, dry air through the freezer. The ice cubes never melt, yet they lose bulk directly as vapor. **Deposition** is the reverse process, when water vapor crystallizes directly as ice.* Frost formed on a cold winter night is an example.

> Water exists in solid, liquid, and gaseous states as ice, water, and water vapor. Latent heat energy is released or absorbed as water changes from one state to another.

---

* *Deposition* is a term used by chemists to describe the deposit of a solid directly from a vapor. Meteorologists generally use the term *sublimation* for both the change from vapor to solid and from solid to vapor, with the difference determined by the context. Since it is clearer to use a separate word for each process, we use the term *deposition* in this text.

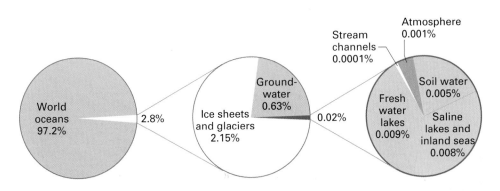

4.2 Volumes of global water in each reservoir of the hydrosphere *Nearly all the Earth's water is contained in the world ocean. Fresh surface and soil water make up only a small fraction of the total volume of global water.*

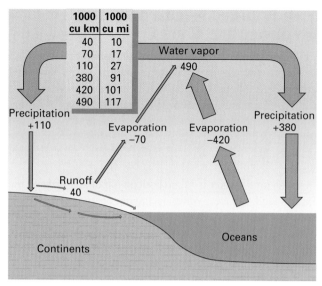

1000 cu km	1000 cu mi
40	10
70	17
110	27
380	91
420	101
490	117

4.3 *The global water balance* *Figures give average annual water flows in and out of world land areas and world oceans. Values are in thousands of cubic kilometers (cubic miles). Global precipitation equals global evaporation. (Based on data of John R. Mather.)*

For now, we will summarize the main features of the hydrologic cycle in the global water balance, shown in Figure 4.3. The cycle begins with evaporation from water or land surfaces, in which water changes state from liquid to vapor and enters the atmosphere. In the case of land, this evaporation is enhanced by transpiration. Total evaporation is still about six times greater over oceans than land, however. This is because the oceans cover most of the planet and because land surfaces are not always wet enough to yield much evaporated water. Once in the atmosphere, water vapor can condense or deposit to form precipitation, which falls to Earth as rain, snow, or hail. Precipitation over the oceans is nearly four times greater than precipitation over land.

Upon reaching the land surface, precipitation has three fates. First, it can evaporate and return to the atmosphere as water vapor. Second, it can sink into the soil and then into the surface rock layers below. As we will see in later chapters, this subsurface water emerges from below to feed rivers, lakes, and even ocean margins. Third, precipitation can run off the land, concentrating in streams and rivers that eventually carry it to the ocean or to a lake in a closed inland basin. This flow of water is known as *runoff*.

The Global Water Balance

Since our planet contains only a fixed amount of water, a global balance must be maintained among flows of water to and from the lands, oceans, and atmosphere. Let's examine this idea in more detail. For our analysis, we will assume that the volume of ocean waters and the overall volume of fresh

water in surface and subsurface water remains constant from year to year. This is probably quite reasonable, unless climate is changing rapidly.

The global water balance is diagrammed in Figure 4.3. Let's consider the ocean first, shown on the right side of the diagram. There are three flows to and from the ocean: evaporation (out), precipitation (in), and runoff (in). For precipitation and runoff, water enters the ocean, while water leaves the ocean by evaporation. Since "in" and "out" must balance, we have

Precipitation + Runoff = Evaporation   (ocean)

Using the values in the figure, which are given as thousands of cubic kilometers, we find that 380 + 40 = 420.

Now let's turn to the continents, shown on the left side of the figure. Again, there are three flows—precipitation (in), evaporation (out), and runoff (out). This time, precipitation is entering (falling onto) the land, while evaporation and runoff are leaving. Thus,

Precipitation = Evaporation + Runoff   (land)

Using the values in the figure, we observe that 110 = 70 + 40.

If we rearrange this last equation to put evaporation first, we have

Evaporation = Precipitation – Runoff   (land)

compared to

Evaporation = Precipitation + Runoff   (ocean)

Taking the land and oceans together and considering only the flow of water from the planet's surface to the atmosphere and the reverse, we can conclude that the runoff terms within the last two equations cancel out. This leaves total evaporation to equal total precipitation, as it must:

Total Evaporation	Total Precipitation
70 (land) + 420 (ocean) cubic kilometers	= 110 (land) + 380 (ocean) cubic kilometers

The point of this simple analysis is not to write equations, but to show that we have constructed a budget that accounts for the flows of water between the reservoirs of the hydrologic cycle. Like a household budget, it must balance. If it does not, then there must be some flow that is not being considered. This is why scientists use budgets as tools to help understand natural processes, like the hydrologic cycle.

For most of this chapter, we will be concerned with one aspect of the hydrologic cycle—the flow of water from the atmosphere to the surface in the form of precipitation. We will begin by examining how the water vapor content of air is measured and how clouds and precipitation form.

The hydrologic cycle describes the global flow of water to and from oceans, land, and atmosphere. Water is exchanged between the atmosphere and oceans and atmosphere and land by evaporation and precipitation. Water moves from land to ocean by runoff.

HUMIDITY

The amount of water vapor present in the air varies widely from place to place and time to time. It ranges from almost nothing in the cold, dry air of arctic regions in winter to as much as 4 or 5 percent of a given volume of air in the warm wet regions near the equator. The general term **humidity** refers to the amount of water vapor present in the air.

Understanding humidity and how the moisture content of air is measured involves an important principle—namely, that the maximum quantity of moisture that can be held at any time in the air is dependent on air temperature. Warm air can hold more water vapor than cold air—a lot more. Air at room temperature (20°C, 68°F) can hold about three times as much water vapor as air at freezing (0°C, 32°F).

> Humidity refers to the amount of water vapor present in air. Warm air can hold much more water vapor than cold air.

Specific Humidity

The actual quantity of water vapor held by a parcel of air is its **specific humidity.** This measure is important because it describes how much water vapor is available for precipitation. Specific humidity is stated as the mass of water vapor contained in a given mass of air and is expressed as grams of water vapor per kilogram of air (g/kg). Figure 4.4 shows the relationship between air temperature and the maximum amount of water vapor that air can hold. We see, for example,

> The dew-point temperature of a mass of air is the temperature at which saturation will occur. The more water vapor in the air, the higher is the dew-point temperature.

that at 20°C (68°F), the maximum amount of water vapor that the air can hold—that is, the maximum specific humidity possible—is about 15 g/kg. At 30°C (86°F), it is nearly doubled—about 26 g/kg. For cold air, the values are quite small. At –10°C (14°F), the maximum is only about 2 g/kg.

Climatologists often use specific humidity to describe the moisture characteristics of a large mass of air. For example, extremely cold, dry air over arctic regions in winter may have a specific humidity as low as 0.2 g/kg. In comparison, the extremely warm, moist air of equatorial regions often holds as much as 18 g/kg. The total natural range on a worldwide basis is very wide. In fact, the largest values of specific humidity observed are from 100 to 200 times as great as the smallest.

Specific humidity is a geographer's yardstick for a basic natural resource—water. It is a measure of the quantity of water in the atmosphere that can be extracted as precipitation. Cold, moist air can supply only a small quantity of rain or snow, but warm, moist air is capable of supplying large amounts.

Figure 4.5 is a set of global profiles showing how specific humidity varies with latitude and how it relates to mean surface air temperature. Both humidity and temperature are measured at the same locations in standard thermometer shelters the world over, and also on ships at sea. Note that the horizontal axis of the graphs has an uneven scale of units—they get closer together toward both left and right edges, which represent the poles. In this case, the units are adjusted so that they reflect the amount of the Earth's surface present at that latitude. That is, since there is much more surface area between 0° and 10° latitude than between 60° and 70°, the region between 0° and 10° is allocated more width on the graph.

Let's look first at specific humidity (upper graph). The curve clearly shows the largest values for the equatorial zones, with values falling off rapidly toward both poles. This curve follows the pattern of insolation quite nicely (see Figure 2.8). More insolation is available at lower latitudes, on average, to evaporate water in oceans or on moist land surfaces. Therefore, specific humidity values are higher at low latitudes than at high latitudes.

The global profile of mean (average) surface air temperature (lower graph) shows a similar shape to the specific humidity profile. We would expect this to be the case, since air temperature and maximum specific humidity vary together, as shown in Figure 4.4.

Another way of describing the water vapor content of air is by its **dew-point temperature.** If air is slowly chilled, it eventually will reach *saturation*. At this temperature, the air holds the maximum amount of water vapor possible. If further cooling continues, condensation will begin and dew will form. The temperature at which satura-

4.4 Specific humidity and temperature *The maximum specific humidity of a mass of air increases sharply with rising temperature.*

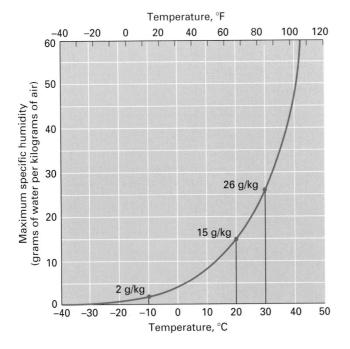

4.5 Global specific humidity and temperature

Pole-to-pole profiles of specific humidity (above) and temperature (below) show similar trends because the ability of air to hold water vapor (measured by specific humidity) is limited by temperature. (Data of J. von Hann, R. Süring, and J. Szava-Kovats as shown in Haurwitz and Austin, Climatology.)

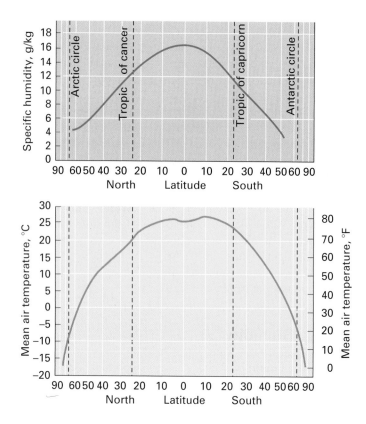

tion occurs is therefore known as the dew-point temperature. Moist air will have a higher dew-point temperature than drier air.

Relative Humidity

The measure of humidity that we encounter every day is **relative humidity.** This measure compares the amount of water vapor present to the maximum amount that the air can hold at that temperature, expressed as a percentage. For example, if the air currently holds half the moisture possible at the present temperature, then the relative humidity is 50 percent. When the humidity is 100 percent, the air holds the maximum amount possible, is saturated, and will be at the dew-point temperature.

A change in relative humidity of the atmosphere can happen in one of two ways. The first is through direct gain or loss of water vapor. For example, if an exposed water surface or wet soil is present, additional water vapor can enter the air, raising the specific humidity. This process is slow because the water vapor molecules must diffuse upward from the surface into the air layer above.

The second way is through a change of temperature. Even though no water vapor is added, a lowering of temperature results in a rise of relative humidity. Recall that the capacity of air to hold water vapor is dependent on temperature. When the air is cooled, this capacity is reduced. The existing amount of water vapor then represents a higher percentage of the total capacity.

> Relative humidity depends on both the water vapor content and the temperature of air. It compares the amount of water held by air to the maximum amount that can be held at that temperature.

An example may help to illustrate these principles (Figure 4.6). Beginning in the early morning hours, at 4 A.M. the temperature is 5°C (41°F), and the relative humidity of the air is 100 percent. That is, the air is saturated and cannot hold any additional water vapor. By 10 A.M., the temperature has risen to 16°C (61°F). The relative humidity has dropped to 50 percent, even though the amount of water vapor in the air remains the same. By 3 P.M., the air has been warmed by the sun to 32°C (90°F). The relative humidity has

4.6 Relative humidity and air temperature

Relative humidity changes with temperature because the capacity of warm air to hold water vapor is greater than that of cold air. In this example, the amount of water vapor stays the same, and only the capacity to hold water changes.

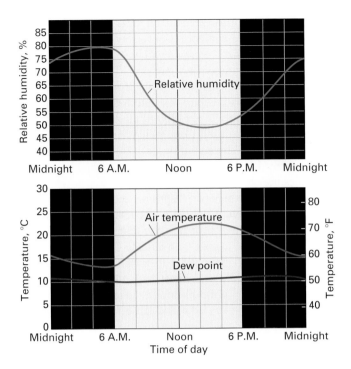

4.7 Daily cycle of relative humidity, air temperature, and dew point *Relative humidity (above) drops during the day as air temperature (below) rises in this example of hourly averages from Washington, D.C., for the month of May. Dew-point temperature remains nearly constant. (Data from National Weather Service.)*

dropped to 20 percent, which is very dry air. The same amount of water vapor is present in the air, but the capacity of the air to hold water vapor has greatly increased. As air temperature falls in the evening, relative humidity again rises, and the cycle repeats itself.

Figure 4.7 illustrates the principle of relative humidity change caused by temperature change. Values shown are averages for May at Washington, D.C. As air temperature rises during the day, relative humidity falls, and vice versa. The dew-point temperature, which indicates the true moisture content, remains nearly constant.

How is humidity measured? A simple method uses two thermometers mounted together side by side in an instrument called a *sling psychrometer* (Figure 4.8). The wet-bulb thermometer bulb is covered with a sleeve of cotton gauze and is wetted. The dry-bulb thermometer remains dry. To operate the sling psychrometer, water is applied to the sleeve covering the wet bulb, and the two thermometers are then whirled in the open air. Evaporation cools the wet bulb. If the air is saturated, then evaporation cannot occur. In this case, no cooling results, and the temperature of the two thermometers is the same. If the air is not saturated, evaporation occurs and the wet-bulb thermometer is cooled. The drier the air, the greater is the cooling. To determine the relative humidity, a special scale is used (Figure 4.8). There are also instruments that read relative humidity directly. One such instrument uses a thin layer of a special material bonded to a metal film. The material absorbs water vapor in an amount depending on the relative humidity. The water vapor affects the ability of the metal film to hold an electric charge. This ability is sensed by an electronic circuit and converted to a direct reading of relative humidity.

THE ADIABATIC PROCESS

Given that ample water vapor is present in a mass of air, how is that related to precipitation? In other words, how is the water vapor turned into liquid or solid particles that can fall to Earth? The answer is by natural cooling of the air. When air is cooled to the dew point and below, some water vapor molecules must change state to form water droplets or ice crystals. Think about a moist sponge. To extract the water, you have to squeeze the sponge—that is, reduce its ability to hold water. Chilling the air is like squeezing the sponge. *GEODISCOVERIES*

How is air chilled sufficiently to produce precipitation? One mechanism for chilling air is nighttime cooling. As we saw in Chapter 3, the ground surface can become quite cold on a clear night through loss of longwave radiation. Thus, still air near the surface can be cooled below the condensation point, producing dew or frost. However, this mechanism

4.8 Sling psychrometer *The sling psychrometer uses paired wet- and dry-bulb thermometers to measure relative humidity. The gauze cover on the wet-bulb thermometer is wetted with water and the thermometers are whirled overhead. Evaporation from the gauze causes the wet-bulb temperature to drop in proportion to the relative humidity of the air. The dry-bulb temperature and the difference in dry- and wet-bulb temperatures are used to find the relative humidity. The scale on the handle is used for this purpose.*

is not sufficient to form precipitation. Precipitation is formed only when a substantial mass of air experiences a steady drop in temperature below the dew point. This happens when an air parcel is uplifted to a higher level in the atmosphere, as we will see in the following section.

Dry Adiabatic Rate

An important principle of phsyics, which we can refer to as the *adiabatic principle,* is that when a gas is allowed to expand, its temperature drops. Conversely, when a gas is compressed, its temperature increases. If you have ever pumped up a bicycle tire using a hand pump, you have observed this latter effect yourself. As you pump vigorously, the metal pump gets hot. This occurs because air inside the pump is being compressed and therefore heated. In the same way, when a small jet of air escapes from a high-pressure hose, it feels cool. Perhaps you have flown on an airplane and noticed a small nozzle overhead that directs a stream of cool air toward you. When the air moves from higher pressure in the air hose that feeds the nozzle to lower pressure in the cabin, it expands and cools. Physicists use the term **adiabatic process** to refer to a heating or cooling process that occurs solely as a result of pressure change. That is, the change in temperature is not caused by heat flowing into or away from a volume of air, but only by a change in pressure on a volume of air.

Given that air cools or heats when the pressure on it changes, how does that relate to uplift and precipitation? The missing link is simply that atmospheric pressure decreases with an increase in altitude. So, if a parcel of air

---

The adiabatic principle states that compression warms a gas, while expansion cools it. In an adiabatic process, the temperature of a gas changes solely as a result of a change in pressure—not from an inflow or outflow of heat energy.

---

is uplifted, atmospheric pressure on the parcel will be lower, and it will expand and cool. Conversely, if a parcel of air descends, atmospheric pressure will be higher, and it will be compressed and warmed (Figure 4.9). The **dry adiabatic lapse rate** describes this behavior for a rising air parcel that has not reached saturation. This rate has a value of about 10°C per 1000 m (5.5°F per 1000 ft) of vertical rise. That is, if a parcel of air is raised 1 km, its temperature will drop by 10°C. Or, in English units, if raised 1000 ft, its temperature will drop by 5.5°F. This is the "dry" rate because condensation does not occur.

Note that in the previous chapter we encountered the environmental temperature lapse rate. The temperature lapse rate is simply an expression of how the temperature of still air varies with altitude. This rate will vary from time to time and from place to place, depending on the state of the atmosphere. It is quite different from the dry adiabatic lapse rate. The dry adiabatic rate applies to a mass of air in vertical motion. It is always constant and is determined by physical laws. No motion of air is implied for the temperature lapse rate.

Wet Adiabatic Rate

Let's continue examining the fate of a parcel of air that is moved upward in the atmosphere (Figure 4.10). We will assume that the parcel starts with a temperature of 20°C (68°F). As the parcel moves upward, its temperature drops at the dry adiabatic rate, 10°C/1000 m (5.5°F/1000 ft). At 500 m (1600 ft), the temperature has dropped by 5°C (9°F) to 15°C (59°F). At 1000 m (3300 ft), the temperature has fallen to 10°C (50°F). As the rising process continues, the air is

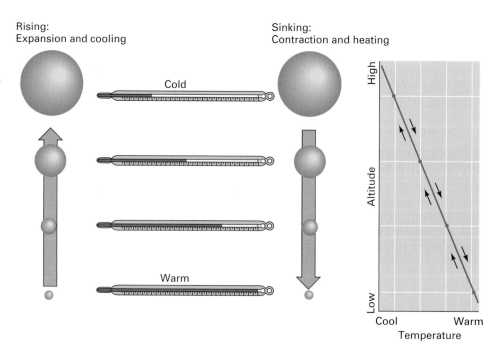

4.9 Adiabatic cooling and heating *A schematic diagram of adiabatic cooling and heating that accompanies the rising and sinking of a mass of air. When air is forced to rise, it expands and its temperature decreases. When air is forced to descend, its temperature increases. (A. N. Strahler.)*

Rising:
Expansion and cooling

Sinking:
Contraction and heating

Cold

Warm

High

Altitude

Low

Cool   Warm
Temperature

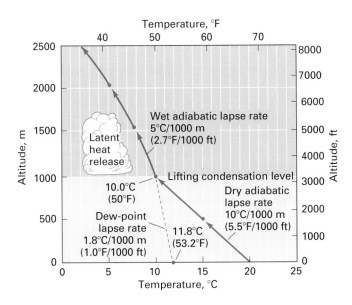

4.10 Adiabatic cooling *Adiabatic decrease of temperature in a rising parcel of air leads to condensation of water vapor into water droplets and the formation of a cloud. (Copyright © A. N. Strahler.)*

cooled to its dew-point temperature, and condensation starts to occur. This is shown on the figure as the *lifting condensation level*.

Note, however, that the dew-point temperature changes slightly with elevation. Instead of remaining constant, it falls at the *dew-point lapse rate* of 1.8°C/1000 m (1.0°F/1000 ft). Suppose the dew-point temperature of the air mass is 11.8°C (53.2°F). Then at 1000 m (3300 ft), the dew-point temperature will be 10°C (50°F). That will also be the temperature of the parcel for the example above, so condensation will begin at that level. The lifting condensation level is thus determined by the initial temperature of the air and its dew point.

If cooling continues, water droplets will form, producing a cloud. If the parcel of saturated air continues to rise, however, a new principle comes into effect—latent heat release. That is, when condensation occurs, latent heat is released and warms the uplifted air. In other words, two effects are occurring at once. First, the uplifted air is being cooled by the reduction in atmospheric pressure. Second, it is being warmed by the release of latent heat from condensation.

Which effect is stronger? As it turns out, the cooling effect is stronger, so the air will continue to cool as it is uplifted. But because of the release of latent heat, the cooling will occur at a lesser rate. This cooling rate for saturated air is called the **wet adiabatic lapse rate** and ranges between 4 and 9°C per 1000 m (2.2 and 4.9°F per 1000 ft). Unlike the

> Rising air cools less rapidly when condensation is occurring, owing to the release of latent heat. This explains why the wet adiabatic cooling rate has a lesser value than the dry adiabatic cooling rate.

> A cloud consists of water droplets, ice particles, or a mixture of both. These form on tiny condensation nuclei, which are normally minute specks of sea salt or dust.

dry adiabatic lapse rate, which remains constant, the wet adiabatic lapse rate is variable because it depends on the temperature and pressure of the air and its moisture content. For most situations, however, we can use a value of 5°C/1000 m (2.7°F/1000 ft). The higher rates apply only to cold, relatively dry air that contains little moisture and therefore little latent heat. In Figure 4.10, the wet adiabatic rate is shown as a slightly curving line to indicate that its value increases with altitude.

CLOUDS

Clouds are frequent features of the atmosphere. Views of the Earth from space show that about half of the Earth is covered by clouds at any given time. We noted earlier that low clouds reflect solar energy, thus cooling the Earth–atmosphere system, while high clouds absorb outgoing longwave radiation, thus warming the Earth–atmosphere system. In Chapter 6 we will return to this topic in more detail. In this chapter, however, we are concerned with clouds as the source of precipitation.

A **cloud** is made up of water droplets, ice particles, or a mixture of both, suspended in air. These particles have a diameter in the range of 20 to 50 μm (0.0008 to 0.002 in.). Recall from Chapter 2 that μm denotes the *micrometer*, or one-millionth of a meter. Each cloud particle is formed on a tiny center of solid matter, called a *condensation nucleus*. This nucleus has a diameter in the range 0.1 to 1 μm (0.000004 to 0.00004 in.).

An important source of condensation nuclei is the surface of the sea. When winds create waves, droplets of spray from the crests of the waves are carried rapidly upward in turbulent air. Evaporation of sea water droplets leaves a tiny residue of crystalline salt suspended in the air. This aerosol strongly attracts water molecules. Another source of nuclei is the heavy load of dust carried by polluted air over cities, which substantially aids condensation and the formation of clouds and fog. But even very clean and clear air contains enough condensation nuclei for the formation of clouds.

In our everyday life at the Earth's surface, liquid water turns to ice when the surrounding temperature falls to the freezing point, 0°C (32°F), or below. However, when water is dispersed as tiny droplets in clouds, it can remain in the liquid state at temperatures far below freezing. Such water is described as *supercooled*. Clouds consist entirely of water droplets at temperatures down to about −12°C (10°F). As cloud temperatures grow colder, a mix of water droplets and ice crystals occurs. The coldest clouds, with temperatures below −40°C (−40°F), occur at altitudes of 6 to

12 km (20,000 to 40,000 ft) and are formed entirely of ice particles.

Cloud Forms

Clouds come in many shapes and sizes, from the small, white puffy clouds often seen in summer to the dark layers that produce a good, old-fashioned rainy day. Meteorologists classify clouds into four families, arranged by height: high, middle, and low clouds, and clouds with vertical development. These are shown in Figures 4.11 and 4.12. Some individual types with their names are also shown.

Clouds are grouped into two major classes on the basis of form—stratiform, or layered clouds, and cumuliform, or globular clouds. *Stratiform* clouds are blanket-like and cover large areas. A common type is *stratus,* which covers the entire sky. Stratus clouds are formed when air layers are forced to rise gradually over large regions. This can happen when one air layer overrides another. As the overriding layer rises and is cooled, condensation occurs over a large area, and a blanket-like cloud forms. If the overriding layer is quite moist and rising continues, dense, thick stratiform clouds result that can produce abundant rain or snow.

Cumuliform clouds are globular masses of cloud that are associated with small to large parcels of rising air. The air parcels rise because they are warmer than the surrounding air. Like bubbles in a fluid, they move upward. Of course, as

> There are two major classes of clouds—stratiform (layered) and cumuliform (globular). Cumulonimbus clouds are dense, tall clouds that produce thundershowers.

they are buoyed upward, they are cooled by the adiabatic process. When condensation occurs, a cloud is formed. The most common cloud of this type is the *cumulus* cloud (Figure 4.12). Sometimes the upward movement yields dense, tall clouds that produce thunderstorms. This form of cloud is the *cumulonimbus.* ("Nimbus" is the Latin word for rain cloud, or storm.)

At the scale of an observer on Earth, cloud shapes and forms are visible clues to air motions that are otherwise invisible to the eye. At the continental or global scale, cloud patterns also serve to mark air motions and reveal major features of global air circulation. One of the most important tools for observing cloud patterns remotely is the GOES series of geostationary satellites. *Focus on Remote Sensing 4.1 • Observing Clouds from GOES* shows some examples of GOES images. GEODISCOVERIES

Fog

Fog is simply a cloud layer at or very close to the surface. In our industrialized world, fog is a major environmental hazard. Dense fog on high-speed highways can cause chain-reaction accidents, sometimes involving dozens of vehicles and causing injury or death to their occupants. At airports, landing delays and shutdowns because of fog bring economic losses to airlines and can inconvenience many thousands of travelers in a single day. For centuries, fog at sea has been a naviga-

4.11 Cloud families and types *Clouds are grouped into families on the basis of height. Individual cloud types are named according to their form.*

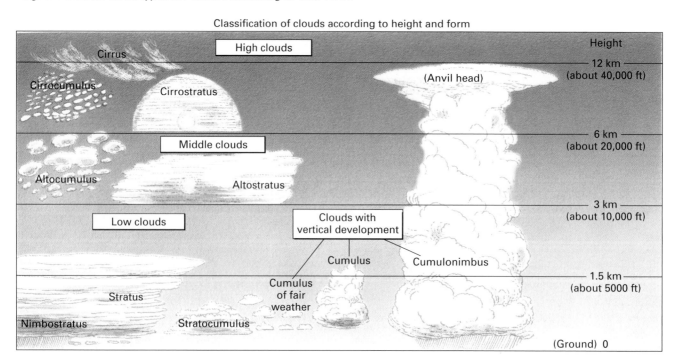

Classification of clouds according to height and form

4.12 Cloud photos *(a) Rolls of altocumulus clouds, also known as a "mackerel sky." (b) High cirrus clouds lie above puffy cumulus clouds in this scene from Holland. (c) Fibrous cirrus clouds are drawn out in streaks by high-level winds to form "mare's tails." (d) Upper layer: lenticular clouds, a type of altostratus formed when moist air is uplifted as it crosses a range of hills. Lower layer: cumulus clouds, also triggered by the uplift.*

tional hazard, increasing the danger of ship collisions and groundings. In addition, polluted fogs, like those of London in the early part of the twentieth century, can injure urban dwellers' lungs and take a heavy toll in lives.

One type of fog, known as *radiation fog,* is formed at night when the temperature of the air layer at the ground level falls below the dew point. This kind of fog is associated with a low-level temperature inversion (Figure 3.12). Another fog type, *advection fog,* results when a warm, moist air layer moves over a cold surface. As the warm air layer loses heat to the surface, its temperature drops below the dew point, and condensation sets in. Advection fog commonly occurs over oceans where warm and cold currents occur side by side. When warm, moist air above the warm current moves over the cold current, condensation occurs. Fogs of the Grand Banks off Newfoundland are formed in this way because here the cold Labrador current comes in contact with the warmer waters of the Gulf Stream.

Sea fog is frequently found along the California coast. It forms within a cool marine air layer in direct contact with the colder water of the California current (Figure 4.13). Similar fogs are found on continental west coasts in the tropical latitude zones where cool, equatorward currents parallel the shoreline. *GEODISCOVERIES*

PRECIPITATION

Clouds are the source of precipitation—the process that provides the fresh water essential for most forms of terrestrial life. Let's look in more detail at how precipitation is formed inside clouds. Precipitation can form in two ways. In warm clouds, fine water droplets condense, collide, and coalesce into larger and larger droplets that can fall as rain. In colder clouds, ice crystals form and grow in a cloud that contains a mixture of both ice crystals and water droplets.

The first process occurs when saturated air rises rapidly, and cooling forces additional condensation. Cloud droplets grow by added condensation and attain a diameter of 50 to 100 μm (0.002 to 0.004 in.). In collisions with one another, the droplets grow to about 500 μm in diameter (about 0.02 in.). This is the size of water droplets in drizzle. Further collisions and coalescence increase drop size and yield *rain.* Average raindrops have diameters of about

> In warm clouds experiencing strong uplift, water droplets grow by condensation, then collide and coalesce to form raindrops. In colder clouds, ice particles grow from collisions with supercooled water droplets that freeze on contact.

Focus on Remote Sensing | 4.1

Observing Clouds from GOES

Some of the most familiar images of Earth acquired by satellite instruments are those of the Geostationary Operational Environmental Satellite (GOES) system. Images from the GOES series of satellites, and its pathfinding predecessors, Synchronous Meterological Satellite (SMS) –1 and –2, have been in constant use by meterologists and weather forecasters since 1974. The primary mission of the GOES series is to view cloud patterns and track weather systems by providing frequent images of the Earth from a consistent viewpoint in space. This capability is of great economic value in forecasting storms and severe weather.

A key feature of the GOES series is its geostationary orbit. Recall from *Geographer's Tools 2.1* • *Remote Sensing for Physical Geography* that a satellite in geostationary orbit rotates above the equator in the same direction as the Earth and at the same angular velocity—that is, it remains constantly above the same point on the equator. From this vantage point, an instrument can acquire a constant stream of images of the Earth beneath it. Thus, a geostationary imager is ideal for viewing clouds and tracking cloud patterns.

Altogether, 10 GOES satellites in three major generations have been placed in orbit since 1975. They have maintained a nearly continuous stream of images from two points on the equator, at latitudes 75°W and 135°W, that bracket North America. From 135°W (GOES-West), Pacific storm systems can be tracked as they approach the continent and move across the western states and provinces. From 75°W (GOES-East), weather systems are observed in the eastern part of the continent as they move in from the west. GOES-East also observes the tropical Atlantic, allowing identification of tropical storms and hurricanes as they form and move eastward toward the Caribbean Sea and southeastern United States.

The present generation of GOES platforms carries two primary instruments—the GOES imager and the GOES Sounder. The imager acquires data in five spectral bands, ranging from the visible red to the thermal infrared. Images of clouds, water vapor, surface temperature, winds, albedo, and fires and smoke are routinely acquired. The sounder uses 19 spectral channels to sound the atmosphere and observe atmospheric profiles of temperature, moisture, ozone, and cloud height and amount.

The image at the right was acquired from the GOES-8 imager at 75°W lat. on September 2, 1994. In this near-noon image, the entire side of the globe nearest to the satellite is illuminated. The red and green primary colors in the image are derived from visible band data. Vegetated areas of the Earth's surface appear green, while semiarid and desert landscapes appear yellow-brown. The blue primary color is derived from a thermal infrared band but is scaled inversely so that cold areas are bright and warm areas are dark. Since clouds are bright white in visible wavelengths

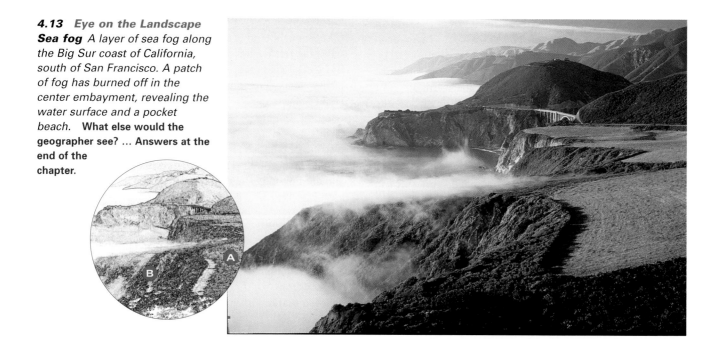

4.13 *Eye on the Landscape*
Sea fog *A layer of sea fog along the Big Sur coast of California, south of San Francisco. A patch of fog has burned off in the center embayment, revealing the water surface and a pocket beach.* **What else would the geographer see? ... Answers at the end of the chapter.**

and are colder than surface features, they appear white. The oceans appear blue because they are dark in the visible bands but are still somewhat warm.

For weather forecasting, an important tool of the latest generation of GOES imagers is the water vapor image. The image at the right shows a global water vapor image for August 11, 2000, constructed from GOES-East and West

Water vapor composite image *Areas of high atmospheric water vapor content are bright in this image, prepared by merging data from four geostationary satellite instruments. (University of Wisconsin Space Science and Engineering Center.)*

Earth from GOES-8 *The GOES-8 geostationary satellite acquired this image on September 2, 1994. (Courtesy NASA. Image produced by M. Jentoft-Nilsen, F. Hasler, D. Chesters, and T. Nielsen.)*

images as well as those of two other geostationary satellites (MeteoSat and GMS). The brightest areas show regions of active precipitation—note the many spots of convective shower patches in the equatorial zone. Dark areas show low water vapor content—the desert band stretching across the Middle East to central Asia and the desert regions of southern Africa are examples. The dark lines moving across North America are frontal systems in which cool, dry air is pushing warmer, moister air southward and eastward. Note also dark tongues of dry air spiraling with light tongues of moist air in cyclonic systems in the southern hemisphere (Chapter 7). A bright patch in the Atlantic off the Carolina coast is Hurricane Alberto. The eye of the storm is faintly visible as a dark spot at the center.

1000 to 2000 μm (about 0.04 to 0.1 in.), but they can reach a maximum diameter of about 7000 μm (about 0.25 in.). Above this value they become unstable and break into smaller drops while falling. This type of precipitation formation occurs in warm clouds typical of the equatorial and tropical zones.

Snow is produced by the second process in cool clouds that are a mixture of ice crystals and supercooled water droplets. As the ice crystals take up water vapor and grow by deposition, the supercooled water droplets lose water vapor by evaporation and shrink. When an ice crystal collides with a droplet of supercooled water, it induces freezing of the droplet. The ice crystals further coalesce to form snow particles, which can become heavy enough to fall from the cloud. Snowflakes formed entirely by deposition can have intricate crystal structures (Figure 4.14), but most particles of snow have endured collisions and coalescence and are simply tiny lumps of ice.

When the underlying air layer is below freezing, snow reaches the ground as a solid form of precipitation. Other-

wise, it melts and arrives as rain. A reverse process, the fall of raindrops through a cold air layer, results in the freezing of rain and produces pellets or grains of ice. These are commonly referred to in North America as *sleet*. (Among the British, sleet refers to a mixture of snow and rain.)

Perhaps you have experienced an *ice storm*. This occurs when the ground is frozen and the lowest air layer is also below freezing. Rain falling through the layer is chilled and freezes onto ground surfaces as a clear glaze. Ice storms cause great damage, especially to telephone and power lines and to tree limbs pulled down by the weight of the ice. In addition, roads and sidewalks are made extremely hazardous by the slippery glaze. Actually, the ice storm is more accurately named an "icing" storm, since it is not ice that is falling but supercooled rain.

Hail, another form of precipitation, consists of lumps of ice ranging from pea- to grapefruit-sized—that is, with a diameter of 5 mm (0.2 in.) or larger. Most hail particles are roughly spherical in shape. The formation of hail will be explained in our discussion of the thunderstorm.

4.14 Snowflakes *These individual snow crystals, greatly magnified, were selected for their beauty. Most snowflakes are lumps of ice without such a delicate structure.*

rainfall, but this ratio may range from 30 to 1 in very loose snow to 2 to 1 in old, partly melted snow.

Precipitation Processes

So far, we have seen how air that is moving upward will be chilled by the adiabatic process to the saturation point and then to condensation, and how eventually precipitation will form. However, one key piece of the precipitation puzzle has been missing—what causes air to move upward? Air can move upward in three ways. First, it can be forced upward as a through-flowing wind. Consider the case of a mass of air moving up and over a mountain range, for example. If the mountain range is high enough and the air is moist enough, then precipitation will occur. This is called **orographic precipitation,** and we'll describe it more fully in the following section.

A second way for air to be forced upward is through *convection.* In this process, a parcel of air is heated, perhaps by a patch of warm ground, so that it is warmer and therefore less dense than the air around it. Like a bubble, it rises. If the air is quite moist and condensation sets in, then the release of latent heat of condensation will ensure that the parcel remains warmer than the surrounding air as it rises. Precipitation that results from this process is **convectional precipitation,** and it will be described after orographic precipitation.

4.15 Rain gauge *This rain gauge consists of two clear plastic cylinders, here partly filled with red-tinted water. The inner cylinder receives rainwater from the funnel top. When it fills, the water overflows into the larger, outer cylinder.*

Precipitation is measured in units of depth of fall per unit of time—for example, centimeters or inches per hour or per day. A centimeter (inch) of rainfall would cover the ground to a depth of 1 cm (1 in.) if the water did not run off or sink into the soil. Rainfall is measured with a *rain gauge.* The simplest rain gauge is a straight-sided, flat-bottomed pan, which is set outside before a rainfall event. (An empty coffee can does nicely.) After the rain, the depth of water in the pan is measured.

A very small amount of rainfall, such as 2 mm (0.1 in.), makes too thin a layer to be accurately measured in a flat pan. To avoid this difficulty, a typical rain gauge is constructed from a narrow cylinder with a funnel at the top (Figure 4.15). The funnel gathers rain from a wider area than the mouth of the cylinder, so the cylinder fills more quickly. The water level gives the amount of precipitation, which is read on a graduated scale.

Snowfall is measured by melting a sample column of snow and reducing it to an equivalent in rainfall. In this way, rainfall and snowfall records may be combined in a single record of precipitation. Ordinarily, a 10-cm (or 10-in.) layer of snow is assumed to be equivalent to 1 cm (or 1 in.) of

A third way for air to be forced upward is through the movement of air masses. As we will see in Chapter 6, air masses are large bodies of air, each with a set of relatively uniform temperature and moisture properties. Air masses move normally from west to east in midlatitudes, and one can overtake another. When this happens, one of them will be forced aloft. Since this overtaking usually occurs in a common type of storm called a cyclone, this type of precipitation is known as *cyclonic precipitation*. We will treat the subject of cyclonic precipitation in Chapter 6.

Three types of precipitation are orographic, convectional, and cyclonic.

Orographic Precipitation

In orographic precipitation, through-flowing winds move moist air up and over a mountain barrier. (The term *orographic* means "related to mountains.") Figure 4.16 shows how this process works. Moist air arrives at the coast after passing over a large ocean surface (1). As the air rises on the windward side of the range, it is cooled at the dry adiabatic rate. When cooling is sufficient, the lifting condensation level is reached, condensation sets in, and clouds begin to form (2). Cooling now proceeds at the wet adiabatic rate. Eventually, precipitation begins. As the air con-

----
In orographic precipitation, moist air is forced up and over a mountain barrier, producing cooling, condensation, and precipitation.
----

tinues up the slope, precipitation continues to fall. GEODISCOVERIES

After passing over the mountain summit, the air begins to descend down the leeward slopes of the range (3). However, as it descends, it is compressed. This means that it is warmed, according to the adiabatic principle. Since cooling is no longer occurring, cloud droplets and ice crystals evaporate or sublimate. Because evaporation and sublimation take up latent heat, at first the warming follows the wet adiabatic rate. Eventually the air clears, and as it continues to descend, it warms further at the dry adiabatic rate. At the base of the mountain on the far side (4), the air is now warmer—and drier, since much of its moisture has been removed in fallen precipitation. This effect creates a *rain shadow* on the far side of the mountain—a belt of dry climate that extends down the leeward slope and beyond. Several of the Earth's great deserts are of this type.

When strong and persistent, the descending flow of hot, dry air is termed a *chinook* wind. Channeled into valleys on the leeward side, it can raise local temperatures very rapidly. The dry air has great evaporating ability, and it can rapidly sublimate a snow cover in winter or dry a brush

4.16 Orographic precipitation *The forced ascent of a warm, moist oceanic air mass over a mountain barrier produces precipitation and a rain shadow desert. As the air moves up the mountain barrier, it loses moisture through precipitation. As it descends the far slope, it is warmed.*

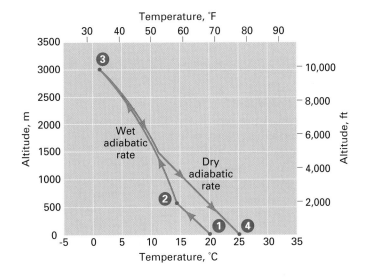

cover to tinder in summer. The chinook wind is one type of local wind, and we will discuss several other local wind types in Chapter 5.

California's rainfall patterns provide an excellent example of orographic precipitation and the rain shadow effect. A map of California (Figure 4.17) shows mean annual precipitation. It uses lines of equal precipitation, called *isohyets*. Focus now on central California and follow the arrow on the map. Prevailing westerly winds bring moist air in from the Pacific Ocean, first over the Coast Ranges of central and northern California, and then, after rain is deposited on the Coast Ranges, the air descends into the broad Central Valley. Here, precipitation is low, averaging less than 25 cm (10 in.) per year.

Next the air continues up and over the great Sierra Nevada, whose summits rise to 4200 m (about 14,000 ft) above sea level. Heavy precipitation, largely in the form of winter snow, falls on the western slopes of these ranges and nourishes rich forests. Passing down the steep eastern face of the Sierra Nevada, the air descends quickly into the Owens Valley, at about 1000 m (3300 ft) elevation. The adiabatic heating warms and dries the air, creating a rain shadow desert. In the Owens Valley, annual precipitation is less than 10 cm (4 in.). In this way, the orographic effect on air moving across the mountains of California produces a part of America's great interior desert zone, extending from eastern California and across Nevada.

Convectional Precipitation

The second process of inducing uplift in a parcel of air is *convection,* the upward motion of a parcel of heated air. In this process, strong updrafts occur within *convection cells*—vertical columns of rising air that are often found above warm land surfaces. Air rises in a convection cell because it is warmer, and therefore less dense, than the surrounding air.

GEODISCOVERIES

The convection process starts when a surface is heated unequally. For example, consider an agricultural field surrounded by a forest. Since the field surface consists largely of bare soil and only a low layer of vegetation, it will be warmer under steady sunshine than the adjacent forest. This means that as the day progresses, the air above the field will grow warmer than the air above the forest.

Now, the density of air depends on its temperature—warm air is less dense than cooler air. The hot-air balloon operates on this principle. The balloon is open at the bottom, and in the basket below a large gas burner forces heated air into the balloon. Because the heated air is less dense than the surrounding air, the balloon rises. The same principle will cause a bubble of air to form over the field, rise, and break free from the surface. Figure 4.18 diagrams this process.

As the bubble of air rises, it is cooled adiabatically and its temperature will decrease as it rises. However, we know that the temperature of the surrounding

4.17 Orographic precipitation effects in California
The effect of mountain ranges on precipitation is strong in the state of California because of the prevailing flow of moist oceanic air from west to east. Centers of high precipitation coincide with the western slopes of mountain ranges, including the coast ranges and Sierra Nevada. To the east, in their rain shadows, lie desert regions.

In convectional precipitation, moist air is warmed at the surface, expands, becomes less dense than surrounding, cooler air, and is buoyed upward. At the lifting condensation level, clouds begin to form.

air will normally decrease with altitude as well. Nonetheless, as long as the bubble is still warmer than the surrounding air, it will be less dense and will therefore continue to rise.

If the bubble remains warmer than the surrounding air and uplift continues, adiabatic cooling chills the bubble below the dew point. Condensation occurs, and the rising air column becomes a

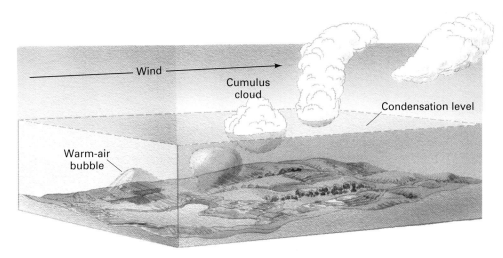

4.18 Formation of a cumulus cloud *A bubble of heated air rises above the lifting condensation level to form a cumulus cloud.*

puffy cumulus cloud. The flat base of the cloud marks the lifting condensation level at which condensation begins. The bulging "cauliflower" top of the cloud is the top of the rising warm-air column, pushing into higher levels of the atmosphere. Normally, the small cumulus cloud will encounter winds aloft that mix it into the local air. After drifting some distance downwind, the cloud evaporates.

UNSTABLE AIR   Sometimes, however, convection continues strongly, and the cloud develops into a dense cumulonimbus mass, or thunderstorm, from which heavy rain will fall. Two environmental conditions encourage the development of

thunderstorm: (1) air that is very warm and moist, and (2) an environmental temperature lapse rate in which temperature decreases more rapidly with altitude than it does for either the dry or wet adiabatic lapse rates. Figure 4.19, which diagrams the convection process in unstable air, shows this type of lapse rate as having a flatter slope than the dry or wet adiabatic rates. Air with these characteristics is referred to as **unstable air.**

For the first condition, recall that there are two adiabatic rates—dry and wet (Figure 4.10). The wet rate is about half the dry rate. While condensation is occurring, the lesser wet rate applies. Thus, air in which condensation is occurring

4.19 Convection in unstable air *When the air is unstable, a parcel of air that is heated sufficiently to rise will continue to rise to great heights.*

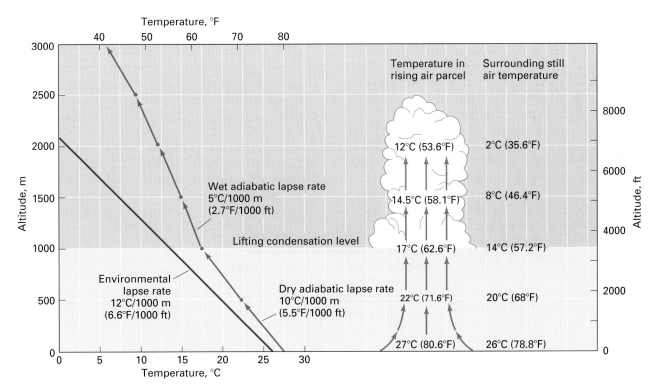

cools less rapidly with uplift. Furthermore, the wet rate is smallest for warm, moist air. This means that the temperature decrease experienced by warm, moist, rising, condensing air is quite small. And since the temperature decrease is small, the rising air is more likely to stay warmer than the surrounding air. Thus, uplift continues.

With regard to the second condition—an environmental temperature lapse rate value that exceeds the adiabatic rates—keep in mind that this rate describes the temperature of the surrounding, still air. Therefore, the temperature of the still air falls quite rapidly with altitude. This means that the condensing air in a rising parcel will stay warmer than the surrounding air. Again, uplift continues.

A simple example may make convection in unstable air clearer. Figure 4.19 shows how temperature changes with altitude for a parcel moving upward by convection in unstable air. The surrounding air is at a temperature of 26°C (78.8°F) at ground level and has a lapse rate of 12°C/1000 m (6.6°F/1000 ft). A parcel of air is heated by 1°C (1.8°F) to 27°C (80.6°F), and it begins to rise. At first, it cools at the dry adiabatic rate. At 500 m (1640 ft), the parcel is at 22°C (71.6°F), while the surrounding air is at 20°C (68°F). Since it is still warmer than the surrounding air, uplift continues. At 1000 m, the lifting condensation level is reached. The temperature of the parcel is 17°C (62.6°F).

As the parcel rises above the condensation level, the wet adiabatic rate applies, here using the value 5°C/1000 m (2.7°F/1000 ft). Now the parcel cools more slowly as it rises. At 1500 m (4920 ft), the parcel is 14.5°C (58.1°F), while the surrounding air is 8°C (46.4°F). Since the parcel is still warmer than the surrounding air, it continues to rise. Note that the difference in temperature between the rising parcel and the surrounding air now actually increases with altitude. This means that the parcel will be buoyed upward ever more strongly, forcing even more condensation and precipitation.

The key to the convectional precipitation process is latent heat. When water vapor condenses into cloud droplets or ice particles, it releases latent heat to the rising air parcel. By keeping the parcel warmer than the surrounding air, this latent heat fuels the convection process, driving the parcel ever higher. When the parcel reaches a high altitude, most of its water will have condensed. As adiabatic cooling continues, less latent heat will be released. As a result, the uplift will weaken. Eventually, uplift will stop, since the energy source, latent heat, is gone. The cell dies and dissipates into the surrounding air.

Unstable air is typical of hot summer air masses in the central and southeastern United States. As we will see later, summer weather patterns sweep warm, humid air from the Gulf of Mexico over the continent. Over a period of days, the intense summer insolation strongly heats the air layer near the ground, producing a steep environmental lapse rate. Thus, both of the conditions that create unsta-

ble air are present. As a result, thunderstorms are very common in these regions during the summer.

Unstable air is also likely to be found in the vast, warm and humid regions of the equatorial and tropical zones. In these regions, convective showers and thundershowers are common. At low latitudes, much of the orographic rainfall is actually in the form of heavy showers and thundershowers produced by convection. In this case, the forced ascent of unstable air up a mountain slope easily produces rapid condensation, which then triggers the convection process.

Thunderstorms

The thunderstorm is an awesome event (Figure 4.20). Intense rain and violent winds, coupled with lightning and thunder, renew our respect for nature's power. What is the anatomy of a thunderstorm?

A **thunderstorm** is an intense local storm associated with a tall, dense cumulonimbus cloud in which there are very strong updrafts of air. A single thunderstorm typically consists of several individual convection cells. A single convection cell is diagrammed in Figure 4.20. Air rises within the cell as a succession of bubble-like air parcels. Intense adiabatic cooling within the bubbles produces precipitation. It can be in the form of water at the lower levels, mixed water and snow at intermediate levels, and snow at high levels where cloud temperatures are coldest.

As the rising air parcels reach high levels, which may be 6 to 12 km (about 20,000 to 40,000 ft) or even higher, the rising rate slows. Strong winds typically present at such high altitudes drag the cloud top downwind, giving the thunderstorm cloud its distinctive shape—resembling an old-fashioned blacksmith's anvil.

Ice particles falling from the cloud top act as nuclei for freezing and deposition at lower levels. Large ice crystals form and begin to sink rapidly. Melting, they coalesce into large, falling droplets. The rapid fall of raindrops adjacent to the rising air bubbles pulls the air downward and feeds a downdraft within the convection cell. Emerging from the cloud base, the downdraft of cool air, laden with precipitation, approaches the surface and spreads out in all directions. Part of the downdraft moves forward, causing more warm, moist surface air to rise and enter the updraft portion of the storm. This effect helps perpetuate the storm. The downdraft creates strong local winds. Wind gusts can sometimes be violent enough to topple trees and raise the roofs of weak buildings.

In addition to powerful wind gusts and heavy rains, thunderstorms can produce hail. Hailstones (Figure 4.21) are formed by the accumulation of ice layers on ice pellets that are suspended in the strong updrafts of the thunderstorm. In this process, they can reach diameters of 3 to 5 cm (1.2 to 2.0 in.). When

> Unstable air—warm, moist, and heated by the surface—can produce abundant convectional precipitation. It is typical of hot summer air masses in the central and southeastern United States.

> A thunderstorm is an intense local storm associated with a tall cumulonimbus cloud in which there are strong updrafts of air. Thunderstorms can produce both hail and cloud-to-ground lightning.

4.20 *Thunderstorms*

a...*Anatomy of a thunderstorm cell* Successive bubbles of moist condensing air push upward in the cell. Their upward movement creates a corresponding downdraft, expelling rain, hail, and cool air from the storm as it moves forward.

b...*Arizona thunderstorm* This massive thunderstorm, photographed in southeastern Arizona by noted storm chaser Warren Faidley, is moving from left to right. The lightning strike caught here on film illuminates the storm's anvil top especially clearly.

4.21 *Hailstones Fresh hailstones, up to about 1 cm (0.5 in.) in diameter.*

they become too heavy for the updraft to support, they fall to Earth. Figure 4.22 shows a hailstorm in progress.

Annual losses from crop destruction caused by hailstorms amount to several hundred million dollars. Damage to wheat and corn crops is particularly severe in the Great Plains, running through Nebraska, Kansas, Missouri, Oklahoma, and northern Texas (Figure 4.23).

Another effect of convection cell activity is to generate lightning. This phenomenon occurs when updrafts and downdrafts cause positive and negative static charges to accumulate within different regions of the cloud. *Lightning* is a great electric arc—a series of gigantic sparks—passing between differently charged parts of the cloud mass or between cloud and ground (see Figure 4.20). During a lightning discharge, a current of as much as 60,000 to 100,000 amperes may develop. This current heats the air intensely, which makes it expand very rapidly—much like an explosion. This expansion sends out sound waves, which we recognize as a thunderclap. Most lightning discharges occur within the cloud, but a significant proportion strike land. In the United States, lightning causes a yearly average of about 150 human deaths and property damage of hundreds

4.22 **Hailstorm in progress, Santa Cruz County, Arizona** *The white streaks in this photo are marble-sized hailstones. Although the hail is falling from a cloud directly overhead, the low afternoon Sun still illuminates the foreground scene.*

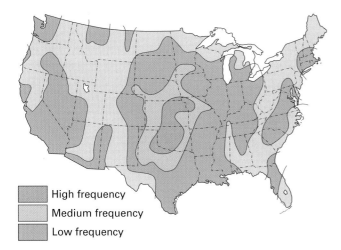

High frequency

Medium frequency

Low frequency

4.23 Frequency of severe hailstorms *As shown in this map of the 48 contiguous United States, severe hailstorms are most frequent in the midwestern plains states of Oklahoma and Kansas. A severe hailstorm is defined as a local convective storm producing hailstones equal to or greater than 1.9 cm (0.75 in.) in diameter. (From R. H. Skaggs, Proc. Assoc. American Geographers, vol. 6, Figure 2. Used by permission.)*

of millions of dollars, including loss by structural and forest fires set by lightning.

Microbursts

The downdraft that accompanies a thunderstorm can sometimes be very intense—so intense that it is capable of causing low-flying aircraft to crash. This type of intense downdraft is called a *microburst*. A diagram of a microburst is shown in Figure 4.24. The downward-moving air flows outward in all directions. It is often, but not always, accompanied by rain.

An aircraft flying through the microburst first encounters strong headwinds. They may cause a bumpy ride but do not interfere with the airplane's ability to fly. However, as it passes through the far side of the microburst, the aircraft encounters a strong tailwind. The lift of the airplane's wings depends on the speed of the air flowing across them, and the tailwind greatly reduces the air speed. This causes a loss of lift. If the tailwind is strong enough, the airplane cannot hold its altitude and may crash. Microbursts can be detected by special radar instruments that measure horizontal wind speeds. They are quite expensive, however, and are installed only at some major airports.

Training procedures for pilots have reduced the incidence of aviation accidents in the United States attributed to microbursts and associated wind shear. However, in 1994 a jet aircraft attempting to land at Charlotte, North Carolina, crashed during a violent thunderstorm, killing 37 and injuring 20 others. A microburst was detected just as the airplane approached the airport, but the warning was not broadcast to the pilots on the radio channel they were following.

Although they attempted to abort the landing, a pilot error occurred, and their aircraft was brought down by the severe tailwind of the microburst.

AIR QUALITY

One of the more important impacts of human activity on our planet is air pollution. Industrial processes and fossil fuel combustion release large quantities of unwanted and unhealthful substances into the air. Air quality is of great concern, not only for the health of ecosystems, but also for human health and comfort. *Eye on the Environment 4.2 ● Air Pollution* provides an overview of air pollution, identifying the most important pollutants and describing some climatic effects of air pollution.

A LOOK AHEAD

Precipitation has been the major focus of this chapter. Keep in mind that no matter what the cause of precipitation, latent heat is always released. Because a significant amount of oceanic evaporation results in precipitation over land, there is a very significant flow of latent heat from oceans to land. And, as we will see in the following chapter, global air circulation patterns move this latent heat poleward, helping to warm the continents and creating the distinctive pattern of climates that differentiates our planet.

4.24 Anatomy of a microburst *(above) Schematic cross section showing a downdraft reaching the ground and producing a horizontal outflow. (below) Schematic profile of an aircraft taking off through a microburst, leading to a loss of lift and eventual crash. (Adapted from diagrams by Research Applications Program, National Center for Atmospheric Research, Boulder, Colorado.)*

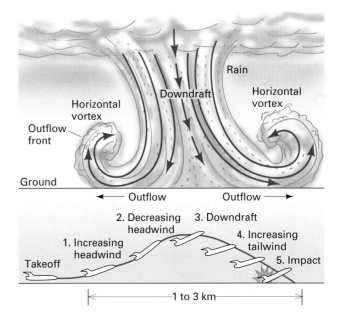

Eye on the Environment | 4.2

Air Pollution

Most people living in or near urban areas have experienced air pollution first hand. Perhaps you've felt your eyes sting or your throat tickle as you drive an urban freeway. Or you've noticed black dust on window sills or window screens and realized that you are breathing in that dust as well. Air pollution is an atmospheric phenomenon that is largely the result of human activity. In this *Eye on the Environment* feature, we provide a short introduction to air pollution.

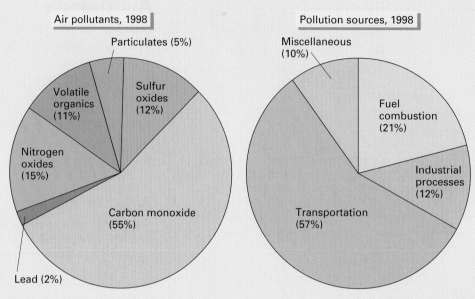

E4.1   *Air pollutants and sources* *These data are for the United States in 1998. (Data from EPA.)*

In preceding chapters, we have discussed two kinds of substances in air—aerosols and gases. *Aerosols* are small bits of matter in the air, so small that they float freely with normal air movements. *Gases* are molecular compounds that are mixed together to form the main body of air. To these two categories, we add a third category, *particulates*—larger, heavier particles that sooner or later fall back to Earth.

An *air pollutant* is an unwanted substance injected into the atmosphere from the Earth's surface by either natural or human activities. Air pollutants come as aerosols, gases, and particulates. Most pollutants generated by human activity arise in two ways. The first is through the day-to-day activities of large numbers of people, for example, in driving automobiles within urban centers. The second is through industrial activities, such as fossil fuel combustion or the smelting of mineral ores to produce metals.

Figure E4.1 shows the relative proportions of the most common air pollutants and their sources for the United States in 1998. The gases are carbon monoxide (CO), sulfur oxides, and nitrogen oxides. The nitrogen oxides can be NO, NO_2, or NO_3; this mixture is usually referred to as NO_x. Similarly, the sulfur oxides SO_2 and SO_3 are referred to as SO_x. Volatile organic compounds include evaporated gasoline, dry cleaning fluids, incompletely combusted fossil fuels, and similar pollutants. They occur as both gases and aerosols. The remaining pollutants shown in the chart are in the form of lead and particulates. Here, particulates include pollution particles that are fine enough to remain in the air as aerosols.

Combustion of fossil fuel, in transportation or as stationary fuel combustion, is the most important source of all these pollutants. As shown in the right half of the figure, it accounts for 78 percent of the emissions. Exhausts from gasoline and diesel engines contribute most of the carbon monoxide, half the volatile organic compounds, and about a third of the nitrogen oxides.

Stationary sources of fuel combustion include power plants that generate electricity and various industrial plants that also burn fossil fuels. They contribute most of the sulfur oxides because they burn coal or lower-grade fuel oils. These fuels are richer in sulfur than gasoline and diesel fuel. Stationary sources also supply most of the particulate matter. Some of it is fly ash—coarse soot particles emitted from smokestacks of power plants. These particles settle out quite quickly within close range of the source. Smaller and finer carbon particles, however, can remain suspended almost indefinitely and become aerosols. Industrial processes, such as the smelting of ore, are also important sources of air pollutants. They are the main source of lead (Figure E4.2).

Smog and Haze

When aerosols and gaseous pollutants are present in considerable density over an urban area, the resultant mixture is known as **smog** (Figure E4.3). This term was coined by combining the words

E4.2 Air pollution from smelters *This satellite image shows the devastating effects of pollutants emitted from smelting operations near Wawa, Ontario, Canada. Healthy vegetation is shown in red to pink tones. The blue streak on the right-hand side of the image is barren of vegetation. The smelting facilities are located at the narrowest point of the blue plume.*

E4.3 Smog in the city *Downtown Los Angeles projects skyward into a layer of smog.*

"smoke" and "fog." Typically, smog allows hazy sunlight to reach the ground, but it may also be dense enough to hide aircraft flying overhead from view. Smog irritates the eyes and throat, and it can corrode structures over long periods of time.

Modern urban smog has three main toxic ingredients: nitrogen oxides, volatile organic compounds, and ozone. Nitrogen oxides and volatile organic compounds are largely automobile pollutants, although significant amounts can be released by industrial power plants and petroleum processing and storage facilities. Ozone is not normally produced directly by pollution sources in the city but forms through a photochemical reaction in the air. In this reaction, nitrogen oxides react with volatile organic compounds in the presence of sunlight to form ozone. Ozone in urban smog can harm plant tissues and eventually kill sensitive plants. It is also harmful to human lung tissue, and it aggravates bronchitis, emphysema, and asthma. This is the same compound that absorbs ultraviolet radiation in the ozone layer within the stratosphere. So, ozone in the stratosphere is desirable, but if close to Earth, it is undesirable indeed. Photochemical reactions can also produce other toxic compounds in smog.

Haze is a condition of the atmosphere in which aerosols. obscure distant objects. Haze builds up naturally in stationary air as a result of human and natural activity. When the air is humid and abundant water vapor is available, water films grow on suspended nuclei. This creates aerosol particles large enough to obscure and scatter light, reducing visibility. Natural haze aerosols include soil dust, salt crystals from the sea surface, hydrocarbon compounds from plants, plant pollen, and smoke from forest and grass fires. Some regions are noted for naturally occurring haze—the Great Smoky Mountains are an example.

Fallout and Washout

Pollutants generated by a combustion process are contained within hot exhaust air emerging from a factory smokestack or an auto tailpipe. Since hot air rises, the pollutants are at first carried aloft by convection. However, the larger particulates soon settle under gravity and return to the surface as *fallout.* Particles too small to settle out are later swept down to Earth by precipitation in a process called *washout.* Through a combination of fallout and washout, the atmosphere tends to be cleaned of pollutants. Although a balance between input and output of pollutants is achieved in the long run, the quantities stored in the air at a given time fluctuate widely.

Pollutants are also eliminated from the air over their source areas by wind. Strong, through-flowing winds will disperse pollutants into large volumes of cleaner air in the downwind direction. Strong winds can quickly sweep away most pollutants from an urban area, but during periods when winds are light or absent, the concentrations can rise to high values.

Inversion and Smog

The concentration of pollutants over a source area rises to its highest levels when vertical mixing (convection) of the air is inhibited. This happens in an inversion—a condition in which the temperature of the air increases with altitude.

Why does the inversion inhibit mixing? Recall that a heated air parcel, perhaps emerging from a smokestack or chimney, will rise as long as it is warmer than the surrounding air. Recall, too, that as it rises, it is cooled according to the adiabatic principle. In an inversion, however, the surrounding air gets warmer, not colder, with altitude. So, the parcel will cool as it rises, while the surrounding air becomes warmer. Thus, the parcel will quickly arrive at the temperature of the surrounding air, and uplift will stop. Under these conditions, heated air will move only a short distance upward. Pollutants in the air parcel will disperse at low levels, keeping concentrations high near the ground.

Two types of inversions are important in causing high air pollutant concentrations—low-level and high-level inversions. When a *low-level temperature inversion* develops over an urban area with many air pollution sources, pollutants are trapped under the "inversion lid." Heavy smog or highly toxic fog can develop. An example is the tragedy that occurred in Donora, Pennsylvania, in late October 1948, when a persistent low-level inversion developed. The city occupies a valley floor hemmed in by steeply rising valley walls. The walls prevented the free mixing of the lower air layer with that of the surrounding region (Figure E4.4) and helped keep the inversion intact. Industrial smoke and gases from factories poured into the inversion layer for five days, increasing the pollution level. A poisonous fog formed and began to take its toll.

Twenty people died, and several thousands were stricken before a change in weather patterns dispersed the polluted layer.

Another type of inversion is responsible for the smog problem experienced in the Los Angeles Basin and other California coastal regions, ranging north to San Francisco and south to San Diego. Here special climatic conditions produce prolonged inversions and smog accumulations (Figure E4.5). Off the California coast is a persistent fair-weather system that is especially strong in the summer. This system produces a layer of warm, dry air at upper elevations. However, a cold current of upwelling ocean bottom water runs along the coast, just offshore. Moist ocean air moves across this cool current and is chilled, creating a cool, marine air layer.

Now, the Los Angeles Basin is a low, sloping plain lying between the Pacific Ocean and a massive mountain barrier on the north and east sides. Weak winds from the south and southwest move the cool, marine air inland over the basin. Further landward movement is blocked by the mountain barrier. Since there is a warm layer above this cool marine air layer, the result is an inversion. Pollutants accumulate in the cool air layer and produce smog. The upper limit of the smog stands out sharply in contrast to the clear air above it, filling the basin like a lake and extending into valleys in the bordering mountains (Figure E4.6). Since this type of inversion persists to a higher level in the atmosphere, we can

Poison fog

E4.4 Low-level temperature inversion *An inversion held air pollutants close to the ground, creating a poison fog accumulation at Donora, Pennsylvania, in October 1948.*

E4.5 High-level temperature inversion on the California coast *A layer of warm, dry, descending air from a persistent fair-weather system rides over a cool, moist marine air layer at the surface to create a persistent temperature inversion.*

Warm, dry descending air

32°C (90°F)
16°C (61°F)

Ocean

Cool, moist air

Cold

Upwelling

W

E

describe it as a *high-level temperature inversion*.

An actual temperature inversion, in which temperature close to the surface increases with altitude, is not essential for building a high concentration of pollutants above a city. Light or calm winds and stable air are all that are required. *Stable air* has a temperature profile that decreases with altitude but at a slow rate. Some convectional mixing occurs in stable air, but the convectional precipitation process is inhibited. At certain times of the year, slow-moving masses of dry, stable air occupy the central and eastern portions of the North American continent. Under these conditions, a broad *pollution dome* can form over a city or region, and air quality will suffer (Figure E4.7). When there is a regional wind, the pollution from a large city will be carried downwind to form a *pollution plume*.

Climatic Effects of Urban Air Pollution

Urban air pollution reduces visibility and illumination. Specifically, a smog layer can cut illumination by 10 percent in summer and 20 percent in winter. Ultraviolet radiation is absorbed by ozone in smog. At times, ultraviolet radiation is completely prevented from reaching the ground. Although this reduces the risk of human skin cancer, it may

E4.6 Eye on the Landscape Smog and haze at Los Angeles *A dense layer of smog and marine haze fills the Los Angeles Basin. The view is from a point over the San Gabriel Mountains, looking southwest.* **What else would the geographer see? ... Answers at the end of the Chapter.**

E4.7 *Pollution dome and pollution plume* *If calm, stable air overlies a major city, a pollution dome can form. When a wind is present, pollutants are carried away as a pollution plume.*

also permit increased viral and bacterial activity at ground level.

Winter fogs are much more frequent over cities than over the surrounding countryside. The fog is enhanced by the abundance of urban aerosols and particulates. Coastal airports, such as those of New York City, Newark, and Boston, suffer from an increased frequency of fog resulting from urban air pollution. Cities also show an increase in cloudiness and precipitation as compared to the surrounding countryside. This increase results from intensified convection generated when the lower air is heated by human activities.

Acid Deposition
Yet another air pollution problem is *acid deposition*. Perhaps you've heard about acid rain killing fish and poisoning trees. Acid rain is part of the phenomenon of acid deposition, which includes not only acid rain but also dry acidic dust particles. Acid rain simply consists of raindrops that have been acidified by air pollutants. Dry acidic particles are dust particles that are acidic in nature. They fall to Earth and coat the surface in a thin dust layer. When wetted by rain or fog, they acidify the water on leaves

and soils. In winter, acid particles can mix with snow as it accumulates. In spring, when the snow melts, a surge of acid water is then released to soils and streams.

Acid deposition is produced by the release of sulfur dioxide (SO_2) and nitric oxide (NO_2) into the air. This occurs when fossil fuels are burned. Once in the air, SO_2 and NO_2 readily combine with oxygen and water in the presence of sunlight and dust particles to form sulfuric and nitric acid aerosols. These aerosols serve as condensation nuclei and acidify the tiny water droplets created. When the droplets coalesce in precipitation, acid raindrops or ice crystals result. Sulfuric and nitric acids can also be formed on dust particles, creating dry acid particles. These can be as damaging to plants, soils, and aquatic life as acid rain.

What are the effects of acid deposition? A primary effect is acidification of lakes and streams, injuring aquatic plants and animals and perhaps eventually sterilizing a lake, pond, or stream. Another problem is soil damage. Through a series of chemical effects, acid deposition can cause soils to lose nutrients, affecting the natural ecosystem within a region.

To measure acidity, scientists use the pH scale. Values on this scale range from 1 to 14, with high values indicating alkalinity and the low values acidity. Pure water has a pH of 7. Normally, rainwater is slightly acid, showing pH values of 5 to 6. In the 1960s European water chemists observed that the pH of rain in northwestern Europe had dropped to values as low as 3 in some samples, similar to household vinegar. Because pH numbers are on a logarithmic scale, these values mean that rain in these samples was 100 to 1000 times more acid than normal.

In North America, scientists studying the chemical quality of rainwater reported in the 1970s and 1980s that rainwater over a large area of the northeastern United States had an average pH of about 4. Values less than pH 4 occurred at times over many heavily industrialized U.S. cities, including Boston, New York, Philadelphia, Birmingham, Chicago, Los Angeles, and San Francisco.

Figure E4.8 is a map of the United States showing the acidity of rainwater averaged over 1999. There has been a significant improvement in the last decade, but values are still low in the Ohio Valley and in the northeast. Compare this with Figure E4.9, showing emissions of sulfur dioxide and nitrogen oxides by states. Although the emissions data are for an earlier year, a strong relationship, especially between SO_2 and the average acidity of rainwater, is clear.

An important factor in the level of impact of acid deposition on the environment is the ability of the soil and surface water to absorb and neutralize acid. This factor ranges widely. In dry climates, acids can be readily neutralized because surface waters are normally somewhat alkaline. Areas where soil water is naturally somewhat acidic are most sensitive to acid deposition. Such areas are associated with moist climates generally and include the eastern United States and Canada, high-

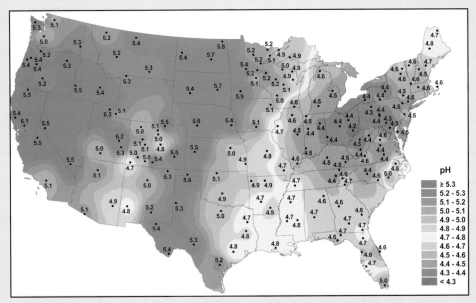

E4.8 Acidity of rainwater for the United States in 1999 *Values are pH units. The northeastern United States shows the lowest, most acid values. (National Atmospheric Deposition Program (NRSP-3)/National Trends Network, Illinois State Water Survey.)*

mountain regions of the western states and provinces, and the Pacific Northwest.

Acid deposition in Europe and North America has severely impacted some ecosystems. In Norway, acidification of stream water has virtually eliminated many salmon runs by inhibiting salmon egg development. Increased fish mortality, attributed to acidification, has been observed

in many lakes in eastern Canada. In 1990, American scientists estimated that 14 percent of Adirondack lakes were heavily acidic, along with 12 to 14 percent of the streams in the Mid-Atlantic states.

Forests, too, have been damaged by acid deposition (Figure E4.10). In western Germany, the impact has been especially severe in the Harz Mountains and the Black Forest. In the 1990s, Europe

continued to experience heavy losses in timber, resulting from fallout of sulfur dioxide in combination with nitrogen oxide emissions from vehicles, industry, and farm wastes. Fallout of sulfur oxides was most heavily concentrated in Germany, the Czech Republic, Slovakia, Poland, and Hungary.

During the 1990s, releases of sulfur oxides, nitrogen oxides, and volatile organic compounds were

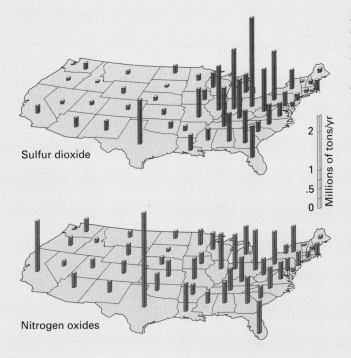

E4.9 Emissions of sulfur dioxide and nitrogen oxides by states for the year 1993 *Sulfur dioxide releases are especially heavy in the states that burn Appalachian coal. Releases of nitrogen oxides are high in Texas, due to oil processing, and in California, due to auto emissions. (Data from EPA, 1994.)*

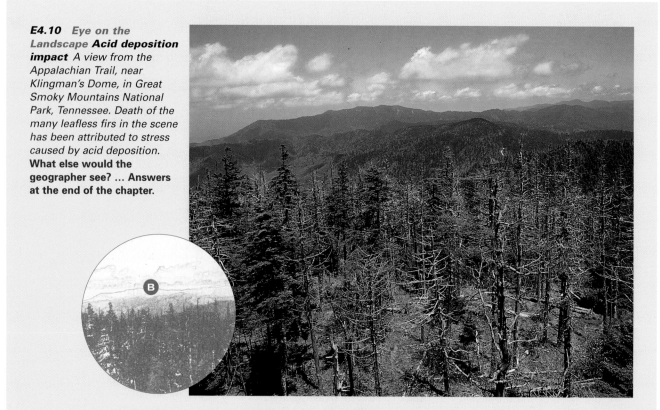

E4.10 *Eye on the Landscape* **Acid deposition impact** *A view from the Appalachian Trail, near Klingman's Dome, in Great Smoky Mountains National Park, Tennessee. Death of the many leafless firs in the scene has been attributed to stress caused by acid deposition.* **What else would the geographer see? ... Answers at the end of the chapter.**

significantly reduced in the United States. These effects were due primarily to improved industrial emission controls. Although the situation is improving, acid deposition is still a very important problem in many parts of the world—especially Eastern Europe and the states of the former Soviet Union. There, air pollution controls have been virtually nonexistent for decades. Reducing pollution levels and cleaning up polluted areas will be a major task for these nations in the next decades.

Air Pollution Control
The United States and Canada have some of the strictest laws in the world limiting air pollution. As a result, many strategies have been developed to reduce emissions. Sometimes these strategies involve trapping and processing pollutants after they are generated. The "smog controls" on automobiles are an example. Sometimes the use of alternative, nonpolluting technology is in order—for example, substituting solar, wind, or

geothermal power for coal burning to generate electricity.

In any event, both as individuals and as a society we can do much to preserve the quality of the air we breathe. Reducing fossil fuel consumption is an example. It will be a considerable challenge to our society and others to preserve and increase global air quality in the face of an expanding human population and increasing human resource consumption.

In Review | Atmospheric Moisture and Precipitation

- **Precipitation** is the fall of liquid or solid water from the atmosphere to reach the Earth's land or ocean surface. Evaporating, condensing, melting, freezing, **sublimation,** and **deposition** describe the changes of state of water.

- Water moves freely between ocean, atmosphere, and land in the **hydrologic cycle.** The global water balance describes these flows. The fresh water in the atmosphere and on land in lakes, streams, rivers, and *ground water* is only a very small portion of the total water in the **hydrosphere.**

- **Humidity** describes the amount of water vapor present in air. The ability of air to hold water vapor depends on temperature. Warm air can hold much more water vapor than cold air.

- **Specific humidity** measures the mass of water vapor in a mass of air, in grams of water vapor per kilogram of air. **Relative humidity** measures water vapor in the air as the percentage of the maximum amount of water vapor that can be held at the given air temperature. The **dew-point temperature,** at which condensation occurs, also measures the moisture content of air.

- The *adiabatic principle* states that when a gas is compressed, it warms, and when a gas expands, it cools. When an air parcel moves upward in the atmosphere, it encounters a lower pressure and so expands and cools.

- The **dry adiabatic lapse rate** describes the rate of cooling with altitude. If the air is cooled below the dew point, condensation or deposition occurs, and latent heat is released. This heat reduces the parcel's rate of cooling with altitude. When condensation or deposition is occurring, the cooling rate is described as the **wet adiabatic lapse rate.**

- Clouds are composed of droplets of water or crystals of ice that form on *condensation nuclei.* Clouds typically occur in layers, as *stratiform* clouds, or in globular masses, as *cumuliform* clouds. *Fog* occurs when a cloud forms at ground level.

- Precipitation from clouds occurs as *rain, hail, snow,* and *sleet.* When *supercooled* rain falls on a surface below freezing, it produces an *ice storm.*

- There are three types of precipitation processes—**orographic, convectional,** and *cyclonic.* In orographic precipitation, air moves up and over a mountain barrier. As it moves up, it is cooled adiabatically and rain forms. As it descends the far side of the mountain, it is warmed, producing a *rain shadow* effect.

- In convectional precipitation, unequal heating of the surface causes an air parcel to become warmer and less dense than the surrounding air. Because it is less dense, it rises. As it moves upward, it cools, and condensation with precipitation may occur. Under conditions of **unstable air, thunderstorms** can form, yielding hail and *lightning.* (Cyclonic precipitation is described in Chapter 6.)

- *Air pollution* is defined as unwanted gases, aerosols, and particulates injected into the air by human and natural activity. Polluting *gases, aerosols,* and *particulates* are generated largely by fuel combustion. **Smog,** a common form of air pollution, contains nitrogen oxides, volatile organic compounds, and ozone. *Inversions* can trap smog and other pollutant mixtures in a layer close to the ground, creating unhealthy air. *Acid deposition* of sulfate and nitrate particles can acidify soils and lakes, causing fish and tree mortality.

Key Terms

precipitation, p. 120
sublimation, p. 120
deposition, p. 120
hydrosphere, p. 120
hydrologic cycle, p. 120
humidity, p. 123
specific humidity, p. 123

dew-point temperature, p. 123
relative humidity, p. 124
adiabatic process, p. 126
dry adiabatic lapse rate, p. 126

wet adiabatic lapse rate, p. 127
cloud, p. 127
orographic precipitation, p. 132

convectional precipitation, p. 132
unstable air, p. 135
thunderstorm, p. 136
smog, p. 140

Review Questions

1. Identify the three states of water and the six terms used to describe possible changes of state.
2. What is the hydrosphere? Where is water found on our planet? in what amounts? How does water move in the hydrologic cycle?
3. Define specific humidity. How is the moisture content of air influenced by air temperature?
4. Define relative humidity. How is relative humidity measured? Sketch a graph showing relative humidity and temperature through a 24-hour cycle.
5. Use the terms *saturation, dew point,* and *condensation* in describing what happens when an air parcel of moist air is chilled.
6. What is the adiabatic process? Why is it important?
7. Distinguish between dry and wet adiabatic lapse rates. In a parcel of air moving upward in the atmosphere, when do they apply? Why is the wet adiabatic lapse rate less than the dry adiabatic rate? Why is the wet adiabatic rate variable in amount?
8. How are clouds classified? Name four cloud families, two broad types of cloud forms, and three specific cloud types.
9. What is fog? Explain how radiation fog and advection fog form.
10. How is precipitation formed? Describe the process for warm and cool clouds.
11. Describe the orographic precipitation process. What is a rain shadow? Provide an example of the rain shadow effect.
12. What is unstable air? What are its characteristics?
13. Describe the convectional precipitation process. What is the energy source that powers this source of precipitation? Explain.

Eye on the Environment 4.2 ● Air Pollution

1. What are the most abundant air pollutants, and what are their sources?
2. What is smog? What important pollutant forms within smog, and how does this happen?
3. Distinguish between low-level and high-level inversions. How are they formed? What is their effect on air pollution? Give an example of a pollution situation of each type.
4. To what type of pollution does the term *acid deposition* refer? What are its causes? What are its effects?

Visualizing Exercises

1. Lay out a simple diagram showing the main features of the hydrologic cycle. Include flows of water connecting land, ocean, and atmosphere. Label the flow paths.
2. Graph the temperature of a parcel of air as it moves up and over a mountain barrier, producing precipitation.
3. Sketch the anatomy of a thunderstorm cell. Show rising bubbles of air, updraft, downdraft, precipitation, and other features.

Essay Questions

1. Water in the atmosphere is a very important topic for understanding weather and climate. Organize an essay or oral presentation on this topic, focusing on the following questions: What part of the global supply of water is atmospheric? Why is it important? What is its global role? How does the capacity of air to hold water vapor vary? How is the moisture content of air measured? Clouds and fog visibly demonstrate the presence of atmospheric water. What are they? How do they form?
2. Compare and contrast orographic and convectional precipitation. Begin with a discussion of the adiabatic process and the generation of precipitation within clouds. Then compare the two processes, paying special attention to the conditions that create uplift. Can convectional precipitation occur in an orographic situation? under what conditions?

Eye on the Landscape

4.13 Sea fog An interesting feature of this dramatic seascape is the presence of gently sloping terrace surfaces **(A)**, used as hay fields, that lie above the marine cliffs. These terrace surfaces were formed by wave abrasion in the near-shore zone and have now been uplifted to well above present sea level. Wave abrasion is covered in Chapter 18; Figure 18.4 shows an abrasion platform that is currently being formed. The uplift is the result of tectonic activity just off the coast, where the Pacific crustal plate and the American plate are moving in different directions (see Chapter 12 for plate tectonics). Another interesting feature of this photo is the gullying and rapid erosion of the cliff **(B)**. The coast ranges of Big Sur receive abundant precipitation, as moist, eastward-moving air is raised over the mountain barrier. (See Orographic Precipitation, this chapter.) The soft sediments are easily eroded by heavy rains, keeping vegetation from stabilizing the slope. Loose sediment falls to the shoreline and is carried offshore by coastal currents. Erosion and gullying are subjects of Chapter 16; Figure 16.4 shows gullies developing on a flat landscape. Coastal erosion is covered in Chapter 18.

E4.6 Los Angeles smog and haze; E4.10 Acid deposition impact Both of these photos show mountain profiles, contrasting the San Gabriel **(A)** and Great Smoky Mountains **(B)**. Note how the profile of the San Gabriels is sharper and more angular. In the photo, you can see steep slopes with chutes of snow-covered rock, devoid of vegetation, that are obviously carved by running water. In the Smokies, the summits and slopes are clothed in forest and are much more rounded. Here, much of the shaping of these ancient mountains takes place by weathering and dissolution of rock acting through a thick layer of vegetation and soil. See fluvial erosion in Chapter 16 and solution weathering in Chapter 14.

5 | WINDS AND GLOBAL CIRCULATION

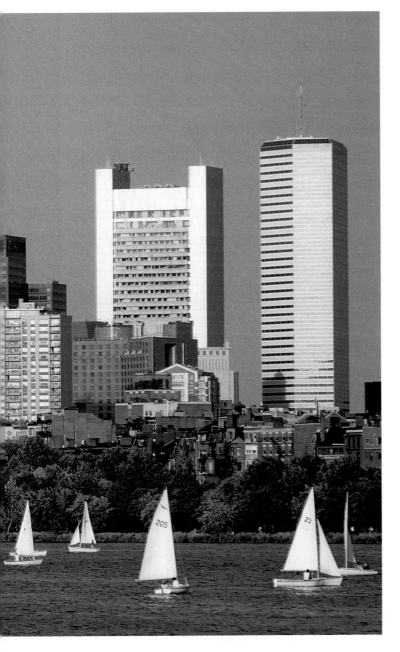

• • • • • • • • • • • • • • • •

PUTTING YOU IN THE PICTURE

Coming about! You push the tiller hard to the opposite side of the boat, and the small craft turns easily into the wind. The sail goes slack and begins to swing across the boat. You duck under the boom, the long pole that holds the bottom of the sail taut, and you step neatly to the other side. The sail now catches the wind from the other side and begins to fill. You pull in the line attached to the sail and correct the rudder for your new course. The sail tightens, and the boat leans away from the wind, gathering speed. You quickly move to the boat's edge and, sitting on the rim, lean out over the water. Your weight is held in check by the strong pull of the sail on the line in your right hand. The boat swings upright, and you are skimming across the water again, thanks to the cool, southeast breeze.

What a marvelous setting for a sail! With the soaring skyscrapers of Boston on the south and the leafy Cambridge campus of MIT on the north, the mouth of the Charles River is the magnificent jewel in the center of the famed necklace of parks, green spaces, and beaches that frames the map of the city. Unfortunately, you're not the only one out for a sail. There are perhaps a dozen other boats sharing the water with you, and since it's still early in the afternoon, the fleet is sure to increase. Watching out for them requires attention.

Sailing on the Charles River in downtown Boston is always something of a challenge. Not only are there other sailors, but also the wind can be a bit fickle, to say the least. Today, the tall buildings stand as obstacles to the southeast breeze, creating currents and eddies in the downstream airflow as it crosses the river mouth. At any moment the breeze can suddenly shift direction. Without the proper adjustment of the sail and the rudder, you can wind up in the water!

Distracted for a moment by the beauty of the city and its setting, you fail to notice a boat quickly approaching your bow. Turn, and fast! You push the tiller over hard, and the boat pivots cleanly away from the oncoming craft. Oh, oh. The wind's shifting, just at the wrong moment. You're going to jibe! Duck! The sail comes flying across the boat, and the line attached to the boom jerks tight in your hand. You'd better move fast, now—your weight's on the wrong side of the boat.

Scrambling to the other side of the boat, you drop the tiller! A major mis-take. Without the control of the rudder, the boat spins, tilting violently in the strong breeze. Water enters the boat as the rim tips below the waterline. Now the boom is in the water! You put all your weight on the opposite rim of the boat, trying to right the boat against the forces aligned to tip it over. Completing the spin, the boat comes to rest pointed into the wind, the unrestrained sail flap-ping noisily overhead. You are crouched in the bottom of the boat, breathless, under the wriggling boom. Ankle-deep water covers your feet. Humbled by the power of the wind, you think, "Welcome to the Charles!"

Wind is the subject of this chapter. Even if you are not a sailor who is constantly concerned with wind strength and direction, wind is part of the everyday weather that we all expe-rience. In the summer, a bit of a breeze can make the difference between a pleasant day and a hot one. In the winter, that same breeze can make the chill of a freezing day more intense. Wind direction is typically determined by the position of weather systems, which are the subject of Chapter 6. In the midlatitudes, strong easterly winds are often associated with wind and rain, while strong west-erly winds typically foretell clear skies.

Winds are caused by unequal heat-ing of the atmosphere. As you know from Chapter 4, heating causes air to expand and become less dense. When this occurs, cooler, denser, surrounding air presses inward to buoy up the lighter, heated air, and a wind along the surface is created. This principle directly explains many types of local winds, such as the sea breeze you might experience on an afternoon at the beach. It is also the cause of global wind motions, when we consider that the equatorial and tropical regions are heated more intensely by the Sun than are the mid- and higher latitudes. ■

WINDS AND GLOBAL CIRCULATION

The air around us is always in motion. From a gentle breeze on a summer's day to a cold winter wind, we experience moving air much of the time. Why does the air move? What are the forces that cause winds to blow? Why do winds blow more often in some directions than others? What is the pat-tern of wind flow in the upper atmosphere, and how does it affect our weather? How does the global wind pattern pro-duce ocean currents? These are some of the questions we will answer in this chapter.

Winds are caused by unequal heating of the Earth's atmosphere. When the atmosphere in one location is heated to a temperature that is warmer than that in another location, a difference in pressure, or pressure gradient, results. As we will see later in more detail, this pressure gradient pushes air at the surface toward the warmer location and pushes air at upper levels away from the warmer location. This process directly explains many types of local winds, such as the sea breeze you might experience on an afternoon at the beach. It is also the cause of global wind motions, since the equato-rial and tropical atmosphere is heated more intensely by the Sun than is the atmosphere at mid- and higher latitudes. In response to this difference in heating, global-scale pressure gradients move vast bodies of warm air poleward, and huge pools of cool air shift equatorward, strongly influencing our day-to-day weather. The direction of these global wind motions is also influenced by the Earth's rotation—a factor that we also discuss in this chapter.

Another feature of the global circulation system is ocean currents, which are largely driven by surface winds. Ocean currents, like global wind motions, act to move heat energy across parallels of latitude and are also affected by the Earth's rotation.

Before we turn to local and global wind patterns, how-ever, we will need to examine atmospheric pressure and how wind develops in response to pressure differences. These are the subjects of the next few pages.

ATMOSPHERIC PRESSURE

The Earth's land surface is actually located at the bottom of a vast ocean of air—the Earth's atmosphere. Like the water in the ocean, the air in the atmosphere is constantly pressing on the solid or liquid surface beneath it. This pressure is exerted in all directions.

Atmospheric pressure exists because air molecules have mass and are constantly being pulled toward the Earth by gravity. As we all know and experience every day, the accel-eration of mass by gravity creates a force that we refer to as weight. Pressure is a force per unit area. For the atmosphere, this force is thus produced by the weight of a column of air above a unit area of surface. At sea level, the force is the weight of about 1 kilogram of mass of air that lies above each square centimeter of surface (1 kg/cm^2) (Figure 5.1). In

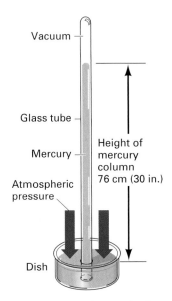

5.2 Mercury barometer *Atmospheric pressure pushes the mercury upward into the tube, balancing the pressure exerted by the weight of the mercury column. As atmospheric pressure changes, the level of mercury in the tube changes. At sea level, the average height of the mercury level is about 76 cm (30 in.).*

5.1 Atmospheric pressure *This figure depicts atmospheric pressure as the weight of a column of air. (a) Metric system. The weight of a column of air 1 cm on a side is balanced by the weight of a mass of about 1 kg. (b) English system. The weight of a column of air 1 in. on a side is balanced by a weight of about 15 pounds.*

the English system, the force is about 15 pounds of weight for each square inch of surface (15 lb/in.²).

In the metric system, the unit of pressure is the *pascal* (Pa). At sea level, the average pressure of air is 101,320 Pa. Although the pascal is the proper metric unit for pressure, many atmospheric pressure measurements are based on the *bar* and the **millibar** (mb) (1 bar = 1000 mb = 10,000 Pa), two former metric units that are still in wide use. In this text, we will use the millibar as the metric unit of atmospheric pressure, following this long tradition. Standard sea-level atmospheric pressure is 1013.2 mb.

Measuring Atmospheric Pressure

Perhaps you know that atmospheric pressure is measured by a **barometer.** But how does a barometer work? The principle is simple, and you have experienced it often. Visualize sipping a soda through a straw. By lowering your jaw and moving your tongue, you expand the volume of your mouth. This creates a partial vacuum, and air pressure forces the soda up through the straw and into your mouth.

A *mercury barometer* works on the same principle (Figure 5.2)—that atmospheric pressure will force a liquid

> **In the mercury barometer, air pressure balances the pull of gravity of a column of mercury about 76 cm (30 in.) high.**

up into a tube when a vacuum is present. To construct the barometer, a glass tube about 1 m (3 ft) long, sealed at one end, is completely filled with mercury. The open end is temporarily held closed. Then the tube is inverted, and the open end is immersed into a dish of mercury. When the opening of the tube is uncovered, the mercury in the tube falls some distance, but then it remains fixed at a level about 76 cm (30 in.) above the surface of the mercury in the dish.

What is happening here? At first, mercury flows out of the tube, under the force of gravity, and into the dish. Since there is no air in the top of the tube, a vacuum is formed. At the same time, atmospheric pressure pushes on the surface of the mercury in the dish, pressing the mercury upward into the tube. Since there is no air in the tube, there is no air pressure to push back against it. The pull of gravity on the mercury column then balances the pressure of the air, and the mercury level stays at a stable height. If air pressure changes, so will the height of the mercury column. So, this device will measure air pressure directly—but as a height in centimeters or inches rather than a pressure. The mercury barometer has become the standard instrument for weather measurements because of its high accuracy.

The widespread use of the mercury barometer has led to the common measurement of air pressure in heights of mercury, as centimeters or inches. On this scale, standard sea-level pressure is 76 cm Hg (29.92 in. Hg). The letters Hg are the chemical symbol for mercury. In this text, we will use in. Hg as the English unit for atmospheric pressure.

5.3 Aneroid barometer *As air pressure varies, the diaphragm moves up and down. This motion is transmitted mechanically to move an indicator hand along a scale.*

More serious effects on human physiology occur at greater elevations. Perhaps you've felt out of breath, or tired more easily, while walking along a high-mountain trail. With decreased air pressure, oxygen moves into lung tissues more slowly, producing shortness of breath and fatigue. These symptoms, along with headache, nosebleed, or nausea, are referred to as mountain sickness and are likely to occur at altitudes of 3000 m (about 10,000 ft) or higher. Most people adjust to the reduced air pressure of a high-mountain environment after several days.

WIND

From a gentle, cooling breeze on a warm summer afternoon to the howling gusts of a major winter storm, wind is a familiar phenomenon. **Wind** is defined as air motion with respect to the Earth's surface. Wind movement is dominantly horizontal. Air motions that are dominantly vertical are not called winds but are known by other terms, such as updrafts or downdrafts.

A more common type of barometer is the *aneroid barometer* (Figure 5.3). It uses a sealed canister of air, much like a small, flat can you might see in a supermarket. The canister has a slight partial vacuum, tensing the lid, or diaphragm, against a spring attached to the lid. The diaphragm flexes as air pressure changes, and through a mechanical linkage this fine movement drives a hand indicator that moves along a scale.

Atmospheric pressures at a location vary from day to day by only a small proportion. On a cold, clear winter day, the barometric pressures at sea level might be as high as 1030 mb (30.4 in. Hg), while in the center of a storm system, pressure might drop to 980 mb (28.9 in. Hg), a difference of about 5 percent. Pressure changes are associated with traveling weather systems, a topic that we will discuss in Chapter 6.

How Air Pressure Changes with Altitude

Figure 5.4 shows how atmospheric pressure decreases with altitude. Moving upward from the Earth's surface, the decrease of pressure with altitude is initially quite rapid. At higher altitudes, the decrease is much slower. Because atmospheric pressure decreases rapidly with altitude near the surface, a small change in elevation will often produce a significant change in air pressure. For example, you may have noticed that your ears sometimes "pop" from a pressure change during an elevator ride in a tall building. You may also have noticed your ears popping on an airplane flight. Aircraft cabins are pressurized to about 800 mb (24 in. Hg), which corresponds to an elevation of about 1800 m (5900 ft).

5.4 Atmospheric pressure and altitude *Atmospheric pressure decreases with increasing altitude above the Earth's surface.*

5.5 Combination cup anemometer and wind vane
The anemometer and wind vane observe wind speed and
direction, which are displayed on the meter below. The
wind vane and anemometer are mounted outside, with a
cable from the instruments leading to the meter, which is
located indoors. Also shown are a barometer and
maximum-minimum thermometer. (Courtesy of Taylor
Instrument Company and Ward's Natural Science
Establishment, Rochester, N. Y.).

Measurement of Wind

The motion of air in wind, like all motion, is characterized
by a direction and velocity (Figure 5.5). Wind direction can
be determined by a *wind vane,* the most common of the
weather instruments. The design of the wind vane keeps it
facing into the wind. Wind direction is always given as the
direction from which the wind is coming. So, a west wind is
one that comes from the west and moves to the east (Figure
5.6).

Wind speed is measured by an *anemometer.* The most
common type is the cup anemometer, which consists of
three funnel-shaped cups mounted on the ends of spokes of
a horizontal wheel. The cups rotate with a speed proportional to that of the wind. One version of the anemometer
turns a small electric generator. The more rapidly the wheel
rotates, the more current is generated. The current is measured by a meter, which is calibrated in units of wind speed.
Units are either meters per second or miles per hour.

Winds and Pressure Gradients

Wind is caused by differences in atmospheric pressure from
place to place. Air tends to move from high to low pressure
until the air pressures are equal. This follows the simple
physical principle that any flowing fluid (such as air) subjected to gravity will move until the pressure at every level
is uniform.

Figure 5.7 illustrates this principle. It is a simple map
showing two locations, taken here as Wichita and Columbus.
At Wichita, the pressure is high (H), and the barometer
(adjusted to sea-level pressure) reads 1028 mb (30.4 in. Hg).
At Columbus, the barometer is low (L) and reads 997 mb
(29.4 in. Hg). Thus, there is a **pressure gradient** between
Wichita and Columbus. The lines on the map around
Wichita and Columbus connect locations of equal pressure
and are called **isobars.** As you move between Wichita and
Columbus, you cross these isobars on the map, encountering
changing pressures as a result of the drop from 1028 to 997
mb (30.4 to 29.4 in. Hg).

Because atmospheric pressure is unequal at Wichita and
Columbus, a *pressure gradient force* will push air from
Wichita toward Columbus. This pressure gradient will produce a wind. The greater the pressure difference between
the two locations, the greater this force will be and the
stronger the wind. However, the wind will not necessarily
move directly from Wichita to Columbus. It will be
deflected by the Coriolis effect, as we will see later in this
chapter.

5.6 Wind direction Winds are designated according to
the compass point from which the wind comes. A west
wind comes from the west, but the air is moving eastward.

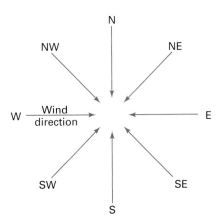

5.7 Isobars and a pressure gradient High pressure is
centered at Wichita, and low pressure is centered at
Columbus.

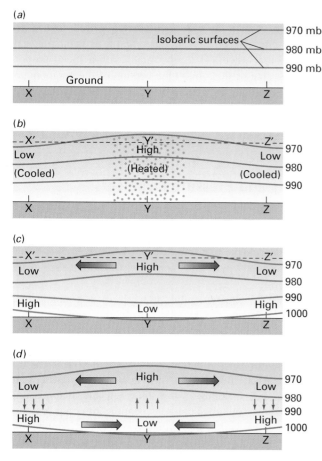

5.8 Formation of a simple convection loop *Heating of the atmosphere over point Y creates a pressure gradient that causes a convection loop.*

A Simple Convective Wind System

How do pressure gradients come about? Pressure gradients develop because of unequal heating of the atmosphere. To understand this process, we will look at a simple convective wind system, shown in Figure 5.8. Part (*a*) depicts a uniform atmosphere above a ground surface. The lines mark isobaric surfaces, seen in cross section, and points *X, Y,* and *Z* mark the left, middle, and right ends of the cross section. Let's now imagine that the air is heated over *Y* (perhaps by sunlight) and cooled over *X* and *Z* (perhaps by radiation loss), as in part (*b*). As the air at *Y* expands in response to heating, the isobaric surfaces move upward. At a constant height in the atmosphere, for example, along line *X'-Y'-Z'*, a pressure gradient is induced. We can see this by noting that at *Y'*, the pressure is between 980 and 970 mb, while at *X'* and *Z'*, it is less than 970 mb. Thus, there will be a pressure gradient from *Y'* to *X'* and *Y'* to *Z'*.

The pressure gradient next causes air to move away from *Y'*, as shown in part (*c*). However, as soon as this air moves, it changes the surface pressure at *X, Y,* and *Z*. Since there is less air above *Y*, pressure at the ground falls. Meanwhile, there is more air above points *X* and *Z*, so pressure there

> Heating of the atmosphere creates a pressure gradient that moves air toward the warm region at low levels and away from the warm region at high levels.

rises. This creates a new pressure gradient at the surface that moves air from *X* and from *Z* to *Y* as shown in part (*d*). The resulting circulation has produced two *convection loops* in which air at the surface converges at the location of heating and moves away, or diverges, at higher altitude. The principle that heating of the air produces a pressure gradient that causes air to move is an important one. We will return to it later in the chapter when we consider upper-level winds.

Sea and Land Breezes

Sea and *land breezes* are simple examples of how heating and cooling of the air produce convection loops to create local winds. If you've ever vacationed at the beach in the summer, you may have noticed that in the afternoons a sea breeze often sets in. This wind brings cool air off the water, dropping temperatures and refreshing beachgoers and residents living close to the beach. Late at night, a land breeze may develop. This wind moves cooler air, which is chilled over land by nighttime radiant cooling, toward the water.

Figure 5.9 shows how these breezes work in more detail. Recall from Chapter 3 that land surfaces heat and cool more rapidly than water, so that the coastline is a location where temperature contrasts can easily develop. During the day, the air over the land is warmed by the land surface, and a convection loop forms as in Figure 5.8*d*. Air moves oceanward aloft, while surface winds bring cool marine air landward at the surface (part *b*). At night, radiation cooling over land creates a reversed convection loop, and an offshore breeze develops (part *c*).

The sea breeze–land breeze convection loop provides an example of surface thermal high- and low-pressure zones. That is, low surface pressure is associated with warm air, and high surface pressure with cool air. We will see this later in the chapter at a more global scale, when we note the tendency for deserts to have low atmospheric pressure and persistent regions of ice and snow to have high pressure.

Local Winds

Sea and land breezes are examples of *local wind* systems that are generated by local effects, such as a difference in heating of the atmosphere by different land surface types. Other types of local winds include mountain and valley winds, drainage winds, and Santa Ana winds.

Mountain winds and *valley winds* are local winds that alternate in direction in a manner similar to the land and sea breezes. During the day, mountain hill slopes are heated intensely by the Sun, causing a convection loop to form. As the surface part of the loop, an air current moves up valleys from the plains below—upward over rising mountain slopes, toward the summits. At night, the hill slopes are chilled by radiation, setting up a reversed convection loop. The cooler, denser hill slope air then moves valleyward, down the hill slopes, to the plain below.

980 mb
990 mb
1000 mb

Sea   Land

(a)   Early morning—calm

(Warmed air)   980
                      990
High   Low   1000

(b)   Afternoon—sea breeze

(Cooled air)   980
                      990
                      1000
Low   High

(c)   Night—land breeze

5.9 *Sea and land breezes* *The contrast between the land surface, which heats and cools rapidly, and the ocean surface, which has a more uniform temperature, induces pressure gradients to create sea and land breezes.*

Another group of local winds are known as *drainage winds,* in which cold, dense air flows under the influence of gravity from higher to lower regions. In a typical situation, cold, dense air accumulates in winter over a high plateau or high interior valley. Under favorable conditions, some of this cold air spills over low divides or through passes, flowing out on adjacent lowlands as a strong, cold wind.

Drainage winds occur in many mountainous regions of the world and go by various local names. The *mistral* of the Rhône Valley in southern France is a well-known example—it is a cold, dry local wind. On the ice sheets of Greenland and Antarctica, powerful drainage winds move down the gradient of the ice surface and are funneled through coastal

> Local winds include land and sea breezes, mountain and valley winds, drainage winds, and Santa Ana winds.

valleys. Picking up loose snow, these winds produce blizzardlike conditions that last for days at a time.

Another type of local wind occurs when the outward flow of dry air from an anticyclone is combined with the local effects of mountainous terrain. An example is the *Santa Ana*—a hot, dry easterly wind that sometimes blows from the interior desert region of southern California across coastal mountain ranges to reach the Pacific coast. This wind is funneled through local mountain gaps or canyon floors where it gains great force. At times the Santa Ana wind carries large amounts of dust. Because this wind is dry, hot, and strong, it can easily fan wildfires in brush or forest out of control (Figure 5.10).

5.10 *Eye on the Landscape* **Brushfire** *Brushfires, such as this one in Oakland, California, are often fanned by strong downslope winds of hot, dry air (Santa Ana winds).* **What else would the geographer see? ... Answers at the end of the chapter**

Eye on the Environment | 5.1

Wind Power, Wave Power, and Current Power

Wind power is an indirect form of solar energy that has been used for centuries. In the Low Countries of Europe, the windmill played a major role in pumping water from the polders as they were reclaimed from tidal land. In these low, flat areas, streams could not be adapted to waterpower, and the windmill was also used to grind grain.

The design of new forms of windmills has intrigued inventors for many decades. The total supply of wind energy is enormous. The World Meteorological Organization has estimated that the combined electrical-generating wind power of favorable sites throughout the world comes to about 20 million megawatts, a figure about 100 times greater than the total electrical generating capacity of the United States. Many problems must be solved, however, before wind power can become a major resource.

Small wind turbines are a promising source of supplementary electric power for individual farms, ranches, and homes. The Darrieus

rotor, with circular blades turning on a vertical axis, is well adapted to small generators—those with less than 50 kilowatts output. Wind turbines presently in operation and being developed are adjusted in scale according to the purposes they serve. Turbines capable of producing power in the range of 50 to 200 kilowatts are already in operation to serve small communities.

Wind turbines with a generating capacity in the range of 50 to 100 kilowatts have been assembled in large numbers at favorable locations to form "wind

Modern windmills *A wind farm in San Gorgonio Pass, near Palm Springs, California. Wind speeds here average 7.5 m/s (17m/hr).*

Still another type of local wind—the *chinook,* which we encountered in Chapter 4—results when strong regional winds pass over a mountain range and descend on the lee side. (See Figure 4.16.) As we know from studying orographic precipitation in Chapter 4, the descending air is warmed and dried. A chinook wind can sublimate snow or dry out soils very rapidly.

Surface winds are generated when pressure gradient forces cause air to move. Energy of motion—kinetic energy—is stored in the air motion. The stored energy can be extracted by windmills or wind turbines to do other work, such as generate electric power. *Eye on the Environment 5.1 • Wind Power, Wave Power, and Current Power* describes this source of energy as well as those of waves and ocean currents.

The Coriolis Effect and Winds

The pressure gradient force tends to move air from high pressure to low pressure. For sea and land breezes, which

are local in nature, this push produces a wind motion in about the same direction as the pressure gradient. For wind systems on a global scale, however, the direction of air motion will be somewhat different. The difference is due to the Earth's rotation, through the Coriolis effect. What is this effect, and how does it come about? GEODISCOVERIES

Through the **Coriolis effect,** an object in motion on the Earth's surface always appears to be deflected away from its course. The object moves as though a force were pulling it sideward. The apparent deflection is to the right in the northern hemisphere and to the left in the southern hemisphere. Deflection is strongest near the poles and decreases to zero at the equator. Figure 5.11 diagrams the effect. Note that the apparent force does not depend on direction of motion—the deflection occurs whether the object is moving toward the north, south, east, or west.

farms." Arranged in rows along ridge crests, the turbines intercept local winds of exceptional frequency and strength. One such locality is the Tehachapi Pass in California, where winds moving southeastward in the San Joaquin Valley are funneled through a narrow mountain gap before entering the Mojave Desert. The San Gorgonio Pass, near Palm Springs, presents a similar situation favorable to the development of wind farms. Another important wind farm locality lies about 80 km (50 mi) east of San Francisco in the Altamont Pass area of Alameda and Contra Costa counties. Here, daytime westerly winds of great persistence develop in response to the buildup of surface low pressure in the Great Valley to the east. A group of wind farms built here with a total of about 3000 turbines provides the Pacific Gas and Electric Company with a yield of over 500 million kilowatt-hours of electricity per year. Larger wind turbines, each capable of generating 2 to 4 megawatts, have been constructed at a number of sites.

Wave energy is another indirect form of solar energy. Nearly all ocean waves are produced by the stress of wind blowing over the sea surface. Water waves are characterized by motion of water particles in vertical orbits. Energy is extracted from wave motion by use of a floating object tethered to the seafloor. As the floating mass rises and falls, a mechanism is operated and drives a generator. Pneumatic systems use the principle of the bellows, operated by the changing pressure of the surrounding water as the water level rises and falls. A number of such devices are in the planning and testing stages.

Harnessing the vast power of a great current stream, such as the Gulf Stream or the Kuroshio, is a prospect that has not gone unnoticed. One rather grandiose plan proposes a large number of current-driven turbines to be tethered to the ocean floor so as to operate below the ocean surface. Each turbine would be 170 m (558 ft) in diameter and capable of generating 83 megawatts of power.

Vertical wind turbines *These vertical wind turbines generate power without needing to rotate to face the wind. Altamont Pass, California.*

An array of 242 such turbines could deliver 10,000 megawatts—a large part of Florida's power requirement and equivalent to the use of about 130 million barrels of crude oil per year.

The Coriolis effect is a result of the Earth's rotation. Imagine that a rocket is launched from the north pole toward New York, following the 74° W longitude meridian (Figure 5.12). However, as it travels toward New York, the Earth rotates beneath its trajectory, moving from west to east. The rocket continues on its straight path, but to an observer at the launch point on the rotating Earth below, the rocket will appear to curve to the right, away from New York and toward Chicago. For the rocket to reach New York, its flight path would have to be adjusted to take the Earth's rotation into account.

What would happen if the rocket were launched from New York heading north along the 74° W meridian toward the pole? Because of the Earth's rotation, a point on the Earth's surface at the latitude of New York (about 40° N) moves eastward at about 360 m/s (about 800 mi/hr). Although the rocket is aimed properly along the meridian, at its launch its motion will have an eastward component of 360 m/s. However, as the rocket continues northward, it passes over land that is moving less swiftly eastward. At 60° N latitude, the eastward velocity of the Earth's surface is only about 225 m/s (about 500 mi/hr). So, the rocket will appear to move more quickly eastward than the land underneath, and its path will appear deflected to the right.

From this analysis, we can see that the Coriolis effect is actually produced because observers on the ground and the geographic grid around them are not fixed but are in motion along with the Earth. Figure 5.13 illustrates this point. Imagine yourself

> **The Coriolis effect causes the apparent motion of winds to be deflected away from the direction of the pressure gradient. Deflection is to the right in the northern hemisphere and to the left in the southern hemisphere as viewed from the starting point.**

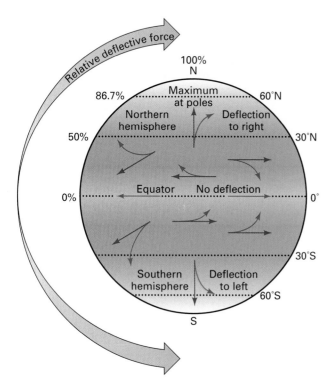

5.11 Coriolis effect direction and strength *The Coriolis effect acts to deflect the paths of winds or ocean currents to the right in the northern hemisphere and to the left in the southern hemisphere as viewed from the starting point. Blue arrows show the direction of initial motion, and red arrows show the direction of motion apparent to the Earth observer.*

5.12 The Coriolis effect *The path of a rocket launched from the north pole toward New York will apparently be deflected to the right by the Earth's rotation and might land near Chicago. If launched from New York toward the north pole, its path will appear to be deflected to the right as well.*

fixed in space above the north pole (*a*). As you watch, the geographic grid slowly turns underneath you. A straight motion, such as that of a rocket, will trace a curving path across the turning grid. Now imagine yourself fixed above New York, viewing the Earth from about the same point in space as in Figure 5.13*b*. As you watch the Earth turn to the right, the intersections of meridians and parallels near New York will appear to rotate counterclockwise. Again, a straight path will appear to curve on this turning grid. Now imagine yourself at the equator (*c*). Here the grid will move eastward but will not rotate. So a straight path does not curve at the equator. This is why the Coriolis effect is not felt at the equator.

In most real situations, we are concerned with analyzing the motions of ocean currents or air masses not from the viewpoint of space, but from the viewpoint of an Earth observer on the geographic grid. As a shortcut, we can treat the Coriolis effect as a sideward-turning force that always acts at right angles to the direction of motion. The strength of this *Coriolis force* increases with the speed of motion but decreases with latitude. If we treat the Coriolis effect this way, it will properly describe motion within the geographic grid.

Considering the Coriolis effect as a sideward-turning force acting on Earthbound motion, how does the Coriolis force affect wind patterns? A parcel of air in motion near the surface is subjected to three forces (Figure 5.14). First is the pressure gradient force that pushes the parcel toward low pressure. Second is the Coriolis force, which always acts at right angles to the direction of motion. Third is a frictional force exerted by the ground surface, which is proportional to

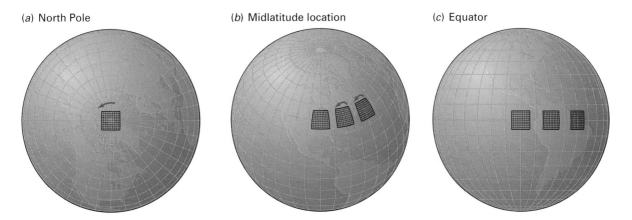

(a) North Pole   (b) Midlatitude location   (c) Equator

5.13 *Rotation of the geographic grid* *The Coriolis effect is produced by Earth's rotation, which turns the geographic grid. This makes straight motion appear curved. Apparent rotation is greatest at the poles and decreases to zero at the equator.*

the wind speed and always acts in the opposite direction to the direction of motion. The sum of these three forces produces a direction of motion that is toward low pressure but at an angle to the pressure gradient.

A parcel of air in motion is subjected to three forces: pressure gradient, Coriolis, and friction. The balance among these forces determines the direction of motion of the air parcel.

Cyclones and Anticyclones

Low- and high-pressure centers are common features of the daily weather map. Think of them as marking vast whirls of air in spiraling motion. In low-pressure centers, known as **cyclones,** air spirals inward and upward. This inward spiraling motion is *convergence.* In high-pressure centers, known as **anticyclones,** air spirals downward and outward. This outward spiraling motion is *divergence.*

How does the inward and outward motion of air in cyclones and anticyclones come about? Figure 5.15 shows surface wind patterns around cyclones and anticyclones in northern and southern hemispheres. When low pressure is at the center (cyclone), the pressure gradient is straight inward (left side of figure). When high pressure is at the cen-

In a cyclone, winds converge, spiraling inward and upward. In an anticyclone, winds diverge, spiraling downward and outward.

ter (anticyclone), the gradient is straight outward (right side). But because of the Coriolis force and friction with the surface, the surface air moves at an angle across the gradient. This creates the spiraling motion. In the northern hemisphere (upper portion), the cyclonic spiral will be counterclockwise because the Coriolis force acts to the right. In the southern hemisphere (lower portion), the cyclonic spiral will be clockwise because the Coriolis force acts to the left. For anticyclones, the situation is reversed.

We know from experience that low-pressure centers (cyclones) are often associated with cloudy or rainy weather and that high-pressure centers (anticyclones) are often associated with fair weather. Why? Recall from Chapter 4 that when air is forced aloft, it is cooled according to the adiabatic principle, and condensation and precipitation can occur. Thus, cloudy and rainy weather is often associated with the inward and upward air motion of cyclones. In the anticyclone, diverging air sinks and spirals outward. When air descends, we know from Chapter 4 that it is warmed according to the adi-

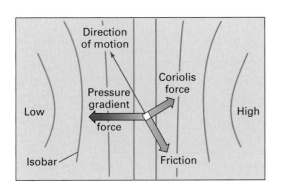

5.14 *Balance of forces on a parcel of surface air* *Although the pressure gradient will push a parcel of air toward low pressure, it will be deflected by the Coriolis force and slowed by friction with the surface. The direction of air motion will be the result of the three forces acting together.*

5.15 Air motion in cyclones and anticyclones
Surface winds spiral inward toward the center of a cyclone but outward and away from the center of an anticyclone. Because the Coriolis effect deflects moving air to the right in the northern hemisphere and to the left in the sourthern hemisphere, the direction of inspiraling and outspiraling reverses from one hemisphere to the other.

abatic principle, and thus condensation cannot occur. So anticyclones are often associated with fair weather.

Cyclones and anticyclones are typically large features of the lower atmosphere—perhaps a thousand kilometers (about 600 mi) across, or more. For example, a fair weather system, which is an anticyclone, may stretch from the Rockies to the Appalachians. Cyclones and anticyclones can remain more or less in one location, or they can move, sometimes rapidly, to create weather disturbances.

Surface Winds on an Ideal Earth

Let's consider wind and pressure patterns on an ideal Earth—one without a complicated pattern of land and water and no seasonal changes. Pressure and winds for such an ideal Earth are shown in Figure 5.16. The surface wind patterns are shown on the globe's surface, and the wind patterns aloft are shown in the cross section on the right. **GEODISCOVERIES**

The most important features for understanding the wind pattern are Hadley cells, shown on the cross section. Recall that when air is heated, a convection loop can be formed. Since insolation is strongest when the Sun is directly overhead, the surface and atmosphere at

In the Hadley cell convection loop, air rises at the intertropical convergence zone and descends in the subtropical high-pressure cells.

the equator will be heated more strongly than other places on this featureless Earth. The result will be two convection loops, the **Hadley cells,** which form in the northern and southern hemispheres. In each Hadley cell, air rises over the equator and is drawn poleward by the pressure gradient. Of course, the air is turned by the Coriolis force and so heads westward as well as poleward. Eventually, it descends at about a 30° latitude, completing a convection loop. (The Hadley cell is named for George Hadley, who first proposed its existence in 1735.)

Since air rises at the equator, a zone of surface low pressure will result. This zone is known as the *equatorial trough.* As shown by the wind arrows, air in both hemispheres moves toward the equatorial trough, where it converges and moves aloft as part of the Hadley cell circulation. Convergence occurs in a narrow zone, named the **intertropical convergence zone (ITCZ).** Because the air is generally rising, winds in the ITCZ are light and variable. In earlier centuries mariners on sailing vessels referred to this region as the *doldrums,* where they were sometimes becalmed for days at a time.

On the poleward side of the Hadley cell circulation, air descends and surface pressures are high. This produces two **subtropical high-pressure belts,** each centered at about 30° latitude. Within the belts, two, three, or four very large and stable anticyclones are formed. At the centers of these anticylones, air descends, and winds are weak. Calm prevails as much as one-quarter of the time. Because of the high frequency of calms, mariners named this belt the *horse latitudes.* The name is said to have originated in colonial times when New England traders carried cargoes of horses to the West Indies. When their ships were stranded for long periods of time and the fresh water supplies ran low, the horses were thrown overboard.

Winds around the subtropical high-pressure centers are outspiraling and move toward equatorial as well as middle latitudes. The winds moving equatorward are the strong and dependable *trade winds.* North of the equator, these are from the northeast and are referred to as the *northeast trades.* To the south of the equator, they are from the southeast and are the *southeast trades.* Poleward of the subtropical highs, air spiraling outward produces southwesterly winds in the northern hemisphere and northwesterly winds in the southern hemisphere.

Between about 30° and 60° latitude, the pressure and wind pattern becomes more complex. This latitudinal belt is a zone of conflict between air bodies with different characteristics. Masses of cool, dry air move into the region, heading eastward and equatorward. These form *polar outbreaks.* The border of the polar outbreak is known as the *polar front.* (We'll have much more to say about air masses and the polar front in Chapter 6.) As a result of this activity, pressures and winds can be quite variable in the midlatitudes from day to day and week to week. On average, however, winds are more often from the west, so the region is said to have *prevailing westerlies.*

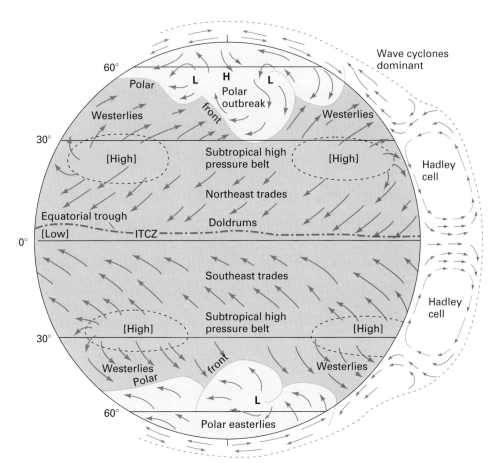

5.16 Global surface winds on an ideal Earth *This schematic diagram of global surface winds and pressures shows the features of an ideal Earth, without the disrupting effect of oceans and continents and the variation of the seasons. Surface winds are shown on the disk of the Earth, while the cross section at the right shows winds aloft.*

At the poles, the air is intensely cold. As a result, high pressure occurs. Outspiraling of winds around a polar anticyclone should create surface winds from a generally easterly direction, known specifically as *polar easterlies.* As we will see in discussing actual pressure and wind patterns, this situation exists only in the south polar region. In the north polar region, winds tend to have an eastward component, but there is too much variation in direction to consider polar easterlies as dominant winds there.

GLOBAL WIND AND PRESSURE PATTERNS

So far, we've discussed the wind pattern for a seasonless, featureless Earth. Let's turn now to actual global surface wind and pressure patterns. Here, we will use the global maps of wind and pressure for January and July, shown in Figure 5.17. The pressures and winds shown are averages over many years for all daily observations in either January or July. They are corrected for the elevation of the recording station, so that pressures are shown for sea level. Average barometric pressure is 1013 mb (29.2 in. Hg). Values greater than this

---

The intertropical convergence zone and subtropical high-pressure belts are prominent features of the global wind and pressure pattern. They shift northward and southward with the seasons.

---

value are "high" (red lines), while values lower are "low" (green lines). The maps make use of both Mercator and polar projections. As we discuss the various features of global pressure and winds, be sure to examine both sets of maps. GEODISCOVERIES

Subtropical High-Pressure Belts

The largest and most prominent features of the maps are the subtropical high-pressure belts, created by the Hadley cell circulation. The high-pressure belt in the southern hemisphere conforms well to the pattern of Figure 5.16. It has three large high-pressure cells, each developed over oceans, that persist year round. A fourth, weaker high-pressure cell forms over Australia in July, as the continent cools during the southern hemisphere winter.

In the northern hemisphere, the situation is somewhat different. The subtropical high-pressure belt shows two large anticyclones centered over oceans—the Hawaiian High in the Pacific and the Azores High in the Atlantic. From January to July, these intensify and move northward. From their positions off the east and west coasts of the United States, they

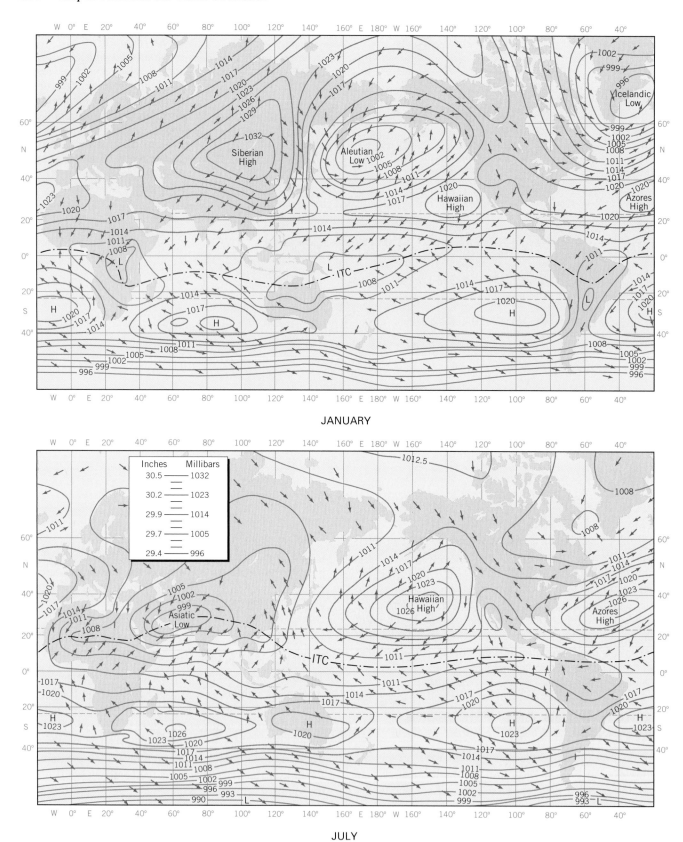

JANUARY

JULY

5.17 **Atmospheric pressure maps** *The maps show mean monthly atmospheric pressure and prevailing surface winds for January and July. Pressure units are millibars reduced to sea level. Many of the wind arrows are inferred from isobars. (Data compiled by John E. Oliver.)*

JANUARY

JULY

Mb	948	952	956	960	964	968	972	976	980	984	988	992	996	1000	1004	1008
In	28.0	28.1	28.2	28.3	28.4	28.5	28.6	28.7	28.8	28.9	29.0	29.1	29.2	29.3	29.4	29.5	29.6	29.7	29.8

Mb	996	1000	1004	1008	1012	1016	1020	1024	1028	1032	1036	1040	1044	1048	1052	1056
In	29.4	29.5	29.6	29.7	29.8	29.9	30.0	30.1	30.2	30.3	30.4	30.5	30.6	30.7	30.8	30.9	31.0	31.1	31.2

5.17 Continued

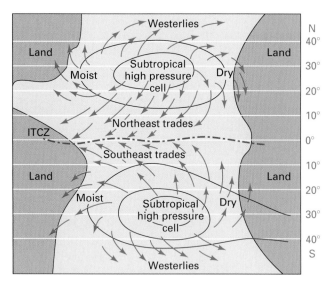

5.18 Subtropical high-pressure cells *Over the oceans, surface winds spiral outward from the subtropical high-pressure cells, feeding the trades and the westerlies. On the eastern side of the cells, air subsides more strongly, producing dry winds. On the western sides, subsidence is not as strong, and a long passage over oceans brings warm, moist air to the continents.*

have a dominant influence on summer weather in North America.

Figure 5.18 is a schematic map of two large high-pressure cells bounded by continents, showing the features of circulation around the cells. As we noted earlier, the outspiraling circulation produces the easterly trade winds in the tropical and equatorial zones, and the westerlies in the subtropical zones and poleward. In these high-pressure cells, air on the east side subsides more intensely. This means that winds spiraling outward on the eastern side are drier. On the west side, subsidence is less strong. In addition, these winds travel long distances across warm, tropical ocean surfaces before reaching land, and so they pick up moisture and heat on their journey.

When the Hawaiian and Azores Highs intensify and move northward in the summer, our east and west coasts feel their effects. On the west coast, dry subsiding air from the Hawaiian High dominates, so fair weather and rainless conditions prevail. On the east coast, warm, moist air from the Azores High flows across the continent from the southeast, producing generally hot, humid weather for the central and eastern United States. In winter, these two anticyclones weaken and move to the south—leaving our weather to the mercy of colder winds and air masses from the north and west.

The ITCZ and the Monsoon Circulation

Recall from Chapter 2 that insolation is most intense when the Sun is directly overhead. Remember, too, that the latitude at which the Sun is directly overhead changes with the seasons, migrating between the tropics of cancer and capricorn. Since the Hadley cell circulation is driven by this heat-

ing, we can expect that the elements of the Hadley cell circulation—ITCZ and subtropical high-pressure belts—will also shift with the seasons. We have already noted this shift for the Hawaiian and Azores Highs, which intensify and move northward as July approaches.

In general, the ITCZ also shifts with the seasons, as we might expect. The shift is moderate in the western hemisphere. The ITCZ moves a few degrees north from January to July over the oceans. In South America, the ITCZ lies across the Amazon in January and swings northward by about 20° to the northern part of the continent. But what happens in Africa and Asia? By comparing the two Mercator maps of Figure 5.17, you can see that there is a huge shift indeed! In January, the ITCZ runs south across eastern Africa and crosses the Indian Ocean to northern Australia at a latitude of about 15° S. In July, it swings north across Africa along the southern side of the Sahara, and then it rises further north to lie along the south rim of the Himalayas, in India, at a latitude of about 25° N. This is a shift of about 40 degrees of latitude!

Why does such a large shift occur in Asia? To answer this question, look at the pressure and wind pattern over Asia. In January, an intense high-pressure system, the Siberian High, is found there. Recall from Chapter 3 that, in winter,

5.19 Monsoon wind patterns *The Asiatic monsoon winds alternate in direction from January to July, responding to reversals of barometric pressure over the large continent.*

JANUARY

JULY

temperatures in northern Asia are very cold, so this high pressure is to be expected. In July, this high-pressure center is absent, replaced instead by a low centered over the Middle Eastern desert region. The Asiatic low is produced by the intense summer heating of the landscape.

The movement of the ITCZ and the change in the pressure pattern with the seasons create a reversing wind pattern in Asia known as the **monsoon** (Figure 5.19). In the winter, there is a strong outflow of dry, continental air from the north across China, Southeast Asia, India, and the Middle East. During this *winter monsoon,* dry conditions prevail. In the summer, warm, humid air from the Indian Ocean and the southwestern Pacific moves northward and north-westward into Asia, passing over India, Indochina, and China. This airflow is known as the *summer monsoon* and is accompanied by heavy rainfall in southeastern Asia.

> **The Asian monsoon wind pattern consists of a cool, dry air flow from the northeast during the low-Sun season, and a warm, moist airflow from the southwest during the high-Sun season.**

North America does not have the remarkable extremes of monsoon winds experienced in Asia. Even so, in summer there is a prevailing tendency for warm, moist air originating in the Gulf of Mexico to move northward across the central and eastern part of the United States. At times, moist air from the Gulf of California also invades the desert Southwest, causing widespread scattered thundershowers and creating the "Arizona monsoon." In winter, the airflow across North America generally reverses, and dry, continental air from Canada moves south and eastward.

Wind and Pressure Features of Higher Latitudes

The northern and southern hemispheres are quite different in their geography. As you can easily see from Figure 5.20, the northern hemisphere has two large continental masses, separated by oceans. An ocean is also at the pole. In the southern hemisphere, we find a large ocean with a cold, glacier-covered continent at the center. These differing land–water patterns strongly influence the development of high- and low-pressure centers with the seasons.

Let's examine the northern hemisphere first (Figure 5.17). Recall from Chapter 3 that continents will be cold in winter and warm in summer, as compared to oceans at the same latitude. But we know from our analysis earlier in the chapter that cold air will be associated with surface high pressure and warm air with surface low pressure. Thus, continents will show high pressure in winter and low pressure in summer.

This pattern is very clear in the northern hemisphere polar maps of Figure 5.17. In winter, the strong Siberian High is found in Asia, and a weaker cousin, the Canadian High, is found in North America. From these high-pressure centers, air spirals outward, bringing cold air to the south. Over the oceans, two large centers of low pressure are striking—the Icelandic Low and the Aleutian Low. These two low-pressure centers are not actually large stable features that we would expect to find on every daily world weather map in January. Rather, they are regions of average low pressure where winter storm systems are spawned.

In summer, the pattern reverses. The continents show generally low surface pressure, while high pressure builds over the oceans. This pattern is easily seen on the northern hemisphere July polar map. The Asiatic Low is strong and intense. Inspiraling winds, forming part of the monsoon cycle, bring warm, moist Indian Ocean air over India and Southeast Asia. A lesser low forms over the deserts of the southwestern United States and northwestern Mexico. The two subtropical highs, the Hawaiian Highs and the Azores Highs, strengthen and dominate the Atlantic and Pacific Ocean regions. Outspiraling winds, following the pattern of Figure 5.18, keep the west coasts of North America and Europe warm and dry, and the east coasts of North America and Asia warm and moist (see Figure 5.17 also).

The higher latitudes of the southern hemisphere present a polar continent surrounded by a large ocean. Since Antarctica is covered by a glacial ice sheet and is cold at all times, a permanent anticyclone, the South Polar High, is centered there. Easterly winds spiral outward from the high-pressure center. Surrounding the high is a band of deep low pressure, with strong, inward-spiraling westerly winds. As early mariners sailed southward, they encountered this band, in which wind strength intensifies toward the pole. Because of the strong prevailing westerlies, they named these southern latitudes the "roaring forties," "flying fifties," and "screaming sixties." *GEODISCOVERIES*

5.20 *Land–ocean contrasts by hemisphere* The *northern and southern hemispheres are quite different in their distributions of lands and oceans. One pole bears a deep ocean, and the other a great continental land mass.*

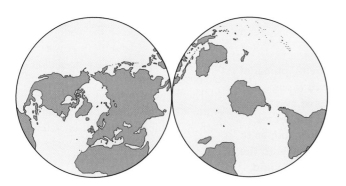

WINDS ALOFT

The surface wind systems we have examined describe airflows at or near the surface. How does air move at the higher levels of the troposphere? Just as we saw for air near the surface, air aloft will move in response to pressure gradients and will be influenced by the Coriolis effect.

5.21 Upper-air pressure gradient *Because the atmosphere is warmer near the equator than at the poles, a pressure gradient force acts to push air poleward. The gradient force increases with altitude.*

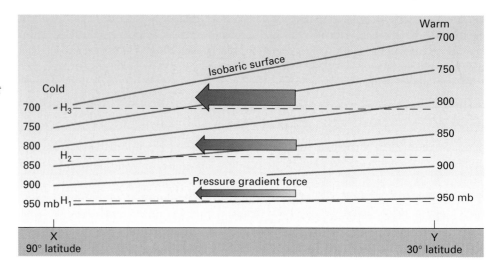

How do pressure gradients arise at upper levels? Before answering this question, we need to restate a simple principle that appeared in our discussion of the convection loop and is shown in Figure 5.8*b*. This principle is that pressure decreases less rapidly with height in warmer air than in colder air. The principle is derived from fundamental physical laws describing the temperature, density, and pressure relationships within a thick layer of gas. Also recall that the Earth's insolation is greatest near the equator and least near the poles. Thus, there will be a temperature gradient from the equator to the poles.

Figure 5.21 shows a diagram of this situation—a simple generalized cross section of the atmosphere from the pole on the left (*X*) to about 30° latitude on the right (*Y*). The isobaric surfaces slope downward from the low latitudes to the pole. This creates a pressure gradient force. For example, at height H_1, the pressure at *X* is about 940 mb, while at *Y* it is about 955 mb, creating a pressure gradient of 15 mb. But note how the pressure difference becomes greater with elevation. At H_2, the polar pressure is about 825, while the low-latitude pressure is about 890. Here the pressure gradient is 65 mb. At H_3, the gradient is about 100 mb. Of course, these values are just examples, but they show that the pressure gradient force increases with altitude, bringing strong winds at high altitudes.

The Geostrophic Wind

Given that a pressure gradient force exists pushing poleward, how will that gradient produce wind and what will be the wind direction? Recall that any wind motion will be subjected to the Coriolis force, which will turn it rightward in the northern hemisphere and leftward in the southern hemisphere. Thus, poleward air motion will be toward the east, creating west winds in both hemispheres.

Figure 5.22*a* shows how the Coriolis force acts on an upper air parcel.

At high altitudes, a moving parcel of air is subjected to two forces—a pressure gradient force and the Coriolis force. When the forces balance, air moves at right angles to the pressure gradient, parallel to the isobars, as the geostrophic wind.

Compared to Figure 5.14, which diagrams the forces on a parcel moving close to the surface, we see that the upper air parcel moves without a friction force, since it is so far from the surface. Thus, there are only two forces on the parcel—the pressure gradient force and the Coriolis force. Imagine a still parcel of air that begins to move poleward in response to the pressure gradient force (*b*). At first it travels in the direction of the pressure gradient, but as it accelerates, the Coriolis force pulls it increasingly toward the right. As its velocity increases, the parcel turns increasingly rightward until the Coriolis force just balances the gradient force. At that point, the sum of forces on the parcel is zero, so its speed and direction remain constant. We use the term *geostrophic wind* to identify this type of airflow. It occurs at upper levels in the atmosphere and is parallel to the isobars.

Figure 5.23 shows an upper-air map for North America on a late June day. The contours show the height of the 500-mb pressure level, and from our analysis of Figure 5.21 above, we can easily see that the 500-mb surface will dip down toward the ground where the upper-air pressure is lower and the air column is cooler. Similarly, the surface will rise up when the upper-air pressure is higher and the air column is warmer. So on this map, low height is low pressure, and high height is high pressure.

The map shows a large, upper-air low centered over the Great Lakes. Since low pressure aloft indicates cold air, this is a mass of cold Canadian air that has moved southward to dominate the eastern part of the continent. A large, but weaker, upper-air high is centered over the southwestern desert. This will be a mass of warm air, heated by intense surface insolation on the desert below.

The figure also shows upper-air wind strengths and directions using the "arrow" symbol employed on scientific weather maps. Note that the winds generally follow the contours well, showing geostrophic flow. They are strongest

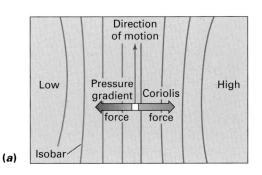

5.22 Geostrophic wind (a) At upper levels in the atmosphere, a parcel of air is subjected to a pressure gradient force and a Coriolis force. (b) As a parcel of air moves in response to a pressure gradient, it is turned progressively sidewards until the gradient force and Coriolis force balance, producing the geostrophic wind.

where the contours are closest together because there the pressure gradient force will be strongest. Note also that the winds around the upper-air low tend to spiral inward and converge on the low. The converging winds descend to the surface, where surface high pressure is present, and then spiral outward. Similarly, the winds around the upper-air high spiral outward and diverge away from the center. At the surface, low pressure causes winds to spiral inward and upward.

Global Circulation at Upper Levels

Figure 5.24 sketches the general pattern of airflows at higher levels in the troposphere. The pattern has four major features—weak equatorial easterlies, tropical high-pressure belts, upper-air westerlies, and a polar low.

We've noted how the general temperature gradient from the tropics to the poles creates a pressure gradient force that generates westerly winds in the

The general pattern of winds aloft is a band of equatorial easterly winds, flanked by tropical high-pressure belts, and a zone of westerly winds to poleward.

upper atmosphere. These *upper-air westerlies* blow in a complete circuit about the Earth, from about 25° lat. almost to the poles. They often exhibit large undulations from their westerly track, as we will describe shortly. At high latitudes the westerlies form a huge circumpolar spiral, circling a great polar low-pressure center. Toward lower latitudes, atmospheric pressure rises steadily, forming a *tropical high-pressure belt* at 15° to 20° N and S lat. This is the high-altitude part of the surface subtropical high-pressure belt, but it is shifted somewhat equatorward. Between the high-pressure ridges is a zone of lower pressure in which the winds are light but generally easterly. These winds are called the *equatorial easterlies*.

Thus, the overall picture of upper-air wind patterns is really quite simple—a band of weak easterly winds in the equatorial zone, belts of high pressure near the tropics of cancer and capricorn, and westerly winds, with some variation in direction, spiraling around polar lows.

5.23 Upper-air wind map *(a) An upper-air map for a day in late June. Lines are height contours for the 500-mb surface. (b) Explanation of wind arrows. 1 knot (nautical mile per hour) = 0.514 meters per second.*

m	ft	m	ft
5450	17,881		
5500	18,045	5750	18,865
5550	18,209	5800	19,029
5600	18,373	5850	19,193
5650	18,537	5900	19,357
5700	18,701	5950	19,521

(a)

(b)

Whole feather = 10 knots (5.2 m/s)
Half feather = 5 knots (2.6 m/s)
Total   = 15 knots (7.7 m/s)

Flag = 50 knots (26 m/s)

Direction of air flow

Shaft

SURFACE WEATHER MAPS

65 knots (33 m/s)   N

280° Bearing

5 (2.6 m/s)

45 (23 m/s)

80 (41 m/s)

90 (46 m/s)

125 (64 m/s)

UPPER AIR CHARTS

Rossby Waves, Jet Streams, and the Polar Front

The smooth westward flow of the upper-air westerlies frequently forms undulations, called **Rossby waves.** Figure 5.25 shows how these waves develop and grow in the northern hemisphere. The waves arise in a zone of contact between cold polar air and warm tropical air, called the *polar front.* The number of Rossby waves ranges from three to seven.

For a period of several days or weeks, the flow may be fairly smooth. Then, an undulation develops, and warm air pushes poleward while a tongue of cold air is brought to the

5.24 *Global upper-level winds* In this generalized plan of global winds high in the troposphere, strong west winds dominate the mid- and high-latitude circulation. They often sweep to the north or south around centers of high and low pressure aloft. In the equatorial region, a weak easterly wind pattern prevails. (Copyright © A. N. Strahler.)

south. Eventually, the tongue is pinched off, leaving a pool of cold air at a latitude far south of its original location. This cold pool may persist for some days or weeks, slowly warming with time. The low in Figure 5.23 is an example of such an air mass. Because of its cold center, it will show low pressure and convergence aloft. Air will descend in its core, and diverge at the surface, creating surface high pressure. Similarly, a warm air pool will contain rising air, with convergence and low pressure at the surface and divergence and high pressure aloft.

Because the Rossby wave circulation brings warm air poleward and cold air equatorward, it is a primary mechanism of poleward heat transport. It is also the reason weather in the midlatitudes is often so variable, as pools of warm, moist air and cold, dry air alternately invade midlatitude land masses.

Jet streams are important features of upper-air circulation. They are narrow zones at a high altitude in which wind streams reach great speeds. They occur where atmospheric pressure gradients are strong. Along a jet stream, pulse-like movements of air follow broadly curving tracks (Figure 5.26). The greatest wind speeds occur in the center of the jet stream, with velocities decreasing away from the center. There are three kinds of jet streams. Two are westerly wind streams, and the third is a weaker jet with easterly winds that develop in Asia as part of the summer monsoon circulation. These are shown in Figure 5.27.

The jet stream begins to undulate.

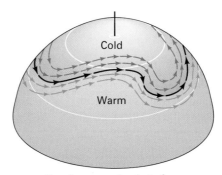

Rossby waves begin to form.

5.25 *Rossby waves*
Rossby waves form in the westerlies of the northern hemisphere, marking the boundary between cold polar air and warm tropical air. This cycle shows the formation of large waves that are pinched off, leaving pools of polar air at lower latitudes. (Copyright © A. N. Strahler.)

Waves are strongly developed. The cold air occupies troughs of low pressure.

When the waves are pinched off, they form cyclones of cold air.

5.26   Polar jet stream *The polar jet stream is shown on this map by lines of equal wind speed. (National Weather Service.)*

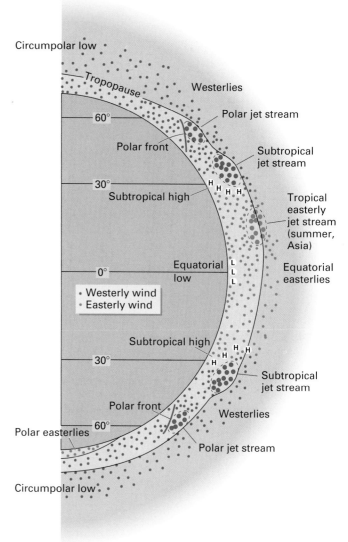

5.27   Upper-level circulation cross section *A schematic diagram of wind directions and jet streams along an average meridian from pole to pole. The four polar and subtropical jets are westerly in direction, in contrast to the single tropical easterly jet. (Copyright © A. N. Strahler.)*

The most poleward type of jet stream is located along the polar front (Figure 5.16). It is designated the *polar-front jet stream* (or simply, the "polar jet"). Located generally between 35° and 65° latitude, it is present in both hemispheres. The polar jet follows the edges of Rossby waves, since these mark the boundary between cold polar air and warm subtropical air. As we know from studying how atmospheric temperature affects pressure aloft, a strong temperature gradient will create a strong pressure gradient, which powers the jet stream.

The polar jet is typically found at altitudes of 10 to 12 km (about 30,000 to 40,000 ft), and wind speeds in the jet range from 75 to as much as 125 m/s (about 170 to 280 mi/hr). In the northern hemisphere, traveling aircraft often use the polar jet to increase ground speed when flying eastward. In the westward direction, flight paths are chosen to avoid the strong winds of the polar jet.

A second type of jet stream forms in the subtropical latitude zone—the *subtropical jet stream.* It occupies a position at the tropopause just above the subtropical high-pressure cells in northern and southern hemispheres (Figure 5.28). Here, westerly wind speeds reach maximum values of 100 to 110 m/s (about 215 to 240 mph). This jet is associated with the increase in velocity, induced by conservation of angular momentum, that occurs as an air parcel moves poleward from the equator. Rotating at the equator along with the Earth, an air parcel has a quantity of angular momentum based on the product of the turning rate of the Earth's rotation and the distance between the parcel and the axis of rotation. As it moves poleward, it gets closer to the Earth's axis of rotation, and by conservation of angular momentum, its turning rate must increase. That translates into an increase in the eastward velocity of the air parcel. This is the same principle that ice skaters utilize when drawing their arms and legs

---

Jet streams are streams of fast-moving air aloft that occur where atmospheric temperature gradients are strong. Each hemisphere normally exhibits westerly polar and subtropical jet streams. An easterly jet occurs in summer over Asia and Africa.

---

inward to increase the speed of a spin. The result is a river of swiftly moving air that piles up as it descends over the subtropical highs. Cloud bands of the subtropical jet stream are pictured in Figure 5.28.

A third type of jet stream is found at even lower latitudes. Known as the *tropical easterly jet stream,* it runs from east to west—opposite in direction to that of the polar-front and subtropical jet streams. The tropical easterly jet occurs only in the summer season and is limited to a northern hemisphere location over Southeast Asia, India, and Africa.

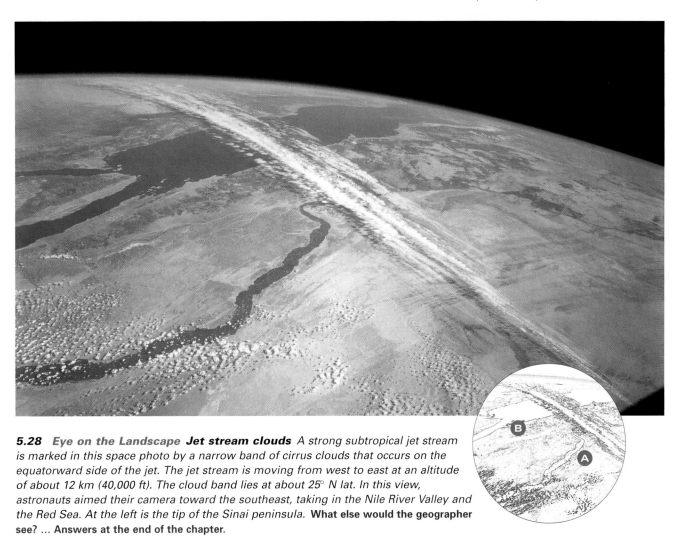

5.28 *Eye on the Landscape* **Jet stream clouds** *A strong subtropical jet stream is marked in this space photo by a narrow band of cirrus clouds that occurs on the equatorward side of the jet. The jet stream is moving from west to east at an altitude of about 12 km (40,000 ft). The cloud band lies at about 25° N lat. In this view, astronauts aimed their camera toward the southeast, taking in the Nile River Valley and the Red Sea. At the left is the tip of the Sinai peninsula.* **What else would the geographer see? ... Answers at the end of the chapter.**

TEMPERATURE LAYERS OF THE OCEAN

As with the atmosphere, the ocean has a layered structure. Ocean layers are recognized in terms of temperature, with temperatures generally highest at the sea surface and decreasing with depth. This trend is not surprising, since the sources of heat that warm the ocean are solar insolation and heat supplied by the overlying atmosphere, both of which act to warm the water at or near the surface.

The ocean's layered temperature structure is shown in Figure 5.29. At low latitudes throughout the year and in middle latitudes in the summer, a warm surface layer develops. Here wave action mixes heated surface water with the water below it to form a warm layer that may be as thick as 500 m (about 1600 ft) with a temperature of 20° to 25°C (68° to 77°F) in oceans of the equatorial belt. Below the warm layer, temperatures drop rapidly in a zone known as the *thermocline*. Below the thermocline is a layer of very cold water extending to the deep ocean floor. Temperatures near the base of the deep layer range from 0°C to 5°C (32° to 41°F).

In arctic and antarctic regions, the warm layer and thermocline are absent, as Figure 5.29 shows.

5.29 *Ocean temperature structure* A schematic north-south cross section of the world ocean shows that the warm surface water layer disappears in arctic and antarctic latitudes, where very cold water lies at the surface. The thickness of the warm layer and the thermocline is greatly exaggerated.

OCEAN CURRENTS

Just as there is a circulation pattern to the atmosphere, so there is a circulation pattern to the oceans. An *ocean current* is any persistent, dominantly horizontal flow of ocean water. Current systems act to exchange heat between low and high latitudes and are essential in sustaining the global energy balance. We can recognize both surface currents and deep currents. Surface currents are driven by prevailing winds. Deep currents are powered by changes in temperature and density occurring in surface waters that cause them to sink.

Surface Currents

The patterns of surface ocean currents are strongly related to prevailing surface winds. Energy is transferred from wind to water by the friction of the air blowing over the water surface. Because of the Coriolis effect, the actual direction of water drift is deflected about 45° from the direction of the driving wind.

Because ocean currents move warm waters poleward and cold waters toward the equator, they are important regulators of air temperatures. Warm surface currents keep winter temperatures in the British Isles from falling much below freezing in winter. Cold surface currents keep weather on the California coast cool, even in the height of summer.

Figure 5.30 shows the general features of the circulation of an idealized ocean between two continental masses. The circulation includes two large circular movements, called **gyres,** that are centered at latitudes of 20° to 30°. These gyres track the movements of air around the subtropical high-pressure cells. An *equatorial current* with westward flow marks the belt of the trade winds. Although the trades blow to the southwest and northwest, at an angle across the parallels of latitude, the water movement follows the parallels. The equatorial currents are separated by an equatorial countercurrent. A slow, eastward movement of water over the zone of the westerlies is named the *west-wind drift*. It covers a broad belt between lat. 35° and 45° in the northern hemisphere and between lat. 30° and 60° in the southern hemisphere.

Figure 5.31 presents a world map that shows surface ocean currents for the month of January. In the equatorial region, ocean currents flow westward, pushed by northeast and southeast trade winds. As these quatorial currents approach land, they are turned poleward along the west sides of the oceans, forming warm currents paralleling the coast. Examples are the Gulf Stream of eastern North America and the Kuroshio Current of Japan.

In the zone of westerly winds, a slow eastward motion of water is named the *west-wind drift*. As west-wind drift waters approach the western sides of the continents, they are deflected equatorward along the coast. The equatorward flows are cool currents. They are often accompanied by *upwelling* along continental margins. In this process, colder water from greater depths rises to the surface. Examples of

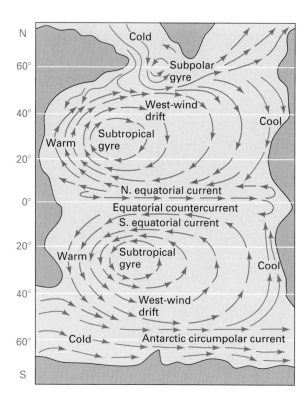

5.30   Ocean current gyres *Two great gyres, one in each hemisphere, dominate the circulation of shallow ocean waters.*

cool currents with upwelling are the Humboldt (or Peru) Current, off the coast of Chile and Peru, the Benguela Current, off the coast of southern Africa, and the California Current.

West-wind drift water also moves poleward to join arctic and antarctic circulations. In the northeastern Atlantic Ocean, the west-wind drift forms a relatively warm current. This is the North Atlantic Drift, which spreads around the British Isles, into the North Sea, and along the Norwegian coast. The Russian port of Murmansk, on the arctic circle, remains ice-free year round because of this warm drift current.

In the northern hemisphere, where the polar sea is largely landlocked, cold currents flow equatorward along the east sides of continents. Two examples are the Kamchatka Current, which flows southward along the Asian coast across from Alaska, and the Labrador Current, which flows between Labrador and Greenland to reach the coasts of Newfoundland, Nova Scotia, and New England.

The strong west winds around Antarctica produce an antarctic circumpolar current of cold water, shown in Figure 5.32. Some of this flow branches equatorward along the west coast of South America, adding to the Humboldt Current (Figure 5.31). Two small gyres lie close to the antarctic continent.

Global surface ocean currents are dominated by huge, wind-driven circular gyres centered near the subtropical high-pressure cells. Upwelling of cold bottom water often occurs on the west coasts of continents in the subtropical zones.

5.31 January ocean currents Surface drifts and currents of the oceans in January. (Based on data from U.S. Navy Oceanographic Office. Redrawn and revised by A. N. Strahler.)

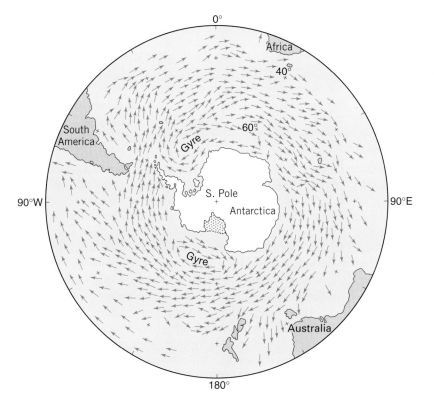

5.32 Currents of the Southern Ocean
Surface water of the Southern Ocean flows continuously from west to east around Antarctica as a broad circumpolar current. This map is greatly simplified from a computer-generated model. (Based on NOAA data as presented by D. Olbers and M. Wenzel in EOS, American Geophysical Union, vol. 71, no. 1.)

5.33 *Eye on the Landscape* **Sea-surface temperatures in the western Atlantic** *A satellite image showing sea-surface temperature for a week in April from data acquired by the NOAA-7 orbiting satellite. Cold water appears in green and blue tones; warm water in red and yellow tones. The image shows the Gulf Stream (GS) and its interactions with cold water of the Continental Slope (SW), brought down by the Labrador Current, and the warm water of the Sargasso Sea (SS). Other features include a meander (M), a warm-core ring (WR), and a cold-core ring (CR). The image was made from data collected by the NOAA-7 orbiting satellite. Data of 35 satellite passes during a week in April were processed together to remove the effects of clouds on single images. (NASA. Courtesy of Otis B. Brown, Robert Evans, and M. Carle, University of Miami, Rosenstiel School of Marine and Atmospheric Science, Florida.)* **What else would the geographer see? ... Answers at the end of the chapter.**

Figure 5.33 shows a satellite image of ocean temperature along the east coast of North America for a week in April. The Gulf Stream stands out as a tongue of red and yellow (warm) color, moving along the coast and heading off from North Carolina in a northeasterly direction. Cooler water from the Labrador Current, in green and blue tones, hugs the northern Atlantic coast. This current heads south and then turns to follow the Gulf Stream to the northeast. Instead of mixing, the two flows remain quite distinct. The boundary between them shows a wavelike flow, much like Rossby waves in the atmosphere. Warm and cold bodies of water are cut off to float freely, forming warm-core and cold-core rings. *GEODISCOVERIES*

Another phenomenon of ocean surface currents is *El Niño.* During an El Niño event, Pacific surface currents shift into an unusual pattern. Pacific upwelling along the Peruvian coast ceases, trade winds weaken, and a weak equatorial eastward current develops. Global patterns of precipitation also change during El Niño events, bringing floods to some regions and droughts to others. In contrast to El Niño is *La Niña,* in which normal Peruvian coastal upwelling is enhanced, trade winds strengthen, and cool water is carried far westward in an equatorial plume.

Figure 5.34 shows two satellite images of sea-surface temperature observed during El Niño and La Niña years. During La Niña conditions (left), the cool water of the Humboldt Current, moving northward along the coast of Chile and Peru, is carried westward into the Pacific in a long plume. Much cold, upwelling water (dark green) is brought to the surface along the Peruvian coast. During an El Niño year (right), the eastward motion of warm water holds the Humboldt Current in check. Some slight upwelling occurs (yellow color), but the amount is greatly reduced compared to normal or La Niña years. *Eye on Global Change 5.2 ● El Niño* provides a fuller description of the El Niño–La Niña phenomenon. *GEODISCOVERIES*

Deep Currents and Thermohaline Circulation

Deep currents move ocean waters in a slow circuit across the floors of the world's oceans. They are generated when surface waters become more dense and slowly sink downward. Coupled with these deep currents are very broad and slow surface currents on which the more rapid surface currents, described above, are superimposed. Figure 5.35 diagrams this slow flow pattern, which links all of the world's oceans. It is referred to as *thermohaline circulation,* since it depends on the temperature and salinity of North Atlantic Ocean waters. These factors act through changes to the density of sea water. Since warm water is slightly less dense than cold water, warm surface waters do not normally mix with cold bottom waters. Follow the figure carefully as we explain the process.

Beginning at *A* in the figure, warm Atlantic surface water slowly moves northward through the equatorial and tropical zones. As this surface layer is warmed, evaporation occurs and it becomes saltier. This increases the density of the layer slightly. As the water reaches the northern boundaries of the Atlantic, it loses heat to the atmosphere, and so it becomes

(a)

(b)

5.34   La Niña and El Niño sea-surface temperatures *This striking image shows sea-surface temperatures in the eastern tropical Pacific during (a) La Niña and (b) El Niño years as measured by the NOAA-7 satellite. Green tones indicate cooler temperatures, while red tones indicate warmer temperatures. (Otis B. Brown/Univ. of Miami and Gene Carl Feldman, NASA/GSFC.)*

colder and still more dense. Eventually, the surface layer becomes dense enough to sink (*B*). This sinking occurs all along the northern boundary of the North Atlantic.

Carried along the bottom, the cold, dense water eventually reaches the Southern Ocean (*C*). Here, upwelling and mixing occur, and the deep waters are carried into the surface waters of the Indian and southern Pacific oceans (*D, E*). A coupled circulation loop moves surface water from the Pacific through the Indonesian seas (*F*) and into the Indian Ocean (*G*). Most of this flow connects with a westward flow link along the Southern Ocean, poleward of the African and Australian continents. However, some of the flow escapes to the west around the southern tip of Africa to enter the South Atlantic (*H*), completing the entire circuit.

Thermohaline circulation plays an important role in the carbon cycle by moving CO_2-rich surface waters into the ocean depths. As noted in Chapter 3 in *Eye on the Environment 3.1 • Carbon Dioxide–On the Increase*, deep ocean circulation provides a

conveyor belt for storage and release of CO_2 in a cycle of about 1500 years' duration. This allows the ocean to moderate rapid changes in atmospheric CO_2 concentration, such as those produced by human activity through fossil fuel burning.

Some scientists have observed that thermohaline circulation could be slowed or stopped by inputs of fresh water into the North Atlantic. Such freshwater inputs could come from the sudden drainage of large lakes formed by melting ice at the close of the last Ice Age. The fresh water would decrease the density of the ocean water, keeping the water from becoming dense enough to sink. Without sinking, circulation would stop. In turn, this would interrupt a major flow pathway for the transfer of heat from equatorial regions to the northern midlatitudes. This mechanism could result in relatively rapid climatic change and is one explanation for the periodic cycles of warm and cold temperatures experienced since the melting of continental ice sheets about 12,000 years ago.

> **In the thermohaline oceanic circulation, dense, cold, salty surface water sinks along the northern boundary of the Atlantic Ocean, generating very slow bottom currents that in turn cause slow upwelling as far away as the Indian Ocean, western Pacific Ocean, and the coast of Antarctica.**

Eye on Global Change | 5.2

El Niño

At intervals of about three to eight years, a remarkable disturbance of ocean and atmosphere occurs. It begins in the eastern Pacific Ocean and spreads its effects widely over the globe. This disturbance lasts more than a year, bringing droughts, heavy rainfalls, severe spells of heat and cold, or a high incidence of cyclonic storms to various parts of the Pacific and its eastern coasts. This phenomenon is called *El Niño*. The expression comes from Peruvian fishermen, who refer to the *Corriente del Niño* or the "Current of the Christ Child," in describing an invasion of warm surface water that occurs

once every few years around Christmas time and greatly depletes their catch of fish. El Niño occurs at irregular intervals and with varying degrees of intensity. Notable El Niño events occurred in 1891, 1925, 1940–1941, 1965, 1972–1973, 1982–1983, 1989–1990, 1991–1992, 1994–1995, and 1997–1998.

Normally, the cool Humboldt (Peru) Current flows northward off the South American coast, and then at about the equator it turns westward across the Pacific as the south equatorial current (see Figure 5.31). The Humboldt Current is fed by upwelling of cold, deep

water, bringing with it nutrients that serve as food for marine life. With the onset of El Niño, upwelling ceases, the cool water is replaced by warm, sterile water from the west, and the abundant marine life disappears.

In an El Niño year, a major change in barometric pressure occurs across the entire stretch of the equatorial zone as far west as southeastern Asia. Normally, low pressure prevails over northern Australia, the East Indies, and New Guinea, where the largest and warmest body of ocean water can be found (see figure below, left). Abundant rainfall occurs in this

El Niño maps Maps of pressures in the tropical Pacific and eastern Indian Ocean in November during normal and El Niño years. In a normal year (a), low pressure dominates in Malaysia and northern Australia. In an El Niño year (b), low pressure moves eastward to the central part of the western Pacific, and sea-surface temperatures become warmer in the eastern Central Pacific. (Copyright © A. N. Strahler.)

(a)

(b)

area during December, which is the high-Sun period in the southern hemisphere.

During an El Niño event, the low-pressure system is replaced by weak high pressure and drought ensues. Pressures drop in the equatorial zone of the eastern Pacific, strengthening the equatorial trough. Rainfall is abundant in this new low-pressure region (right part of figure). The shift in barometric pressure patterns is known as the *Southern Oscillation.*

Surface winds and currents also change with this change in pressure. During normal conditions, the strong, prevailing trade winds blow westward, causing very warm ocean water to move to the western Pacific and to "pile up" near the western equatorial low. This westward motion causes the normal upwelling along the South American coast, as bottom water is carried up to replace the water dragged to the west. During an El Niño event, the easterly trade winds die with the change in atmospheric pressure. A weak westerly wind flow sometimes occurs, completely reversing the normal wind direction. Without the pressure of the trade winds to hold them back, warm waters surge eastward. Sea-surface temperatures and actual sea levels rise off the tropical western coasts of the Americas.

The major change in sea-surface temperatures that accompanies an El Niño can also shift weather patterns across large regions of the globe. The winter of 1997–1998 experienced one of the strongest El Niños of the century. Torrential rains drenched Peruvian and Ecuadorean coast ranges, produc-ing mudflows, debris avalanches, and extensive river flooding. Large portions of Australia and the East Indies went rainless for months, and forest fires burned out of control in Sumatra, Borneo, and Malaysia. In east Africa, Kenya experienced rainfall 100 cm (80 in.) above normal. In North America, a series of powerful winter storms lashed the Pacific coast, doing extensive damage in California. Monstrous tornadoes ripped through Florida, killing over 40 people and destroying more than 800 homes. Meanwhile, mild winter conditions east of the Rockies saved vast amounts of fossil fuel while generating an ice storm that left 4 million people without power in Quebec and the northeastern United States. All told, the deranged weather of the El Niño of 1997–1998 did property damage estimated at $33 billion and killed an estimated 2100 people.

A somewhat rarer phenomenon, also capable of altering global weather patterns, is La Niña (the girl child), a condition roughly opposite to El Niño. During a La Niña period, sea-surface temperatures in the central and western Pacific Ocean fall to lower than average levels. This happens because the South Pacific subtropical high becomes very strongly developed during the high-Sun season. The result is abnormally strong southeast trade winds. The force of these winds drags a more-than-normal amount of warm surface water westward, which enhances upwelling along western continental coasts. Figure 5.34 provides a pair of satellite images of sea-surface temperature that contrast El Niño and La Niña conditions.

The El Niño of 1997–1998 was rapidly followed by the La Niña of 1998–1999. The result was heavier monsoon rains in India and more rain in Australia. In North America, winter conditions were colder than normal in the Northwest and upper Midwest. The eastern Atlantic region endured drought through spring and early summer. The hurricane season of 1998, spawning the monster storm Hurricane Mitch, was the deadliest in the past two centuries.

What causes the El Niño/Southern Oscillation phenomenon (also known as ENSO)? At first, meteorologists tried to explain the changes in winds and pressures by the changes in ocean surface currents, while oceanographers tried to explain the changes in surface currents by changes in winds and pressures. It wasn't until the 1950s that scientists realized that the two phenomena were linked. One view is that the cycle is simply a natural oscillation caused by the way in which the atmosphere and oceans are coupled by energy exchange. In any event, scientists now have good computer models that accept sea-surface temperature along with air temperature and pressure data and can predict El Niño events reasonably well some months before they occur.

El Niño and its alter ego La Niña show how dynamic our planet really is. As a grand-scale, global phenomenon, El Niño–La Niña shows how the circulation patterns of the ocean and atmosphere are linked and interact to provide teleconnections capable of producing extreme events affecting millions of people throughout the world.

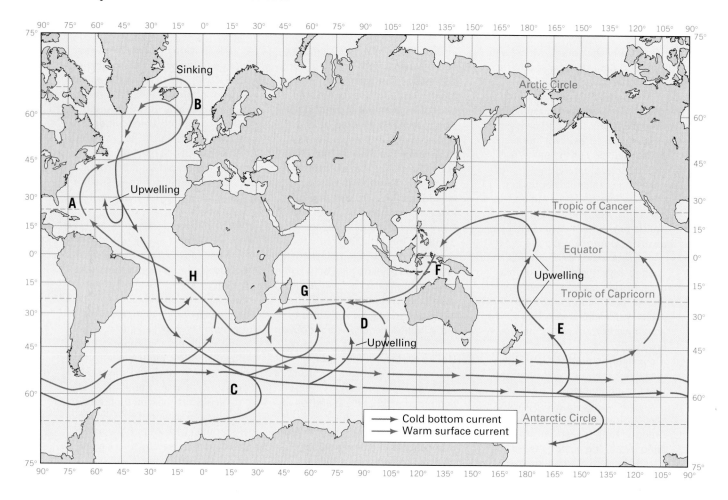

5.35 Ocean circulation *Deep ocean currents, generated by the sinking of cold, salty water in the northern Atlantic, circulate sea water in slowly moving coupled loops involving the Atlantic, Pacific, Indian, and Southern oceans. Actual flows are much broader than the narrow flow paths shown here. (After A. J. Gordon, Nature 382:399–400, August, 1996. Used by permission.)*

A LOOK AHEAD The global circulation of winds and currents paves the way for our next subject—weather systems. Recall from Chapter 4 that when warm, moist air rises, precipitation can occur. This happens in the centers of cyclones, where air converges. Although cyclones and anticyclones are generally large surface features, they move from day to day and are steered by the global pattern of winds. Thus, your knowledge of global winds and pressures will help you to understand how weather systems and storms develop and migrate. Your knowledge will also be very useful for the study of climate, which we take up in Chapter 7.

In Review | Winds and Global Circulation

- The term *atmospheric pressure* describes the weight of air pressing on a unit of surface area. Atmospheric pressure is measured by a **barometer.** Atmospheric pressure decreases rapidly as altitude increases.

- Wind occurs when air moves with respect to the Earth's surface. Air motion is produced by **pressure gradients** that are formed when air in one location is heated to a temperature that is warmer than another. Heating creates high pressure aloft, which moves high-level air away from the area of heating. This motion induces low pressure at the surface, pulling surface air toward the area of heating, and a convection loop is formed.

- *Sea* and *land breezes* are examples of *convection loops* formed from unequal heating and cooling of the land surface as compared to a nearby water surface.

- *Local winds* are generated by local pressure gradients. The sea and land breezes, as well as *mountain* and *valley winds,* are examples caused by local surface heating. Other local winds include *drainage winds, Santa Ana,* and *chinook winds.*

- The Earth's rotation strongly influences atmospheric circulation through the **Coriolis effect.** The *Coriolis force* deflects wind motion, producing circular or spiraling flow paths around **cyclones** (centers of low pressure and *convergence*) and **anticyclones** (centers of high pressure and *divergence*).

- Because the equatorial and tropical regions are heated more intensely than the higher latitudes, two convection loops develop—the **Hadley cells.** These loops drive the *northeast* and *southeast trade winds,* the convergence and lifting of air at the **intertropical convergence zone (ITCZ),** and the sinking and divergence of air in the **subtropical high-pressure belts.**

- The most persistent features of the global pattern of atmospheric pressure are the subtropical high-pressure belts, which are generated by the Hadley cell circulation. They intensify during their high-Sun season.

- The **monsoon** circulation of Asia responds to a reversal of atmospheric pressure over the continent with the seasons. A *winter monsoon* flow of cool, dry air from the northeast alternates with a *summer monsoon* flow of warm, moist air from the southwest.

- In the midlatitudes and poleward, westerly winds prevail. In winter, continents develop high pressure, and intense oceanic low-pressure centers are found off the Aleutian Islands and near Iceland in the northern hemisphere. In the summer, the continents develop low pressure as oceanic subtropical high-pressure cells intensify and move poleward.

- Winds aloft are dominated by a global pressure gradient force between the tropics and pole in each hemisphere that is generated by the hemispheric temperature gradient from warm to cold. Coupled with the Coriolis force, the gradient generates strong westerly *geostrophic winds* in the upper air. In the equatorial region, weak easterlies dominate the upper-level wind pattern.

- **Rossby waves** develop in the *upper-air westerlies,* bringing cold, polar air equatorward and warmer air poleward. The *polar-front* and *subtropical* **jet streams** are concentrated westerly wind streams with high wind speeds. The *tropical easterly jet stream* is weaker and limited to Southeast Asia, India, and Africa.

- Oceans show a warm surface layer, a *thermocline,* and a deep cold layer. Near the poles, the warm layer and thermocline are absent.

- *Ocean* surface *currents* are dominated by huge **gyres** that are driven by the global surface wind pattern. *Equatorial currents* move warm water westward and then poleward along the east coasts of continents. Return flows bring cold water equatorward along the west coasts of continents.

- *El Niño* events occur when an unusual flow of warm water in the equatorial Pacific moves eastward to the coasts of Central and South America, suppressing the normal northward flow of the Humboldt Current. Upwelling along the Peruvian coast is greatly reduced. El Niño events normally occur on a three- to eight-year cycle and affect global patterns of precipitation in many regions. The causes of El Niño are not well understood.

- Slow, deep ocean currents are driven by the sinking of cold, salty water in the northern Atlantic. This *thermohaline circulation* pattern involves nearly all the Earth's ocean basins, and also acts to moderate the buildup of atmospheric CO_2 by moving CO_2-rich surface waters to ocean depths.

Key Terms

millibar, p. 153	isobar, p. 155	intertropical convergence zone (ITCZ), p. 162	monsoon, p. 167
barometer, p. 153	Coriolis effect, p. 158		Rossby waves, p. 170
wind, p. 154	cyclones, p. 161		jet streams, p. 171
pressure gradient, p. 155	anticyclones, p. 161	subtropical high-pressure belts, p. 162	gyres, p. 174
	Hadley cells, p. 162		

Review Questions

1. Explain atmospheric pressure. Why does it occur? How is atmospheric pressure measured and in what units? What is the normal value of atmospheric pressure at sea level? How does atmospheric pressure change with altitude?

2. Describe a simple convective wind system, explaining how air motion arises from a pressure gradient force induced by heating of the atmosphere.

3. Describe land and sea breezes. How do they illustrate the concepts of pressure gradient and convection loop?

4. What are drainage winds? What local names are applied to them?

5. What is the Coriolis effect, and why is it important? What produces it? How does it influence the motion of wind and ocean currents in the northern hemisphere? in the southern hemisphere?

6. Define cyclone and anticyclone. How does air move within each? What is the direction of circulation of each in the northern and southern hemispheres? What type of weather is associated with each and why?

7. What is the Asian monsoon? Describe the features of this circulation in summer and winter. How is the ITCZ involved? How is the monsoon circulation related to the high- and low-pressure centers that develop seasonally in Asia?

8. Compare the winter and summer patterns of high and low pressure that develop in the northern hemisphere with those that develop in the southern hemisphere.

9. How does global-scale heating of the atmosphere create a pressure gradient force that increases with altitude?

10. What is the geostrophic wind, and what is its direction with respect to the pressure gradient force?

11. Describe the basic pattern of global atmospheric circulation at upper levels.

12. What are Rossby waves? Why are they important?

13. Identify five jet streams. Where do they occur? In which direction do they flow?

14. What is the general pattern of ocean surface current circulation? How is it related to global wind patterns?

15. How does thermohaline circulation induce deep ocean currents?

Eye on the Environment 5.1 • Wind Power, Wave Power, and Current Power

1. Discuss wind power as a source of energy. Include indirect use in tapping waves and ocean currents for energy.

Eye on Global Change 5.2 • El Niño

1. Compare the normal pattern of wind, pressure, and ocean currents in the equatorial Pacific with the pattern during an El Niño event.

2. What are some of the weather changes reported for El Niño events?

3. What is La Niña, and how does it compare with the normal pattern?

Visualizing Exercises

1. Sketch an ideal Earth (without seasons or ocean—continent features) and its global wind system. Label the following on your sketch: doldrums, equatorial trough, Hadley cell, ITCZ, northeast trades, polar easterlies, polar front, polar outbreak, southeast trades, subtropical high-pressure belts, and westerlies.

2. Draw four spiral patterns showing outward and inward flow in clockwise and counterclockwise directions. Label each as appropriate to cyclonic or anticyclonic circulation in the northern or southern hemisphere.

Essay Questions

1. An airline pilot is planning a nonstop flight from Los Angeles to Sydney, Australia. What general wind conditions can the pilot expect to find in the upper atmosphere as the airplane travels? What jet streams will be encountered? Will they slow or speed the aircraft on its way?

2. You are planning to take a round-the-world cruise, leaving New York in October. Your vessel's route will take you through the Mediterranean Sea to Cairo, Egypt, in early December. Then you will pass through the Suez Canal and Red Sea to the Indian Ocean, calling at Bombay, India, in January. From Bombay, you will sail to Djakarta, Indonesia, and then go directly to Perth, Australia, arriving in March. Rounding the southern coast of Australia, your next port of call is Auckland, New Zealand, which you will reach in April. From Auckland, you head directly to San Francisco, your final destination, arriving in June. Describe the general wind and weather conditions you will experience on each leg of your journey.

Eye on the Landscape

Chapter Opener Sailboats on the Charles River, Boston An urban landscape In a large city, such as Boston, very little of the landscape is in a natural state. Although the skyscrapers (A) are anchored firmly to Boston's bedrock, much of the city is built on fill brought in near the turn of the twentieth century. The Charles River, once a tidal estuary, is dammed and now amounts to a shallow freshwater lake (B). The vegetation on the Esplanade that borders the river is nearly all planted with culvated trees chosen for their appearance and utility. In short, just about everything in this scene is a product of human activity.

5.10 Brushfire The dense, shrubby vegetation on this slope is chaparral, a mix of tough woody shrubs with leathery leaves (A) that is often found on steep slopes in Mediterranean climates (covered in Chapter 7). To reduce moisture loss during the hot, dry summer, many plants of the chaparral have waxy, oily leaves that can prove very flammable under the right conditions. Chaparral is a vegetation type that depends on burning at regular intervals to maintain itself, so the fire in this picture is merely part of a natural process.

5.28 Jet stream clouds A desert landscape seen from space. Large expanses of the Middle East are essentially devoid of vegetation, and so a space photo such as this shows variation in soil and surface color. Many of the white areas (A) are wind-blown sand sheets (see Dunes, Chapter 18), while the medium tones are weathered rock materials that have not been carried as far by wind or water from their origin (see Chapters 16 and 18 for sediment transportation by water and wind). Ranges of hills and low mountains (B) appear darker due to a sparse vegetation cover. The Nile River Valley cuts a wide swath through the region, with its broad band of irrigated agriculture (see Chapter 15 on rivers). The overall blue tint is produced by atmospheric scattering (Chapter 2), which is more intense in the blue region of the spectrum.

5.33 Sea-surface temperatures This image shows a temperature pattern over the land (A) as well as the water. The general gradient from yellow and red tones (hot) to dark tones (cold) shows land surface temperatures decreasing strongly from Florida to the Maritime Provinces, as we might expect for a week in April. Industrial and residential regions, such as the Washington to New York corridor, Hudson River Valley, and Connecticut River Valley show warmer temperatures due to human activity (see Chapter 3 on urban-rural surface temperature contrasts). Note the dark color of the Appalachians, which are colder because of their elevation (see Chapter 3 for how surface temperature varies with elevation).

6 | WEATHER SYSTEMS

● ● ● ● ● ● ● ● ● ● ● ● ● ● ●

PUTTING YOU IN THE PICTURE

Our day-to-day weather may occasionally be stormy, bringing the inconvenience of rain or snow to our daily activities. But rare storms, with high winds, high coastal ocean waters, and torrential rains or blizzards of snow can cause great destruction and loss of life. What is it like to experience a severe storm?

On August 23, 1992, Hurricane Andrew approached the coast of south Florida, near Miami. Following the path of the storm was veteran storm chaser and photographer, Warren Faidley, who decided to ride out the storm with two friends, Steve and Mike. For their shelter, they selected a sturdy parking garage, built of concrete and steel, in a coastal community south of Miami. Here is part of Faidley's account of the terrifying night of the storm.

At sunset, my friends and I gathered at the garage. Throughout the evening we monitored radio and television broadcasts. With ominous accuracy, Robert Sheets, director of the National Hurricane Center, gave hourly updates on Andrew, now a Category 4 … Like sentries awaiting an advancing army, we took turns scanning the Miami skyline from our borrowed fortress, searching for early signs of Andrew. Bright flashes from downed power lines on the horizon marked Andrew's imminent arrival.

Steve and I held an anemometer through the opening on the fourth floor. "Fifty-five miles per hour!" I yelled over the howling. "A gust to 65!" The meter almost flew out of my hand. The electrical explosions across the city grew more intense, yet the fury of the wind swallowed up all sound save the car alarms whining inside the garage. Eventually even the alarms were overwhelmed by the whistling wind.

Around 4 A.M. all hell broke loose. I have experienced severe storms with winds in excess of 75 m.p.h., but these gusts were blowing at well over 140 m.p.h. The anemometer and later damage surveys indicated even stronger gusts. By now only an occasional thud—some building collapsing or losing a roof—would punctuate the noise. Windows from the surrounding buildings imploded, scattering glass everywhere. I looked down at my arm and saw blood, not knowing when I'd been cut. Putting on my reading glasses to protect my eyes, I made a mental note: next time, bring goggles.

My senses were on red alert, a self-preservation mode I adopt only in the most extreme danger. Everything I see, hear, feel—even taste—goes by so quickly that my mind simply slows it all down to preserve rational thought. The alternative is sheer panic.

The infamous "hurricane wail" people talk about is in many ways unreal—"the scream of the devil," Steve and I agreed. The piercing sound alone is wicked enough, but mixed with breaking glass and crashing debris, it rattled me inside. I will never forget that sound…

In the garage, not all was well. The elevator shafts had begun to flood; water poured out between the doors and cascaded down the emergency stairwells. The lights in the garage flickered on and off and the fire alarm blared, yet it was hardly audible in the din. … I trusted the structural integrity of our fortress, but under Andrew's full force, the garage began to oscillate. After two hours of this swaying, I began to worry one of the floors might collapse…

For more than five hours Andrew pounded the garage and the rest of South Florida with everything it had. As the building shook and the storm raged all around, I felt as though I was inside a tornado and riding an earthquake at the same time.*

Hurricane Andrew was the most costly natural disaster to strike the United States, with some damage estimates as high as $30 billion. *Eye on the Environment 6.2 • Hurricane Andrew—Killer Cyclone* provides more information on this devastating storm. Hurricanes, or tropical cyclones, as they are termed by meteorologists, are but one of a number of types of weather systems that are the subject of this chapter. ∎

* From *Weatherwise,* vol. 45, no. 6, Copyright © Warren Faidley. Used by permission.

WEATHER SYSTEMS

As we saw in Chapter 5, the Earth's atmosphere is in constant motion, driven by the planet's rotation and its uneven heating by the Sun. The horizontal motion of the wind moves air from one place to another, allowing air to acquire characteristics of temperature and humidity in one region and then carry those characteristics into another region. The vertical motion affects clouds and precipitation. When air is lifted, it is cooled, enabling clouds and precipitation to form. When air descends, it is warmed, retarding the formation of clouds and precipitation. In this way, the Earth's wind system influences the weather we experience from day to day—the temperature and humidity of the air, cloudiness, and the amount of precipitation.

Some patterns of wind circulation occur commonly and so present recurring patterns of weather. For example, traveling low-pressure centers (cyclones) of converging, inspiraling air often bring warm, moist air in contact with cooler, drier air, with clouds and precipitation as the result. We recognize these recurring circulation patterns and their associated weather as **weather systems.**

Weather systems range in size from a few kilometers, in the case of the tornado, to a thousand kilometers or more, in the case of a large traveling anticyclone. They may last for hours or weeks, depending on their size and strength. Some forms of weather systems—tornadoes and hurricanes, for example—involve high winds and heavy rainfall and can be very destructive to life and property.

AIR MASSES

Weather systems are often associated with the motion of air masses. An **air mass** is a large body of air with fairly uniform temperature and moisture characteristics. It can be several thousand kilometers or miles across, and it can extend upward to the top of the troposphere. A given air mass is characterized by a distinctive combination of surface temperature, environmental temperature lapse rate, and surface-specific humidity. Air masses range widely in temperature—from searing hot to icy cold—as well as in moisture content.

Air masses acquire their characteristics in *source regions.* In a source region, air moves slowly or stagnates, which allows the air to acquire temperature and moisture characteristics from the region's surface. For example, an air mass with warm temperatures and a high water vapor content develops over a warm equatorial ocean. Over a large tropical desert, slowly subsiding air forms a hot air mass with low humidity. A very cold air mass with a low water vapor content is generated over cold, snow-covered land surfaces in the arctic zone in winter.

Air masses move from one region to another under the influence of pressure gradients and upper-level wind patterns and are sometimes pushed or blocked by high-level jet stream winds. When an air mass moves to a new area, its properties will begin to change because it is influenced by the new surface environment. For example, the air mass may lose heat or take up water vapor.

Air masses are classified on the basis of the latitudinal position and the nature of the underlying surface of their source regions. Latitudinal position primarily determines surface temperature and the environmental temperature lapse rate of the air mass, while the nature of the underlying surface—continent or ocean—usually determines the moisture content. For latitudinal position, five types of air masses are distinguished, as shown in the following table.

Air Mass	Symbol	Source Region
Arctic	A	Arctic ocean and fringing lands
Antarctic	AA	Antarctica
Polar	P	Continents and oceans, lat. 50–60° N and S
Tropical	T	Continents and oceans, lat. 20–35° N and S
Equatorial	E	Oceans close to equator

For the type of underlying surface, two subdivisions are used:

Air Mass	Symbol	Source Region
Maritime	m	Oceans
Continental	c	Continents

Combining these two types of labels produces a list of six important types of air masses, shown in Table 6.1. The table also gives some typical values of surface temperature and specific humidity, although these can vary widely, depending on season. Air mass temperature can range from –46°C (–51°F) for continental arctic (cA) air masses to 27°C (81°F) for maritime equatorial air masses (mE). Specific humidity shows a very high range—from 0.1 g/kg for the cA air mass to as much as 19 g/kg for the mE air mass. In other words,

> Air masses are large bodies of air with fairly uniform temperature and moisture characteristics that are acquired from a source region. They are classified by latitudinal position (e.g., arctic, tropical), which largely determines temperature, and by underlying surface type (maritime, continental), which largely determines moisture content.

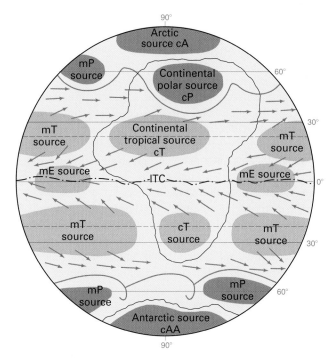

6.1 Global air masses and source regions *An idealized continent, producing continental (c) air masses, is shown at the center. It is surrounded by oceans, producing maritime air masses (m). Tropical (T) and equatorial (E) source regions provide warm or hot air masses, while polar (P), arctic (A), and antarctic (AA) source regions provide colder air masses of low specific humidity.*

maritime equatorial air can hold about 200 times as much moisture as continental arctic air.

Figure 6.1 shows schematically the global distribution of source regions of these air masses. An idealized continent is shown in the center of the figure surrounded by ocean. Note from the figure that the polar air masses (mP, cP) originate in the subarctic latitude zone, not in the polar latitude zone. Recall that in Chapter 2 we defined "polar" as a region containing one of the poles. However, meteorologists use the word "polar" to describe air masses from the subarctic

Table 6.1 | Properties of Typical Air Masses

Air Mass	Symbol	Source Region	Properties	Temperature °C	Temperature (°F)	Specific Humidity (gm/kg)
Maritime equatorial	mE	Warm oceans in the equatorial zone	Warm, very moist	27°	(81°)	19
Maritime tropical	mT	Warm oceans in the tropical zone	Warm, moist	24°	(75°)	17
Continental tropical	cT	Subtropical deserts	Warm, dry	24°	(75°)	17
Maritime polar	mP	Midlatitude oceans	Cool, moist (winter)	4°	(39°)	4.4
Continental polar	cP	Northern continental interiors	Cold, dry (winter)	–11°	(12°)	1.4
Continental arctic (and continental antarctic)	cA (cAA)	Regions near north and south poles	Very cold, very dry (winter)	–46°	(–51°)	0.1

and subantarctic zones, and we will follow their usage when referring to air masses.

The maritime tropical air mass (mT) and maritime equatorial air mass (mE) originate over warm oceans in the tropical and equatorial zones. They are quite similar in temperature and water vapor content. With their high values of specific humidity, both are capable of very heavy yields of precipitation. The continental tropical air mass (cT) has its source region over subtropical deserts of the continents. Although this air mass may have a substantial water vapor content, it tends to be stable and has low relative humidity when highly heated during the daytime.

The maritime polar air mass (mP) originates over midlatitude oceans. Since the quantity of water vapor it holds is not as large as maritime tropical air masses, the mP air mass yields only moderate precipitation. Much of this precipitation is orographic and occurs over mountain ranges on the western coasts of continents. The continental polar air mass (cP) originates over North America and Eurasia in the subarctic zone. It has low specific humidity and is very cold in winter. Last is the continental arctic (and continental antarctic) air mass type (cA, cAA), which is extremely cold and holds almost no water vapor.

North American Air Masses

The air masses that form in and near North America have a strong influence on North American weather. The source regions of these air masses are shown in Figure 6.2. The continental polar (cP) air mass of North America originates over north-central Canada. This air mass forms tongues of cold, dry air that periodically extend south and east from the source region to produce anticyclones accompanied by cool or cold temperatures and clear skies. The arctic air mass (cA) that develops over the Arctic Ocean and its bordering lands of the arctic zone in winter is extremely cold and stable. When this air mass moves southward, it produces a severe cold wave.

The maritime polar air mass (mP) originates over the North Pacific and Bering Strait, in the region of the persistent Aleutian low-pressure center. This air mass is characteristically cool and moist, with a tendency in winter to become unstable, giving heavy precipitation over coastal ranges. Another maritime polar air mass of the North American region originates over the North Atlantic Ocean. It, too, is cool and moist, and is felt especially in Canada's Maritime Provinces in winter.

Of the tropical air masses, the most common visitor to the central and eastern United States is the maritime tropical air mass (mT) from the Gulf of Mexico. It moves northward, bringing warm, moist, unstable air over the eastern part of the country. In the summer, particularly, this air mass brings hot, sultry weather to the central and eastern United States. It also produces many thunderstorms. Closely related is a maritime tropical air mass from the Atlantic Ocean east of Florida, over the Bahamas.

Over the Pacific Ocean, a source region of another maritime tropical air mass (mT) lies in the cell of high pressure

6.2 North American air mass source regions and trajectories *Air masses acquire temperature and moisture characteristics in their source regions, then move across the continent. (Data from U.S. of Dept. Commerce.)*

located to the southwest of lower California. Occasionally, in summer, this moist unstable air mass penetrates the southwestern desert region, bringing severe thunderstorms to southern California and southern Arizona. In winter, a tongue of mT air frequently reaches the California coast, bringing heavy rainfall that is intensified when forced to rise over coastal mountain ranges.

A hot, dry continental tropical air mass (cT) originates over northern Mexico, western Texas, New Mexico, and Arizona during the summer. This air mass does not travel widely but governs weather conditions over the source region.

Cold, Warm, and Occluded Fronts

A given air mass usually has a sharply defined boundary between itself and a neighboring air mass. This boundary is termed a **front.** We saw an example of a front in the contact between polar and tropical air masses, shown in Figures 5.16, 5.25, and 5.27. This feature is the *polar front,* and it is located below the axis of the jet stream in the upper-air waves. *GEODISCOVERIES*

Figure 6.3 shows the structure of a front along which a cold air mass invades a zone occupied by a warm air mass. A front of this type is called a **cold front.** Because the colder air mass is denser, it remains in contact with the ground. As it moves forward, it forces the warmer air mass to rise above it. If the warm air is unstable, severe thunderstorms may develop.

6.3 Cold front *At a cold front, a cold air mass lifts a warm air mass aloft. The upward motion sets off a line of thunderstorms. The frontal boundary is actually much less steep than is shown in this schematic drawing. (Drawn by A. N. Strahler.)*

Thunderstorms near a cold front often form a long line of massive clouds stretching for tens of kilometers (Figure 6.4).

Figure 6.5 diagrams a **warm front** in which warm air moves into a region of colder air. Here, again, the cold air

Fronts are boundaries between air masses. When cold air invades warmer air, the boundary is a cold front. When warm air invades colder air, the boundary is a warm front.

mass remains in contact with the ground because it is denser. The warm air mass is forced to rise on a long ramp over the cold air below. The rising motion causes stratus clouds to form, and precipitation often follows. If the warm air is stable, the precipitation will be steady (Figure

6.4 Cold front cumulus *A line of cumulus clouds marks the advance of a cold front, moving from left to right. The cold air pushes warmer, moister air aloft, triggering cloud formation.*

6.5 **Warm front** *In a warm front, warm air advances toward cold air and rides up and over the cold air. A notch of cloud is cut away to show rain falling from the dense stratus cloud layer. (Drawn by A. N. Strahler.)*

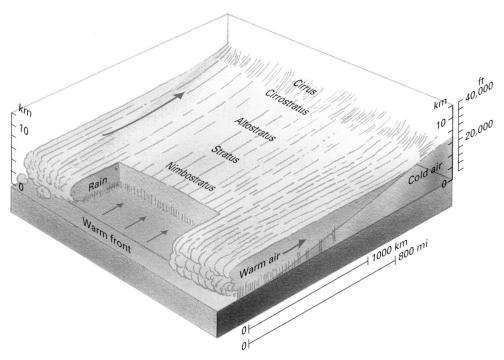

6.5). If the warm air is unstable, convection cells can develop, producing cumulonimbus clouds with heavy showers or thunderstorms (not shown in the figure).

Cold fronts normally move along the ground at a faster rate than warm fronts. Thus, when both types are in the same neighborhood, a cold front can overtake a warm front. The result is an **occluded front,** diagrammed in Figure 6.6. ("Occluded" means closed or shut off.) The colder air of the fast-moving cold front remains next to the ground, forcing both the warm air and the less cold air ahead to rise over it. The warm air mass is lifted completely free of the ground.

> **In an occluded front, a cold front overtakes a warm front, lifting a pool of warm, moist air upward. The result is a region of clouds and precipitation.**

TRAVELING CYCLONES AND ANTICYCLONES

Air masses are set in motion by wind systems—typically, cyclones and anticyclones that involve masses of air moving in a spiraling motion. As we saw in Chapter 5, air spirals inward and converges in a *cyclone,* while air spirals outward and diverges in an *anticyclone.* Most types of cyclones and anticyclones are large features that move slowly across the Earth's surface, bringing changes in the weather as they move. These are referred to as *traveling cyclones* and *traveling anticyclones.*

In a cyclone, convergence and upward motion cause air to rise and be cooled adiabatically. If the air is moist, condensation or deposition can occur. This is **cyclonic precipitation.** Many cyclones are weak and pass overhead

> **In middle and higher latitudes, traveling cyclones and anticyclones bring changing weather systems. In cyclones, convergence and uplift typically cause condensation and precipitation, while in anticyclones divergence and subsidence cause air to be warmed, producing clear conditions.**

with little more than a period of cloud cover and light precipitation. However, when pressure gradients are steep and the inspiraling motion is strong, intense winds and heavy rain or snow can accompany the cyclone. In this case, the disturbance is called a **cyclonic storm.**

Traveling cyclones fall into three types. First is the wave cyclone of midlatitude, arctic, and antarctic zones. This type of cyclone ranges in intensity from a weak disturbance to a powerful storm. Second is the tropical cyclone of tropical and subtropical zones. This type of cyclone ranges in intensity from a mild disturbance to the highly destructive hurricane or typhoon. A third type is the tornado, a small, intense cyclone of enormously powerful winds. The tornado is much, much smaller in size than other cyclones, and it is related to strong convectional activity.

In an anticyclone, divergence and downward motion cause air to descend and be warmed adiabatically. Thus, condensation does not occur. Skies are fair, except for occasional puffy cumulus clouds that sometimes develop in a moist surface air layer. Because of these characteristics, anticyclones are often termed *fair-weather systems.* Toward the center of an anticyclone, the pressure gradient is weak, and winds are light and variable. Traveling anticyclones are found in the midlatitudes. They are typically associated with ridges or domes of clear, dry air that move eastward and equatorward. Figure 6.7 is a geostationary satellite image of eastern North America. A large anticyclone is centered over the area, bringing fair weather and cloudless skies.

6.6 ***Occluded front*** *In an occluded front, a warm front is overtaken by a cold front. The warm air is pushed aloft, and it no longer contacts the ground. Abrupt lifting by the denser cold air produces precipitation. (Drawn by A. N. Strahler.)*

6.7 ***Eye on the Landscape*** ***Picture of an anticyclone*** *This geostationary satellite image of eastern North America shows a large anticyclone centered over the area, bringing fair weather and cloudless skies. The boundary between clear sky and clouds running across the Gulf of Mexico and the Florida peninsula delineates the leading edge of the cool, dry air mass. The cloud edge at the top of the photo marks the cold fronts of two cold air masses advancing eastward and southward.* **What else would the geographer see? ... Answers at the end of the chapter.**

6.8 Conditions for formation of a wave cyclone

Two anticyclones, one with warm subtropical air and the other with cold polar air, are in contact on the polar front. The shaded area is shown in block (a) as the early stage of development of a wave cyclone.

Wave Cyclones

In middle and high latitudes, the dominant form of weather system is the **wave cyclone,** a large inspiral of air that repeatedly forms, intensifies, and dissolves along the polar front. Figure 6.8 shows a situation favorable to the formation of a wave cyclone. Two large anticyclones are in contact on the polar front. One contains a cold, dry polar air mass, and the other a warm, moist maritime air mass. Airflow converges from opposite directions on the two sides of the front, setting up an unstable situation. The wave cyclone will begin to form between the two high-pressure cells in a zone of lower pressure referred to as a *low-pressure trough.*

How does a wave cyclone form, grow, and eventually dissolve? Figure 6.9 shows the life history of a wave cyclone. In block (*a*) (early stage), the polar-front region shows a

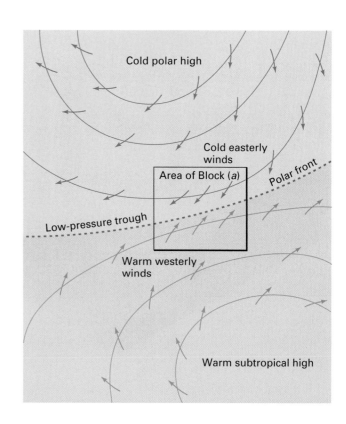

6.9 Development of a wave cyclone
In (a), a wave motion begins at a point along the polar front. The wave along the cold and warm fronts deepens and intensifies in (b). In (c), the cold front overtakes the warm front, producing an occluded front in the center of the cyclone. Later, the polar front is reestablished with a mass of warm air isolated aloft (d).

wave beginning to form. Cold air is turned in a southerly direciton and warm air in a northerly direction, so that each advances on the other. As these frontal motions develop, precipitation will form.

In block (b) (open stage), the wave disturbance along the polar front has deepened and intensified. Cold air actively pushes southward along a cold front, and warm air actively moves northeastward along a warm front. The zones of precipitation along the two fronts are now strongly developed but are wider along the warm front than along the cold front.

In block (c) (occluded stage), the cold front has overtaken the warm front, producing an occluded front. The warm air mass at the center of the inspiral is forced off the ground, intensifying precipitation. Eventually, the polar front is reestablished (block (d), dissolving stage), but a pool of warm, moist air remains aloft. As the moisture content of the pool is reduced, precipitation dies out, and the clouds gradually dissolve.

Keep in mind that a wave cyclone is quite a large feature—1000 km (about 600 mi) or more across. Also, the cyclone normally moves eastward as it develops, propelled by prevailing westerlies aloft. Therefore, blocks (a)–(d) are like three-dimensional snapshots taken at intervals along an eastbound track.

Weather Changes within a Wave Cyclone

How does weather change as a wave cyclone passes through a region? Figure 6.10 shows two simplified weather maps of the eastern United States depicting conditions on successive days. The structure of the storm is defined by the isobars, labeled in millibars. The three kinds of fronts are shown by special line symbols. Areas of precipitation are shown in gray.

The map on the left shows the cyclone in an open stage, similar to Figure 6.9, block (b). The isobars show that the cyclone is a low-pressure center with inspiraling winds. The cold front is pushing south and east, supported by a flow of cold, dry continental polar air from the northwest filling in behind it. Note that the wind direction changes abruptly as

6.10 Simplified surface weather maps and cross sections through a wave cyclone *In the open stage (left), cold and warm fronts pivot around the center of the cyclone. In the occluded stage (right), the cold front has overtaken the warm front, and a large pool of warm, moist air has been forced aloft.*

the cold front passes. There is also a sharp drop in temperature behind the cold front as cP air fills in. The warm front is moving north and somewhat east, with warm, moist maritime tropical air following. The precipitation pattern includes a broad zone near the warm front and the central area of the cyclone. A thin band of precipitation extends down the length of the cold front. Cloudiness generally prevails over much of the cyclone.

A cross section along the line A–A' shows how the fronts and clouds are related. Along the warm front is a broad layer of stratus clouds. These take the form of a wedge with a thin leading edge of cirrus. (See Figures 4.11 and 4.12 for cloud types.) Westward, this wedge thickens to altostratus, then to stratus, and finally to nimbostratus with steady rain. Within the sector of warm air, the sky may partially clear with scattered cumulus. Along the cold front are cumulonimbus clouds associated with thunderstorms. These yield heavy rains but only along a narrow belt.

The weather map on the right shows conditions 24 hours later. The cyclone has moved rapidly northeastward, its track shown by the red line. The center has moved about 1600 km (1000 mi) in 24 hours—a speed of just over 65 km (40 mi)

per hour. The cold front has overtaken the warm front, forming an occluded front in the central part of the disturbance. A high-pressure area, or tongue of cold polar air, has moved in to the area west and south of the cyclone, and the cold front has pushed far south and east. Within the cold air tongue, the skies are clear. A cross section below the map shows conditions along the line B–B', cutting through the occluded part of the storm. Notice that the warm air mass is lifted well off the ground and yields heavy precipitation.

Cyclone Tracks and Cyclone Families

Wave cyclones tend to form in certain areas and travel common paths until they dissolve. Figure 6.11 is a world map showing common paths of wave cyclones and tropical cyclones. (We will discuss tropical cyclones in detail later in this chapter.) The western coast of North America commonly receives wave cyclones arising in the North Pacific Ocean. Wave cyclones also originate over land, shown by the tracks starting in Alaska, the Pacific Northwest, the south-central United States, and along the Gulf coast. Most of these tracks converge toward the northeast and pass into the

6.11 Paths of tropical cyclones and wave cyclones *This world map shows typical paths of tropical cyclones (red) and midlatitude wave cyclones (blue). (Based on data of S. Pettersen, B. Haurwitz and N. M. Austin, J. Namias, M. J. Rubin, and J-H. Chang.)*

6.12 **Daily world weather map** *A daily weather map of the world for a given day during July or August might look like this map, which is a composite of typical weather conditions. (After M. A. Garbell.)*

Legend:
- Cold front
- Warm front
- Occluded front
- ITCZ
- (H) High
- (L) Low
- (◎) Hurricane

North Atlantic, where they tend to concentrate in the region of the Icelandic Low.

In the northern hemisphere, wave cyclones are heavily concentrated in the neighborhood of the Aleutian and Icelandic Lows. These cyclones commonly form in a succession, traveling as a chain across the North Atlantic and North Pacific oceans. Figure 6.12, a world weather map, shows several such cyclone families. Each wave cyclone moves northeastward, deepening in low pressure and occluding to form an upper-air low. For this reason, intense cyclones arriving at the western coasts of North America and Europe are usually occluded.

In the southern hemisphere, storm tracks are more nearly along a single lane, following the parallels of latitude. Three such cyclones are shown in Figure 6.12. This track is more uniform because of the uniform pattern of ocean surface circling the globe at these latitudes. Only the southern tip of South America projects southward to break the monotonous expanse of ocean.

The Tornado

A **tornado** is a small but intense cyclonic vortex in which air spirals at tremendous speed. It is associated with thunderstorms spawned by fronts in midlatitudes of North America. Tornadoes can also occur inside tropical cyclones (hurricanes).

The tornado appears as a dark funnel cloud hanging from the base of a dense cumulonimbus cloud (Figure 6.13). At its lower end, the funnel may be 100 to 450 m (about 300 to 1500 ft) in diameter. The base of the funnel appears dark because of the density of condensing moisture, dust, and debris swept up by the wind. Wind speeds in a tornado exceed speeds known in any other storm. Estimates of wind speed run as high as 100 m/s (about 225 mi/hr). As the tornado moves across the countryside, the funnel writhes and twists. Where it touches the ground, it can cause the complete destruction of almost anything in its path. GEODISCOVERIES

Tornadoes occur as parts of cumulonimbus clouds traveling in advance of a cold front. They seem to originate where turbulence is greatest. They are most common in the spring and summer but can occur in any month. Where a cold front of maritime polar air lifts warm, moist maritime tropical air, conditions are most favorable for tornadoes. As shown in Figure 6.14, they occur in greatest numbers in the central and southeastern states and are rare over mountainous and forested regions. They are almost unknown west of the Rocky Mountains and are relatively less frequent on the eastern seaboard. Tornadoes are a typically American phenomenon, being most frequent and violent in the United States. They also occur in Australia in substantial numbers and are occasionally reported from other midlatitude locations.

Devastation from a tornado is often complete within the narrow limits of its path (Figure 6.15). Only the strongest buildings constructed of concrete and steel can withstand the extremely violent winds. The National Weather Service

> **The tornado is a small but intense cyclonic vortex with very high wind speeds. Tornados occur within cumulonimbus clouds traveling in advance of a cold front, and are most common in spring and summer.**

6.13 Tornado *This tornado touched down near Clearwater, Kansas, on May 16, 1991. Surrounding the funnel is a cloud of dust and debris carried into the air by the violent winds.*

maintains a tornado forecasting and warning system. Whenever weather conditions favor tornado development, the danger area is alerted, and systems for observing and reporting a tornado are set in readiness. **GEODISCOVERIES**

6.14 Frequency of occurrence of observed tornadoes in the 48 contiguous United States *The data shown in this map span a 30-year record, 1960–1989. (Courtesy of Edward W. Ferguson, National Severe Storms Forecast Center, National Weather Service.)*

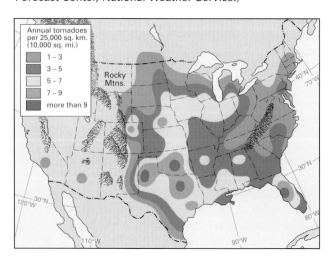

TROPICAL AND EQUATORIAL WEATHER SYSTEMS

So far, the weather systems we have discussed are those of the midlatitudes and poleward. Weather systems of the tropical and equatorial zones show some basic differences from those of the midlatitudes. Upper-air winds are often weak, so air mass movement is slow and gradual. Air masses are warm and moist, and tend to have similar characteristics. Thus, clearly defined fronts and large, intense wave cyclones are missing. On the other hand, intense convectional activity occurs because of the high moisture content of low-latitude maritime air masses. In these very moist air masses, even slight convergence and uplifting can be sufficient to trigger precipitation.

Clouds and precipitation in tropical and equatorial regions have recently become the object of a remote sensing mission. Of special interest is the intensity and duration of rainfall over oceans, which is not well monitored. *Focus on Remote Sensing 6.1 ● TRMM—The Tropical Rainfall Monitoring Mission* provides more information.

Easterly Waves and Weak Equatorial Lows

One of the simplest forms of tropical weather systems is an *easterly wave*, a slowly moving trough of low pressure within the belt of tropical easterlies (trades). These waves occur in latitudes 5° to 30° N and S over oceans, but not over the equator itself. Figure 6.16 is a simplified upper-air map of an easterly wave showing wind patterns, the axis of the wave, isobars, and the zone of showers. At the surface, a zone of weak low pressure underlies the axis of the wave. The wave travels westward at a rate of 300 to 500 km (about 200 to 300 mi) per day. Surface airflow converges on the eastern, or rear, side of the wave axis. This convergence causes the moist air to be lifted, producing scattered show-

6.16 Easterly wave *An easterly wave passing over the West Indies. (Data from H. Riehl, Tropical Meteorology, New York: McGraw-Hill.)*

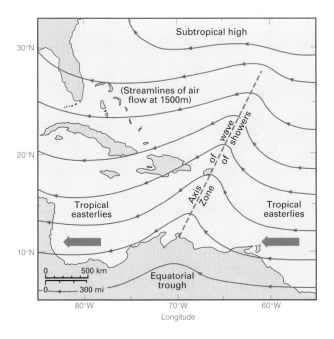

ers and thunderstorms. The rainy period may last a day or two as the wave passes.

Another related weather system is the *weak equatorial low,* a disturbance that forms near the center of the equatorial trough. Moist equatorial air masses converge on the center of the low, causing rainfall from many individual convectional storms. Several such weak lows are shown on the world weather map (Figure 6.12), lying along the ITCZ. Because the map is for a day in July or August, the ITCZ is shifted well north of the equator. During this season, the rainy monsoon is in progress in Southeast Asia. It is marked by a weak equatorial low in northern India.

Polar Outbreaks

Another distinctive feature of low-latitude weather is the occasional penetration of powerful tongues of cold polar air from the midlatitudes into very low latitudes. These tongues are known as *polar outbreaks.* The leading edge of a polar outbreak is a cold front with squalls, which is followed by unusually cool, clear weather with strong, steady winds. The polar outbreak is best developed in the Americas. Outbreaks that move southward from the United States into the Caribbean Sea and Central America are called "northers" or "nortes," while those that move north from Patagonia into tropical South American are called "pamperos." One such outbreak is shown over South America on the world weather map (Figure 6.12). A severe polar outbreak may bring subfreezing temperatures to the highlands of South America and severely damage such essential crops as coffee.

Tropical Cyclones

The most powerful and destructive type of cyclonic storm is the **tropical cyclone,** which is known as the *hurricane* in the

Focus on Remote Sensing | **6.1**

TRMM—The Tropical Rainfall Monitoring Mission

As we saw when studying the Earth's global radiation balance in Chapter 2, the latitude belt between the Earth's two tropics receives the major portion of the Sun's energy. It is the region of the Hadley cell circulation, covered in Chapter 5, in which warm, moist air converges near the equator (ITCZ) and rises in a giant convection loop that moves heat from the equatorial region to the tropics. As the air rises, convectional precipitation occurs. In the condensation process, latent heat is released to the surrounding atmosphere. The effect is to gather up solar energy by evaporation of water from large areas of ocean or moist land surface and then release that energy to the atmosphere over a much smaller region as precipitation forms.

Mathematical models of atmospheric circulation have long identified this heating of middle levels of the atmosphere in the intertropical convergence zone (ITCZ) as an important process in atmospheric dynamics. However, scientists have lacked specific knowledge about where and when intertropical convectional precipitation occurs and how rainfall formation, and therefore energy release, occurs within rain clouds. Although weather radars have allowed the study of convectional precipitation over land, there has been no such source of data over oceans.

This lack of data, and its importance for weather and climate modeling, led NASA and the National Space Development Agency of Japan (NASDA) to devise and fly the Tropical Rainfall Measuring Mission (TRMM), which monitors clouds and pre-

cipitation in the intertropical zone. This satellite platform, launched from the Tanegashima, Japan, spaceport in November 1997, has a unique orbit. Instead of the polar Sun-synchronous orbit of a typical Earth imager (see *Geographer's Tools 2.1 • Remote Sensing for Physical Geography*), TRMM's orbit is strongly inclined so that the platform revolves from tropic to tropic, covering only the region from 35° N to 35° S latitudes. With a low altitude and a short orbital period, the orbit allows the platform's instru-

ments to observe precipitation over land and ocean surfaces in this region at all times of day and night. In this way, scientists can accumulate accurate statistics about rainfall frequency, intensity, duration, and energy released for this portion of the globe. This, in turn, will allow mathematical models of atmospheric circulation to predict global energy transport, winds, and precipitation with much better accuracy.

The TRMM satellite platform carries five instruments, of which three are used for observations

TRMM observes Hurricane Floyd *This data swath from TRMM instruments shows Hurricane Floyd as it passed north and east of Cuba and the Greater Antilles in September 1999. Cross sections show the concentration of raindrops and ice particles inside the clouds. (NASA)*

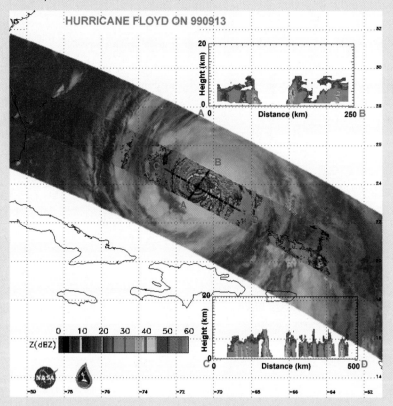

of clouds and precipitation: the precipitation radar, passive microwave imager, and visible and infrared scanner. The precipitation radar provides three-dimensional maps of storm structure, including the intensity and distribution of rain, precipitation type, height of storm clouds, and height at which snow melts into rain. It works by sending a pulse of radio waves toward the Earth and then listening for echoes that are scattered back by water droplets and ice particles. The timing and intensity of the echoes are used to reconstruct a profile of precipitation from the surface to the top of the storm cloud.

The passive microwave imager uses a different principle to monitor water vapor, cloud water, and rainfall intensity. The principle is that water droplets emit more energy in the microwave portion of the spectrum than does a water surface. This allows the measurement of precipitation over oceans. The emitted radiation is very small, however, and the measuring instrument must therefore be quite accurate. There are also methods for estimating precipitation over land using microwave emissions, but they are less effective.

The third imager is a visible and infrared scanner. It acquires images in five spectral regions, ranging from the visible to the thermal infrared. Its visible band tracks clouds by imaging them as bright objects. Thermal bands measure cloud-top temperature, which indicates cloud height. This imager is also quite similar to a number of instruments on other satellites, so it provides a way of linking the unique observations of the precipitation radar and microwave imager to data collected by other platforms at other times. The TRMM platform also carries a CERES

Global rainfall *This image shows precipitation as observed over a two-day period by TRMM instruments. (NASA)*

instrument (see *Focus on Remote Sensing 2.3 • CERES—Clouds and the Earth's Radiant Energy System*) and a lightning imaging sensor.

The image on the facing page shows TRMM data obtained while passing over Hurricane Floyd on September 13, 1999. The inward-spiraling cloud pattern of this tropical cyclone is clearly shown by the wide track of the visible and infrared scanner. The narrow track shows precipitation as imaged by the other instruments. The two cross sections, A–B and C–D, show the structure of the storm. Precipitation is most intense in the yellow and red areas in the eye-wall of the storm, but some of the outer precipitation bands visible in cross section C–D also show very intense precipitation.

The image above is a composite of precipitation measured over a two-day period. Again, yellow and red tones indicate intense rainfall. Note the tropical cyclone present off the west coast of Mexico. Also visible are lines of frontal precipitation across the United States at the top of the image and across the southern Pacific from the equator southeast to southern Chile.

Like many other new orbiting platforms of instruments observing the Earth from space, TRMM is rapidly improving our scientific knowledge of the Earth as a system as well as our ability to model that system and to predict how human activity will alter it.

western hemisphere, the *typhoon* in the western Pacific off the coast of Asia, and the *cyclone* in the Indian Ocean. This type of storm develops over oceans in 8° to 15° N and S latitudes but not closer to the equator. The exact mechanism of formation is not known, but typically the tropical cyclone originates as an easterly wave or weak low, which then intensifies and grows into a deep, circular low. High sea-surface temperatures, over 27°C (81°F), are required for tropical cyclones to form. Once formed, the storm moves westward through the trade-wind belt, often intensifying as it travels. It can then curve northwest, north, and northeast, steered by westerly winds aloft. Tropical cyclones can penetrate well into the midlatitudes, as many residents of the southern and eastern coasts of the United States can attest.

GEO***DISCOVERIES***

An intense tropical cyclone is an almost circular storm center of extremely low pressure. Because of the very strong pressure gradient, winds spiral inward at high speed. Convergence and uplift are intense, producing very heavy rainfall (Figure 6.17). The storm gains its energy through the release of latent heat as the intense precipitation forms. The storm's diameter may be 150 to 500 km (about 100 to 300 mi). Wind speeds can range from 30 to 50 m/s (about 65 to 135 mi/hr) and sometimes much higher. Barometric pressure in the storm center commonly falls to 950 mb (28.1 in. Hg) or lower.

A characteristic feature of the tropical cyclone is its central eye, in which clear skies and calm winds prevail (Figure 6.18). The eye is a cloud-free vortex produced by the intense spiraling of the storm. In the eye, air descends from high altitudes and is adiabatically warmed. As the eye passes over a site, calm prevails, and the sky clears. Passage of the eye may take about half an hour, after which the storm strikes with renewed ferocity, but with winds in the opposite direction. Wind speeds are highest along the cloud wall of the eye.

The intensity of tropical cyclones is rated on the Simpson–Saffir scale, shown in Table 6.2. This scale ranks storms based on the central pressure of the storm, mean wind speed, and height of accompanying storm surge. Category 1 storms are weak, while category 5 storms are devastating.

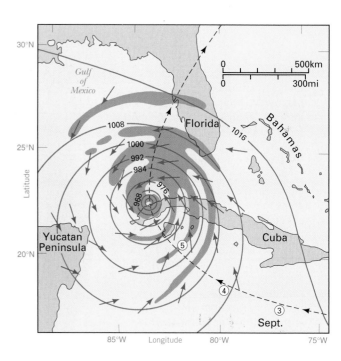

6.17 Hurricane weather map *A simplified weather map of a hurricane passing over the western tip of Cuba. Daily locations, beginning on September 3, are shown as circled numerals. The recurving path will take the storm over Florida. Shaded areas show dense rain clouds as seen in a satellite image.*

Tropical cyclones (hurricanes or typhoons) are the most powerful and destructive of cyclonic storms, with wind speeds ranging from 30 to 50 m/s (about 65 to 135 mi/hr) and above. In the central eye of the storm, barometric pressure may fall to 950 mb (28.1 in. Hg) or lower.

Refer to Figure 6.11, presented earlier, which shows typical paths of both wave cyclones and tropical cyclones. As you can see, tropical cyclones always form over oceans. In the western hemisphere, hurricanes originate in the Atlantic off the west coast of Africa, in the Caribbean Sea, or off the west coast of Mexico. Curiously, tropical cyclones do not form in the South Atlantic or southeast Pacific regions. As a result, South America is never threatened by these severe storms. In the Indian Ocean, cyclones originate both north and south of the equator, moving north and east to strike India,

6.18 Anatomy of a hurricane *In this schematic diagram, cumulonimbus (Cb) clouds in concentric rings rise through dense stratiform clouds. Cirrus clouds (Ci) fringe out ahead of the storm. Width of diagram represents about 1000 km (about 600 mi). (Redrawn from NOAA, National Weather Service.)*

Table 6.2	Simpson–Saffir Scale of Tropical Cyclone Intensity		
Category	Central Pressure mb (in. Hg)	Storm Surge m (ft)	Mean Wind m/s (mph)
1 Weak	>980 (>29.0)	1.2–1.7 (4–5)	33–42 (74–95)
2 Moderate	965–979 (28.5–29.0)	1.8–2.6 (6–8)	43–49 (96–110)
3 Strong	945–964 (27.9–28.5)	2.7–3.8 (9–12)	50–58 (111–130)
4 Very Strong	920–944 (27.2–27.9)	3.9–5.6 (13–18)	59–69 (131–155)
5 Devastating	<920 (<27.2)	>5.6 (>18)	>69 (>155)

Pakistan, and Bangladesh, as well as south and west to strike the eastern coasts of Africa and Madagascar. Typhoons of the western Pacific also form both north and south of the equator, moving into northern Australia, Southeast Asia, China, and Japan.

Tracks of tropical cyclones of the North Atlantic are shown in detail in Figure 6.19. Most of the storms originate at 10° to 20° N latitude, travel westward and northwestward through the trades, and then turn northeast at about 30° to 35° N latitude into the zone of the westerlies. Here their intensity lessens, especially if they move over land. In the tradewind belt, the cyclones travel 10 to 20 km (6 to 12 mi) per hour. In the zone of the westerlies, their speed is more variable.

Tropical cyclones are now tracked using satellite images. They are often

> **The strong onshore winds of a tropical cyclone, coupled with high tides and a favorable offshore sea-bottom configuration, can create a devastating storm surge that inundates wide areas near the coast. Heavy rains, while recharging water supplies, can also produce damaging floods.**

easy to identify from their distinctive pattern of inspiraling bands of clouds and a clear central eye. Figure 6.20 provides a gallery of satellite images of tropical cyclones.

Tropical cyclones occur only during certain seasons. For hurricanes of the North Atlantic, the season runs from May through November, with maximum frequency in late summer or early autumn. In the southern hemisphere, the season is roughly the opposite. These periods follow the annual migrations of the ITCZ to the north and south with the seasons, and correspond to periods when ocean temperatures are warmest.

For convenience, tropical cyclones are given names as they are tracked by weather forecasters. Male and female names are alternated in an alphabetical sequence renewed each season. Two sets of names are used—one for hurricanes of the Atlantic and one for typhoons of the Pacific. Names are reused, but the names of storms that cause significant damage or destruction are retired from further use. GEODISCOVERIES

Impacts of Tropical Cyclones

Tropical cyclones can be tremendously destructive storms. Islands and coasts feel the full force of the high winds and flooding as tropical cyclones move onshore. For example, Hurricane Andrew, striking the Florida coast near Miami, was the most devastating storm ever to occur in the United States, claiming as much as $25 billion in property damage and 43 lives. *Eye on the Environment 6.2 • Hurricane Andrew—Killer Cyclone* provides more information about this costly natural disaster.

The most serious effect of tropical cyclones is usually coastal destruction by storm waves and very high tides. Since the atmospheric pressure at the center of the cyclone is so low, sea level rises toward the center of the storm. High winds create a damaging surf and push water toward the coast, raising sea level even higher. Waves attack the shore at points far inland of the normal tidal range. Low pressure, winds, and the underwater shape of a bay floor can combine to produce a sudden rise of water level, known as a **storm surge.** During surges, ships are lifted by the high waters and can be stranded far inland.

If high tide accompanies the storm, the limits reached by inundation are even higher. This flooding can create enormous death tolls. At Galveston, Texas, in 1900, a sudden storm surge generated by a severe hurricane flooded the low coastal city and drowned about 6000 persons. Low-lying coral atolls of the western Pacific may be entirely swept over by wind-driven sea water, washing away palm trees and houses and drowning the inhabitants.

Also important is the large amount of rainfall produced by tropical cyclones. For some coastal regions, these storms provide much of the summer rainfall. Although this rainfall is a valuable water resource, it can also produce freshwater flooding, raising rivers and streams out of their banks. On

6.19 Tracks of typical hurricanes occurring during August The storms arise in warm tropical waters and move northwest. On entering the region of prevailing westerlies, the storms change direction and move toward the northeast.

6.20 *Tropical Cyclone Gallery*

a...*Hurricane Mitch* *on October 26, 1998, off the coast of Honduras, as it approached the Yucatan Peninsula. Data from NOAA's AVHRR instrument. (NOAA.)*

b...*Hurricane Linda* *approaching Baja California on September 12, 1997. Data are from the NOAA GOES-9 satellite. (NASA. Images and rendering by Marit Jentoft-Nilsen.)*

c...**Typhoon Bilis,** a category 5 tropical cyclone, approaching Taiwan on August 22, 2000. Data are from the SeaWiFS (Sea Wide-Field Scanner) instrument. (NASA, SeaWiFS Project Office.)

d ... **Hurricane Michael,** a category 2 hurricane, seen on October 18, 2000. This moderate hurricane made landfall on Newfoundland the following day, where wind gusts to nearly 50 m/s (112 mi/hr) were observed. It was only the second hurricane to hit Canada in 25 years. Damage was minor.

Eye on the Environment | 6.2

Hurricane Andrew—Killer Cyclone

Tropical cyclones, or hurricanes, are among the most destructive of natural phenomena. With sustained high winds and rain, they can wreak a pattern of destruction across vast areas of land. Coastal areas are particularly vulnerable. Since hurricanes develop and reach their maximum intensity over ocean waters, coastal regions experience the full brunt of the storm. Storm surges caused by strong onshore winds can raise sea levels and bring storm-ravaged surf inland to batter coastal structures well back from the beach. Low coastal islands can be completely submerged, drowning their inhabitants.

Hurricane Andrew, striking the south Florida Peninsula on August 23–24, 1992, and moving on to ravage the Louisiana coast on August 26, was the most damaging natural disaster ever to hit the United States. In the path of Andrew in south Dade Country, Florida, was a population of 355,000. By the time the storm was over, 80,000 dwellings were demolished or destroyed. Another 55,000 dwellings were seriously damaged. Estimates of property loss ranged between $20 and $30 billion. The death toll was small, however—the storm claimed only 43 victims. Evacuations and emergency warnings reduced the toll significantly. In Louisiana, Andrew struck a rural area of agricultural fields, marshlands, and small towns, so damage was less intense. Another 15 people died, and additional damage was estimated at about $2 billion.

Hurricane Andrew, like many other tropical cyclones, began as a

tropical easterly wave of showers off the African coast near the Cape Verde Islands on August 14. Moving westward, it became a tropical depression with winds of 18 m/s (40 mi/hr) on August 17 and was named Andrew—the first tropical cyclone of the 1992 season. By August 20, the storm had moved westward but gained little strength. Opposing the westward surface motion of the storm was an upper airflow from west to east, creating a high-level wind shear that kept the storm from developing a strong vertical structure. However, by August 21, an unusual easterly upper airflow took hold, eliminating the wind shear. This allowed Andrew to gain strength. By 11 P.M. on the 21st, Andrew's winds had increased to 27 m/s (60mi/hr).

Andrew continued to intensify. On Saturday morning, August 22,

the storm became a category 1 hurricane on the Simpson–Saffir scale, and later in the day, it attained category 2 status. The storm continued its rapid development, becoming a category 3, then a category 4 storm on Sunday, August 23. Bearing winds of 72 m/s (160 mi/hr) and a central low pressure of 924 mb (27.33 in. Hg), the hurricane swept through the southern Bahamas that evening, killing four and leaving 1700 homeless. Between 5:00 and 6:00 A.M. Monday morning, Andrew's eye crossed the coast of Florida just south of Miami. Sustained winds were in the range of 65 m/s (145 mi/hr). A few miles to the north of the eye, the anemometer atop the National Hurricane Center measured a peak of 73 m/s (164 mi/hr) before breaking at the height of the storm. Other anemometers meas-

Andrew's wrath *Hurricane Andrew blew the roofs from these homes in the suburbs of Miami, Florida, exposing their contents to wind and water damage.*

steep slopes, soil saturation and high winds can topple trees and produce disastrous mudslides and landslides.

Tropical cyclone activity varies from year to year and decade to decade. In the Atlantic Basin, 13 strong hurricanes of category 3 or higher struck the eastern United States or Florida Peninsula from 1947 to 1969, while only one hurri-

cane struck the same region in the period 1970–1987. Although a number of strong storms, including category 4 or 5 hurricanes Gilbert (1988), Hugo (1989), and Andrew (1992) occurred in 1988–1992, Atlantic hurricane activity remained depressed until 1994. However, the 1995 season provided 19 named tropical storms—two fewer than the ured gusts of about 78 m/s (175 mi/hr). The storm surge brought sea level up to 5.2 m (16.9 ft) above normal.

Crossing Florida at a speed of 7 to 8 m/s (16 to 18 mi/hr) and entering the Gulf of Mexico, Andrew headed west and slightly north. Intensifying over the warm Gulf water, the hurricane reattained category 4 status and struck the Louisiana coast on Wednesday, August 26. Low barrier islands were overtopped, eroded, and cut with new channels. Marshes and shrimping grounds were filled with sand and debris. Sugar cane crops were flattened. Many dwellings were severely damaged. Moving inland and weakening, Andrew headed northeast, finally disintegrating into a region of showers in the Ohio River Valley on August 28.

Although Andrew's winds in south Florida were measured in the 60 to 74 m/s (135 to 165 mi/hr) range, an analysis of the damage patterns afterward indicated that winds were still higher in some locations. These winds may have resulted from a particularly intense convectional cell that developed on the north side of the storm as it crossed the coastline.

Another possibility, according to Professor Ted Fujita of the University of Chicago, is a "suction vortex"—a swirl of faster winds that occurs within the storm's eye-wall. Although the vortex normally achieves wind speeds of only 9 to 13 m/s (20 to 30 mi/hr), the vortex can be sucked upward by a strong local updraft, causing it to narrow and spin more intensely. In fact, the vortex can attain wind speeds of up to 35 m/s (80 mi/hr). Where the swirling motion reinforces the overall storm wind of perhaps 55 m/s (120 mi/hr), total wind speed can approach 90 m/s (200 mi/hr).

Andrew's harvest *The entrance to Pinelands, in the Everglades National Park, is blocked by fallen trees brought down by the high winds of Hurricane Andrew.*

No normal structure can survive such winds for more than a minute or two. Fortunately, these winds are restricted to a narrow, tornado-like swath of up to 100 m (about 300 ft) in width.

The damage was not restricted to human settlements. Hurricane Andrew delivered a solid blow to the Everglades as well. Most "hammocks"—tree-covered islands in the watery grassland—were stripped bare of leaves, their trees broken and splintered. Mangrove swamps on the western edge of the Everglades were also defoliated.

Although the Everglades ecosystem has endured the periodic assaults of hurricanes for centuries and is well-adjusted to surviving them, ecologists are concerned about the regrowth process. A number of imported species threaten to take over, crowding out the native ones. It is not certain how the presence of these exotic plants will influence the recovery of the hammocks and other areas opened up by Andrew's high winds and waters.

Andrew is an example of the brutal power of the tropical cyclone to change the face of the Earth. Only a massive earthquake in a major urban area is capable of doing more damage. Although little can be done to stop such disasters from occurring, much can be done to reduce the damage they cause. Warnings and evacuations have reduced death rates to very low numbers. What is needed is better attention to building structures that are more wind-resistant and have such simple protections as storm shutters. Preserving shallow bays and mangrove swamps in natural states can also help by brunting the force of storm surges. Keeping beach-front development in check can reduce the number and size of structures exposed to the full force of hurricane winds and waves. Although Andrew was the first major hurricane to hit south Florida in nearly 30 years, the next hurricane will probably hit much sooner. Let's hope that this region will fare better next time, thanks to the lessons learned from Andrew.

record of 1933—and 11 hurricanes—only one short of the record set in 1969. During August 27 and 28, 1995, a record of five tropical cyclones were active simultaneously in the Atlantic Basin. The 1996 season brought a respite with only two moderate hurricanes—Bertha and Fran—striking nearly the same region of the North Carolina coast. The following

year also showed reduced activity, with only weak Hurricane Danny making landfall on the Louisiana–Alabama coast.

But hurricanes returned with a vengeance in 1998, with two killer storms—Georges and Mitch. Hurricane Georges, a category 4 storm in the central Atlantic, weakened somewhat before making landfalls in the Lesser Antilles islands of Antigua, St. Kitts, and Nevis, and then proceeded along the Greater Antilles chain from Puerto Rico to Hispaniola and Cuba. After grazing Key West, Florida, Georges finally made continental landfall near Biloxi, Mississippi, on September 28 with winds of about 50 m/s (112 mi/hr).

About a month later, Hurricane Mitch, one of the deadliest Atlantic tropical cyclones in history, struck the heart of Central America. Claiming a toll of 9086, Mitch killed most of its victims with its torrential rains and their aftermath. The monster category 5 storm attained a central low pressure of 905 mb with winds of 77.2 m/s (172 mi/hr) in the western Caribbean on October 26, 1998. This central pressure is the fourth lowest ever measured in an Atlantic hurricane and ties Hurricane Mitch with Hurricane Camille of 1969. Moving at an average speed of only 2.2 m/s (5 mi/hr), Mitch hovered over the seven-nation Central American isthmus for nearly a week, dumping about 60 cm (24 in.) of rain at most locations. As much as 91 cm (36 in.) was reported at one mountain weather station, and it is likely that higher rainfalls went unrecorded. Countless mudflows and debris avalanches coursed down the rugged slopes of the western ranges of Honduras and Nicaragua, burying entire villages and villagers alike. Major rivers quickly flooded, sweeping floodplain settlements away on roiling waters. The damage was truly devastating. Honduras and Nicaragua suffered losses of about half their annual gross national products. Guatemala and El Salvador were also hard hit.

The most recent forecasts suggest that we may be entering an active period for Atlantic hurricanes, similar to that of 1947–1969. If so, we can expect the cost to human society to increase substantially, given the buildup in coastal development that has occurred in hurricane-prone regions.

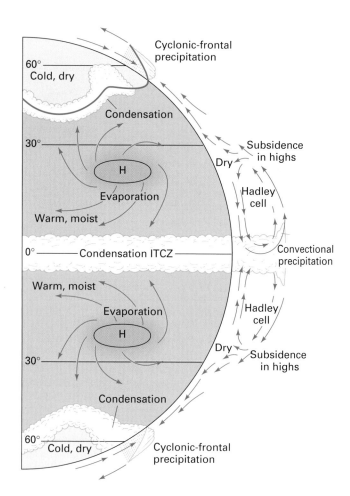

6.21 Global atmospheric transport of heat and moisture *Major mechanisms of heat transport in the atmosphere include intertropical convergence, the Hadley cell circulation, and cyclonic-frontal precipitation in the midlatitudes.*

POLEWARD TRANSPORT OF HEAT AND MOISTURE

Recall from Chapter 2 that vast global heat flows extend poleward from the equatorial and tropical regions, where the Sun's energy is concentrated and net radiation is positive, to the middle and higher latitudes, where net radiation is negative. These heat flows are referred to as *poleward heat transport*. Weather systems such as cyclones are byproducts of this heat transport, which sends warm, moist air poleward to mix with cooler, drier air, thus creating cyclonic precipitation. With our knowledge of the global circulation patterns of the atmosphere and oceans, and our knowledge of precipitation processes, we can now examine poleward heat transport in more detail.

Atmospheric Heat and Moisture Transport

Figure 6.21 is a schematic diagram of global heat and moisture flow within the atmosphere. An important feature is the Hadley cell circulation—a global convection loop in which moist air converges and rises in the intertropical conver-

gence zone (ITCZ) while subsiding and diverging in the subtropical high-pressure belts.

The Hadley cell convection loop acts to pump heat from warm equatorial oceans poleward to the subtropical zone. Near the surface, air flowing toward the ITCZ picks up water vapor evaporated by sunlight from warm ocean surfaces, increasing the moisture and latent heat content of the air. With convergence and uplift of the air at the ITCZ, the latent heat is released as sensible heat in the condensation that occurs. Air traveling poleward in the return circulation aloft retains much of this heat, although some is lost to space by radiant cooling. When the air descends in the subtropical high-pressure belts, the sensible heat becomes available at the surface. The net effect is thus to gather heat from tropical and equatorial zones and release the heat in the subtropical zone, where it can be conveyed further poleward by the motion of mT and cT air masses into the midlatitudes.

In the mid- and high latitudes, poleward heat transport is also produced by the Rossby wave mechanism (see Figure 5.25). Lobes of cold, dry polar or arctic air (cP, cA, and cAA air masses) plunge toward the equator, while tongues of warmer, moister air (mT and mP air masses) flow toward the poles. At their margins, cyclones develop, releasing latent heat in precipitation and providing a heat flow that warms the mid-and higher latitudes well beyond the capabilities of the Sun.

Oceanic Heat Transport

Just as atmospheric circulation plays a role in moving heat from one region of the globe to another, so does oceanic circulation. Chapter 5 documented the thermohaline circulation of the ocean, in which the sinking of cold, salty water in the North Atlantic is linked to a slow-moving system of deep ocean currents, upwelling, and surface currents (Figure 5.35). Figure 6.22 diagrams the key features of this flow. At the surface, it involves slow drifting of huge volumes of Atlantic surface water from the South Atlantic to the North Atlantic. Exposed to intense insolation in the equatorial and tropical zones, the water is warmed. Evaporation also concentrates salt in the

Mechanisms of poleward heat transport include the Hadley cells, which move equatorial heat to the tropics, Rossby waves, which move tongues of warm, moist air poleward and cold, dry air equatorward in midlatitudes, and the Atlantic thermohaline circulation, which moves warm ocean surface water northward where it ultimately warms Europe.

surface layer, making the water more dense. As it travels northward, the water cools, losing heat and warming the atmosphere above it. Finally, the water is chilled almost to its freezing point. Because of its cold temperature and saltiness, the North Atlantic water is dense enough to sink to the bottom of the Atlantic Basin. This cold, Atlantic bottom water then heads slowly southward.

Note that this circulation pattern is also a type of sinking convection loop powered by the loss of heat and increase in the density of the surface water as it moves poleward. In effect, this loop acts like a heat pump in which sensible heat is acquired in tropical and equatorial regions and is moved northward into the North Atlantic, where it is transferred to the air. Since wind patterns move air eastward at higher latitudes, this heat ultimately warms Europe. The amount of heat released is quite large. In fact, a recent calculation shows that it is equal to about 35 percent of the total solar energy received by the Atlantic Ocean north of 40° latitude! This type of circulation does not occur in the Pacific or Indian oceans.

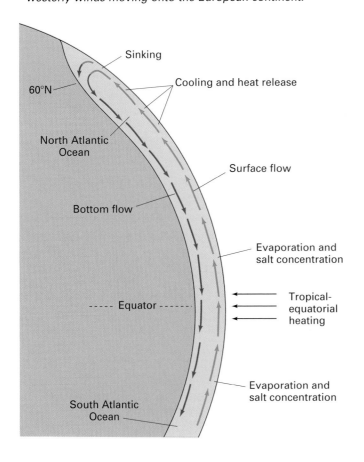

6.22 Thermohaline convection loop *Warm surface waters flow into the North Atlantic, cool, and sink to the deep Atlantic Basin, creating a circulation that warms westerly winds moving onto the European continent.*

CLOUD COVER, PRECIPITATION, AND GLOBAL WARMING

Thus far in this text, we have examined climatic change from several different perspectives. Let's now focus on how global climate might be influenced by an increase in clouds and precipitation. Recall that global temperatures have been rising over the last 20 years. As part of this global warming, satellite data have detected a rise in temperature of the global ocean surface of about 1°C (1.8°F) over the past decade. Any rise in sea-surface temperature increases the rate of evaporation. And the increase in evaporation will raise the average atmospheric content of water vapor. What effect will this have on climate?

Water has several roles in global climate. First, in its vapor state it is one of the greenhouse gases. That is, it absorbs and emits longwave radiation, thus enhancing the warming effect of the atmosphere above the Earth's surface. In fact, water is more important than CO_2 in creating the greenhouse effect. Thus, one result of an increase in global water vapor in the atmosphere should enhance warming.

Second, water vapor can condense or deposit, forming clouds. Will more clouds increase or decrease global temperatures? That question is still being debated by the scientists who study and model the atmosphere. Clouds can have two different effects on the surface radiation balance. As large, white bodies, they can reflect a large proportion of incoming shortwave radiation back to space, thus acting to cool global temperatures. But cloud droplets and ice particles also absorb longwave radiation from the ground and return that emission as counterradiation. This absorption is an important part of the greenhouse effect, and it is much stronger for water as cloud droplets or ice particles than as water vapor. Thus, clouds also act to warm global temperatures by enhancing longwave reradiation from the atmosphere to the surface.

Which effect, longwave warming or shortwave cooling, will dominate? The best information at present, obtained from satellite measurements, is that the average flow of shortwave energy reflected by clouds back to space is about 50 W/m^2, while the greenhouse warming effect of clouds amounts to about 30 W/m^2. Thus, the net effect of clouds is a cooling of the planet by an energy flow of about 20 W/m^2.

What will happen when surface temperature increases, water vapor in the atmosphere increases, and more clouds form? Computer models of global climate generally agree that longwave warming will be enhanced more than shortwave cooling. This means that the net effect of clouds will be a positive feedback that enhances the greenhouse effect and so accentuates the surface warming. After a small increase, the effect of clouds will still be to cool the planet but not at so great a rate. Shortwave cooling will still prevail but to a lesser degree. Another way to put it is to say that with global warming, the cooling effects of clouds on climate will be reduced somewhat, leading to temperatures that are even warmer.

What about precipitation? This is the third role of water in global climate. With more water vapor and more clouds in the air, more precipitation should result. Think, however, about what might happen if precipitation increases in arctic and subarctic zones. In this case, more of the Earth's surface

---

While low, thick clouds tend to cool the Earth and high, thin clouds tend to warm it, in balance clouds presently act to cool the Earth. As increasing surface temperatures from global warming cause more water to evaporate into the atmosphere, the net effect will be more high clouds, thus reducing the cooling effect and enhancing warming.

---

could be covered by snow and ice. Since snow is a good reflector of solar energy, this would increase the Earth's albedo, thus tending to reduce global temperatures. Another effect might be to increase the depth of snow, thus tying up more water in snow packs and reducing runoff to the oceans. Reducing runoff would reduce the rate at which sea level has been rising, presumably as a result of climate warming (see Chapter 18).

At this time, scientists are unsure how the global climate system will respond to global warming induced by the CO_2 increases predicted for the twenty-first century. Perhaps increased cloud cover and enhanced precipitation will slow the warming trend—or perhaps they will enhance it. As time goes by, however, our understanding of global climate and our ability to predict its changes are certain to increase.

A LOOK AHEAD   With a knowledge of the global circulation patterns of the atmosphere and oceans, as well as an understanding of weather systems and how they produce precipitation, the stage is set for global climate—which is the topic of Chapter 7. As we will see, the annual cycles of temperature and precipitation that most regions experience are quite predictable, given the changes in wind patterns, air mass flows, and weather systems that occur with the seasons. Our description of the world's climates grows easily and naturally from the principles you have mastered thus far in your study of physical geography.

In Review | Weather Systems

- A **weather system** is a recurring atmospheric circulation pattern with its associated weather. Weather systems include wave cyclones, traveling anticyclones, tornadoes, easterly waves, weak equatorial lows, and tropical cyclones.

- **Air masses** are distinguished by the latitudinal location and type of surface of their *source regions*. Air masses influencing North America include those of continental and maritime source regions, and of arctic, polar, and tropical latitudes.

- The boundaries between air masses are termed **fronts.** These include **cold** and **warm fronts,** where cold or warm air masses are advancing. In the **occluded front,** a cold front overtakes a warm front, pushing a pool of warm, moist air mass above the surface.

- *Traveling cyclones* include wave cyclones, tropical cyclones, and the tornado. The *traveling anticyclone* is typically a fair-weather system.

- **Wave cyclones** form in the midlatitudes at the boundary between cool, dry air masses and warm, moist air masses. In the wave cyclone, a vast inspiraling motion produces cold and warm fronts, and eventually an occluded front. Precipitation normally occurs with each type of front.

- **Tornadoes** are very small, intense cyclones that occur as a part of thunderstorm activity. Their high winds can be very destructive.

- Tropical weather systems include *easterly waves* and *weak equatorial lows.* Easterly waves occur when a weak low-pressure trough develops in the easterly wind circulation of the tropical zones, producing convergence, uplift, and shower activity. Weak equatorial lows occur near the intertropical convergence zone. In these areas of low pressure, convergence triggers abundant convectional precipitation.

- **Tropical cyclones** can be the most powerful of all storms. They develop over very warm tropical oceans and can intensify to become vast inspiraling systems of very high winds with very low central pressures. As they move onto land, they bring heavy surf and storm surges of very high waters. Tropical cyclones have caused great death and destruction in coastal regions.

- Global air and ocean circulation provides the mechanism for *poleward heat transport* by which excess heat moves from the equatorial and tropical regions toward the poles. In the atmosphere, the heat is carried primarily in the movement of warm, moist air poleward, which releases its latent heat when precipitation occurs. In the oceans, a global circulation moves warm surface water northward through the Atlantic Ocean. Heated in the equatorial and tropical regions, the surface water loses its heat to the air in the North Atlantic and sinks to the bottom. These heat flows help make northern and southern climates warmer than we might expect based on solar heating alone.

- Because global warming, produced by increasing CO_2 levels in the atmosphere, will increase the evaporation of surface water, atmospheric moisture levels will increase. This will tend to enhance the greenhouse effect. But more clouds are likely to form, and this should cool the planet. Increased moisture could also reduce temperatures by increasing the amount and duration of snow cover. Further research on the effect of global warming on climate is needed.

Key Terms

Review Questions

1. Define air mass. What two features are used to classify air masses?
2. Compare the characteristics and source regions for mP and cT air mass types.
3. Describe a tornado. Where and under what conditions do tornadoes typically occur?
4. Identify three weather systems that bring rain in equatorial and tropical regions. Describe each system briefly.
5. Describe the structure of a tropical cyclone. What conditions are necessary for the development of a tropical cyclone? Give a typical path for the movement of a tropical cyclone in the northern hemisphere.
6. Why are tropical cyclones so dangerous?
7. How does the global circulation of the atmosphere and oceans provide poleward heat transport?
8. How does water, as vapor, clouds, and precipitation, influence global climate? How might water in these forms act to enhance or retard climatic warming?

Eye on Environment 6.2 • Hurricane Andrew—Killer Cyclone

1. Where did Hurricane Andrew first form, and how did it develop and move?
2. What damage was sustained by south Florida and coastal Louisiana from Hurricane Andrew?
3. What were the maximum wind speeds of Andrew? How were they achieved?
4. How did Hurricane Andrew affect the Everglades?

Visualizing Exercises

1. Identify three types of fronts and draw a cross section through each. Show the air masses involved, the contacts between them, and the direction of air mass motion.
2. Sketch two weather maps, showing a wave cyclone in open and occluded stages. Include isobars on your sketch. Identify the center of the cyclone as a low. Lightly shade areas where precipitation is likely to occur.

Essay Questions

1. Compare and contrast midlatitude and tropical weather systems. Be sure to include the following terms or concepts in your discussion: air mass, convectional precipitation, cyclonic precipitation, easterly wave, polar front, stable air, traveling anticyclone, tropical cyclone, unstable air, wave cyclone, and weak equatorial low.
2. Prepare a description of the annual weather patterns that are experienced through the year at your location. Refer to the general temperature and precipitation pattern as well as the types of weather systems that occur in each season.

Eye on the Landscape

Chapter Opener Hurricane Andrew Remote sensing can sometimes reveal ocean features, especially in coastal regions **(A)**. Here the brown tones in Florida Bay and to the east, north of Andros Island, Bahamas, indicate waters rich in terrestrial sediment and organic matter. Some of this material may have been washed into these shallow waters by the torrential rains of the hurricane. Between the two land areas, along the eastern coast of south Florida, the clear (dark) waters of the Florida current flow northward as the start of the Gulf Stream. (See Chapter 5 for ocean currents.) The Everglades appears as the southernmost dark spot on the Florida Peninsula. Lake Okeechobee is to the north.

6.7 Picture of an anticyclone On this clear satellite image, it is easy to see some of the major features of the eastern North American landscape. At **(A)**, note the difference in soil color along the Mississippi River. These are alluvial soils, deposited over millennia by the Mississippi during the Ice Ages. At **(B)**, note the curving arc that marks the boundary between soft coastal plain sediments to the south and the weathered rocks and soils of the piedmont, to the north. (Figure 17.8 diagrams this location.) Actually, this boundary can be traced all the way to New York City. Further inland lie the Appalachians **(C)**, showing puffy orographic cumulus clouds. (We covered orographic precipitation in Chapter 4.)

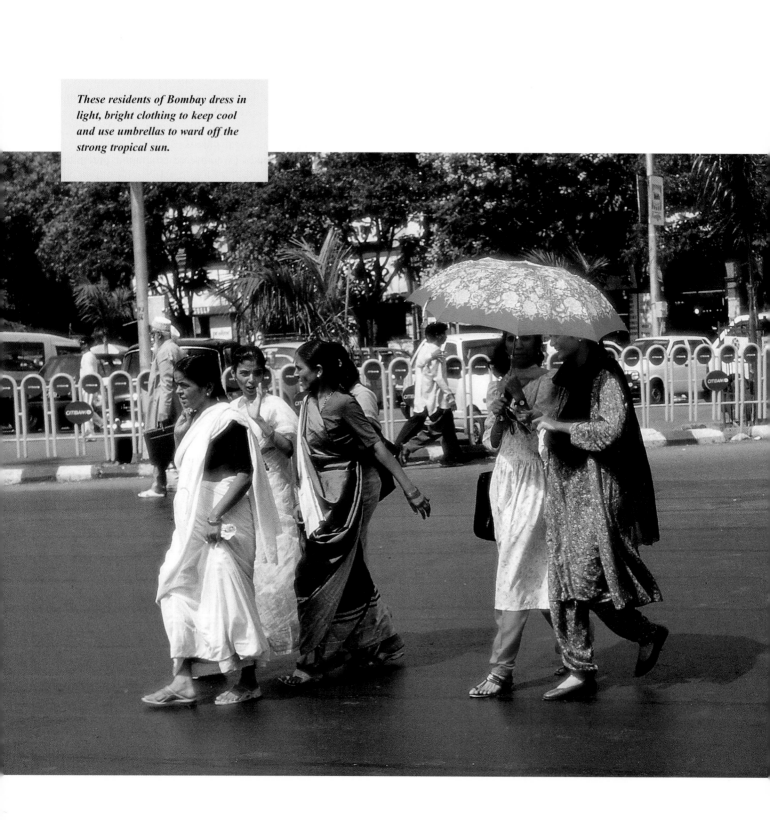

These residents of Bombay dress in light, bright clothing to keep cool and use umbrellas to ward off the strong tropical sun.

7 | GLOBAL CLIMATES

. . . .
PUTTING YOU IN THE PICTURE

London, Rome, Cairo, Delhi, Singapore, Tokyo, Honolulu, and Anchorage! An around-the-world tour! It was hard to believe that you were the winner of the sweepstakes drawing sponsored by the credit card company, but now that you've picked up your first-class air ticket, it's finally hitting home. Although you'll only spend a few days in each of these famous cities, it will be a real thrill to experience them first-hand.

But what kind of weather should you expect? Will it be hot or cold, sunny or rainy? Should you pack your overcoat and sweaters? Or will your lightweight clothes be more appropriate? Will you need an umbrella in any of these cities? You'll be leaving in January, so you know what to expect for your trip to the airport, but what happens after that? Fortunately, you still have your college physical geography textbook to refer to. That evening, you brush up on the climates you will encounter.

"Let's see, London. Oh yes, that's a marine west-coast climate type. Lots of occluded wave cyclones in January. Hmm. I'd better have my raincoat and umbrella. Wool would be good. Perhaps that lightweight gray sweater will fit in my carry-on bag. Then on to Rome. That's the Mediterranean climate. Well, winter is the wet season for the Mediterranean climate, but the rainfall is so much lighter than in London that I'm likely to get away with just an umbrella. Still, I'll need some kind of light coat, since it won't be very warm. The raincoat will probably do fine.

"Cairo—now, that's the dry subtropical climate. Not much chance of rain in that climate, and the hottest temperatures aren't till later in the year, when the sun moves to the tropic of cancer and is nearly overhead. Better have something light and cool. And I'll need my swimsuit for lounging by the hotel pool. Then there's Delhi. It's dry tropical there as well. Hmm. With the winter monsoon bringing dry, continental air down the slope of the Himalayas, it should be pretty clear, with warm days and cool nights.

"Then on from Delhi to Singapore. Oh oh, Singapore is right on the equator, so that would be the wet equatorial climate. I'd better make sure my umbrella isn't buried too deeply in my suitcase. Those convectional showers don't last long, but they can be pretty intense. It'll certainly be hot and steamy. Now, Tokyo is in the moist continental climate, at about 35° N latitude. That's the same as New York. In January? Looks like I'll need those wool clothes again. And I'd better keep that umbrella handy.

"Now, Honolulu! That's the wet equatorial climate, the same as Singapore. But Hawaii's at about 20° N, so the ocean will be cooler and the daily temperature should be just perfect! Last of all is Anchorage. That's the boreal forest climate—and in January! I'll certainly need something warm. But I hate to lug my heavy winter coat around the world just for the last few days. Here's an idea! I've been putting off buying that new ski suit for a couple of years, but how about buying it in Anchorage? And skis, too. What a way to end the trip! Swimming in the warm Pacific and skiing in Alaska!"

Who said learning physical geography wouldn't come in handy some day? After finishing this chapter, you'll be set to pack your clothes for any destination, any time!

As we examine the Earth's climates, you will find that your earlier studies of air temperatures, precipitation, global winds and circulation patterns, air masses, and weather systems will all combine to help you understand the nature of global climates as they are located over the globe. Thus, this chapter will help you to bring together much of the knowledge of physical geography that you have already acquired in a focus on the breadth and diversity of the world's climatic regions. ∎

GLOBAL CLIMATES

What do we mean by climate? In its most general sense, **climate** is the average weather of a region. To describe the average weather of a region, we could use many of the measures describing the state of the atmosphere that we have already encountered. These might include daily net radiation, barometric pressure, wind speed and direction, cloud cover and type, presence of fog, precipitation type and intensity, incidence of cyclones and anticyclones, frequency of frontal passages, and other such items of weather information. However, observations as detailed as these are not made regularly at most weather stations around the world. To study climate on a worldwide basis, we must turn to the two simple measurements that are made daily at every weather station—temperature and precipitation. We will be concerned with both their average values for each month of the year and their variation across the months of the year.

Note that temperature and precipitation strongly influence the natural vegetation of a region—for example, forests occur generally in moist regions, and grasslands in dry regions. The natural vegetation cover is often a distinctive feature of a climatic region and typically influences the human use of the area. Temperature and precipitation are also important factors in the cultivation of crop plants—a necessary process for human survival. And the development of soils, as well as the types of processes that shape landforms, are partly dependent on temperature and precipitation. For these reasons, we will find that climates defined on the basis of temperature and precipitation also help set apart many features of the environment, not just climate alone. This is why the study of global climates is such an important part of physical geography.

KEYS TO CLIMATE

A few simple principles discussed in earlier chapters are very helpful in understanding the global scope of climate. First, recall from Chapter 3 that two major factors influence the annual cycle of air temperature experienced at a station: latitude and coastal versus continental location.

- *Latitude.* The annual cycle of temperature at a station depends on its latitude. Near the equator, temperatures are warmer and the annual range is low. Toward the poles, temperatures are colder and the annual range is greater. These effects are produced by the annual cycle of insolation, which varies with latitude.
- *Coastal-continental location.* Coastal stations show a smaller annual variation in temperature, while the variation is larger for stations in continental interiors. This effect occurs because ocean surface temperatures vary less with the seasons than do temperatures of land surfaces.

214

Air temperature also has an important effect on precipitation. Recall this principle from Chapter 4:

- *Warm air can hold more moisture than cold air.* This means that colder regions generally have lower precipitation than warmer regions. Also, precipitation will tend to be greater during the warmer months of the temperature cycle.

Another key idea associated with climate is the *time cycle,* which we discussed in our Introduction. The primary driving force for weather, as we have seen, is the flow of solar energy received by the Earth and atmosphere. Since that energy flow varies on daily cycles with the planet's rotation and on annual cycles with its revolution in orbit, it imposes these cycles on temperature and precipitation. Other time cycles appear in climate as well.

Keeping these key ideas in mind as you read this chapter and the two that follow will help make the climates easy to understand and explain.

Temperature Regimes

Let's look in more detail at the influence of latitude and location on the annual temperature cycle of a station. Figure 7.1 shows some typical patterns of mean monthly temperatures observed at stations around the globe. We can refer to these patterns as *temperature regimes*—distinctive types of annual temperature cycles related to latitude and location. In the figure, each regime has been labeled according to its latitude zone: equatorial, tropical, midlatitude, and subarctic. Some labels also describe the location of the station in terms of its position on a landmass—"continental" for a continental interior location, and "west coast" or "marine" for a location close to the ocean.

The equatorial regime (Douala, Cameroon, 4° N) is uniformly very warm. Temperatures are close to 27°C (81°F) year-round. There are no temperature seasons because insolation is nearly uniform through the year. In contrast, the tropical continental regime (In Salah, Algeria, 27° N) shows a very strong temperature cycle. Temperatures change from very hot when the Sun is high, near one solstice, to mild at the opposite solstice. However, the situation is quite different at Walvis Bay, Southwest Africa (23° S), which is at nearly the same latitude and so has about the same insolation cycle as In Salah. At Walvis Bay, we find the tropical west-coast regime, which has only a weak annual cycle and no extreme heat. The difference, of course, is due to the moderating effects of the maritime location of Walvis Bay. This moderating effect persists poleward, as shown in the two temperature graphs for the midlatitude west-coast regime—Monterey, California (36° N) and Sitka, Alaska (57° N).

In continental interiors, however, the annual temperature cycle remains strong. The midlatitude continental regime of Omaha, Nebraska (41° N), and the subarctic continental regime of Fort Vermilion, Alberta (58° N), show annual variations in mean monthly temperature of about 30°C (54°F) and 40°C (72°F), respectively. The ice sheet regime of Greenland (Eismitte, 71° N) is in a class by itself, with severe cold all year.

Other regimes can be identified, and, because they grade into one another, the list could be expanded indefinitely. However, what is important is that (1) annual variation in insolation, which is determined by latitude, provides the basic control on temperature patterns, and (2) the effect of location—maritime or continental—moderates that variation.

The contrast in temperatures with season at a location depends on both latitude and location. At higher latitudes and in continental interiors, the contrast is greater.

Global Precipitation

Global precipitation patterns are largely determined by air masses and their movements, which in turn are produced by global air circulation patterns. To see this, let's look at the world map of precipitation, Figure 7.2, in more detail. This map of mean annual precipitation shows **isohyets**—lines drawn through all points having the same annual precipitation. Analyzing the map, we can recognize seven global precipitation regions as follows. (Note that for regions where all or most of the precipitation is rain, we use the word "rainfall." For regions where snow is a significant part of the annual total, we use the word "precipitation.") Table 7.1 summarizes the observations below.

1. **Wet equatorial belt.** This zone of heavy rainfall, over 200 cm (80 in.) annually, straddles the equator and includes the Amazon River Basin in South America, the Congo River Basin of equatorial Africa, much of the African coast from Nigeria west to Guinea, and the East Indies. Here the prevailing warm temperatures and high-moisture content of the mE air masses favor abundant convectional rainfall. Thunderstorms are frequent year-round.

2. **Trade-wind coasts.** Narrow coastal belts of high rainfall, 150 to 200 cm (about 60 to 80 in.), and locally even more, extend from near the equator to latitudes of about 25° to 30° N and S on the eastern sides of every continent or large island. Examples include the eastern coast of Brazil, Central America, Madagascar, and northeastern Australia. The rainfall of these coasts is supplied by moist mT air masses from warm oceans, brought over the land by the trade winds. As they encounter coastal hills and mountains, these air masses produce heavy orographic rainfall.

3. **Tropical deserts.** In striking contrast to the wet equatorial belt astride the equator are the two zones of vast tropical deserts lying approximately on the tropics of cancer and capricorn. These are hot, barren deserts, with less than 25 cm (10 in.) of rainfall annually and in many places with less than 5 cm (2 in.). They are located under and are caused by the large, stationary subtropical cells of high pressure, in which the subsiding cT air mass is adiabatically warmed and dried.

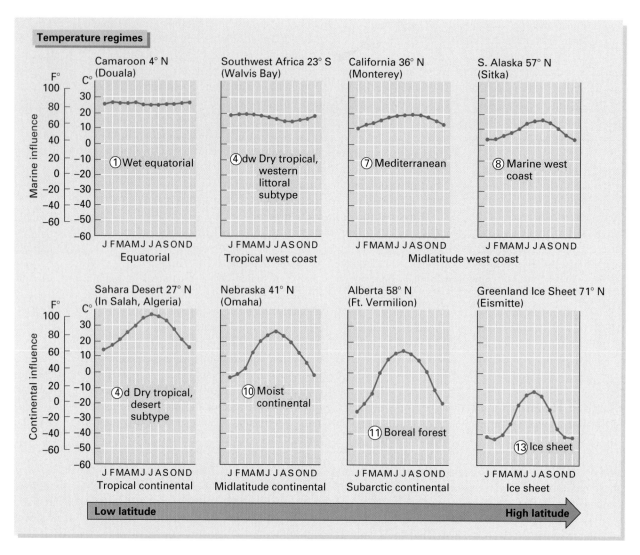

7.1   *Temperature regimes* *(above and right) Some important temperature regimes, represented by annual cycles of air temperature. (Based on the Goode Base Map.)*

These deserts extend off the west coasts of the lands and out over the oceans. Rain here is largely convectional and extremely unreliable.

4. **Midlatitude deserts and steppes.** Farther northward, in the interiors of Asia and North America between lat. 30° and lat. 50°, are great deserts, as well as vast expanses of semiarid grasslands known as *steppes.* Annual precipitation ranges from less than 10 cm (4 in.) in the driest areas to 50 cm (20 in.) in the moister steppes. Dryness here results from remoteness from ocean sources of moisture.

Located in regions of prevailing westerly winds, these arid lands typically lie in the rain shadows of coastal mountains and highlands. For example, the Cordilleran Ranges of Oregon, Washington, British Columbia, and Alaska shield the interior of North America from moist mP air masses originating in the

At lower latitudes, annual rainfall is heavy in the wet equatorial belt and trade-wind coasts. In contrast, tropical deserts are very dry.

Pacific. Upon descending into the intermountain basins and interior plains, the mP air masses are warmed and dried. Similarly, mountains of Europe and the Scandinavian Peninsula obstruct the flow of moist mP air masses from the North Atlantic into western Asia. The great southern Asiatic ranges also prevent the entry of moist mT and mE air masses from the Indian Ocean.

The southern hemisphere has too little land in the midlatitudes to produce a true continental desert, but the dry steppes of Patagonia, lying on the lee side of the Andean chain, are roughly the counterpart of the North American deserts and steppes of Oregon and northern Nevada.

5. **Moist subtropical regions.** On the southeastern sides of the continents of North America and Asia, in lat. 25° to 45° N, are the moist subtropical regions, with 100 to 150 cm (about 40 to 60 in.) of rainfall annually. Smaller

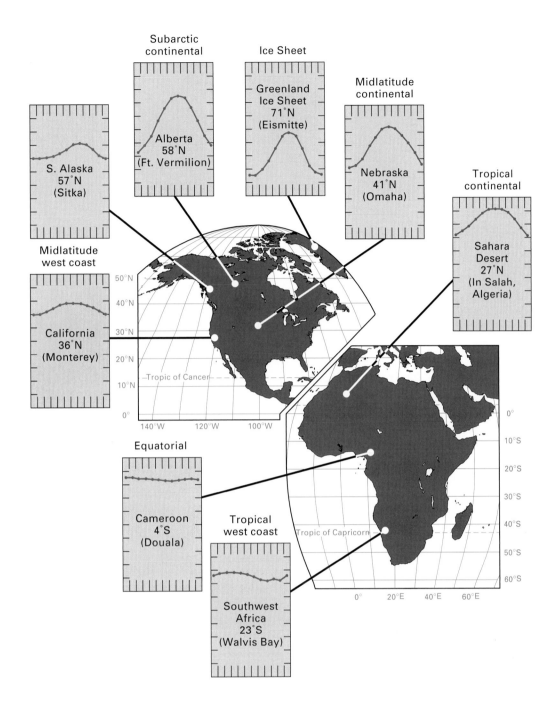

areas of the same kind are found in the southern hemisphere in Uruguay, Argentina, and southeastern Australia. These regions are positioned on the moist western sides of the oceanic subtropical high-pressure centers. As a result, the lands receive moist mT air masses from the tropical ocean that are carried poleward over the adjoining land. Commonly, too, these areas receive heavy rains from tropical cyclones.

6. **Midlatitude west coasts.** Another distinctive wet location is on midlatitude west coasts of all continents and

At higher latitudes, east and west coasts have higher precipitation, while continental interiors and arctic and polar regions are drier.

large islands lying between about 35° and 65° in the region of prevailing westerly winds. In these zones, abundant orographic precipitation occurs as a result of forced uplift of mP air masses. Where the coasts are mountainous, as in Alaska and British Columbia, southern Chile, Scotland, Norway, and South Island of New Zealand, the annual precipitation is over 200 cm (79 in.). During the Ice Age, this precipitation fed alpine glaciers that descended to the coast, carving the picturesque deep bays (fiords) that are so typically a part of the scenery there.

7.2 World precipitation *Mean annual precipitation of the world. Isohyets are labeled in cm (in.)*

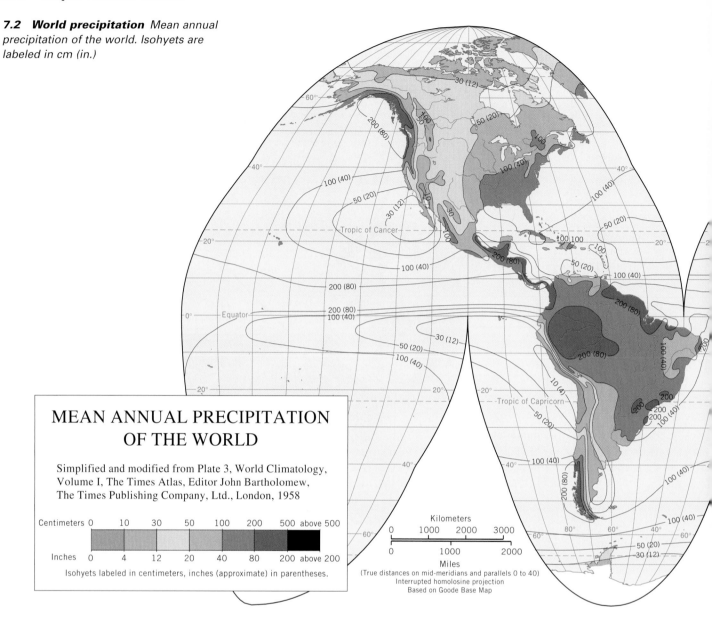

MEAN ANNUAL PRECIPITATION
OF THE WORLD

Simplified and modified from Plate 3, World Climatology,
Volume I, The Times Atlas, Editor John Bartholomew,
The Times Publishing Company, Ltd., London, 1958

Centimeters 0   10   30   50   100   200   500 above 500

Inches 0   4   12   20   40   80   200 above 200

Isohyets labeled in centimeters, inches (approximate) in parentheses.

Kilometers
0   1000   2000   3000

0   1000   2000
Miles

(True distances on mid-meridians and parallels 0 to 40)
Interrupted homolosine projection
Based on Goode Base Map

Table 7.1 | **World Precipitation Regions**

Name	Latitude Range	Continental Location	Prevailing Air Mass	Annual Precipitation Centimeters	Inches
1. Wet equatorial belt	10° N to 10° S	Interiors, coasts	mE	Over 200	Over 80
2. Trade-wind coasts (windward tropical coasts)	5–30° N and S	Narrow coastal zones	mT	Over 150	Over 60
3. Tropical deserts	10–35° N and S	Interiors, west coasts	cT	Under 25	Under 10
4. Midlatitude deserts and steppes	30–50° N and S	Interiors	cT, cP	10–50	4–20
5. Moist subtropical regions	25–45° N and S	Interiors, coasts	mT (summer)	100–150	40–60
6. Midlatitude west coasts	35–65° N and S	West coasts	mP	Over 100	Over 40
7. Arctic and polar deserts	60–90° N and S	Interiors, coasts	cP, cA	Under 30	Under 12

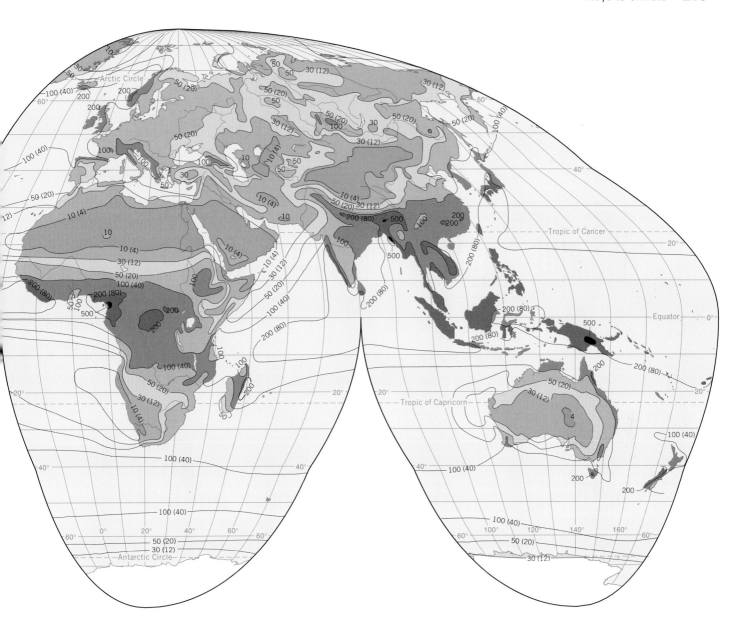

7. **Arctic and polar deserts.** A seventh precipitation region is formed by the arctic and polar deserts. Northward of the 60th parallel, annual precipitation is largely under 30 cm (12 in.), except for the west-coast belts. Cold cP and cA air masses cannot hold much moisture, and consequently, they do not yield large amounts of precipitation. At the same time, the relative humidity is high and evaporation rates are low.

Total annual precipitation is a useful quantity in establishing the character of a climate type, but it does not account for the seasonality of precipitation. The variation in monthly precipitation through the annual cycle is a very important factor in climate description. If there is a pattern of alternating dry and wet seasons instead of a uniform dis-

tribution of precipitation throughout the year, we can expect that the natural vegetation, soils, crops, and human use of the land will all be different. It also makes a great deal of difference whether the wet season coincides with a season of higher temperatures or with a season of lower temperatures. If the warm season is also wet, growth of both native plants and crops will be enhanced. If the warm season is dry, the stress on growing plants will be great and irrigation will be required for crops.

Monthly precipitation patterns can be described largely by three types: (1) uniformly distributed precipitation; (2) a precipitation maximum during the summer (or season of high Sun), in which insolation is at its peak; and (3) a precipitation maximum during the winter or cooler season (season of low Sun), when insolation is least. Note that the

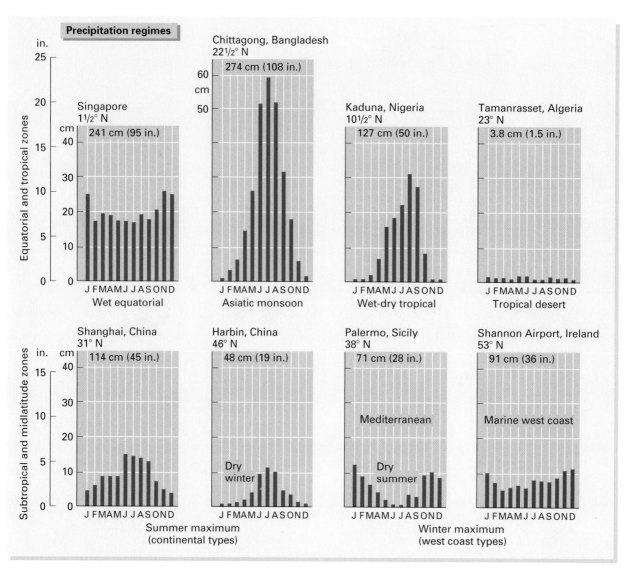

7.3 *Seasonal precipitation patterns* (above and right) Eight precipitation types selected to show various seasonal patterns. (Based on the Goode Base Map.)

uniform type of pattern can include a wide range of possibilities from little or no precipitation in any month to abundant precipitation in all months.

Figure 7.3 shows a set of monthly precipitation diagrams selected to illustrate the major types that occur over the globe. Two stations show the uniformly distributed pattern described above—Singapore, a wet equatorial station near the equator ($1\frac{1}{2}°$ N), and Tamanrasset, Algeria, a tropical desert station very near the tropic of cancer at 23° N. At Singapore, rainfall is abundant in all months, but some months have somewhat more than others. Tamanrasset has so little rain in any month that it scarcely shows on the graph.

Chittagong, Bangaladesh ($22\frac{1}{2}°$ N), and Kaduna, Nigeria ($10\frac{1}{2}°$ N), both show patterns of the second type—that is, a wet season at the time of high Sun (summer solstice) and a dry season at the time of low Sun (winter solstice).

> **Monthly patterns of precipitation are typically of three types— uniformly distributed, summer or high-Sun maximum, and winter or low-Sun maximum.**

Chittagong is an Asian monsoon station, with a very large amount of precipitation falling during the high-Sun season. Kaduna, an African station with about half the total annual precipitation, shows a similar pattern and is also of the wet-dry tropical type. Both of these stations experience their wet season when the intertropical convergence zone (ITCZ) is nearby and their dry season when the ITCZ has retreated to the other hemisphere.

The summer precipitation maximum also occurs at higher latitudes on the eastern sides of continents. Shanghai, China (31° N), shows this pattern nicely in the subtropical zone. The same summer maximum persists into the midlatitudes. For example, Harbin, in eastern China (46° N), has a long, dry winter with a marked period of summer rain.

In contrast to these patterns are cycles with a winter precipitation maximum. Palermo, Sicily (38° N), is an

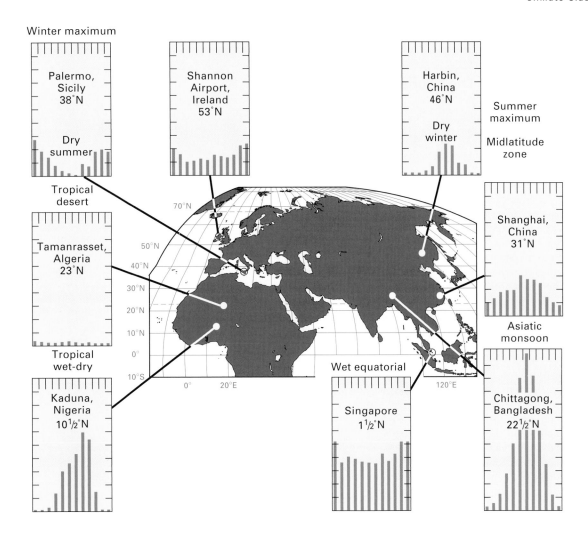

example of the Mediterranean type, named for its prevalence in the lands surrounding the Mediterranean Sea. This type experiences a very dry summer but has a moist winter. Southern and central California are also regions of this climate type. In Mediterranean climates, summer drought is produced by subtropical high-pressure cells, which intensify and move poleward during the high-Sun season. They extend into the regions of Mediterranean climate, providing the hot dry weather of their associated cT air masses and blocking the passage of other, moister, air masses. In the low-Sun season, the subtropical high-pressure cells move equatorward and weaken, allowing frontal and cyclonic precipitation to penetrate Mediterranean climate regions.

The dry-summer, moist-winter cycle is carried into higher midlatitudes along narrow strips of west coasts. Shannon Airport, Ireland (53° N), shows this marine west-coast type, although the difference between summer and winter rainfall is not as marked. Summers have less rainfall here for two reasons. First, the blocking effects of subtropical high-pressure cells tend to extend poleward into these regions, keeping moister air masses and cyclonic storms away. Second, the cyclonic storms that produce much of the winter precipitation are reduced in intensity during the high-Sun season. This occurs because the temperature and mois-

ture contrasts between polar and arctic air masses and tropical air masses are weaker in summer, owing to increased high-latitude insolation.

CLIMATE CLASSIFICATION

Mean monthly values of air temperature and precipitation can describe the climate of a weather station and its nearby region quite accurately. To study climates from a global viewpoint, climatologists classify these values into distinctive climate types. This requires developing a set of rules to use in examining monthly temperature and precipitation values. By applying the rules, the climatologist can use each station's data to determine the climate to which it belongs. This textbook recognizes 13 distinctive climate types that are designed to be understood and explained by air mass movements and frontal zones. The types follow quite naturally from the principles governing temperature and precipitation that we have discussed above.

An alternative climate classification is that devised by the Austrian climatologist Vladimir Köppen in 1918 and modified by Geiger and Pohl in 1953. It uses a system of letters to label climates. It is presented in *Special Supplement 7.1 • The Köppen Climate System.*

The Köppen Climate System

Air temperature and precipitation data have formed the basis for several climate classifications. One of the most important of these is the Köppen climate system, devised in 1918 by Dr. Vladimir Köppen of the University of Graz in Austria. For several decades, this system, with various later revisions, was the most widely used climate classification among geographers. Köppen was both a climatologist and plant geographer, so his main interest lay in finding climate boundaries that coincided approximately with boundaries between major vegetation types.

Under the Köppen system, each climate is defined according to assigned values of temperature and precipitation, computed in terms of annual or monthly values. Any given station can be assigned to its particular climate group and subgroup solely on the basis of the records of temperature and precipitation at that place.

Note that mean annual temperature refers to the average of 12 monthly temperatures for the year. Mean annual precipitation refers to the average of the entire year's precipitation as observed over many years.

The Köppen system features a shorthand code of letters designating major climate groups, subgroups within the major groups, and further subdivisions to distinguish particular seasonal characteristics of temperature and precipitation. Five major climate groups are designated by capital letters as follows:

A *Tropical rainy climates*
Average temperature of every month is above 18°C (64.4°F).

These climates have no winter season. Annual rainfall is large and exceeds annual evaporation.

B *Dry climates*
Evaporation exceeds precipitation on the average throughout the year. There is no water surplus; hence, no permanent streams originate in B climate zones.

C *Mild, humid (mesothermal) climates*
The coldest month has an average temperature of under 18°C (64.4°F) but above –3°C (26.6°F); at least one month has an average temperature above 10°C (50°F). The C climates thus have both a summer and a winter.

D *Snowy-forest (microthermal) climates*
The coldest month has an average temperature of under –3°C (26.6°F). The average temperature of the warmest month is above 10°C (50°F). (Forest is not generally found where the warmest month is colder than 10°C (50°F).)

E *Polar climates*
The average temperature of the warmest month is below 10°C (50°F). These climates have no true summer.

Note that four of these five groups (A, C, D, and E) are defined by temperature averages, whereas one (B) is defined by precipitation-to-evaporation ratios. Groups A, C, and D have sufficient heat and precipitation for the growth of forest and woodland vegetation. Figure S7.1 shows the boundaries of the five major climate groups, and Figure S7.2 is a world map of Köppen climates.

Subgroups within the five major groups are designated by a second

letter according to the following code.

S Semiarid (steppe)
W Arid (desert)

(The capital letters S and W are applied only to the dry B climates.)

f Moist, adequate precipitation in all months, no dry season. This modifier is applied to A, C, and D groups.
w Dry season in the winter of the respective hemisphere (low-Sun season).
s Dry season in the summer of the respective hemisphere (high-Sun season).
m Rainforest climate, despite short, dry season in monsoon type of precipitation cycle. Applies only to A climates.

From combinations of the two letter groups, 12 distinct climates emerge:

A *Tropical rainforest climate*
The rainfall of the driest month is 6 cm (2.4 in.) or more.

Am *Monsoon variety of Af*
The rainfall of the driest month is less than 6 cm (2.4 in.). The dry season is strongly developed.

Aw *Tropical savanna climate*
At least one month has rainfall less than 6 cm (2.4 in.). The dry season is strongly developed.

Figure S7.3 shows the boundaries between Af, Am, and Aw climates as determined by both annual rainfall and rainfall of the driest month.

BS *Steppe climate*
A semiarid climate characterized by grasslands, it occupies an intermediate position between the

S7.1 Generalized Köppen climate map *Highly generalized world map of major climate regions according to the Köppen classification. Highland areas are in black. (Based on Goode Base Map.)*

desert climate (BW) and the more humid climates of the A, C, and D groups. Boundaries are determined by formulas given in Figure S7.4.

BW   Desert climate

Desert has an arid climate with annual precipitation of usually less than 40 cm (15.7 in.). The boundary with the adjacent steppe climate (BS) is determined by formulas given in Figure S7.4.

C   Mild humid climate with no dry season

Precipitation of the driest month averages more than 3 cm (1.2 in.).

Cw   Mild humid climate with a dry winter

The wettest month of summer has at least 10 times the precipitation of the driest month of winter. (Alternative definition: 70 percent or more of the mean annual precipitation falls in the warmer six months.)

Cs   Mild humid climate with a dry summer

Precipitation of the driest month of summer is less than 3 cm (1.2 in.). Precipitation is at least three times as much as the driest month of summer. (Alternative definition: 70 percent or more of the mean annual precipitation falls in the six months of winter.)

Df   Snowy-forest climate with a moist winter

No dry season.

Dw   Snowy-forest climate with a dry winter

ET   Tundra climate

The mean temperature of the warmest month is above 0°C (32°F) but below 10°C (50°F).

EF   Perpetual frost climate

In this ice sheet climate, the mean monthly temperatures of all months are below 0°C (32°F).

To denote further variations in climate, Köppen added a third let-

ter to the code group. The meanings are as follows:

a With hot summer; warmest month is over 22°C (71.6°F); C and D climates.

b With warm summer; warmest month is below 22°C (71.6°F); C and D climates.

c With cool, short summer; less than four months are over 10°C (50°F); C and D climates.

d With very cold winter; coldest month is below −38°C (−36.4°F); D climates only.

h Dry-hot; mean annual temperature is over 18°C (64.4°F); B climates only.

k Dry-cold; mean annual temperature is under 18°C (64.4°F); B climates only.

As an example of a complete Köppen climate code, BWk refers to a cool desert climate, and Dfc refers to a cold, snowy-forest climate with cool, short summer.

S7.2 Köppen climates of the world *World map of climates according to the Köppen–Geiger–Pohl system.*

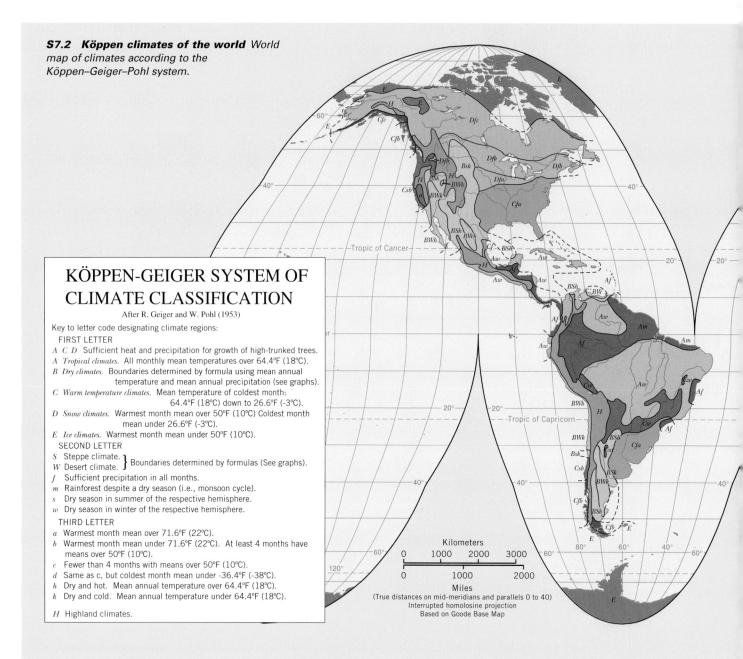

KÖPPEN-GEIGER SYSTEM OF CLIMATE CLASSIFICATION

After R. Geiger and W. Pohl (1953)

Key to letter code designating climate regions:

FIRST LETTER

A C D Sufficient heat and precipitation for growth of high-trunked trees.

A Tropical climates. All monthly mean temperatures over 64.4°F (18°C).

B Dry climates. Boundaries determined by formula using mean annual temperature and mean annual precipitation (see graphs).

C Warm temperature climates. Mean temperature of coldest month: 64.4°F (18°C) down to 26.6°F (-3°C).

D Snow climates. Warmest month mean over 50°F (10°C) Coldest month mean under 26.6°F (-3°C).

E Ice climates. Warmest month mean under 50°F (10°C).

SECOND LETTER

S Steppe climate.
W Desert climate. } Boundaries determined by formulas (See graphs).

f Sufficient precipitation in all months.

m Rainforest despite a dry season (i.e., monsoon cycle).

s Dry season in summer of the respective hemisphere.

w Dry season in winter of the respective hemisphere.

THIRD LETTER

a Warmest month mean over 71.6°F (22°C).

b Warmest month mean under 71.6°F (22°C). At least 4 months have means over 50°F (10°C).

c Fewer than 4 months with means over 50°F (10°C).

d Same as c, but coldest month mean under -36.4°F (-38°C).

h Dry and hot. Mean annual temperature over 64.4°F (18°C).

k Dry and cold. Mean annual temperature under 64.4°F (18°C).

H Highland climates.

Kilometers
0 1000 2000 3000

0 1000 2000
Miles

(True distances on mid-meridians and parallels 0 to 40)
Interrupted homolosine projection
Based on Goode Base Map

S7.3 Boundaries of the A climates

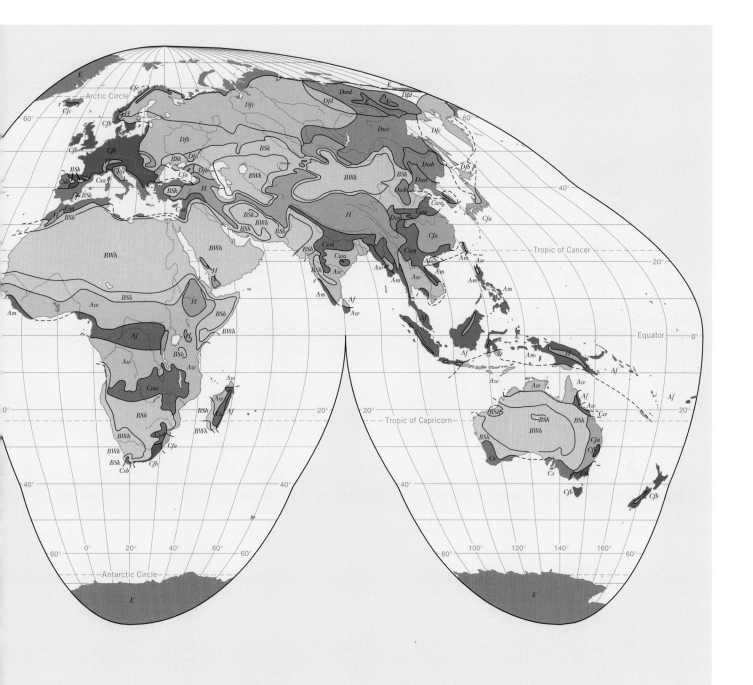

S7.4 Boundaries of the B climates *Upper figures: metric system. Lower figures: English system*

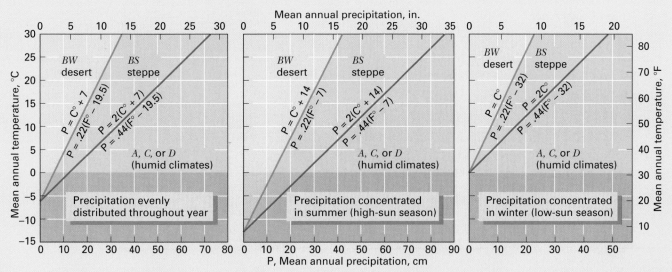

Recall from Chapter 6 that air masses are classified according to the general latitude of their source regions and their surface type—land or ocean—within that region (see Figure 6.1). The latitude determines the temperature of the air mass, which can also depend on the season. The kind of surface, land or ocean, controls the moisture content. Since the air mass characteristics control the two most important climate variables—temperature and precipitation—we can explain climates using air masses as a guide.

We also know from Chapter 6 that frontal zones are regions in which air masses are in contact. And when unlike air masses are in contact, cyclonic precipitation is likely to develop. However, the position of frontal zones changes with the seasons. For example, the polar-front zone lies generally across the middle latitudes of the United States in winter, but it retreats northward to Canada during the summer. The seasonal movements of frontal zones therefore influence annual cycles of temperature and precipitation.

Figure 7.4 shows our schematic diagram of air mass source regions from Chapter 6. We have subdivided this diagram into global bands that contain three broad groups of climates: low-latitude (Group I), midlatitude (Group II), and high-latitude (Group III). They are described briefly as follows.

- *Group I: Low-Latitude Climates.* The region of low-latitude climates (Group I) is dominated by the source regions of continental tropical (cT), maritime tropical (mT), and maritime equatorial (mE) air masses. These

source regions are related to the three most obvious atmospheric features that occur within their latitude band—the two subtropical high-pressure belts and the equatorial trough at the intertropical convergence zone (ITCZ). Air of polar origin occasionally invades regions of low-latitude climates. Easterly waves and tropical cyclones are important weather systems in this climate group.

- *Group II: Midlatitude Climates.* The region of midlatitude climates (Group II) lies in the polar-front zone—a zone of intense interaction between unlike air masses. Here tropical air masses moving poleward and polar air masses moving equatorward are in conflict. Wave cyclones are normal features of the polar front, and this zone may contain as many as a dozen wave cyclones around the globe.

- *Group III: High-Latitude Climates.* The region of high-latitude climates (Group III) is dominated by polar and arctic (including antarctic) air masses. In the arctic belt of the 60th to 70th parallels, continental polar air masses meet arctic air masses along an *arctic-front zone,* creating a series of eastward-moving wave cyclones. In the southern hemisphere, there are no source regions in the subantarctic belt for continental polar air—just a great single oceanic source region for maritime polar (mP) air masses. The pole-centered continent of Antarctica provides a single great source of the extermely cold, dry antarctic air mass (cAA). These two air masses interact along the *antarctic-front zone.*

7.4 Climate groups and air mass source regions *Using the map of air mass source regions, we can identify five global bands associated with three major climate groups. Within each group is a set of distinctive climates with unique characteristics that are explained by the movements of air masses and frontal zones.*

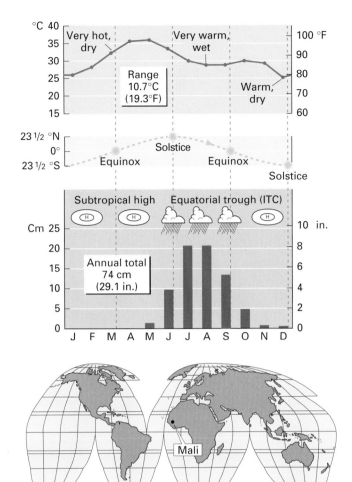

7.5 Typical climograph *Kayes, Mali, lat. 14° N, is located in western Africa, just south of the Sahara Desert. Here, the climate is nearly rainless for seven months of the year. During these months, dry subsiding air from subtropical high-pressure cells dominates. Rainfall occurs when the intertropical convergence zone moves northward and reaches the vicinity of Kayes in June or July. Temperatures are hottest in April and May, before the rainy season begins.*

Within each of these three climate groups are a number of climate types (or simply, climates)—four low-latitude climates (Group I), six midlatitude climates (Group II), and three high-latitude climates (Group III)—for a total of 13 climate types. In this textbook, the climates are numbered for ease of identification on maps and diagrams. We will refer to each climate by name because the names describe the general nature of the climate and also suggest its global location. For convenience, we include the climate number next to its name in the text.

In presenting the climates, we will make use of a pictorial device called a **climograph.** It shows the annual cycles of monthly mean air temperature and monthly mean precipitation for a location, along with some other useful information. Figure 7.5 is an example of a climograph. The data plotted are for Kayes, a station in the African country of

Mali, at 14° N. At the top of the climograph, the mean monthly temperature is plotted as a line graph. At the bottom, the mean monthly precipitation is shown as a bar graph. The annual range in temperature and the total annual precipitation are stated on every climograph as well. Most climographs also display dominant weather features, which are shown using picture symbols. For Kayes, the two dominant features are the subtropical high and equatorial trough (ITCZ). Many of our climographs also include a small graph of the Sun's declination in order to help show when solstices and equinoxes occur.

The world map of climates, Figure 7.6, shows the actual distribution of climate types on the continents. This map, based on data collected at a large number of observing stations, is simplified because the climate boundaries are uncertain in many areas where observing stations are thinly distributed.

Overview of the Climates

Although we will provide a thorough examination of the 13 climates in the remainder of this chapter, here is a quick and informal description of each.

LOW-LATITUDE CLIMATES (GROUP I)

- *Wet equatorial* ①. Warm to hot with abundant rainfall, this is the steamy climate of the Amazon and Congo basins.
- *Monsoon and trade-wind coastal* ②. This warm to hot climate has a very wet rainy season. It occurs in coastal regions that are influenced by trade winds or a monsoon circulation. The climates of Vietnam and Bangladesh are good examples.
- *Wet-dry tropical* ③. A warm to hot climate with very distinct wet and dry seasons. The monsoon region of India falls into this type, as does much of the Sahel region of Africa.
- *Dry tropical* ④. The climate of the world's hottest deserts—extremely hot in the high-Sun season, a little cooler in the low-Sun season, with little or no rainfall. The Sahara Desert, Saudi Arabia, and the central Australian desert are of this climate.

MIDLATITUDE CLIMATES (GROUP II)

- *Dry subtropical* ⑤. Another desert climate, but not quite as hot as the dry tropical climate ④, since it is found farther poleward. This type includes the hottest part of the American Southwest desert.
- *Moist subtropical* ⑥. The climate of the southeastern regions of the United States and China—hot and humid summers, with mild winters and ample rainfall year-round.
- *Mediterranean* ⑦. Hot, dry summers and rainy winters mark this climate. Southern and central California, as well as the lands of the Mediterranean region—Spain, southern Italy, Greece, and the coastal regions of Lebanon and Israel—are prime examples.

7.6 *Climates of the world* Compiled from station data by A. N. Strahler.

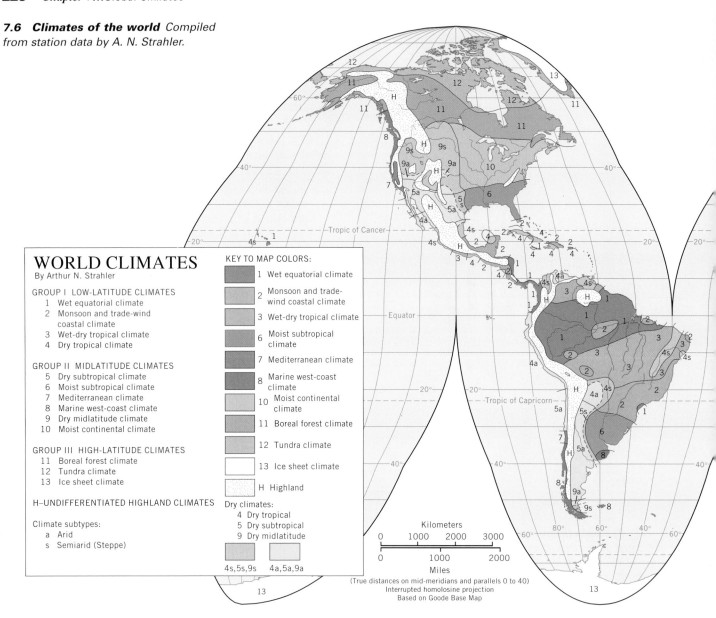

WORLD CLIMATES
By Arthur N. Strahler

GROUP I LOW-LATITUDE CLIMATES
1 Wet equatorial climate
2 Monsoon and trade-wind coastal climate
3 Wet-dry tropical climate
4 Dry tropical climate

GROUP II MIDLATITUDE CLIMATES
5 Dry subtropical climate
6 Moist subtropical climate
7 Mediterranean climate
8 Marine west-coast climate
9 Dry midlatitude climate
10 Moist continental climate

GROUP III HIGH-LATITUDE CLIMATES
11 Boreal forest climate
12 Tundra climate
13 Ice sheet climate

H—UNDIFFERENTIATED HIGHLAND CLIMATES

Climate subtypes:
a Arid
s Semiarid (Steppe)

KEY TO MAP COLORS:
1 Wet equatorial climate
2 Monsoon and trade-wind coastal climate
3 Wet-dry tropical climate
6 Moist subtropical climate
7 Mediterranean climate
8 Marine west-coast climate
10 Moist continental climate
11 Boreal forest climate
12 Tundra climate
13 Ice sheet climate
H Highland

Dry climates:
4 Dry tropical
5 Dry subtropical
9 Dry midlatitude

4s,5s,9s   4a,5a,9a

Kilometers
0   1000   2000   3000

Miles
0   1000   2000

(True distances on mid-meridians and parallels 0 to 40)
Interrupted homolosine projection
Based on Goode Base Map

- *Marine west-coast* ⑧. The climate of the Pacific Northwest—coastal Oregon, Washington, and British Columbia. Warm summers and cool winters, with more rainfall in winter, are characteristics of this climate.
- *Dry midlatitude* ⑨. This dry climate is found in midlatitude continental interiors. The steppes of central Asia and the Great Plains of North America are familiar locales with this climate—warm to hot in summer, cold in winter, and with low annual precipitation.
- *Moist continental* ⑩. This is the climate of the eastern United States and lower Canada—cold in winter, warm in summer, with ample precipitation through the year.

HIGH-LATITUDE CLIMATES (GROUP III)

- *Boreal forest* ⑪. Short, cool summers and long, bitterly cold winters characterize this snowy climate. Northern Canada, Siberia, and central Alaska are regions of boreal forest climate.

- *Tundra* ⑫. Although this climate has a long, severe winter, temperatures on the tundra are somewhat moderated by proximity to the Arctic Ocean. This is the climate of the coastal arctic regions of Canada, Alaska, Siberia, and Scandinavia.
- *Ice sheet* ⑬. The bitterly cold temperatures of this climate, restricted to Greenland and Antarctica, can drop below −50°C (−58°F) during the sunless winter months. Even during the 24-hour days of summer, temperatures remain well below freezing.

Dry and Moist Climates

All but two of the 13 climate types introduced above are classified as either dry climates or moist climates. Dry climates are those in which total annual evaporation of moisture from the soil and from plant foliage exceeds the annual precipitation by a wide margin. Generally speaking, the dry

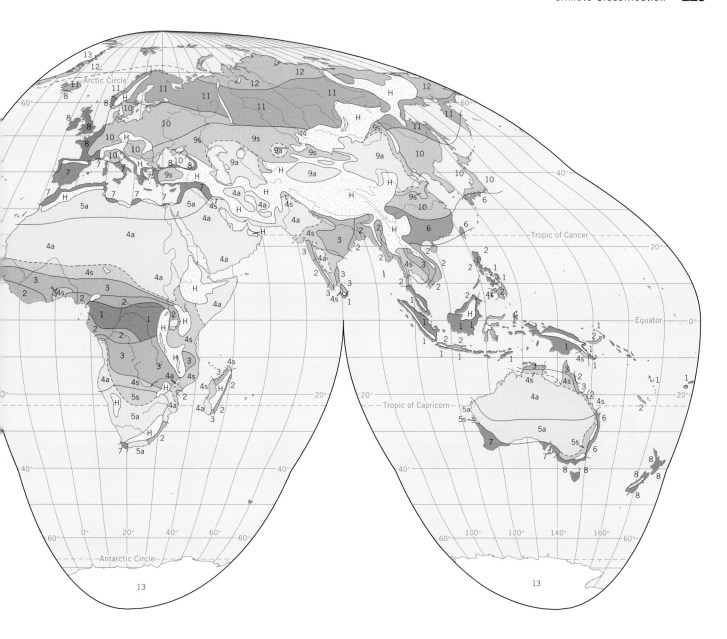

climates do not support permanently flowing streams. The soil is dry much of the year, and the land surface is clothed with sparse plant cover—scattered grasses or shrubs—or simply lacks a plant cover. Moist climates are those with sufficient rainfall to maintain the soil in a moist condition through much of the year and to sustain the year-round flow of the larger streams. Moist climates support forests or prairies of dense tall grasses.

Within the dry climates there is a wide range of degree of aridity, ranging from very dry deserts nearly devoid of plant life to moister regions that support a partial cover of grasses or shrubs. To recognize this diversity, we will refer to two dry climate subtypes: (1) semiarid (or steppe) and (2) arid. The semiarid (steppe) subtype, designated by the letter

> **Dry climates are those where evaporation and transpiration exceed precipitation by a wide margin. In moist climates, precipitation maintains moist soils and active streamflow for most or all of the year. Wet-dry climates alternate through the year between wet and dry states.**

s, has enough precipitation to support sparse grasses and shrubs. This subtype is found adjacent to moist climates. The arid subtype, indicated by the letter *a,* ranges from extremely dry to transitional with semiarid.

Two of our 13 climates cannot be accurately described as either dry or moist climates. These are the wet-dry tropical ③ and Mediterranean ⑦ climate types. Instead, they show a seasonal alteration between a very wet season and a very dry season. This striking contrast in seasons gives a special character to the two climates, and thus, we have singled them out for special recognition as wet-dry climates. In Figure 7.3 they are designated as having the wet-dry tropical and the Mediterranean precipitation patterns. Table 7.2 sum-

Table 7.2 | Moist, Dry, and Wet-Dry Climate Types

	Climate Type		
Climate Group	**Moist**	**Dry**	**Wet-Dry**
I: Low-latitude Climates	Wet equatorial ① Monsoon and trade-wind coastal ②	Dry tropical ④ (*s*, steppe; *a*, arid)	Wet-dry tropical ③
II: Midlatitude Climates	Moist subtropical ⑥ Marine west-coast ⑧ Moist continental ⑩	Dry subtropical ⑤ (*s, a*) Dry midlatitude ⑨ (*s, a*)	Mediterranean ⑦
III: High-latitude Climates	Boreal forest ⑪ Tundra ⑫	Ice sheet ⑬	

marizes moist, dry, wet-dry climates as they occur within the three main climate groups.

With our introduction to global climates now complete, we can describe each of the climates and their associated environments in turn. Keep in mind that the annual cycles of temperature and precipitation for each climate region can be easily related to the latitude and location (coastal or interior) of the region, as well as to the characteristic movements of air masses and fronts through the region as they are controlled by global-scale patterns of atmospheric circulation. Our description of the environments associated with the climate types will stress the natural vegetation covers, agricultural practices, and other climate-related human activities that occur in the regions of each climate type. GEODISCOVERIES

LOW-LATITUDE CLIMATES

The *low-latitude climates* lie for the most part between the tropics of cancer and capricorn. In terms of world latitude zones, the low-latitude climates occupy all of the equatorial zone (10° N to 10° S), most of the tropical zone (10–15° N and S), and part of the subtropical zone. In terms of prevailing pressure and wind systems, the region of low-latitude climates includes the equatorial trough of the intertropical convergence zone (ITCZ), the belt of tropical easterlies (northeast and southeast trades), and large portions of the oceanic subtropical high-pressure belt (Figure 7.4).

Figure 7.7 shows climographs for the four low-latitude climates. These climates range from extremely moist—the wet equatorial climate ①—to extremely dry—the dry tropical climate ④. They also vary strongly in the seasonality of their rainfall. In the wet equatorial climate ①, rainfall is abundant year-round. But in the wet-dry tropical climate ③, rainfall is abundant for only part of the year. During the remainder of the year, little or no rain falls. The

---

The wet equatorial climate ① is dominated by warm, moist tropical and equatorial maritime air masses yielding abundant rainfall year-round. The monsoon and trade-wind coastal climate ② has a strong wet season occurring when the ITCZ is nearby.

---

seasonal temperature cycle also varies among these climates. In the wet equatorial climate ①, temperatures are nearly uniform throughout the year. In the dry tropical climate ④, there is a strong annual temperature cycle. Table 7.3 summarizes some of the characteristics of these climates.

The Wet Equatorial Climate ① (Köppen: *Af*)

The **wet equatorial climate** ① is a climate of the intertropical convergence zone (ITCZ), which is nearby for most of the year. The climate is dominated by warm, moist maritime equatorial (mE) and maritime tropical (mT) air masses that yield heavy convectional rainfall. Precipitation is plentiful in all months, and the annual total often exceeds 250 cm (about 100 in.). However, there is usually a seasonal pattern to the rainfall, so that rainfall is greater during some part of the year. This period of heavier rainfall occurs when the ITCZ migrates into the region. Remarkably uniform temperatures prevail throughout the year. Both mean monthly and mean annual temperatures are typically close to 27°C (81°F).

Figure 7.8 shows the world distribution of the wet equatorial climate ①. This climate is found in the latitude range 10° N to 10° S. Its major regions of occurrence include the Amazon lowland of South America, the Congo Basin of equatorial Africa, and the East Indies, from Sumatra to New Guinea.

Figure 7.9 is a climograph for Iquitos, Peru (located in Figure 7.8), a typical wet equatorial station situated close to the equator in the broad, low basin of the upper Amazon River. Notice the very small annual range in temperature and the very large annual rainfall total. Monthly air temperatures are extremely uniform in the wet equatorial climate ①. Typically, mean monthly air temperature will range between 26° and 29°C (79° and 84°F) for stations at low elevation in the equatorial zone.

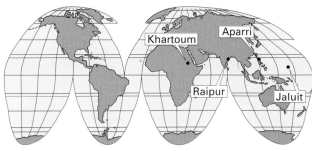

7.7 Low-latitude climographs *These four climographs show the key features of the four low-latitude climates.*

The Monsoon and Trade-Wind Coastal Climate ② (Köppen: *Af, Am*)

Like the wet equatorial climate ①, the **monsoon and trade-wind coastal climate** ② has abundant rainfall. But unlike the wet equatorial climate ①, the rainfall of the monsoon and trade-wind coastal climate ② always shows a strong seasonal pattern. This seasonal pattern is due to the migration of the intertropical convergence zone (ITCZ). In the high-Sun season ("summer," depending on the hemisphere), the ITCZ is nearby and monthly rainfall is greater. In the low-Sun season, when the ITCZ has migrated to the other hemisphere, subtropical high pressure dominates and monthly rainfall is less. Figure 7.8 shows the global distribution of the monsoon and trade-wind coastal climate ②. The climate occurs over latitudes from 5° to 25° N and S.

As the name of the monsoon and trade-wind coastal climate ② suggests, this climate type is produced by two somewhat different situations. On trade-wind coasts, rainfall is produced by moisture-laden maritime tropical (mT) and maritime equatorial (mE) air masses. These are moved onshore onto narrow coastal zones by trade winds or by

In the wet periods of the monsoon and trade-wind coastal climate ②, equatorial east coasts receive warm moist air masses from easterly trades while tropical south and west coasts receive moist air from southwesterly monsoon winds.

monsoon circulation patterns. As the warm, moist air passes over coastal hills and mountains, the orographic effect touches off convectional shower activity. Shower activity is also intensified by easterly waves, which are more frequent when the ITCZ is nearby. The east coasts of land masses experience this trade-wind effect because the trade winds blow from east to west. Trade-wind coasts are found along the east sides of Central and South America, the Caribbean Islands, Madagascar (Malagasy), Southeast Asia, the Philippines, and northeast Australia (see Figure 7.8).

The coastal precipitation effect also applies to the summer monsoon of Asia, when the monsoon circulation brings mT air onshore. However, the onshore monsoon winds blow from southwest to northeast, so it is the western coasts of land masses that are exposed to this moist airflow. Western India and Myanmar (formerly Burma) are examples. Moist air also penetrates well inland in Bangladesh, providing the very heavy monsoon rains for which the region is well known.

In central and western Africa, and southern Brazil, the monsoon pattern shifts the intertropical convergence zone

7.8 World map of wet equatorial ① and monsoon and trade-wind coastal climates ② (Based on Goode Base Map.)

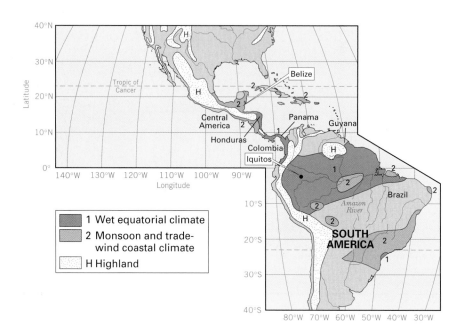

1 Wet equatorial climate
2 Monsoon and trade-wind coastal climate
H Highland

Table 7.3 | **Low-Latitude Climates**

Climate	Temperature	Precipitation	Explanation
Wet equatorial ①	Uniform temperatures, mean near 27 °C (81°F).	Abundant rainfall, all months, from mT and mE air masses. Annual total may exceed 250 cm (100 in.).	The ITCZ dominates this climate, with abundant convectional precipitation generated by convergence in weak equatorial lows. Rainfall is heaviest when the ITCZ is nearby.
Monsoon and trade-wind coastal ②	Temperatures show an annual cycle, with warmest temperatures in the high-Sun season.	Abundant rainfall but with a strong seasonal pattern.	Trade-wind coastal: Rainfall from mE and mT air masses is heavy when the ITCZ is nearby, lighter when the ITCZ moves to the opposite hemisphere. Asian monsoon coasts: dry air flowing southwest in low-Sun season alternates with moist oceanic air flowing northeast, producing a seasonal rainfall pattern on west coasts.
Wet-dry tropical ③	Marked temperature cycle, with hottest temperatures before the rainy season.	Wet high-Sun season alternates with dry low-Sun season.	Subtropical high pressure moves into this climate in the low-Sun season, bringing very dry conditions. In the high-Sun season, the ITCZ approaches and rainfall occurs. Asian monsoon climate: alternation of dry continental air in low-Sun season with moist oceanic air in high-Sun season brings a strong pattern of dry and wet seasons.
Dry tropical ④	Strong temperature cycle, with intense hot temperatures during high-Sun season.	Low precipitation. Sometimes rainfall occurs when the ITCZ is near.	This climate is dominated by subtropical high pressure, which provides clear, stable air for much or all of year. Insolation is intense during high-Sun period.

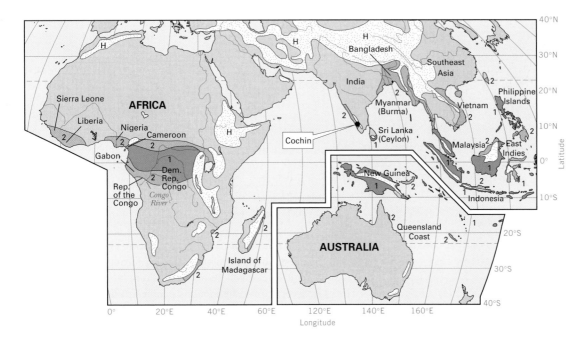

over 20° of latitude, or more (see Figure 5.19). Here, heavy rainfall occurs in the high-Sun season, when the ITCZ is nearby. Drier conditions prevail in the low-Sun season, when the ITCZ is far away.

Temperatures in the monsoon and trade-wind coastal climate ②, though warm throughout the year, also show an annual cycle. Warmest temperatures occur in the high-Sun

season, just before arrival of the ITCZ brings clouds and rain. Minimum temperatures occur at the time of low Sun.

Figure 7.10 is a climograph for the city of Belize, in the Central American country of Belize (Figure 7.8). This east-coast city, located at lat. 17° N, is exposed to the tropical easterly trade winds. Rainfall is abundant from June through November, when the ITCZ is nearby. Easterly waves are

7.9 Wet equatorial climate ① *Iquitos, Peru, lat. 3° S, is located in the upper Amazon lowland, close to the equator. Temperatures differ very little from month to month, and there is abundant rainfall throughout the year.*

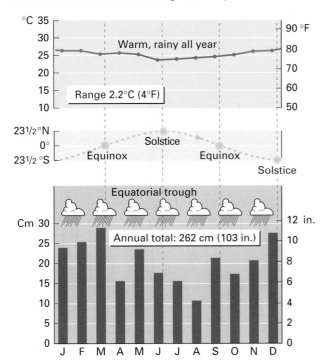

7.10 Trade-wind coastal climate ② *This climograph for Belize, a Central American east-coast city at lat. 17° N, shows a marked season of low rainfall following the period of low Sun. For the remainder of the year, precipitation is high, produced by warm, moist northeast trade winds.*

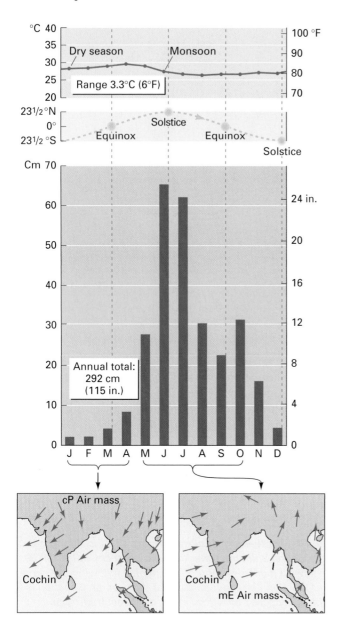

7.11 Monsoon coastal climate ② *Cochin, India, on a windward coast at lat. 10° N, shows an extreme peak of rainfall during the rainy monsoon, contrasting with a short dry season at time of low Sun.*

India, provides an example (Figure 7.11). Located at lat. 10° N on the west coast of lower peninsular India (see Figure 7.8), Cochin receives the warm, moist southwest winds of the summer monsoon. In this season, monthly rainfall is extreme in both June and July. A strongly pronounced season of low rainfall occurs at time of low Sun—December through March. Air temperatures show only a very weak annual cycle, cooling a bit during the rains, so the annual range is small at this low latitude.

THE LOW-LATITUDE RAINFOREST ENVIRONMENT Our first two climates—wet equatorial ① and monsoon and trade-wind coastal ②—are quite uniform in temperature and have a high annual rainfall. These factors create a special environment—the *low-latitude rainforest environment* (Figure 7.12). In the rainforest, streams flow abundantly throughout most of the year, and river channels are lined along the banks with dense forest vegetation. Indigenous peoples of the rainforest once traveled the rivers in dugout canoes, although today most enjoy easier travel powered by outboard motor. Over a century ago, larger shallow-draft river craft turned the major waterways into the main arteries of trade, connecting towns and cities on the river banks. Aircraft have added a new dimension in mobility, criss-crossing the almost trackless green sea of forest to find landings in clearings or on broad reaches of the larger rivers.

Low-latitude rainforests, unlike higher latitude forests, possess a great diversity of plant and animal species. The low-latitude rainforest can contain as many as 3000 different tree species in an area of only a few square kilometers, whereas midlatitude forests possess fewer than one-tenth that number. The number of types of animals found in the rainforest is also very large. A 16 km² (6 mi²) area in Panama near the Canal, for example, contains about 20,000 species of insects, while all of France has only a few hundred.

Many products of the rainforest have economic value. Rainforest lumber, such as mahogany, ebony, or balsawood, is an important export. Quinine, cocaine, and other drugs come from the bark and leaves of tropical plants. Cocoa is derived from the seed kernel of the cacao plant. Natural rubber is made from the sap of the rubber tree. The tree comes from South America, where it was first exploited. Rubber trees also are widely distributed through the rainforest of Africa. Today, the principal production is from plantations in Indonesia, Malaysia, Thailand, Vietnam, and Sri Lanka (Ceylon).

common in this season, and on occasion a tropical cyclone will bring torrential rainfall. Following the December solstice, rainfall is greatly reduced, with minimum values in March and April. At this time, the ITCZ lies farthest away, and the climate is dominated by subtropical high pressure. Air temperatures show an annual range of 5° C (9° F) with maximum in the high-Sun months.

The Asiatic monsoon shows a similar pattern, but there is an extreme peak of rainfall during the high-Sun period and a well-developed dry season with two or three months of only small rainfall amounts. The climograph for Cochin,

The natural environment of the wet equatorial ① and monsoon and trade-wind coastal climates ② is a rainforest of broadleaf evergreen trees. The rainforest is complex and diverse and yields many useful products.

In many areas of the rainforest and its fringes, agriculture is practiced by the slash-and-burn method, in which small patches of forest are cut down and then burned on the site. The wood ashes contain nutrients, and when spread over the soil, they serve as a simple fertilizer. Crops are then grown in the soil for a few years, until the nutrients are gone. The now unproductive plot is abandoned and eventually returns to rainforest. *Eye on the Environment 9.1 ● Exploita-*

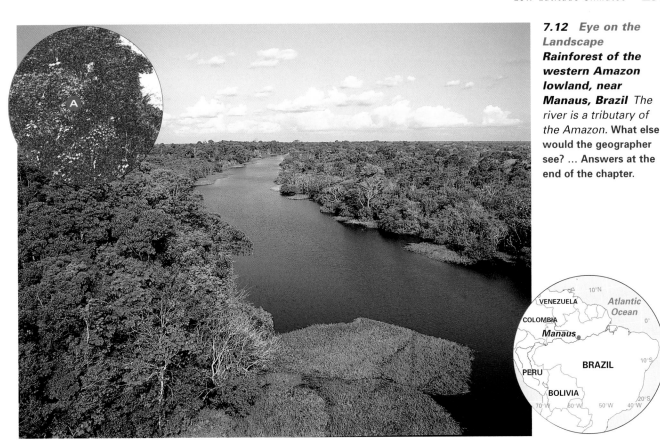

7.12 *Eye on the Landscape* **Rainforest of the western Amazon lowland, near Manaus, Brazil** *The river is a tributary of the Amazon.* **What else would the geographer see? ... Answers at the end of the chapter.**

tion of the Low-Latitude Rainforest Environment provides more detail on this process and discusses the effects of large-area clearing of the rainforest for mechanized agriculture and grazing.

The Wet-Dry Tropical Climate ③ (Köppen: *Aw, Cwa*)

In the monsoon and trade-wind coastal climate ②, we noted that the movements of the intertropical convergence zone into and away from the climate region produce a seasonal cycle of rainfall and temperature. As we move farther poleward, this cycle becomes stronger, and the monsoon and trade-wind coastal climate ② grades into the **wet-dry tropical climate ③**.

The wet-dry tropical climate ③ is distinguished by a very dry season at low Sun that alternates with a very wet season at high Sun. During the low-Sun season, when the equatorial trough is far away, dry continental tropical (cT) air masses prevail. In the high-Sun season, when the ITCZ is nearby, moist maritime tropical (mT) and maritime equatorial (mE) air masses dominate. Cooler temperatures accompany the dry season but give way to a very hot period before the rains begin.

Figure 7.13 shows the global distribution of the wet-dry tropical climate ③. It is found at latitudes of 5° to 20° N and

> The wet-dry tropical climate ③ has a very dry season alternating with a very wet season. A typical vegetation cover in this climate is savanna woodland, a sparse cover of trees over grassland.

S in Africa and the Americas, and at 10° to 30° N in Asia. In Africa and South America, the climate occupies broad bands poleward of the wet equatorial and monsoon and trade-wind coastal climates. Because these regions are farther away from the ITCZ, less rainfall is triggered by the ITCZ during the rainy season, and subtropical high pressure can dominate more strongly during the low-Sun season. In central India and Indochina, the regions of wet-dry tropical climate ③ are somewhat protected by mountain barriers from the warm, moist mE and mT air flows provided by trade and monsoon winds. These barriers create a rain shadow effect, so that even less rainfall occurs during the rainy season and the dry season is drier still.

Figure 7.14 is a climograph for Timbo, Guinea, at lat. 10° N in West Africa (Figure 7.13). Here the rainy season begins just after the March equinox and reaches a peak in August, about two months following the June solstice. At this time, the ITCZ has migrated to its most northerly position, and moist mE air masses flow into the region from the ocean lying to the south. Monthly rainfall then decreases as the low-Sun season arrives and the ITCZ moves to the south. Three months—December through February—are practically rainless. During this season, subtropical high pressure dominates the climate, and stable, subsiding continental tropical (cT) air pervades the region. The temperature cycle is closely linked to both the solar cycle and the pre-

7.13 World map of the wet-dry tropical climate ③ (Based on Goode Base Map.)

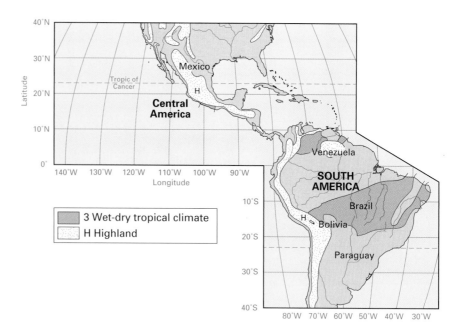

cipitation pattern. In February and March, insolation increases, and air temperature rises sharply. A brief hot season occurs. As soon as the rains set in, the effect of cloud cover and evaporation of rain causes the temperature to decline. By July, temperatures have resumed an even level.

The native vegetation of the wet-dry tropical climate ③ must survive alternating seasons of very dry and very wet

7.14 Wet-dry tropical climate ③ *Timbo, Guinea, at lat. 10° N, is located in West Africa. A long wet season at time of high Sun alternates with an almost rainless dry season at time of low Sun.*

weather. Most plants enter a dormant phase during the dry period, then burst forth into leaf and bloom with the coming of the rains. For this reason, the native plant cover can be described as *rain-green vegetation.*

Rain-green vegetation consists of two basic types. First is *savanna woodland* (Figure 7.15). By *woodland* we mean an open forest in which trees are widely spaced apart. In the savanna woodland, coarse grasses occupy the open space between the trees, which are often coarse-barked and thorny. Large expanses of grassland may also be present. In the dry season, the grasses turn to straw, and many of the tree species shed their leaves to cope with the drought. The second vegetation type is found in the more arid parts of the climate region. Here, small thorny trees and large shrubs form dense patches. This type of vegetation cover is referred to as *thorntree-tall-grass savanna.* (Further details are given in Chapter 9.) Because of the prevalence of the savanna vegetation in the wet-dry tropical climate ③, it can be identified as the savanna environment.

In the savanna environment, river channels that are not fed by nearby moist mountain regions are nearly or completely dry in the low-Sun dry season. In the rainy season, these river channels become filled to their banks with swiftly flowing, turbid water. The rains are not reliable, and agriculture without irrigation is hazardous at best. When the rains fail, a devastating famine can ensue. The Sahel region of Africa, discussed further in *Eye on Global Change 7.2 • Drought and Land Degradation in the African Sahel,* is a region well-known for such droughts and famines.

The Dry Tropical Climate ④ (Köppen: *BWh, BSh*)

The **dry tropical climate** ④ is found in the center and east sides of subtropical high-pressure cells. Here, air strongly subsides, warming adiabatically and inhibiting condensa-

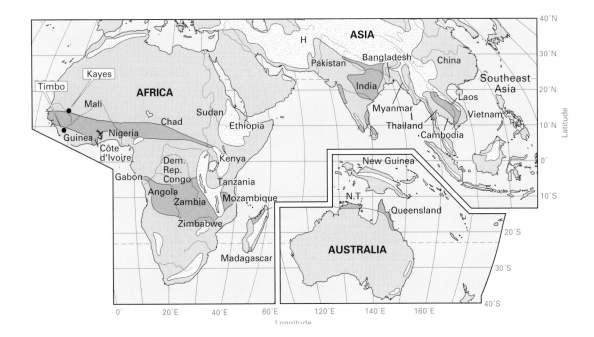

tion. Rainfall is very rare and occurs only when unusual weather conditions move moist air into the region. Since skies are clear most of the time, the Sun heats the surface intensely, keeping air temperatures high. During the high-Sun period, heat is extreme. During the low-Sun period, temperatures are cooler. Given the dry air and lack of cloud cover, the daily temperature range is very large.

The driest areas of the dry tropical climate ④ are near the tropics of cancer and capricorn. As we travel from the tropics toward the equator, we find that somewhat more rain falls. Continuing in this direction, we encounter a short rainy season at the time of year when the ITCZ is near, and the climate grades into the wet-dry tropical ③ type.

Figure 7.16 shows the global distribution of the dry tropical climate ④. Nearly all of the dry tropical climate ④ areas lie in the latitude range 15° to 25° N and S. The largest region is the Sahara–Saudi Arabia–Iran–Thar desert belt of North Africa and southern Asia. This vast desert expanse includes some of the driest regions on Earth. Another large region of dry tropical climate ④ is the desert of central Aus-

7.15 Savanna woodland of the Serengeti Plains, Tanzania, East Africa *Acacia trees with flattened crowns are scattered on this grassy landscape.*

Eye on Global Change | **7.2**

Drought and Land Degradation in the African Sahel

The wet-dry tropical climate ③ is subject to years of devastating drought as well as to years of abnormally high rainfall that can result in severe floods. Climate records show that two or three successive years of abnormally low rainfall (a drought) typically alternate with several successive years of average or higher than average rainfall. This variability is a permanent feature of the wet-dry tropical climate ③, and the plants and animals inhabiting this region have adjusted to the natural variability in rainfall, with one exception: the human species.

The wet-dry tropical climate ③ of West Africa, including the adjacent semiarid southern belt of the dry tropical climate ④ to the north, provides a lesson on the human impact on a delicate ecological system. Countries of this perilous belt, called the *Sahel,* or *Sahelian zone,* are shown in the figure at the right. From 1968 through 1974, all these countries were struck by a severe drought. Both nomadic cattle herders and grain farmers share this zone. During the drought, grain crops failed and foraging cattle could find no food to eat. In the worst stages of the Sahel drought, nomads were forced to sell their remaining cattle. Because the cattle were their sole means of subsistence, the nomads soon starved. Some 5 million cattle perished, and it has been estimated that 100,000 people died of starvation and disease in 1973 alone.

The Sahelian drought of 1968–1974 was associated with a special phenomenon, which at that time was called *desertification*—the permanent transformation of the land surface by human activities to resemble a desert, largely through the destruction of grasses, shrubs, and trees by grazing animals and

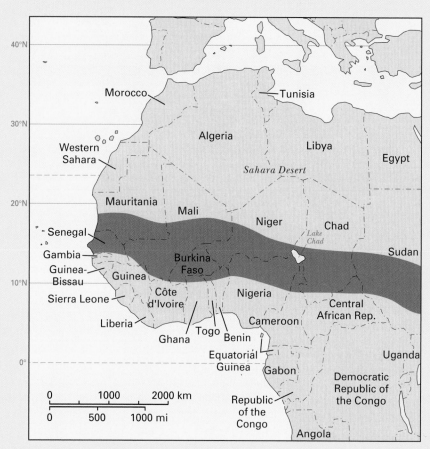

The African Sahel *The Sahel, or Sahelian zone, shown in color, lies south of the great Sahara Desert of North Africa.*

fuel wood harvesting. That term has now been abandoned in favor of *land degradation.* This degradation accelerates the effects of soil erosion, such as gullying of slopes and accumulations of sediment in stream channels. The removal of soil by wind is also intensified.

Periodic droughts throughout past decades are well documented in the Sahel, as they are in other world regions of wet-dry tropical climate ③. Presented at right is a graph showing the percentage of departures from the long-term mean of each year's rainfall in the western Sahel from 1901 through

1995. Note the wide year-to-year variation. Since about 1950, the durations of periods of continuous departures both above and below the mean seem to have increased substantially. The period of sustained high-rainfall years in the 1950s contrasts sharply with a series of severe drought episodes starting in 1971. To obtain an earlier record, scientists have examined fluctuations in the level of Lake Chad. In times of drought, the lake's shoreline retreats, while in times of abundant rainfall, it expands. The changes document periods of rainfall deficiency and

Sahelian drought *At the height of the Sahelian drought, vast numbers of cattle had perished and even the goats were hard pressed to survive. Trampling of the dry ground prepared the region for devastating soil erosion in the rains that eventually ended the drought.*

excess—1820–1840, below normal; 1870–1895, above normal; 1895–1920, below normal.

The sustained drought periods of the past show that droughts and wet periods are a normal phenomenon in the Sahel. In prior eras, the landscape was able to recover from the droughts during periods of abundant rainfall. Today, the pressure of increased populations of humans and cattle keeps the land degraded. As long as these populations remain high, the land degradation will be permanent.

The effect of an accelerated global warming in the coming decades on regions of wet-dry tropical climate ③ is difficult to predict. Some regions may experience greater swings between drought and surplus precipitation. At the same time, these regions may migrate poleward as a result of intensification of the Hadley cell circulation. Thus, the Sahel region may move northward into the desert zone of the Sahara. However, such changes are highly speculative. At this time, there is no scientific consensus on how climatic change will affect the Sahel or other regions of wet-dry tropical climate ③.

Rainfall in the Sahel *Rainfall fluctuations for stations in the western Sahel, 1901–1995, expressed as a percent departure from the long-term mean. (Courtesy of Sharon E. Nicholson, Department of Meteorology, Florida State University, Tallahassee.)*

7.16 World map of the dry tropical ④, dry subtropical ⑤, and dry midlatitude ⑨ climates
The latter two climates are poleward and eastward extensions of the dry tropical climate ④ with cooler temperatures. (Based on Goode Base Map.)

tralia. The west coast of South America, including portions of Ecuador, Peru, and Chile, also exhibits the dry tropical climate ④. However, temperatures there are moderated by a cool marine air layer that blankets the coast.

Figure 7.17 is a climograph for a dry tropical station in the heart of the North African desert. Wadi Halfa, Sudan (Figure 7.16), lies at lat. 22° N, almost on the tropic of cancer. The temperature record shows a strong annual cycle with a very hot period at the time of high Sun, when three consecutive months average 32°C (90°F). Daytime maximum air temperatures are frequently between 43° and 48°C (about 110° to 120°F) in the warmer months. There is a comparatively cool season at the time of low Sun, but the coolest month averages a mild 16°C (61°F), and freezing temperatures are rarely recorded. No rainfall bars are shown on the climograph because precipitation averages less than 0.25 cm (0.1 in.) in all months. Over a 39-year period, the maximum rainfall recorded in a 24-hour period at Wadi Halfa was only 0.75 cm (0.3 in.).

Although the world's dry climates consist largely of extremely arid deserts, there are in addition broad zones at the margins of the desert that are best described as *semiarid*. These zones have a short wet season that supports the growth of grasses on which animals (both wild and domestic) graze. Geographers also call these semiarid regions **steppes.** Nomadic tribes and their herds of animals visit these areas during and after the brief moist period. In Figures 7.6 (world climate map) and 7.16 (world map of dry climates), the two subdivisions of dry climates are distinguished with the letters *a* (arid) and *s* (semiarid).

The dry tropical climate ④ lies in the belt of persistent subtropical high pressure, so rainfall is rare. Temperatures are very hot during the high-Sun season but are significantly cooler during the low-Sun season.

An example of the semiarid dry tropical climate ④ is that of Kayes, Mali, which we presented earlier as a sample climograph (Figure 7.5). Located in the Sahel region of Africa, this station has a distinct rainy season that occurs when the intertropical convergence zone moves north in the high-Sun season. This precipitation pattern shows the semiarid subtype as a transition between the arid dry tropical climate ④ and the wet-dry tropical climate ③.

The Earth's desert landscapes are actually quite varied (Figure 7.18). Much of the arid desert consists of barren areas of drifting sand or sterile salt flats. However, in semiarid regions, thorny trees and shrubs are often abundant, since the climate includes a small amount of regular rainfall. Figure 7.18*a* shows a scene from the semiarid part of the Kalahari Desert in southwest Africa (Figure 7.16), featuring the strange-looking baobab tree. Baja California also supports a sparse collection of plants especially adapted to semiarid desert conditions.

An important variation of the dry tropical climate ④ occurs in narrow coastal zones along the western edge of continents. These regions are strongly influenced by cold ocean currents and the upwelling of deep, cold water, which occurs just offshore. The cool water moderates coastal zone temperatures and reduces the seasonality of the temperature cycle. Figure 7.19 shows a climograph for Walvis Bay, a port city on the west coast of Namibia (southwest Africa), at lat. 23° S (Figure 7.16). (In the figure, the yearly cycle begins with July because this is a southern hemisphere station.) For a location nearly on the tropic of capricorn (23 1/2° S), the monthly temperatures are remarkably cool, with the

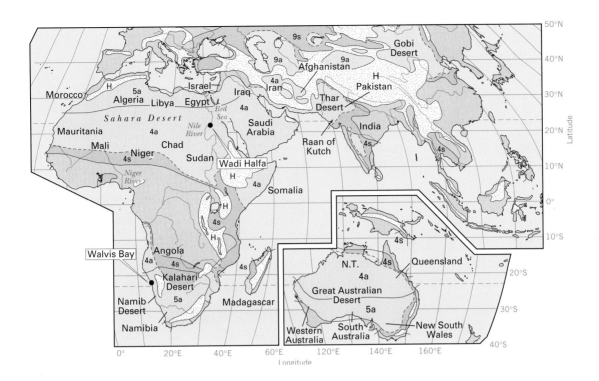

warmest monthly mean only 20°C (68°F) and the coolest monthly mean 14°C (57°F). This provides an annual range of only 5°C (9°F). Coastal fog is a persistent feature of this climate. Another important occurrence of this western coastal desert subtype is along the western coast of South America in Peru and Chile, where it is known as the Atacama Desert.

MIDLATITUDE CLIMATES

The *midlatitude climates* almost fully occupy the land areas of the midlatitude zone and a large proportion of the subtropical latitude zone. Along the western fringe of Europe, they extend into the

> The midlatitude climates lie in a zone of interaction between tropical and polar air masses that is marked by traveling cyclones, traveling anticyclones, and frontal boundaries. They range from very wet to very dry and usually show a strong variation in temperature and/or precipitation through the year.

subarctic latitude zone as well, reaching to the 60th parallel. Unlike the low-latitude climates, which are about equally distributed between northern and southern hemispheres, nearly all of the midlatitude climate area is in the northern hemisphere. In the southern hemisphere, the land area poleward of the 40th parallel is so small that the climates are dominated by a great southern ocean and do not develop the continental characteristics of their counterparts in the northern hemisphere.

The midlatitude climates of the northern hemisphere lie in a broad zone of intense interaction between two groups of very unlike air masses (Figure 7.4). From the subtropical zone, tongues of

7.17 Dry tropical climate ④, dry desert *Wadi Halfa is a city on the Nile River in Sudan at lat. 22° N, close to the Egyptian border. Too little rain falls to be shown on the graph. Air temperatures are very high during the high-Sun months.*

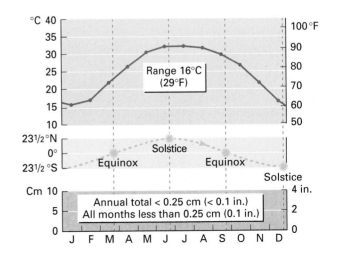

7.18 Desert Landscapes

a...*Eye on the Landscape* *The baobab* This strange-looking baobab tree is a common inhabitant of the thorntree semidesert of Botswana, in the Kalahari region of southern Africa. **What else would the geographer see?** ... Answers at the end of the chapter.

b...*Great Australian Desert* Red colors dominate this desert scene from the Rainbow Valley, south of Alice Springs.

c...*Sahara Desert* This sandy plain in Algeria is dotted with date palms that tap ground water near the surface.

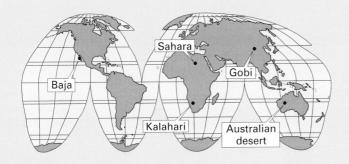

d...Eye on the Landscape
Gobi Desert *Dry, wind-swept basins alternate with rock outcrops and ranges in the high Gobi Desert of Mongolia.* **What else would the geographer see? ... Answers at the end of the chapter.**

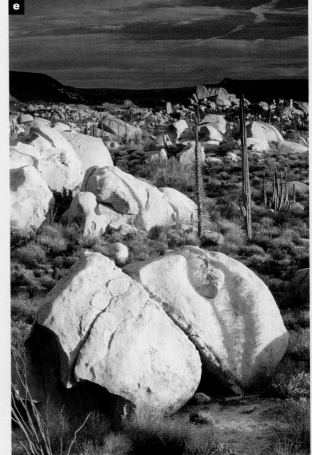

e...Baja California Desert
Rounded granite boulders and many plants adapted to dry desert conditions are visible in this scene from the Catavina Desert of the northern Baja California Pennisula, Mexico.

7.19 Dry tropical climate ④, western coastal desert subtype *Walvis Bay, Namibia (southwest Africa), is a desert station on the west coast of Africa at lat. 23° S. Air temperatures are cool and remarkably uniform throughout the year.*

maritime tropical (mT) air masses enter the midlatitude zone. There, they meet and conflict with tongues of maritime polar (mP) and continental polar (cP) air masses along the discontinuous and shifting polar-front zone.

In terms of prevailing pressure and wind systems, the midlatitude climates include the poleward halves of the great subtropical high-pressure systems and much of the belt of prevailing westerly winds (see Figure 5.16). As a result, weather systems, such as traveling cyclones and their fronts, characteristically move from west to east. This dominant global eastward airflow influences the distribution of climates from west to east across the North American and Eurasian continents.

The six midlatitude climate types range from two that are very dry to three that are extremely moist (Figure 7.20). The midlatitude climates span the range from those with strong wet and dry seasons—the Mediterranean climate ⑦—to those with precipitation that is more or less uniformly distributed through the year. Temperature cycles are also quite varied. Low annual ranges are seen along the windward west coasts. In contrast, annual ranges are large in the continental interiors. Table 7.4 summarizes the important features of these climates.

7.20 Climographs for the six midlatitude climates

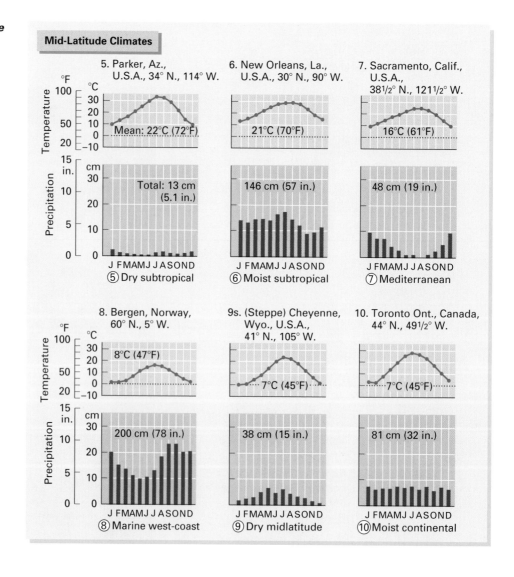

Table 7.4 | Midlatitude Climates

Climate	Temperature	Precipitation	Explanation
Dry subtropical ⑤	Distinct cool or cold season at low-Sun period.	Precipitation is low in nearly all months.	This climate lies poleward of the subtropical high-pressure cells and is dominated by dry cT air most of the year. Rainfall occurs when moist mT air reaches the region, either in summer monsoon flows or in winter frontal movements.
Moist subtropical ⑥	Temperatures show strong annual cycle, but with no winter month below freezing.	Abundant rainfall, cyclonic in winter and convectional in summer. Humidity generally high.	The flow of mT air from the west sides of subtropical high-pressure cells provides moist air most of the year. cP air may reach this region during the winter.
Mediterranean ⑦	Temperature range is moderate, with warm to hot summers and mild winters.	Unusual pattern of wet winter and dry summer. Overall, drier when nearer to subtropical high pressure.	The poleward migration of subtropical high-pressure cells moves clear, stable cT air into this region in the summer. In winter, cyclonic storms and polar frontal precipitation reach the area.
Marine west-coast ⑧	Temperature cycle is moderated by marine influence.	Abundant precipitation but with a winter maximum.	Moist mP air, moving inland from the ocean to the west, dominates this climate most of the year. In the summer, subtropical high pressure reaches these regions, reducing precipitation.
Dry midlatitude ⑨	Strong temperature cycle with large annual range. Summers warm to hot, winters cold to very cold.	Precipitation is low in all months but usually shows a summer maximum.	This climate is dry because of its interior location, far from mP source regions. In winter, cP dominates. In summer, a local dry continental air mass develops.
Moist continental ⑩	Summers warm, winters cold with three months below freezing. Very large annual temperature range.	Ample precipitation, with a summer maximum.	This climate lies in the polar-front zone. In winter, cP air dominates, while mT invades frequently in summer. Precipitation is abundant, cyclonic in winter and convectional in summer.

The Dry Subtropical Climate ⑤ (Köppen: *BWh, BWk, BSh, BSk*)

The **dry subtropical climate** ⑤ is simply a poleward extension of the dry tropical climate ④, caused by somewhat similar air mass patterns. A point of difference is in the annual temperature range, which is greater for the dry subtropical climate ⑤. The lower latitude portions have a distinct cool season, and the higher latitude portions, a cold season. The cold season, which occurs at a time of low Sun, is due in part to invasions of cold continental polar (cP) air masses from higher latitudes. Precipitation occurring in the low-Sun season is produced by midlatitude cyclones that occasionally move into the subtropical zone. As in the dry tropical climate ④, both arid and semiarid subtypes are recognized.

Figure 7.16 shows the global distribution of the dry subtropical climate ⑤. A broad band of this climate type is found in North Africa, connecting with the Near East. Southern Africa and southern Australia also contain regions of dry subtropical climate ⑤ poleward of the dry tropical climate ④. In South America, a band of dry subtropical climate ⑤ occupies Patagonia, a region east of the Andes in Argentina. In North America, the Mojave and Sonoran

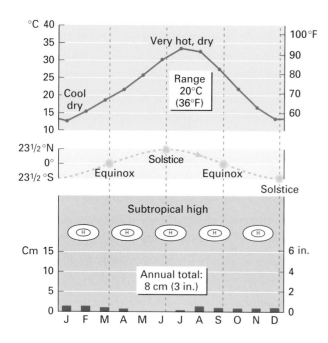

7.21 Dry subtropical climate ⑤ *Yuma, Arizona, lat. 33° N, has a strong seasonal temperature cycle. Compare with Wadi Halfa (Figure 7.17).*

the great summer heat of the low-elevation regions of the Mojave Desert is comparable to that experienced in the Sahara Desert, the low Sun brings a winter season that is not found in the tropical deserts. Here, cyclonic precipitation can occur in most months, including the cool low-Sun months.

In the Mojave Desert and adjacent Sonoran Desert, plants are often large and numerous, in some places giving the appearance of an open woodland. One example is the occurrence of forest-like stands of the tall, cylindrical saguaro cactus. Another is the woodland of Joshua trees found in higher parts of the Mojave Desert (Figure 7.22). Other distinctive plants include the prickly pear cactus, the ocotillo plant, and the creosote bush.

Both plants and animals of deserts are adapted to the dry environment. As in the dry tropical environment, many annual plants remain dormant as seeds during long dry periods, then spring to life, flower, and bloom very quickly when rain falls. Certain invertebrate animals adopt the same life pattern. For example, the tiny brine shrimp of the North American desert may wait many years in dormancy until normally dry lakebeds fill with water, an event that occurs perhaps three or four times per century. The shrimp then emerge and complete their life cycles before the lake evaporates.

deserts of the American Southwest and northwest Mexico are of the dry subtropical type.

Figure 7.21 is a climograph for Yuma, Arizona, a city within the arid subtype of the dry subtropical climate ⑤, close to the Mexican border at lat. 33° N. The pattern of monthly temperatures shows a strong seasonal cycle with a dry, hot summer. A cold season brings monthly means as low as 13°C (55°F). Freezing temperatures—0°C (32°F) and below—can be expected at night in December and January. The annual range is 20°C (36°F). Precipitation, which totals about 8 cm (3 in.), is small in all months but has peaks in late winter and late summer. The August maximum is caused by the invasion of maritime tropical (mT) air masses, which bring thunderstorms to the region. Higher rainfalls from December through March are produced by midlatitude wave cyclones following a southerly path. Two months, May and June, are nearly rainless.

The environment of the dry subtropical climate ⑤ is similar to that of the dry tropical climate ④ in that both are very dry. The boundary between these two climate types is gradational. But if we were to travel northward in the subtropical climate zone of North America, arriving at about 34° N in the interior Mojave Desert of southeastern California, we would encounter environmental features significantly different from those of the low-latitude deserts of tropical Africa, Arabia, and northern Australia. Although

---

The dry subtropical climate ⑤ resembles the dry tropical climate ④ but has cooler temperatures during the low-Sun season, when continental polar air masses invade the region.

The Moist Subtropical Climate ⑥ (Köppen: *Cfa*)

Recall that circulation around the subtropical high-pressure cells provides a flow of warm, moist air onto the eastern side of continents (see Figure 5.18). This flow of maritime tropical (mT) air dominates the **moist subtropical climate** ⑥. Summer in this climate sees abundant rainfall, much of it convectional. Occasional tropical cyclones further enhance summer precipitation. In Southeast Asia, this climate is characterized by a strong monsoon effect, with summer rainfall much increased above winter rainfall. Summer temperatures are warm, with persistent high humidity.

Winter precipitation in the moist subtropical climate ⑥ is also plentiful, produced in midlatitude cyclones. Invasions of continental polar (cP) air masses are frequent in winter, bringing spells of subfreezing weather. No winter month has a mean temperature below 0°C (32°F).

Figure 7.23 presents a global map of the moist subtropical climate ⑥ It is found on the eastern sides of continents in the latitude range 20° to 35° N and S. In South America, it includes parts of Uruguay, Brazil, and Argentina. In Australia, it consists of a narrow band between the eastern coastline and the eastern interior ranges. Southern China, Taiwan, and southernmost Japan are regions of the moist subtropical climate ⑥ in Asia. In the United States, the moist subtropical climate ⑥ covers most of the Southeast, from the Carolinas to east Texas.

7.22   A Mojave Desert landscape *The strange-looking Joshua tree, shown in the foreground, is abundant at higher elevations in the Mojave Desert.*

Figure 7.24 is a climograph for Charleston, South Carolina, located on the eastern seaboard at lat. 33° N. In this region, a marked summer maximum of precipitation is typical. Total annual rainfall is abundant—120 cm (47 in.)—and ample precipitation falls in every month. The annual temperature cycle is strongly developed, with a large annual range of 17°C (31°F). Winters are mild, with the January mean temperature well above the freezing mark.

Much of the natural vegetation in the North American area of this moist climate type consists of *broadleaf deciduous forest.* (In a deciduous forest, leaf shedding occurs annually, and the trees are bare throughout the cold or dry season.) Broadleaf deciduous forest can be found on the interior

> The moist subtropical climate ⑥, found on the eastern sides of continents in the midlatitudes, has abundant precipitation. In summer, warm, moist maritime tropical air masses provide convectional showers, while in winter, wave cyclones provide rain and occasional snow.

uplands of the Carolinas and across Tennessee. An entirely different forest type occurs farther south and east on the low, sandy coastal plain of the Gulf states, on the Florida Peninsula, and on the Atlantic coastal plain of Georgia and the Carolinas. It is the *southern pine forest,* which is well adapted to sandy soils. Over a large part of southern China and the south island of Japan, the native vegetation was formerly a *broadleaf evergreen* forest (Chapter 9). Today this forest is gone from China, but some small areas in Japan remain. On the North American continent, in Louisiana and farther east on the Gulf coast, the evergreen broadleaf forest is represented by a few small areas of Evangeline oak and magnolia (Figure 7.25).

7.23 World map of the moist subtropical climate *(Based on Goode Base Map.)*

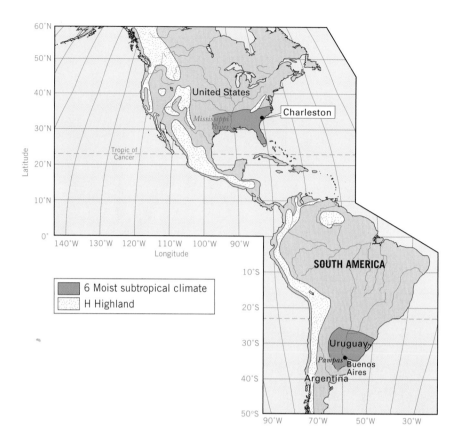

6 Moist subtropical climate

H Highland

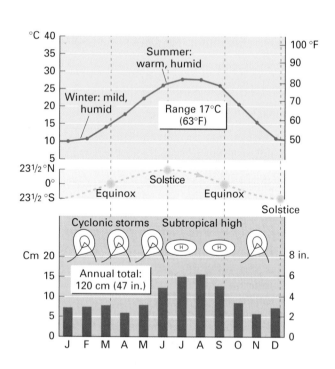

7.24 Moist subtropical climate ⑥ *Charleston, South Carolina, lat. 33° N, has a mild winter and a warm summer. There is ample precipitation in all months but a definite summer maximum.*

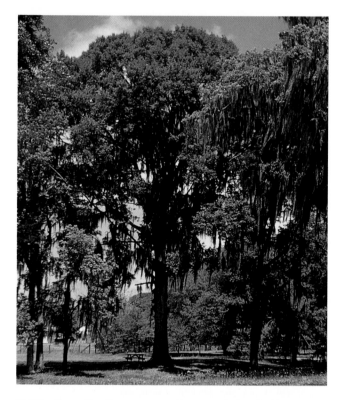

7.25 Broadleaf evergreen forest of the Gulf coast *A typical species of the broadleaf evergreen forest is the Evangeline oak, here bearing Spanish "moss"—an epiphyte that forms long beard-like streamers. The ground beneath is maintained as a lawn. Evangeline State Park, Bayou Teche, Louisiana. (A. N. Strahler.)*

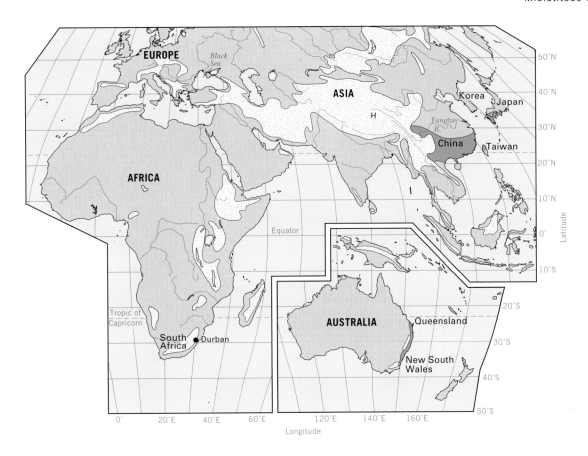

The Mediterranean Climate ⑦ (Köppen: *Csa, Csb*)

The **Mediterranean climate** ⑦ is unique among the climate types because its annual precipitation cycle has a wet winter and a very dry summer. The reason for this precipitation cycle lies in the poleward movement of the subtropical high-pressure cells during the summer season. The Mediterranean climate ⑦ is located along the west coasts of continents, just poleward of the dry, eastern side of the subtropical high-pressure cells (see Figure 5.18). When the subtropical high-pressure cells move poleward in summer, they enter the region of this climate. Dry continental tropical (cT) air then dominates, producing the dry summer season. In winter, the moist mP air mass invades with cyclonic storms and generates ample rainfall.

In terms of total annual rainfall, the Mediterranean climate ⑦ spans a wide range from arid to humid, depending on location. Generally, the closer an area is to the tropics, the stronger the influence of subtropical high pressure will be, and thus the drier the climate. The temperature range is moderate, with warm to hot summers and mild winters. Coastal zones between lat. 30° and 35° N and S, such as southern California, show a smaller annual range, with very mild winters.

Figure 7.26 is a climograph for Moneterey, California, a Pacific coastal city at lat. 36° N. The annual temperature cycle is very weak. The small annual range and cool summer

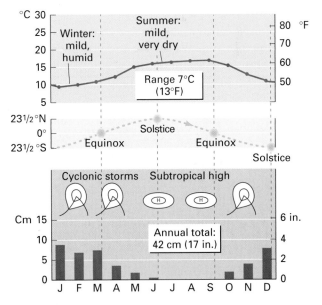

7.26 Mediterranean climate ⑦ *Monterey, California, lat. 36° N, has a very weak annual temperature cycle because of its closeness to the Pacific Ocean. The summer is very dry.*

7.27 World map of the Mediterranean ⑦ and marine west-coast ⑧ climates *(Based on Goode Base Map.)*

reflect the strong control of the cold California current and its cool, marine air layer. Fogs are frequent. This temperature regime is found only in a narrow coastal zone. Rainfall drops to nearly zero for four consecutive summer months but rises to substantial amounts in the rainy winter season. In California's Central Valley, the dry summers persist, but the annual temperature range is greater, so that daily high temperatures match those of the adjacent dry climates.

A global map of the Mediterranean climate ⑦ is shown in Figure 7.27. It is found in the latitude range 30° to 45° N and S. In the southern hemisphere, it occurs along the coast of Chile, in the Cape Town region of South Africa, and along the southern and western coasts of Australia. In North America, it is found in central and southern California. In Europe, this climate type surrounds the Mediterranean Sea, which gives the climate its distinctive name.

The native vegetation of the Mediterranean climate environment is adapted to survival through the long summer drought. Shrubs and trees are typically equipped with small, hard or thick leaves that resist water loss through transpiration. These plants are called *sclerophylls;* the prefix scler, from the Greek for "hard," is combined with phyllo, which is Greek for "leaf." Sclerophylls of the

> The Mediterranean climate ⑦, found along midlatitude west coasts, is distinguished by its dry summer and wet winter. In summer, dry subtropical high pressure blocks rainfall, while in winter, wave cyclones produce ample precipitation.

Mediterranean environment are typically evergreen, retaining their leaves through the entire yearly cycle. Examples are the evergreen oaks, of which there are several common species in California, and the cork oak of the Mediterranean lands (Figure 7.28a). Another example is the olive tree, native to the Mediterranean lands. In Australia, the thick-leafed eucalyptus tree is the dominant sclerophyll. Oak woodland of California bears a ground cover of grasses that turn to straw in the summer (Figure 7.28b). At higher elevations, drought-resistant conifers often occur.

Another form of native vegetation in this environment is a cover of drought-resistant shrubs, including sclerophylls and spiny-leafed species. In the Mediterranean lands, this scrub vegetation goes under the name of maquis or garrigue. In California, where it is called *chaparral,* it clothes steep hill and mountain slopes too dry to support oak woodland or oak forest (Figure 7.28c).

Wildfire is an integral part of the Mediterranean environment of California. Chaparral is extremely flammable during the long summer fire season (Figure 5.10). Brushfires rage through chaparral and oak forests and leave the soil surface bare and unprotected. When torrential rains occur in winter, large quantities of coarse mineral debris are swept

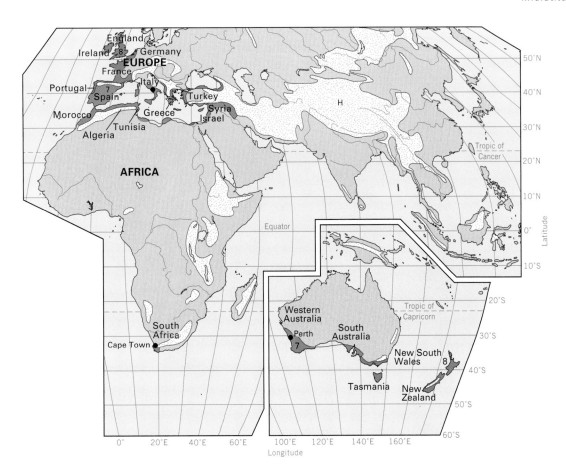

downslope and carried long distances by streams in flood. Mudflows and debris floods (usually called "mudslides" in the news media) are particularly destructive to communities on canyon floors (see Chapter 14).

The Marine West-Coast Climate ⑧ (Köppen: *Cfb, Cfc*)

The **marine west-coast climate** ⑧ occupies midlatitude west coasts. These locations receive the prevailing wester-lies from over a large ocean and experience frequent cyclonic storms involving cool, moist mP air masses. Where the coast is mountainous, the orographic effect causes a very large annual precipita-tion. In this moist climate, precipita-tion is plentiful in all months, but there is often a distinct winter maximum. In summer, subtropical high pressure extends poleward into the region, reducing rainfall. The annual tempera-ture range is comparatively small for midlatitudes. The marine influence keeps winter temperatures mild, as compared with inland locations at equivalent latitudes.

The global map of the marine west-coast climate ⑧ (Figure 7.27) shows the areas in which this climate occurs.

> The marine west-coast climate ⑧ features frequent cyclonic storms that provide abundant precipitation, especially when enhanced by an orographic effect. Summers are normally drier as subtropical high pressure moves poleward, blocking storm tracks.

In North America, the climate occupies the western coast from Oregon to northern British Columbia. In Western Europe, the British Isles, Portugal, and much of France fall into the marine west-coast climate. New Zealand and the southern tip of Australia, as well as the island of Tasmania, are marine west-coast climate regions found in the south-ern hemisphere, as is the Chilean coast south of 35° S. The general latitude range of this climate is 35° to 60° N and S.

Figure 7.29 is a climograph for Vancouver, British Columbia, just north of the U.S.–Canadian border. The annual precipitation is very great, and most of it falls dur-ing the winter months. Notice the greatly reduced rainfall in the summer months. The temperature cycle shows a remarkably small range for this lati-tude. Even the winter months have averages above the freezing mark.

Forest is the native vegetation of this climate. Dense needleleaf forests of redwood, fir, cedar, hemlock, and spruce (Figure 7.30) flourish in the wet mountainous areas of the northern Pacific coast. Under the lower precipi-tation regime of Ireland, southern England, France, and the Low Countries, a broadleaf deciduous forest was the native vegetation, but much of it disappeared many centuries ago under cultivation. Only scattered forest plots or groves

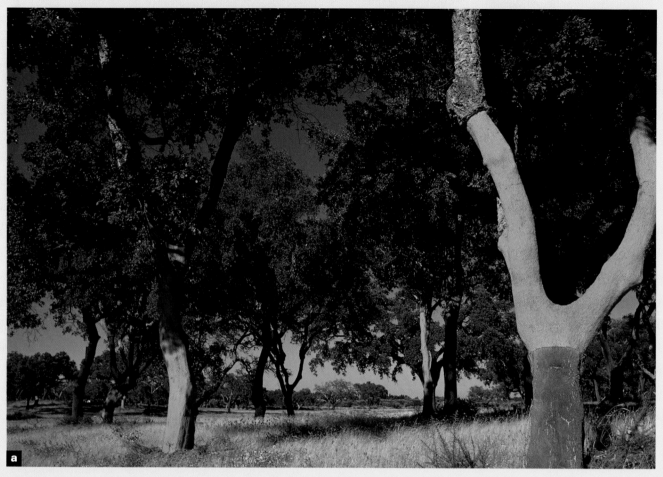

a...Cork oak *The cork oak, Quercus suber, has tough, thick outer bark that is harvested for wine corks. The underlying live bark is reddish in color. Algarve Province, Portugal.*

b...California oak woodland in summer *Blue oaks (Quercus douglasii) are scattered over a cover of dry grasses in the inner Coast Ranges, near Williams, California.*

c...*California chaparral*
Wiry shrubs cloak the
steep slopes of a coast
range near Acton,
California.

d...*Orchard agriculture*
Groves of lemon, orange,
and avocado trees
surround these homes and
stables in Montecito,
California. Chaparral
covers the steep slopes of
the Santa Ynez Mountains
in the background. (A. N.
Strahler.)

°C 30 / 80 °F

Winter: cool, moist

Summer: mild, dry

Range 15°C (27°F)

23½°N
0°
23½°S

Equinox · Solstice · Equinox · Solstice

Cyclonic storms · Subtropical high · Cyclonic storms

Cm 25 / 10 in.

Annual total: 146 cm (57 in.)

J F M A M J J A S O N D

7.29 Marine west-coast climate ⑧ *Vancouver, British Columbia, lat. 49° N, has a large annual total precipitation but with greatly reduced amounts in the summer. The annual temperature range is small, and winters are very mild for this latitude.*

remain (see Figure 7.31). Sometimes called "summer-green" deciduous forest, it is dominated by tall broadleaf trees that provide a continuous and dense canopy in summer but shed their leaves completely in winter.

The Dry Midlatitude Climate ⑨ (Köppen: *BWk, BSk*)

The **dry midlatitude climate** ⑨ is limited almost exclusively to interior regions of North America and Eurasia, where it lies within the rain shadow of mountain ranges on the west or south. Maritime air masses are effectively blocked out much of the time, so that the continental polar (cP) air mass dominates the climate in winter. In summer, a dry continental air mass of local origin is dominant. Summer rainfall is mostly convectional and is caused by occasional invasions of maritime air masses. The annual temperature cycle is strongly developed, with a large annual range. Summers are warm to hot, but winters are cold to very cold.

The largest expanse of the dry midlatitude climate ⑨ is in Eurasia, stretching from the southern republics of the former Soviet Union to the Gobi Desert and northern China (Figure 7.16). In the central portions of this region lie true deserts of the arid climate subtype, with very low precipitation. Extensive areas of highlands occur here as well. In North America, the dry western interior regions, including the Great Basin, Columbia Plateau, and the Great Plains, are of the semiarid subtype. A small area of dry midlatitude climate ⑨

7.30 Needleleaf forest *Sitka spruce and hemlock populate the Hoh rainforest, Olympic National Park, Washington. Ferns and mosses provide a lush ground cover in this very wet environment.*

7.31 Marine west-coast landscape *This lush landscape near Devon, England, is located in the domain of the summer-green deciduous forest. After many centuries of human occupation and cultivation, the forest is now reduced to small patches and scattered individual trees in hollows and along fence rows.*

is found in southern Patagonia, near the tip of South America. The latitude range of this climate is 35° to 55° N.

Figure 7.32 is a climograph for Pueblo, Colorado, a semiarid station located at lat. 38° N, just east of the

The dry midlatitude climate ⑨ occupies continental interiors in rain shadows or far from oceanic moisture sources. Precipitation is low and the annual temperature variation is large.

Rocky Mountains. Total annual precipitation is 31 cm (12 in.). Most of this precipitation is in the form of convectional summer rainfall, which occurs when moist maritime tropical (mT) air masses invade from the south and produce thunderstorms. In winter, snowfall is light and yields only small monthly precipitation averages. The temperature cycle has a large annual range, with warm summers and cold winters. January, the coldest winter month, has a mean temperature just below freezing.

The low precipitation and cold winters of this semiarid climate produce a steppe landscape dominated by hardy perennial short grasses. In North America, this cover is termed *short-grass prairie* rather than steppe. Wheat is a major crop of the semiarid, dry midlatitude steppelands, but wheat harvests are at the mercy of rainfall variations from year to year. With good spring rains, there is a good crop, but if spring rains fail, so does the wheat crop.

The Moist Continental Climate ⑩ (Köppen: *Dfa, Dfb, Dwa, Dwb*)

The **moist continental climate** ⑩ is located in central and eastern parts of North America and Eurasia in the midlatitudes. This climate lies in the polar-front zone—the battleground of polar and tropical air masses. Seasonal temperature contrasts are strong, and day-to-day weather is highly variable. Ample precipitation throughout the year is increased in summer by invading maritime tropical (mT) air masses. Cold winters are dominated by continental polar (cP) and continental arctic (cA) air masses from subarctic source regions.

7.32 Dry midlatitude climate ⑨ *Pueblo, Colorado, lat. 38° N, shows a marked summer maximum of rainfall in the summer months. Note also the cold, dry winter season. Figure 7.16 shows the location of this station.*

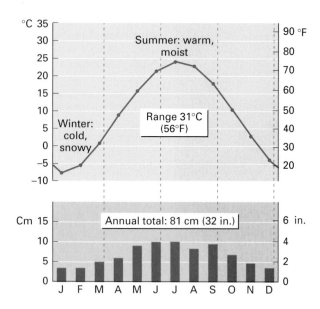

7.33 Moist continental climate ⑩ *Madison, Wisconsin, lat. 43° N, has cold winters and warm summers, making the annual temperature range very large.*

⑩ lies in a higher latitude belt (45° to 60° N) and receives precipitation from mP air masses coming from the North Atlantic.

Madison, Wisconsin, lat. 43° N, in the American Midwest (Figure 7.33), provides an example of the moist continental climate ⑩. The annual temperature range is very large. Summers are warm, but winters are cold, with three consecutive monthly means well below freezing. Precipitation is ample in all months, and the annual total is large. There is a summer maximum of precipitation when the maritime tropical (mT) air mass invades, and thunderstorms are formed along moving cold fronts and squall lines. Much of the winter precipitation is in the form of snow, which remains on the ground for long periods.

Figure 7.34 shows the global locations of the moist continental climate ⑩. It is restricted to the northern hemisphere, occurring in latitudes 30° to 55° N in North America and Asia, and in latitudes 45° to 60° N in Europe. In Asia, it is found in northern China, Korea, and Japan. Most of central and eastern Europe has a moist continental climate, as does most of the eastern half of the United States from Tennessee to the north, as well as the southernmost strip of eastern Canada.

Throughout most of the moist continental climate ⑩, forests are the dominant natural vegetation cover. However, where the climate grades into drier climates, such as the dry midlatitude climate ⑨, tall, dense grasses may be the natural cover. The *tall-grass prairie* of Illinois, Iowa, and Nebraska is an example. These lands are now the heart of the American corn belt. In Europe, lands of the moist continental climate ⑩ have been in cultivation for many centuries, providing a multitude of crops. Forests are planted and managed for timber yield. The result is a landscape that shows human influence everywhere (Figure 7.35).

In eastern Asia—China, Korea, and Japan—the seasonal precipitation pattern shows more summer rainfall and a drier winter than in North America. This is an effect of the monsoon circulation, which moves moist maritime tropical (mT) air across the eastern side of the continent in summer and dry continental polar southward through the region in winter. In Europe, the moist continental climate

---

The moist continental climate ⑩ lies in the polar-front zone where day-to-day weather is highly variable. Ample frontal precipitation is enhanced in summer by maritime tropical air masses. Winter temperatures fall well below freezing.

7.34 World map of the moist continental climate ⑩ *(Based on Goode Base Map.)*

**7.35 Eye on the Landscape *Moist continental
landscape*** *The Mosel River in its winding gorge through the
Rhineland-Pfalz province of western Germany. On the steep
valley wall at the right are vineyards and forested land. Abundant
rainfall, fertile soils, and mild summer temperatures provide an environment well
suited to agriculture.* **What else would the geographer see? … Answers at the end of
the chapter.**

7.36 Climographs of the three high-latitude climates

High-Latitude Climates

11. Yatkutsk, Russia
61° N., 130° E.

Mean: −11°C
(12.5°F)

Total: 19 cm
(7.5 in.)

J FMAMJ JASOND
⑪Boreal forest

12. Barrow Point, Alaska U.S.A.,
71° N., 157° W.

−12°C (10°F)

10 cm (4 in.)

J FMAMJ JASOND
⑫Tundra

13. Amundsen-Scott Sta.
S. Pole,
90° N.

−48°C (−54°F)

J FMAMJ JASOND
⑬Ice-sheet

HIGH-LATITUDE CLIMATES

By and large, the *high-latitude climates* are climates of the northern hemisphere (Figure 7.36). They occupy the northern subarctic and arctic latitude zones, but they also extend southward into the midlatitude zone as far south as about the 47th parallel in eastern North America and eastern Asia. One of these, the ice sheet climate ⑬, is present in both hemispheres in the polar zones. Table 7.5 provides an overview of the three high-latitude climates.

The high-latitude climates coincide closely with the belt of prevailing westerly winds that circles each pole (see Figure 5.16). In the northern hemisphere, this circulation sweeps maritime polar (mP) air masses, formed over the northern oceans, into conflict with continental polar (cP) and continental arctic (cA) air masses on the continents. Rossby waves form in the westerly flow, bringing lobes of warmer, moister air poleward into the region in exchange for colder, drier air moved equatorward. The result of these processes is frequent wave cyclones, produced along a discontinuous and constantly fluctuating arctic-front zone. In summer, tongues of maritime tropical air masses (mT) reach

the subarctic latitudes to interact with polar air masses and yield important amounts of precipitation.

The Boreal Forest Climate ⑪
(Köppen: *Dfc, Dfd, Dwc, Dwd*)

The **boreal forest climate** ⑪ is a continental climate with long, bitterly cold winters and short, cool summers. It occupies the source region for cP air masses, which are cold, dry, and stable in the winter. Invasions of the very cold cA air mass are common. The annual range of temperature is greater than that of any other climate and is greatest in Siberia. Precipitation increases substantially in summer, when maritime air masses penetrate the continent with traveling cyclones, but the total annual precipitation is small. Although much of the boreal forest climate is moist, large areas in western Canada and Siberia have low annual precipitation and are therefore cold and dry.

The global extent of the boreal forest climate ⑪ is presented in Figure 7.37. In North America, it stretches from central and western Alaska, across the Yukon and Northwest

7.37 World map of the boreal forest climate ⑪ *(Based on Goode Base Map.)*

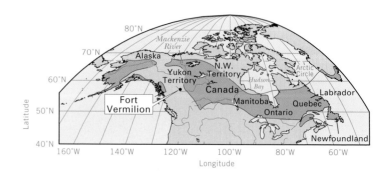

11 Boreal forest climate

Table 7.5 | **High-Latitude Climates**

Climate	Temperature	Precipitation	Explanation
Boreal forest ⑪	Short, cool summers and long, bitterly cold winters. Greatest annual range of all climates.	Annual precipitation small, falling mostly in summer months.	This climate falls in the source region for cold, dry stable cP air masses. Traveling cyclones, more frequent in summer, bring precipitation from mP air.
Tundra ⑫	No true summer but a short mild season. Otherwise, cold temperatures.	Annual precipitation small, falling mostly during mild season.	The coastal arctic fringes occupied by this climate are dominated by cP, mP, and cA air masses. The maritime influence keeps winter temperatures from falling to the extreme lows of interiors.
Ice sheet ⑬	All months below freezing, with lowest global temperatures on Earth experienced during Antarctic winter.	Very low precipitation, but snow accumulates since temperatures are always below freezing.	Ice sheets are the source regions of cA and cAA air masses. Temperatures are intensely cold.

Territories to Labrador on the Atlantic coast. In Europe and Asia, it reaches from the Scandinavian Peninsula eastward across all of Siberia to the Pacific. In latitude, this climate type ranges from 50° to 70° N.

Figure 7.38 is a climograph for Fort Vermilion, Alberta, at lat. 58° N. The very great annual temperature range shown here is typical for North America. Monthly mean air temperatures are below freezing for seven consecutive months. The summers are short and cool. Precipitation shows a marked annual cycle with a summer maximum, but the total annual precipitation is small. Although precipitation in winter is small, a snow cover remains over solidly frozen ground through the entire winter. On the same climograph, temperature data are shown for Yakutsk, a Siberian city at lat. 62° N. The enormous annual range is evident, as well as the extremely low means in winter months. January reaches a mean of about –42°C (–44°F). Except for the ice sheet inte-

The boreal forest climate ⑪ is a climate of long, bitterly cold winters and short, cool summers. For most of the year, cold, dry continental arctic and polar air masses dominate. In summer, occasional maritime air masses provide moisture for precipitation.

riors of Antarctica and Greenland, this region is the coldest on Earth. Precipitation is not shown for Yakutsk, but the annual total is very small.

The land surface features of much of the region of boreal forest climate were shaped beneath the great ice sheets of the last ice age. Severe erosion by the moving ice exposed hard bedrock over vast areas and created numerous shallow rock basins. Bouldery rock debris mantles the rock surface in many places. Many of the shallower rock basins have been filled by organic bog materials.

The dominant upland vegetation of the boreal forest climate ⑪ region is boreal forest, consisting of *needleleaf trees*. In North America and Europe, these are evergreen needleleaf trees, mostly pine, spruce, and fir. Figure 9.15 is a composite map showing the extent of the boreal forest. In central and eastern Siberia, the boreal forest is dominated by the larch, which sheds its needles in winter and is thus a deciduous

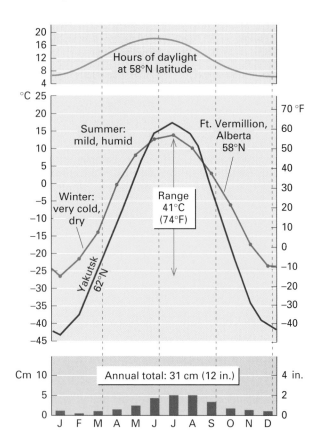

7.38 **Boreal forest climate** ⑪ *Extreme winter cold and a very great annual range in temperature characterize the climate of Fort Vermilion, Alberta. The temperature range of Yakutsk, Siberia, is even greater.*

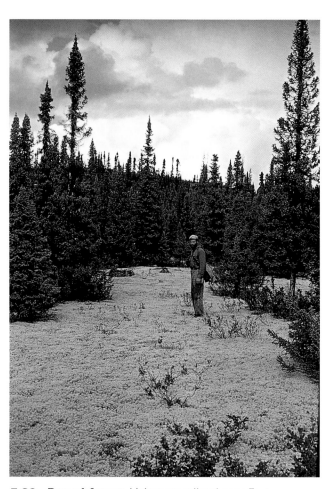

7.39 **Boreal forest** *Lichen woodland near Fort McKenzie, lat. 57° N, in northern Quebec. The trees are black spruce. Between the trees is a carpet of lichen. (R. N. Drummond.)*

tree. Associated with the needleleaf trees are stands of aspen, balsam poplar, willow, and birch. Along the northern fringe of boreal forest lies a zone of woodland in which low trees, such as black spruce, are spaced widely apart. The open areas are covered by a surface layer of lichens and mosses (Figure 7.39). This cold woodland is referred to as *taiga*.

Although the growing season in the boreal forest climate ⑪ is short, crop farming is still possible. It is largely limited to lands surrounding the Baltic Sea, bordering Finland and Sweden. Crops grown in this area include barley, oats, rye, and wheat. Along with dairying, these crops primarily supply food for subsistence. The needleleaf forests provide paper, pulp, cellulose, and construction lumber.

The Tundra Climate ⑫ (Köppen: *ET*)

The **tundra climate** ⑫ occupies arctic coastal fringes and is dominated by polar (cP, mP) and arctic (cA) air masses. Winters are long and severe. A moderating influence of the nearby ocean water prevents winter temperatures from falling to the extreme lows

> The tundra climate ⑫ occupies arctic coastal fringes. Although the climate is very cold, the maritime influence keeps winter temperatures from falling to the levels of the boreal forest climate ⑪. A mild season provides a few months of thaw.

found in the continental interior. There is a very short mild season, which many climatologists do not recognize as a true summer.

The world map of the tundra climate ⑫ (Figure 7.40) shows the tundra ringing the Arctic Ocean and extending across the island region of northern Canada. It includes the Alaskan north slope, the Hudson Bay region, and the Greenland coast in North America. In Eurasia, this climate type occupies the northernmost fringe of the Scandinavian Peninsula and Siberian coast. The Antarctic Peninsula (not shown in Figure 7.40) belongs to the tundra climate ⑫. The latitude range for this climate is 60° to 75° N and S, except for the northern coast of Greenland, where tundra occurs at latitudes greater than 80° N.

Figure 7.41 is a climograph for Upernavik, located on the west coast of Greenland at lat. 73° N. A short mild period, with above-freezing temperatures, is equivalent to a summer season in lower latitudes. The long winter is

7.40 World map of the tundra climate ⑫

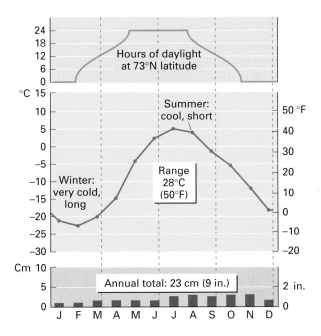

7.41 Tundra climate ⑫ *Upernavik, Greenland, lat. 73° N, shows a smaller annual range than Fort Vermilion (Figure 7.38).*

very cold, but the annual temperature range is not as large as that for the boreal forest climate to the south. Total annual precipitation is small. Increased precipitation beginning in July is explained by the melting of the sea-ice cover and a warming of ocean water temperatures. This increases the moisture content of the local air mass, allowing more precipitation.

The term *tundra* describes both an environmental region and a major class of vegetation (Figure 7.42). (An equivalent climatic environment—called *alpine tundra*—prevails in many global locations in high mountains above the timberline.) Soils of the arctic tundra are poorly developed and consist of freshly broken mineral particles and varying amounts of humus (finely divided, partially decomposed plant matter). Peat bogs are numerous. Because soil water is solidly and permanently frozen not far below the surface, the summer thaw brings a condition of water saturation to the soil.

Vegetation of the tundra consists of a cover of scattered grasses, sedges, and lichens, along with shrubs of willow. Vegetation is scarce on dry, exposed slopes and summits. Here the surface cover is often a rocky pavement of angular rock fragments. Trees exist in the tundra only as small, shrub-like plants. They are stunted because of the seasonal damage to roots by freeze and thaw of the soil layer and to branches exposed to the abrading action of wind-driven snow. In some places, a distinct tree line separates the forest and tundra. It coincides approximately with the 10°C (50°F)

7.42 *Tundra landscape* Caribou migration across the arctic tundra of northern Alaska.

isotherm of the warmest month and has been used by geographers as a boundary between boreal forest and tundra.

Because of the cold temperatures experienced in the tundra and northern boreal forest climate zones, the ground is typically frozen to great depth. This perennially frozen ground, or **permafrost,** prevails over the tundra region and a wide bordering area of boreal forest climate. Normally, a top layer of the ground will thaw each year during the mild season. This active layer of seasonal thaw is from 0.6 to 4 m (2 to 13 ft) thick, depending on latitude and the nature of the ground.

Continuous permafrost, which extends without gaps or interruptions under all surface features, coincides largely with the tundra climate, but also includes a large part of the boreal forest climate in Siberia. *Discontinuous permafrost,* which occurs in patches separated by frost-free zones under lakes and rivers, occupies much of the boreal forest climate zone of North America and Eurasia. Sporadic occurrence of permafrost in small patches extends into the southern limits of the boreal forest climate. Chapter 14 provides more detail on permafrost in arctic and boreal regions.

> The ice sheet climate ⑬ has the lowest temperatures found on Earth. No month shows mean temperatures above freezing, and winter mean monthly temperatures can fall to 40°C (–40°F) and below.

The Ice Sheet Climate ⑬ (Köppen: *EF*)

The **ice sheet climate** ⑬ coincides with the source regions of arctic (A) and antarctic (AA) air masses, situated on the vast, high ice sheets of Greenland and Antarctica and over polar sea ice of the Arctic Ocean. Mean annual temperature is much lower than that of any other climate, with no monthly mean above freezing. Strong temperature inversions, caused by radiation loss from the surface, develop over the ice sheets. In Antarctica and Greenland, the high surface altitude of the ice sheets intensifies the cold. Strong cyclones with blizzard winds are frequent. Precipitation, almost all occurring as snow, is very low but accumulates because of the continuous cold. The latitude range for this climate is 65° to 90° N and S.

Figure 7.43 shows temperature graphs for several representative ice sheet stations. The graph for Eismitte, a research station on the Greenland ice cap, shows the northern hemisphere temperature cycle, whereas the other four examples are all from Antarctica. Temperatures in the interior of Antarctica have proved to be far lower than those at any other place on Earth. A Russian meteorological station at Vos-

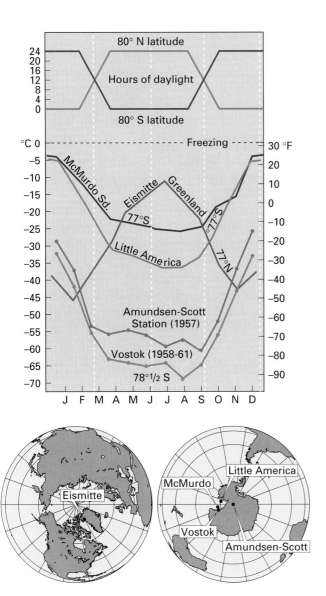

7.43 Ice sheet climate ⑬ *Temperature graphs for five ice sheet stations.*

tok, located about 1300 km (about 800 mi) from the south pole at an altitude of about 3500 m (11,500 ft), may be the world's coldest spot. Here a low of –88.3°C (–127°F) was observed in 1958. At the pole itself (Amundsen-Scott Station), July, August, and September of 1957 had averages of about –60°C (–76°F). Temperatures are considerably higher, month for month, at Little America in Antarctica because it is located close to the Ross Sea and is at a low altitude.

A LOOK AHEAD The climates of the Earth are remarkably diverse, ranging from the hot, humid wet equatorial climate ① at the equator to the bitterly cold and dry ice sheet climate ⑬ at the poles. Between these extremes are the other climates, each with distinctive features—such as dry summers or dry winters, or uniform or widely varying temperatures. The global environments associated with each climate type are also highly varied, from the lush, equatorial rainforest to the stunted willows of the tundra. Some of these environmental regions are more hospitable to human habitation and use than are others. As we have shown in this chapter and its predecessor, climate exerts strong controls on vegetation and soils, especially at the global level. We turn to a more detailed look at vegetation and soils in the next three chapters.

In Review | Global Climates

- **Climate** is the average weather of a region. Because temperature and precipitation are measured at many stations worldwide, we can use the combined annual patterns of monthly averages of temperature and precipitation to assign climate types.

- *Temperature regimes* are typical patterns of annual variation in temperature. They depend on latitude, which determines the annual pattern of insolation, and on location—continental or maritime—which enhances or moderates the annual insolation cycle.

- Global precipitation patterns are largely determined by air masses and their movements, which in turn are produced by global air circulation patterns. The main features of the global pattern of rainfall are:

1. a wet equatorial belt produced by convectional precipitation around the ITCZ;
2. trade-wind coasts that receive moist flows of mT air from trade winds as well as tropical cyclones;
3. tropical deserts located under subtropical high-pressure cells;
4. midlatitude deserts and *steppes,* which are dry because they are far from maritime moisture sources;
5. moist subtropical regions that receive westward flows of moist mT air in the summer and eastward-moving wave cyclones in winter;
6. midlatitude west coasts, which are subjected to eastward flows of mP air and occluded wave cyclones by prevailing westerly winds; and

7. polar and arctic deserts, where little precipitation falls because the air is too cold to hold much moisture.

- Annual patterns of precipitation fall into three patterns: uniform (ranging from abundant to scarce); high-Sun (summer) maximum; and low-Sun (winter) maximum.

- There are three groups of climate types, arranged by latitude. Low-latitude climates (Group I) are dominated by mE, mT, and cT air masses, and are largely related to the global circulation patterns that produce the ITCZ, trade winds, and subtropical high-pressure cells. Midlatitude climates (Group II) lie in the polar-front zone and are strongly influenced by eastward-moving wave cyclones in which mT, mP, and cP air masses interact. High-latitude climates (Group III) are dominated by polar and arctic (antarctic) air masses. Wave cyclones mixing mP and cP air masses along the *arctic-front zone* provide precipitation in this region.

- Dry climates are those in which precipitation is largely evaporated from soil surfaces and transpired by vegetation, so that permanent streams cannot be supported. Within dry climates, there are two subtypes: arid (driest) and semiarid or steppe (a little wetter). In moist climates, precipitation exceeds evaporation and transpiration, providing for sustained year-round stream flow. In wet-dry climates, strong wet and dry seasons alternate.

- In the **wet equatorial climate** ①, the ITCZ is always nearby, and so rainfall is abundant throughout the year. The annual temperature cycle is very weak, so that the daily range greatly exceeds the annual range.

- The **monsoon and trade-wind coastal climate** ② shows a dry period when the ITCZ has migrated toward the opposite tropic. With the return of the ITCZ, monsoon circulation and enhanced easterly waves provide increased rainfall in this climate type. Temperatures are highest in the dry weather before the onset of the wet season.

- The equatorial climate ① and monsoon and trade-wind coastal climate ② provide the *low-latitude rainforest environment,* which is one of tall, broadleaved evergreen trees and great species diversity. The rainforest provides many products of economic value, including hardwoods, pharmaceuticals, and rubber.

- The **wet-dry tropical climate** ③ has a very dry period at the time of low Sun and a wet season at the time of high Sun, when the ITCZ is near. Temperatures peak strongly just before the onset of the wet season. Associated with this climate type is the savanna environment, including *savanna woodland* and *thorntree-tall-grass savanna.*

- The **dry tropical climate** ④ is dominated by subtropical high pressure and is often nearly rainless. Temperatures are very high during the high-Sun season. While the semiarid subtype of this climate has a short wet season and supports **steppes** of sparse grasses, the arid subtype provides a desert environment.

- The **dry subtropical climate** ⑤ is dominated by subtropical high pressure and resembles the dry tropical climate ④, but it has a larger annual temperature range and a distinct cool season. The Mojave and Sonoran deserts, which occur in this climate type, are distinguished by a cover of specialized plants, such as the Joshua tree and saguaro cactus.

- Abundant rainfall with a summer maximum is a characteristic of the **moist subtropical climate** ⑥. The temperature cycle of this type includes cool winters with spells of subfreezing weather and warm, humid summers. The natural vegetation cover is forest. However, agriculture has largely replaced natural forests throughout this climate type.

- The **Mediterranean climate** ⑦ is unique because its annual precipitation cycle has a wet winter and a dry summer. The temperature range is moderate, with warm to hot summers and mild winters. The vegetation of the Mediterranean climate environment includes many evergreen sclerophylls, such as the cork oak, olive tree, and eucalyptus. Oak woodlands are also typical. Steep slopes are clothed with the drought-resistant shrubs of the chaparral.

- The **marine west-coast climate** ⑧, like the Mediterranean climate ⑦, has a winter precipitation maximum. The marine influence keeps temperature mild with a narrowed annual range. Forest is the native vegetation of the marine west-coast environment—dense needleleaf forest on the northern Pacific coast and broadleaf deciduous forest on the coast of western Europe. In Europe, lands of this climate have been under intensive cultivation for many centuries, and little forest remains.

- The **dry midlatitude climate** ⑨ has warm to hot summers and cold to very cold winters. The semiarid subtype typically has fertile soils with a sparse cover of grasses, which is termed *short-grass* prairie in North America and steppe in Eurasia. Wheat is the dominant crop. Because rainfall is highly variable, drought is a recurring event.

- The **moist continental climate** ⑩ has ample precipitation with a summer maximum. Summers are warm and winters cold. The moist continental environment typically sustains a cover of forest. In North America, a region of this climate in the Middle West once supported *tall-grass prairie.* This region is now the corn belt, an area of legendary crop productivity.

- The **boreal forest climate** ⑪ has long, bitterly cold winters. The boreal forest consists of needleleaf trees of pine, spruce, fir, and larch. Patches of deciduous aspen, poplar, willow, and birch also occur. Dairying and limited crop farming are economic activities of the boreal forest region.

- The **tundra climate** ⑫ occupies arctic coastal fringes. Because of the marine influence, winter temperatures are not as bitterly cold as those of the boreal forest climate ⑪. The tundra is a vegetation cover of scattered grasses, sedges, lichens, and dwarf shrubs, with patches of rock fragments devoid of plants. **Permafrost** underlies much of the tundra, which presents special problems for construction of buildings and roads.

- The **ice sheet climate** ⑬ is the coldest of all climates, with no monthly mean temperature above freezing. The severity of the climate prohibits nearly all human habitation.

Key Terms

climate, p. 214
isohyets, p. 215
climograph, p. 227
wet equatorial climate ①, p. 230
monsoon and trade-wind coastal climate ②, p. 231

wet-dry tropical climate ③, p. 235
dry tropical climate ④, p. 236
steppes, p. 240
dry subtropical climate ⑤, p. 245

moist subtropical climate ⑥, p. 246
Mediterranean climate ⑦, p. 249
marine west-coast climate ⑧, p. 251
dry midlatitude climate ⑨, p. 254

moist continental climate ⑩, p. 255
boreal forest climate ⑪, p. 258
tundra climate ⑫, p. 260
permafrost, p. 262
ice sheet climate ⑬, p. 262

Review Questions

1. Discuss the use of monthly records of average temperature and precipitation to characterize the climate of a region. Why are these measures useful?

2. Why are latitude and location (maritime or continental) important factors in determining the annual temperature cycle of a station?

3. How does air temperature, as a climatic variable, influence precipitation?

4. Describe three temperature regimes and explain how they are related to latitude and location.

5. Identify seven important features of the global map of precipitation and describe the factors that produce each.

6. The seasonality of precipitation at a station can generally be described as following one of three patterns. Identify them, explaining how each pattern can arise and providing an example.

7. What are the important global circulation patterns and air masses that influence low-latitude (Group I) climates? What are their effects?

8. What air masses and circulation patterns influence the midlatitude (Group II) climates and how?

9. Identify the air masses and frontal zones that are important in determining high-latitude climates (Group III) and explain their effects.

10. Why is the annual temperature cycle of the wet equatorial climate ① so uniform?

11. What are the main features of the low-latitude rain forest environment? Name some typical plant products of the rainforest.

12. The wet-dry tropical climate ③ has two distinct seasons. What factors produce the dry season? the wet season? Identify two main types of natural vegetation associated with the tropical wet-dry climate ③.

13. Why the dry tropical climate ④ dry? How do the arid (a) and semiarid (s) subtypes of this climate differ?

14. How does the dry subtropical climate ⑤ differ from the dry tropical climate ④? Describe some of the features of the Mojave Desert environment.

15. Both the moist subtropical ⑥ and moist continental ⑩ climates are found on eastern sides of continents in the midlatitudes. What are the major factors that determine their temperature and precipitation cycles? How do these two climates differ?

16. Both the Mediterranean ⑦ and marine west-coast ⑧ climates are found on the west coasts of continents. Why do they experience more precipitation in winter than in summer? How do the two climates differ?

17. Identify the characteristic features of the natural vegetation of the Mediterranean climate environment. Contrast these with the natural vegetation types and human use of the land in marine west-coast environments.

18. Describe the monthly temperature and precipitation cycles of the dry midlatitude climate ⑨. What is the natural vegetation of this climate type?

19. What vegetation covers characterize the moist continental climate ⑩?

20. Both the boreal forest ⑪ and tundra ⑫ climate are climates of the northern regions, but the tundra is found fringing the Arctic Ocean and the boreal forest is located further inland. Compare these two climates from the viewpoint of coastal-continental effects.

21. The boreal forest environment includes vast areas of land that are little influenced by human activity. Describe the natural vegetation of the boreal forest environment.

22. What is the coldest climate on Earth? How is the annual temperature cycle of this climate related to the cycle of insolation?

**Eye on Global Change 7.2 • Drought and Land Degrada-
tion in the African Sahel**

1. What is meant by the term *land degradation?* Provide an example of a region in which land degradation has occurred.

2. Examine the Sahel rainfall graph carefully. Compare the pattern of rainfall fluctuations for the periods 1901–1950 and 1951–present. How do they differ?

Visualizing Exercises

1. Sketch the temperature and rainfall cycles for a typical station in the monsoon and trade-wind coastal climate ②. What factors contribute to the seasonality of the two cycles?

2. Sketch climographs for the Mediterranean climate ⑦ and the dry midlatitude climate ⑨. What are the essential differences between them? Explain why they occur.

3. Suppose South America were turned over. That is, imagine that the continent was cut out and flipped over end-for-end so that the southern tip was at about 10° N latitude and the northern end (Venezuela) was positioned at about 55° S. The Andean chain would still be on the west side, but the shape of the land mass would now be quite different. Sketch this continent and draw possible climate boundaries, using your knowledge of global air circulation patterns, frontal zones, and air mass movements.

Essay Questions

1. The intertropical convergence zone (ITCZ) moves north and south with the seasons. Describe how this movement affects the four low-latitude climates.

2. Discuss the role of the polar front and the air masses that come in conflict in the polar-front zone in the temperature and precipitation cycles of the midlatitude and high-latitude climates.

3. What climate is your home in? Compare it to the climate of another location with which you are familiar. In your comparison, stress the factors that determine the thermal and precipitation regimes, including the global circulation patterns, air masses, and frontal zones that influence each climate. Construct a possible climograph for each.

Eye on the Landscape

7.12 Amazon rainforest The rainforest is a very diverse and complex environment **(A)**. In this aerial view, you can distinguish many different types of trees based on leaf color and texture, height, and branching pattern. Note that some very large tree crowns project above the others—these are *emergents,* which are characteristic of mature rainforest and play a special role in the ecology of the rainforest. See Chapter 9, Global Biogeography, Low-latitude Rainforest.

7.18a The baobab Note the rounded rocks around the base of the baobab **(A)**. In desert climates, chemical weathering breaks apart some types of crystalline rocks grain by grain, producing the rounded forms such as these. See Chapter 14, granular disintegration and chemical weathering.

7.18d Gobi Desert This scene gives the impression of ancient weathered rocks **(A)** emerging from a sea of rock fragments **(B)**. The gently sloping surfaces are fans built of weathered rock material that has been carried away from the mountains in runoff from infrequent, but severe, rainstorms (see Chapter 16, Alluvial Fans). In some cases, the sloping surface may be underlain by eroded rock with a thin veneer of gravel (Chapter 16, pediments). The surfaces are smoothed by wind action, which removes fine material and leaves coarse sand and gravel mantling the surface (see Chapter 18, desert pavement).

7.35 Moist continental landscape The Mosel, like many larger rivers, carves large loops in its path that are termed *meanders* **(A)**. As the loop expands outward, it cuts strongly into the adjacent terrain, creating a steep slope **(B)**. As it migrates outward, the loop leaves gentle slopes mantled by river sediments **(C)**. River meanders are discussed in Chapter 16.

8 | BIOGEOGRAPHIC PROCESSES

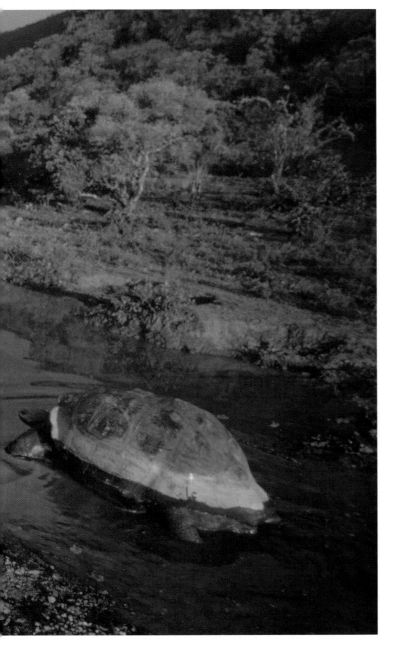

. . . .
PUTTING YOU IN THE PICTURE

The Earth's biological realm is large and diverse, including millions of species of organisms found from the ocean's abyssal depths to the land's highest peaks. This diversity has arisen through evolution, a biogeographic process in which natural selection acts on variation that occurs in reproducing populations of species to permute and transform strains of organisms, creating new species even as old strains die out.

The observation that natural selection might be responsible for evolution is credited to Charles Darwin, who published his ideas on *The Origin of Species by Means of Natural Selection* in 1859. One of Darwin's pivotal experiences in developing his theory of evolution was a visit to the Galápagos Islands, made in 1835 as the naturalist aboard the British survey vessel *HMS Beagle.* Impressed by "the amount of creative force displayed on these small, barren and rocky islands," he documented a variety of life-forms that he would later hypothesize to have evolved from common ancestors and become specialized and adjusted to the habitats of each island. Although human activity has influenced the islands in the many years since his visit, much of the striking character of these islands remains. Imagine yourself on a modern-day visit to the Galápagos…

The guide cuts the motor, and the rubber boat runs aground on the small crescent of beach in Shipton Cove, on the eastern shore of Isla Isabela, largest of the Galápagos Islands. Your small group steps neatly to the sand, having practiced this landing before on the many earlier stops of your two-week-long boat tour of the islands. This is the high point of your trip—a two-day round-trip trek to the summit of Alcedo Volcano, one of five volcanoes on this nearly uninhabited island. You'll climb from sea level to nearly 1100 m (3600 ft) in about 12 km (7.5 mi), reaching the summit camp by midafternoon. The views are supposed to be truly spectacular. And you are sure to see many of the famous Galápagos tortoises in their natural habitat.

"Look, a flightless cormorant!" The guide alerts your to one of the rarer and more unusual birds of the Galápagos. Swimming through the water with only its brown head and neck visible, the bird flashes its turquoise eyes in your direction. Leaping upward and then diving downward in a single motion, it heads

off in search of its favorite bottom-living prey, the octopus.

You clamber up the steep path around the ocean cliff, barely noticing the basking marine iquanas that at first seemed so strange and by now are routine. These land reptiles have acquired the habit of grazing on underwater algae and seaweed. Like nearly all the other native animals of the Galápagos, they have no fear of humans and regard you with their typical disdain.

At first, the trail winds through an arid zone of prickly-pear and candelabra cacti alternating with stands of spindly palo santo trees and acacias. Suddenly you come upon several land iguanas chewing away on the pads of a prickly pear. Although resembling their marine cousins, they are tan to brownish in color, with shorter spines along their backbones, and are considerably rarer. Both species are descended from a common ancestor, probably the green iquana now common in mainland Ecuador.

As you climb, the vegetation becomes more and more lush, and you enter a cloud forest of odd-looking trees festooned with mosses, ferns, and even orchids. Your guide points out a woodpecker finch, a small gray-green bird perched on a dead shrub near the trail. Holding a cactus spine in its stout beak, it probes holes in the dead wood for insects. According to your guide, it is

one of the 13 finch species unique to the Galápagos that were studied by Darwin and used as evidence for his theory of evolution by means of natural selection.

Some hours later, you reach the rim of Alcedo Volcano's crater. The view is indeed spectacular. A vast expanse of green shrub forest sprawls below you, with a brown network of seasonal pools on the crater floor. In the far distance, the peaks of the other four volcanoes on Isabela are visible, two to the left and two to the right.

The next morning, you arise early in camp and set off for the crater, anxious to see the famous giant tortoises in their natural habitat. By ones and twos, the giant tortoises appear near the trail, extending their long necks to graze on the lower branches of the shrubs. As your guide reports, these are members of the *vandenberghi* subspecies, which is restricted to this volcano. In fact, each of the five volcanoes on Isabela has its own unique subspecies of giant tortoise. Finally, you reach a large group of tortoises wallowing in a seasonal pool. To the clicking of cameras, you watch as they frolic in their communal mud bath. What an odd and curious place this is, for reptiles such as tortoises and iguanas to be so common! No wonder the Galápagos Islands have captured the imagination of naturalists like Charles Darwin for almost two centuries. ■

BIOGEOGRAPHIC PROCESSES

This chapter is the first of two chapters that are concerned with biogeography. **Biogeography** is a branch of geography that focuses on the distribution of plants and animals—the *biota*—over the Earth. It attempts to identify and describe the processes that influence plant and animal distribution patterns at varying scales of space and time. We can think of biogeography as encompassing two major views or themes. *Ecological biogeography* is concerned with how the distribution patterns of organisms are affected by the environment, including both the physical environment experienced by organisms and the biological environment created by interaction with other organisms. *Historical biogeography* focuses on how spatial distribution patterns of organisms arise over time and space. Evolution of species, migration of organisms and their

methods of dispersal, and extinction of species are processes of interest in historical biogeography.

In this chapter, we identify the key processes of ecological and historical biogeography that determine how the distribution patterns of plants and animals have arisen and how they change with time. Our next chapter inventories the global pattern of vegetation and animal life, stressing how climate influences that pattern.

ENERGY AND MATTER FLOW IN ECOSYSTEMS

We begin our first chapter by examining some ideas from the domain of **ecology,** which is the study of the interactions between life-forms and their environment. These ideas focus on how organisms live and interact as ecosystems and how energy and matter are cycled by ecosystems.

We can define an **ecosystem** as a group of organisms and the environment with which they interact. Ecosystems have inputs of matter and energy that plants and animals use to grow, reproduce, and maintain life. Matter and energy are also exported from ecosystems. An ecosystem tends to achieve a set of balances of the various processes and activities that occur within it. Although many of these balances are robust and self-regulating, some are quite sensitive and can be easily upset or destroyed.

The Food Web

A salt marsh provides a good example of an ecosystem (Figure 8.1). A variety of organisms are present—algae and aquatic plants, microorganisms, insects, snails, and crayfish, as well as such larger organisms as fishes, birds, shrews, mice, and rats. Inorganic components will be found as well—water, air, clay particles and organic sediment, inorganic nutrients, trace elements, and light energy. Energy transformations in

> The food web describes how food energy flows from organism to organism within an ecosystem. Primary producers support primary, secondary, and higher-level consumers. Decomposers feed on dead plant and animal matter from all levels.

the ecosystem occur by means of a series of steps or levels, referred to a **food chain** or **food web.**

The plants and algae in the food web are the **primary producers.** They use light energy to convert carbon dioxide and water into carbohydrates (long chains of sugar molecules) and eventually into other biochemical molecules needed for the support of life. This process of energy conversion is called *photosynthesis,* and we will return to it in more detail shortly. Organisms engaged in photosynthesis form the base of the food web.

The primary producers support the **consumers**—organisms that ingest other organisms as their food source. At the lowest level of consumers are the *primary consumers* (the snails, insects, and fishes). At the next level are the *secondary consumers* (the mammals, birds, and larger fishes), which feed on the primary consumers. Still higher levels of feeding occur in the salt-marsh ecosystem as marsh hawks and owls consume the smaller animals below them in the food web. The **decomposers** feed on *detritus,* or decaying organic matter, derived from all levels. They are largely microscopic organisms (microorganisms) and bacteria.

The food web is really an energy flow system, tracing the path of solar energy through the ecosystem. Solar

8.1 Energy flow diagram of a salt-marsh ecosystem in winter *The arrows show how energy flows from the Sun to producers, consumers, and decomposers. (Food chain after R. L. Smith, Ecology and Field Biology, Harper and Row, New York.)*

8.2 Energy loss *Percentage of energy passed up the steps of the food chain, assuming 90 percent is lost energy at each step.*

energy is absorbed by the primary producers and stored in the chemical products of photosynthesis. As these organisms are eaten and digested by consumers, chemical energy is released. This chemical energy is used to power new biochemical reactions, which again produce stored chemical energy in the bodies of the consumers.

At each level of energy flow in the food web, energy is lost to *respiration*. Respiration can be thought of as the burning of fuel to keep the organism operating. It will be discussed in more detail in the next section. Energy expended in respiration is ultimately lost as waste heat and cannot be stored for use by other organisms higher up in the food chain. This means that, generally, both the numbers of organisms and their total amount of living tissue must decrease greatly up the food chain. In general, only 10 to 50 percent of the energy stored in organic matter at one level can be passed up the chain to the next level. Normally, there are about four levels of consumers.

Figure 8.2 is a bar graph showing the percentage of energy passed up the chain when only 10 percent moves from one level to the next. The horizontal scale is in powers of 10. In ecosystems of the lands, the mass of organic matter and the number of individuals of the consuming animals decrease with each upward step. In the food chain shown in Figure 8.1, there are only a few marsh hawks and owls in the third level of consumers, while countless individuals are found in the primary level.

For individual species, the number of individuals of a species present in an ecosystem ultimately depends on the level of resources available to support the species population. If these resources provide a steady supply of energy, the population size will normally attain a steady level. In some cases, however, resources vary with time, for example, in an annual cycle. The population size of a species depending on these resources may then also fluctuate in a corresponding cycle.

> Photosynthesis is the process in which plants combine water, carbon dioxide, and solar energy to form carbohydrate. Respiration is the reverse process, in which carbohydrate is oxidized in living tissues to yield the energy that sustains life.

Photosynthesis and Respiration

Stated in the simplest possible terms, **photosynthesis** is the production of carbohydrate. *Carbohydrate* is a general term for a class of organic compounds consisting of the elements carbon, hydrogen, and oxygen. Carbohydrate molecules are composed of short chains of carbon bonded to one another. Also bonded to each carbon are hydrogen (H) atoms and hydroxyl (OH) molecules. We can symbolize a single carbon atom with its attached hydrogen atom and hydroxyl molecule as –CHOH–. The leading and trailing dashes indicate that the unit is just one portion of a longer chain of connected carbon atoms.

Photosynthesis of carbohydrate requires a series of complex biochemical reactions using water (H_2O) and carbon dioxide (CO_2) as well as light energy. A simplified chemical reaction for photosynthesis can be written as follows:

$$H_2O + CO_2 + \text{light energy} \rightarrow \text{–CHOH–} + O_2$$

Oxygen in the form of gas molecules (O_2) is a byproduct of photosynthesis. Photosynthesis is also referred to as *carbon fixation*, since in the process gaseous carbon as CO_2 is "fixed" to a solid form in carbohydrate.

Respiration is the process opposite to photosynthesis in which carbohydrate is broken down and combined with oxygen to yield carbon dioxide and water. The overall reaction is as follows:

$$\text{–CHOH–} + O_2 \rightarrow CO_2 + H_2O + \text{chemical energy}$$

As in the case of photosynthesis, the actual reactions are far from simple. The chemical energy released is stored in several types of energy-carrying molecules in living cells and used later to synthesize all the biological molecules necessary to sustain life.

Because both photosynthesis and respiration occur simultaneously in a plant, the amount of new carbohydrate placed in storage is less than the total carbohydrate being synthesized. We must thus distinguish between gross photosynthesis and net photosynthesis. **Gross photosynthesis** is the total amount of carbohydrate produced by photosynthesis. **Net photosynthesis** is the amount of carbohydrate remaining after respiration has broken down sufficient carbohydrate to power the plant. Stated as an equation,

Net photosynthesis = Gross photosynthesis – Respiration

Because both photosynthesis and respiration occur in the same cell, gross photosynthesis cannot be measured readily. Instead, we will deal with net photosynthesis. In most cases, respiration will be held constant, so use of the net instead of the gross will show the same trends.

The rate of net photosynthesis is strongly dependent on the intensity of light energy available, up to a limit. Figure

8.3 shows this principle. The rate of net photosynthesis is indicated on the vertical axis by the rate at which a plant takes up carbon dioxide. On the horizontal axis, light intensity increases from left to right. At first, net photosynthesis rises rapidly as light intensity increases. The rate then slows and reaches a maximum value, shown by the plateau in the curve. Above this maximum, the rate falls off because the incoming light is also causing heating. This heating increases the rate of respiration, which offsets gross production by photosynthesis and decreases the net.

Light intensity sufficient to allow maximum net photosynthesis is only 10 to 30 percent of full summer sunlight for most green plants. Additional light energy is simply ineffective. Duration of daylight then becomes the important factor in the rate at which products of photosynthesis accumulate as plant tissues. On this subject, you can draw on your knowledge of the seasons and the changing angle of the Sun's rays with latitude. Figure 8.4 shows the duration of the daylight period with changing seasons for a wide range of latitudes in the northern hemisphere. At low latitudes, days are not far from the average 12-hour length throughout the year. At high latitudes, days are short in winter but long in summer. The seasonal contrast in day length increases with latitude. In subarctic latitudes, photosynthesis can go on in summer during most of the 24-hour day, a factor that can compensate significantly for the shortness of the growing season.

The rate of photosynthesis also increases as air temperature increases, up to a limit. Figure 8.5 shows the results of a laboratory experiment in which sphagnum moss was grown under constant illumination. Gross photosynthesis increased rapidly to a maximum at about 20°C (68°F), then leveled off. Respiration increased quite steadily to the limit of the experiment. Net photosynthesis, which is the difference between the values in the two curves, peaked at about 18°C (64°F), then fell off rapidly.

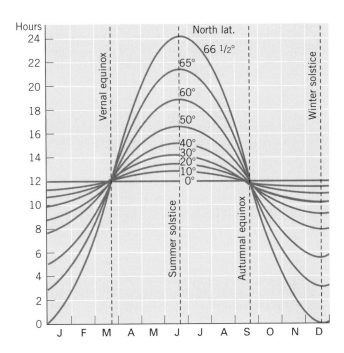

8.4 Day length variation *Duration of the daylight period at various latitudes throughout the year. The vertical scale gives the number of hours the Sun is above the horizon.*

Net Primary Production

Plant ecologists measure the accumulated net production by photosynthesis in terms of the **biomass,** which is the dry weight of organic matter. This quantity could, of course, be stated for a single plant or animal, but a more useful measurement is the biomass per unit of surface area within the ecosystem—that is, grams of biomass per square meter or

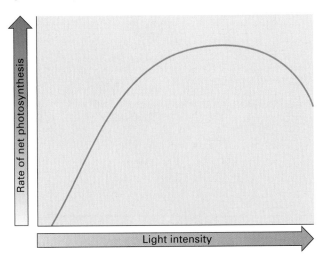

8.3 Net photosynthesis *The curve of net photosynthesis shows a steep initial rise, then levels off as light intensity rises.*

8.5 Temperature and energy flow *Respiration and gross and net photosynthesis vary with temperature. (Data of Stofelt, in A. C. Leopold, Plant Growth and Development, McGraw-Hill, New York.)*

Table 8.1 | **Net Primary Production for Various Ecosystems**

	Grams per Square Meter per Year	
	Average	**Typical Range**
Lands		
Rainforest of the equatorial zone	2000	1000–5000
Freshwater swamps and marshes	2500	800–4000
Midlatitude forest	1300	600–2500
Midlatitude grassland	500	150–1500
Agricultural land	650	100–4000
Lakes and streams	500	100–1500
Extreme desert	3	0–10
Oceans		
Algal beds and reefs	2000	1000–3000
Estuaries (tidal)	1800	500–4000
Continental shelf	360	300–600
Open ocean	125	1–400

(metric) tons of biomass per hectare (1 hectare = 10^4 m²). Of all ecosystems, forests have the greatest biomass because of the large amount of wood that the trees accumulate through time. The biomass of grasslands and croplands is much smaller in comparison. For freshwater bodies and the oceans,

---

Net primary production measures the rate of accumulation of carbohydrate by primary producers. Equatorial rainforests and freshwater swamps and marshes are among the most productive ecosystems, while deserts are least productive.

---

the biomass is even smaller—on the order of one-hundredth that of the grasslands and croplands.

Although the amount of biomass present per unit area is an important indicator of the amount of photosynthetic activity, it can be misleading. In some ecosystems, biomass is broken down very quickly by consumers and decomposers, so the amount maintained is less. From the viewpoint of ecosystem productivity, what is important is the annual yield of useful energy produced by the ecosystem, or the *net primary production.*

Table 8.1 provides the net primary production of various ecosystems in units of grams of dry organic matter produced annually from one square meter of surface. The figures are rough estimates, but they are nevertheless highly meaningful. Note that the highest values are in two quite unlike environments: forests and wetlands (estuaries). Agricultural land compares favorably with grassland, but the range is very large in agricultural land, reflecting many factors such as availability of soil water, soil fertility, and use of fertilizers and machinery.

Productivity of the oceans is generally low. The deep water oceanic zone, which comprises about 90 percent of the world ocean area, is the least productive of the marine ecosystems. Continental shelf areas are a good deal more productive and support much of the world's fishing industry (Figure 8.6).

Upwelling zones are also highly productive. Upwelling of cold water from ocean depths brings nutrients to the surface and greatly increases the growth of microscopic floating plants known as *phytoplankton.* These, in turn, serve as food sources for marine animals in the food chain. Consequently, zones of

---

8.6 Distribution of World fisheries *Coastal areas and upwelling areas together supply over 99 percent of world production. (Compiled by the National Science Board, National Science Foundation.)*

upwelling near habitable coastlines are highly productive fisheries. An example is the Peru Current off the west coast of South America. Here, countless individuals of a single species of small fish, the anchoveta, provide food for larger fish and for birds. The birds, in turn, excrete their wastes on the mainland coast of Peru. The accumulated deposit, called guano, is a rich source of nitrate fertilizer that is now severely depleted.

Within the past decade or so, remote sensing has come into use as a tool for measuring and mapping primary productivity on a global scale. Figure 8.7 shows a polar view of the northern hemisphere, displaying in image format satellite measurements of phytoplankton pigment concentration, which is an index of photosynthetic activity near the ocean surface. The red and yellow colors, marking high pigment concentrations, emphasize the importance of the polar seas in ocean productivity. The data were collected over a period of years and show averages. It is important to realize that the productivity is highly seasonal because it is limited by light. Thus, polar productivity falls to near zero during the polar winter.

8.7 *Ocean productivity as inferred using remote sensing.* *This polar image shows northern hemisphere phytoplankton pigment concentrations averaged over the period 1978 to 1986. They were derived by processing data obtained from the Coastal Zone Color Scanner instrument on the Nimbus-7 satellite. Pigment concentrations are indicated by colors: red (highest) through yellow, green, and blue, to purple (lowest).*

Net Production and Climate

What climatic factors control net primary productivity? We have already identified light intensity and duration, as well as temperature, as influencing net photosynthesis. Another important factor is the availability of water. A shortage or surplus of soil water might be the best climatic factor to examine, but data are not available. Ecologists have related net annual primary production to mean annual precipitation, as shown in Figure 8.8. The production values are for plant structures above the ground surface. Although the productivity increases rapidly with precipitation in the lower range from desert through semiarid to subhumid climates, it levels off in the humid range.

> **Day length, air and soil temperature, and water availability are the most important climatic factors that control net primary productivity.**

Combining the effects of light intensity, temperature, and precipitation, we can assign rough values of productivity to each of the climates as follows (units are grams of carbon per square meter per year):

Highest (over 800)	Wet equatorial ①
Very high (600–800)	Monsoon and trade-wind coastal ②
	Wet-dry tropical ③
High (400–600)	Wet-dry tropical ③ (Southeast Asia)
	Moist subtropical ⑥
	Marine west coast ⑧
Moderate (200–400)	Mediterranean ⑦
	Moist continental ⑩
Low (100–200)	Dry tropical, semiarid ④s
	Dry midlatitude, semiarid ⑨s
	Boreal forest ⑩

Very low (0–100)	Dry tropical, desert ④d
	Dry midlatitude, desert ⑨d
	Boreal forest ⑩
	Tundra ⑫

Remote sensing can also be used to monitor net primary productivity of the Earth's land surface. Figure 8.9 shows oceanic phytoplankton pigment concentration (as in Figure 8.7), in addition to which a land vegetation productivity index has been overlain on the continents. The dark-green colors are assigned to the most productive areas, while the yellow-brown tones indicate the least productive. The world's deserts show clearly and present an obvious contrast to equatorial rainforest regions. Tropical savanna and midlatitude grassland are intermediate in productivity, whereas boreal forest and tundra are low.

Biomass as an Energy Source

Net primary production represents a source of renewable energy derived from the Sun that can be exploited to fill human energy needs. The use of biomass as an energy source involves releasing solar energy that has been fixed in plant tissues through photosynthesis. This process can take place in a number of ways—the simplest is direct burning of plant matter as fuel, as in a campfire or a wood-burning stove. Other approaches involve the generation of intermediate fuels from plant matter—methane gas, charcoal, and alcohol, for example. Biomass energy conversion is not highly energy efficient. Typical values of net annual primary production of plant communities range from 1 to 3 percent of available solar energy. However, the abundance of terrestrial biomass is so great that biomass utilization could provide the energy equivalent to 3 million barrels of oil per day for the United States with proper development.

One important use of biomass energy is the burning of firewood for cooking (and some space heating) in developing nations. The annual growth of wood in the forest of developing countries totals about half the world's energy production—plenty of firewood is thus available. However, fuelwood use exceeds production in many areas, creating local shortages and severe strains on some forest ecosystems. The forest–desert transition areas of thorntree, savanna, and desert scrub in central Africa south of the Sahara Desert are examples.

Even in closed stoves, wood burning is not very efficient, ranging from 10 to 15 percent for cooking. However, the conversion of wood to charcoal or gas can boost efficiencies to values as high as 70 to 80 percent with appropriate technology. In this process, termed *pyrolysis,* controlled partial burning in an oxygen-deficient environment reduces carbohydrate to free carbon (charcoal) and yields flammable gases such as carbon monoxide and hydrogen. Charcoal is more energy efficient than wood, burns more cleanly, and is easier to transport. As an added advantage, charcoal can be made

8.8 Precipitation and net production *Net primary production increases rapidly with increasing precipitation but levels off in the higher values. Observed values fall mostly within the shaded zone. (Data of Whittaker, 1970.)*

8.9 Global productivity as inferred using remote sensing *Oceans: From NOAA's Coastal Zone Color Scanner instrument as described for Figure 8.7. Lands: Displayed is an annual vegetation greenness index obtained from the Advanced Very High Resolution Radiometer (AVHRR) instrument aboard the NOAA-7 spacecraft. The most productive regions are shown in dark green tones, ranging through medium and light green tones to yellow-brown tones (least productive).*

from waste fibers and agricultural residues that would normally be discarded. Thus, charcoal is an efficient fuel that can help extend the firewood supply in areas where wood is in high demand.

A second method of extracting energy from biomass uses anaerobic digestion to produce *biogas.* In this process, animal and human wastes are fed into a closed digesting chamber, where anaerobic bacteria break down the waste to produce a gas that is a mixture of methane and carbon dioxide. The biogas can be easily burned for cooking or heating, or it may be used to generate electric power. The digested residue is a sweet-smelling fertilizer. China now maintains a vigorous program of construction of biogas digesters for the use of small family units. The benefits include better sanitation and reduced air pollution, as well as more efficient fuel usage.

Another use of biomass that is increasing in importance is the conversion of agricultural wastes to alcohol. In this process, yeast microorganisms are used to convert the carbohydrate to alcohol through fermentation. An advantage of alcohol is that it can serve as a substitute and extender for gasoline. Gasohol, a mixture of up to 10 percent alcohol in gasoline, can be burned in conventional engines without adjustment.

Brazil, a country without adequate petroleum production, has relied heavily on alcohol fuel derived from sugarcane. In a recent year, for example, alcohol provided 63 percent of Brazil's automotive fuel needs. Distillation of alcohol, how-

ever, requires heating, thus greatly reducing the net energy yield. Alcohol, charcoal, and firewood are all alternatives to fossil fuels that will become increasingly important as petroleum becomes scarcer and more costly in the coming decades.

Relying on biomass energy can also yield important benefits in reducing carbon dioxide emissions. However, burning biomass does not reduce the CO_2 flow to the atmosphere directly. The burning of biomass quickly releases CO_2 that would normally be released more slowly, as the biomass decays. But the energy obtained in the biomass burning will in all likelihood substitute for some fossil fuel burning. Because this fossil fuel is not burned, its CO_2 is not released to the atmosphere, thus reducing overall carbon dioxide emissions.

Biomass burning is not always controlled. In some ecosystems, such as grasslands, savannas, and woodlands, wildfires are frequent, and ecosystems are well-adjusted to burning. In others, such as forest, wildfires may be infrequent but severe and damaging. Global monitoring of wildfires is now possible using remote sensing. *Focus on Remote Sensing 8.1 • Remote Sensing of Fires* shows how several satellite sensors can be used to track wildfires.

The Carbon Cycle

We have seen how energy of solar origin flows through ecosystems, passing from one part of the food chain to the

Focus on Remote Sensing | **8.1**

Remote Sensing of Fires

Wildfires occur frequently on the Earth's land surface, and biomass burning has important effects on both local and global ecosystems. Biomass burns inefficiently, releasing not only carbon dioxide and water, but also a number of other greenhouse gases that absorb outgoing longwave radiation and enhance the greenhouse effect. Aerosols are another byproduct of inefficient combustion that can effect atmospheric processes. Burning mobilizes such nutrients as nitrogen, phosphorus, and sulfur in ash so that they become available for a new generation of plants, but it also carries them upward in smoke and as gases. Fire affects ecosystems by changing species composition and creating a patchy structure of diversity on the landscape. It can also stimulate runoff and soil erosion where a significant layer of vegetation is lost to fire. For these reasons and more, monitoring of fires by remote sensing is a topic of intense interest to global change researchers.

Fires can be remotely sensed in several ways. Thermal imagers detect active fires as bright spots because they emit more heat energy than normal surfaces. However, fires may be obscured under clouds. Smoke plumes can also identify the location of fires but are hard to distinguish from clouds in some images. Fire scars can also be detected, especially using mid-infrared bands.

Image (*a*) shows the Earth as imaged by NASA's Moderate Resolution Imaging Spectroradiometer (MODIS) in true color on August 23, 2000. Superimposed on the true color browse image are red squares indicating high temperatures as sensed by MODIS thermal bands. The inset images highlight fires in the Amazon Basin, southern Africa, and Australia. Image (*b*) is a MODIS image of bush fires in northern Australia on October 2, 2000. Inset into the Gulf of Carpentaria is a closeup of one of the

(a)

MODIS fire browse image *Global MODIS coverage for August 23, 2000, displayed in true color. (MODIS Land Team/C. Justice/L. Giglio/J. Descloitres/NASA.)*

larger ones. The red outlines identify pixels where fire has been detected. The dark burn scar and smoke plumes are readily visible.

Images (*c*) and (*d*) show Los Alamos, New Mexico, from Landsat-7 before and after the Cerro Grande Fire of May, 2000. Over

Australian bush fires from MODIS *Fires in north-central Australia, shown in true color. Red outlines identify pixels flagged as burning by thermal bands. (MODIS Land Team/C. Justice/L. Giglio/J.Descloitres/NASA)*

(b)

Los Alamos on April 14, 2000 (Robert Simmon, NASA GSFC.)

Los Alamos on June 17, 2000, after the fire (Robert Simmon, NASA GSFC.)

20,000 people were forced to flee their homes as this disastrous fire progressed, and more than 200 homes were consumed by the flames. The images are composites of mid-infrared, near-infrared, and green Landsat bands assigned colors of red, green, and blue, respectively. Green colors show vegetation, with the brightest green identifying the dense grass of golf courses. Blue and purplish tones are developed areas. The strong red color in the June image identifies the burned area, which is very dark in green and near-infrared bands, but bright in the mid-infrared. The blue stripes on the left edge of the April image (*c*) are ski runs covered with snow. They appear bright green due to their grass cover in June.

Like MODIS, the SeaWiFS instrument produces a wide view of the Earth's surface that can detect the smoke from fires. Image (*e*) is a true color image of the southeastern United States acquired by the instrument from a track over west Texas on February 5, 2001. The view is to the east, along the Gulf coast from Texas to Florida. Dozens of smoke plumes are clearly visible. The Mississippi River Valley crosses the center of the image from left to right as an area of light-colored fallow soils. Louisiana's Mississippi Delta is an obvious coastal feature, with a plume of fresh sediment extending out into the Gulf of Mexico.

Southeastern U.S. from SeaWiFS
This oblique view of the Gulf Coast from the SeaWiFS imager aboard the OrbView-2 satellite shows numerous smoke plumes. (Image courtesy SeaWiFS Project, NASA/Goddard Space Flight Center, and ORBIMAGE.)

next, until it is ultimately lost from the biosphere as energy radiated to space. Matter also moves through ecosystems, but because gravity keeps surface material Earthbound, matter cannot be lost in the global ecosystem. As molecules are formed and reformed by chemical and biochemical reactions within an ecosystem, the atoms that compose them are not changed or lost. Thus, matter is conserved within an ecosystem, and atoms and molecules can be used and reused, or cycled, within ecosystems.

Atoms and molecules move through ecosystems under the influence of both physical and biological processes. The pathways of a particular type of matter through the Earth's ecosystem comprise a **biogeochemical cycle** (sometimes referred to as a *material cycle* or *nutrient cycle*).

The major features of a biogeochemical cycle are diagrammed in Figure 8.10. Any area or location of concentration of a material is a **pool**. There are two types of pools: *active pools,* where materials are in forms and places easily accessible to life processes, and *storage pools,* where materials are more or less inaccessible to life. A system of pathways of material flows connects the various active and storage pools within the cycle. Pathways can involve the movement of material in all three states of matter—gas, liquid, and solid. For example, carbon moves freely in the atmosphere as carbon dioxide gas and freely in water as dissolved CO_2 and as carbonate ion (CO_3^-). It also takes the form of a solid in deposits of limestone and dolomite (calcium and magnesium carbonate).

Ecologists have studied and documented biogeochemical cycles for many elements, including carbon, oxygen, nitrogen, sulfur, and phosphorus. Of these, the **carbon cycle** is probably the most important for two reasons. First, all life is composed of carbon compounds of one form or another. Second, human activities are modifying the carbon cycle in important ways.

Some details of the *carbon cycle* are shown in a schematic diagram, Figure 8.11. In the gaseous portion of the cycle, carbon moves largely as carbon dioxide (CO_2), which is a free gas in the atmosphere and a dissolved gas in fresh and saltwater. In the sedimentary portion of its cycle, carbon resides in carbohydrate molecules in organic matter, as hydrocarbon compounds in rock (petroleum, coal), and as mineral carbonate compounds such as calcium carbonate ($CaCO_3$). The world supply of atmospheric carbon dioxide is represented in Figure 8.11 by a box. It is a small portion of the carbon in active pools, constituting less than 2 percent. This atmospheric pool is supplied by plant and animal respiration in the oceans and on the lands. Under natural conditions, some new carbon enters the atmosphere each year from volcanoes by outgasing in the form of CO_2 and carbon monoxide (CO). Industry injects substantial amounts of carbon into the atmosphere through combustion of fossil fuels. This increment from fuel combustion and its probable effects on global air temperatures were discussed in Chapter 3.

Carbon dioxide leaves the atmospheric pool to enter the oceans, where it is used in photosynthesis by phytoplankton. These organisms are primary producers in the ocean ecosystem and are consumed by marine animals in the food chain. Phytoplankton also build skeletal structures of calcium carbonate. This mineral matter settles to the ocean floor to accumulate as sedimentary strata, an enormous storage pool not available to organisms until released later by rock weathering. Organic compounds synthesized by phytoplankton also settle to the ocean floor and eventually are transformed into the hydrocarbon compounds making up petroleum and natural gas. On the lands, plant matter accumulating over geologic time forms layers of peat that are ultimately transformed into coal. Petroleum, natural gas, and coal comprise the fossil fuels, and these represent huge storage pools of carbon. GEODISCOVERIES

Human activity is presently affecting the carbon cycle very significantly. Through the burning of fossil fuels, CO_2 is being released to the atmosphere at a rate far beyond that of any natural process. *Eye on Global Change 8.2 • Human Impact on the Carbon Cycle* documents how human activity has influenced the major flows within the carbon cycle. GEODISCOVERIES

> The carbon cycle is a biogeochemical cycle in which carbon flows among storage pools in the atmosphere, ocean, and on the land. Human activity has affected the carbon cycle, causing carbon dioxide concentrations in the atmospheric storage pool to increase.

8.10 *General features of a biogeochemical cycle*

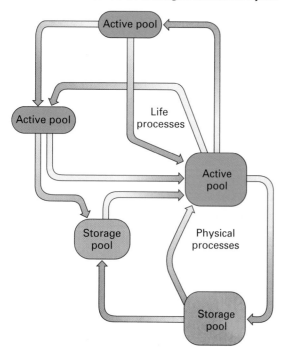

ECOLOGICAL BIOGEOGRAPHY

In the preceding section, we viewed ecosystems from the perspective of energy and showed how the biosphere fixes

8.11   The carbon cycle *(Copyright © A. N. Strahler.)*

sunlight to power global cycles of elements, such as the carbon cycle. But ecosystems are composed of individual organisms that utilize and interact with their environment in different ways. From fungi digesting organic matter on a forest floor to ospreys fishing in a coastal estuary, each organism has a range of environmental conditions that limits its survival as well as a set of characteristic adaptations that it exploits to obtain the energy it needs to live. **Ecological biogeography** examines the distribution patterns of plants and animals from the viewpoint of ecological factors and interrelationships between species.

We begin our discussion of ecological biogeography with the relationship between organisms and their physical environment. As we travel through a hilly, wooded area, it is easy to see that the ecosystems we encounter are strongly influenced by landform and soil. For example, the upland forest of oak and hickory on the Blue Ridge Mountains of Virginia gives way to hemlock, birch, and maple in small valleys and low places. Upland soils are thick, stony, and well-drained, while the soils of the valleys and swales are finer, richer in organic matter, and wetter more of the time. Forest communities are also strikingly different in form on rocky ridges and on steep cliffs, where pines and scrubby oaks abound. Here water drains away rapidly, and soil is thin or largely absent. These variations illustrate the concept of the **habitat**—a subdivision of the environment according to the needs and preferences of organisms or groups of organisms. Figure 8.12 presents an example taken from a Canadian boreal forest. Here, there are six distinctive habitats: upland, bog, bottomland, ridge, cliff, and active sand dune. Each habitat supports a different type of ecosystem.

A concept related to the habitat, the *ecological niche,* includes the functional role played by an organism as well as the physical space it inhabits. If the habitat is the individual's "address," then the niche is its "profession," including how and where it obtains its energy and how it influences other species and the environment around it. Included in the ecological niche are the organism's tolerances and responses to changes in moisture, temperature, soil chemistry, illumination, and other factors. Although many different species may occupy the same habitat, only a few of these species will ever share the same ecological niche, for, as we'll see shortly, evolution will tend to separate them.

As we move from habitat to habitat, we find that each is the home of a group of organisms, each occupying different but interrelated ecological niches. We can define a *community* as an assemblage of organisms that live in a particular habitat and interact with one another. Although every organ-

> The *habitat* of a species describes the physical environment that harbors its activities. The *ecological niche* of a species describes how it obtains its energy and how it influences other species and its own environment. The *community* is a group of interacting organisms that occupy a particular habitat.

Eye on Global Change | 8.2

Human Impact on the Carbon Cycle

Carbon is an element that is abundant at the Earth's surface and is also essential for life. As noted earlier in this chapter, carbon cycles continuously among the land surface, atmosphere, and ocean in many complex pathways. However, these flows are now strongly influenced by human activity. The most important human impact on the carbon cycle is the burning of fossil fuels. Another important human impact lies in changing the Earth's land covers— for example, in clearing forests or abandoning agricultural areas. Let's look at these impacts in more detail.

The figure below shows a simple diagram of the major flows within the carbon cycle for the period 1989–1998. The magnitudes of the annual flows are shown in gigatons (Gt) of carbon per year (1

gigaton = 10^9 metric tons = 10^{12} kg = 1.1×10^9 English tons = 1.1 English gigatons). These flows are estimates, and a second value after each value indicates its uncertainty. For example, fossil fuel burning liberates 6.3 ± 0.6 Gt/yr, which we can interpret as a flow that is most likely to be in the range $6.3 - 0.6 = 5.7$ to $6.3 + 0.6 = 6.9$ Gt/yr.

By comparing the flows, we see that about half of the output of carbon by fossil fuel burning is taken up by the atmosphere (3.3 ± 0.2 Gt/yr). Of the remaining amount, about 2/3 (2.3 ± 0.8 Gt/yr) is absorbed by the oceans. This leaves unaccounted an amount of about $6.3 - 3.3 - 2.3 = 0.7$ Gt/yr. Since there are no other significant pathways, this carbon must be flowing into the biosphere. In other words, ecosystems are a *sink* for

CO_2, accepting about 0.7 Gt/yr of carbon.

Ecosystems cycle carbon in photosynthesis, respiration, decomposition, and combustion. Photosynthesis and respiration are basic physiological processes that fix and release CO_2. Decomposition is the process in which bacteria and fungi digest dead organic matter, and is actually a form of respiration. Combustion refers to uncontrolled combustion, as when an ecosystem burns. It is very hard to account for these processes globally in such a way as to know their net effect, but by applying the logic of budgeting, we know that they must sum to the value of 0.7 Gt/yr of carbon buildup in land ecosystems mentioned above.

If the value of 0.7 Gt/yr in carbon uptake is correct, the amount

The global carbon cycle Values are in gigatons of carbon per year. (Data from UN/IPCC.)

of terrestrial biomass must be increasing at that rate. However, forests are presently diminishing in area as they are logged or converted to farmland or grazing land. This conversion is primarily occurring in tropical and equatorial regions, and it is estimated to release about 1.6 ± 0.8 Gt/yr of carbon to the atmosphere. Since this release is included in the net land ecosystem uptake of 0.7 Gt/yr, the remainder of the world's forests must be taking up the 1.6 Gt/yr loss from deforestation as well as an additional 0.7 Gt/yr. So we can estimate that mid- and high-latitude forests are increasing in area or biomass to fix carbon at a rate of 1.6 + 0.7 = 2.3 Gt/yr.

Independent evidence seems to confirm this conclusion. In Europe, for example, forest statistics show an increase of growing stock—the volume of living trees—of about 25 percent from 1970 to 1990. This increase has been sustained in spite of damage to forests by air pollution, especially in eastern Europe. In North America, forest areas are increasing in many regions as agricultural production has abandoned marginal areas to natural forest growth. New England is a good example of this trend. A century ago, only a small portion of New England was forested. Now only a small portion is cleared.

Some of the increase in global biomass may also be the effect of enhancement of photosynthesis by warmer temperatures and increased CO_2 concentrations (discussed in more detail shortly). Another factor proposed to account for increased ecosystem productivity is nitrogen fertilization of soils by washout of nitrogen pollutant gases in the atmosphere.

Some foresters have observed that harvesting mature forests and replacing them with young, fast-growing timber should increase the rate of withdrawal of CO_2 from the atmosphere. Since the lumber of the mature forests goes into semipermanent storage in dwellings and structures where it is protected from decay and oxidation to CO_2, it represents a withdrawal of CO_2 from the atmosphere. The young forests that replace the mature ones grow quickly, fixing carbon at a much faster rate than the older, mature forest, in which annual growth has slowed.

A report of research scientists at the College of Forestry of Oregon State University showed, however, that the conversion of old-growth forests to young, fast-growing forests will not significantly decrease atmospheric CO_2. They calculated that while 42 percent of the harvested timber goes into comparatively long-term storage (greater than five years) in building structures, much of the remainder is directly discarded on the logging site where it is burned or rapidly decomposes. In addition, some biomass becomes waste in factory processing of the lumber, where sawdust and scrap are burned as fuel. Similarly, the manufacture of paper also results in short-term conversion of a large proportion of the harvested trees to CO_2. In sum, harvesting of old-growth forests as now practiced actually contributes substantially to atmospheric CO_2.

Some environmentalists have advocated increased tree planting as a way of enhancing CO_2 fixation. To take up the quantity of carbon now being released by fossil fuel burning would require some 7 million square kilometers of new closed-crown broadleaved deciduous forest—an area about the size of Australia. To absorb the net increase in atmospheric carbon would require about half that area. This would be a daunting task at best.

Another factor is that increased CO_2 concentration in the atmosphere might enhance photosynthesis and thus increase the rate of carbon fixation. The enhancement of photosynthesis by increased CO_2 concentrations has been observed for many plants and demonstrated as a way of increasing yields of some crops. However, CO_2 is only one factor in photosynthesis—light, temperature, nutrients, and water are also needed, and restrictions in any of these will reduce photosynthesis. In one research study, forest photosynthesis under enriched CO_2 conditions was stimulated, but no net increase in carbon storage was observed. On the other hand, the CO_2 enrichment seemed to enhance root growth and nitrogen fixation by root nodules, thus making the trees more able to withstand stress.

While the dynamics of forests are important in the global carbon cycle, soils may be even more important. Recent inventories estimate that about four times as much carbon resides in soils than in above-ground biomass. The largest reservoir of soil carbon is in the boreal forest. In fact, there is about as much carbon in boreal forest soils as in all above-ground vegetation. This soil carbon has accumulated over thousands of years under cold conditions that have retarded its decay. However, there is now great concern that global warming, which is acting more strongly at high latitudes, will increase the rate of decay of this vast carbon pool and that boreal forests, which are presently a sink for CO_2, will become a source.

Reducing the rate of carbon dioxide buildup in the atmosphere is a matter of great international concern. As we noted in Chapter 3, an international treaty limiting emissions of CO_2 and other greenhouse gases was signed at the Rio de Janeiro Earth Summit in 1992 by nearly 150 nations. Since that time, the world's nations have been struggling with the implementation of a plan to control these emissions. While much good progress has been made, more work is needed. An effective global commitment to reduction of CO_2 releases and control of global warming still awaits us.

8.12 Habitats within the Canadian boreal forest
Habitats of the Canadian boreal forest are quite varied and include moving dune, bottomland, ridge, bog, and upland. (After P. Dansereau.)

Deciduous forest

Needleleaf forest

Moving dune

Bottomland

Ridge

Canadian forest habitats

Bog

Upland

Cliff

ism must adjust to variations in the environment on its own, we find that similar habitats often contain similar assemblages of organisms. Biogeographers and ecologists recognize specific types of communities, called *associations,* in which typical organisms are likely to be found together. Sometimes these associations are defined by species, as in the beech-birch-maple forest that is found in the Great Lakes region and that stretches to New England. Other times they are defined more generally by the life-form of the vegetation cover, as in the boreal forest biome (sometimes facetiously termed the "spruce-moose" biome), which consists of the broad circumpolar band of coniferous forest found in the northern hemisphere and includes many similar and related species of plants, animals, and microbes. (The *biome* is the largest division of ecosystems. (We'll say more about biomes in Chapter 9.)

What physical environmental factors are most important in determining where organisms, as individuals and species, are found? In general, moisture and temperature are most important. Although organisms are sometimes present under conditions of extreme temperatures or dryness as spores or cysts, nearly all organisms have limits that are exceeded at least somewhere on Earth at some time. At the global scale, temperature and moisture patterns translate into climate. For this reason, as we will see in Chapter 9, there is a very strong relationship between climate and vegetation.

Other environmental factors that influence plant and animal distribution patterns and life cycles are light and wind. Light varies within communities, as, for example, between the top and bottom of a forest canopy, as well as with time, as in the changes in daylight length that occur with the seasons. Wind exposes plants to drying and can cause plants to be stunted on the windward side. Let's look at environmental factors in more detail.

Water Need

Both plants and animals show a variety of adaptations that enable them to cope with the abundance or scarcity of water. In plants, many of these adaptations affect the transpiration mechanism. Evaporation at the leaf is controlled by specialized leaf pores, which provide openings in the outer layer of cells. When soil water is depleted, the pores close and evap-

oration is greatly reduced. Plants that are adapted to drought conditions are termed *xerophytes.* The word "xerophyte" comes from the Greek roots *xero-,* meaning "dry," and *phyton,* meaning "plant." Some xerophytes are adapted to habitats that dry quickly following rapid drainage of precipitation—for example, sand dunes, beaches, and bare rock surfaces. Others are adapted to habitats in which rainfall is simply scarce, such as deserts.

In some xerophytes, water loss is reduced by a thick layer of wax or wax-like material on leaves and stems. The wax helps to seal water vapor inside the leaf or stem. Still other xerophytes adapt to a desert environment by greatly reducing their leaf area or by bearing no leaves at all. Needlelike leaves, or spines in place of leaves, are also adaptations of plants to conserve water. In cactus plants, the foliage leaf is not present, and transpiration is limited to thickened, water-filled stems that store water for use during long, dry periods (Figure 8.13).

Adaptations of plants to water-scarce environments also include improved abilities to obtain and store water. Roots may extend deeply to reach soil moisture far from the surface. In cases where the roots reach to the ground water zone, a steady supply of water is assured. Plants drawing from ground water are termed *phreatophytes* and may be found along dry stream channels and valley floors in desert

8.13 Prickly pear cactus (Opuntia) *The prickly pear cactus is widely distributed in the western hemisphere. This clump of cactus is in the Sonoran Desert, Arizona.*

regions. In these environments, ground water is usually near the surface. Other desert plants produce a widespread, but shallow, root system. This system enables them to absorb water from short desert downpours that saturate only the uppermost soil layer. Commonly, leaves and stems of desert plants are greatly thickened by a spongy tissue in which much water can be stored. Plants employing this adaptation are called *succulents.*

Another adaptation to extreme aridity is a very short life cycle. Many small desert plants will germinate from seed, then leaf out, bear flowers, and produce seed in the few weeks immediately following a heavy rain shower. In this way, they complete their life cycle when soil moisture is available, and they survive the dry period as dormant seeds that require no moisture.

Certain climates, such as the wet-dry tropical climate ③ and the moist continental climate ⑩, have a yearly cycle with one season in which water is unavailable to plants because of lack of precipitation or because the soil water is frozen. This season alternates with one in which there is abundant water. Plants adapted to such regimes are called *tropophytes,* from the Greek word *trophos,* meaning "change" or "turn." Tropohytes respond to this pattern by dropping their leaves at the close of the moist season and becoming dormant during the dry season. When water is again available, they leaf out and grow at a rapid rate. Trees and shrubs that shed their leaves seasonally are termed *deciduous,* while *evergreen* plants retain most of their leaves in a green state through one or more years.

The Mediterranean climate ⑦ also has a strong seasonal wet-dry alternation, with dry summers and wet winters. Plants in this climate are often xerophytic and characteristically have hard, thick, leathery leaves. An example is the live oak, which

> Xerophytes are plants that are adapted to a dry and sometimes hot environment. Examples are phreatophytes, which have deep roots, succulents, which store water internally in spongy tissues, and a deciduous habit, in which leaf drop limits water loss from transpiration.

holds most of its leaves through the dry season (Figure 8.14). As we saw in Chapter 7, such hard-leaved evergreen trees and woody shrubs are called *sclerophylls.* (The prefix *scler-* is from the Greek root for "hard" and is combined with the Greek word for leaf, *phyllon.*) Plants that hold their leaves through a dry or cold season have the advantage of being able to resume photosynthesis immediately when growing conditions become favorable, whereas the deciduous plants must grow a new set of leaves.

To cope with water shortages, **xeric animals** have evolved methods that are somewhat similar to those used by the plants. Many of the invertebrates exhibit the same pattern as ephemeral annuals—evading the dry period in dormant states. When rain falls, they emerge to take advantage of the new and short-lived vegetation that often results. For example, many species of birds regulate their behavior to nest only when the rains occur, the time of most abundant food for their offspring. The tiny brine shrimp of the Great Basin may wait many years in dormancy until normally dry lakebeds fill with water, an event that occurs perhaps three or four times a century. The shrimp then emerge and complete their life cycles before the lake evaporates.

Mammals are by nature poorly adapted to desert environments, but many survive through a variety of mechanisms that enable them to avoid water loss. Just as plants reduce transpiration to conserve water, so many desert mammals do not sweat through skin glands. Instead they rely on other methods of cooling, such as avoiding the Sun and becoming active only at night. In this respect, they are joined by most of the rest of the desert fauna, spending their days in cool burrows in the soil and their nights foraging for food (Figure 8.15).

8.14 Live oak This California live oak is an example of a sclerophyll—a plant with thick, leathery leaves that is adapted to an environment with a very dry season.

8.15 Desert reptile *The noctural activity of the night snake, Hypsiglena torquata, is captured in this flash photo from Baja, Mexico.*

Temperature

The temperature of the air and soil, the second of the important climatic factors in ecology, acts directly on organisms through its influence on the rates at which physiological processes take place in plant and animal tissues. In general, each plant species has an optimum temperature associated with each of its functions, such as photosynthesis, flowering, fruiting, or seed germination. There are also limiting lower and upper temperatures for these individual functions and for the total survival of the plant itself.

Temperature can also act indirectly on plants and animals. For example, higher air temperatures reduce the relative humidity of the air (Chapter 4), thus enhancing transpiration from plant leaves as well as increasing direct evaporation of soil water.

In general, the colder the climate, the fewer the species that are capable of surviving. A large number of tropical

> Temperature affects physiological processes occurring in plant and animal tissues. In general, colder climates have fewer plant and animal species.

plant species cannot survive below-freezing temperatures for more than a few hours. In the severely cold arctic and alpine environments of high latitudes and high altitudes, only a few plant and animal species are found. This principle explains why an equatorial rainforest contains such a diverse array of plants and animals, whereas a forest of the subarctic zone will be dominated by only a few. Tolerance to cold in plants is closely tied to the physical disruption that accompanies the growth of ice crystals inside cells. Cold-tolerant plant species can expel excess water from cells to spaces between cells, where freezing does no damage.

The effects of temperature variations on animals are moderated by their physiology and by their ability to seek sheltered environments. Most animals lack a physiological mechanism for internal temperature regulation. These animals, including reptiles, invertebrates, fish, and amphibians, are *cold-blooded animals*—their body temperature passively follows the environment. With a few exceptions (notably fish and some social insects), these animals are active only during the warmer parts of the year. They survive the cold weather of the midlatitude zone winter by becoming dormant. Some vertebrates enter a dormant state termed *hibernation* in which metabolic processes virtually stop and body temperatures closely parallel those of the surroundings. Most hibernators seek out burrows, nests, or other environments where winter temperatures do not reach extremes or fluctuate rapidly. Because the annual range of soil temperatures is greatly reduced below the uppermost layers, soil burrows are particularly suited to hibernation.

Other animals maintain their tissues at a constant temperature by internal metabolism. This group includes the birds and mammals. These *warm-blooded animals* possess a variety of adaptations to maintain a constant internal temperature. Fur, hair, and feathers act as insulation by trapping dead air spaces next to the skin surface, reducing heat loss to the surrounding air or water. A thick layer of fat will also

8.16 Alaskan brown bear
This brown bear, snacking on a salmon, is well-insulated against winter cold by a heavy coat and a layer of fat.

8.17 *Eye on the Landscape* **Grazing caribou** *A herd of caribou grazes on tundra in the foothills of the Brooks Range, Alaska.* **What else would the geographer see? ... Answers at the end of the chapter.**

provide excellent insulation (Figure 8.16). Other adaptations are for cooling—for example, sweating or panting uses the high latent heat of vaporization of water to remove heat. Heat loss is also facilitated by exposing blood-circulating tissues to the cooler surroundings. The seal's flippers and bird's feet serve this function.

Adaptations to temperature extremes can allow a species to exploit a potentially harsh environment. The caribou (Figure 8.17) is but one of several species of large grazing mammals that can endure very cold temperatures. During the mild season, caribou graze the far northern tundra, then migrate southward to pass the winter in a less hostile environment.

Other Climatic Factors

The factor of light is also important in determining local plant distribution patterns. Some plants are adapted to bright sunlight, while others require shade (Figure 8.18). The amount of light available to a plant will depend in large part on the plant's position. Tree crowns in the upper layer of a forest receive maximum light but correspondingly reduce the amount available to lower layers. In extreme cases, forest trees so effectively cut off light that the forest floor is almost free of shrubs and smaller plants. In certain decidu-

ous forests of midlatitudes, the period of early spring, before the trees are in leaf, is one of high light intensity at ground level, permitting the smaller plants to go through a rapid growth cycle. In summer these plants will largely disappear as the leaf canopy is completed. Other low plants in the same habitat require shade and do not appear until the leaf canopy is well developed.

Treated on a global basis, the factor of light available for plant growth varies by latitude. Duration of daylight in summer increases rapidly with higher latitude and reaches its maximum poleward of the arctic and antarctic circles, where the Sun may be above the horizon for 24 hours (see Figure 8.4). Although the growing season for plants is greatly shortened at high latitudes by frost, the rate of plant growth in the short frost-free summer is greatly accelerated by the prolonged daylight.

In midlatitudes, where vegetation is of a deciduous type, the annual rhythm of increasing and decreasing periods of daylight determines the timing of budding, flowering, fruiting, leaf shedding, and other vegetation activities. As to the importance of light intensity itself, even on overcast days there is sufficient light to permit most plants to carry out photosynthesis at their maximum rates.

Light also influences animal behavior. The day–night cycle controls the activity patterns of many animals. Birds,

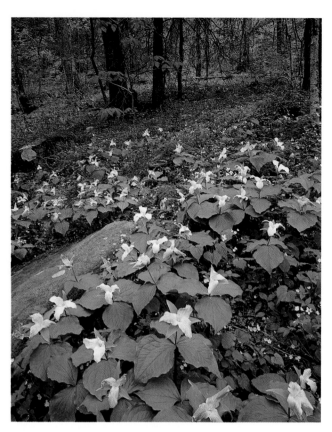

8.18 *Sun-loving and shade-loving plants* *Sun-loving spring flowers, including baby blue-eyes and California poppies, bloom in this open field near San Luis Opisbo, California (left). In contrast, the forest floor in this scene in the Great Smoky Mountains National Park is carpeted with shade-loving trillium flowers (right).*

for example, are generally active during the day, whereas small foraging mammals, such as weasels, skunks, and chipmunks, are more active at night. Light also controls seasonal activity through *photoperiod,* or daylight length, in midlatitudes. As autumn days grow shorter and shorter, squirrels and other rodents hoard food for the coming winter season. Later, increasing photoperiod will trigger such activities as mating and reproduction in the spring.

Wind is also an important environmental factor in the structure of vegetation in highly exposed positions. Close to the timberline in high mountains and along the northern limits of tree growth in the arctic zone, trees are deformed by wind in such a way that the branches project from the lee side of the trunk only (flag shape) (see Figure 9.24). Some trees may show trunks and branches bent to near-horizontal attitude, facing away from the prevailing wind direction. The effect of wind is to cause excessive drying, damaging the exposed side of the plant. The tree limit on mountainsides thus varies in elevation with degree of exposure of the slope to strong prevailing winds and will extend higher on lee slopes and in sheltered pockets.

In addition to temperature and moisture, ecological factors of light intensity, length of the daylight period, length of the growing season, and wind duration and intensity act to determine plant and animal distribution patterns.

Bioclimatic Frontiers

Taken separately or together, climatic factors of moisture, temperature, light, and wind can act to limit the distribution of plant and animal species. Biogeographers recognize that there is a critical level of climatic stress beyond which a species cannot survive and that there will be a geographic boundary marking the limits of the potential distribution of a species. Such a boundary is sometimes referred to as a **bioclimatic frontier.** Although the frontier is usually marked by a complex of climatic elements, it is sometimes possible to single out one climatic element related to soil water or temperature that coincides with it.

The distribution of ponderosa pine *(Pinus ponderosa)* in western North America provides an example (Figure 8.19). In this mountainous region, annual rainfall varies sharply with elevation. The 50 cm (20 in.) isohyet (rainfall contour) of annual total precipitation encloses most of the upland areas having the yellow pine. It is the parallelism of the isohyet with forest boundary, rather than actual degree of coincidence, that is significant. The sugar maple *(Acer saccharum)* is a somewhat more complex case

8.19 Distribution of ponderosa pine in western North America *Areas of ponderosa pine (Pinus ponderosa) are shown in black. The edge of the shaded area marks the rainfall contour (isohyet) of 50 cm (20 in.).*

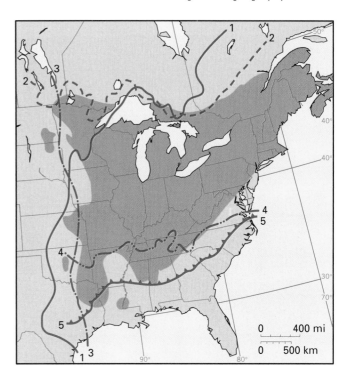

8.20 Bioclimatic limits of sugar maple *The shaded area shows the distribution of sugar maple (Acer saccharum) in eastern North America. Line 1, 75 cm (30 in.) annual precipitation. Line 2, −40° C (−40° F) mean annual minimum temperature. Line 3, eastern limit of yearly boundary between arid and humid climates. Line 4, 25 cm (10 in.) mean annual snowfall. Line 5, −10° C (16° F) mean annual minimum temperature.*

(Figure 8.20). Here the boundaries on the north, west, and south coincide roughly with selected values of annual precipitation, mean annual minimum temperature, and mean annual snowfall.

Although bioclimatic limits must exist for all species, no plant or animal need necessarily be found at its frontier. Many other factors may act to keep the spread of a species in check. A species may be limited by diseases or predators found in adjacent regions. In another example, a species (especially a plant species) may migrate slowly and may still be radiating outward from the location in which it evolved. (More details will be given later.) Or a species may be dependent on another species and therefore be limited by the latter's distribution.

> Geomorphic (landform) factors influencing plant and animal distributions include slope angle, slope aspect, and relief. Edaphic (soil) factors include soil particle size (e.g., coarse sand or fine clay) and amount and nature of organic matter in the soil.

Geomorphic Factors

Geomorphic factors (landform factors) influencing ecosystems include such elements as slope steepness (the angle that the ground surface makes with respect to the horizontal), slope aspect (the orientation of a sloping ground surface with respect to geographic north), and relief (the difference in elevation of divides and adjacent valley bottoms). In a much broader sense, geomorphic factors include the entire sculpturing of the landforms of a region by processes of erosion, transportation, and deposition via streams, waves, wind, and ice, and by forces of volcanism and mountain building. These are topics covered in detail in Chapters 13 through 19.

Slope steepness acts indirectly by influencing the rate at which precipitation is drained from the surface. On steep slopes, surface runoff is rapid. On gentle slopes, much of the precipitation can penetrate the soil and be retained. More rapid erosion on steep slopes may result in thin soil, whereas that on gentler slopes is thicker. Slope aspect has a direct influence on plants by increasing or decreasing the exposure to sunlight and prevailing winds. Slopes facing the Sun have a warmer, drier environment than slopes that face away from the Sun and therefore lie in shade for much longer periods of the day. In midlatitudes, these slope-aspect contrasts may be so strong as to produce quite different biotic communities on north-facing and south-facing slopes (Figure 8.21).

Geomorphic factors are in part responsible for the dryness or wetness of the habitat within a region that has essentially the same overall climate. Each community has its own microclimate. On divides, peaks, and ridge crests, the soil tends to dryness because of rapid drainage and because the

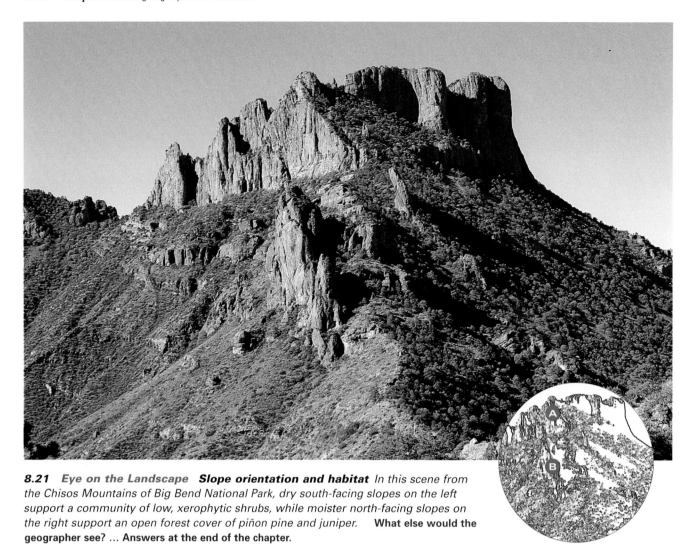

8.21 *Eye on the Landscape* ***Slope orientation and habitat*** *In this scene from the Chisos Mountains of Big Bend National Park, dry south-facing slopes on the left support a community of low, xerophytic shrubs, while moister north-facing slopes on the right support an open forest cover of piñon pine and juniper.* **What else would the geographer see? ... Answers at the end of the chapter.**

surfaces are more exposed to sunlight and drying winds. By contrast, the valley floors are wetter because surface runoff over the ground and into streams causes water to converge there. In humid climates, the ground water table in the valley floors may lie close to or at the ground surface to produce marshes, swamps, ponds, and bogs (see Chapter 15).

Edaphic Factors

Edaphic factors are those related to the soil. In terms of biogeography, we can look at soils from two perspectives. First, the broad pattern of soil types that we will recognize in Chapter 10 is largely related to climate, as is the broad pattern of distribution of plants and animals that we will see in Chapter 9. Thus, there is a strong coincidence between climate, soil, and biota at the global level. We'll return to this topic in those chapters.

A second viewpoint is in terms of habitats—the fine-scale mosaic of place-to-place variations of the Earth's surface. Soils can vary widely from one small area to the next, thus influencing the local distribution of plants and animals. For example, sandy soils store less water than soils with abundant silt and clay, and thus often display a vegetation

cover with xeric adaptations. A high content of organic matter in the soil often means nutrients are abundant rather than scarce. Such a rich soil usually harbors more plant species and allows them to grow more rapidly. We will return to some of the soil factors that are important to plants and animals in Chapter 10.

While soil conditions often affect the types of plants and animals found at a locality, the biota can also act to change the soil conditions. Given a barren habitat recently formed by some geologic event, such as the outpouring of lava or the emergence of a coastal zone from beneath the sea, the soil undergoes a gradual evolution with the occupance of the habitat by a succession of biotic communities. The plants alter the soil by such processes as contributing organic matter or producing acids that act on the mineral matter. Animal life, feeding on the plant life, also makes its contribution to the physical and chemical processes of soil evolution.

Interactions among Species

Species interactions can also be important factors in determining the distribution patterns of plants and animals. Two species that are part of the same ecosystem can interact with

8.22 Predation *This giant anteater enjoys a lunch of Brazilian termites.*

one another in three ways: interaction may be negative to one or both species; or the two species may be neutral, not affecting each other; or interaction may be positive, benefiting at least one of the species.

Competition, a negative interaction between species, occurs whenever two species require a common resource that is in short supply. Because neither species has full use of the resource, both populations suffer, showing growth rates lower than those when only one of the species is present. Sometimes one species will win the competition and crowd out its competitor. At other times, the two species may remain in competition indefinitely.

Competition between species is an unstable situation. If a genetic strain within one of the populations emerges that uses a substitute resource for which there is no competition, its survival rate will be higher than that of the remaining strain, which still competes. The original strain may become extinct. In this way, evolutionary mechanisms tend to reduce competition among species.

Predation and parasitism are other negative interactions between species. *Predation* occurs when one species feeds on another (Figure 8.22). The benefits are obviously positive to the predator species, which obtains energy for survival, and negative to the prey species. *Parasitism* occurs when one species gains nutrition from another, typically when the parasite organism invades or attaches to the body of the host in some way.

Although we tend to think of predation and parasitism as negative processes that benefit one species at the expense of the other, it may well be that these interactions are really beneficial in the long term to the host or prey populations. A classic example is the growth of the deer herd on the Kaibab Plateau north of the Grand Canyon in Arizona (Figure 8.23). Initially at a population of about 4000, in an area of 283,000

hectares (700,000 acres), the herd grew to nearly 100,000 in the short span of 1907–1924 in direct proportion to a government predator control and game protection program. Wolves became extinct in the area, and populations of coyotes and mountain lions were greatly reduced. The huge deer population, however, proved too much for the land, and overgrazing led to a population crash. In one year, half the animals starved to death; by the late 1930s, the population had declined to a stable level near 10,000. Thus, predation maintained the deer population at levels that were in harmony with the supportive ability of the environment. In addition to maintenance of equilibrium population levels, predation and parasitism differentially remove the weaker individuals and can improve the genetic composition of the species.

8.23 Rise and fall of the Kaibab deer herd *This graph plots the population size of the deer herd in the Kaibab National Forest, Arizona, as it rose in response to the abolition of predators and then crashed when the population ran out of food. (After D. I. Rasmussen, Ecological Monographs, vol. 11, 1941, p. 237.)*

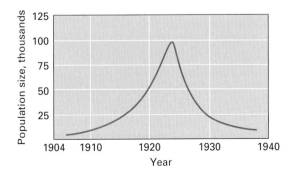

A third type of predation is *herbivory,* in which grazing of plants by animals reduces the viability of the plant species population. Some species are well-adapted to grazing and can maintain themselves well in the face of grazing pressure. Others may be quite sensitive to this process. When overgrazing occurs, these differing sensitivities can produce significant changes in the structure and composition of plant communities.

A fourth type of negative interaction is *allelopathy,* a phenomenon of the plant kingdom in which chemical toxins produced by one species serve to inhibit the growth of others. As an example, sage, a common shrub species in the California chaparral, produces leaves rich in volatile toxins (cineole and camphor). As the leaves fall and accumulate in the soil, the allelopathic toxins build to a level sufficient to inhibit the growth of herbaceous plants, such as grasses, which are thus only found in adjacent areas. Still other chaparral shrubs produce water-soluble antibiotics that also inhibit the growth of nearby grasses. These chemical defenses, however, are broken down by periodic fires, which are events essential to maintenance of the chaparral ecosystem. The fires destroy the toxins and also trigger the germination of seeds of many species of annual herbaceous plants by breaking their seed coats. The annuals then dominate the area until the shrubs grow and force them out by allelopathy, beginning the cycle anew.

The term **symbiosis** includes three types of positive interactions between species: commensalism, protocooperation, and mutualism. In commensalism, one of the species is benefited and the other is unaffected. Examples of commensals include the epiphytic plants—such as orchids or Spanish moss—which live on the branches of larger plants (Figure 9.7). These *epiphytes* depend on their hosts for physical support only. In the animal kingdom, small commensal crabs or fishes seek shelter in the burrows of sea worms; or the commensal remora fish attaches itself to a shark, feeding on bits of leftover food as its host dines.

When the relationship benefits both parties but is not essential to their existence, it is termed protocooperation. The attachment of a stinging coelenterate to a crab is an example of protocooperation. The crab gains camouflage and an additional measure of defense, while the coelenterate eats bits of stray food which the crab misses.

Where protocooperation has progressed to the point that one or both species cannot survive alone, the result is mutualism. A classic example is the association of the nitrogen-fixing bacterium *Rhizobium* with the root tissue of certain types of plants (legumes), in which the action of the bacteria converts nitrogen gas to a form directly usable by the plant. The association is mutualistic because *Rhizobium* cannot survive alone.

> Negative interactions among species include competition, predation, parasitism, herbivory, and allelopathy. Positive interaction includes commensalism, protocooperation, and mutualism, which are three forms of symbiosis.

> In the process of ecological succession, an ecosystem proceeds through seral stages to reach a climax. Primary succession occurs on new soil, while secondary succession occurs where disturbance has removed or altered existing communities.

ECOLOGICAL SUCCESSION

The phenomenon of change in plant and animal communities through time is a familiar one. A drive in the country reveals patches of vegetation in many stages of development—from open, cultivated fields through grassy shrublands to forests. Clear lakes, gradually filled with sediment from the rivers that drain into them, become bogs. These kinds of changes, in which biotic communities succeed one another on the way to a stable end point, are referred to as **ecological succession.**

GEODISCOVERIES

In general, succession leads to formation of the most complex community of organisms possible in an area, given its physical controlling factors of climate, soil, and water. The series of communities that follow one another on the way to the stable stage is called a *sere.* Each of the temporary communities is referred to as a *seral stage.* The stable community, which is the end point of succession, is the **climax.** If succession begins on a newly constructed deposit of mineral sediment, it is termed *primary succession.* If succession occurs on a previously vegetated area that has been recently disturbed by such agents as fire, flood, windstorm, or humans, it is referred to as *secondary succession.*

A new site on which primary succession occurs may have one of several origins—for example, a sand dune, a sand beach, the surface of a new lava flow or freshly fallen layer of volcanic ash, or the deposits of silt on the inside of a river bend that is gradually shifting. Such a site will not have a well-developed soil. Rather, it may perhaps be little more than a deposit of coarse mineral fragments. In other cases, such as that of floodplain silt deposits, the surface layer may represent redeposited soil endowed with substantial amounts of organic matter and nutrients.

The first stage of a succession is a *pioneer stage.* It includes a few plant and animal species unusually well adapted to adverse conditions of rapid water drainage and drying of soil and to excessive exposure to sunlight, wind, and extreme ground and lower-air temperatures. As pioneer plants grow, their roots penetrate the soil, and their subsequent decay adds organic matter directly to the soil. Fallen leaves and stems add an organic layer to the ground surface. Bacteria and animals begin to live in the soil in large numbers. Grazing mammals feed on the small plants. Birds forage the newly vegetated area for seeds and grubs.

Soon conditions are favorable for other species that invade the area and displace the pioneers. The new arrivals may be larger plant forms providing more extensive cover of foliage over the ground. In this case the climate near the ground is considerably altered toward one of less extreme air and soil temperatures, high humidities, and less intense insolation. Still other species now invade and thrive in the modified environment. When the succession has finally run its course, a climax community of plant and animal species in a more or less stable composition will exist.

The colonization of a sand dune provides an example of primary succession. Growing foredunes bordering the ocean or lake shore present a sterile habitat. The dune sand—usually largely quartz, feldspar, and other common rock-forming minerals—lacks such important nutrients as nitrogen, calcium, and phosphorus, and its water-holding ability is very low. Under the intense solar radiation of the day, the dune surface is a hot, drying environment. At night, radiation cooling in the absence of moisture produces low surface temperatures.

One of the first pioneers of this extreme environment is beachgrass (Figure 8.24, left). This plant reproduces by sending out rhizomes (creeping underground stems), and the plant thus slowly spreads over the dune. Beachgrass is well adapted to the eolian environment; it does not die when buried by moving sand, but instead puts up shoots to reach the new surface.

After colonization, the shoots of beachgrass act to form a baffle that suppresses movement of sand, and thus the dune becomes more stable. With increasing stabilization, plants that are more adapted to the dry, extreme environment but cannot withstand much burial begin to colonize the dune. Typically, these are low, matlike woody shrubs, such as beach wormwood or false heather.

On older beach and dune ridges of the central Atlantic coastal plain, the species that follow matlike shrubs are typically larger woody plants and such trees as beach plum, bayberry, poison ivy, and choke cherry. These species all have one thing in common—their fruits are eaten by birds. The seeds are excreted as the birds forage among the low dune shrubs, thereby sowing the next stage of succession. As the scrubby bushes and small trees spread, they shade out the mat-like shrubs and any remaining beachgrass. Pines may also enter at this stage.

At this point, the soil begins to accumulate a significant amount of organic matter. No longer dry and sterile, it now possesses organic compounds and nutrients, and it has accumulated enough fine particles to hold water for longer intervals. These soil conditions encourage the growth of such broadleaf species as red maple, hackberry, holly, and oaks, which shade out the existing shrubs and small trees (Figure 8.24, right). Once the forest is established, it tends to reproduce itself; the species of which it is composed are tolerant to shade, and their seeds can germinate on the organic forest floor. Thus, the climax is reached. The stages through which the ecosystem has developed constitute the sere, progressing from beachgrass to low shrubs to higher shrubs and small trees to forest.

Although this example has stressed the changes in plant cover, animal species are also changing as succession proceeds. Table 8.2 shows how some typical invertebrates appear and disappear through succession on the Lake Michigan dunes. Note that the seral stages shown in the table for these inland dunes are somewhat different from those described for the coastal environment.

Where disturbance alters an existing community, secondary succession can occur. *Old-field succession,* taking place on abandoned farmland, is a good example of secondary succession. In the eastern United States, the first stages of the sere often depend on the last use of the land before abandonment. If row crops were last cultivated, one set of pioneers, usually annuals and biennials, will appear. If small grain crops were cultivated, the pioneers are often perennial herbs and grasses. If pasture is abandoned, those pioneers that were not grazed will have a head start. Where mineral soil was freshly exposed by plowing, pines are often important following the first stages of succession

8.24 Dune succession at Sandy Hook, New Jersey *(left) Beachgrass is a pioneer on beach dunes and helps stabilize the dune against wind erosion. (right) The climax forest on dunes. Holly (Ilex opaca), seen on the left, is an important constituent of the climax forest. Note the leaves and decaying organic matter on the forest floor. (Alan Strahler)*

Table 8.2 | **Invertebrate Succession on the Lake Michigan Dunes**

	Successional Stages				
Invertebrate	Beachgrass–Cottonwood	Jack Pine Forest	Black Oak Dry Forest	Oak and Oak–Hickory Moist Forest	Beech–Maple Forest Climax
White tiger beetle	x				
Sand spider	x				
Long-horn grasshopper	x	x			
Burrowing spider	x	x			
Bronze tiger beetle		x			
Migratory locust		x			
Ant lion			x		
Flatbug			x		
Wireworms			x	x	x
Snail			x	x	x
Green tiger beetle				x	x
Camel cricket				x	x
Sowbugs				x	x
Earthworms				x	x
Woodroaches				x	x
Grouse locust					x

Source: V. E. Shelford, as presented in E. P. Odum, *Fundamentals of Ecology* (Philadelphia: W. B Saunders Co.), p. 259.

because pine seeds favor disturbed soil and strong sunlight for germination. Although slower-growing than other pioneers, the pines will eventually shade the others out and become dominant. Their dominance is only temporary, however, because their seeds cannot germinate in shade and litter on the forest floor. Seeds of hardwoods such as maples and oaks can germinate under these conditions, and as the pines die, hardwood seedlings grow quickly to fill the holes produced in the canopy. The climax, then, is the hardwood forest. Figure 8.25 is a schematic diagram showing one example of old-field succession.

Succession, Change, and Equilibrium

The successional changes we have described result from the actions of the plants and animals themselves. One set of inhabitants paves the way for the next. As long as nearby populations of species provide colonizers, the changes lead

8.25 Old-field succession *A typical old-field succession sequence for the southeastern United States, following abandonment of cultivated fields. This is a pictorial graph of continuously changing plant composition spanning about 150 years.*

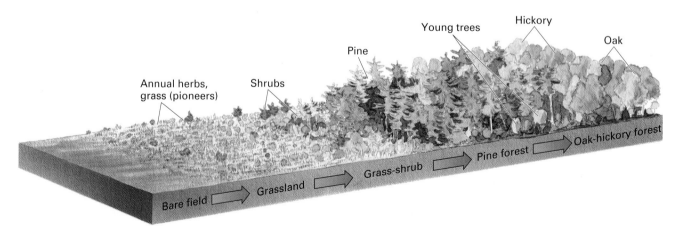

automatically from bare soil or fallow field to climax forest. This type of succession is often termed *autogenic* (self-producing) *succession.*

In many cases, however, autogenic succession does not run its full course. Environmental disturbances, such as wind, fire, flood, or renewed clearing for agriculture, may occur, diverting the course of succession temporarily or even permanently. For example, autogenic succession on seaside dunes may proceed for some years, but sooner or later a severe storm will generate waves and wind that reduce the developing community to a sand bar once again. Or a mature forest may be destroyed by fire, inviting the succession of plants and animals that are specifically adapted to burned environments. (See *Eye on the Environment 8.3 ● The Great Yellowstone Fire.*) In addition, habitat conditions such as site exposure, unusual bedrock, or impeded drainage can hold back or divert the course of succession so successfully that the climax is never reached.

Introduction of new species can also greatly alter existing ecosystems and successional pathways. The parasitic chestnut blight, introduced from Europe to New York in about 1910, decimated populations of the American chestnut tree within a period of about 40 years. This tree species, which may have accounted for as many as one-fourth of the mature trees in eastern forests, is now found only as small blighted stems sprouting from old root systems.

These examples show that while succession is a reasonable model to explain many of the changes that we see in ecosystems with time, it may be more realistic to view the pattern of ecosystems on the landscape as reflecting a spatial dynamic equilibrium between autogenic forces of self-induced change and external forces that reverse or rechannel autogenic change temporarily or permanently. From this viewpoint, the biotic landscape is thus a mosaic of patches of distinctive biotic communities with different biological potentials and different histories.

This view of the landscape assumes that all successional species are always available to colonize new space or invade seral communities. The nature of the biotic communities that we see will then be determined by varying environmental and ecological factors as they act within new space created by physical and human processes. But what about the continental and global scale, taken over longer spans of time? Here, not all species are available to colonize new space. On this scale, the processes of migration, dispersal, evolution, and extinction are more important in determining the spatial patterns we observe. These processes are within the realm of historical biogeography, to which we turn next.

> We can view the pattern of plant and animal communities on the landscape as a dynamic equilibrium between succession, in which the community modifies its own habitat and composition, and environmental disturbance, such as wind, flood, fire, or logging.

HISTORICAL BIOGEOGRAPHY

Historical biogeography focuses on how spatial distribution patterns of organisms arise over space and time. It examines four key processes: evolution, speciation, extinction, and dispersal. Let's look at each of these in turn, beginning with evolution.

Evolution

An astonishing number of organisms exist on Earth, each adapted to the ecosystem in which it carries out its life cycle. About 40,000 species of microorganisms, 350,000 species of plants, and 2.2 million species of animals, including some 800,000 insect species, have been described and identified. However, many organisms remain unclassified. Estimates suggest that species of plants will ultimately number about 540,000. Of an estimated 2 million insect species, only slightly more than one-third have been classified.

How has life gained this astonishing diversity? Through the process of **evolution,** the environment itself has acted on organisms to create this diversity of life-forms, even as organic processes have been a prime factor in shaping the environment. You are probably familiar with the name of Sir Charles Darwin, whose monumental work, *The Origin of Species by Means of Natural Selection,* was published in 1859. Through exhaustive studies, Darwin showed that all life possessed *variation,* the differences that arise between parent and offspring. The environment acts on variation in organisms, Darwin observed, in much the same way that a plant or animal breeder does, selecting for propagation only those individuals with the best qualities, those best suited to their environment. Darwin termed this survival and reproduction of the fittest **natural selection.** He saw that variation could, when acted upon by natural selection through time, bring about the formation of new species whose individuals differed greatly from their ancestors. Thus, Darwin viewed the formation of new species as a product of variation acted upon by natural selection.

The weakness in Darwin's theory lay not in the origin of species as products of natural selection, but in the process of variation. He was at a loss to explain why variation occurred, and he simply accepted it as a natural and an automatic property of life. We know now that variation results from the interaction of two sources: *mutation,* which alters the genetic material and consequently the physical characteristics of organisms; and *recombination,* the pairing of parental genetic material in offspring in unique assortments.

Mutation results when the genetic material (DNA, or deoxyribonucleic acid) of a reproductive cell is changed. This requires breaking and reassembling chemical bonds in the DNA of chromosomes, which can come about when the cell is exposed to heat, ionizing radiation, or certain types of chemical agents. Most mutations either have no effect or are deleterious, but a small proportion may have a positive effect on the genetic makeup of the individual. If that positive effect makes the individual organism more likely to survive and reproduce, then the altered

Eye on the Environment | 8.3

The Great Yellowstone Fire

Yellowstone is the largest and oldest American national park. Thanks in part to the magnificent photographs of its wondrous geysers and travertine terraces, taken by the pioneer photographer, William Henry Jackson, and brought back to Congress in Washington, it was declared a park in 1872. Almost a square in outline, its 900,000 hectares (2.2 million acres) occupy the northwest corner of Wyoming within the Rocky Mountain region. Volcanic rocks dominate the geology, and lava flows occupy much of the area, forming a high plateau. Because an ancient caldera underwent collapse here, geothermal activity continues to this day in many localities in the park. (See Chapter 13 and Figures 13.10 and 13.11.) Its rich, well-watered forests and parks support populations of bison, elk, and grizzly bear, along with many other mammal and bird species—a powerful attraction for millions of tourists. All in all, we have here a priceless forest ecosystem, little disturbed by natural catastrophe or human interference for at least the past two centuries.

Through August and September of 1988, many forest fires burned out of control in Yellowstone Park, casting a smoke pall over large areas of surrounding states. Of the 45 fires, most were started by lightning strikes, and this number was not unusual for the area. It had, however, been the driest summer for more than a century of record, and temperatures were persistently high. Once started, the fires spread rapidly and uncontrollably under the driving force of strong winds that in one fire reached 160 km (100 mi) per hour. Firefighters were powerless to check the spread of the flames. Dense forests of lodgepole pine that had not burned for 250 years were killed by flames that raced through the forest canopy. Fires at ground level spread rapidly through areas of older forest with ample fuel supplies accumulated as fallen trees and branches. The fires also consumed patches of young forest that rarely ignite and have typically served well as natural firebreaks.

Contrary to greatly exaggerated stories in the news media, a later study of air photographs showed that only 20 percent of the park land had been burned over in some degree. Fears of massive extermination of wildlife were also unfounded. Most of the large animals survived: only 9 bison of the herd of 2500 and 350 elk of the herd of 30,000 perished in the fires. Mammals large and small, along with insects and birds, returned quickly to the burned areas, with the herbivores taking advantage of grasses that grew vigorously among the ashes of the forest floor.

For a full century since the establishment of Yellowstone Park in 1872, the National Park Service had practiced its "no-burn" policy of not allowing any fire to spread unchecked. This policy was based on the concept that national parks must be maintained in a pristine condition, their plants being permitted to grow and die unaffected by any human activity or by fires. In 1972, this practice was reversed to a "let-burn" policy under which naturally occurring fires were to be allowed to spread unchecked. As expected, the Park Service immediately came under strong criticism for having practiced their let-burn policy by standing aside and allowing the July blazes to spread to mammoth proportions. By late July, park officials decided to suppress all fires, but a massive effort with national support was to no avail. In retrospect, however, it is question-

---

gene is likely to survive as well and be passed on to offspring.

Recombination describes the process by which an offspring receives two slightly different copies, or *alleles,* of each gene from its parents. One allele may be dominant and suppress the other, or the two alleles may act simultaneously. Because each individual receives two alleles of each gene, and there are typically tens of thousands of genes in an organism, the possible number of genetic combinations is very large. Thus, recombination provides a constant source of variation that acts to make every offspring slightly different from the next.

Mutations act to change the nature of species through time. But just what is a species? For our purposes here, we

> **Natural selection acts on the variation that occurs within populations of species to produce evolution. Variation is produced by mutation, in which genetic material is altered, and recombination, in which new combinations of existing genetic material arise.**

can define a **species** (plural, *species*) as all individuals capable of interbreeding to produce fertile offspring. A *genus* (plural, *genera*) is a collection of closely related species that share a similar genetic evolutionary history. Note that each species has a scientific name, composed of a generic name and a specific name in combination. Thus, red oak, a common deciduous tree of eastern North America, is *Quercus rubra*. The related white oak is *Quercus alba* (Figure 8.26).

Although the true test of the species is the ability of all of its individuals to reproduce successfully with one another, this criterion is not always easily applied. Instead, the species is usually defined by *morphology*—the outward form and appearance of its individuals. The *phenotype* of an individual is the morphological

Yellowstone fire and recovery *The fires that ravaged Yellowstone National Park in 1988 completely destroyed many stands of trees (left). After 10 years, however, natural regeneration had started new forests to take their place (right).*

able whether immediate action to put out the first fires could have made much difference over the vast area of the park. Another carefully drawn conclusion following study of the fire by forest ecologists was that the additional fuel accumulated in the 100-year "no-burn" period made only a minor difference in the extent and intensity of the 1988 fire.

Ecologists view the occurrence of a season of great fires in Yellowstone Park about every 250 to 300 years as part of a natural cycle of fire-driven ecological succession. Following such a fire, saplings of lodgepole pine occupy the burned area, growing tall and closely set to form a dense crown canopy, enduring between 150 and 250 years after the initial fire. During this phase, large canopy burns can occur, restarting the succession. If no fire occurs, the succession continues with a dying out of the pines and the growth of a new generation of other tree species, such as the spruce and fir. By this time—more than 300 years from the start—the forest floor becomes highly flammable and can be readily swept by fire to restart the succession.

Recent studies have established that the natural succession described above affects isolated patches of the entire forested region, resulting in a mosaic of patches in different stages of succession. In this mosaic, the younger patches serve as natural firebreaks to limit the spread of a fire occurring in adjacent older patches. Wildlife is thus able to escape the fires and then reinhabits the burned area, where a new ground cover of grasses and shrubs provides food for the grazing herbivores. This natural process of fire and renewal is important to the maintenance of other types of ecosystems as well, including the tall-grass prairies of Kansas, the pine forests of the southeastern United States, and much of the Canadian boreal forest.

8.26 Red and white oaks *Although similar in general appearance, these two species of oak are easily separated. Red oak acorns have a flat cap and stubby nut, with pointed, bristle tips on the leaf lobes (left). White oak acorns have a deeper cap and a pointed nut, with rounded leaf-lobe tips (right).*

expression of its gene set, or *genotype,* and includes all the physical aspects of its structure that are readily perceivable. Species, then, are usually defined by a characteristic phenotype or range of phenotypes.

Speciation

Biogeographers use the term **speciation** to refer to the process by which species are differentiated and maintained. Actually, speciation is not the result of a single process but arises from a number of component processes acting together through time. One of these component processes is mutation, which we have already described. The rate of mutation depends on the organism's exposure to mutagenic agents as well as the organism's sensitivity to those agents. Another process is natural selection, which as we have seen acts to favor individuals with genotypes that produce successful traits for a particular environment.

A third speciation process is *genetic drift.* In this process, chance mutations without any particular beneficial effect change the genetic composition of a breeding population until it diverges from other populations. Genetic drift is a weak factor in large populations, but in small populations, such as a colony of a few pioneers in a new habitat, random mutations are more likely to be preserved. *Gene flow* is an opposite process in which evolving populations exchange alleles as individuals move among populations. This process helps maintain the uniformity of the gene pool of the species by moving new alleles into all populations.

Speciation often occurs when populations become isolated from one another. This **geographic isolation** can happen in several ways. For example, plate tectonics may uplift a mountain range that separates a population into two different subpopulations by a climatic barrier. Or a chance long-distance dispersal may establish a new population far from the main one. These are examples of *allopatric speciation,* in which populations are geographically isolated and gene flow between the populations does not take place. As genetic drift and natural selection proceed, the populations gradually diverge and eventually lose the ability to interbreed successfully.

The evolution of finch species on the Galápagos Islands is an example of allopatric speciation created by isolation and natural selection (Figure 8.27). This cluster of five major volcanic islands and nine lesser ones, located about 800 km (500 mi) from the coast of Ecuador, was visited by Charles Darwin and provided much raw material for his ideas about evolution. As the story has been reconstructed, the islands were first colonized by a single original finch species, the blue-black grassquit. Over time, individual populations became adapted to conditions on particular islands through natural selection, and, enhanced by their isolation on different islands, evolved into different species. Later, some of these species successfully reinvaded other islands, continuing the

speciation and evolution process. Today 5 genera and 13 species of finches are found on the Galápagos, specializing in diets of seeds, buds, and insects.

Another example of allopatric speciation from the Galápagos is the giant tortoise (see chapter opener photo). Each of the larger islands bears at least one distinctly different population of these large reptiles. Like the finches, they are believed to be evolved from a single ancestral stock that colonized the island chain and then diverged into unique types.

In contrast to allopatric speciation is *sympatric speciation,* in which speciation occurs within a larger population. Imagine a species that has two different primary food sources. Eventually, mutations will arise that favor each food source at the expense of the other—for example, two different lengths or shapes of a bird's beak that facilitate feeding on two different types of fruit or seeds. As these mutations are exposed to natural selection, they will produce two different populations, each adapted to its own food source. Eventually, the populations may become spearate species. For example, hundreds of species of cichlid fishes have evolved from a few founding stocks in the lakes of the African Rift Valley. Lake Victoria alone, which was dry as recently as about 12,000 years before present, supports more than 300 of these species. Although these populations have all been in contact in the same shallow lake, they have specialized and diverged through natural selection to the point of becoming individual species. This type of evolution, in which a new environment provides the opportunity for the formation of many new species adjusted to different habitats is termed *adaptive radiation.*

Another mechanism of sympatric speciation that is quite important in plants is **polyploidy.** Normal organisms have two sets of genes and chromosomes—that is, they are *diploid.* Through accidents in the reproduction process, two closely related species can cross in such a way that the offspring has both sets of genes from both parents. These *tetraploids* are fertile but cannot reproduce with the populations from which they arose, and so they are instantly isolated as new species. By some estimates, 70 to 80 percent of higher plant species have arisen in this fashion.

> Speciation is the process by which species are differentiated and maintained. Component processes affecting speciation include mutation, natural selection, genetic drift, and gene flow. Isolation, in which breeding populations are separated, enhances speciation.

Extinction

Over geologic time, the fate of all species is **extinction.** When conditions change more quickly than populations can evolve new adaptations, population size falls. When that occurs, the population becomes increasingly more vulnerable to chance occurrences, such as a fire, a rare climatic event, or an outbreak of disease. Ultimately, the population succumbs, and the species becomes extinct.

Some extinctions are very rapid, particularly those induced by human activity. A classic example is that of the passenger pigeon, which was a dominant bird of eastern North America in the late nineteenth century. Flying in huge flocks and feeding on seeds and fruits such as beechnuts and

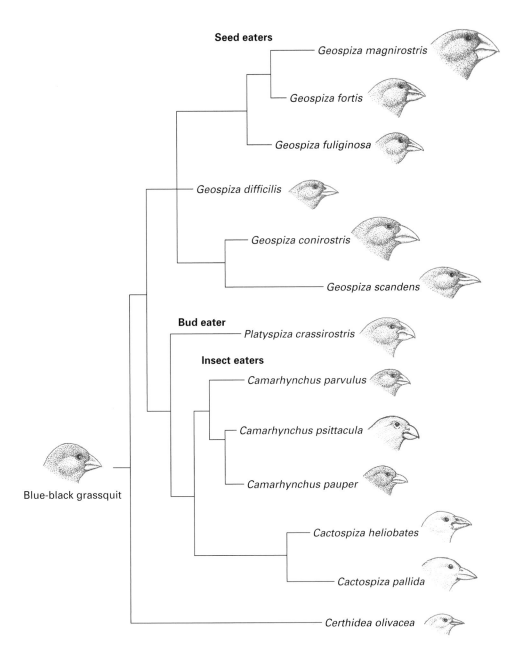

Seed eaters

Geospiza magnirostris

Geospiza fortis

Geospiza fuliginosa

Geospiza difficilis

Geospiza conirostris

Geospiza scandens

Bud eater

Platyspiza crassirostris

Insect eaters

Camarhynchus parvulus

Camarhynchus psittacula

Camarhynchus pauper

Cactospiza heliobates

Cactospiza pallida

Certhidea olivacea

Blue-black grassquit

*8.27 **Adaptive radiation of finch species of the Galápagos Islands*** Five genera and 13 species of finch evolved from a single ancestral population of the blue-black grassquit. Beak shapes are adapted to the primary food source: seeds, buds, and insects. (From J. H. Brown and M. V. Lomolino, Biogeography, second ed., 1998, Sinauer, Sunderland, Massachusetts, used by permission.)

acorns, these birds were easily captured in nets and shipped to markets for food. By 1890, they were virtually gone, and the last known passenger pigeon died in the Cincinnati Zoo in 1914.

Rare but extreme events can also cause extinctions. Many lines of evidence have converged to document that the Earth was struck by an asteroid about 65 million years ago. The impact, which occurred on the continental shelf near the Yucatan Peninsula, raised a global dust cloud that blocked sunlight from the surface, cooling the Earth's climate intensely for a period of perhaps several years. Dinosaurs and many other groups of terrestrial and marine organisms were wiped out. Less affected were organisms that were less sensitive to a brief, but intense, period of cold. These included birds and mammals, which have internal metabolic temperature regulation, as well as seed plants and insects that pass part of their life cycle in a dormant state.

Extinction occurs when all individuals of a species die. If the environment changes too rapidly for a population to adjust by evolution, it dies out. Extreme events, such as the collision of an asteroid with the Earth about 65 million years ago, can cause mass extinctions.

Dispersal

Nearly all types of organisms have some mechanism of **dispersal**—that is, a capacity to move from a location of origin to new sites. Often dispersal is confined to one life stage, as in the dispersal of higher plants as seeds. Even in animals that are inherently mobile, there is often a developmental stage when movement from one site to the next is more likely to occur.

8.28 Cattle egret *This small, white heron migrated from Africa to South America in the late 1880s. It follows grazing animals, such as this Texas steer, and feeds on insects and small invertebrates flushed by the grazing action.*

Normally, dispersal does not change the geographic range of a species. Seeds fall near their sources, and animals seek out nearby habitats to which they are adjusted. Dispersal is thus largely a method for gene flow that helps to encourage the cross-breeding of organisms throughout a population. When land is cleared or new land is formed, dispersal moves colonists into the new environment. We documented this role of dispersal earlier in this chapter as part of succession. Species also disperse by *diffusion*, the slow extension of range from year to year.

A rare, long-distance dispersal event can be very significant in establishing biogeographic patterns. We have already noted how a single ancestral colony of finches invaded the Galápagos Islands and underwent adaptive radiation to populate all of the islands successfully.

Some species have modes of propagation that are especially well-adapted to long-distance dispersal. Mangrove species, which line coastal estuaries in equatorial and tropical regions, have seeds that are carried thousands of miles on winds and ocean currents to populate far distant shores. Another example of a plant well adapted to oceanic dispersal is the coconut palm. Its large seed, housed in a floating husk, has made it a universal occupant of island beaches. Among the animals, birds, bats, and insects are frequent long-distance travelers. Generally, nonflying mammals, freshwater fishes, and amphibians are less likely to make long leaps, with rats and tortoises the exceptions.

The case of the cattle egret demonstrates both long-distance dispersal and diffusion (Figure 8.28). This small heron crossed the Atlantic, arriving in northeastern South America from Africa in the late 1880s. A hundred years later, it had

> **Dispersal is the capacity to move from a location of origin to new sites. In diffusion, species extend their range slowly from year to year. In long-distance dispersal, unlikely events establish breeding populations at remote locations.**

become one of the most abundant herons of the Americas. Figure 8.29 maps its range extension over much of this period. Another example of diffusion is the northward colonization of the British Isles by oaks following the retreat of continental glaciers after the close of the Ice Age. As shown in Figure 8.30, oaks required about 3500 years, from about

8.29 Diffusion of the cattle egret *After long-distance dispersal to northeastern South America, the cattle egret spread to Central and North America as well as to coastal regions of western South America. (From J. H. Brown and M. V. Lomolino, Biogeography, second ed., 1998, Sinauer, Sunderland, Massachusetts, used by permission.)*

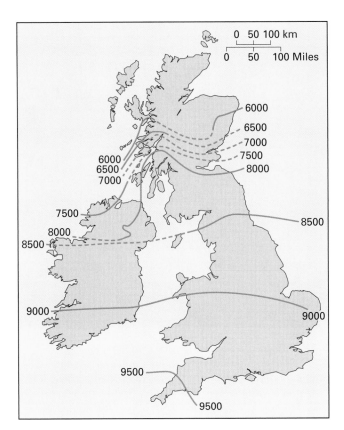

8.30 Diffusion of oaks *Following the retreat and melting of continental glaciers at the close of the Ice Age, oak species diffused northward across the British Isles. Contours indicate northern border at times in years before present. Dashed lines are less certain. (From H. J. B. Birks, J. Biogeography, vol. 16, pp. 503–540. Used by permission.)*

example, the Bering Strait region between Alaska and easternmost Siberia was dry land during the early Cenozoic Era (about 60 million years ago) and during the Ice Age, when sea level dropped by more than 100 m (325 ft). Many plant and animal species of Asia are known to have crossed this bridge and then spread southward into the Americas. One notable migrant species of the last continental glaciation was the aboriginal human, and there is substantial evidence to support the hypothesis that the skilled hunters who crossed the Bering land bridge were responsible for the extinction of many of the large animals, including wooly mammoths and ground sloths, that disappeared from the Americas about 10,000 years ago (Figure 8.31).

Distribution Patterns

The processes of evolution, speciation, extinction, and dispersal have over time produced many spatial patterns of species distribution on today's Earth. One of the simplest patterns is that of the **endemic** species, which is found in one region or location and nowhere else. An endemic distribution can arise in one of two ways—as the result of a contraction of a broader range or as the location of origin of a species that has not spread widely beyond the endemic region. Some endemic species are ancient relics of biological strains that have otherwise gone extinct. An example is the gingko tree

8.31 Wooly mammoth *A reconstruction of the wooly mammoth, a huge tusked mammal that inhabited North America throughout the Ice Age and became extinct about 10,000 years ago, most likely from hunting by prehistoric humans.*

9500 years before present to 6000 years before present, to reach their northern limit.

Dispersal often means surmounting *barriers*—that is, regions a species is unable to colonize or perhaps even occupy for a short period of time. Long-distance dispersal can mean bridging such obvious barriers as an ocean of salt water or an ice sheet by an unlikely accident. But other barriers are not so obvious. For example, the basin and range country of Utah, Nevada, and California presents a sea of desert with islands of forest to some diffusing species. While birds and bats may have no difficulty moving from one island to the next, a small mammal may never cross the desert sea, at least under its own power. In this case, the barrier is one in which the physiological limits of the species are exceeded. But there may be ecological barriers as well—for example, a zone of intense predation or a region occupied by strongly and successfully competing species.

Just as there are barriers to dispersal, so there are corridors that facilitate dispersal. Central America forms a present-day land bridge connecting North and South America, for example. It has been in place for about 3.5 million years. Other corridors of great importance to present-day species distribution patterns have existed in the recent past. For

8.32 Gingko tree *The gingko tree is an eastern Chinese endemic that has survived millions of years with little evolutionary change. Thanks to human activity, it is now much more widely distributed. Shown here in an urban habitat (Brooklyn, New York), the gingko has a distinctive shape with long branches of closely-held leaves (left). The fan-shaped leaves show an uncommon arrangement of parallel veins radiating from the leaf stem (right).*

(Figure 8.32), which was widespread throughout the Mesozoic Era (245 to 60 million years ago) but until recently was restricted to a small region in eastern China. It is now widely planted as an urban street tree, known for its hardiness.

In contrast to endemics are *cosmopolitan* species, which are found very widely. Among these is the human species, *Homo sapiens*. Bats and peregrine falcons are also very widely dispersed, cosmopolitan species. Very small organisms, or organisms with very small propagating forms, are often of

> Cosmopolitan species are found very widely. Endemic species are restricted to a single region or location. Disjunctions occur when closely related species are found in widely separated regions.

cosmopolitan distribution, distributed widely by atmospheric and oceanic circulations.

Another interesting pattern is **disjunction,** in which one or more closely related species are found in widely separated regions. An example is the distribution of the tinamous and flightless ratite birds, which includes such oddities as the ostrich, emu, cassowary, and kiwi (Figure 8.33). This disjunct pattern is thought to result from an ancestral species that was widespread across the ancient continent of Gondwana (see Chapter 12). As plate tectonics split Gond-

8.33 Disjunct distribution *The distribution pattern of the ratite birds and tinamous shows disjunctions resulting from isolation by continental motions. (From J. H. Brown and M. V. Lomolino, Biogeography, second ed., 1998, Sinauer, Sunderland, Massachusetts, used by permission.)*

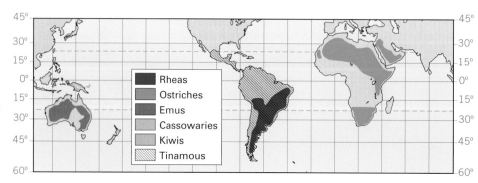

Rheas
Ostriches
Emus
Cassowaries
Kiwis
Tinamous

wana apart into North and South America, Africa, Australia, and New Zealand, isolation and evolution differentiated the ancestral lineage into the diverse array of related species that now inhabit these continents. Another disjunct pattern that is less well understood is the occurrence of closely related species of such common desert plants of the North American Southwest as mesquite, creosote bush, and paloverde in the desert environments of Chile and Argentina.

Biogeographic Regions

As spatial distributions of species are examined on a global scale, certain common patterns emerge. Although closely related species tend to be nearby or to occupy similar regions, larger groups of organisms, such as families and orders, tend to show similar patterns that are often disjunct. For example, the South America–Africa–Australia–New Zealand pattern described for the ratite birds fits the distribution of many other ancient families of plants and animals and is a reflection of common ancestry on Gondwana. This example emphasizes the importance of history in determining these global patterns. Another important determinant of spatial patterns is global climate. Often, members of the same lineage have similar adaptations to environment and so are found in similar climatic regions.

As a result of these factors, biogeographers have been able to define **biogeographic regions** in which the same or closely related plants and animals tend to be found together. That is, when we cross the boundary between two biogeographic regions, we pass from one group of distinctive plants and animals to another. Figure 8.34 shows the major biogeographic regions recognized for plants and animals. Note that many of the boundaries on the two maps are very close, indicating that at the global scale, plants and animals have similar and related histories of evolution and environmental affinity.

> **Biodiversity expresses the variety of biological life on Earth. Human activity has reduced biodiversity by modifying natural habitats and causing extinctions.**

BIODIVERSITY

From the principles of ecological and historical biogeography presented in this chapter, we can see that **biodiversity**—the variety of biological life on Earth—depends both on the variety of the Earth's environments and the processes of evolution, dispersal, and extinction as they have acted through geologic time. At the present time, the Earth's biodiversity is rapidly decreasing as a result of human activity. Our species, *Homo sapiens,* has come to utilize some 20 to 40 percent of global primary productivity as well as exploit 70 percent of its marine fisheries to provide its food. In this process, it has doubled the natural rate of nitrogen fixation, used more than half of the Earth's supply of surface water, and transformed more than 40 percent of the land surface.

Recent human activity has ushered in a wave of extinctions unlike any that has been seen for millions of years. In the last 40 years, biologists have documented the disappearance of several hundred land animal species, including 58 mammals, 115 birds, 100 reptiles, and 64 amphibians. Aquatic species have also been severely impacted, with 40 species or subspecies of freshwater fish lost in North America alone in the last few decades. In the plant kingdom, botanists estimate that over 600 species have become extinct in the past four centuries. Figure 8.35 shows the status of some important groups of plant and animal species of the United States. Many species are extinct or imperiled. Only about a third of the species of crayfishes and freshwater mussels are considered secure, as are about half of the species of amphibians and freshwater fishes.

These documented extinctions may be only the tip of the iceberg. Many species remain as yet undiscovered and so may become extinct without ever being described. Many of these species will likely be important to a better understanding of evolutionary history, if recent experience is any guide, because they will include representatives of new biological divisions of the highest level. In the last decade alone, three new families of flowering plants and two new phyla of animals were described. (The *phylum,* plural *phyla,* is the highest division of higher animal life.) Among animal groups, insects, spiders, and other invertebrates are the most poorly known. Fungi and microbes are other groups in which a large proportion of species have yet to be determined.

How has human activity created extinctions? One way has been to disperse new organisms that outcompete or predate existing organisms. Island populations have especially suffered from this process. Developing in isolation, island species have often not evolved defense mechanisms to protect themselves from predators, including humans. Many islands were subjected to waves of invading species, ranging from rats to weeds, brought first by prehistoric humans and later by explorers and conquerors. Hunting by prehistoric humans alone was sufficient to exterminate many species, not only from islands but from large continental regions as well. Another mechanism is the use of fire. As humans learned to use fire in hunting and to clear and maintain open land, large areas became subject to periodic burning. Habitat alteration and fragmentation is yet another cause of extinctions. By isolating plant and animal populations and altering their environment, human activities cause populations to shrink, making extinction more likely.

Biodiversity is not uniform over the Earth's surface. In general, tropical and equatorial regions have more species and more variation in species composition between different habitats. In areas of isolation, such as islands or mountaintops, species diversity tends to increase with the size of the isolated area and decrease with the degree of isolation from surrounding sources of colonists. Much of this island diversity is contributed by endemic species. Geographic areas in which biodiversity is especially high are referred to as *hotspots.* An important strategy for preservation of global biodiversity is to identify hotspots and take conservation

(a)

8.34 Biogeographic regions *(a) Biogeographic regions of land plants. (After Goode, 1974.) (b) Biogeographic regions of land animals. (After Wallace, 1876.)*

measures to protect them. In this way, conservation efforts can be most effective.

Why is biodiversity important? Nature, operating over millennia, has provided an incredibly rich array of organisms that interact with one another in a seamless web of organic life. Humans, for all their cleverness, are still part of that web. When we cause the extinction of a species, we break a link in the web. Ultimately, the web will thin, with important consequences for both the human species and continued life on Earth. We have no way of knowing which organisms future humans will rely on and how. It seems prudent to keep as many of them around for as long as possible.

A LOOK AHEAD   In this chapter we've focused on the processes that determine the spatial patterns of biota on Earth. We've examined how energy and matter flow in ecosystems and how different ecosystems have different levels of productivity. We have seen how organisms adjust to their environments and how natural selection works in response to environmental pressures. We have also seen how evolution, dispersal, and extinction generate patterns of species distribution and determine biodiversity. Our next chapter takes a more functional view of the life layer by inventorying the global biomes—major divisions of ecosystems that are based largely on the dominant life-form of their vegetation covers.

8.35 Degree of endangerment of plant and animal groups in the United States *These bar graphs show the proportion of species within each group as secure, vulnerable, imperiled, or extinct. (Data from Nature Conservancy.)*

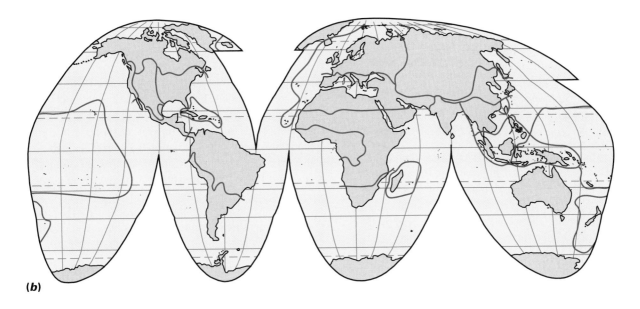

(b)

In Review | Biogeographic Processes

- **Biogeography** focuses on the distribution of plants and animals over the Earth. *Ecological biogeography* examines how relationships between organisms and environment help determine when and where organisms are found. *Historical biogeography* examines how, where, and when species have evolved and how they are distributed over longer times and broader scales.

- **Ecology** is the science of interactions among organisms and their environment. Its focus is the **ecosystem,** which by interaction among components provides pathways for flows of energy and cycles of matter. The **food web** of an ecosystem details how food energy flows from **primary producers** through **consumers** and on to **decomposers.** Because energy is lost at each level, only a relatively few top-level consumers are normally present.

- **Photosynthesis** is the production of carbohydrate from water, carbon dioxide, and light energy by primary producers. **Respiration** is the opposite process, in which carbohydrate is broken down into carbon dioxide and water to yield chemical energy and thus power organisms. **Net photosynthesis** is the amount of carbohydrate remaining after respiration has reduced **gross photosynthesis.** Net photosynthesis increases with increasing light and temperature, up to a point.

- **Biomass** is the accumulated net production of an ecosystem. Forests and wetlands are ecosystems with high rates of *net primary production,* while grasslands and agricultural lands are generally lower. Oceans are most productive in coastal and upwelling zones near continents. Among climate types, those with abundant rainfall and warm temperatures are most productive.

- Biomass is an attractive form of solar-powered energy. Charcoal, *biogas,* and alcohol are biomass products that can be used as fuels.

- **Biogeochemical cycles** consist of *active pools* and *storage pools* linked by flow paths. The **carbon cycle** includes an active pool of biospheric carbon and atmospheric atmospheric CO_2, with a large storage pool of carbonate in sediments. Human activities have provided a pathway from storage to active pools by the burning of fossil fuel. The oxygen cycle features an active pool of atmospheric CO_2, which is increased by photosynthesis and reduced by respiration, combustion, and mineral oxidation.

- A *community* of organisms occupies a particular environment, or **habitat.** *Associations* are often defined by characteristic species or vegetation life-forms. Environmental actors influencing the distribution patterns of organisms include moisture, temperature, light, and wind.

- Organisms require water to live, and so they are limited by the availability of water. **Xerophytes** are adapted to arid habitats. They reduce water loss by having waxy leaves, spines instead of leaves, or no leaves at all. *Phreatophytes* have deep roots; *succulents* store water in spongy tissues. *Tropophytes* are deciduous during the dry or cold season. *Sclerophylls* have thick, leathery leaves that resist drying during the Mediterranean summer climate. **Xeric animals** include vertebrates that are nocturnal and have various adaptations to conserve water. Invertebrates such as brine shrimp can adjust their life cycle to prolonged drought.

- Temperature acts on plants to trigger and control stages of plant growth as well as to limit growth at temperature extremes. Survival below freezing requires special adaptations, and so only a small proportion of plants are frost tolerant. *Cold-blooded animals* have body temperatures that follow the environment, but they can moderate these temperatures by seeking warm or cool places. Mammals and birds are *warm-blooded animals* that maintain constant internal temperatures through a variety of adaptation mechanisms.

- The light available to a plant depends on its position in the structure of the community. Duration and intensity of light vary with latitude and season and serve as a cue to initiate growth stages in many plants. The day–night cycle regulates much of animal behavior, as does *photoperiod* (daylight length). Wind deforms plant growth by dessicating buds and young growth on the windward side of the plant.

- A bioclimatic frontier marks the potential distribution boundary of a species. Other factors may also limit the distribution of a species.

- Geomorphic factors of slope steepness and orientation affect both the moisture and temperature environment of the habitat and serve to differentiate the microclimate of each community. Soil, or *edaphic, factors* can also limit the distribution patterns of organisms or affect community composition.

- Species interact in a number of ways, including **competition,** *predation* and *parasitism,* and *herbivory.* The population explosion and crash of the Kaibab Plateau deer herd provide an example of the beneficial nature of some predator–prey relationships. In *allelopathy,* plant species literally poison the soil environment against competing species. Positive (beneficial) interactions between species are termed **symbiosis.**

- **Ecological succession** comes about as ecosystems change in predictable ways through time. A series of stable communities follows a *sere* to a **climax.** *Primary succession* occurs on new soil substrate, while *secondary succession* occurs on disturbed habitats. Succession on coastal sand dunes follows a series of stages from dune grass to an oak and holly forest. *Old-field succession,* which occurs on abandoned farmland, also leads to a deciduous forest climax. Although succession is a natural tendency for ecosystems to change with time, it is opposed by natural disturbances and limited by local environmental conditions.

- *Historical biogeography* focuses on evolution, speciation, extinction, and dispersal as they influence the distribution patterns of species.

- Life has attained its astonishing diversity through **evolution.** In this process, **natural selection** acts on variation to produce populations that are progressively better adjusted to their environment. Variation arises from *mutation* and *recombination.*

- A **species** is best defined as a population of organisms that are capable of interbreeding successfully, but instead it is usually defined by a typical *morphology* or *phenotype.*

- **Speciation** is the process by which species are differentiated and maintained. It includes mutation, natural selection, *genetic drift,* and *gene flow.* **Geographic isolation** acts to isolate subpopulations of a species, allowing genetic divergence and speciation to occur. The finches of the Galápagos Islands provide an example of *allopatric speciation* by geographic isolation. In *sympatric speciation,* adaptive pressures force a breeding population to separate into different subpopulations that may become species. Sympatric speciation of plants has included **polyploidy,** which is an important mechanism of evolution for higher plants.

- **Extinction** occurs when populations become very small and thus are vulnerable to chance occurrences of fire, disease, or climate anomaly. Rare but very extreme events can cause mass extinctions. An example is the asteroid impact that the Earth suffered about 65 million years ago. The global dust cloud that lingered for several years caused extreme cold temperatures, which wiped out many important lineages of plants and animals.

- Species expand and contract their ranges by **dispersal.** Plants are generally dispersed by seeds, while animals often disperse under their own power. Since most dispersal happens within the range of a species, it acts primarily to encourage gene flow between subpopulations. Long-distance dispersal, though very rare, may still be very important in establishing biogeographic patterns. *Barriers,* often climatic or topographic, inhibit dispersal and induce geographic isolation. Corridors provide pathways that facilitate dispersal.

- **Endemic** species are found in one region or location and nowhere else. They arise by either a contraction of the range of a species or by a recent speciation event. *Cosmopolitan* species are widely dispersed and nearly universal. **Disjunction** occurs when one or more closely related species appear in widely separated regions.

- **Biogeographic regions** capture patterns of occurrence in which the same or closely related plants and animals tend to be found together. They result because their species have common histories and similar environmental preferences.

- **Biodiversity** is rapidly decreasing as human activity progressively affects the Earth. Extinction rates for many groups of plants and animals are as high or higher today than they have been at any time in the past. Humans act to disperse predators, parasites, and competitors widely, disrupting long-established evolutionary adjustments of species with their environments. Hunting and burning have exterminated many species. Habitat alteration and fragmentation also produce extinctions. Preservation of global biodiversity includes a strategy of protecting *hotspots* where diversity is greatest.

Key Terms

biogeography, p. 270
ecology, p. 270
ecosystem, p. 271
food chain, p. 271
food web, p. 271
primary producers, p. 271
consumers, p. 271
decomposers, p. 271
photosynthesis, p. 272
respiration, p. 272
gross photosynthesis,
    p. 272

net photosynthesis, p. 272
biomass, p. 273
biogeochemical cycle,
    p. 280
pool, p. 280
carbon cycle, p. 280
ecological biogeography,
    p. 281
habitat, p. 281
xerophytes, p. 284
xeric animals, p. 285
bioclimatic frontier, p. 288

competition, p. 291
symbiosis, p. 292
ecological succession,
    p. 292
climax, p. 292
historical biogeography,
    p. 295
evolution, p. 295
natural selection, p. 295
species, p. 296
speciation, p. 298

geographic isolation,
    p. 298
polyploidy, p. 298
extinction, p. 298
dispersal, p. 299
endemic, p. 301
disjunction, p. 302
biogeographic region,
    p. 303
biodiversity, p. 303

Review Questions

1. Define the terms *biogeography, ecology,* and *ecosystem.*
2. What is a food web or food chain? What are its essential components? How does energy flow through the food web of an ecosystem?
3. Compare and contrast the processes of photosynthesis and respiration. What classes of organisms are associated with each?
4. How is net primary production related to biomass? Identify some types of terrestrial ecosystems that have a high rate of net primary production and some with a low rate.
5. Which areas of oceans and land are associated with high net primary productivity? How is net primary production on land related to climate?
6. Why is biomass energy a desirable energy source? Identify and describe the most useful forms of biomass energy as fuel.
7. What is a *biogeochemical cycle?* What are its essential features?
8. What are the essential features and flow pathways of the carbon cycle? How have human activities impacted the carbon cycle?
9. What is a *habitat?* What are some of the characteristics that differentiate habitats? Compare habitat with *niche.*
10. Contrast the terms *ecosystem, community,* and *association.*
11. Although water is a necessity for terrestrial life, many organisms have adapted to arid environments. Describe some of the adaptations that plants and animals have evolved to cope with the desert.
12. Terrestrial temperatures vary widely. How does the annual variation in temperature influence plant growth, development, and distribution? How do animals cope with variation in temperature?
13. How does the ecological factor of light affect plants and animals? How does wind affect plants?
14. What is meant by the term *bioclimatic frontier?* Provide an example.
15. How do geomorphic and edaphic factors influence the habitat of a community?

16. Contrast the terms used to describe interactions among species. Provide an example of positive predation.
17. Describe ecological succession using the terms *sere, seral stage, pioneer,* and *climax.* Use dune succession as an example.
18. How do primary succession and secondary succession differ? Describe old-field succession as an example of secondary succession.
19. How does the pattern of ecosystems on the landscape reflect an equilibrium?
20. Explain Darwin's theory of evolution by means of natural selection. What key point was Darwin unable to explain?
21. What two sources of variation act to differentiate offspring from parents?
22. What is *speciation?* Identify and describe four component processes of speciation.
23. What is the effect of geographic isolation on speciation? Provide an example of allopatric speciation.
24. How does sympatric speciation differ from allopatric speciation? Provide an example of sympatric speciation.
25. What is *extinction?* Provide some examples of extinctions of species.
26. Describe the process of dispersal. Provide a few examples of plants and animals suited to long-distance dispersal.
27. Contrast barriers and corridors in the dispersal process.
28. How does an endemic distribution pattern differ from a cosmopolitan pattern? What is a *disjunction?*
29. How are biogeographic regions differentiated?
30. What is biodiversity? How has human activity impacted biodiversity?

**Eye on the Environment 8.3 • The Great Yellowstone
Fire**

1. Describe the great fire that struck Yellowstone National Park in the summer of 1988. How much of the park actually burned? What types of habitats were affected?
2. What is the natural fire succession process for the park?
3. Would a "no-burn" policy have stopped the fires? Did 100 years of no-burn policy contribute significantly to the fires of 1988?

Visualizing Exercises

1. Sketch a graph showing the relationship among gross photosynthesis, net photosynthesis, respiration, and temperature. How is net photosynthesis obtained from gross photosynthesis and respiration for each temperature?

2. Diagram the general features of a biogeochemical cycle in which storage pools and active pools are linked by life processes and physical processes.

3. Draw a timeline illustrating either dune succession or old-field succession. Between the stages, indicate the environmental changes that occur.

4. Carefully compare parts *(a)* and *(b)* of Figure 8.34 Which boundaries are similar, and which are different? Speculate on possible reasons for the similarities and differences.

Essay Questions

1. Suppose atmospheric carbon dioxide concentration doubles. What will be the effect on the carbon cycle? How will flows change? Which pools will increase? decrease?

2. Select three distinctive habitats for plants and animals that occur nearby and with which you are familiar. Organize a field trip (real or imaginary) to visit the habitats. Compare their physical environments and describe the basic characteristics of the ecosystems found there.

3. Imagine yourself as a biogeographer discovering a new group of islands. Select a global location for your island group, including climate and proximity to nearby continents or land masses. What types of organisms would you expect to find within your island group and why?

Eye on the Landscape

Chapter Opener Giant tortoises of the Galápagos Islands Clouds and fog **(A)** ring the edge of the crater, induced by orographic lifting of warm moist air as the trade winds cross the islands. The Galápagos Islands, situated on the Equator, have two distinct seasons. In the warm and wet season (December to May), southeast trade winds bring conditions that are often clear but with occasional heavy showers. In the dry or misty season (May to December), northeast trade winds bring frequent clouds with mist or fog at higher elevations, but rainfall is lower. Note also the shape of the crater's rim **(B)**, with its uniform height and gentle curving circular shape, seen here from the perspective of a wide-angle photo. Craters are formed when the underlying magma in the central vent of a volcano is withdrawn and the surface volcanic deposits subside.

8.17 Grazing caribou Notice the "blobs" of soil on this slope **(A)**, distinguished by color and shape. These are solifluction lobes, produced in the short arctic summer when the snow melts and the surface layer of soil thaws. The saturated soil creeps downhill in lobes until it dries out sufficiently and its motion stops. Chapter 14 describes this process as well as other surface processes characteristic of the arctic environment. Also note the snow patches **(B)**. In some years, these will persist through the summer.

8.21 Slope orientation and habitat The vertical faces of these cliffs **(A)** and pinnacles **(B)** are most likely joint planes—planes of fractures in the rock resulting from cooling or from release of the pressure of overlying rock removed by erosion. Joints are discussed in Chapter 14.

A train streaks across the Mojave Desert at sunset. What else would the geographer see? ... Answers at the end of the chapter.

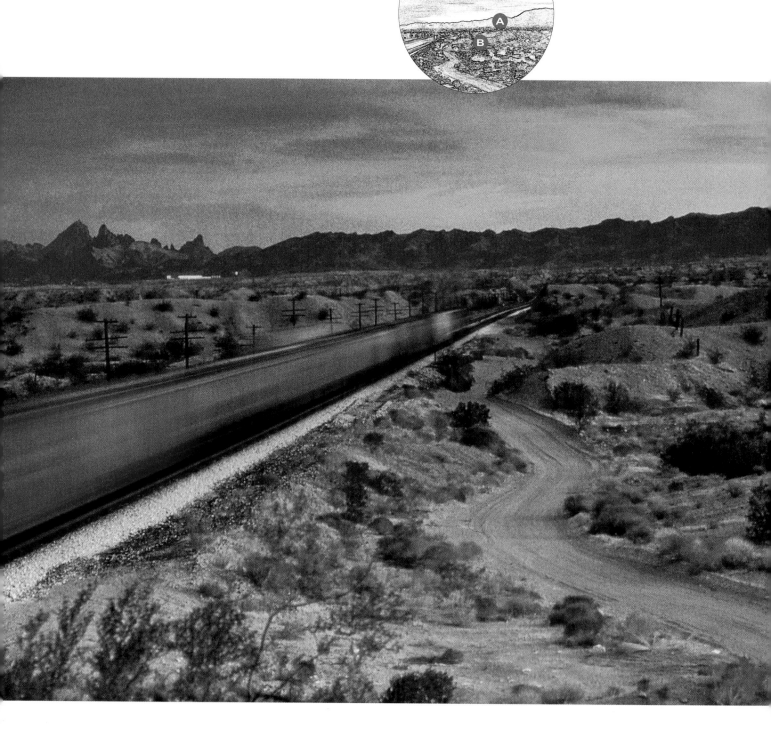

9 | GLOBAL BIOGEOGRAPHY

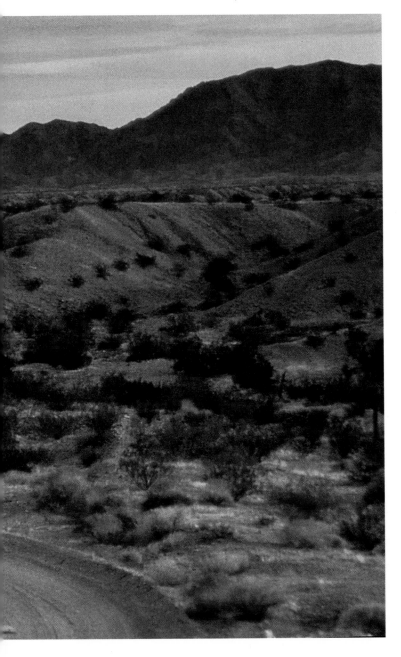

PUTTING YOU IN THE PICTURE

We've seen how the processes of ecological and historical biogeography can influence the patterns of plant and animal species on the landscape. But how do these processes actually play out at the global scale? This chapter surveys the characteristics and distribution patterns of the major units of biotic life on land.

As an introduction to the scale of these patterns, imagine yourself on a train trip across the American West, sampling a transect of the many diverse biogeographic landscapes of this unique region…

All aboard! The conductor's cry resounds through Union Station, Los Angeles, and at precisely 11:50 A.M., your train begins to move. At last your journey is underway! You're headed for Chicago on a three-day train ride that will take you through some of the most breathtaking scenery in America.

An hour later, you are relaxing in the comfortable armchair of your roomette, gazing out the large picture window. The Mojave Desert flashes by. The strange forms of Joshua trees contrast against the gray and tan colors of the desert earth. Overhead a bird with a large wingspan glides effortlessly. You wonder if it could be a California condor, far from its coastal habitat. The coach rushes across a dry lakebed of white salt sparkling in the Sun. Sharp peaks loom in the far distance, their angular foothills terminating the broad plains of grayish creosote bush that sweep toward the mountain fronts.

Early that evening, the train pulls out of Las Vegas, headed up toward the high desert of the Colorado Plateau. As the Sun slowly sets, the reds and browns of the rocks and soils intensify. Soon the desert sky darkens to a blue-black. Sagebrush carpets the landscape, looking soft and inviting in the moonlight. Reluctantly, you draw the shade and turn in for the night.

"More coffee?" The steward expertly refills your breakfast coffee cup, swaying with the motion of the dining car. During the night, the train has slowly gained elevation to enter the rugged upper reaches of the Colorado Basin. Twisted crowns of piñon pines and junipers dot the hills. A frightened squirrel gathering pine nuts leaps away quickly. Gray-green sagebrush blankets the lowlands, with green, willowy salt-bush lining the dry creek beds.

By early afternoon, you are well into your crossing of the Rocky Mountains. The track follows the narrow canyon of the upper Colorado. Steep, conifer-clad slopes enclose the rushing train. Soon you are in the high country. Ragged, windswept peaks jut skyward, spread with a green and white quilt of snowfields, conifer forests, and aspen groves. You glimpse a pair of mule deer, half hidden in a small stand of firs. Suddenly the train plunges into the Moffat Tunnel for the long passage under the continental divide. Emerging on the east side, you descend steeply toward the high plains and the skyscrapers of Denver.

Later that evening, you sit alone in your darkened compartment, gazing out the window. The short-grass prairie sweeps by, its moonlit surface like a table spread with faded lace. A prairie dog town flashes past, with burrows like blisters on the brown soil. Soon, the rocking motion of the train lulls you to sleep.

Noises in the station at Omaha awaken you early in the morning. After breakfast, you watch Iowa roll by. Maturing corn and soybean crops form a dense, green blanket on the Earth. Cattle graze stubbly green fields where hay was harvested earlier in the season. As the train heads eastward, the density of settlement increases. By late afternoon, you reach the Chicago area, and precisely at 4:15 P.M., your train pulls into the station. Regretfully, you hoist your suitcase and step off the train. Your trip is over, but you will long remember the magnificent landscapes and scenic vistas of the American West. ∎

GLOBAL BIOGEOGRAPHY

Our last chapter focused on biogeography, examining the principles and processes that determine the distribution of plants and animals on the Earth's land surface. In this chapter, we survey the broad global distribution patterns of ecosystems, emphasizing the characteristics of the vegetation cover. We emphasize the vegetation because it is a visible and obvious part of the landscape and because humans depend on plants for many life essentials, including food, medicines, fuel, clothing, and shelter. Also, the largest division of ecosystems—the biome—is defined primarily on the characteristics of the vegetation cover. Note that vegetation is often strongly related to climate, so we will find that the occurrence of major biomes coincides in many cases with broad climate types. Thus, your knowledge of climate from Chapter 7 will come in handy as you study the world's vegetation. GEODISCOVERIES

NATURAL VEGETATION

Over the last few thousand years, human societies have come to dominate much of the land area of our planet. In many regions, humans have changed the natural vegetation—sometimes drastically, other times subtly. What do we mean by natural vegetation? **Natural vegetation** is a plant cover that develops with little or no human interference. It is subject to natural forces of modification and destruction, such as storms or fires. Natural vegetation can still be seen over vast areas of the wet equatorial climate, although the rainforests there are being rapidly cleared. Much of the arctic tundra and the boreal forest of the subarctic zones is in a natural state.

In contrast to natural vegetation is *human-influenced vegetation*, which is modified by human activities. Much of the land surface in midlatitudes is totally under human control, through intensive agriculture, grazing, or urbanization. Some areas of natural vegetation appear to be untouched but are actually dominated by human activity in a subtle manner. For example, most national parks and national forests have been protected from fire for many decades. When lightning starts a forest fire, the firefighters put out the flames as fast as possible. However, periodic burning is part of the natural cycle in many regions. One vital function of fire is to release nutrients that are stored in plant tissues. When the vegetation burns, the ashes containing the nutrients remain. These enrich the soil for the next cycle of vegetation cover. In recent years, managers of some parks and forests have stopped suppressing wildfires, allowing the return of more natural periodic burning. *(See Eye on the Environment 8.3 • The Great Yellowstone Fire.)*

Our species has influenced vegetation in yet another way—by moving plant species from their original habitats to

---

Most areas of the Earth's land surface are influenced subtly or strongly by human activity. Human influence includes such factors as clearing for agriculture and grazing, fire suppression, and introduction of alien plants and animals.

---

foreign lands and foreign environments. The eucalyptus tree is a striking example. From Australia, various species of eucalyptus have been transplanted to such far-off lands as California, North Africa, and India. Sometimes exported plants thrive like weeds, forcing out natural species and becoming a major nuisance. Few of the grasses that clothe the coastal ranges of California are native species, yet a casual observer might think that these represent native vegetation.

Even so, all plants have limited tolerance to the environmental conditions of soil water, heat and cold, and soil nutrients. Consequently, the structure and outward appearance of the plant cover conforms to basic environmental controls, and each vegetation type is associated with a characteristic geographical region—whether forest, grassland, or desert.

STRUCTURE AND LIFE-FORM OF PLANTS

Plants come in many types, shapes, and sizes. Botanists recognize and classify plants by species. However, the plant geographer is less concerned with individual species and more concerned with the plant cover as a whole. In describing the plant cover, plant geographers refer to the **life-form** of the plant—its physical structure, size, and shape. Although the life-forms go by common names and are well understood by almost everyone, we will review them to establish a uniform set of meanings.

Both trees and shrubs are erect, woody plants (Figure 9.1). They are *perennial,* meaning that their woody tissues endure from year to year. Most have life spans of many years. *Trees* are large, woody perennial plants having a single upright main trunk, often with few branches in the lower part but branching in the upper part to form a crown. *Shrubs* are woody perennial plants that have several stems branching from a base near the soil surface, so as to place the mass of foliage close to ground level.

Lianas are also woody plants, but they take the form of vines supported on trees and shrubs. Lianas include not only the tall, heavy vines of the wet equatorial and tropical rainforests, but also some woody vines of midlatitude forests. Poison ivy and the tree-climbing form of poison oak are familiar North American examples of lianas.

Herbs comprise a major class of plant life-forms. They lack woody stems and so are usually small, tender plants. They occur in a wide range of shapes and leaf types. Some are *annuals,* living only for a single season—while others are perennials—living for multiple seasons. Some herbs are broad-leaved, and others are narrow-leaved, such as grasses. Herbs as a class share few characteristics in common except that they usually form a low layer as compared with shrubs and trees.

Forest is a vegetation structure in which trees grow close together. Crowns are in contact, so that the foliage largely shades the ground. Many forests in moist climates show at least three layers of life-forms (Figure 9.1). Tree crowns form the uppermost layer, shrubs an intermediate layer, and herbs a lower layer. There is sometimes a fourth, lowermost layer that consists of mosses and related very small plants. In **woodland,** crowns of trees are separated by open areas that usually have a low herb or shrub layer.

Lichens are another life-form seen in a layer close to the ground. They are plant forms in which algae and fungi live together to form a single plant structure. In some alpine and arctic environments, lichens grow in profusion and dominate the vegetation (Figure 9.2).

TERRESTRIAL ECOSYSTEMS— THE BIOMES

From the viewpoint of human use, ecosystems are natural resource systems. Food, fiber, fuel, and structural material are products of ecosystems and are manufactured by organisms using energy derived from the Sun. Humans harvest that energy by using ecosystem products. The products and

9.1 *Layers of a beech-maple-hemlock forest* *In this schematic diagram, the vertical dimensions of the lower layers are greatly exaggerated. (After P. Dansereau.)*

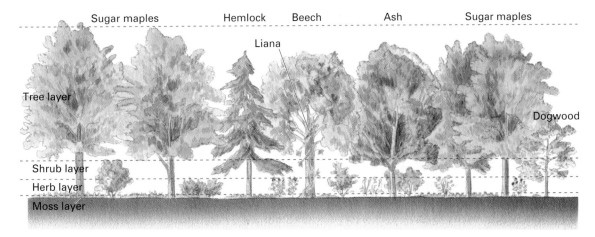

9.2 Lichen *Reindeer moss, a white variety of lichen, seen here on rocky tundra of Alaska.*

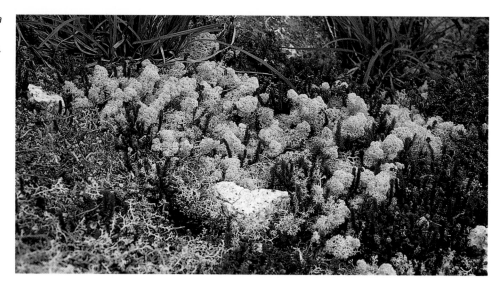

productivity of ecosystems depend to a large degree on climate. Where temperature and rainfall cycles permit, ecosystems provide a rich bounty for human use. Where temperature or rainfall cycles restrict ecosystems, human activities can also be limited. Of course, humans, too, are part of the ecosystem. A persistent theme of human geography is the study of how human societies function within ecosystems, utilizing ecosystem resources and modifying ecosystems for human benefit.

Ecosystems fall into two major groups—aquatic and terrestrial. *Aquatic ecosystems* include marine environments and the freshwater environments of the lands. Marine ecosystems include the open ocean, coastal estuaries, and coral reefs. Freshwater ecosystems include lakes, ponds, streams, marshes, and bogs. Our survey of physical geography will not include these aquatic environments. Instead, we will focus on the **terrestrial ecosystems,** which are dominated by land plants spread widely over the upland surfaces of the continents. The terrestrial ecosystems are directly influenced by climate and interact with the soil. In this way, they are closely woven into the fabric of physical geography.

Within terrestrial ecosystems, the largest recognizable subdivision is the **biome.** Although the biome includes the total assemblage of plant and animal life interacting within the life layer, the green plants dominate the biome physically because of their enormous biomass, as compared with that of other organisms. Plant geographers concentrate on the characteristic life-form of the green plants within the biome. These life-forms are principally trees, shrubs, lianas, and herbs, but other life-forms are important in certain biomes.

There are five principal biomes. The **forest biome** is dominated by trees, which form a closed or nearly closed canopy. Forest requires an abundance of soil water, so forests are found in moist climates. Temperatures must also be suitable, requiring at least a warm season, if not warm temperatures the year round. The **savanna biome** is transitional between forest and grassland. It exhibits an open cover of trees with grasses and herbs underneath. The **grassland biome** develops in regions with moderate shortages of soil water. The semiarid regions of the dry tropical, dry subtropical, and dry midlatitude climates are the home of the grassland biome. Temperatures must also provide adequate warmth during the growing season. The **desert biome** includes organisms that can survive a moderate to severe water shortage for most, if not all, of the year. Temperatures can range from very hot to cool. Plants are often xerophytes, showing adaptations to the dry environment.

The **tundra biome** is limited by cold temperatures. Only small plants that can grow quickly when temperatures warm above freezing in the warmest month or two can survive.

Biogeographers break the biomes down further into smaller vegetation units, called *formation classes,* using the life-form of the plants. For example, at least four and perhaps as many as six kinds of forests are easily distinguished within the forest biome. At least three kinds of grasslands are easily recognizable. Deserts, too, span a wide range in terms of the abundance and life-form of plants. The formation classes described in the remaining portion of this chapter are major, widespread types that are clearly associated with specific climate types. Figure 9.3 is a generalized world map of the formation classes. It simplifies the very complex patterns of natural vegetation to show large uniform regions in which a given formation class might be expected to occur. *GEODISCOVERIES*

Biogeographers recognize five principal biomes: forest, grassland, savanna, desert, and tundra. Formation classes are subdivisions of biomes based on vegetation structure and life-form.

Forest Biome

Within the forest biome, we can recognize six major formations: low-latitude rainforest, monsoon forest, subtropical evergreen forest, midlatitude deciduous forest, needleleaf forest, and sclerophyll forest.

LOW-LATITUDE RAINFOREST *Low-latitude rainforest,* found in the equatorial and tropical latitude zones, consists of tall, closely set trees. Crowns form a continuous canopy of foliage and provide dense shade for the ground and lower layers (Figure 9.4). The trees are characteristically smooth-barked and unbranched in the lower two-thirds. Tree leaves are large and evergreen—thus, the equatorial rainforest is often described as *broadleaf evergreen forest.*

Crowns of the trees of the low-latitude rainforest tend to form two or three layers (Figure 9.5). The highest layer consists of scattered "emergent" crowns that protrude from the closed canopy below, often rising to 40 m (130 ft). Some emergent species develop wide buttress roots, which aid in their physical support (Figure 9.6). Below the layer of emergents is a second, continuous layer, which is 15 to 30 m (about 50 to 100 ft) high. A third, lower layer consists of small, slender trees 5 to 15 m (about 15 to 50 ft) high with narrow crowns.

Typical of the low-latitude rainforest are thick, woody lianas supported by the trunks and branches of trees. Some are slender, like ropes, while others reach thicknesses of 20 cm (8 in.). They climb high into the trees to the upper canopy, where light is available, and develop numerous branches of their own. *Epiphytes* ("air plants") are also common in low-latitude rainforest. These plants attach themselves to the trunk, branches, or foliage of trees and lianas. Their host is used solely as a means of physical support. Epiphytes include plants of many different types—ferns, orchids, mosses, and lichens (Figure 9.7).

A particularly important characteristic of the low-latitude rainforest is the large number of species of trees that coexist. In equatorial regions of rainforest, as many as 3000 species may be found in a few square kilometers. Individuals of a given species are often widely separated. Many species of plants and animals in this very diverse ecosystem still have not been identified or named by biologists.

Equatorial and tropical rainforests are not jungles of impenetrable plants thickets. Rather, the floor of the low-latitude rainforest is usually so densely shaded that plant foliage is sparse close to the ground. This gives the forest an open aspect, making it easy to travel within its interior. The ground surface is covered only by a thin litter of leaves. Dead plant matter rapidly decomposes because the warm temperatures and abundant moisture promote its breakdown by bacteria. Nutrients released by decay are quickly absorbed by roots. As a result, the soil is low in organic matter.

> Low-latitude rainforests are very diverse, containing large numbers of plant and animal species. Broadleaf evergreen trees dominate the vegetation cover. The rainforest climate is wet year round or has a short dry season.

Large herbivores are uncommon in the low-latitude rainforest. They include the African okapi and the tapir of South America and Asia. Most herbivores are climbers, and they include many primates—monkeys and apes. Toucans, parrots, and tinamous join fruit-eating bats as flying grazers of the forest. Tree sloths spend their lifetimes hanging upside down as they browse the forest canopy. There are few large predators. Notable are the leopards of African and Asian forests and the jaguars and ocelots of the South American forests.

Low-latitude rainforest develops in a climate that is continuously warm, frost-free, and has abundant precipitation in all months of the year (or, at most, has only one or two dry months). These conditions occur in the wet equatorial climate ① and the monsoon and trade-wind coastal climate ②. In the absence of a cold or dry season, plant growth goes on continuously throughout the year. In this uniform environment, some plant species grow new leaves continuously, shedding old ones continuously as well. Still other plant species shed their leaves according to their own seasons, responding to the slight changes in daylight length that occur with the seasons.

World distribution of the low-latitude rainforest is shown in Figure 9.8. A large area of rainforest lies astride the equator. In South America, this equatorial rainforest includes the Amazon lowland. In Africa, the *equatorial rainforest* is found in the Congo lowland and in a coastal zone extending westward from Nigeria to Guinea. In Indonesia, the island of Sumatra, on the west, bounds a region of equatorial rainforest that stretches eastward to the islands of the western Pacific.

The low-latitude rainforest extends poleward through the tropical zone (lat. 10° to 25° N and S) along monsoon and trade-wind coasts. The monsoon and trade-wind coastal climate in which this *tropical-zone rainforest* thrives has a short dry season. However, the dry season is not intense enough to deplete soil water.

In the northern hemisphere, trade-wind coasts that have rainforests are found in the Philippine Islands and also along the eastern coasts of Central America and the West Indies. These highlands receive abundant orographic rainfall in the belt of the trade winds. A good example of tropical-zone rainforest is the rainforest of the eastern mountains of Puerto Rico. In Southeast Asia, tropical-zone rainforest is extensive in monsoon coastal zones and highlands that have heavy rainfall and a very short dry season. These occur in Vietnam and Laos, southeastern China, and on the western coasts of India and Myanmar. In the southern hemisphere, belts of tropical-zone rainforest extend down the eastern Brazilian coast, the Madagascar coast, and the coast of northeastern Australia.

Within the regions of low-latitude rainforest are many island-like highland regions where climate is cooler and

9.3   *Natural vegetation of the world*

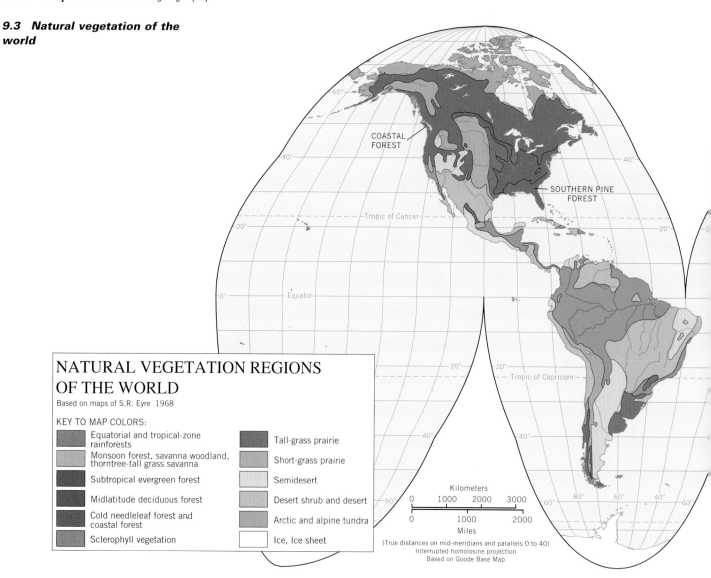

NATURAL VEGETATION REGIONS
OF THE WORLD
Based on maps of S.R. Eyre 1968

KEY TO MAP COLORS:

- Equatorial and tropical-zone rainforests
- Monsoon forest, savanna woodland, thorntree-tall grass savanna
- Subtropical evergreen forest
- Midlatitude deciduous forest
- Cold needleleaf forest and coastal forest
- Sclerophyll vegetation
- Tall-grass prairie
- Short-grass prairie
- Semidesert
- Desert shrub and desert
- Arctic and alpine tundra
- Ice, Ice sheet

Kilometers
0   1000   2000   3000

Miles
0   1000   2000

(True distances on mid-meridians and parallels 0 to 40)
Interrupted homolosine projection
Based on Goode Base Map

9.4   *Aerial view of low-latitude rainforest* *Middle Mazaruni River near Bartica, Guyana.*

9.5 Rainforest layers
This diagram shows the
typical structure of
equatorial rainforest. (After
J. S. Beard, The Natural
Vegetation of Trinidad,
Clarendon Press, Oxford.)

9.6 Buttress roots at the base of a large rainforest tree *From Daintree National Park, Queensland, Australia.*

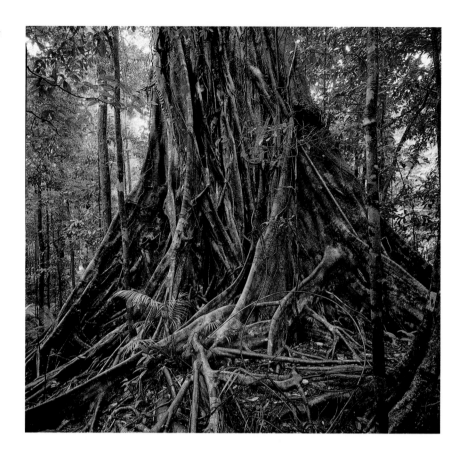

9.7 Epiphytes *In this photo from the El Yunque rainforest, Caribbean National Forest, Puerto Rico, red-flowering epiphytes adorn the trunks of sierra palms.*

rainfall is increased by the orographic effect. Here, rainforest extends upward on the rising mountain slopes. Between 1000 and 2000 m (about 3300 and 6500 ft), the rainforest gradually changes in structure and becomes *montane forest* (Figure 9.8). The canopy of montane forest is more open, and tree heights are lower, than in the rainforest. With increasing elevation, the forest canopy height becomes even lower. Tree ferns and bamboos are numerous, and epiphytes are particularly abundant. As elevation increases, mist and fog become persistent, giving high-elevation montane forest the name *cloud forest.*

Low-latitude rainforest is under increasing human pressure. Slowly but surely, the rainforest is being conquered by logging, clearcutting, and conversion to grazing and farming. *Eye on the Environment 9.1 • Exploitation of the Low-Latitude Rainforest Ecosystem* describes this process in more detail.

MONSOON FOREST *Monsoon forest* of the tropical latitude zone differs from tropical rainforest because it is deciduous, with most of the trees of the monsoon forest shedding their leaves during the dry season. Shedding of leaves results from the stress of a long dry season that occurs at the time of low Sun and cooler temperatures. In the dry season, the forest resembles the deciduous forests of the midlatitudes during their leafless winter season.

Monsoon forest is typically open. It grades into woodland, with open areas occupied by shrubs and grasses (Figure 9.9). Because of its open nature, light easily reaches the lower layers of the monsoon forest. As a result, these lower layers are better developed than in the rainforest. Tree heights are also lower. Typically, many tree species are present—as many as 30 to 40 species in a small tract—although the rainforest has many more. Tree trunks are massive, often with thick, rough bark. Branching starts at a comparatively low level and produces large, round crowns.

Figure 9.10 is a world map of the monsoon forest and closely related formation classes. Monsoon forest develops in the wet-dry tropical climate ③ in which a long rainy season alternates with a dry, rather cool season. These conditions, though most strongly developed in the Asiatic monsoon climate, are not limited to that area. The typical regions of monsoon forest are in Myanmar, Thailand, and Cambodia. In the monsoon forest of southern Asia, the teakwood tree was once abundant and was widely exported to the Western world to make furniture, paneling, and decking. Now this great tree is logged out, and the Indian elephant, once trained to carry out this logging work, is unemployed. Large areas of monsoon forest also occur in south-central Africa and in Central and South America, bordering the equatorial and tropical rainforests.

> **Monsoon forest consists of an open cover of deciduous trees that shed their leaves during the pronounced dry season. It is found in regions of wet-dry tropical climate ③ and ranges from South America to Africa and southern Asia.**

SUBTROPICAL EVERGREEN FOREST *Subtropical evergreen forest* is generally found in regions of moist subtropical climate ⑥, where winters are mild and there is ample rainfall throughout the year. This forest occurs in two forms: broadleaf and needleleaf. The *subtropical broadleaf evergreen forest* differs from the low-latitude rainforests, which are also broadleaf evergreen types, in having relatively few species of trees. Trees are not as tall as in the low-latitude rainforests. Their leaves tend to be smaller and more leathery, and the leaf canopy less dense. The subtropical broadleaf evergreen forest often has a well-developed lower layer of vegetation. Depending on the location, this layer may include tree ferns, small palms, bamboos, shrubs, and herbaceous plants. Lianas and epiphytes are abundant.

Figure 9.11 is a map of the subtropical evergreen forests of the northern hemisphere. Here subtropical evergreen forest consists of broad-leaved trees such as evergreen oaks and trees of the laurel and magnolia families. The name "laurel forest" is applied to these forests, which are associated with the moist subtropical climate ⑥ in the southeastern United States, southern China, and southern Japan. However, these regions are under intense crop cultivation because of their favorable climate. In these areas, the land has been cleared of natural vegetation for centuries, and little natural laurel forest remains.

The *subtropical needleleaf evergreen forest* occurs only in the southeastern United States (Figure 9.11). Here it is referred to as the southern pine forest, since it is dominated by species of pine. It is found on the wide belt of sandy soils that fringes the Atlantic and Gulf coasts. Because the soils are sandy, water drains away quickly, leaving the soils quite dry. In infrequent drought years, these forests may burn. Since pines are well adapted to droughts and fires, they form a stable vegetation cover for large areas of the region. Timber companies have taken advantage of this natural preference for pines, creating many plantations yielding valuable lumber and pulp (Figure 9.12).

> **The subtropical evergreen forest includes both broadleaf and needleleaf evergreen trees and is found in moist subtropical climate ⑥ regions of southeastern North America and Southeast Asia. Most of this formation class has been lost to cultivation.**

MIDLATITUDE DECIDUOUS FOREST *Midlatitude deciduous forest* is the native forest type of eastern North America and Western Europe. It is dominated by tall, broadleaf trees that provide a continuous and dense canopy in summer but shed their leaves completely in the winter. Lower layers of small trees and shrubs are weakly developed. In the spring, a lush layer of lowermost herbs quickly develops but soon fades after the trees reach full foliage and shade the ground.

Figure 9.13 is a map of midlatitude deciduous forests, which are found almost entirely in the northern hemisphere. Throughout much of its range, this forest type is associated with the moist continental climate ⑩. Recall from Chapter 7

Eye on Global Change | 9.1

Exploitation of the Low-Latitude Rainforest Ecosystem

Many of the world's equatorial and tropical regions are home to the rainforest ecosystem. This ecosystem is perhaps the most diverse on Earth. That is, it possesses more species of plants and animals than any other. Very large tracts of rainforest still exist in South America, south Asia, and some parts of Africa. Ecologists regard this ecosystem as a genetic reservoir of many species of plants and animals. But as human populations expand and the quest for agricultural land continues, low-latitude rainforests are being threatened with clearing, logging, cultivation of cash crops, and animal grazing.

In the past, low-latitude rainforests were farmed by native peoples using the *slash-and-burn* method—cutting down all the vegetation in a small area, then burning it (see photo at right). In a rainforest ecosystem, most of the nutrients are held within living plants rather than in the soil. Burning the vegetation on the site releases the trapped nutrients, returning a portion of them to the soil. Here, the nutrients are available to growing crops. The supply of nutrients derived from the original vegetation cover is small, however, and the harvesting of crops rapidly depletes the nutrients. After a few seasons of cultivation, a new field is cleared, and the old field is abandoned. Rainforest plants reestablish their hold on the abandoned area. Eventually, the rainforest returns to its original state. This cycle shows that primitive, slash-and-burn agriculture is fully compatible with maintenance of the rainforest ecosystem.

On the other hand, modern intensive agriculture uses large areas of land and is not compatible with the rainforest ecosystem. When large areas are abandoned, seed sources are so far away that the original forest species cannot take hold. Instead, secondary species dominate, often accompanied by species from other vegetation types. These species are good invaders, and once they enter an area, they tend to stay. The dominance of these secondary species is permanent, at least on the human time scale. Thus, we can regard the rainforest ecosystem as a resource that, once cleared, will never return. The loss of low-latitude rainforest will result in the disappearance of thousands of species of organisms from the rainforest environment—a loss of millions of years of evolution, together with the destruction of the most complex ecosystem on Earth.

In Amazonia, transformation of large areas of rainforest into agricultural land uses heavy machinery to carve out major highways, such as the Trans-Amazon Highway in Brazil, and innumerable secondary roads and trails. Large fields for cattle pasture or commercial crops are created by cutting, bulldozing, clearing, and burning the vegetation. In some regions, the great broadleaved rainforest trees are removed for commercial lumber.

What are the effects of large-scale clearing? A recent study using a computer model by scientists of the University of Maryland and the Brazilian Space Research Institute indicated that when the Amazon rainforest is entirely removed and replaced with pasture, surface and soil temperatures will be increased 1 to 3°C (2 to 5°F).

9.8 Low-latitude rainforest *World map of low-latitude rainforest, showing equatorial and tropical rainforest types. (Data source same as Figure 9.3. Based on Goode Base Map.)*

Slash-and-burn clearing *This rainforest in Maranhão, Brazil, has been felled and burned in preparation for cultivation.*

parable to the original one may be impossible to achieve.

According to a report issued by the United Nations Food and Agriculture Organization, about 0.6 percent of the world's rainforest is lost annually by conversion to other uses. More rainforest land, 2.2 million hectares (about 8500 mi^2), is lost annually in Asia than in Latin America and the Caribbean, where 1.9 million hectares (about 7300 mi^2) are converted every year. Africa's loss of rainforest was estimated at about 470,000 hectares per year (1800 mi^2). Among individual countries, Brazil and Indonesia are the loss leaders, accounting for nearly half of the rainforest area converted to other uses. Note that these values do not include even larger losses of moist deciduous forests in these regions. Deforestation in low-latitude dry deciduous forests and hill and montane forests is also very serious.

Although deforestation rates are very rapid in some regions, many nations are now working to reduce the rate of loss of rainforest environment. However, because the rainforest can provide agricultural land, minerals, and timber, the pressure to allow deforestation continues.

Precipitation in the region will decline by 26 percent and evaporation by 30 percent. The deforestation will change weather and wind patterns so that less water vapor enters the Amazon Basin from outside sources, making the basin even drier. In areas where a marked dry season occurs, that season will be lengthened. Although such models contain simplifications and are subject to error, the results confirm the pessimistic conclusion that once large-scale deforestation has occurred, artificial restoration of a rainforest com-

9.9 Monsoon woodland *This woodland is in the Bandipur Wild Animal Sanctuary in the Nilgiri Hills of southern India. The scene is taken in the rainy season, with trees in full leaf.*

that this climate receives adequate precipitation in all months, normally with a summer maximum. There is a strong annual temperature cycle with a cold winter season and a warm summer.

Common trees of the deciduous forest of eastern North America, southeastern Europe, and eastern Asia are oak, beech, birch, hickory, walnut, maple, elm, and ash. A few needleleaf trees are often present as well—hemlock, for example (Figure 9.14). Where the deciduous forests have been cleared by lumbering, pines readily develop as second-growth forest.

In Western Europe, the midlatitude deciduous forest is associated with the marine west-coast climate ⑧. Here the dominant trees are mostly oak and ash, with beech in cooler and moister areas. In Asia, the midlatitude deciduous forest occurs as a belt between the boreal forest to the north and steppelands to the south. A small area of deciduous forest is found in Patagonia, near the southern tip of South America.

The deciduous forest includes a great variety of animal life, much of it stratified according to canopy layers. As many as five layers can be distinguished: the upper canopy, lower canopy, understory, shrub layer, and ground layer.

9.10 Monsoon forest *World map of monsoon forest and related types—savanna woodland and thorntree-tall grass savanna. (Data source same as Figure 9.3. Based on Goode Base Map.)*

9.11 *Subtropical evergreen forest* *Northern hemisphere map of subtropical evergreen forests, including the southern pine forest. (Data source same as Figure 9.3. Based on Goode Base Map.)*

Because the ground layer presents a more uniform environment in terms of humidity and temperature, it contains the largest concentration of organisms and the greatest diversity of species. Many small mammals burrow in the soil for shelter or food in the form of soil invertebrates. Among this burrowing group are ground squirrels, mice, and shrews, as well as some larger animals—foxes, woodchucks, and rabbits. Most of the larger mammals feed on ground and shrub layer vegetation, except for some, such as the brown bear, which are omnivorous and prey upon the small animals as well.

Even though birds possess the ability to move through the layers at will, most actually restrict themselves to one or more layers. For example, the wood peewee is found in the lower canopy, and the red-eyed vireo is found in the understory. Above them are the scarlet tanagers and blackburnian warblers, which are upper canopy dwellers. Below are the ground dwellers, such as grouse, warblers, and ovenbirds. Flying insects often show similar stratification patterns.

Among the large herbivores that graze in the deciduous forest are the red deer and roe deer of Eurasia and the white-tailed deer of North America. Smaller herbivores include voles, mice, and squirrels. Predators include bears, wildcats, lynx, wolves, foxes, weasels, and, among birds, owl species.

9.12 Pine plantation *This plantation of longleaf pine grows on the sandy soil of the southeastern coastal plain. Near Waycross, Georgia.*

NEEDLELEAF FOREST *Needleleaf forest* refers to a forest composed largely of straight-trunked, cone-shaped trees with relatively short branches and small, narrow, needle-like leaves. These trees are conifers. Most are evergreen, retaining their needles for several years before shedding them. When the needleleaf forest is dense, it provides continuous and deep shade to the ground. Lower layers of vegetation are sparse or absent, except for a thick carpet of mosses that may occur. Species are few—in fact, large tracts of needleleaf forest consist almost entirely of only one or two species.

Boreal forest is the cold-climate needleleaf forest of high latitudes. It occurs in two great continental belts, one in North America and one in Eurasia (Figure 9.15). These belts span their land masses from west to east in latitudes 45° N to 75° N, and they closely correspond to the region of boreal forest climate ⑪. The boreal forest of North America, Europe, and western Siberia is composed of such evergreen conifers as spruce and fir (Figure 9.16). The boreal forest of north-central and eastern Siberia is dominated by larch. The larch tree sheds its needles in winter and is thus a deciduous needleleaf tree. Broadleaf deciduous trees, such as aspen,

9.13 Midlatitude deciduous forest *Northern hemisphere map of midlatitude deciduous forests. (Data source same as Figure 9.3. Based on Goode Base Map.)*

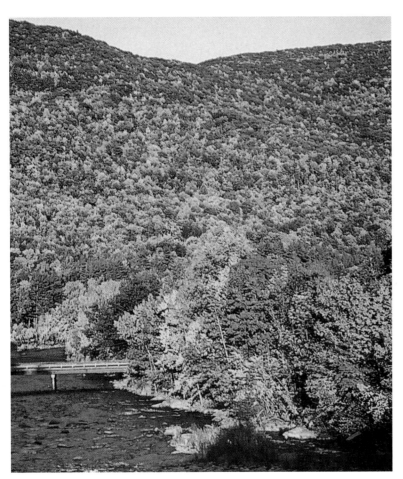

9.14 *Deciduous forest* *In this deciduous forest of the Catskill Mountains, New York, some needleleaf trees—pine and hemlock—are also present.*

balsam poplar, willow, and birch, tend to take over rapidly in areas of needleleaf forest that have been burned over. These species can also be found bordering streams and in open places. Between the boreal forest and the midlatitude deciduous forest lies a broad transition zone of mixed boreal and deciduous forest.

Needleleaf evergreen forest extends into lower latitudes wherever mountain ranges and high plateaus exist. For

example, in western North America this formation class extends southward into the United States on the Sierra Nevada and Rocky Mountain Ranges and over parts of the higher plateaus of the southwestern states (Figure 9.17, *left*). In Europe, needleleaf evergreen forests flourish on all the higher mountain ranges.

In its northernmost range, the boreal needleleaf forest grades into cold woodland. This form of vegetation is lim-

9.15 **Needleleaf forest** *Northern hemisphere map of cold-climate needleleaf forests, including coastal forest. (Data source same as Figure 9.3. Based on Goode Base Map.)*

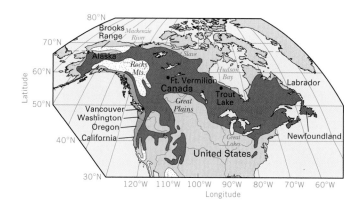

ited to the northern portions of the boreal forest climate ⑪ and the southern portions of the tundra climate ⑫. Trees are low in height and spaced well apart. A shrub layer may be well developed. The ground cover of lichens and mosses is distinctive. Cold woodland is often referred to as *taiga*. It is transitional into the treeless tundra at its northern fringe.

Mammals of the boreal needleleaf forest in North America include deer, moose, elk, black bear, marten, mink, wolf, wolverine, and fisher. Common birds include jays, ravens, chickadees,

---

Needleleaf forest includes boreal and coastal forest. Boreal forest stretches across the northern reaches of North America and Eurasia. Coastal forest is restricted to the steep slopes of the coast ranges in the Pacific Northwest region.

---

nuthatches, and a number of warblers. The caribou, lemming, and snowshoe rabbit inhabit both needleleaf forest and the adjacent tundra biome. The boreal forest often experiences large fluctuations in animal species populations, a result of the low diversity and highly variable environment.

Coastal forest is a distinctive needleleaf evergreen forest of the Pacific Northwest coastal belt, ranging in latitude from northern California to southern Alaska (Figure 9.15). Here, in a band of heavy orographic precipitation,

9.16 **Boreal forest** *A view of the boreal forest, Denali National Park, Alaska, pictured here just after the first snowfall of the season. At this location near the northern limits of the boreal forest, the tree cover is sparse. The golden leaves of aspen mark the presence of this deciduous species.*

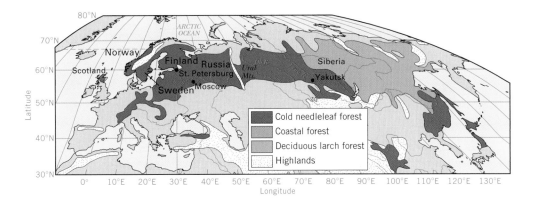

9.17 *Two kinds of needleleaf forest of the western United States* *(left) Open forest of western yellow pine (ponderosa pine), in the Kaibab National Forest, Arizona. (right) A grove of great redwood trees in Humboldt State Park, California.*

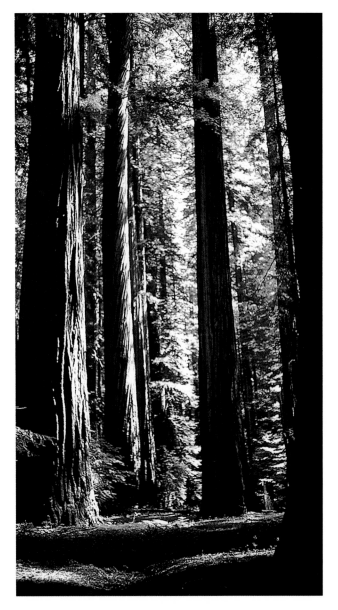

9.18 World map of sclerophyll vegetation (Data source same as Figure 9.3. Based on Goode Base Map.)

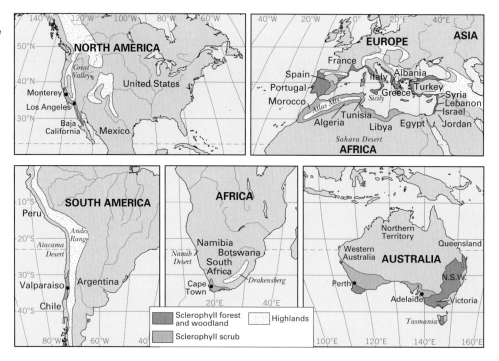

mild temperatures, and high humidity, are perhaps the densest of all conifer forests, with the world's largest trees (Figure 9.17, *right*). Individual redwood trees attain heights of over 100 m (about 330 ft) and girths of over 20 m (about 65 ft).

SCLEROPHYLL FOREST The native vegetation of the Mediterranean climate ⑦ is adapted to survival through the long summer drought. Shrubs and trees that can survive such drought are characteristically equipped with small, hard, or thick leaves that resist water loss through transpiration. As we noted earlier in the chapter, these plants are called sclerophylls.

Sclerophyll forest consists of trees with small, hard, leathery leaves. The trees are often low-branched and gnarled, with thick bark. The formation class includes *sclerophyll woodland,* an open forest in which only 25 to 60 percent of the ground is covered by trees. Also included are extensive areas of *scrub,* a plant formation type consisting of shrubs covering somewhat less than half of the ground area. The trees and shrubs are evergreen, retaining their thickened leaves despite a severe annual drought.

Our map of sclerophyll vegetation, Figure 9.18, includes forest, woodland, and scrub types. Sclerophyll forest is closely associated with the Mediterranean climate ⑦ and is narrowly limited to west coasts between 30° and 40° or 45° N and S latitude. In the Mediterranean lands, the sclerophyll forest forms a narrow, coastal belt ringing the Mediterranean Sea. Here, the Mediterranean forest consists of such trees as cork oak, live oak, Aleppo pine, stone pine, and olive. Over the centuries, human activity has reduced the sclerophyll

> Sclerophyll forest is dominated by low trees with thick, leathery leaves that are well-adapted to the long summer drought of the Mediterranean climate ⑦. Southern California's chaparral, found on coast-range slopes, is a form of sclerophyll scrub.

forest to woodland or destroyed it entirely. Today, large areas of this former forest consist of dense scrub.

The other northern hemisphere region of sclerophyll vegetation is the California coast ranges. Here, the sclerophyll forest or woodland is typically dominated by live oak and white oak. Grassland occupies the open ground between the scattered oaks (see Figure 7.28b). Much of the remaining vegetation is sclerophyll scrub or "dwarf forest," known as *chaparral.* It varies in composition with elevation and exposure. Chaparral may contain wild lilac, manzanita, mountain mahogany, poison oak, and live oak.

In central Chile and in the Cape region of South Africa, sclerophyll vegetation has a similar appearance, but the dominant species are quite different. Important areas of sclerophyll forest, woodland, and scrub are also found in southeast, south-central, and southwest Australia, including many species of eucalyptus and acacia.

Savanna Biome

The savanna biome is usually associated with the tropical wet-dry climate ③ of Africa and South America. It includes vegetation formation classes ranging from woodland to grassland. In *savanna woodland,* the trees are spaced rather widely apart because soil moisture during the dry season is not sufficient to support a full tree cover. The open spacing permits development of a dense lower layer, which usually consists of grasses. The woodland has an open, park-like appearance. Savanna woodland usually lies in a broad belt adjacent to equatorial rainforest.

In the tropical savanna woodland of Africa, the trees are of medium height. Tree crowns are flattened or umbrella-shaped, and the trunks have thick, rough bark (see Figure 7.15). Some species of trees are xerophytic forms with small leaves and thorns. Others are broad-leaved deciduous species that shed their leaves in the dry season. In this respect, savanna woodland resembles monsoon forest.

Fire is a frequent occurrence in the savanna woodland during the dry season, but the tree species of the savanna are particularly resistant to fire. Many geographers hold the view that periodic burning of the savanna grasses maintains the grassland against the invasion of forest. Fire does not kill the underground parts of grass plants, but it limits tree growth to individuals of fire-resistant species. Many rainforest tree species that might otherwise grow in the wet-dry climate regime are prevented by fires from invading. The browsing of animals, which kills many young trees, is also a factor in maintaining grassland at the expense of forest.

The regions of savanna woodland are shown in Figure 9.10, along with monsoon forest (discussed earlier). In Africa, the savanna woodland grades into a belt of *thorntree-tall-grass savanna,* a formation class transitional to the desert biome. The trees are largely of thorny species. Trees are more widely scattered, and the open grassland is more extensive than in the savanna woodland. One characteristic tree is the flat-topped acacia, seen in Figure 7.15. Elephant grass is a common species. It can grow to a height of 5 m (16 ft) to form an impenetrable thicket.

The thorntree-tall-grass savanna is closely identified with the semiarid subtype of the dry tropical and subtropical climates (④s, ⑤s). In the semiarid climate, soil water storage is adequate for the needs of plants only during the brief rainy season. The onset of the rains is quickly followed by the greening of the trees and grasses. For this reason, vegetation of the savanna biome is described as rain-green, an adjective that also applies to the monsoon forest.

The African savanna is widely known for the diversity of its large grazing mammals, which include more than a dozen antelopes. Careful studies have shown that each species has a particular preference for different parts of the grasses they consume—blade, sheath, and stem. Grazing stimulates the grasses to continue to grow, and so the ecosystem is more productive when grazed than when left alone. With these grazers comes a large variety of predators—lions, leopards, cheetahs, hyenas, and jackals. Elephants are the largest animals of the savanna and adjacent woodland regions.

The savanna biome is adapted to a strong wet-dry annual cycle. Grazing by large mammals and periodic burning in the dry season help to maintain the openness of the savanna by suppressing tree seedlings.

Grassland Biome

The grassland biome includes two major formation classes that we will discuss here—tall-grass prairie and steppe (Figure 9.19). *Tall-grass prairie* consists largely of tall grasses.

Forbs, which are broad-leaved herbs, are also present. Trees and shrubs are absent from the prairie but may occur in the same region as narrow patches of forest in stream valleys. The grasses are deeply rooted and form a thick and continuous turf (Figure 9.20).

Prairie grasslands are best developed in regions of the midlatitude and subtropical zones with well-developed winter and summer seasons. The grasses flower in spring and early summer, and the forbs flower in late summer. Tall-grass prairies are closely associated with the drier areas of moist continental climate ⑩. Here, soil water is in short supply during the summer months.

When European settlers first arrived in North America, the tall-grass prairies were found in a belt extending from the Texas Gulf coast northward to southern Saskatchewan (Figure 9.19). A broad peninsula of tall-grass prairie extended eastward into Illinois, where conditions are somewhat more moist. Since the time of settlement, these prairies have been converted almost entirely to agricultural land. Another major area of tall-grass prairie is the Pampa region of South America, which occupies parts of Uruguay and eastern Argentina. The Pampa region falls into the moist subtropical climate ⑥, with mild winters and abundant precipitation.

Recall from Chapter 7 that **steppe,** also called *short-grass prairie,* is a vegetation type consisting of short grasses occurring in sparse clumps or bunches. Scattered shrubs and low trees may also be found in the steppe. The plant cover is poor, and much bare soil is exposed. Many species of grasses and forbs occur. A typical grass of the American steppe is buffalo grass. Other typical plants are the sunflower and loco weed. Steppe grades into semidesert in dry environments and into prairie where rainfall is higher.

Our map of the grassland biome (Figure 9.19) shows that steppe grassland is concentrated largely in the midlatitude areas of North America and Eurasia. The only southern hemisphere occurrence shown on the map is the "veldt" region of South Africa, a highland steppe surface in Orange Free State and Transvaal.

Steppe grasslands correspond well with the semiarid subtype of the dry continental climate ⑨. Spring rains nourish the grasses, which grow rapidly until early summer. By midsummer, the grasses are usually dormant. Occasional summer rainstorms cause periods of revived growth.

The animals of the grassland are distinctive. Before the exploitation of the North American grassland for cattle ranching, large grazing mammals were abundant. These included the pronghorn antelope, the elk, and the buffalo, which ranged widely from tall-grass prairie to steppe. Now nearly extinct, the buffalo *(Bison)* once numbered 60 million and roamed the

The grassland biome includes tall-grass prairie and short-grass prairie (steppe). Tall-grass prairie provides rich agricultural land suited to cultivation and cropping. Short-grass prairie occupies vast regions of semidesert and is suited to grazing.

9.19 Grassland *World map of the grassland biome in subtropical and midlatitude zones. (Data source same as Figure 9.3. Based on Goode Base Map.)*

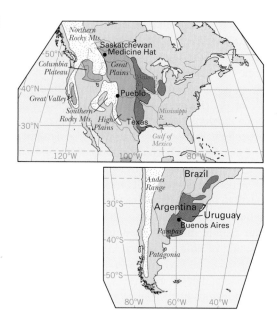

grasslands from the Rockies to the Shenandoah Valley of Virginia. By 1889, however, this herd had been reduced to 800 individuals, or approximately the same number that are alive today. Most of the present buffalo population is confined to Yellowstone National Park. Today, rodents, and rabbits join cattle as the major grazers in the grasslands ecosystem.

The grassland ecosystem supports some rather unique adaptations to life. A common adaptive mechanism is jumping or leaping locomotion, assuring an unimpeded view of the surroundings. Jackrabbits and jumping mice are examples of jumping rodents. The pronghorn combines the leap with great speed, which allows it to avoid predators and fire. Burrowing is also another common life habit, for the soil

9.20 Tall-grass prairie, Iowa *In addition to grasses, tall-grass prairie vegetation includes many forbs, such as the flowering species shown in this photo.*

provides the only shelter in the exposed grasslands. Examples are burrowing rodents, including prairie dogs, gophers, and field mice. Rabbits exploit old burrows, using them for nesting or shelter. Invertebrates also seek shelter in the soil, and many are adapted to living with the burrows of rodents, where extremes of moisture and temperature are substantially moderated.

Desert Biome

The desert biome includes several formation classes that are transitional from grassland and savanna biomes into vegetation of the arid desert. Here we recognize two basic formation classes: semidesert and dry desert. Figure 9.21 is a world map of the desert biome.

Semidesert is a transitional formation class found in a wide latitude range—from the tropical zone to the midlatitude zone. It is identified primarily with the arid subtypes of all three dry climates. Semidesert consists of sparse xerophytic shrubs. One example is the sage-brush vegetation of the middle and southern Rocky Mountain region and Colorado Plateau (Figure 9.22). Recently, as a result of overgrazing and trampling by livestock, semidesert shrub vegetation seems to have expanded widely into areas of the western United States that were formerly steppe grasslands.

Thorntree semidesert of the tropical zone consists of xerophytic trees and shrubs that are adapted to a climate with a very long, hot dry season and only a very brief, but intense, rainy season. These conditions are found in the semiarid and arid subtypes of the dry tropical ④ and dry subtropical ⑤ climates. The thorny trees and shrubs are known locally as thorn forest, thornbush, or thornwoods

---

The desert biome includes semidesert and dry desert and occupies the dry tropical ④, dry subtropical ⑤, and dry midlatitude ⑨ climate zones. Desert plants vary widely in appearance and in adaptation to the dry environment.

(see Figure 7.18*a*). Many of these are deciduous plants that shed their leaves in the dry season. The shrubs may be closely intergrown to form dense thickets. Cactus plants are present in some localities.

Dry desert is a formation class of xerophytic plants that are widely dispersed over only a very small proportion of the ground. The visible vegetation of dry desert consists of small, hard-leaved, or spiny shrubs, succulent plants (such as cactus), or hard grasses. Many species of small annual plants may be present but appear only after a rare, but heavy, desert downpour. Much of the world map area assigned to desert vegetation has no plant cover at all because the surface consists of shifting dune sands or sterile salt flats.

Desert plants differ greatly in appearance from one part of the world to another. In the Mojave and Sonoran deserts of the southwestern United States, plants are often large, giving the appearance of a woodland (Figure 9.23). Examples are the tree-like saguaro cactus, the prickly pear cactus, the ocotillo, creosote bush, and smoke tree.

Desert animals, like the plants, are typically adapted to the dry conditions of the desert. We have already mentioned some adaptions of xeric animals to the water scarcity of the desert. Important herbivores in American deserts include kangaroo rats, jackrabbits, and grasshopper mice. Insects are abundant, as are insect-eating bats and birds such as the cactus wren. Reptiles, especially lizards, are also common.

Tundra Biome

Arctic tundra is a formation class of the tundra climate ⑫. (See Figure 7.40 for a polar map of the arctic tundra.) In this

9.21 Desert *World map of the desert biome, including desert and semidesert formation classes. (Data source same as Figure 9.3. Based on Goode Base Map.)*

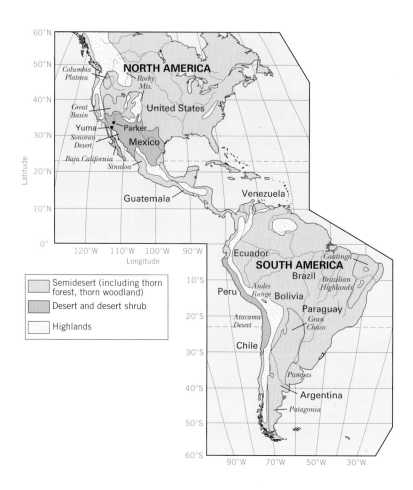

Semidesert (including thorn forest, thorn woodland)

Desert and desert shrub

Highlands

9.22 Eye on the Landscape Sagebrush semidesert *Sagebrush dominates the landscape near Monument Valley, Utah—a region of mesas, buttes, and pinnacles on the Colorado Plateau.* **What else would the geographer see? ... Answers at the end of the chapter.**

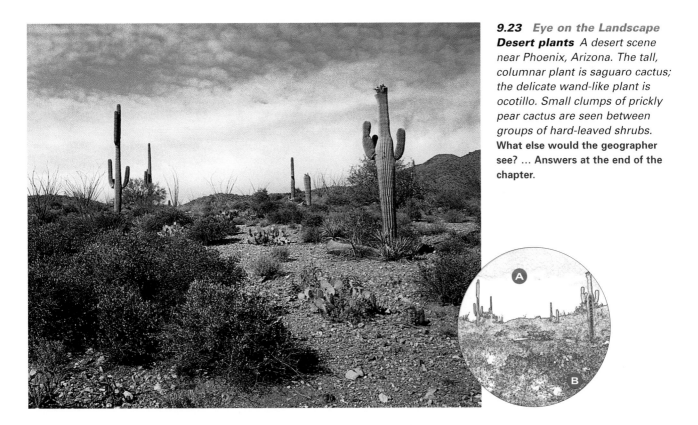

9.23 Eye on the Landscape
Desert plants A desert scene near Phoenix, Arizona. The tall, columnar plant is saguaro cactus; the delicate wand-like plant is ocotillo. Small clumps of prickly pear cactus are seen between groups of hard-leaved shrubs. **What else would the geographer see? ... Answers at the end of the chapter.**

climate, plants grow during the brief summer of long days and short (or absent) nights. At this time, air temperatures rise above freezing, and a shallow surface layer of ground ice thaws. The permafrost beneath, however, remains frozen, keeping the meltwater at the surface. These conditions create a marshy environment for at least a short time over wide areas. Because plant remains decay very slowly within the cold meltwater, layers of organic matter can build up in the marshy ground. Frost action in the soil fractures and breaks large roots, keeping tundra plants small. In winter, wind-driven snow and extreme cold also injure plant parts that project above the snow.

Plants of the arctic tundra are mostly low herbs, although dwarf willow, a small woody plant, occurs in places. Sedges, grasses, mosses, and lichens dominate the tundra in a low layer (Figure 7.42). Typical plant species are ridge sedge, arctic meadow grass, cotton grasses, and snow lichen. There are also many species of forbs that flower brightly in

---

The tundra biome includes low plants that are adapted to survival through a harsh, cold winter. They grow, bloom and set seed during a short summer thaw.

---

the summer. Tundra composition varies greatly as soils range from wet to well drained. One form of tundra consists of sturdy hummocks of plants with low, water-covered ground between. Some areas of arctic scrub vegetation composed of willows and birches are also found in tundra.

Tundra vegetation is also found at high elevations. This *alpine tundra* develops above the limit of tree growth and below the vegetation-free zone of bare rock and perpetual snow (Figure 9.24). Alpine tundra resembles arctic tundra in many physical respects.

As is most often true in particularly dynamic environments, species diversity in the tundra is low, but the abundance of individuals is high. Among the animals, vast herds of caribou in North America and reindeer (their Eurasian relatives) roam the tundra, lightly grazing the lichens and plants and moving constantly. A smaller number of musk-oxen are also primary consumers of the tundra vegetation. Wolves and wolverines, as well as arctic foxes

9.24 Alpine tundra *Flag-shaped spruce trees (on left side of photo), shaped by prevailing winds, mark the upper limit of tree growth. Near the summit of the Snowy Range, Wyoming.*

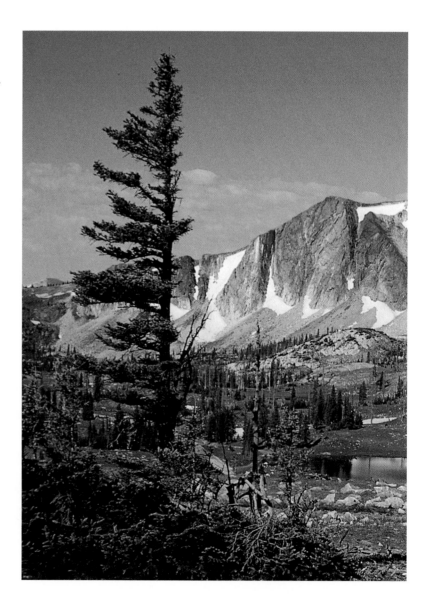

and polar bears, are predators. Among the smaller mammals, snowshoe rabbits and lemmings are important herbivores. Invertebrates are scarce in the tundra, except for a small number of insect species. Black flies, deerflies, mosquitoes, and "no-see-ums" (tiny biting midges) are all abundant and can make July on the tundra most uncomfortable. Reptiles and amphibians are also rare. The boggy tundra, however, presents an ideal summer environment for many migratory birds such as waterfowl, sandpipers, and plovers.

The food web of the tundra ecosystem is simple and direct. The important producer is reindeer moss, the lichen *Cladonia rangifera.* In addition to the caribou and reindeer, lemmings, ptarmigan (arctic grouse), and snowshoe rabbits are important lichen grazers. The important predators are the fox, wolf, and lynx, although all these animals may feed directly on plants as well. During the summer, the abundant insects help support the migratory waterfowl populations.

Elevations range from about 700 m (about 2300 ft) at the bottom of the Grand Canyon to 3844 m (12,608 ft) at the top of San Francisco Peak. The vegetation cover and rainfall range are shown on the left. The vegetation zonation includes desert shrub, grassland, woodland, pine forest, Douglas fir forest, Engelmann spruce forest, and alpine meadow. Annual rainfall ranges from 12 to 25 cm (about 5 to 10 in.) in the desert shrub vegetation type to 80–90 cm (about 30–35 in.) in the Engelmann spruce forest.

Biogeographers have used the vegetation zonation to set up a series of *life zones,* which are also shown on the figure. These range from the lower Sonoran life zone, which includes the desert shrubs typical of the Sonoran Desert, to the arctic-alpine life zone, which includes the alpine meadows at the top of San Francisco Peak. Other life zones take their names from typical regions of vegetation cover. For example, the Hudsonian zone, 2900 to 3500 m (about 9500 to 11,500 ft), bears a needleleaf forest quite similar to needleleaf boreal forest of the subarctic zone near Hudson's Bay, Canada.

Altitude Zones of Vegetation

In earlier chapters, we described the effects of increasing elevation on climatic factors, particularly air temperature and precipitation. We noted that, with elevation, temperatures decrease and precipitation generally increases. These changes produce systematic changes in the vegetation cover as well, yielding a sequence of vegetation zones related to altitude.

The vegetation zones of the Colorado Plateau region in northern Arizona and adjacent states provide a striking example of this altitude zonation. Figure 9.25 is a diagram showing a cross section of the land surface in this region.

Because climate factors of temperature and precipitation vary with elevation and over space, vegetation patterns often show zonation with altitude and systematic variation on long transects.

Climatic Gradients and Vegetation Types

In discussing the major formation classes of vegetation, we have emphasized the importance of climate. As climate changes with latitude or longitude, vegetation will also change. Figure 9.26 shows three transects across portions of continents that illustrate this principle. (For these transects, we will ignore the effects of mountains or highland regions on climate and vegetation.)

The upper transect stretches from the equator to the tropic of cancer in Africa. Across this region, climate ranges

9.25 Altitude zone of vegetation in the arid southwestern United States *The profile shows the Grand Canyon-San Francisco Mountain district of northern Arizona. (Based on data of G. A. Pearson, C. H. Merriam, and A. N. Strahler.)*

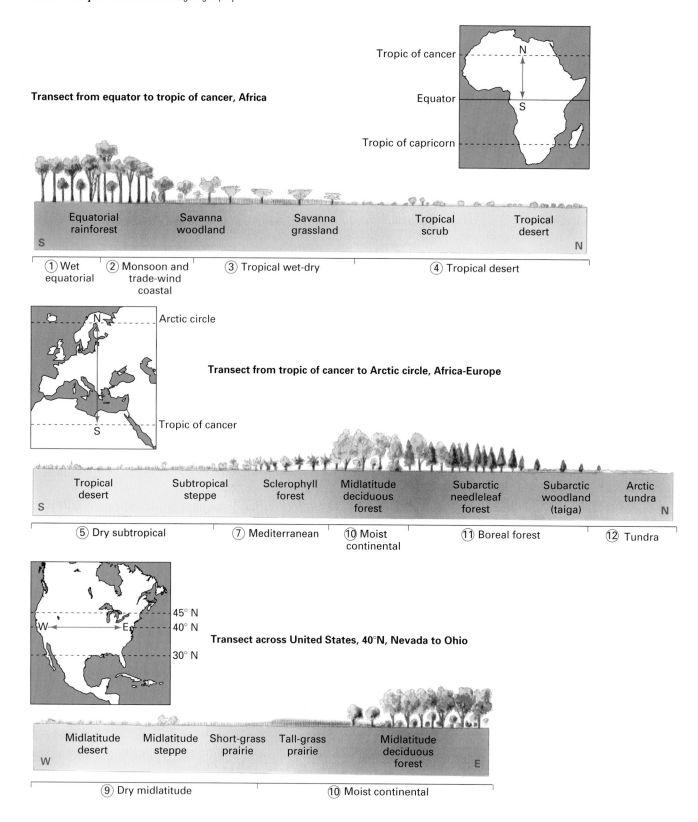

Transect from equator to tropic of cancer, Africa

Tropic of cancer
N
Equator
S
Tropic of capricorn

| Equatorial rainforest | Savanna woodland | Savanna grassland | Tropical scrub | Tropical desert |

S — N

① Wet equatorial ② Monsoon and trade-wind coastal ③ Tropical wet-dry ④ Tropical desert

Arctic circle
N

Transect from tropic of cancer to Arctic circle, Africa-Europe

Tropic of cancer
S

| Tropical desert | Subtropical steppe | Sclerophyll forest | Midlatitude deciduous forest | Subarctic needleleaf forest | Subarctic woodland (taiga) | Arctic tundra |

S — N

⑤ Dry subtropical ⑦ Mediterranean ⑩ Moist continental ⑪ Boreal forest ⑫ Tundra

45° N
40° N
W — E
30° N

Transect across United States, 40°N, Nevada to Ohio

| Midlatitude desert | Midlatitude steppe | Short-grass prairie | Tall-grass prairie | Midlatitude deciduous forest |

W — E

⑨ Dry midlatitude ⑩ Moist continental

9.26 Vegetation transects *Three continental transects showing the succession of plant formation classes across climatic gradients.*

through all four low-latitude climates: wet equatorial ①, monsoon and trade-wind coastal ②, wet-dry tropical ③, and dry tropical ④. Vegetation grades from equatorial rainforest, savanna woodland, and savanna grassland to tropical scrub and tropical desert.

The middle transect is a composite from the tropic of cancer to the arctic circle in Africa and Eurasia. Climates include many of the mid- and high-latitude types: dry subtropical ⑤, Mediterranean ⑦, moist continental ⑩, boreal forest ⑪, and tundra ⑫. The vegetation cover grades from tropical desert through subtropical steppe to sclerophyll forest in the Mediterranean. Further north is the midlatitude deciduous forest in the region of moist continental climate ⑩, which grades into boreal needleleaf forest, subarctic woodland, and finally tundra.

The lower transect ranges across the United States, from Nevada to Ohio. On this transect, the climate begins as dry midlatitude ⑨. Precipitation gradually increases eastward, reaching moist continental ⑩ near the Mississippi River. The vegetation changes from midlatitude desert and steppe to short-grass prairie, tall-grass prairie, and midlatitude deciduous forest.

The changes on these transects are largely gradational rather than abrupt. Yet, the global maps of both vegetation and climate show distinct boundaries from one region to the next. Which is correct? The true situation is gradational rather than abrupt. Maps must necessarily have boundaries to communicate information. But climate and vegetation know no specific boundaries. Instead, they are classified into specific types for convenience in studying their spatial patterns. When studying any map of natural features, keep in mind that boundaries are always approximate and gradational.

A LOOK AHEAD   As we have seen in this chapter, climate has an important influence on the Earth's biomes and their global distribution pattern. The key ingredients of climate are temperature, moisture, and the variation of temperature and moisture through the year. These same factors are important in the formation of soils, which is the subject of the next chapter. The vegetation cover also influences soil formation. For example, soils developed on grasslands are very different from those developed under conifer forests. Other important determinants of soil formation are the nature of the soil's parent material as it is derived from weathered rock and the time allowed for soil formation to proceed.

In Review | Global Biogeography

- **Natural vegetation** is a plant cover that develops with little or no human interference. Although much vegetation appears to be in a natural state, humans influence the vegetation cover by fire suppression and introduction of new species.

- The **life-form** of a plant refers to its physical structure, size, and shape. Life-forms include *trees, shrubs, lianas, herbs,* and *lichens.*

- The largest unit of terrestrial ecosystems is the **biome:** forest, grassland, savanna, desert, and tundra. The **forest biome** includes a number of important forest formation classes. The *low-latitude rainforest* exhibits a dense canopy and open floor with a very large number of species. *Subtropical evergreen forest* occurs in *broadleaf* and *needleleaf* forms in the moist subtropical climate ⑥. *Monsoon forest* is largely deciduous, with most species shedding their leaves after the wet season.

- *Midlatitude deciduous forest* is associated with the moist continental climate ⑩. Its species shed their leaves before the cold season. *Needleleaf forest* consists largely of evergreen conifers. It includes the *coastal forest* of the Pacific Northwest, the *boreal forest* of high latitudes, and needle-leaved mountain forests. *Sclerophyll forest* is comprised of trees with small, hard, leathery leaves and is found in the Mediterranean climate ⑦ region.

- The **savanna biome** consists of widely spaced trees with an understory, often of grasses. Dry-season fire is frequent in the savanna biome, limiting the number of trees and encouraging the growth of grasses.

- The **grassland biome** of midlatitude regions includes *tall-grass prairie* in moister environments and *short-grass prairie,* or **steppe,** in semiarid areas.

- Vegetation of the **desert biome** ranges from thorny shrubs and small trees to dry desert vegetation comprised of drought-adapted species.

- **Tundra biome** vegetation is limited largely to low herbs that are adapted to the severe drying cold experienced on the fringes of the Arctic Ocean.

- Since climate changes with altitude, vegetation typically occurs in altitudinal zones. Climate also changes gradually with latitude, and so biome changes are typically gradual, without abrupt boundaries.

Key Terms

natural vegetation, p. 312

life form, p. 313

forest, p. 313

woodland, p. 313

terrestrial ecosystems, p. 314

biome, p. 314

forest biome, p. 314

savanna biome, p. 314

grassland biome, p. 314

desert biome, p. 314

tundra biome, p. 314

steppe, p. 329

Review Questions

1. What is natural vegetation? How do humans influence vegetation?

2. Plant geographers describe vegetation by its overall structure and by the life-forms of individual plants. Define and differentiate the following terms: forest, woodland, tree, shrub, herb, liana, perennial, deciduous, evergreen, broadleaf, and needleleaf.

3. What are the five main biome types that ecologists and biogeographers recognize? Describe each briefly.

4. Low-latitude rainforests occupy a large region of the Earth's land surface. What are the characteristics of these forests? Include forest structure, types of plants, diversity, and climate in your answer.

5. Monsoon forest and midlatitude deciduous forest are both deciduous but for different reasons. Compare the characteristics of these two formation classes and their climates.

6. Subtropical broadleaf evergreen forest and tall-grass prairie are two vegetation formation classes that have been greatly altered by human activities. How was this done and why?

7. Distinguish among the types of needleleaf forest. What characteristics do they share? How are they different? How do their climates compare?

8. Which type of forest, with related woodland and scrub types, is associated with the Mediterranean climate? What are the features of these vegetation types? How are they adapted to the Mediterranean climate?

9. How do traditional agricultural practices in the low-latitude rainforest compare to present-day practices? What are the implications for the rainforest environment?

10. What are the effects of large-scale clearing on the rainforest environment?

11. Describe the formation classes of the savanna biome. Where is this biome found and in what climate types? What role does fire play in the savanna biome?

12. Compare the two formation classes of the grassland biome. How do their climates differ?

13. Describe the vegetation types of the desert biome.

14. What are the features of arctic and alpine tundra? How does the cold tundra climate influence the vegetation cover?

15. How does elevation influence vegetation? Provide an example of how vegetation zonation is related to elevation.

Visualizing Exercises

1. Forests often contain plants of many different life-forms. Sketch a cross section of a forest including typical life-forms, and identify them with labels.

2. How does elevation influence vegetation? Sketch a hypothetical mountain peak in the southwestern U.S. desert that rises from a plain at about 500 m (about 1600 ft) elevation to a summit at about 4000 m (about 13,000 ft) and label the vegetation zones you might expect to find on its flanks.

Essay Questions

1. Figure 9.26 presents a vegetation transect from Nevada to Ohio. Expand the transect on the west so that it begins in Los Angeles. On the east, extend it northeast from Ohio through Pennsylvania, New York, western Massachusetts, and New Hampshire, to end in Maine. Sketch the vegetation types in your additions and label them, as in the diagram. Below your vegetation transect, draw a long bar subdivided to show the climate types.

2. Construct a similar transect of climate and vegetation from Miami to St. Louis, Minneapolis, and Winnipeg.

Eye on the Landscape

Chapter Opener A train streaks across the Mojave Desert Angular mountains **(A)** ascend from gravel-covered plains **(B)** in this sunset view of the Mojave Desert. The steep-sided mountain slopes are carved by running water from desert storms. Although such storms are infrequent, their effects are intense because of the lack of vegetation cover to hold the soil in place. As debris-laden streams drain into the lowlands, they deposit gravelly sediment in fans and plane off the underlying rock to form pediments. Stream erosion is the subject of Chapter 16, and it includes a section on landforms of mountainous deserts.

9.22 Sagebrush semidesert These tall columnar landforms are buttes **(A)**. Mesas are larger, isolated rock platforms **(B)**. Here in Monument Valley they are formed by a thick and uniform rock layer that weathers into rectangular blocks along joint planes. The blocks fall away, leaving a vertical cliff face behind. Buttes and mesas are discussed in Chapter 17, and joints are covered in Chapter 14.

9.23 Desert plants This sky of puffy altocumulus clouds and layered altostratus **(A)** indicates moisture at higher levels in the troposphere. The photo, taken in July, reflects conditions of the local "monsoon" season, in which moist air from the Gulf of California moves into southern Arizona, often generating intense thunderstorms and flash floods. The gravel-covered ground surface **(B)** is characteristic of deserts, where wind and storm runoff remove fine particles and leave coarser rock fragments behind. Clouds are covered in Chapter 4, and the Arizona monsoon is mentioned in Chapter 5. Erosion by runoff and wind are treated in Chapters 16 and 18.

EYE ON THE LANDSCAPE

Piedmont farmland in southern Georgia. The ancient soils of this region are highly productive with proper treatment. What else would the geographer see? … Answers at the end of the chapter.

10 | GLOBAL SOILS

● ●
PUTTING YOU IN THE PICTURE

This chapter is about soil—the loose surface layer of Earth that supports the growth of vegetation. North American soils provide the base for the most efficient and productive agriculture in the world. But this has not always been the case.

At the close of the eighteenth century, the fertility of the agricultural lands of the Southeast was declining rapidly. Romantically, we picture George Washington, happily retired on his Virginia estate, Mount Vernon, reaping a richly deserved bounty from expansive farmlands under his personal supervision. This vision is far from the truth. In fact, agricultural yields at Mount Vernon, and elsewhere in the Southeast, were dropping. In 1834, 35 years after Washington's death, a visitor to Mount Vernon declared that "a more widespread and perfect agricultural ruin could not be imagined."

The mystery of failing agricultural fertility was solved by Edmund Ruffin (1794–1865), who owned lands on the coastal plain of Virginia that were rapidly declining in the early 1800s. Quite by chance, Ruffin obtained a copy of Sir Humphry Davy's *Agricultural Chemistry,* published in 1813. Despite Ruffin's lack of formal education in science, he was quick to grasp the significance of one statement: "any acid matter … may be ameliorated by application of quicklime."

So it came about that on a February morning in 1818, Ruffin directed his fieldhands to haul marl from pits in low areas of his lands. Marl is a soft, lime-rich mud that occurs widely on the eastern coastal plain. The workers spread 200 bushels of marl over several acres of newly cleared ridge land of poor quality. In the spring, Ruffin planted this area in corn to test the effect of the marl. In the words of historian Avery Craven, this is what happened: "Eagerly he waited. As the season advanced, he found reason for joy. From the very start the plants on marled ground showed

341

marked superiority, and at harvest time they yielded an advantage of fully forty per cent. The carts went back to the pits. Fields took on fresh life. A new era in agricultural history of the region had dawned."*

Why did the application of lime enhance soil fertility so dramatically? As we will see in this chapter, nutrients are held in soils on the surfaces of fine particles. In an acid environment, these surfaces do not hold nutrients well, and so downward-moving water from precipitation washes nutrients out of the reach of plant roots. Lime reduces acidity and restores the ability of soil particles to hold nutrients, so that they remain available to plants.

Why didn't American farmers understand the importance of liming their fields? The colonists who settled this region were largely from England, where soils are formed on freshly ground mineral matter left by the great ice sheets. The breakdown products of this fresh mineral matter act like lime to reduce soil acidity, so that liming is not needed. In contrast, upland soils of the eastern seaboard from Virginia to Georgia have been continuously exposed to a mild, moist climate for tens of thousands of years. Fresh mineral matter in these soils has long been broken down, and the residual mineral matter that remains is highly acid. Thus, the colonists had no experience with these fundamentally different southeastern soils.

Soils are influenced by the rock material from which they develop, by the climate of the region (especially precipitation and temperature), by the vegetation cover that develops on them, and by the length of time that the soil has been in place. Your new knowledge of global climate and vegetation, acquired in the last two chapters, will be quite helpful in understanding soils and the processes that form them. ■

* Avery Craven, *Edmund Ruffin, Southerner* (New York; Appleton, 1932), p. 55.

GLOBAL SOILS

This chapter is devoted to soil systems, including the processes that form soils and give them their distinctive characteristics, to soil classification, and to global distribution of soil types. **Soil** is the uppermost layer of the land surface that plants use and depend on for nutrients, water, and physical support. Soils can vary greatly from continent to continent, from region to region, and even from field to field. This is because they are influenced by factors and processes that can vary widely from place to place. For example, a field near a large river that floods regularly may acquire a layer of nutrient-rich silt, making its soil very productive. A nearby field at a higher elevation, without the benefit of silt enrichment, may be sandy or stony and require the addition of fertilizers to grow crops productively.

Vegetation is an important factor in determining soil qualities. For example, some of America's richest soils developed in the Middle West under a cover of thick grass sod. The deep roots of the grass, in a cycle of growth and decay, deposited nutrients and organic matter throughout the thick soil layer. In the Northeast, conifer forests provided a surface layer of decaying needles that kept the soil quite acid. This acidity allowed nutrients to be washed below root depth, out of the reach of plants. When farmed today, these soils need applications of lime to reduce their acidity and enhance their fertility.

> Soil is the uppermost layer of the land surface. Plants depend on soil for nutrients, water, and physical support. Vegetation, climate, and time are important factors in soil development.

Climate, measured by precipitation and temperature, is also an important determiner of soil properties. Precipitation controls the downward movement of nutrients and other chemical compounds in soils. If precipitation is abundant, water tends to wash soluble compounds, including nutrients, deeper into the soil and out of reach of plant roots.

Temperature acts to control the rate of decay of organic matter that falls to the soil from the plant cover or that is provided to the soil by the death of roots. When conditions are warm and moist, decay organisms work efficiently, consuming organic matter readily. Thus, organic matter and nutrients in soils of the tropical and equatorial zones are generally low. Where conditions are cooler, decay proceeds more slowly, and organic matter is more abundant in the soil. Of course, if the climate is very dry or desert-like, then vegetation growth is slow or absent. No matter what desert temperatures are like, organic matter will be low.

Time is also an important factor. The characteristics and properties of soils require time for development. For example, a fresh deposit of mineral matter, like the clean, sorted sand of a dune, may require hundreds to thousands of years to acquire the structure and properties of a sandy soil.

Geographers are keenly interested in the differences in soils from place to place over the globe. The ability of the soils and climate of a region to produce food largely determines the size of the population it will support. In spite of the growth of cities, most of the world's inhabitants still live close to the soil that furnishes their food. And many of those same inhabitants die prematurely when the soil does not furnish enough food for all.

THE NATURE OF THE SOIL

Soil, as the term is used in soil science, is a natural surface layer that contains living matter and can support plants. The soil consists of matter in all three states—solid, liquid, and gas. It includes both *mineral matter* and *organic matter.* Mineral matter is largely derived from rock material, whereas organic matter is of biological origin and may be living or dead. Living matter in the soil consists not only of plant roots, but also of many kinds of organisms, including microorganisms.

Soil scientists use the term *humus* to describe finely divided, partially decomposed organic matter in soils. Some humus rests on the soil surface, and some is mixed through the soil. Humus particles of the finest size are gradually carried downward to lower soil layers by rainfall that sinks in and moves through the soil. When abundant, humus particles can give the soil a brown or black coloration.

Both air and water are found in soil. Water may tend to contain high levels of dissolved substances, such as nutrients. Air in soils may have high levels of such gases as carbon dioxide or methane, and low levels of oxygen.

The solid, liquid, and gaseous matter of the soil are constantly changing and interacting through chemical and physical processes. This makes the soil a very dynamic layer. Because of these processes, soil science, often called *pedology,* is a highly complex body of knowledge.

Although we may think of soil as occurring everywhere, large expanses of continents possess a surface layer that cannot be called soil. For example, dunes of moving sand, bare rock surfaces of deserts and high mountains, and surfaces of fresh lava near active volcanoes do not have a soil layer.

The characteristics of soils are developed over a long period of time through a combination of many processes acting together. Physical processes act to break down rock fragments of regolith into smaller and smaller pieces. Chemical processes alter the mineral composition of the original rock, producing new minerals. Taken together, these physical and chemical processes are referred to as *weathering,* which we will describe in Chapter 14. Weathering occurs in soils and is part of the process by which soils develop their properties and characteristics.

In most soils, the inorganic material of the soil consists of fine particles of mineral matter. The term **parent material** describes all forms of mineral matter that are suitable for transformation into soil. Parent material may be derived

Vegetation
Soil
Regolith
Bedrock
Forest litter

10.1 A cross section through the land surface *In this cross section, vegetation and forest litter lie atop the soil. Below is regolith, produced by the breakup of the underlying bedrock.*

from the underlying *bedrock,* which is solid rock below the soil layer (Figure 10.1). Over time, weathering processes soften, disintegrate, and break bedrock apart, forming a layer of *regolith,* or residual mineral matter. Regolith is one of the common forms of parent material. Other kinds of regolith consist of mineral particles transported to a place of rest by the action of streams, glaciers, waves and water currents, or winds. For example, dunes formed of sand transported by wind are a type of regolith on which soil may be formed.

> Soil is a complex mixture of solids, liquids, and gases. Solid matter includes both mineral and organic matter. Watery solutions and atmospheric gases are also found in soils.

Soil Color and Texture

The most obvious feature of a soil or soil layer is probably its color. Some color relationships are quite simple. For example, the soils of the Midwest prairies have a black or dark brown color because they contain abundant particles of humus. Red or yellow colors often mark the soils of the Southeast. These colors are created by the presence of iron-containing oxides.

In some areas, soil color may be inherited from the mineral parent material, but, more generally, soil color is generated by soil-forming processes. For example, a white surface layer in soils of dry climates often indicates the presence of mineral salts brought upward by evaporation. A pale, ashgray layer near the top of soils of the boreal forest climate results when organic matter and various colored minerals are washed downward, leaving only pure, light-colored mineral matter behind. As we explain soil-forming processes and describe the various classes of soils, soil color will take on more meaning.

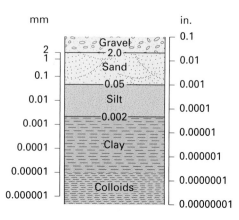

10.2 Mineral particle sizes *Size grades, which are names like sand, silt, and clay, refer to mineral particles within a specific size range. They are defined using the metric system. English equivalents are also shown.*

The mineral matter of the soil consists of individual mineral particles that vary widely in size. The term **soil texture** refers to the proportion of particles that fall into each of three size grades— *sand, silt,* and *clay.* The diameter range of each of these grades is shown in Figure 10.2. Millimeters are the standard units. Each unit on the scale represents a power of ten, so that clay particles of 0.000,001 millimeter diameter are one-millionth the size of sand grains 1 mm in diameter. The finest of all soil particles are termed *colloids.* In measuring soil texture, gravel and larger particles are eliminated, since these play no important role in soil processes.

Soil texture is described by a series of names that emphasize the dominant particle size, whether sand, silt, or clay (including colloids). Figure 10.3 gives examples of five soil textures with typical percentage compositions. A *loam* is a

> Soil texture refers to the proportions of sand, silt, and clay found in a soil. Soil color is usually determined by soil-forming processes and varies widely.

mixture containing a substantial proportion of each of the three grades. Loams are classified as sandy, silty, or clayrich when one of these grades is dominant.

Why is soil texture important? Texture largely determines the ability of the soil to retain water. Coarse-textured (sandy) soils have many small passages between touching mineral grains that quickly conduct water through to deeper layers. If the soil consists of fine particles, passages and spaces are much smaller. Thus, water will penetrate more slowly and also tend to be retained. We will return to the important topic of the water-holding ability of soils in a later section.

Soil Colloids

Soil colloids consist of particles smaller than one hundred-thousandth of a millimeter (0.000,01 mm, 0.000,000,4 in.). Like other soil particles, some colloids are mineral, while others are organic. Mineral colloids are usually very fine particles of clay minerals. If you examine mineral colloids under a microscope, you will find that they consist of thin, plate-like bodies (Figure 10.4). When well mixed in water, particles this small remain suspended indefinitely, giving the water a murky appearance. Organic colloids are tiny bits of organic matter that are resistant to decay.

Soil colloids are important because their surfaces attract soil nutrients, which are in the form of ions dissolved in soil water. Figure 10.5 diagrams a colloidal particle, showing this effect. Colloid surfaces tend to be negatively

10.4 Mineral colloids *Seen here enlarged about 20,000 times are tiny flakes of clay minerals of colloidal dimensions. These particles have settled from suspension in San Francisco Bay.*

10.3 Soil textures *These diagrams show the proportion of sand, silt, and clay in five different soil texture classes. For description of soil texture, clay includes colloids.*

10.5 Schematic diagram of a colloid particle *The thin, flat colloidal particle has negative surface charges and a layer of positively charged ions held to the surface.*

charged because of their molecular structure, and thus attract and hold positively charged ions. Among the many ions in soil water, one important group consists of **bases,** which are ions of four elements: calcium (Ca^{++}), magnesium (Mg^{++}), potassium (K^+), and sodium (Na^+). Because plants require these elements, they are among the *plant nutrients*—ions or chemical compounds that are needed for plant growth. Colloids hold these ions, but also give them up to plants when in close contact with root membranes. Without this ion-holding ability of soil colloids, most of the vital nutrients would be carried out of the soil by percolating water and would be taken out of the region in streams, eventually reaching the sea. This leaching process goes on continually in moist climates, but loss is greatly retarded by the ion-holding capacity of soil colloids.

> Colloids are the finest particles in the soil. The surfaces of colloids attract base ions, which plants use as nutrients.

> Soil acidity, as measured by pH, varies widely. Soils of cool, moist regions are generally acid, while soils of arid climates are alkaline. Acid soils often have a low base status.

Soil Acidity and Alkalinity

The soil solution also contains hydrogen (H^+) and aluminum (Al^{+++}) ions. But unlike the bases, they are not considered to be plant nutrients. The presence of these acid ions in the soil solution tends to make the solution acid in chemical balance.

An important principle of soil chemistry is that the acid ions have the power to replace the nutrient bases clinging to the surfaces of the soil colloids. As acid ions accumulate, the bases are released to the soil solution. They are gradually washed downward below rooting level, reducing soil fertility. When this happens, the soil acidity is increased.

The degree of acidity or alkalinity of a solution is designated by the *pH value.* The lower the pH value, the greater the degree of acidity. A pH value of 7 represents a neutral state—for example, pure water has a pH of 7. Lower values are in the acid range, while higher values are in the alkaline range.

Table 10.1 shows the natural range of acidity and alkalinity found in soils. High soil acidity is typical of cold, humid climates. In arid climates, soils are typically alkaline. Acidity can be corrected by the application of lime, a compound of calcium, carbon, and oxygen ($CaCO_3$), which removes acid ions and replaces them with the base calcium.

Soil Structure

Soil structure refers to the way in which soil grains are grouped together into larger masses, called *peds.* Peds range in size from small grains to large blocks. They are bound together by soil colloids. Small peds, roughly shaped like spheres, give the soil a granular structure or crumb structure (see Figure 10.6). Larger peds provide an angular, blocky structure. Peds form when colloid-rich clays shrink in volume as they dry out. Shrinkage results in formation of soil cracks, which define the surfaces of the peds.

Soils with a well-developed granular or blocky structure are easy to cultivate. This is an important agricultural factor

Table 10.1 | Soil Acidity and Alkalinity

pH	4.0	4.5	5.0	5.5	6.0	6.5	6.7	7.0	8.0	9.0	10.0	11.0
Acidity	Very strongly acid		Strongly acid	Moderately acid		Slightly acid		Neutral	Weakly alkaline	Alkaline	Strongly alkaline	Excessively alkaline
Lime requirements	Lime needed except for crops requiring acid soil		Lime needed for all but acid-tolerant crops			Lime generally not required		No lime needed				
Occurrence	Rare	Frequent	Very common in cultivated soils of humid climates				Common in subhumid and arid climates				Limited areas in deserts	

Source: Based on data of C. E. Millar, L. M. Turk, and H. D. Foth, *Fundamentals of Soil Science,* John Wiley & Sons, New York.

10.6 Soil texture *This soil shows a granular texture. The grains are referred to as peds.*

in lands where primitive plows, drawn by animals, are still widely used. Soils with a high clay content can lack peds. These soils are sticky and heavy when wet and are difficult to cultivate. When dry, they become too hard to be worked.

Minerals of the Soil

Soil scientists recognize two classes of minerals abundant in soils: primary minerals and secondary minerals. The *primary minerals* are compounds present in unaltered rock. These are mostly silicate minerals—compounds of silicon and oxygen, with varying proportions of aluminum, calcium, sodium, iron, and magnesium. (The silicate minerals are described more fully in Chapter 11.) Primary minerals form a large fraction of the solid matter of many kinds of soils, but they play no important role in sustaining plant or animal life.

When primary minerals are exposed to air and water at or near the Earth's surface, they are slowly altered in chemical composition. This process is part of *mineral alteration,* a chemical weathering process that is explained in more detail in Chapter 11. The primary minerals are altered into **secondary minerals,** which are essential to soil development and to soil fertility.

In terms of the properties of soils, the most important secondary minerals are the *clay minerals.* They form the majority of fine mineral particles in soils. From the viewpoint of soil fertility, the ability of a clay mineral to hold base ions is its most important property. This ability varies with the particular type of clay mineral; some hold bases tightly, and others loosely.

The nature of the clay minerals in a soil determines its *base status.* If the clay minerals can hold abundant base ions, the soil is of *high base status* and generally will be highly fertile. If the clay minerals hold a smaller supply of

---

Primary minerals in soils are minerals that remain from unaltered rock. Secondary minerals are formed by mineral alteration. Clay minerals and sesquioxides are important secondary minerals.

---

bases, the soil is of *low base status* and is generally less fertile. Humus colloids have a high capacity to hold bases so that the presence of humus is usually associated with potentially high soil fertility.

Mineral oxides are secondary minerals of importance in soils. They occur in many kinds of soils, particularly those that remain in place in areas of warm, moist climates over very long periods of time (hundreds of thousands of years). Under these conditions, minerals are ultimately broken down chemically into simple oxides, compounds in which a single element is combined with oxygen.

Oxides of aluminum and iron are the most important oxides in soils. Two atoms of aluminum are combined with three atoms of oxygen to form the *sesquioxide* of aluminum (Al_2O_3). (The prefix *sesqui-* means "one and a half" and refers to the chemical composition of one and one-half atoms of oxygen for every atom of aluminum.) In soils, aluminum oxide forms the mineral *bauxite,* which is a combination of aluminum sesquioxide and water molecules bound together. It occurs as hard, rock-like lumps and layers below the soil surface. Where bauxite layers are thick and uniform, they are sometimes strip-mined as aluminum ore.

Sesquioxide of iron (Fe_2O_3), again held in combination with water molecules, is *limonite,* a yellowish to reddish mineral that supplies the typical reddish to chocolate-brown colors of soils and rocks. Some shallow accumulations of limonite were formerly mined as a source of iron. Limonite and bauxite occur in close association in soils of warm, moist climates in low latitudes.

Soil Moisture

Besides providing nutrients for plant growth, the soil layer serves as a reservoir for the moisture that plants require. Soil moisture is a key factor in determining how the soils of a region support vegetation and crops.

The soil receives water from rain and from melting snow. Where does this water go? First, some of the water can run off the soil surface and not sink in. Instead, it flows into brooks, streams, and rivers, eventually reaching the sea. What of the water that sinks into the soil? Some of this water is returned to the atmosphere as water vapor. This happens when soil water evaporates and when transpiration by plants lifts soil water from roots to leaves, where it evaporates. Taken together, we can term these last two losses *evapotranspiration.* Some water can also flow completely through the soil layer to recharge supplies of ground water at depths below the reach of plant roots.

When precipitation infiltrates the soil, the water wets the soil layer. This process is called *soil water recharge.* Eventually, the soil layer holds the maximum possible quantity of water, even though the larger pores may remain filled with air. Water movement then continues downward.

Suppose now that no further water enters the soil for a time. Excess soil water continues to drain downward, but some water clings to the soil particles. This water resists the pull of gravity because of the force of *capillary tension.* To understand this force, think about a droplet of condensation that has formed on the cold surface of a glass of ice water. The water droplet seems to be enclosed in a "skin" of surface molecules, drawing the droplet together into a rounded shape. The "skin" is produced by capillary tension. This force keeps the drop clinging to the side of the glass indefinitely, defying the force of gravity. Similarly, tiny films of water adhere to soil grains, particularly at the points of grain contacts. They remain until they evaporate or are absorbed by plant rootlets.

When a soil has first been saturated by water and then allowed to drain under gravity until no more water moves downward, the soil is said to be holding its *storage capacity* of water. For most soils, drainage takes no more than two or three days. Most excess water is drained out within one day.

Storage capacity is measured in units of depth, usually centimeters or inches, as with precipitation. It depends largely on the texture of the soil, as shown in Figure 10.7. Finer textures hold more water than coarser textures. This occurs because fine particles have a much larger surface area in a unit of volume than coarse particles. Thus, a sandy soil has a small storage capacity, while a clay soil has a large storage capacity.

The figure also shows the *wilting point,* the water storage level below which plants will wilt. The wilting point depends on soil texture. Because fine particles hold water more tightly, it is more difficult for plants to extract moisture from fine soils. Thus, plants can wilt in fine-textured soils even though more soil water is present than in coarse-textured soils. The difference between the storage capacity of a soil and its wilting point is the *available water capacity*—that is, the maximum amount of water available

10.7 Storage capacity and wilting point according to soil texture *Finer textured soils hold more water. They also hold water more tightly, so that plants wilt more quickly.*

to plants when the soil is at storage capacity. The available water capacity is greatest in loamy soils.

THE SOIL WATER BALANCE

Water in the soil is a critical resource needed for plant growth. The amount of water available at any given time is determined by the *soil water balance,* which includes the gain, loss, and storage of soil water. Figure 10.8 is a pictorial flow diagram that illustrates the components of the balance. Water held in storage in the soil water zone is increased by recharge during precipitation but decreased by use through evapotranspiration. Surplus water is disposed of by downward percolation to the ground water zone or by overland flow.

To proceed, we must recognize two ways to define evapotranspiration. First is *actual evapotranspiration (Ea),* which is the true or real rate of water vapor return to the atmosphere from the ground and its plant cover. Second is *potential evapotranspiration (Ep),* representing the water vapor loss under an ideal set of conditions. One condition is that there is present a complete (or closed) cover of uniform vegetation consisting of fresh green leaves and no bare ground exposed through that cover. A second condition is that there is an adequate water supply, so that the storage capacity of the soil is maintained at all times. This condition can be fulfilled naturally by abundant and frequent precipitation, or artificially by irrigation.

To simplify the ponderous terms we have just defined, they may be transformed as follows:

> Soil moisture clings to soil particles by capillary action. Soil storage capacity, which depends on soil texture, measures the ability of a soil to hold moisture after excess water has drained away.

Evapotranspiration (E)

Precipitation
(P)

Soil
water
zone

Soil water
storage

Recharge
(+G)

Utilization
(−G)

Surplus (R)
as overland
flow

Inter-
mediate
zone

Surplus (R)
as ground water
recharge

Total
runoff
(R)

Ground water
zone

Ground water
outflow

10.8 Schematic diagram of the soil water balance in a soil column *(Copyright © A. N. Strahler.)*

- Actual evapotranspiration *(Ea)* is **water use.**
- Potential evapotranspiration *(Ep)* is **water need.**

The word "need" signifies the quantity of soil water needed if plant growth is to be maximized for the given conditions of solar radiation and air temperature and the available supply of nutrients. The most important factor in determining water need is temperature. In warmer months, water need will be greater, while in cooler months, it will be less.

The difference between water use and water need is the *soil water shortage,* or *deficit.* This is the quantity of water that must be furnished by irrigation to achieve maximum crop growth within an agricultural system.

A Simple Soil Water Budget

We now turn to a simple accounting of the monthly and annual quantities of the components of the soil water balance. The numerical accounting is called a *soil water budget,* and it involves only simple addition and subtraction of monthly mean values for a given observing station. All terms of the soil water budget are stated in centimeters of water depth, the same as for precipitation.

A simplified soil water budget is shown in Figure 10.9. The seven terms we need for a complete budget are listed as follows, along with abbreviations used on the graph:

Precipitation, *P*
Water need, *Ep*

> **The soil water budget measures the balance between water need, water use, and precipitation for each month. Depending on the balance, soil moisture may be lowered or recharged.**

Water use, *Ea*
Storage withdrawal, −*G*
Storage recharge, +*G*
Soil water shortage, *D*
Water surplus, *R*

Points on the graph represent average monthly values. They are connected by smooth curves to enhance annual cycles of change. In our example, precipitation *(P)* is much the same in all months, with no strong annual cycle. In contrast, water need *(Ep)* shows a strong seasonal cycle, with low values in winter and a high summer peak. This example would fit a typical moist midlatitude climate with mild winters.

At the start of the year, precipitation greatly exceeds water use and a large water surplus *(R)* exists. This surplus is disposed of by runoff. By May, water use exceeds precipitation, and a water deficit occurs. In this month plants begin to withdraw soil water from storage.

Storage withdrawal (−G) is represented by the difference between the water-use curve and the precipitation curve. As storage withdrawal continues, however, plants draw soil water only with increasing difficulty. Thus, water use *(Ea)* is less than water need *(Ep)* during this period. Storage withdrawal continues throughout the summer. The deficit period lasts through September. The area labeled soil water shortage *(D)* is the difference between water need and water use. It represents the total quantity of water needed by irrigation to ensure maximum growth throughout the deficit period.

In October, precipitation *(P)* again begins to exceed water need *(Ep)*, but the soil must first absorb an amount equal to the summer storage withdrawal. So there follows a period of *storage recharge (+G)* that lasts through November. In December the soil reaches its full storage capacity, arbitrarily fixed at 30 cm

10.9 A simplified soil water budget *This soil water budget is typical of a moist climate in middle latitudes.*

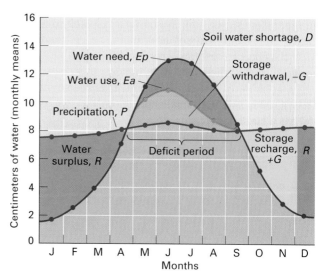

(11.8 in.). Now, a water surplus *(R)* again sets in, lasting through the winter.

The soil water budget was developed by C. Warren Thornthwaite, a distinguished climatologist and geographer who was concerned with practical problems of crop irrigation. He developed the calculation of the soil water budget in order to place irrigation on a precise, accurate basis. Only in the equatorial zone and in a few parts of the tropical and midlatitude zones is precipitation ample to fulfill the water need during the growing season. In a world beset by severe and prolonged food shortages, the Thornthwaite concepts and calculations are of great value in assessing the benefits to be gained by increased irrigation.

Soil Development

How do soils develop their distinctive characteristics? Let's turn to the processes that act to form soils and soil layers. We begin with soil horizons.

Soil Horizons

Most soils possess **soil horizons**—distinctive horizontal layers that differ in physical composition, chemical composition, or organic content or structure (Figure 10.10). Soil

> **Soil horizons are distinctive layers found in soils that differ in physical or chemical composition, organic content, or structure. The soil profile refers to the display of horizons on a cross section through the soil.**

horizons are developed by the interactions through time of climate, living organisms, and the configuration of the land surface. Horizons usually develop by either selective removal or accumulation of certain ions, colloids, and chemical compounds. The removal or accumulation is normally produced by water seeping down through the soil profile from the surface to deeper layers. Horizons are often distinguished by their color. The display of horizons on a cross section through the soil is termed a **soil profile.** GEODISCOVERIES

Let's review briefly the main types of horizons and their characteristics. For now, this discussion will apply to the types of horizons and processes that are found in moist forest climates. These are shown in Figure 10.11. Soil horizons are of two types: organic and mineral. Organic horizons, designated by the capital letter *O*, overlie the mineral horizons and are formed from accumulations of organic matter derived from plants and animals. Soil scientists recognize two possible layers. The upper O_i horizon

10.10 Soil horizons *A column of soil will normally show a series of horizons, which are horizontal layers with different properties.*

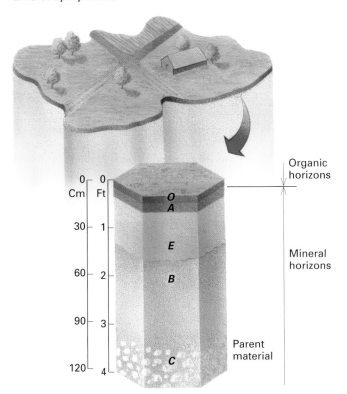

10.11 Soil horizons of moist forest climates *A sequence of horizons that might appear in a forest soil developed under a cool, moist climate. (Soil Conservation Service, U.S. Department of Agriculture.)*

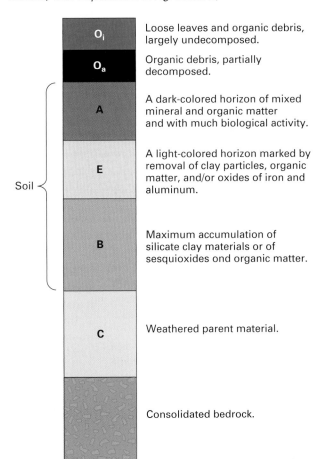

O_i Loose leaves and organic debris, largely undecomposed.

O_a Organic debris, partially decomposed.

A A dark-colored horizon of mixed mineral and organic matter and with much biological activity.

E A light-colored horizon marked by removal of clay particles, organic matter, and/or oxides of iron and aluminum.

B Maximum accumulation of silicate clay materials or of sesquioxides ond organic matter.

C Weathered parent material.

Consolidated bedrock.

contains decomposing organic matter that is recognizable as leaves or twigs. The lower O_a horizon contains material that is broken down beyond recognition by eye. This material is humus, which we mentioned earlier.

Mineral horizons lie below the organic horizons. Four main horizons are important—*A, E, B,* and *C.* Plant roots readily penetrate *A, E,* and *B* horizons and influence soil development within them. Soil scientists limit the term *soil* to refer to the *A, E,* and *B* horizons. The *A horizon* is the uppermost mineral horizon. It is rich in organic matter, consisting of numerous plant roots and downwashed humus from the organic horizons above. Next is the *E horizon.* Clay particles and oxides of aluminum and iron are removed from the *E horizon* by downward-seeping water, leaving behind pure grains of sand or coarse silt.

The *B horizon* receives the clay particles, aluminum, and iron oxides, as well as organic matter washed down from the *A* and *E* horizons. It is made dense and tough by the filling of natural spaces with clays and oxides. Transitional horizons are present at some locations.

Beneath the *B horizon* is the *C horizon.* It consists of the parent mineral matter of the soil and has been described earlier in this chapter as regolith (see Figure 10.1). Below the regolith lies bedrock or sediments of much older age than the soil.

Soil-Forming Processes

There are four classes of soil-forming processes. The first includes processes of *soil enrichment,* which add material to the soil body. For example, inorganic enrichment occurs when sediment is brought from higher to lower areas by overland flow. Stream flooding also deposits fine mineral particles on low-lying soil surfaces. Wind is another source of fine material that can accumulate on the soil surface. Organic enrichment occurs when humus accumulating in *O* horizons is carried downward to enrich the *A* horizon below.

The second class includes processes that remove material from the soil body. This *removal* occurs when surface erosion carries sediment away from the uppermost layer of soil. Another important process of loss is *leaching,* in which seeping water dissolves soil materials and moves them to deep levels or to ground water.

The third class of soil-forming processes involves *translocation,* in which materials are moved within the soil body, usually from one horizon to another. Two processes of translocation that operate simultaneously are eluviation and illuviation. **Eluviation** consists of the downward transport of fine particles, particularly the clays and colloids, from the uppermost part of the soil. Eluviation leaves behind grains of sand or coarse silt, forming the *E* horizon. **Illuviation** is the accumulation of materials that are brought down downward, normally from the *E* horizon to the *B* horizon. The materials that accumulate may be

10.12 *A forest soil profile on outer Cape Cod* *The pale grayish E horizon overlies a reddish B horizon. A thin layer of wind-deposited silt and dune sand (pale brown layer) has been deposited on top.*

> The four classes of soil-forming processes are enrichment, removal, translocation, and transformation. In translocation, fine particles are transported downward by eluviation and accumulate in lower horizons by illuviation.

clay particles, humus, or sesquioxides of iron and aluminum.

Figure 10.12 is a soil profile developed under a cool, humid forest climate. It shows the effects of both soil enrichment and translocation processes. The topmost layer of the soil is a thin deposit of wind-blown silt and dune sand, which has enriched the soil profile. In translocation processes, eluviation has removed colloids and sesquioxides from the whitened *E* horizon. Illuviation has added them to the *B* horizon, which displays the orange-red colors of iron sesquioxide.

The translocation of calcium carbonate is another important process. In pure form, this secondary mineral is calcite ($CaCO_3$). In many areas, the parent material of the soil contains a substantial proportion of calcium carbonate derived from the disintegration of limestone, a common variety of bedrock. Carbonic acid, which forms when carbon dioxide gas dissolves in rainwater or soil water, readily reacts with calcium carbonate. The products of this reaction remain dissolved in solution as ions.

In moist climates, a large amount of surplus soil water moves downward to the ground water zone. This water movement leaches calcium carbonate from the entire soil in a process called *decalcification.* Soils that have lost most of their calcium are also usually acid in chemical balance and so are low in bases. Addition of lime or pulverized limestone not only corrects the acid condition, but also restores the calcium, which is used as a plant nutrient.

In dry climates, calcium carbonate is dissolved in the upper layers of the soil during periods of rain or snowmelt when soil water recharge is taking place. The dissolved carbonate matter is carried down to the *B* horizon, where water penetration reaches its limits. Here, the carbonate matter is precipitated (deposited in crystalline form) in the *B* horizon, a process called *calcification.* Calcium carbonate deposition takes the form of white or pale-colored grains, nodules, or plates in the *B* or *C* horizons.

A last process of translocation occurs in desert climates. In some low areas, a layer of ground water lies close to the surface, producing a flat, poorly drained area. Evaporation of water at or near the soil surface draws up a continual flow of ground water by capillary tension, much like a cotton wick draws oil upward in an oil lamp. Moreover, the ground water is often rich in dissolved salts. When evaporation occurs, the salts precipitate and accumulate as a distinctive *salic horizon.* This process is called *salinization.* Most of the salts are compounds of sodium, of which ordinary table salt (sodium chloride, or halite, NaCl) is a familiar example. Sodium in large amounts is associated with highly alkaline conditions and is toxic to many kinds of plants. When salinization occurs in irrigated lands in a desert climate, the soil can be ruined for further agricultural use. *Eye on the Environment 10.1 ● Death of a Civilization* shows how the ancient cultures of Babylon and Sumer declined as their irrigated desert soils deteriorated from salinization and waterlogging.

The last class of soil-forming processes involves the *transformation* of material within the soil body. An example is the conversion of minerals from primary to secondary types, which we have already described. Another example is decomposition of organic matter to produce humus, a process termed *humification.* In warm, moist climates, humification can decompose organic matter completely to yield carbon dioxide and water, leaving virtually no organic matter in the soil.

Soil Temperature

Soil temperature is another important factor in determining the chemical development of soils and the formation of horizons. Temperature acts as a control over biologic activity and also influences the intensity of chemical processes affecting soil minerals. Below 10°C (50°F), biological activity is slowed, and at or below the freezing point (0°C, 32°F), biological activity stops. Chemical processes affecting minerals are inactive. The root growth of most plants and germination of their seeds require soil temperatures above 5°C (41°F). For plants of the warm, wet low-latitude climates, germination of seeds may require a soil temperature of at least 24°C (75°F).

The temperature of the uppermost soil layer and the soil surface strongly affects the rate at which organic matter is decomposed by microorganisms. Thus, in cold climates, where decomposition is slow, organic matter in the form of fallen leaves and stems tends to accumulate to form a thick *O* horizon. As we have already described, this material becomes humus, which is carried downward to enrich the *A* horizon.

In warm, moist climates of low latitudes, the rate of decomposition of plant material is rapid, so that nearly all the fallen leaves and stems are disposed of by bacterial activity. Under these conditions, the *O* horizon may be missing, and the entire soil profile will contain very little organic matter.

> **Temperature has an important influence on soil development. In cold climates, decomposition of organic matter is slow and organic matter accumulates. In warm climates, organic matter decomposes rapidly and soil organic matter is scarce.**

SOIL AND SURFACE CONFIGURATION The configuration, or shape, of the ground surface is an important factor in soil formation. Configuration includes the steepness of the ground surface, or *slope,* as well as its compass orientation, or *aspect.* Generally speaking, soil horizons are thick on gentle slopes but thin on steep slopes. This is because the soil is more rapidly removed by erosion processes on the steeper slopes.

Aspect acts to influence soil temperatures and the soil water regime. Slopes facing away from the Sun are sheltered from direct insolation and tend to have cooler, moister soils. Slopes facing toward the Sun are exposed to direct solar rays, raising soil temperatures and increasing evapotranspiration.

BIOLOGICAL PROCESSES IN SOIL FORMATION The presence and activities of living plants and animals, as well as their nonliving organic products, have an important influence on soil. We have already noted the role that organic matter as humus plays in soil fertility. The colloidal structure of humus holds bases, which are needed for plant growth. In this way, humus helps keep nutrients cycling through plants and soils, and ensures that nutrients will not be lost by leaching to ground water. Humus also helps bind the soil into crumbs and clumps. This structure allows water and air to penetrate the soil freely, providing a healthy environment for plant roots.

Animals living in the soil include many species and come in many individual sizes—from bacteria to burrowing mammals. The total role of animals in soil formation is extremely important in soils that are warm and moist enough to support large animal populations. For example, earthworms continually rework the soil not only by burrowing, but also by passing soil through their intestinal tracts. They ingest large amounts of decaying leaf matter, carry it down from the surface, and incorporate it into the mineral soil horizons. Many forms of insect larvae perform a similar function. Tube-like openings are made by larger animals—moles, gophers, rabbits, badgers, prairie dogs, and other species.

Eye on the Environment | 10.1

Death of a Civilization

What causes a civilization to collapse after flourishing for thousands of years? Wars and conquests come to mind, but sometimes the cause is less obvious. A case in point is the decline of the civilization of Sumer, and its successor, Babylon. Perhaps you have already learned something of these ancient Middle Eastern cultures, which were nourished by the Tigris and Euphrates rivers thousands of years ago. These two famous rivers drain the Zagros and Taurus Mountains of Turkey and Iran, and flow out through a broad valley across the deserts of Iraq to the Persian Gulf.

The Sumerian civilization evolved in the lower part of the Tigris–Euphrates Valley, beginning with a village culture dating to about five millennia B.C. By 3000 B.C., the Sumerian civilization was well established. The agriculture that supported Sumer depended on a system of irrigation canals utilizing the waters of the Tigris and Euphrates. Supported by this highly productive agricultural base, urban culture progressed to a high level, including the development of a written language. Skilled crafts-

men produced pottery, jewelry, and ornate weapons of copper, silver, and gold (see figure at right).

Trouble set in for the Sumerians in about 2400 B.C. It came in the form of a deterioration of their croplands caused by an accumulation of salt in the soil. In this process, today called salinization, a small increment of salt is added to the soil each year by the evaporation of irrigation water containing a low concentration of salts. Over time, the salt content of the soil increases until it reaches levels high enough to affect crop plants. Wheat, which was a staple crop of Sumer, is quite sensitive to salt, and its yield declines sharply as the salt concentration rises in the soil. Barley, another staple crop of that era, is somewhat less sensitive to salt.

Archaeologists and historians have inferred the onset of salinization in Sumerian agricultural lands by a shift in the proportions of wheat and barley in the national agricultural output. It seems that around 3500 B.C. these two grains were grown in about equal amounts. By 2500 B.C., however, wheat accounted for only about one-sixth of the total grain production. Evidently, the more salt-toler-

ant barley had replaced wheat over much of the area.

At this point in time, the Sumerian grain yield also began to decline seriously. Ancient records show that in 2400 B.C., around the Sumerian city of Girsu, fields were yielding an average of about 300 kg of grain per hectare (270 lb/acre)—a high rate, even by modern standards. By 2100 B.C. the grain yield had declined to 180 kg per hectare (160 lb/acre), and by 1700 B.C., to a mere 100 kg/ha (90 lb/acre). Cities so declined in population that they finally dwindled into villages. As Sumerian civilization withered in the south, Babylonia, in the northern part of the valley, became more important. Soon, political control of the region passed from Sumer to Babylon.

Historians chronicle an event that also seems to have played a part in the decline of the Sumerian civilization. A group of Sumerian cities along the Euphrates River needed more irrigation water than they had. After a long and fruitless series of conflicts with upstream cities to gain additional water, the problem was solved by running a canal across the valley from the Tigris River. Now there was plenty of water. However, it was used so abundantly that the ground water table began to rise. As the water table came closer to the ground surface, capillary tension drew the

---

Human activity also influences the physical and chemical nature of the soil. Large areas of agricultural soils have been cultivated for centuries. As a result, both the structure and composition of these agricultural soils have undergone great changes. These altered soils are often recognized as distinct soil classes that are just as important as natural soils.

THE GLOBAL SCOPE OF SOILS

An important aspect of soil science for physical geography is the classification of soils into major types and

subtypes that are recognized in terms of their distribution over the Earth's land surfaces. Geographers are particularly interested in the linkage of climate, parent material, time, biologic process, and landform with the distribution of types of soils. Geographers are also interested in the kinds of natural vegetation associated with each of the major soil classes. The geography of soils is thus essential in determining the quality of environments of the globe. It is important because soil fertility, along with availability of fresh water, is a basic measure of the ability of an environmental region to produce food for human consumption.

An ancient enameled panel commorating the victory in war of a king in the first dynasty of Ur, Mesopotamia. Note the rich agricultural spoils depicted in the second and third rows—cattle, sheep, fishes, sacks of grain.

ground water to the surface, where it evaporated rapidly in the hot sun. The result was an even faster accumulation of salt. When the salty water table came up still further, it saturated the roots of the barley plants, devastating the crops. Unknowingly, in their quest for water the Sumerians destroyed their own land.

The rising civilization of Babylon was doomed to fall in turn. Along the Tigris River, east of the modern city of Baghdad, an extremely elaborate system of irrigation canals was built. This system was begun in a pre-Babylonian period, 3000 to 2400 B.C. After a long history of abandonments and reconstructions, it was superseded between A.D. 200 and 500 by a final irrigation system based on a central canal, the Nahrwan Canal.

A problem soon developed, however. The irrigation system featured long, branching channels

that proved to be traps for silt, which settled out of muddy river water. Frequent cleaning of silt from the channels was required, and silt was piled up in great embankments and mounds beside the channels. Gradually, silt from the mounds was washed by rains into nearby fields, adding layer upon layer to the land surface. According to one estimate, about 1 m (3 ft) of silt accumulated over the fields in a 500-year period. This rise in level of the farmland by silting made its irrigation all the more difficult. Until the strong central government authority collapsed, things went along well enough. Afterward, however, the entire system gradually broke down, and by the twelfth century it was abandoned completely. Not long after, Mongol hordes invaded the valley. By then, little remained of the once prosperous civilization that had dominated the region for over 4000 years.

Salinization, waterlogging, and silt accumulation have affected nearly every major irrigation system developed in a desert climate. Today, Pakistan struggles with salinization and waterlogging in a vast irrigation system in the Indus Valley. The lower Colorado River lands, including areas in Mexico, have suffered from salinization. Israel faces the threat of salinization of vital agricultural lands that have only recently been placed in irrigation. Some new methods may help. For example, soaking the ground through tubes, while keeping the soil covered under a plastic sheet, can reduce the buildup of salt.

Small wonder, then, that new proposals for desert irrigation systems arouse little enthusiasm. Perhaps a record of nearly total failure, spread over the entire span of civilization, is beginning to get its message across.

The soils of the world have been classified according to a system developed by scientists of the U.S. Soil Conservation Service, in cooperation with soil scientists of many other nations. Here we are concerned only with the two highest levels of this classification system. The top level contains 11 **soil orders** summarized in Table 10.2. The second level consists of *suborders*, of which we need to mention only a few.

Soil orders and suborders are often distinguished by the presence of a diagnostic horizon. Each diagnostic horizon

Soils are classified by soil order and suborder. These divisions are largely distinguished by the presence of diagnostic horizons with unique physical or chemical properties.

has some unique combination of physical properties (color, structure, texture) or chemical properties (minerals present or absent). The two basic kinds of diagnostic horizons are (1) a horizon formed at the surface and called an *epipedon* (from the Greek, word *epi*, meaning "over" or "upon") and (2) a subsurface horizon formed by the removal or accumulation of matter. In our descriptions of soil orders, we will refer to a number of diagnostic horizons.

We can recognize three groups of soil orders. The largest group includes seven orders with well-developed horizons

Table 10.2 | Soil Orders

Group I

Soils with well-developed horizons or with fully weathered minerals, resulting from long-continued adjustment to prevailing soil temperature and soil water conditions.

Oxisols	Very old, highly weathered soils of low latitudes, with a subsurface horizon of accumulation of mineral oxides and very low base status.
Ultisols	Soils of equatorial, tropical, and subtropical latitude zones, with a subsurface horizon of clay accumulation and low base status.
Vertisols	Soils of subtropical and tropical zones with high clay content and high base status. Vertisols develop deep, wide cracks when dry, and the soil blocks formed by cracking move with respect to each other.
Alfisols	Soils of humid and subhumid climates with a subsurface horizon of clay accumulation and high base status. Alfisols range from equatorial to subarctic latitude zones.
Spodosols	Soils of cold, moist climates, with a well-developed *B* horizon of illuviation and low base status.
Mollisols	Soils of semiarid and subhumid midlatitude grasslands, with a dark, humus-rich epipedon and very high base status.
Aridisols	Soils of dry climates, low in organic matter, and often having subsurface horizons of accumulation of carbonate minerals or soluble salts.

Group II

Soils with a large proportion of organic matter.

Histosols	Soils with a thick upper layer very rich in organic matter.

Group III

Soils with poorly developed horizons or no horizons, and capable of further mineral alteration.

Entisols	Soils lacking horizons, usually because their parent material has accumulated only recently.
Inceptisols	Soils with weakly developed horizons, having minerals capable of further alteration by weathering processes.
Andisols	Soils with weakly developed horizons, having a high proportion of glassy volcanic parent material produced by erupting volcanoes.

or fully weathered minerals. A second group includes a single soil order that is very rich in organic matter. The last group includes three soil orders with poorly developed horizons or no horizons.

The world soils map presented in Figure 10.13 shows the major areas of occurrence of the soil orders. The Alfisols (Table 10.2) have been subdivided into four important suborders that correspond well to four basic climate zones. The map is quite general, indicating those areas where a given soil order is likely to be found. The map does not show many important areas of Entisols, Inceptisols, Histosols, and Andisols (Table 10.2). These orders are largely of local occurrence, since they are found on recent deposits such as floodplains, glacial landforms, sand dunes, marshlands, bogs, or volcanic ash deposits. The map also shows areas of highlands. In these regions, the soil patterns are too complex to show at a global scale.

Soil Orders

Table 10.3 explains the names of the soil orders. The formative element is a syllable used in the names of suborders and lower groups. Although each order has several suborders, we will refer to only a few. **GEODISCOVERIES**

Three soil orders dominate the vast land areas of low latitudes: Oxisols, Ultisols, and Vertisols. Soils of these orders have developed over long time spans in an environment of warm soil temperatures and soil water that is abundant in a wet season or lasts throughout the year. We will discuss these orders first.

OXISOLS Oxisols have developed in equatorial, tropical, and subtropical zones on land surfaces that have been stable over long periods of time. During soil development, the climate has been moist, with a large water surplus. Oxisols have developed over vast areas of South America and Africa in the wet equatorial climate ①. Here the native vegetation is rainforest. The wet-dry tropical climate ③ with its large seasonal water surplus is also associated with Oxisols in South America and Africa.

Oxisols usually lack distinct horizons, except for darkened surface layers. Soil minerals are weathered to an extreme degree and are dominated by stable sesquioxides of aluminum and iron. Red, yellow, and yellowish-brown colors are normal (Figures 10.14a and 10.15). The base status of the Oxisols is very low, since nearly all the bases required by plants have been removed from the soil profile. A small store of nutrient bases occurs very close to the soil surface. The soil is quite easily broken apart and allows easy penetration by rainwater and plant roots.

Oxisols and Ultisols develop over long time periods in warm, moist climates. Oxisols have substantial accumulations of iron and aluminum sesquioxides. Ultisols have a horizon of clay accumulation.

ULTISOLS Ultisols are quite closely related to the Oxisols in outward appearance and environment of origin. Ultisols are reddish to yellowish in color (Figures 10.14b and 10.16). They have a subsurface horizon of clay accumulation, called an *argillic horizon,* which is not found in the Oxisols. It is a B horizon and has developed through accumulation of clay in the process of illuviation. Although forest is the characterstic native vegetation, the base status of the Ultisols is low. As in the Oxisols, most of the bases are found in a shallow surface where they are released by the decay of plant matter. They are quickly taken up and recycled by the shallow roots of trees and shrubs.

In a few areas, the Ultisol profile contains a subsurface horizon of sesquioxides. This horizon is capable of hardening to a rock-like material if it becomes exposed at the surface and is subjected to repeated wetting and drying. This material is referred to as *plinthite* (from the Greek word *plinthos,* meaning "brick"). In the hardened state, plinthite is referred to as *laterite* (from the Latin *later,* or brick). In Southeast Asia, plinthite is quarried and cut into building blocks (Figure 10.17). These blocks harden into laterite blocks when they are exposed to the air.

Ultisols are widespread throughout Southeast Asia and the East Indies. Other important areas are in eastern Australia, Central America, South America, and the southeastern United States. Ultisols extend into the lower midlatitude zone in the United States, where they correspond quite closely in extent with the area of moist subtropical climate ⑥. In lower latitudes, Ultisols are identified with the wet-dry tropical climate ③ and the monsoon and trade-wind coastal climate ②. Note that all these climates have a dry season, even though it may be short.

Both Oxisols and Ultisols of low latitudes were used for centuries under shifting agriculture prior to the advent of modern agricultural technology. This primitive agricultural method, known as slash-and-burn, is still widely practiced. Without fertilizers, these soils can sustain crops on freshly cleared areas for only two or three years, at most, before the nutrient bases are exhausted and the garden plot must be abandoned. Substantial use of lime, fertilizers, and other industrial inputs is necessary for high, sustained crop yields. Furthermore, the exposed soil surface of the Ultisols is vulnerable to devastating soil erosion, particularly on steep hill slopes.

Vertisols develop on certain rock types in wet-dry climates under grassland and savanna vegetation. They expand and contract strongly on wetting and drying, creating deep cracks that allow the soil to "swallow itself" and remain constantly mixed.

VERTISOLS Vertisols have a unique set of properties that stand in sharp contrast to the Oxisols and Ultisols. Vertisols are black in color and have a high clay content (Figures 10.14c and 10.18). Much of the clay consists of a particular mineral that shrinks and swells greatly with seasonal changes in soil water content. Wide, deep vertical cracks develop in the soil during the dry season. As the dry soil blocks are wetted and softened by rain, some fragments of surface soil drop into the cracks before they close, so that the soil "swallows itself" and is constantly being mixed.

Vertisols typically form under grass and savanna vegetation in subtropical and tropical climates with a pronounced dry season. These climates include the semiarid subtype of the dry tropical steppe climate ④ and the wet-dry tropical climate ③. Because Vertisols require a particular clay mineral as a parent material, the major areas of occurrence are scattered and show no distinctive pattern on the world map. An important region of Vertisols is the Deccan Plateau of western India, where basalt, a dark variety of igneous rock, supplies the silicate minerals that are altered into the necessary clay minerals.

Vertisols are high in base status and are particularly rich in such nutrient bases as calcium and magnesium. The soil solution is nearly neutral in pH, and a moderate content of organic matter is distributed through the soil. The soil retains large amounts of water because of its fine texture, but much of this water is held tightly by the clay particles and is not available to plants. Where soil cultivation depends on human or animal power, as it does in most of the developing nations where the soil occurs, agricultural yields are low. This is because the moist soil becomes highly plastic and is difficult to till with primitive tools. For this reason, many areas of Vertisols have been left in grass or shrub cover, providing grazing for cattle. Soil scientists think that the use of modern technology, including heavy farm machinery, could result in substantial production of food and fiber from Vertisols that are not now in production.

ALFISOLS The **Alfisols** are soils characterized by an argillic horizon, produced by illuviation. Unlike the argillic horizon of the Ultisols, this B horizon is enriched by silicate clay minerals that have an adequate capacity to hold bases such as calcium and magnesium. The base status of the Alfisols is therefore generally quite high.

Above the B horizon of clay accumulation is a horizon of pale color, the E horizon, which has lost some of the original bases, clay minerals, and sesquioxides by the process of eluviation. These materials have become concentrated by illuviation in the B horizon. Alfisols also have a gray, brownish, or reddish surface horizon.

10.13 Soils of the world

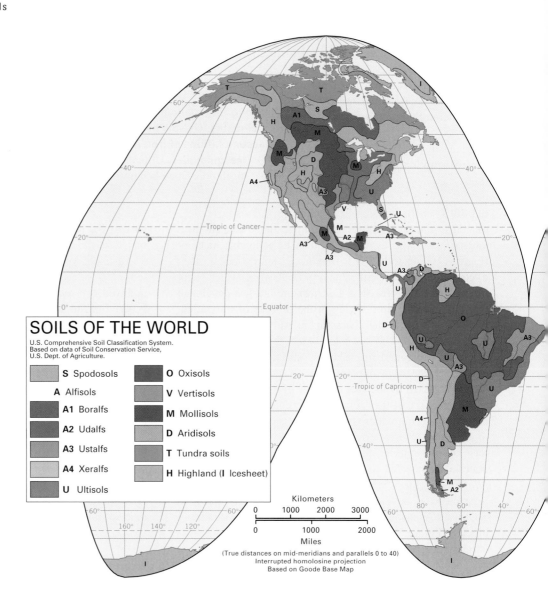

SOILS OF THE WORLD

U.S. Comprehensive Soil Classification System.
Based on data of Soil Conservation Service,
U.S. Dept. of Agriculture.

S Spodosols   **O** Oxisols

A Alfisols   **V** Vertisols

A1 Boralfs   **M** Mollisols

A2 Udalfs   **D** Aridisols

A3 Ustalfs   **T** Tundra soils

A4 Xeralfs   **H** Highland (**I** Icesheet)

U Ultisols

Kilometers
0 1000 2000 3000

0 1000 2000
Miles

(True distances on mid-meridians and parallels 0 to 40)
Interrupted homolosine projection
Based on Goode Base Map

Table 10.3 | Formative Elements in Names of Soil Orders

Name of Order	Formative Element	Derivation of Formative Element	Pronunciation of Formative Element
Entisol	ent	Meaningless syllable	recent
Inceptisol	ept	L. *inceptum,* beginning	inept
Histosol	ist	Gr. *histos,* tissue	histology
Oxisol	ox	F. *oxide,* oxide	ox
Ultisol	ult	L. *ultimus,* last	ultimate
Vertisol	ert	L. *verto,* turn	invert
Alfisol	alf	Meaningless syllable	alfalfa
Spodosol	od	Gr. *spodos,* wood ash	odd
Mollisol	oll	L. *mollis,* soft	mollify
Aridisol	id	L. *aridus,* dry	arid
Andisol	and	Eng. *andesite,* a volcanic rock type	and

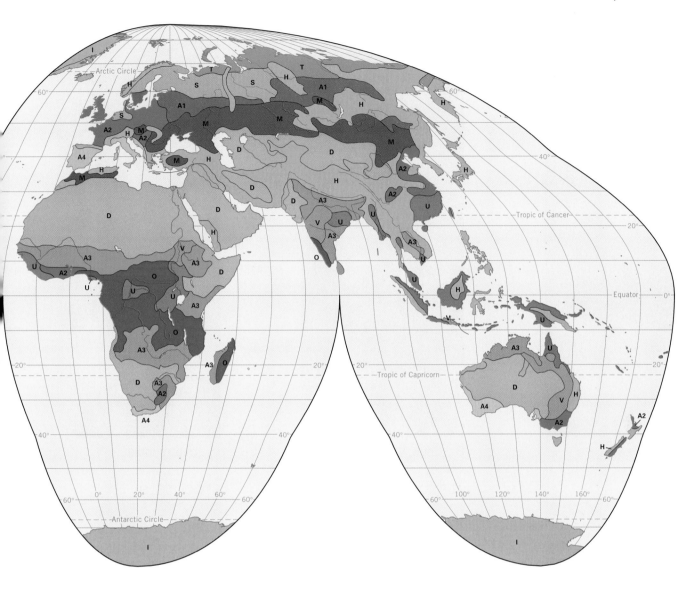

The world distribution of Alfisols is extremely wide in latitude (see Figure 10.13). Alfisols range from latitudes as high as 60° N in North America and Eurasia to the equatorial zone in South America and Africa. Obviously, the Alfisols span an enormous range in climate types. For this reason, we need to recognize four of the important suborders of Alfisols, each with its own climate affiliation.

Boralfs are Alfisols of cold (boreal) forest lands of North America and Eurasia. They have a gray surface horizon and a brownish subsoil. *Udalfs* are brownish Alfisols of the midlatitude zone. They are closely associated with the moist continental climate ⑩ in North America, Europe, and eastern Asia (Figure 10.14d).

Ustalfs are brownish to reddish Alfisols of the warmer climates (Figure 10.14e). They range from the subtropical zone to the equator and are associated with the wet-dry tropical climate ③ in Southeast Asia, Africa, Australia, and

> **Alfisols have horizons of eluviation and illuviation of clays. They have a high base status and can be very productive.**

South America. *Xeralfs* are Alfisols of the Mediterranean climate ⑦, with its cool moist winter and dry summer. The Xeralfs are typically brownish or reddish in color.

SPODOSOLS Poleward of the Alfisols in North America and Eurasia lies a great belt of soils of the order ***Spodosols,*** formed in the cold boreal forest climate ⑪ beneath a needleleaf forest. Spodosols have a unique property—a *B* horizon of accumulation of reddish mineral matter with a low capacity to hold bases (Figure 10.14f). This horizon is called the spodic horizon (Figure 10.19). It is made up of a dense mixture of organic matter and compounds of aluminum and iron, all brought downward by eluviation from an overlying *E* horizon. Because of the intensive removal of matter from the *E* horizon, it has a bleached, pale gray to white appearance (Figure 10.12). This conspicuous feature led to the naming of the soil as *podzol* (ash-soil) by Russian peasants. In modern terminology, this

OXISOLS

A   Torrox, Hawaii

ULTISOLS

B   Udult, Virginia

VERTISOLS

C   Ustert, India

ALFISOLS

D   Udalf, Michigan

ALFISOLS

E   Ustalf, Texas

SPODOSOLS

F   Orthod, France

10.14 Soil profiles of several soil orders

MOLLISOLS

G   Boroll, USSR

MOLLISOLS

H   Udoll, Argentina

MOLLISOLS

I   Ustoll, Colorado

MOLLISOLS

J   Rendoll, Argentina

ARIDISOLS

K   Argid, Colorado

HISTOSOLS

L   Fibrist, Minnesota

10.15 An Oxisol in Hawaii *Sugarcane is being cultivated here.*

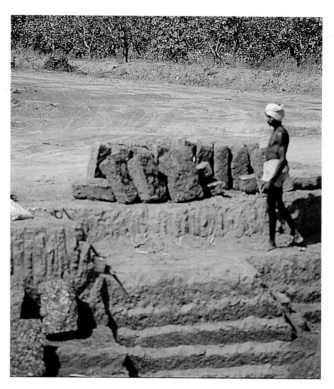

10.17 Laterite in India *Laterite, formed by hardening of plinthite, is being quarried for building stone in this scene from India.*

10.16 Ultisol profile in North Carolina *The thin, pale layer at the top is an E horizon, showing the effects of removal of materials by eluviation. Below the base of the thick, reddish B horizon is a blotchy-colored zone of plinthite.*

10.18 A Vertisol in Texas *The clay minerals that are abundant in Vertisols shrink when they dry out, producing deep cracks in the soil surface.*

10.19 Diagram of a Spodosol profile *The spodic horizon, a dense mixture of organic matter, aluminum, and iron oxides, and the gray to white albic horizon, are diagnostic features of a Spodosol profile.*

and magnesium. They are also low in humus. Although the base status of the Spodosols is low, forests of pine and spruce are supported through the process of recycling of the bases.

Spodosols are closely associated with regions recently covered by the great ice sheets of the Pleistocene Epoch. These soils are therefore very young. Typically, the parent material is coarse sand consisting largely of the mineral quartz. This mineral cannot weather to form clay minerals.

Spodosols are naturally poor soils in terms of agricultural productivity. Because they are acid, application of lime is essential. Heavy applications of fertilizers are also required. With proper management and the input of the required industrial products, Spodosols can be highly productive, if the soil texture is favorable. High yields of potatoes from Spodosols in Maine and New Brunswick are examples. Another factor unfavorable to agriculture is the shortness of the growing season in the more northerly parts of the Spodosol belt.

HISTOSOLS   Throughout the northern regions of Spodosols are countless patches of **Histosols.** This unique soil order has a very high content of organic matter in a thick, dark upper layer (Figure 10.14*l*). Most Histosols go by such common names as peats or mucks. They have formed in shallow lakes and ponds by accumulation of partially decayed plant matter. In time, the water is replaced by a layer of organic matter, or *peat,* and becomes a *bog* (Figure 10.20). Peat bogs are used extensively

pale layer is an *albic horizon.* The *O* horizon, a thin, very dark layer of organic matter, overlies the *A horizon.*

Spodosols are strongly acid and are low in plant nutrients such as calcium

---

Spodisols have a light-colored (albic) horizon of eluviation and a dense horizon of illuvation (spodic horizon). They develop under cold needleleaf forests.

10.20 Eye on the Landscape Peat bog *This peat bog in Galway, Ireland, has been trenched to reveal a Histosol profile. Peat blocks, seen to the sides of the trench, are drying for use as fuel.*   **What else would the geographer see? ... Answers at the end of the chapter.**

10.21 A Histosol in cultivation *Garden crops cultivated on a Histosol of a former glacial lakebed, Southern Ontario, Canada.*

for cultivation of cranberries (cranberry bogs). Sphagnum peat from bogs is dried and baled for sale as a mulch for use on suburban lawns and shrubbery beds. For centuries, Europe has used dried peat from bogs of glacial origin as a low-grade fuel.

Some Histosols are *mucks*—organic soils composed of fine black materials of sticky consistency. These are agriculturally valuable in midlatitudes, where they occur as beds of former lakes in glaciated regions. After appropriate drainage and application of lime and fertilizers, these mucks are remarkably productive for garden vegetables (Figure 10.21). Histosols are also found in low latitudes, where conditions of poor drainage have favored thick accumulations of plant matter.

ENTISOLS Entisols have in common the combination of a mineral soil and the absence of distinct horizons. Entisols are soils in the sense that they support plants, but they may be found in any climate and under any vegetation. Entisols lack distinct horizons for two reasons. It may be the result of a parent material, such as quartz sand, in which horizons do not readily form. Or it may be the result of lack of time for horizons to form in recent deposits of alluvium or on actively eroding slopes.

Histosols are organic soils, often termed peats or mucks. They are typically formed in cool or cold climates in areas of poor drainage.

Entisols lack horizons, often because they are only recently deposited. They may occur in any climate or region.

Inceptisols have only weakly developed horizons. Inceptisols of river floodplains and deltas are often very productive where they are regularly enriched by flood borne silt.

Entisols occur from equatorial to arctic latitude zones. From the standpoint of agricultural productivity, Entisols of the subarctic zone and tropical deserts (along with arctic areas of Inceptisols) are the poorest of all soils. In contrast, Entisols and Inceptisols of floodplains and delta plains in warm and moist climates are among the most highly productive agricultural soils in the world because of their favorable texture, ample nutrient content, and large soil water storage.

INCEPTISOLS Inceptisols are soils with horizons that are weakly developed, usually because the soil is quite young. These areas occur within some of the regions shown on the world map as Ultisols and Oxisols. Especially important are the Inceptisols of river floodplains and delta plains in Southeast Asia that support dense populations of rice farmers.

In these regions, annual river floods cover low-lying plains and deposit layers of fine silt. This sediment is rich in primary minerals that yield bases as they weather chemically over time. The constant enrichment of the soil explains the high soil fertility in a region where uplands develop only Ultisols of low fertility. Inceptisols of these floodplain and delta lands are of a suborder called *Aquepts*—Inceptisols of wet places. Much closer to home is another prime example of Aquepts within the domain of the Ultisols—the lower Mississippi River floodplain and delta plain.

ANDISOLS **Andisols** are soils in which more than half of the parent mineral matter is volcanic ash, spewed high into the air from the craters of active volcanoes and coming to rest in layers over the surrounding landscape. The fine ash particles are glass-like shards. A high proportion of carbon, formed by the decay of plant matter, is also typical, so that the soil usually appears very dark in color. Andisols form over a wide range of latitudes and climates. They are for the most part fertile soils, and in moist climates they support a dense natural vegetation cover.

> Andisols are unique soils that form on volcanic ash of relatively recent origin. They are dark in color and typically fertile.

Andisols do not appear on our world map because they are found in small patches associated with individual volcanoes that are located mostly in the "Ring of Fire"—the chain of volcanic mountains and islands that surrounds the great Pacific Ocean. Andisols are also found on the island of Hawaii, where volcanoes are presently active.

MOLLISOLS **Mollisols** are soils of grasslands that occupy vast areas of semiarid and subhumid climates in midlatitudes. Mollisols are unique in having a very thick, dark brown to black surface horizon called the *mollic epipedon* (Figures 10.14*g* to *j* and 10.22). This layer lies within the *A* horizon and is always more than 25 cm (9.8 in.) thick. The soil has a loose,

> Mollisols are soils of grasslands in subhumid to semiarid climates. They have a thick, dark brown surface layer termed a mollic epipedon. Their loose texture and high base status place them among the most productive soil types in the world.

granular structure (Figure 10.6) or a soft consistency when dry. Other important qualities of the Mollisols are the dominance of calcium among the bases of the *A* and *B* horizons and the very high base status of the soil.

Most areas of Mollisols are closely associated with the semiarid subtype of the dry midlatitude climate ⑨ and the adjacent portion of the moist continental climate ⑩. In North America, Mollisols dominate the Great Plains region, the Columbia Plateau, and the northern Great Basin. In South America, a large area of Mollisols covers the Pampa region of Argentina and Uruguay. In Eurasia, a great belt of Mollisols stretches from Romania eastward across the steppes of Russia, Siberia, and Mongolia. Russians refer to the Mollisols as *chernozems,* a term that has gained widespread use throughout the Western world as well.

Because of their loose texture and very high base status, Mollisols are among the most naturally fertile soils in the world. They now produce most of the world's commercial grain crop. Most of these soils have been used for crop production only in the last century. Prior to that time, they were used mainly for grazing by nomadic herds. The Mollisols have favorable properties for growing cereals in large-scale mechanized farming and are relatively easy to manage. Production of grain varies considerably from one year to the next because seasonal rainfall is highly variable.

A brief mention of four suborders of the Mollisols as they occur in the United States and Canada will help you to understand important regional soil differences related to climate. *Borolls,* the cold-climate suborder of the Mollisols, are found in a large area extending on both sides of the U.S.-Canadian border east of the Rocky Mountains (Figure 10.14*g*). *Udolls* are Mollisols of a relatively moist climate, as compared with the other suborders. Formerly, the Udolls supported tall-grass prairie, but today they are closely identified with the corn belt in the American Midwest (Figure 10.14*h*).

Ustolls are Mollisols of the semiarid subtype of the dry midlatitude climate ⑨, with a substantial soil water shortage in the summer months (Figure 10.14*i*). The Ustolls underlie much of the short-grass prairie region east of the Rockies (Figure 10.23). *Xerolls* are Mollisols of the Mediterranean climate ⑦, with its tendency to cool, moist winters and rainless summers.

Desert and Tundra Soils

Desert and tundra soils are soils of extreme environments. Aridisols characterize the desert climate. As might be expected, they are low in organic matter and high in salts. Tundra soils are poorly developed because they are formed on very recent parent material, left in place by glacial activ-

10.22 Schematic diagram of a Mollisol profile *The thick, dark brown to black mollic epipedon is a diagnostic feature of the Mollisol.*

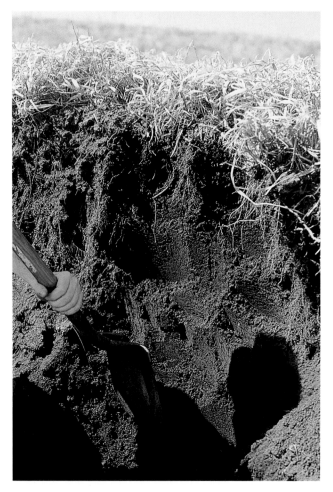

10.23 A Mollisol developed on loess *Dry prairie grasses are seen on the surface.*

10.24 Salic horizon *Appearing as a white layer, a salic horizon lies close to the surface in this Aridisol profile in the Nevada desert. The scale is marked in feet (1 ft = 30.5 cm).*

ity during the Ice Age. Cold temperatures have also restricted soil development in tundra regions.

ARIDISOLS Aridisols, soils of the desert climate, are dry for long periods of time. Because the climate supports only a very sparse vegetation, humus is lacking and the soil color ranges from pale gray to pale red (Figure 10.14*k*). Soil horizons are weakly developed, but there may be important subsurface horizons of accumulated calcium carbonate (petrocalcic horizon) or soluble salts (salic horizon) (Figure 10.24). The salts, of which sodium is a dominant constituent, give the soil a very high degree of alkalinity. The Aridisols are closely correlated with the arid subtype of the dry tropical ④, dry subtropical ⑤, and dry midlatitude ⑨ climates.

Most Aridisols are used for nomadic grazing, as they have been through the ages. This use is dictated by the limited rainfall, which is inadequate for crops

---

Aridisols are desert soils with weakly developed horizons. They often exhibit subsurface layers of accumulation of calcium carbonate or soluble salts. With irrigation and proper management, they are quite fertile.

Tundra soils are wet and cold Inceptisols. They are often underlain by permafrost.

---

without irrigation. Locally, where water supplies from mountain streams or ground water permit, Aridisols can be highly productive for a wide variety of crops under irrigation (Figure 10.25). Great irrigation systems, such as those of the Imperial Valley of the United States, the Nile Valley of Egypt, and the Indus Valley of Pakistan, have made Aridisols highly productive, but not without problems of salt buildup and water-logging.

TUNDRA SOILS Soils of the arctic tundra fall largely into the order of Inceptisols, soils with weakly developed horizons that are usually associated with a moist climate. Inceptisols of the tundra climate belong to the suborder of Aquepts, which as we noted earlier are Inceptisols of wet places. More specifically, the tundra soils can be assigned to the *Cryaquepts,* a subdivision within the Aquepts. The prefix *cry* is derived from the Greek word *kryos,* meaning "icy cold." We may refer to these soils simply as *tundra soils.*

10.25 *Eye on the Landscape* **Irrigated Aridisols** *Cotton farming along the Colorado River, near Parker, Arizona, is shown in this oblique air photo.* **What else would the geographer see? ... Answers at the end of the chapter.**

Tundra soils are formed largely of primary minerals ranging in size from silt to clay that are broken down by frost action and glacial grinding. Layers of peat are often present between mineral layers. Beneath the tundra soil lies persenially frozen ground (permafrost), described in Chapter 14. Because the annual summer thaw affects only a shallow surface layer, soil water cannot easily drain away. Thus, the soil is saturated with water over large areas. Repeated freezing and thawing of this shallow surface layer disrupts plant roots, so that only small, shallow-rooted plants can maintain a hold.

A Midcontinental Transect from Aridisols to Alfisols

Figure 10.26 provides a diagram showing generalized changes in soil profile and soil type as they might occur on a transect from a cool, dry desert to a cool moist climate,

10.26 *Soil profile transect A schematic diagram of the changing soil profile from a cool dry desert on the West to a cool moist climate on the East. (After C. E. Millar, L. M. Turk, and H. D. Foth, Fundamentals of Soil Science, John Wiley and Sons, New York.)*

starting in the northern desert of the Great Basin and crossing the Midwest. Starting with the Aridisols in the West, horizons are thin but show an accumulation of clay minerals in the *B* horizon. As precipitation increases, grasses dominate and Mollisols are formed. The *B* horizon thickens and includes calcium accumulation. In the east, where alfisols occur, enhanced precipitation generates an *E* horizon that takes its place between the *A* and *B* horizons.

A LOOK AHEAD

This chapter completes a sequence of chapters describing three interrelated topics—climate, vegetation, and soils. The climate of a region is determined by the characteristics of air masses and weather systems that it experiences through the year. The nature of the vegetation cover depends largely on the climate, varying from desert to tundra biomes. Soils depend on both climate and vegetation to develop their unique characteristics. But soils can also influence vegetation development. Similarly, the climate at a location can be influenced by its vegetation cover. Thus, there is a strong interaction between climate, soils, and vegetation.

The next three chapters also comprise a set that are strongly related. They deal with the lithosphere—the realm of the solid Earth, as we described in the Prologue. Our survey of the lithosphere will begin with the nature of Earth materials, then move to a discussion of how continents and ocean basins are formed and how they are continually changing, even today. Finally, we will discuss landforms occurring within continents that are produced by such lithospheric processes as volcanic activity and earthquake faulting.

In Review | Global Soils

- The **soil** layer is a complex mixture of solid, liquid, and gaseous components. It is derived from **parent material,** or *regolith,* that is produced from rock by *weathering.* The major factors influencing soil and soil development are parent material, climate, vegetation, and time.

- **Soil texture** refers to the proportions of *sand, silt,* and *clay* that are present. *Colloids* are the finest particles in soils and are important because they help retain nutrient ions, or **bases,** that are used by plants. Soils show a wide range of *pH values,* from acid to alkaline. Soils with granular or blocky structures are most easily cultivated.

- In soils, *primary minerals* are chemically altered to **secondary minerals,** which include oxides and *clay minerals.* The nature of the clay minerals determines the soil's base status. If *base status* is *high,* the soil retains nutrients; if *low,* the soil can lack fertility. When a soil is fully wetted by heavy rainfall or snowmelt and allowed to drain, it reaches its *storage capacity.* Evaporation from the surface and transpiration from plants draws down the soil water store until precipitation occurs again to recharge it.

- The *soil water balance* describes the gain, loss, and storage of soil water. It depends on **water need** *(potential evapotranspiration),* **water use** *(actual evapotranspiration),* and precipitation. Monitoring these values on a monthly basis provides a *soil water budget* for the year.

- Most soils possess distinctive horizontal layers called **horizons.** These layers are developed by processes of *enrichment, removal, translocation,* and *transformation.* In downward translocation, materials such as humus, clay particles, and mineral oxides are removed by **eluviation** from an upper horizon and accumulate by **illuviation** in a lower one. In *salinization,* salts are translocated upward by evaporating water to form a *salic horizon.* In *humification,* a transformation process, organic matter is broken down by bacterial decay. Where soil temperatures are warm, this process can be highly effective, leaving a soil low in organic content. Animals, such as earthworms, can be very important in soil formation where they are abundant.

- Global soils are classified into 11 **soil orders,** often by the presence of one of more diagnostic horizons. **Oxisols** are old, highly weathered soils of low latitudes. They have a horizon of mineral oxide accumulation and a low base status. **Ultisols** are also found in low latitudes. They have a horizon of clay accumulation and are also of low base status. **Vertisols** are rich in a type of clay mineral that expands and contracts with wetting and drying, and has a high base status. **Alfisols** have a horizon of clay accumulation like Ultisols, but they are of high base status. They are found in moist climates from equatorial to subarctic zones. **Spodosols,** found in cold, moist climates, exhibit a horizon of illuviation and low base status. **Histosols** have a thick upper layer formed almost entirely of organic matter.

- Three soil orders have poorly developed horizons or no horizons—**Entisols, Inceptisols,** and **Andisols.** Entisols are composed of fresh parent material and have no horizons. The horizons of Inceptisols are only weakly developed. Andisols are weakly developed soils occurring on young volcanic deposits.

- **Mollisols** have a thick upper layer rich in humus. They are soils of midlatitude grasslands. **Aridisols** are soils of arid regions, marked by horizons of accumulation of carbonate minerals or salts.

- *Tundra soils* are largely wet, cold Inceptisols.

Key Terms

soil, p. 342
parent material, p. 343
soil texture, p. 344
soil colloids, p. 344
bases, p. 345
secondary minerals,
    p. 346

water use, p. 348
water need, p. 348
soil horizons, p. 349
soil profile, p. 349
eluviation, p. 350
illuviation, p. 350
soil orders, p. 353

Oxisols, p. 354
Ultisols, p. 355
Vertisols, p. 355
Alfisols, p. 355
Spodosols, p. 357
Histosols, p. 361
Entisols, p. 362

Inceptisols, p. 362
Andisols, p. 363
Mollisols, p. 363
Aridisols, p. 364

Review Questions

1. Which important factors condition the nature and development of the soil?
2. Soil color, soil texture, and soil structure are used to describe soils and soil horizons. Identify each of these three terms, showing how they are applied.
3. Explain the concepts of acidity and alkalinity as they apply to soils.
4. Identify two important classes of secondary minerals in soils and provide examples of each class.
5. How does the ability of soils to hold water vary, and how does this ability relate to soil texture?
6. Define water need (potential evapotranspiration) and water use (actual evapotranspiration). How are they used in the soil water balance?
7. Identify the following terms as used in the soil water budget: storage, withdrawal, storage recharge, soil water shortage, water surplus.
8. What is a soil horizon? How are soil horizons named? Provide two examples.
9. Identify four classes of soil-forming processes and describe each.
10. What are translocation processes? Identify and describe four translocation processes.
11. How many soil orders are there? Try to name them all.
12. Name three soil orders that are especially associated with low latitudes. For each order, provide at least one distinguishing characteristic and explain it.
13. Compare Alfisols and Spodosols. What features do they share? What features differentiate them? Where are they found?
14. Where are Mollisols found? How are the properties of Mollisols related to climate and vegetation cover? Name four suborders within the Mollisols.
15. Desert and tundra are extreme environments. Which soil order is characteristic of each environment? Briefly describe desert and tundra soils.

Eye on the Environment 10.1 • Death of a Civilization

1. Describe the processes of salinization, waterlogging, and siltation and their impact on agriculture.
2. A technological "fix" for siltation is to build a dam upstream of an irrigation canal complex, so that silt will be trapped behind the dam. What other effects might building such a dam produce, both desirable and undesirable?

Visualizing Exercises

1. Sketch the profile of a Spodosol, labeling *O, A, E, B,* and *C* horizons. Diagram the movement of materials from the zone of eluviation to the zone of illuviation.

2. Examine the world soils map (Figure 10.13) and identify three soil types that are found near your location. Develop a short list of characteristics that would help you tell them apart.

Essay Questions

1. Document the important role of clay particles and clay mineral colloids in soils. What is meant by the term *clay?* What are colloids? What are their properties? How does the type of clay mineral influence soil fertility? How does the amount of clay influence the water-holding capacity of the soil? What is the role of clay minerals in horizon development?

2. Using the world maps of global soils and global climate, compare the pattern of soils on a transect along the 20° E longitude meridian with the patterns of climate encountered along the same meridian. What conclusions can you draw about the relationship between soils and climate? Be specific.

Eye on the Landscape

Chapter Opener   Piedmont farm-land in southern Georgia This aerial oblique view shows some of the typical diversity seen in an American agricultural landscape. Fallow fields (**A**), some furrowed and others disk-harrowed, show red soil colors, typical of the Ultisols of this region. Light brown-toned fields (**B**) are probably small grain crops ready for harvest. The lighter green fields (**C**) may be a cover crop, planted to keep the soil stabilized against erosion. A stand of planted pines (**D**) lies in the foreground, while a patch of deciduous forest (**E**) is in the distance. Ultisols are treated in this chapter.

10.25   Irrigated   Aridisols The broad floodplain of the Colorado River (right) shows how productive Aridisols can be when properly managed and irrigated. A consistent climate with abundant sunlight, coupled with abundant river water, makes for high productivity and crop yields. Fallow fields (**A**) show typical Aridisol colors of gray to brown, while cotton fields are vibrant green (**B**). Note the network of canals (**C**) serving the fields. Aridisols are covered in this chapter.

10.20   Peat bog Peat bogs are typical of glacial terrain. Moving ice sheets remove and take up loose soil and rock, then leave sediment behind as they melt. The low, hummocky terrain (**A**) at the edge of this bog may be part of a moraine, a landform that occurs at the melting edge of an ice sheet as wasting ice leaves piles of sediment behind in a chaotic and disorganized fashion. The isolated angular boulders (**B**) may be erratics—large rocks that are carried in the ice and moved long distances from their sources. Landforms of glacial ice sheets are described in Chapter 19.

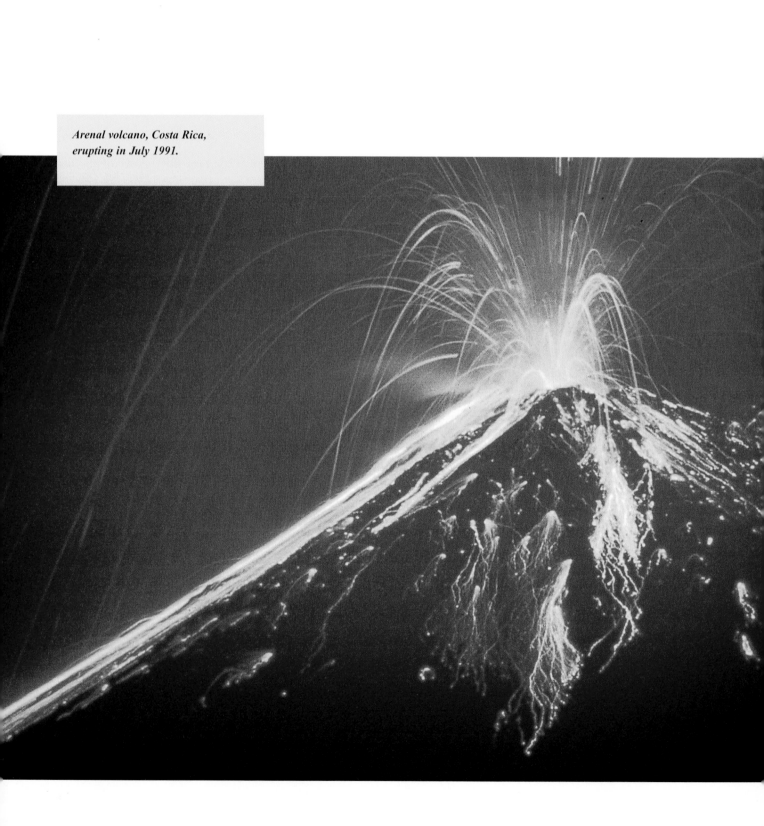

Arenal volcano, Costa Rica, erupting in July 1991.

11 | EARTH MATERIALS

PUTTING YOU IN THE PICTURE

Perhaps from collecting pebbles along a beach or stones from the bed of a stream, you may be aware that there are many different kinds of rocks. The primary objective of this chapter is to describe the more important types of rocks and minerals and how they are formed. Although most processes of rock formation are slow and steady, some are rapid and even a bit dangerous to witness. Imagine yourself on a journey to a pit crater on the Hawaiian volcano of Kilauea to watch the formation of new rock from molten lava…

The "Visitor Parking" sign emerges quickly from the fog, and you nose the car to the left, pulling into a vacant space in the parking lot. You expected better weather, but the whole eastern half of the "Big Island"—Hawaii—is covered by trade-wind clouds. At this elevation, about 1200 m (4000 ft), you are in the middle of the wet, gray, cloud layer. You open the trunk to get the sweatshirt from your backpack and then change into your sturdy hiking boots. Your destination is Mauna Ulu, a nearly circular pit crater that contains a pool of molten lava.

At first, the trail winds through a scrubby forest of low trees and shrubs. The tough rock underfoot is broken into angular blocks, separated by cracks and fissures. The footing is difficult. Each step requires attention to keep your boot from becoming wedged into a crevice. Soon you emerge from the woodland onto a surface of fresh lava. The trail is marked by small flags every 20 feet or so. The fog thickens a bit. As you walk carefully from one flag to the next, you seem to be inside a dim, gray bubble on a strange planet. There are no landmarks—nothing familiar is within sight— as you pick your way across the rough terrain.

You gradually become aware of a noise, like a distant waterfall. As you continue along the trail, it gets louder and you begin to feel the ground shaking beneath your feet. The fog takes on a redish glow. The acrid scent of sulfur catches in your nose and tightens your throat. Dimly, something emerges ahead. You are nearing the rim of the crater. You make out a wooden platform on the edge, with perhaps a half-dozen people clustered toward the back. As you reach the platform, the noise builds to a great, lowpitched roar. You mount the trembling platform and approach the edge. The heat from the lava strikes your face with an unexpected intensity.

The sides of the crater drop away steeply. At the bottom of the great pit is a black surface crust of congealed lava, covering the hot magma. Long cracks gash the crust, glowing red from deep within. Suddenly, a huge rock mass falls from the side of the crater into the pit, fracturing the black surface into jagged pieces that are tossed outward by the impact. Molten rock splashes slowly upward, spattering the rock wall with red and white spray. The heat of the fresh molten rock is unbearable, and you turn away.

After watching the incredible spectacle for as long as you can endure the heat, you leave the platform and head back. As you walk, the roaring fades, and the fog loses its red glow. The lava landscape seems mysterious rather than ominous and threatening. The scrubby woods seems somehow welcoming, its twisted trees beckoning as outposts of Earthly life on your return journey to familiar ground. You feel small and insignificant, humbled by the immense physical power of the Earth.

The cooling of molten lava that is released at the Earth's surface is one way in which new rock is formed. This chapter explores other ways that are perhaps not so dramatic, but equally important. Our objective will be to lay the groundwork necessary for understanding how the Earth's crustal processes control landforms and landscapes, as well as provide the raw materials for the land-forming agents of water, wind, and glacial ice that we discuss in later chapters. ■

EARTH MATERIALS

This chapter is the first in a sequence of three chapters that focus on the solid mineral realm of our planet. The first chapter deals with the materials that compose the Earth's outer rock layer—minerals and rocks that are formed in various ways both deep within the Earth and near or at its surface. In the second chapter, we will see that the outermost layer of the solid Earth is a hardened crust of rock that is fractured, broken into large plates, and moved by forces deep within the Earth. In the third chapter, we will show how surface features reveal the history of the crust and the forces that have shaped it. The topics in these three chapters are drawn from *geology,* the science of the solid Earth and its history.

Why are physical geographers concerned with the solid Earth? The Earth's outer layer serves as a platform, or base, for life on the lands. It provides the continental surfaces that are carved into landforms by moving water, wind, and glacial ice. Landforms, in turn, influence the distribution of ecosystems and exert strong controls over human occupation of the lands. Landforms made by water, wind, and ice are the subject of the closing chapters of this book.

We begin our study of the solid Earth by examining the minerals and rocks that are found at or near the Earth's surface. There are a great number of minerals and rocks, but we will focus only on the few that are most important for a broad understanding of the continents, ocean basins, and their varied features. We will also examine the rock cycle, in which Earth materials are constantly formed and reformed over geologic time in a cycle powered largely by heat from the Earth's interior.

THE CRUST AND ITS COMPOSITION

The thin, outermost layer of our planet is the *Earth's crust.* This skin of varied rocks and minerals ranges from about 8 to 40 km (about 5 to 25 mi) thick and contains the continents and ocean basins. It is the source of soil on the lands, of salts of the sea, of gases of the atmosphere, and of all the water of the oceans, atmosphere, and lands.

The Earth's crust is the outermost layer of the solid Earth. It contains the rocks of the continents and ocean basins, and is largely composed of oxygen, silicon, aluminum, and iron.

Figure 11.1 displays the eight most abundant elements of the Earth's crust in terms of percentage by weight. Oxygen, the predominant element, accounts for a little less than half the total weight. Second is silicon, which accounts for a little more than a quarter. Together they account for 75 percent of the crust, by weight.

Aluminum accounts for approximately 8 percent and iron for about 5 percent of the Earth's crust. These metals are very important to our industrial civilization, and, fortunately, they are relatively abundant. Four other metallic elements—calcium, sodium, potassium, and magnesium—

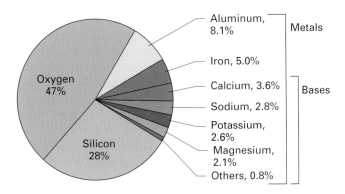

11.1 Crustal composition *The eight most abundant elements in the Earth's crust, measured by percentage of weight. Oxygen and silicon dominate, with aluminum and iron following.*

make up the remaining 12 percent. All four occur at about the same order of abundance (2 to 4 percent). These elements are essential nutrients for plant and animal life and are described in Chapter 8 as plant nutrients. Because of this role, they are also important determiners of soil fertility (Chapter 10).

Of the remaining chemical elements composing the Earth's crust, a few are very important from the viewpoint of powering the cycles of the solid Earth. These are the radioactive forms of elements that, through slow radioactive decay, provide a nearly infinite source of heat that seeps slowly outward from the Earth's interior. We will return to the topic of radioactive decay and radiogenic heating in Chapter 12.

Rocks and Minerals

The elements of the Earth's crust are usually bonded with other elements to form chemical compounds. We recognize these compounds as minerals. A **mineral** is a naturally occurring, inorganic substance that usually possesses a definite chemical composition and characteristic atomic structure. Most minerals have a crystalline structure—gemstones such as diamonds or rubies are examples, although their crystalline shape is enhanced by the stonecutter. Quartz is an example of a very common crystalline mineral. It usually occurs as a clear, six-sided prism (Figure 11.2).

Minerals are combined into **rock,** which we can broadly define as an assemblage of minerals in the solid state. Rock comes in a very wide range of compositions, physical characteristics, and ages. A given variety of rock is usually composed of two or more minerals, and often many different minerals are present. However, a few rock varieties consist almost entirely of one mineral. Most rock of the Earth's crust is extremely old by human standards, with the age of

11.2 Quartz crystals *These large crystals of quartz form six-sided, translucent columns.*

formation often ranging back many millions of years. But rock is also being formed at this very hour as active volcanoes emit lava that solidifies on contact with the atmosphere or ocean.

Rocks of the Earth's crust fall into three major classes. (1) **Igneous rocks** are solidified from mineral matter in a high-temperature molten state. (2) **Sedimentary rocks** are formed from layered accumulations of mineral particles derived mostly by weathering and erosion of preexisting rocks. (3) **Metamorphic rocks** are formed from igneous or sedimentary rocks that have been physically or chemically changed, usually by application of heat and pressure during crustal mountain-making.

The three classes of rocks are constantly being transformed from one to another in a continuous process through which the crustal minerals have been recycled during many millions of years of geologic time. Figure 11.3 diagrams these transformations. In the process of melting, preexisting rock of any class is melted and then later cools to form igneous rock. In weathering and erosion, preexisting rock is broken down and accumulated in layers that become sedimentary rock. Heat and pressure convert igneous and sedimentary rocks to metamorphic rock. We will return to this cycle at the end of the chapter, after we have taken a more detailed look at rocks, minerals, and their formation processes.

Rocks are composed of minerals, which are naturally occurring inorganic substances. The three major classes of rocks are igneous, sedimentary, and metamorphic. Earth processes transform rocks from one class to another.

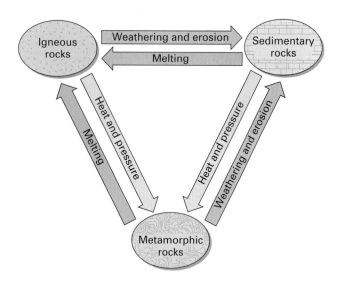

11.3 Rock transformation *The three classes of rock are transformed into one another by weathering and erosion, melting, and exposure to heat and pressure.*

IGNEOUS ROCKS

Igneous rocks are formed when molten material moves from deep within the Earth to a position within or atop the crust. There the molten material cools, forming rocks composed of mineral crystals. Table 11.1 presents the more important igneous rock types that we will refer to in this chapter.

Most igneous rock consists of *silicate minerals,* chemical compounds that contain silicon and oxygen atoms. Most of the silicate minerals also have one, two, or more of the metallic elements listed in Figure 11.1—that is, aluminum, iron, calcium, sodium, potassium, and magnesium. Although there are many silicate minerals, we will only be concerned with seven of them, which are shown in Figure 11.4.

Among the most common minerals of all rock classes is **quartz,** mentioned earlier, which is silicon dioxide (SiO_2) (Figure 11.2). It is quite hard and

resists chemical breakdown. Two silicate-aluminum minerals called *feldspars* follow. One type, *potash feldspar,* contains potassium as the dominant metal besides aluminum. A second type, *plagioclase feldspar,* is rich in sodium, calcium, or both. Quartz and feldspar form a silicate mineral group described as **felsic** ("fel" for feldspar; "si" for silicate).

Quartz and the feldspars are light in color (white, pink, or grayish) and lower in density than the other silicate minerals. By *density,* we mean the mass of matter contained in a unit volume. Figure 11.5 shows the density of four Earth materials: water, quartz, the mineral olivine (discussed below), and pure iron.

The next three silicate minerals are actually groups of minerals, with a number of mineral varieties in each group. They are the *mica, amphibole,* and *pyroxene* groups. All three are silicates containing aluminum, magnesium, iron, and potassium or calcium. The seventh mineral, *olivine,* is a silicate of only magnesium and iron that lacks aluminum. Altogether, these minerals are described as **mafic** ("ma" for magnesium; "f" from the chemical symbol for iron, Fe). The mafic minerals are dark in color (usually black) and are denser than the felsic minerals. *GEODISCOVERIES*

Common Igneous Rocks

Igneous rocks solidify from rock in a hot, molten state, known as **magma.** From pockets a few kilometers below the Earth's surface, magma makes its way upward through fractures in older solid rock and eventually solidifies as igneous rock. No single igneous rock is made up of all seven silicate minerals listed in Figure 11.4. Instead, a given rock variety contains three or four of those minerals as the major ingredients. The mineral grains in igneous rocks are very tightly interlocked, and the rock is normally very strong.

The column in the center of Figure 11.4 shows four common igneous rocks. Each rock is connected by arrows to the principal minerals it contains. These four rocks are carefully selected to be used in our later explanation of features of the Earth's crust.

> Igneous rocks form when molten rock cools, forming mineral crystals. Silicate minerals in igneous rocks are classed as felsic (light colored, less dense) and mafic (dark colored, more dense). The felsic mineral quartz is very common.

Table 11.1 | Some Common Igneous Rock Types

Subclass	Rock Type	Composition
Intrusive (cooling at depth, producing coarse crystal texture)	Granite	Felsic minerals, typically quartz, feldspars, and mica
	Diorite	Felsic minerals without quartz, usually including plagioclase feldspar and amphibole
	Gabbro	Mafic minerals, typically plagioclase feldspar, pyroxene, and olivine
	Peridotite	An ultramafic rock of pyroxene and olivine
Extrusive (cooling at the surface, producing fine crystal texture)	Rhyolite	Same as granite
	Andesite	Same as diorite
	Basalt	Same as gabbro

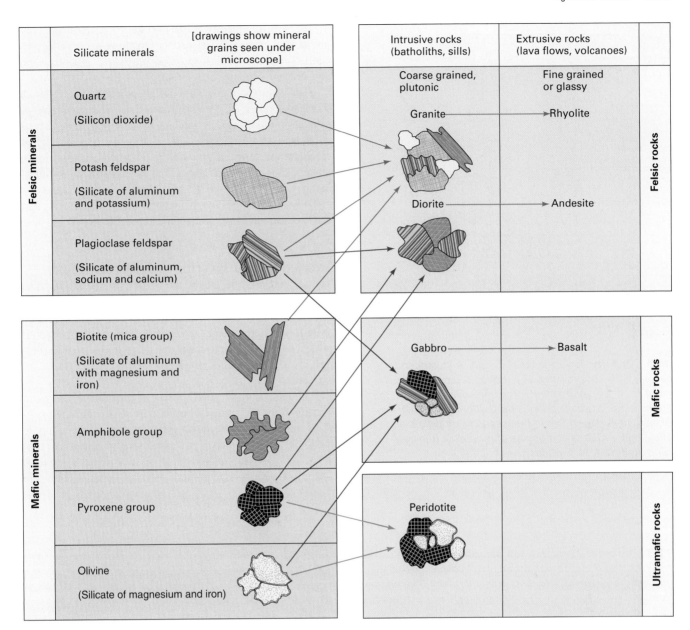

11.4 *Silicate minerals and igneous rocks* *Only the most important silicate mineral groups are listed, along with four common igneous rock types. The patterns shown for mineral grains indicate their general appearance through a microscope, and do not necessarily indicate the relative amounts of the minerals within the rocks.*

The first igneous rock is **granite.** The bulk of granite consists of quartz (27 percent), potash feldspar (40 percent), and plagioclase feldspar (15 percent). The remainder is mostly biotite and amphibole. Because most of the volume of granite is of felsic minerals, we classify granite as a *felsic igneous rock.* Granite is a mixture of white, grayish or pinkish, and black grains, but the overall appearance is a light gray or pink color (Figure 11.6*a*).

Diorite, the second igneous rock on the list, lacks quartz. It consists largely of plagioclase feldspar (60 percent) and secondary amounts of amphibole

> Granite, diorite, gabbro, and peridotite are igneous rocks ranging from felsic to mafic and ultramafic in composition.

and pyroxene. Diorite is a light-colored felsic rock that is only slightly denser than granite. When its grains are coarse, it may have a speckled appearance (Figure 11.6*b*).

The third igneous rock is *gabbro,* in which the major mineral is pyroxene (60 percent). A substantial amount of plagioclase feldspar (20 to 40 percent) is present, and, in addition, there may be some olivine (0 to 20 percent). Since the mafic minerals pyroxene and olivine are dominant, gabbro is classed as a *mafic igneous rock.* It is dark in color and denser than the felsic rocks.

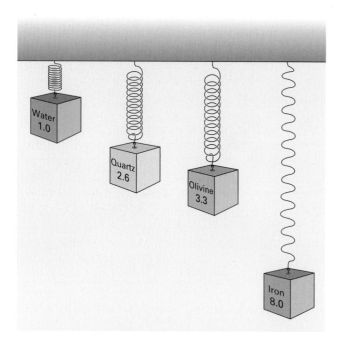

11.5 *Density* *The concept of density is illustrated by several cubes of the same size, but of different materials, hung from coil springs under the influence of gravity. The stretching of the coil spring is proportional to the density of each material, shown in thousands of kilograms per cubic meter. Quartz and olivine are common minerals. (Copyright © A. N. Strahler.)*

The fourth igneous rock, *peridotite,* is dominated by olivine (60 percent). The rest is mostly pyroxene (40 percent). Peridotite is classed as an *ultramafic igneous rock,* denser even than the mafic types.

These four common varieties of igneous rock show an increasing range of density, from felsic, through mafic, to ultramafic types. This arrangement according to density is duplicated on a grand scale in the principal rock layers that comprise the solid Earth, with the least dense layer (mostly felsic rocks) near the surface and the densest layer (ultramafic rocks) deep in the Earth's interior. We will stress this layered arrangement again in describing the Earth's crust and the deeper interior zones in Chapter 12.

Intrusive and Extrusive Igneous Rocks

Magma that solidifies below the Earth's surface and remains surrounded by older, preexisting rock is called **intrusive igneous rock.** The process of injection into existing rock is *intrusion.* Where magma reaches the surface, it emerges as **lava,** which solidifies to form **extrusive igneous rock** (Figure 11.7). The process of release at the surface is called *extrusion.*

Although both intrusive rock and extrusive rock can solidify from the same original body of magma, their outward appearances are quite different when you compare freshly broken samples of each. Intrusive igneous rocks cool very slowly—over hundreds or thousands of years—and, as a result, develop large mineral crystals—that is, they are

11.6 *Intrusive igneous rock samples* *(a) A coarse-grained granite. Dark grains are amphibole and biotite; light grains are feldspars; clear grains are quartz. (b) A coarse-grained diorite. Feldspar grains are light; amphibole and pyroxene grains are dark. Compare with granite in (a)—clear quartz grains are lacking.*

(a)

(b)

idly, the individual crystals are very small. They can only be seen through a microscope. Rocks with such small crystals are termed *fine-textured*. If the lava contains dissolved gases, the gases expand as the lava cools, forming a rock with a frothy, bubble-filled texture, called *scoria* (Figure 11.8*b*). Sometimes a lava cools to form a shiny natural volcanic glass (Figure 11.8*a*). Most lava solidifies simply as a dense, uniform rock with a dark, dull surface.

Since the outward appearance of intrusive and extrusive rocks formed from the same magma are so different, they are named differently. The igneous rock types we have discussed so far—granite, diorite, gabbro, and peridotite—are intrusive rocks. Except for peridotite, all have counterparts as extrusive rocks. They are named in the right column of Figure 11.4. Each one has the same mineral composition as the intrusive rock named at the left. *Rhyolite* is the name for lava of the same composition as granite; **andesite** is lava with the mineral composition of diorite; and **basalt** is lava of the composition of gabbro. Andesite and basalt are the two most common types of lavas. Rhyolite and andesite are pale grayish or pink in color, whereas basalt is black. Lava flows, along with particles of solidified lava blown explosively from narrow vents, often accumulate as isolated hills or mountains that we recognize as volcanoes. They are described in Chapter 13.

A body of intrusive igneous rock is called a **pluton.** Granite typically accumulates in enormous plutons, called *batholiths*. Figure 11.9 shows the relationship of a batholith to the overlying rock. As the hot fluid magma rises, it melts and incorporates the older rock lying above it. A single batholith extends down several kilometers and may occupy an area of several thousand square kilometers.

Figure 11.9 shows other common forms of plutons. One is a *sill*, a plate-like layer formed when magma forces its way between two preexisting rock layers. In the example

11.7 Fresh lava *This fresh black lava shows a glassy surface and flow structures. Chain of Craters Road, Hawaii Volcanoes National Park, Hawaii.*

coarse-textured. The granite and diorite samples in Figure 11.6 are good examples of coarse texture.

In an intrusive igneous rock, individual mineral crystals are visible with the unaided eye or with the help of a simple magnifying lens. In an extrusive rock, which cools very rap-

11.8 Extrusive igneous rock samples *(a) Obsidian, or volcanic glass. The smooth, glassy appearance is acquired when a gas-free lava cools very rapidly. This sample shows red and black streaks caused by minor variations in composition. (b) A specimen of scoria, a form of lava containing many small holes and cavities produced by gas bubbles.*

(a)

(b)

11.9 *Volcanic rock formations* This block diagram illustrates various forms of intrusive igneous rock plutons as well as an extrusive lava flow.

shown, the sill has lifted the overlying rock layers to make room. A second kind of pluton is the *dike,* a wall-like body formed when a vertical rock fracture is forced open by magma. Commonly, these vertical fractures conduct magma to the land surface in the process of extrusion. Figure 11.10 shows a dike of mafic rock cutting across layers of older rock. The dike rock is fine-textured because of its rapid cooling. Magma entering small, irregular, branching fractures in the surrounding rock solidifies in a branching network of thin *veins.*

> Intrusive igneous rocks solidify below the Earth's surface. Because they cool slowly, they develop mineral crystals visible to the eye. Extrusive igneous rocks cool very rapidly on the land surface or ocean bottom and thus show crystals of only microscopic size.

11.10 *Exposure of a dike* A dike of mafic igneous rock with nearly vertical parallel sides, cutting across flat-lying sedimentary rock layers. Arrows mark the contact between igneous rock and sedimentary rock. Spanish Peaks region, Colorado.

The formation of extrusive igneous rock can be witnessed today where volcanic processes are active. *Eye on the Environment 11.1 • Battling Iceland's Heimaey Volcano* describes the struggle of courageous Icelanders to save their volcanic island village and harbor from the onslaught of a volcanic eruption.

Chemical Alteration of Igneous Rocks

The minerals in igneous rocks are formed at high temperatures, and often at high pressures, as magma cools. When igneous rocks are exposed at or near the Earth's surface, the conditions are quite different. Temperatures and pressures are low. Also, the rocks are exposed to soil water and ground water solutions that contain dissolved oxygen and carbon dioxide. In this new environment, the minerals within an igneous rock may no longer be stable. Instead, most of these minerals undergo a slow chemical change that weakens their structure. Chemical change in response to this alien environment is called **mineral alteration.** It is a process of *weathering,* which we will define in Chapter 15 as the physical and chemical processes that break down rocks.

Weathering also includes the physical forces of disintegration that break up igneous rock into small fragments and separate the component minerals, grain by grain. This breakup, or fragmentation, is essential for the chemical reactions of mineral alteration. The reason is that fragmentation results in a great increase in the mineral surface area that is exposed to chemically active solutions in soils. (We will take up the processes of physical disintegration of rocks in Chapter 14.)

Another chemical alteration process is *oxidation.* This process occurs when oxygen dissolved in soil or ground water reacts with minerals. Oxidation is the normal fate of most silicate minerals exposed at the surface. With oxidation, the silicate minerals are converted to *oxides,* in which silicon and the metallic elements—such as calcium, magnesium, and iron—each bond completely with oxygen. Oxides are very stable. Quartz, with the composition silicon dioxide (SiO_2), is a common mineral oxide. It is very long-lasting and is found abundantly in many types of rocks and sediments. As we will see shortly, quartz is a major constituent of sedimentary rocks.

Silicate minerals do not generally dissolve in water, but some react chemically with water in a chemical alteration process called *hydrolysis.* This process is not merely a soaking or wetting of the mineral, but a true chemical change that produces a different mineral compound. The products of hydrolysis are stable and long-lasting, as are the products of oxidation.

Some of the alteration products of silicate minerals are clay minerals, described in Chapter 10. These minerals are produced by a combination of oxidation and hydrolysis acting on silicates. A **clay mineral** is one that has plastic properties when moist, because it consists of very small, thin flakes that become lubricated by layers of water molecules. Clay minerals formed by mineral alteration are abundant in common types of sedimentary rocks.

Solution is another mechanism by which minerals are altered. Most minerals do not dissolve directly in water. One important exception is rock salt (sodium chloride), which is a constituent of sea water. When carbon dioxide dissolves in water, a weak acid—*carbonic acid*—is formed. Carbonic acid can dissolve certain minerals, especially calcium carbonate. In addition, where decaying vegetation is present, soil water contains many complex organic acids that are capable of reacting with minerals.

Mineral alteration occurs when the minerals in igneous rocks are transformed chemically into new minerals that are more stable at or near the Earth's surface. Processes of mineral alteration include oxidation, hydrolysis, and solution.

SEDIMENTS AND SEDIMENTARY ROCKS

We can now turn to the second great rock class, the *sedimentary rocks.* The mineral particles in sedimentary rocks can be derived from preexisting rock of any of the three rock classes as well as from newly formed organic matter. However, igneous rock is the most important original source of the inorganic mineral matter that makes up sedimentary rock. For example, a granite can weather to yield grains of quartz and particles of clay minerals derived from feldspars, thus contributing sand and clay to a sedimentary rock. Sedimentary rocks include rock types with a wide range of physical and chemical properties. We will only touch on a few of the most important kinds of sedimentary rocks, which are shown in Table 11.2.

In the process of mineral alteration, solid rock is weakened, softened, and fragmented, yielding particles of many sizes and mineral compositions. When transported by a fluid medium—air, water, or glacial ice—these particles are known collectively as **sediment.** Used in its broadest sense, sediment includes both inorganic and organic matter. Dissolved mineral matter in solution must also be included.

Streams and rivers carry sediment to lower land levels, where sediment can accumulate. The most favorable sites of sediment accumulation are shallow seafloors bordering continents. But sediments also accumulate in inland valleys, lakes, and marshes. Thick accumulations of sediment may become deeply buried under newer (younger) sediments. Wind and glacial ice also transport sediment but not necessarily to lower elevations or to places suitable for accumulation. Over long spans of time, the sediments can undergo physical or chemical changes, becoming compacted and hardened to form sedimentary rock.

There are three major classes of sediment. First is **clastic sediment,** which consists of inorganic rock and mineral fragments, called *clasts.* Examples are the materials in a sand bar of a river bed or on a ocean beach. Second is **chemically precipitated sediment,** which consists of inorganic mineral compounds precipitated from a saltwater solution or as hard parts of organisms. In the process of chemical pre-

Table 11.2 | **Some Common Sedimentary Rock Types**

Subclass	Rock Type	Composition
Clastic (composed of rock and/or mineral fragments)	Sandstone	Cemented sand grains
	Siltstone	Cemented silt particles
	Conglomerate	Sandstone containing pebbles of hard rock
	Mudstone	Silt and clay, with some sand
	Claystone	Clay
	Shale	Clay, breaking easily into flat flakes and plates
Chemically precipitated (formed by chemical precipitation from sea water or salty inland lakes)	Limestone	Calcium carbonate, formed by precipitation on sea or lake floors
	Dolomite	Magnesium and calcium carbonates, similar to limestone
	Chert	Silica, a microcrystalline form of quartz
	Evaporites	Minerals formed by evaporation of salty solutions in shallow inland lakes or coastal lagoons
Organic (formed from organic material)	Coal	Rock formed from peat or other organic deposits; may be burned as a mineral fuel
	Petroleum (mineral fuel)	Liquid hydrocarbon found in sedimentary deposits; not a true rock but a mineral fuel
	Natural gas (mineral fuel)	Gaseous hydrocarbon found in sedimentary deposits; not a true rock but a mineral fuel

Eye on the Environment | 11.1

Battling Iceland's Heimaey Volcano

Normally the processes of rock formation are very slow, but when a volcanic eruption occurs, large volumes of fresh rock can be created very quickly. In 1973, inhabitants of Heimaey, a small island very near the southern coast of Iceland, found themselves witnessing the birth of new rock from a vantage point that was a little too close.

It was around 2:00 A.M., early on a January morning, that the eruption began. First, a fissure split the eastern side of the island from one coast to the other, sending up a curtain of fire in a pyrotechnical display nearly 2 km (1.2 mi) long. Soon, however, the spraying fountains of volcanic debris became restricted to a small area not far from Helgafell, an older volcanic

cone. The lava and lava fragments of ash and cinders, called *tephra*, poured out at a rate of 100 m³/sec (about 3500 ft³/sec), building a cone that soon reached a height of about 100 m (about 300 ft) above sea level. It was dubbed Kirkjufell after a farmstead, Kirkjubaer, which lay beneath the debris.

It wasn't long before strong easterly winds set in, and Iceland's

Vestmannaeyjar after the eruption *Lava (right) invaded the village of Vestmannaeyjar, while fine black cinders (now swept from roofs and roads) covered the town.*

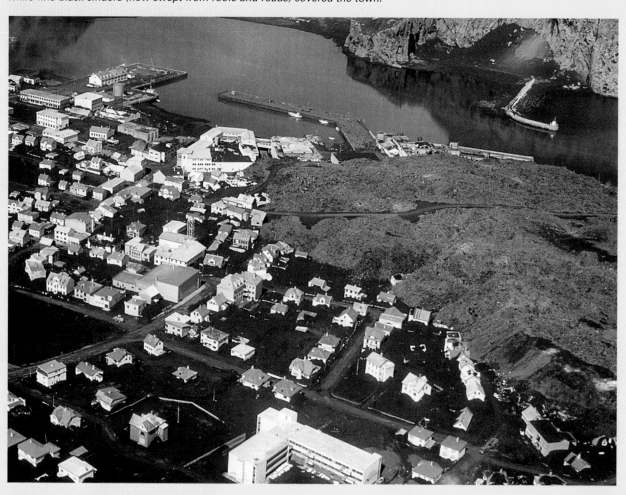

main fishing port, Vestmannaeyjar, located within a kilometer (about a half-mile) from the eruption, began to receive a "snowfall" of fine, black cinders. Houses on the east side of the town were buried under the tephra. Many of the houses collapsed under the weight of the hot ash, while others were set afire. Lava flows soon reached the village, burying and burning still other buildings (see photo).

Over the passing weeks, the emissions of tephra ceased, but lava continued to flow from the cone, which soon reached 200 m (about 600 ft) in height. Unfortunately, the lava began to flow northward, narrowing the harbor and threatening the future livelihood of the town and its evacuated inhabitants. Fishing is a major industry in Iceland, accounting for nearly 80 percent of Iceland's foreign exchange. Vestmannaeyjar normally lands and processes 20 percent of the nation's fish catch.

Because the harbor and fishing port are important to the nation's economy, Icelanders embarked on a bold plan to save the harbor by altering the course of the lava, diverting the flow eastward. This was to be accomplished by cooling the flow on its north edge with water streams, creating a natural wall to channel the flow alongside the harbor instead of across it. Within a few weeks, the first pumping of water onto the flowing lava began. By early March, a pump ship came into operation in the harbor, providing a steady flow of sea water to cool the slowly moving flows. It was joined in April by as many as 47 high-capacity pumps floating on barges, delivering in total as much as 1 cu m of sea water per second (35 ft³/sec).

The most effective technique for slowing the lava at a particular location began with cooling the edge

Map of Heimaey Island, Iceland *The map shows the extent of lava flows from the eruptions of the Kirkjufell crater. (Modified from U.S. Geological Survey.)*

and nearby surface of the flow with water from hoses. This allowed bulldozers to build a crude road up and over the slowly moving flow, using nearby tephra as the road material. Large plastic pipes were then laid along the road and across the flow, with small holes spraying water on hotspots. As long as sea water was flowing in the plastic pipes, they remained cool enough not to melt. After a day or so, the flow typically began to slow. Pumping usually continued for about two weeks, until the lava stopped steaming at spray points. The result of the water applications was to build a broad wall of cool, lava rubble with thickening lava behind it.

The huge undertaking successfully stabilized the northern front of the lava flow, keeping the harbor

from becoming closed (see map). Although the lava had indeed reached the harbor, it had merely narrowed its entrance. In fact, the new flow improved the harbor's ability to shelter the boats within it (see map). Within about five months, the eruption was over, and the digging out began in earnest. Within a year, life was back to normal on Heimaey.

The diversion of the lava and reconstruction of the village of Vestmannaeyjar were extremely costly. To pay for the cost, Iceland passed a special tax increase, requiring the average Icelandic family to pay about 10 percent of its annual income for one year. Generous foreign aid also helped cover the enormous expenses borne by the tiny nation.

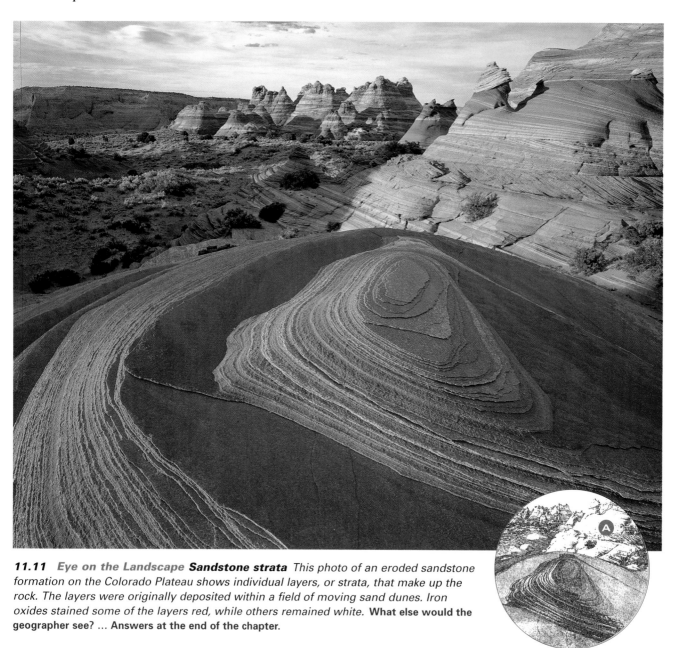

11.11 *Eye on the Landscape* **Sandstone strata** *This photo of an eroded sandstone formation on the Colorado Plateau shows individual layers, or strata, that make up the rock. The layers were originally deposited within a field of moving sand dunes. Iron oxides stained some of the layers red, while others remained white.* **What else would the geographer see?** *... Answers at the end of the chapter.*

cipitation, ions in solution combine to form solid mineral matter separate from the solution. A layer of rock salt, such as that found in dry lakebeds in arid regions, is an example. A third class is **organic sediment** which consists of the tissues of plants and animals, accumulated and preserved after the death of the organism. An example is a layer of peat in a bog or marsh.

Sediment accumulates in more-or-less horizontal layers, called **strata**, or simply "beds" (Figure 11.11). Individual strata are separated from those below and above by surfaces called stratification planes or bedding planes. These separation surfaces allow one layer to be easily removed from the next. Strata of widely different compositions can occur alternately, one above the next. **GEODISCOVERIES**

Sedimentary rocks are composed of sediment, which may be clastic, chemically precipitated, or organic. Layers of sediment are termed strata.

Clastic Sedimentary Rocks

Clastic sediments are derived from clasts of any and all of the rock groups—igneous, sedimentary, metamorphic—and thus may include a very wide range of minerals for sedimentary rock formation. Silicate minerals are the most important, both in original form and as altered by oxidation and hydrolysis. Quartz and feldspar usually dominate. Because quartz is hard and is immune to alteration, it is the most important single component of the clastic sediments (Figure 11.12). Second in abundance are fragments of unaltered fine-grained parent rocks, such as tiny pieces of lava rock. Feldspar and mica are also commonly present. Clay minerals are major constituents of the finest clastic sediments.

11.12   Rounded quartz grains from an ancient sandstone *The grains average about 1 mm (0.039 in.) in diameter.*

The range of particle sizes in a clastic sediment determines how easily and how far the particles are transported by water currents. The finer the particles, the more easily they are held suspended in the fluid. On the other hand, the coarser particles tend to settle to the bottom of the fluid layer. In this way, a separation of size grades, called *sorting,* occurs. Sorting determines the texture of the sediment deposit and of the sedimentary rock derived from that sediment. The finest clay particles do not settle out unless they are made to clot together into larger clumps. This clotting process, called *flocculation,* normally occurs when river water carrying clay mixes with the saltwater of the ocean.

When sediments accumulate in thick sequences, the lower strata are exposed to the pressure produced by the weight of the sediments above them. This pressure compacts the sediments, squeezing out excess water. Cementation occurs as dissolved minerals recrystallize where grains touch and in the spaces between mineral particles. Silicon dioxide (quartz, SiO_2) is very slightly soluble in water, and so the cement is often a form of quartz, called *silica,* which lacks a true crystalline form. Calcium carbonate ($CaCO_3$) is another common material that cements clastic sedimentary rocks. Compaction and cementation produce sedimentary rock.

The important varieties of clastic sedimentary rock are distinguished by the size of their particles. They include sandstone, conglomerate, mudstone, claystone, and shale. **Sandstone** is formed from fine to coarse sand (Figure 11.13). The cement may be silica or calcium carbonate. The sand grains are commonly of quartz, such as those shown in Figure 11.12. (Refer to Figure 10.2 for the names and diameters of the various grades of sediment particles.) Sandstone containing numerous rounded pebbles of hard rock is called *conglomerate* (Figure 11.14).

A mixture of water with particles of silt and clay, along with some sand grains, is called *mud.* The sedimentary rock hardened from such a mixture is called *mudstone.* Compacted and hardened clay layers become *claystone.* Sedimentary rocks of mud composition are commonly layered in such a way that they easily break apart into small flakes and plates. The rock is then described as being *fissile* and is given the name **shale.** Shale,

---

Important varieties of clastic sedimentary rocks include sandstone, conglomerate, mudstone, claystone, and shale. They are formed when sediments are compressed and cemented by solutions containing silica or calcium carbonate.

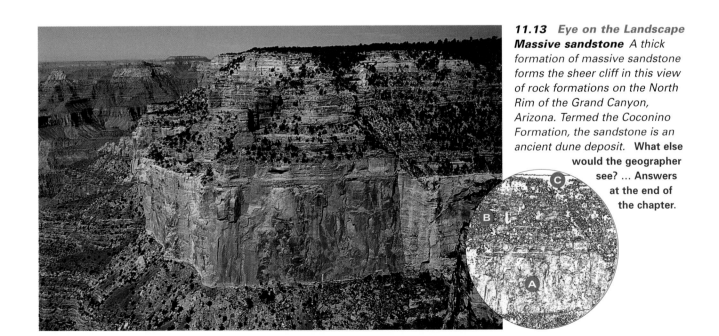

11.13 *Eye on the Landscape* **Massive sandstone** *A thick formation of massive sandstone forms the sheer cliff in this view of rock formations on the North Rim of the Grand Canyon, Arizona. Termed the Coconino Formation, the sandstone is an ancient dune deposit.* **What else would the geographer see? ... Answers at the end of the chapter.**

11.14 Conglomerate section *A piece of quartzitic conglomerate, cut through and polished, reveals rounded pebbles of quartz (clear and milky colors) and chert (grayish). It is about 12 cm (4.7 in.) in diameter.*

stone, composed largely of the mineral calcite. *Calcite* is calcium carbonate ($CaCO_3$). Marine limestones—limestone strata formed on the seafloor—accumulated in thick layers in many ancient seaways in past geologic eras (Figure 11.16). A closely related rock, also formed by chemical precipitation, is *dolomite,* composed of calcium-magnesium carbonate. Limestone and dolomite are grouped together as the *carbonate rocks.* They are dense rocks with white, pale gray, or even black color.

Sea water also yields sedimentary layers of silica in a hard, noncrystalline form called *chert.* Chert is a variety of sedimentary rock, but it also commonly occurs combined with limestone (cherty limestone).

Shallow water bodies acquire a very high level of salinity where evaporation is sustained and intense. One type of shallow water body is a bay or estuary in a coastal desert region. Another type is the salty lake of inland desert basins (see Figure 16.32). Sedimentary minerals and rocks deposited from such concentrated solutions are called **evaporites.** Ordinary rock salt, the mineral halite (sodium chloride), has accumulated in this way in thick sedimentary rock layers. These layers are mined as major commercial sources of halite.

the most abundant of all sedimentary rocks, is formed largely of clay minerals. The compaction of the mud to form mudstone and shale involves a considerable loss of volume as water is driven out of the clay. Figure 11.15 shows an eroded shale formation in southern Utah. Here some of the shale beds have distinctive colors, so the bedding is especially prominent.

> Limestone, dolomite, and chert are rocks formed by chemical precipitation in a marine environment. Evaporites form when salty water evaporates, leaving layers of soluble minerals behind.

Chemically Precipitated Sedimentary Rocks

Under favorable conditions, mineral compounds are deposited from the salt solutions of sea water and of salty inland lakes in desert climates. One of the most common sedimentary rocks formed by chemical precipitation is **lime-**

Hydrocarbon Compounds in Sedimentary Rocks

Hydrocarbon compounds (compounds of carbon, hydrogen, and oxygen) form a most important type of organic sediment—one on which human society increasingly depends for fuel. These substances occur both as solids (peat and coal) and as liquids and gases (petroleum and natural gas). Only coal qualifies physically as a rock. *Peat* is a soft, fibrous substance of brown to black color. (See Figure 10.20.) It accumulates in a bog environment where the continual presence of water inhibits the decay of plant remains.

At various times and places in the geologic past, plant remains accumulated on a large scale, accompanied by

11.15 Shale butte *These flat-lying strata consist of layers of different colored shales that are easily eroded. Near Glen Canyon City in southern Utah. (A. N. Strahler.)*

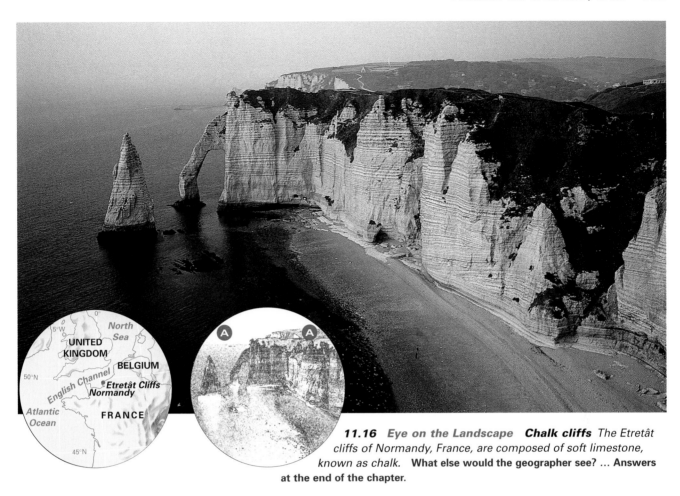

11.16 *Eye on the Landscape*   ***Chalk cliffs*** *The Etretât cliffs of Normandy, France, are composed of soft limestone, known as chalk.*   **What else would the geographer see? ... Answers at the end of the chapter.**

sinking of the area and burial of the compacted organic matter under thick layers of inorganic clastic sediments. *Coal* is the end result of this process (Figure 11.17). Individual coal seams are interbedded with shale, sandstone, and limestone strata. *Petroleum* (or *crude oil,* as the liquid form is often called) includes many hydrocarbon compounds. *Natural gas,* which is found in close association with accumulations of liquid petroleum, is a mixture of gases. The principal gas is methane (marsh gas, CH_4).

Geologists generally agree that petroleum and natural gas are of organic origin, but the nature of their formation is not fully understood. They are classed not as minerals but rather separately as mineral fuels.

Natural gas and petroleum commonly occupy open interconnected pores in a thick sedimentary rock layer—a porous sandstone, for example. The simplest arrangement of strata favorable to trapping petroleum and natural gas is an uparching of the type shown in Figure 11.18. Shale forms an

11.17 *Strip mine* *A coal seam near Sheridan, Wyoming, being strip mined by heavy equipment.*

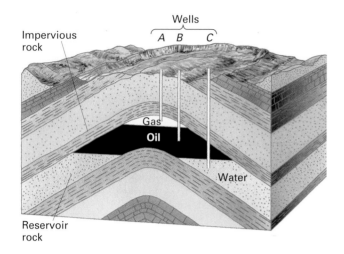

Wells
A B C

Impervious rock

Gas

Oil

Water

Reservoir rock

11.18 Trapping of oil and gas *Idealized cross section of an oil pool on a dome structure in sedimentary strata. Well A will draw gas; well B will draw oil; and well C will draw water. The cap rock is shale; the reservoir rock is sandstone. (Copyright © A. N. Strahler.)*

impervious cap rock. A porous sandstone beneath the cap rock serves as a reservoir. Natural gas occupies the highest position, with the oil below it, and then water, in increasing order of density.

Yet another form of occurrence of hydrocarbon fuels is *bitumen,* a variety of petroleum that behaves much as a solid, although it is actually a highly viscous liquid. Bitumen goes by other common names, such as tar, asphalt, or pitch. In some localities, bitumen occupies pore spaces in layers of sand or porous sandstone. It remains immobile in the enclosing sand and will flow only when heated. Outcrops of *bituminous sand (oil sand)* exposed to the Sun will show bleeding of the bitumen. Perhaps the best known of the great bituminous sand deposits are those occurring in Alberta, Canada. Where exposed along the banks of the Athabasca River, the oil sand is extracted from surface mines.

Hydrocarbon compounds in sedimentary rocks are important because they provide an energy resource on which modern human civilization depends. These **fossil fuels,** as they are called collectively, have required millions of years to accumulate. However, they are being consumed at a very rapid rate by our industrial society. These fuels are nonrenewable resources. Once they are gone, there will be no more because the quantity produced by geologic processes even in a thousand years is scarcely measurable in comparison to the quantity stored through geologic time.

GEODISCOVERIES

Table 11.3	Some Common Metamorphic Rock Types
Rock Type	**Description**
Slate	Shale exposed to heat and pressure that splits into hard flat plates
Schist	Shale exposed to intense heat and pressure that shows evidence of shearing
Quartzite	Sandstone that is "welded" by a silica cement into a very hard rock of solid quartz
Marble	Limestone exposed to heat and pressure, resulting in larger, more uniform crystals
Gneiss	Rock resulting from the exposure of clastic sedimentary or intrusive igneous rocks to heat and pressure

pany the mountain-building processes of the Earth's crust. The result is a rock so changed in texture and structure as to be reclassified as **metamorphic rock.** Mineral components of the parent rock are, in many cases, reconstituted into different mineral varieties. Recrystallization of the original minerals can also occur. Our discussion of metamorphic rocks will mention only five common types—slate, schist, quartzite, marble, and gneiss (Table 11.3).

Slate is formed from shale that is heated and compressed by mountain-making forces. This fine-textured rock splits neatly into thin plates, which are familiar as roofing shingles and as patio flagstones. With application of increased heat and pressure, slate changes into **schist,** representing the most advanced stage of metamorphism. Schist has a structure called foliation, consisting of thin but rough, irregularly curved planes of parting in the rock (Figure 11.19). These are evidence of *shearing*—a stress that pushes the layers sideways, like a deck of cards pushed into a fan with the sweep of a palm. Schist is set apart from slate by the coarse texture of the mineral grains, the abundance of mica, and

> Hydrocarbons in sedimentary rocks include coal, petroleum, natural gas, and peat. These mineral fuels power modern industrial society.

11.19 Schist sample *A specimen of foliated schist.*

METAMORPHIC ROCKS

Any type of igneous or sedimentary rock may be altered by the tremendous pressures and high temperatures that accom-

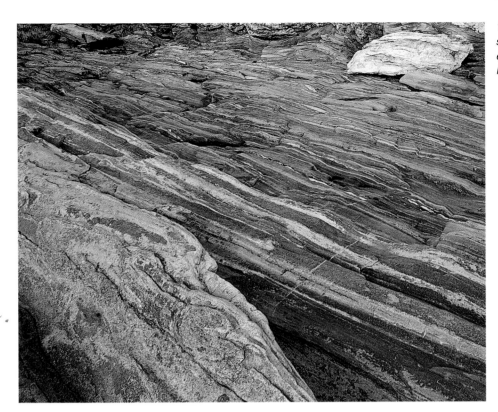

11.20 Banded gneiss *This surf-washed rock surface exposes banded gneiss, Pemaquid Point, Maine.*

occasionally the presence of scattered large crystals of newly formed minerals, such as garnet.

The metamorphic equivalent of conglomerate, sandstone, and siltstone is **quartzite,** which is formed by the addition of silica to fill completely the open spaces between grains. This process is carried out by the slow movement of underground waters carrying silicate into the rock, where it is deposited.

Limestone, after undergoing metamorphism, becomes *marble,* a rock of sugary texture when freshly broken. During the process of internal shearing, calcite in the limestone re-forms into larger, more uniform crystals than before. Bedding planes are obscured, and masses of mineral impurities are drawn out into swirling bands.

Finally, the important metamorphic rock **gneiss** may be formed either from intrusive igneous rocks or from clastic sedimentary rocks that have been in close contact with intrusive magmas. A single description will not fit all gneisses because they vary considerably in appearance, mineral composition, and structure. One variety of gneiss is strongly banded into light and dark layers or lenses (Figure 11.20), which may be bent into wavy folds. These bands have differing mineral compositions. They are thought to be relics of sedimentary strata, such as shale and sandstone, to which new mineral matter has been added from nearby intrusive magmas.

Metamorphic rocks are formed from preexisting rocks by intense heat and pressure, which alter rock structure and chemical composition. Shale is transformed to slate or schist, sandstone to quartzite, and limestone to marble. Gneiss forms when an intrusive magma cools next to igneous or sedimentary rocks.

One of the more exciting developments in remote sensing in the past few years has been the launch of the ASTER instrument (Advanced Spaceborne Thermal Emission and Reflection Spectroradiometer) on NASA's Terra satellite platform. It has a unique ability to aid in geologic mapping and in identification of areas with potentially valuable ore deposits. *Focus on Remote Sensing 11.2 • Geologic Mapping with ASTER* tells this interesting story. *GEODISCOVERIES*

THE CYCLE OF ROCK CHANGE

The processes that form rocks, when taken together, constitute a single system that cycles and recycles Earth materials over geologic time from one form to another. The **cycle of rock change,** shown in Figure 11.21, describes this system. There are two environments—a surface environment of low pressures and temperatures and a deep environment of high pressures and temperatures. The surface environment is the site of rock alteration and sediment deposition. In this environment, igneous, sedimentary, and metamorphic rocks are uplifted and exposed to air and water. Their minerals are altered chemically and broken free from the parent rock, yielding sediment. The sediment accumulates in basins, where deeply buried sediment layers are compressed and cemented into sedimentary rock.

Focus on Remote Sensing | 11.2

Geologic Mapping with ASTER

A unique imaging instrument built by Japan that is of special interest to geologists is aboard NASA's Terra satellite platform, launched in December 1999. ASTER, the Advanced Spaceborne Thermal Emission and Reflection Spectroradiometer, acquires multispectral images in three different waveband regions—visible and near-infrared, shortwave infrared, and thermal infrared. Recall from *Geographer's Tools 2.1 • Remote Sensing for Physical Geography* that human eyes are sensitive only to the visible light portion of the spectrum, from about 0.4 to 0.65 μm, while the Sun provides enough energy for good imaging from about 0.3 to 3.0 μm. This means that surface substances can be imaged in these wavelengths and have "colors" that are unknown in human experience. We refer to these colors as spectral signatures. As it happens, minerals and rocks often have distinctive spectral signatures in near- and shortwave infrared domains, and so near- and shortwave infrared images are particularly useful for geologic mapping.

Another spectral region of particular interest is the thermal infrared, especially the wavelengths from 8.0 to 12.0 μm. Here the energy for sensing is the radiant energy emitted by substances as a function of their temperature. However, some types of rocks and minerals emit more energy of a given wavelength at a given temperature than others. In this way, we can think of these substances as having spectral signatures in the thermal infrared, too.

The two graphs below illustrate this principle. They show reflectance plotted against wavelength for two common minerals—quartz and calcite. Also shown are the spectral locations of the ASTER wavebands. In the left graph, we can see that both quartz and calcite are quite bright in ASTER's two visible bands (bands 1 and 2, green and red). They appear white to the eye, with calcite, which is slightly brighter in the red band, appearing to be a "warmer" white. In the shortwave infrared, however, calcite is significantly darker in ASTER band 8 than in surrounding bands 7 and 9. The right graph shows spectral reflectance in the thermal infrared spectral region. Calcite is brightest in ASTER band 13 while

being darkest in the adjacent band 14. Quartz has a distinctive signature as well, being about 40 percent brighter in bands 10–12 than in 13–14. Thus, both minerals have unique spectral signatures in this spectral region, too.

Color images in the three waveband regions, shown at right, illustrate how ASTER can help identify different types of rocks and minerals by their spectral signatures. The images are from the Saline Valley, a dry semidesert region in southeastern California near Death Valley, and were acquired on March 30, 2000. Trending diagonally across the center of the image is the valley itself, which is the now-nearly-dry bed of a lake that was filled with water during the Ice Age. Lying to the southwest is the Inyo mountain chain, with a number of steep valleys cutting into its flanks. In the lower left corner, we see a small part of the Owens Valley. To the northeast of Saline Valley lies a complex geologic region at the northern end of the Panamint Range.

Image *(a)* is a color infrared display in which true colors of green, red, and near-infrared are shown

Reflectance spectra for quartz and calcite *Wavebands in which ASTER acquires images are identified as numbered bands. Left, the visible and near-infrared spectral region; right, the thermal infrared spectral region.*

Multispectral ASTER images of Saline Valley, California (a) visible and near-infrared image; (b) shortwave infrared image; (c) thermal infrared image. (NASA/GSFC/MITI/ERSDAC/JAROS and U.S./Japan ASTER Science Team.)

as blue, green, and red in the image. The forests of the Inyos appear in a distinctive red color because vegetation is very bright in the near-infrared region. Snow along the summit of the Inyos is white, as are the salty dry basins of the valleys. Many different rock types appear, in tones ranging from brown, gray, and yellow to blue. Image *(b)* displays ASTER shortwave infrared bands 8, 6, and

4 as blue, green and red. These bands distinguish carbonate, sulfate, and clay minerals. Limestones are yellow-green, while the clay mineral kaolinite appears in purplish tones. In *(c)*, ASTER bands 10, 12, and 13 are shown in blue, green, and red, respectively. Here, rocks rich in quartz appear in red tones, while carbonate rocks are green. Mafic volcanic rocks are visible in purple tones.

By comparing the three images, it is easy to see how the spectral information acquired by ASTER has the ability to make geologic mapping faster and easier. By using ASTER data, geologists can not only come to understand better the geologic history of a region, but also identify the geologic settings in which valuable mineral ores may be found.

11.21 The cycle of rock change
This cycle links sediment with sedimentary, metamorphic, intrusive igneous, and extrusive igneous rocks in processes of rock formation and destruction.

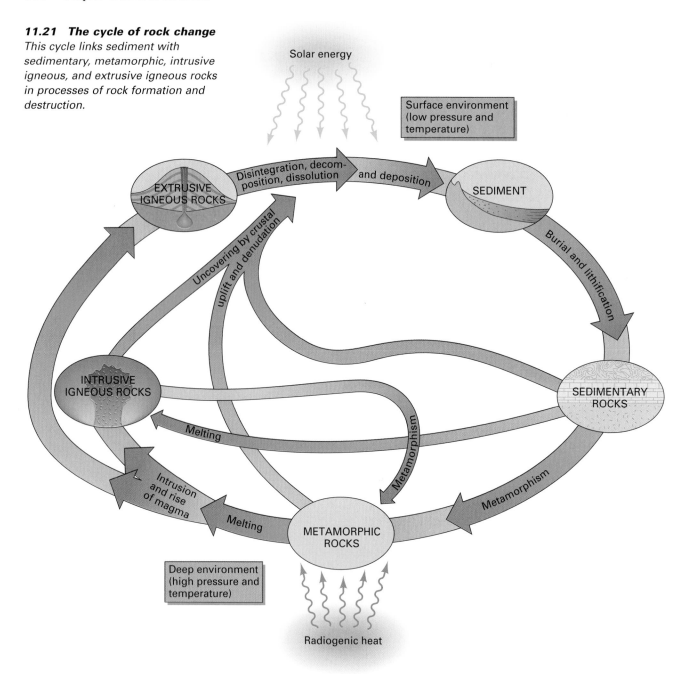

Sedimentary rock, entering the deep environment, is heated by the slow radioactive decay of elements and comes into a zone of high confining pressure. Here, it is transformed into metamorphic rock. Pockets of magma are formed in the deep environment and move upward, melting and incorporating surrounding rock as they rise. Upon reaching a higher level, magma cools and solidifies, becoming intrusive

The cycle of rock change describes how Earth materials are cycled and recycled by Earth processes over geologic time. In the surface environment, rocks weather into sediment. In the deep environment, heat and pressure transform sediment into rock that is eventually exposed at the surface.

igneous rock. Or it may emerge at the surface to form extrusive igneous rock. Either way, the cycle is completed.

The cycle of rock change has been active since our planet became solid and internally stable, continuously forming and re-forming rocks of all three major classes. Not even the oldest igneous and metamorphic rocks found thus far are the "original" rocks of the Earth's crust, for they were recycled eons ago.

A LOOK AHEAD This chapter has described the minerals and rocks of the Earth's surface and the processes of their formation. As we look further into the Earth's outermost layers in the next chapter, we will see that these processes do not occur everywhere. Instead, there is a grand plan that organizes the formation and destruction of rocks and distributes the processes of the cycle of rock change in a geographic pattern. This pattern is controlled by the pattern in which the solid Earth's brittle outer layer is fractured into great plates that split and separate and also converge and collide. The grand plan is plate tectonics, the scheme for understanding the dynamics of the Earth's crust over millions of years of geologic time.

In Review | Earth Materials

- The elements oxygen and silicon dominate the *Earth's crust.* Metallic elements, which include aluminum, iron, and the base elements, account for nearly all the remainder.

- **Minerals** are naturally occurring, inorganic substances. Each has an individual chemical composition and atomic structure.

- *Silicate minerals* make up the bulk of **igneous rocks.** They contain silicon and oxygen together with some of the metallic elements.

- There are three broad classes of igneous rocks, depending on their mineral content. **Felsic** rocks contain mostly felsic minerals and are least dense; **mafic** rocks, containing mostly mafic minerals, are denser; and *ultramafic* rocks are most dense. Because felsic rocks are least dense, they are generally found in the upper layers of the Earth's crust. Mafic and ultramafic rocks are more abundant in the layers below.

- If **magma** erupts on the surface to cool rapidly as lava, the rocks formed are **extrusive** and have a fine crystal texture. If the magma cools slowly below the surface as a **pluton,** the rocks are **intrusive** and the crystals are larger. **Granite** (felsic, intrusive), **andesite** (felsic, extrusive), and **basalt** (mafic, extrusive) are three very common igneous rock types.

- Most silicate minerals found in igneous rocks undergo **mineral alteration** when exposed to air and moisture at the Earth's surface. Mineral alteration occurs through *oxidation, hydrolysis,* or *solution.* **Clay minerals** are commonly produced by mineral alteration.

- **Sedimentary rocks** are formed in layers, or **strata. Clastic sedimentary** rocks are composed of fragments of rocks and minerals that usually accumulate on ocean floors. As the layers are buried more and more deeply, water is pressed out and particles are cemented together. **Sandstone** and **shale** are common examples.

- **Chemical precipitation** also produces sedimentary rocks, such as **limestone.** Coal, petroleum, and natural gas are hydrocarbon compounds occurring in sedimentary rocks that are used as mineral fuels.

- **Metamorphic rocks** are formed when igneous or sedimentary rocks are exposed to heat and pressure. Shale is altered to *slate* or **schist,** sandstones become **quartzite,** and intrusive igneous rocks or clastic sediments are metamorphosed into **gneiss.**

- In the **cycle of rock change,** rocks are exposed at the Earth's surface, and their minerals are broken free and altered to form sediment. The sediment accumulates in basins, where the layers are compressed and cemented into sedimentary rock. Deep within the Earth, the heat of radioactive decay melts preexisting rock into magma, which can move upward into the crust to form igneous rocks that cool at or below the surface. Rocks deep in the crust are exposed to the heat and pressure, forming metamorphic rock. Mountain-building forces move deep igneous, sedimentary, and metamorphic rocks upward to the surface, providing new material for surface alteration and breakup and completing the cycle.

Key Terms

mineral, p. 373
rock, p. 373
igneous rocks, p. 373
sedimentary rocks, p. 373
metamorphic rocks,
   p. 373
quartz, p. 374
felsic, p. 374
mafic, p. 374
magma, p. 374

granite, p. 375
intrusive igneous rock,
   p. 376
lava, p. 376
extrusive igneous rock,
   p. 376
andesite, p. 377
basalt, p. 377
pluton, p. 377
mineral alteration, p. 378

clay mineral, p. 379
carbonic acid, p. 379
sediment, p. 379
clastic sediment, p. 379
chemically precipitated
   sediment, p. 379
organic sediment, p. 382
strata, p. 382
sandstone, p. 383
shale, p. 383

limestone, p. 384
evaporites, p. 384
fossil fuels, p. 386
metamorphic rock,
   p. 386
schist, p. 386
quartzite, p. 387
gneiss, p. 387
cycle of rock change,
   p. 387

Review Questions

1. What is the Earth's crust? What elements are most abundant in the crust?
2. Define the terms *mineral* and *rock.* Name the three major classes of rocks.
3. What are silicate minerals? Describe two classes of silicate minerals.
4. Name four types of igneous rocks and arrange them in order of density.
5. How do igneous rocks differ when magma cools (a) at depth and (b) at the surface?
6. How are igneous rocks chemically altered? Identify and describe three processes of chemical alteration.
7. What is sediment? Define and describe three types of sediments.
8. Describe two processes that produce sedimentary rocks, and identify at least three important varieties of clastic sedimentary rocks.
9. How are sedimentary rocks formed by chemical precipitation?
10. What types of sedimentary deposits consist of hydrocarbon compounds? How are they formed?
11. What are metamorphic rocks? Describe at least three types of metamorphic rocks and how they are formed.

Eye on the Environment 11.1 ● Battling Iceland's
Heimaey Volcano

1. What were the initial effects of the eruption of Heimaey's volcano?
2. What important industry was threatened by the volcanic activity and why?
3. How was sea water used to keep lava from blocking the harbor?

Visualizing Exercises

1. Sketch a cross section of the Earth showing the following features: batholith, sill, dike, veins, lava, and volcano.
2. Sketch the cycle of rock change and describe the processes that act within it to form igneous, sedimentary, and metamorphic rocks.

Essay Question

1. A granite is exposed at the Earth's surface, high in the Sierra Nevada mountain range. Describe how mineral grains from this granite might be released, altered, and eventually become incorporated in a sedimentary rock. Trace the route and processes that would incorporate the same grains in a metamorphic rock.

Eye on the Landscape

11.11 Sandstone strata Note how these fine sedimentary beds (**A**) are crossed at angles by other beds. This cross-bedding is characteristic of dune sands. As the sand dune moves forward, sand layers accumulate on its sloping forward surface. Later, the wind erodes the dune, forming a new surface across the beds that cuts them at an angle. That surface is then covered by more sand layers. The result is the distinctive pattern of sloping sand beds cut by other beds at low angles that we see in this rock formation. Sand dune formation and movement is covered in Chapter 18.

11.13 Massive sandstone The "stairstep" topography of the Grand Canyon is a unique product of uplift of the Colorado Plateau and deep downcutting by the through-flowing Colorado River. The steep canyon walls carved by this process form vertical faces ("risers") (**A**) where massive, hard, rock formations, such as the Coconino Formation, are exposed. Weaker beds of sandstone and shale form sloping surfaces ("steps") (**B**) with a cover of rock fragments and sediment. Although the region is very dry, there is enough rainfall at higher elevations to support an open stand of pinyon pine and juniper forest atop the plateau (**C**) as well as a sparse cover on the slopes of the canyon walls. You can read more about landforms of horizontal rock strata in Chapter 17. Altitudinal zonation of vegetation in the Grand Canyon region was discussed in Chapter 9.

11.16 Chalk cliffs Note the hazy air (**A**) that is apparent in this photo. Marine aerosols are particularly rich in tiny particles of sea salt. Like ordinary table salt during humid weather, these particles attract water vapor. The vapor accumulates around the salt particles until tiny droplets are formed. These droplets absorb and scatter sunlight, creating the marine haze that is so evident in this photo. Recall that haze was covered in Chapter 4 in *Eye on the Environment 4.2 • Air Pollution.*

EYE ON THE LANDSCAPE

Sinai Peninsula, seen from space. Red Sea, with Gulfs of Suez and Aqaba to left and right.
What else would the geographer see? … Answers at the end of the chapter.

12 | THE LITHOSPHERE AND PLATE TECTONICS

• • • • • • • • • • • • • • • • • • • •

PUTTING YOU IN THE PICTURE

Perhaps it was in that eighth grade geography class when you began to stare at the wall map of the world that hung in the front of the room, and you noticed that some parts of the map looked like a puzzle coming apart. The upper right corner of Africa was particularly suspect. Twisting the continent toward the northeast would close the Red Sea and Persian Gulf, making the Arabian Peninsula join Africa and the Middle East in a continuous broad land bridge.

Then you noticed the Mediterranean. With a little squeezing, it could shut quite nicely, uniting Europe and Africa. Greece would have to tuck in under Turkey, which would move northward to fill in the Black Sea. Italy and its islands would have to move south, filling in the obvious gap in the African coast of Libya and Tunisia. And then there was that funny bulge on the east coast of South America that would lie rather nicely along the west coast of Africa, with the Guianas butting against the west African republics from Liberia to Nigeria.

North America was a lot tougher, with those messy Caribbean islands and Central America to worry about. Could North and South America have been closer? If the Yucatan and Honduran-Nicaraguan peninsulas curled north and west, they would tuck in along the Gulf coast. Then Florida would nestle up against Panama, with Cuba, Jamaica, and Hispaniola lying along the Venezuelan coast. Now, with North America closer to South America, the east coast would fit against the bulge of western Africa rather nicely.

Perhaps, you wondered, all these lands actually did once fit together in a single world continent. Fun to think about but utterly fantastic.

Or was it? The idea that the continents were once united and then broken apart is an old one. In fact, in 1668, a Frenchman interpreted the matching coastlines of eastern

395

South America and western Africa as proof that the two continents were separated during the biblical flood!

Within the last 30 years, the slow movement of continents over long spans of geologic time has become widely accepted by geologists and physical geographers. Evidence for the fracturing of the Earth's outer rock layer and movement of the resulting rock plates has come from many sources—earthquake analysis, studies of ancient collections of plants and animals as they may have existed on united continents, and the matching of geologic structures, such as mountain chains and rock sequences, from conti-

nent to continent. Most recently, the rates of movement of continents have actually been measured using laser beams bounced from continents to orbiting satellites and back. Thus, many lines of evidence now converge on *plate tectonics*—the body of knowledge describing the movements and changes through time of lithospheric plates carrying continents and ocean basins.

Plate tectonics provides the plan of the Earth's major relief features, from the locations of mountain ranges and basins to the positions of oceans and inland seas. An understanding of plate tectonics is thus important to the understanding of the global patterns of the Earth's landscapes—including climate, soils, vegetation, landforms,

and, ultimately, human activity. In other words, the processes of physical geography are superimposed on a global geography given by plate tectonics. For this reason, the motions and histories of continental plates— splitting and separating to form ocean basins, closing and colliding to form mountain ranges—can be taken as a grand plan to organize a vast body of geographic and geologic knowledge. That grand plan—and the forces that have shaped it—are the subjects of this chapter. ■

THE LITHOSPHERE AND PLATE TECTONICS

On the globes and maps we've seen since early childhood, the outline of each continent is so unique that we would never mistake one continent for another. But why are no two continents even closely alike? As we will see in this chapter, the continents have had a long history. In fact, in each of the continents some regions of metamorphic rocks date back more than 2 billion years. As part of that history, the continents have been fractured and split apart, as well as pushed together and joined.

Perhaps you've visited an old New England farmhouse that was constructed over hundreds of years—first the small two-room house, to which the kitchen shed was added at the back, and then the parlor wing at the side. Later the roof was raised for a second story, the tool house and barn were joined to the main house, the carriage house was built, and so it went. The Earth's continents have that kind of history. They are composed of huge masses of continental crust that have been assembled at different times in each continent's history. The theory describing the motions and changes through time of the continents and ocean basins, and the processes that fracture and fuse them, is called *plate tectonics.*

Briefly stated, plate tectonic theory maintains that the Earth's outermost solid layer consists of huge plates that are in constant but slow motion, powered by energy sources deep within the Earth. Floating on a layer of plastic rock below, these rigid plates can both collide with other plates and split apart to form new plates. When crustal plates collide, one plate may be forced under the other in a process called *subduction.* In this case, part of the subducted plate can melt, creating pockets of magma that move upward to

the surface and create volcanic mountain chains and island arcs. Plates can also collide without subduction, forming mountain chains of rocks that are fractured, folded, and compressed into complex mountain structures. Where crustal plates split apart, creating rift valleys and ultimately ocean basins, lava moves upward to fill the gap. Thus, the Earth's surface arrangement of continents and oceans is constantly but slowly changing. Ocean basins expand with rifting and contract with subduction, while continents grow by collision and lose parts by splitting.

Because plate motions are powered by internal Earth forces, they are independent of surface conditions. So we find volcanoes erupting in the cold desert of Antarctica as well as near the equator in African savannas. An alpine mountain range has been pushed up in the cold subarctic zone of Alaska, where it runs east-west. Yet another range lies astride the equator in South America and runs north-south. Both ranges lie in belts of crustal collision, where many strong earthquakes occur.

Although internal tectonic processes operate independently, the processes the determine climate, vegetation, and soils are dependent on the major relief features and Earth materials provided by the tectonic setting. Thus, an understanding of plate tectonics is important to our understanding of the global patterns of the Earth's landscapes—including its climate, soils, vegetation, and, ultimately, human activity.

In this chapter we will survey the major geologic features of our planet, starting with the layered structure of its deep interior. We will then examine the crust and compare the crust of the continents with the crust of the ocean basins. Lastly, we will turn to plate tectonics and describe how plate movements have created broad regions of igneous, sedimentary, and metamorphic rocks. The motions and histories of these vast rock plates, splitting and separating to form ocean basins, closing and colliding to form mountain ranges, can be taken as an overarching framework to organize a great body of geologic and geographic knowledge. That is why this revolutionary framework—plate tectonic theory—ranks with the theory of evolution as one of the great milestones in scientific study of the Earth.

THE STRUCTURE OF THE EARTH

What lies deep within the Earth? From studies of earthquake waves, reflected from deep Earth layers, scientists have discovered that our Earth is far from uniform from its outer crust to its center. Instead, it consists of a central core with several layers, or shells, surrounding it. The densest matter is at the center, and each layer above it is increasingly less dense. This structure is inherited from the Earth's earliest history as it was formed by accretion from a mass of gas and dust orbiting the Sun. We will begin our examination of the Earth's inner structure at the center and then work outward.

The Earth's Interior

Figure 12.1 is a cutaway diagram of the Earth showing its interior. The Earth as a whole is an almost spherical body

The layers of the Earth's interior include the crust, mantle, liquid outer core, and solid inner core. Continental crust has both felsic and mafic rock zones, while oceanic crust has only mafic rock.

approximately 6400 km (about 4000 mi) in radius. The center is occupied by the **core,** which is about 3500 km (about 2200 mi) in radius. Because earthquake waves suddenly change behavior upon reaching the core, scientists have concluded that the outer core has the properties of a liquid. However, the innermost part of the core is in the solid state. Based on earthquake waves (and other kinds of data), it has long been inferred that the core consists mostly of iron, with some nickel. The core is very hot, but its temperature is not known very well. It is estimated to lie somewhere between 3000°C and 5000°C (about 5400°F to 9000°F).

Enclosing the metallic core is the **mantle,** a rock shell about 2900 km (about 1800 mi) thick. Earthquake waves indicate that mantle rock is composed of mafic minerals similar to olivine (a silicate of magnesium and iron) and may resemble the ultramafic igneous rock peridotite, which is found exposed here and there on the continental surface (see Figure 11.4 and Table 11.1). Temperatures in the mantle range from about 2800°C (about 5100°F) near the core to about 1800°C (about 3300°F) near the crust.

The outermost and thinnest of the Earth shells is the **crust,** a layer normally about 8 to 40 km (about 5 to 25 mi) thick (Figure 12.2). It is formed largely of igneous rock, but it also contains substantial proportions of metamorphic rock and a comparatively thin upper layer of sedimentary rock. The base of the crust is sharply defined where it contacts the mantle. This contact is detected by the way in which earth-

12.1 The Earth's interior *This cutaway diagram of the Earth shows the inner core of iron, which is solid, surrounded by a liquid outer core. The mantle, which surrounds the core, is a thick inner layer of ultramafic rock. The crust is too thin to show to correct scale. Thicknesses of interior layers are rounded to the nearest 100 units.*

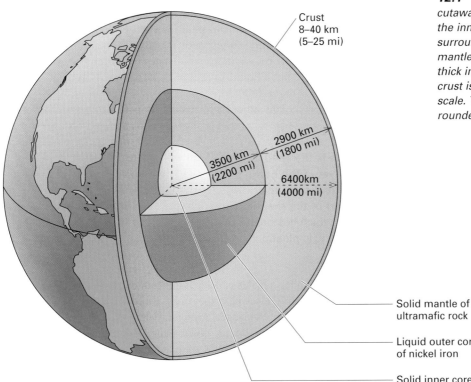

Crust
8–40 km
(5–25 mi)

3500 km
(2200 mi)

2900 km
(1800 mi)

6400 km
(4000 mi)

Solid mantle of
ultramafic rock

Liquid outer core
of nickel iron

Solid inner core

(a)

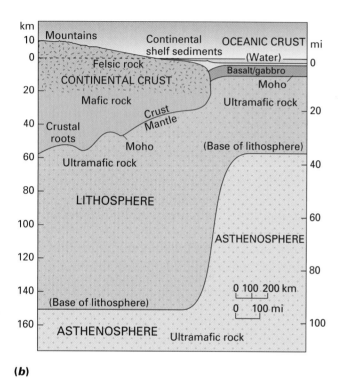

(b)

12.2 Outer layers of the Earth *(a) Idealized cross section of the Earth's crust and upper mantle. (b) Details of the crust and mantle at the edge of a continent, including the types of rocks found there. Also shown are the lithosphere and asthenosphere. (Copyright © A. N. Strahler.)*

has a chemical composition similar to that of granite, it is commonly described as being *granitic rock.* Much of the granitic rock is metamorphic rock. There is no sharply defined separation between the felsic and mafic zones.

While continental crust typically has two layers of different composition, **oceanic crust** has only one. It consists almost entirely of mafic rocks with the composition of basalt and gabbro. Basalt, as lava, forms an upper zone, whereas gabbro, an intrusive rock of the same composition, lies beneath the basalt.

Another key distinction between continental and oceanic crust is that the crust is much thicker beneath the continents than beneath the ocean floors (Figure 12.2). While 35 km (22 mi) is a good average figure for crustal thickness beneath the continents, 7 km (4 mi) is a typical figure for thickness of the basalt and gabbro crust beneath the deep ocean floors. Later in this chapter we will relate this difference in thickness to the processes that form the two types of crust.

The Lithosphere and Asthenosphere

Geologists use the term **lithosphere** to mean an outer Earth zone, or shell, of rigid, brittle rock. The lithosphere includes not only the crust, but also the cooler, upper part of the mantle that is composed of brittle rock. The lithosphere ranges in thickness from 60 to 150 km (40 to 95 mi). It is thickest under the continents and thinnest under the ocean basins.

Some tens of kilometers deep in the Earth, the brittle condition of the lithospheric rock gives way gradually to a plastic, or "soft," layer named the **asthenosphere** (Figure 12.2). (This word is derived from the Greek root *asthenes,* meaning "weak.") However, at still greater depth in the mantle, the strength of the rock material again increases. Thus, the asthenosphere is a soft layer sandwiched between the "hard" lithosphere above and a "strong" mantle rock layer below. In terms of states of matter, the asthenosphere is not a liquid, even though its temperature reaches 1400°C (about 2600°F).

The lithosphere resting atop the asthenosphere can be thought of as a hard, brittle shell resting on a soft, plastic underlayer. Because the asthenosphere is soft and plastic, the rigid lithosphere can easily move over it. The lithospheric shell consists of large pieces called **lithospheric plates.** A single plate can be as large as a continent and can move independently of the plates that surround it. Like great slabs of floating ice on the polar sea, lithospheric plates can separate from one another at one location, while elsewhere they may collide in crushing impacts that raise great ridges. Along these collision ridges, one plate can be found diving down beneath the edge of its neighbor. These varied sorts of plate movements will be described shortly.

quake waves abruptly change velocity at that level. The boundary surface between crust and mantle is called the *Moho,* a simplification of the name of the scientist, Andrija Mohorovicic, who discovered it in 1909.

The continental crust is quite different from the crust beneath the oceans. From studies of earthquake waves, geologists have concluded that the **continental crust** consists of two continuous zones—a lower, continuous rock zone of mafic composition, which is more dense, and an upper, continuous zone of felsic rock, which is less dense (Figure 12.2). Because the felsic portion

The lithosphere is the solid, brittle outermost layer of the Earth. It includes the crust and the cooler, brittle upper part of the mantle. The asthenosphere, which lies below the lithosphere, is plastic.

THE GEOLOGIC TIME SCALE

To place the great movements of lithospheric plates in their correct positions in historical sequence, we will need to refer to some major units in the scale of geologic time. Table 12.1 lists the major geologic time divisions. All time older than 570 million years (m.y.) before the present is *Precambrian time.* Three *eras* of time follow: *Paleozoic, Mesozoic,* and *Cenozoic.* These eras saw the evolution of life-forms in the oceans and on the lands. The geologic eras are subdivided into *periods.* Their names, ages, and durations are also given in Table 12.1.

The Cenozoic Era is particularly important in terms of the continental surfaces because nearly all landscape features seen today have been produced in the 65 million years since that era began. The Cenozoic Era is comparatively short in duration, scarcely more than the average duration of a single period in older eras. It is subdivided directly into

The geologic time scale includes the Paleozoic, Mesozoic, and Cenozoic Eras. The Cenozoic Era is most recent, and nearly all the landscape features visible today have been formed within that era.

seven lesser time units called *epochs.* (Note that the terms *Tertiary Period* and *Quaternary Period* are sometimes applied to the Cenozoic Epochs Paleocene through Pliocene, and Pleistocene, respectively.) Details of the Pleistocene and Holocene epochs are given in Chapter 19.

The human genus *Homo* evolved during the late Pliocene Epoch and throughout the Pleistocene Epoch. As you can see, the period of human occupation of the Earth's surface—a few million years at best—is but a fleeting moment in the vast duration of our planet's history.

MAJOR RELIEF FEATURES OF THE EARTH'S SURFACE

Before we begin our discussion of plate tectonics, it is important to examine the major relief features of the Earth—its continents and ocean basins—since they are pro-

Table 12.1 | Geologic Time Scale

Era	Period	Epoch	Duration (millions of years)	Age (millions of years)	Orogenies	Evolution of Life-Forms
Cenozoic		Holocene	(10,000 yr)			
		Pleistocene	2	2		Human genus
		Pliocene	3	5		
		Miocene	19	24		Hominoids
		Oligocene	13	37		Whales
		Eocene	21	58		Bats
		Paleocene	8	66	Cordilleran	Mammals
Mesozoic	Cretaceous		78	144		Flowering plants
	Jurassic		64	208	Allegheny, or Hercynian	Dinosaurs (extinct)
	Triassic		37	245		Turtles
Paleozoic	Permian		41	286		Frogs
	Carboniferous		74	360		Conifers; higher fishes
	Devonian		48	408	Caledonian	Vascular plants; primitive fishes
	Silurian		30	438		
	Ordovician		67	505		
	Cambrian		65	570		Invertebrates

Precambrian Time (Extends to oldest known rocks, about 4 billion years)
Age of Earth as a planet: 4.6 to 4.7 billion years.
Age of universe: 17 to 18 billion years.

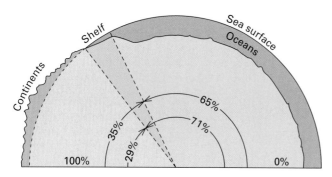

12.3 Continent and ocean areas compared *Actual global percentages of land and ocean areas compared with percentages if sea level were to drop 180 m (about 600 ft), exposing the continental shelves.*

duced by plate tectonic processes. Detailed global maps show that about 29 percent of the Earth's surface is land and 71 percent oceans (Figure 12.3). If the seas were to drain away, however, we would see that broad sloping areas lie close to the continental shores. These continental shelves are covered by shallow water, less than 150 m (500 ft) deep. From these relatively shallow continental shelves, the ocean floor drops rapidly to depths of thousands of meters. In other

words, the ocean basins seem to be more than "brim full" of water—the oceans have spread over the margins of ground that would otherwise be assigned to the continents. If the ocean level were to drop by 150 m (about 500 ft), the shelves would be exposed, adding about 6 percent to the area of the continents. The surface area of continents would then be increased to 35 percent, and the ocean basin area decreased to 65 percent. These revised figures represent the true relative proportions of continents and oceans.

Figure 12.4 is a map showing some of the major relief features of continents and oceans, including mountain arcs, island arcs, ocean trenches, and the midoceanic ridge. We will refer to this map as we discuss the relief features of continents and oceans in turn.

Relief Features of the Continents

Broadly viewed, the continents consist of two basic subdivisions: active belts of mountain-making and inactive regions of old, stable rock. The mountain ranges in the active belts grow through one of two very different geologic processes. First is *volcanism,* the formation of massive accumulations of volcanic rock by extrusion of magma. Many lofty mountain ranges consist of chains of volcanoes built of extrusive igneous rocks.

12.4 Tectonic features of the world *Principal mountain arcs, island arcs, and trenches of the world and the midoceanic ridge. (Midoceanic ridge map copyright © A. N. Strahler.)*

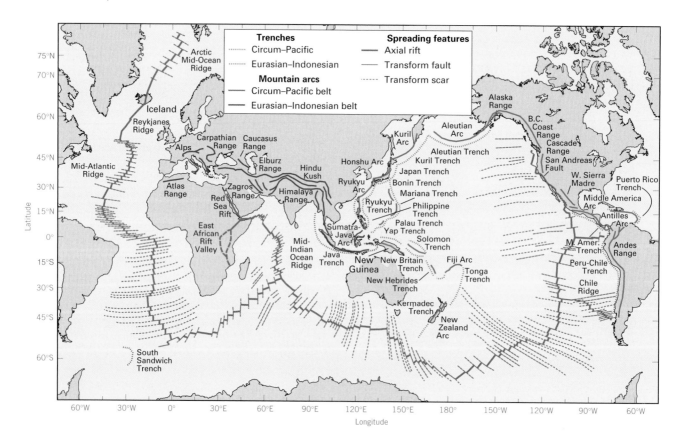

The second mountain-building process is *tectonic activity,* the breaking and bending of the Earth's crust under internal Earth forces. This tectonic activity usually occurs when great lithospheric plates come together in titanic collisions. (We will return to this topic later in this chapter.) Crustal masses that are raised by tectonic activity form mountains and plateaus. In some instances, both volcanism and tectonic activity have combined to produce a mountain range. Tectonic activity can also lower crustal masses to form depressions that may be occupied by ocean embayments or inland seas. Landforms produced by volcanic and tectonic activity are the subject of Chapter 13.

ALPINE CHAINS Active mountain-making belts are narrow zones that are usually found along the margins of lithospheric plates. These belts are sometimes referred to as *alpine chains* because they are characterized by high, rugged mountains, such as the Alps of Central Europe. These mountain belts were formed in the Cenozoic Era by volcanism or tectonic activity or a combination of both. Alpine mountain-building continues even today in many places.

The alpine chains are characterized by broadly curved lines on the world map (Figure 12.4). Each curved section of an alpine chain is referred to as a **mountain arc.** The arcs are linked in sequence to form two principal mountain belts. One is the *circum-Pacific belt* (shown in green), which rings the Pacific Ocean Basin. In North and South America, this belt is largely on the continents and includes the Andes and Cordilleran Ranges. In the western part of the Pacific Basin, the mountain arcs lie well offshore from the continents and take the form of **island arcs.** Partly submerged, they join the Aleutians, Kurils, Japan, the Philippines, and other smaller islands. These island arcs are the result of volcanic activity. Between the larger islands, the arcs are represented by volcanoes rising above the sea as small, isolated islands.

The second chain of major mountain arcs forms the *Eurasian-Indonesian belt,* shown in blue in Figure 12.4. It starts in the west at the Atlas Mountains of North Africa and continues through the European Alps and the ranges of the Near East and Iran to join the Himalayas. The belt then continues through Southeast Asia into Indonesia, where it abruptly meets the circum-Pacific belt. Later, we will return to these active belts of mountain-making and explain them in terms of subduction of lithospheric plates.

CONTINENTAL SHIELDS Belts of recent and active mountain-making account for only a small portion of the continental crust. The remainder consists of comparatively

The two basic subdivisions of continental masses are active belts of mountain-making and inactive regions of old, stable rock. Mountains are built by volcanism and tectonic activity.

Alpine chains include mountain arcs and island arcs. They occur in two prominent belts—the circum-Pacific belt and the Eurasian-Indonesian belt.

Inactive continental regions of stable rocks include continental shields and ancient mountain roots. Continental shields are low-lying areas of old igneous and metamorphic rock and may be exposed or covered by sediments.

inactive regions of much older rock. Within these stable regions we can recognize two types of crustal structure—continental shields and mountain roots. **Continental shields** are low-lying continental surfaces beneath which lie igneous and metamorphic rocks in a complex arrangement. Figure 12.5 is a generalized map showing the shield areas of the continents. Two classes of shields are shown: exposed shields and covered shields.

Exposed shields include very old rocks, mostly from Precambrian time, and have had a very complex geologic history. An example of an exposed shield is the Canadian Shield of North America. Exposed shields are also extensive in Scandinavia, South America, Africa, Asia, peninsular India, and Australia. The exposed shields are largely regions of low hills and low plateaus, although there are some exceptions where large crustal blocks have been recently uplifted. Many thousands of meters of rock have been eroded from these shields throughout the past half-billion or more years.

Covered shields are areas of the continental shields that are covered by younger sedimentary layers, ranging in age from Paleozoic through Cenozoic eras. These strata accumulated at times when the shields were inundated by shallow seas. Marine sediments were laid down on the ancient shield rocks in thicknesses ranging from hundreds to thousands of meters. These shield areas were then broadly arched upward to became land surfaces again. Erosion has since removed large sections of their sedimentary cover, but it still remains intact over vast areas. The covered shields are shown in Figure 12.5.

Some core areas of the shields are composed of rock as old as early Precambrian time, dating back to a time period called the Archean Eon, 2.5 to 3.5 billion years ago. On our map, these ancient areas are shown encircled by bold lines. The ancient cores are exposed in some areas but covered in others.

ANCIENT MOUNTAIN ROOTS Remains of older mountain belts lie within the shields in many places. These *mountain roots* are mostly formed of Paleozoic and early Mesozoic sedimentary rocks that have been intensely bent and folded, and in some locations changed into metamorphic rocks—slate, schist, and quartzite, for example. Thousands of meters of overlying rocks have been removed from these old tectonic belts, so that only the lowermost structures remain. Roots appear as chains of long, narrow ridges, rarely rising over a thousand meters above sea level.

One important system of mountain roots was formed in the Paleozoic Era, during a great collision between two enormous lithospheric plates that took place about 400 million years

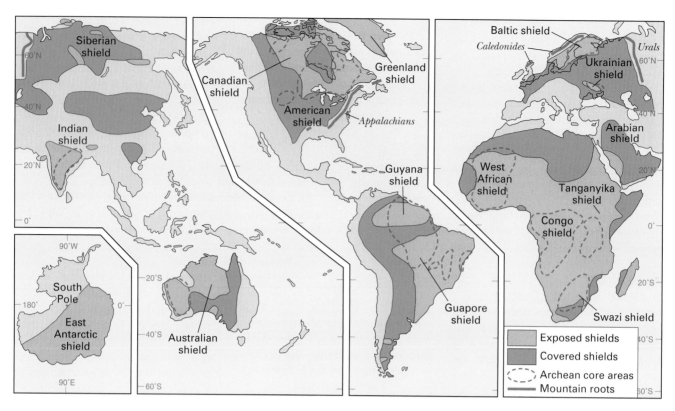

12.5 Continental shields *Generalized world map of continental shields, exposed and covered. Continental centers of early Precambrian age (Archean) lie with the areas encircled by a broken red line. Heavy brown lines show mountain roots of Caledonian and Hercynian (Appalachian) orogenies. (Based in part on data of R. E. Murphy, P. M. Hurley, and others. Copyright © A. N. Strahler.)*

ago. This collision created high alpine mountain chains that have since been worn down to belts of subdued mountains and hills. Today these roots, called Caledonides, form a highland belt across the northern British Isles and Scandinavia. They are also present in the Maritime Provinces of eastern Canada and in New England. A second, but younger, root system was formed during another great collision of plates near the close of the Paleozoic Era, about 250 million years ago. In North America, this highland system is represented by the Appalachian Mountains. The Caledonides and Appalachians are shown as mountain roots in Figure 12.5.

> Mountain roots mark ancient collisions of lithospheric plates. They are now eroded into regions of low highlands and ridges.

Relief Features of the Ocean Basins

The relief features of ocean basins are quite different from those of the continents. Crustal rock of the ocean floors consists almost entirely of basalt, which is covered over large areas by a comparatively thin accumulation of sediments. Age determinations of the basalt and its sediment cover show that the oceanic crust is quite young, geologically speaking. Much of the oceanic crust is less than 60 million years old, although some large areas have ages of about 65 to 135 million years. When we consider that the great bulk of the continental crust is of Precambrian age—mostly over

1 billion years old—the young age of the oceanic crust is quite remarkable. However, we will soon see how plate tectonic theory explains this young age.

THE MIDOCEANIC RIDGE AND OCEAN BASIN FLOOR Figure 12.6 shows the important relief features of ocean basins. The ocean basins are characterized by a central ridge structure that divides the basin about in half. This *midoceanic ridge* consists of submarine hills that rise gradually to a rugged central zone. Precisely in the center of the ridge, at its highest point, is the *axial rift,* which is a narrow, trench-like feature. The location and form of this rift suggest that the crust is being pulled apart along the line of the rift.

The midoceanic ridge and its principal branches can be traced through the ocean basins for a total distance of about 60,000 km (about 37,000 mi). Figure 12.4 shows the extent of the ridge. Beginning in the Arctic Ocean, it then divides the Atlantic Ocean Basin from Iceland to the South Atlantic. Turning east, it enters the Indian Ocean, where one branch penetrates Africa. The other branch continues east between Australia and Antarctica, and then swings across the South Pacific. Nearing South America, it turns north and reaches North America at the head of the Gulf of California.

On either side of the midoceanic ridge are broad, deep plains that belong to the *ocean basin floor* (Figure 12.6). Their

12.6 Ocean basins *This schematic block diagram shows the main features of ocean basins. It applies particularly well to the North and South Atlantic oceans.*

average depth below sea level is about 5000 m (about 16,000 ft). These large, flat expanses of ocean floor are called *abyssal plains.* They are extremely smooth because they have been built up of fine sediment that has settled slowly and evenly from ocean water above. Some abyssal regions include *abyssal hills,* small hills rising to heights from a few tens of meters to a few hundred meters above the ocean floor.

Many details of the ocean basins and their submarine landforms are shown in Figure 12.7. This image of the ocean floor was constructed from data of a U.S. Navy satellite that measured the surface height of the ocean very precisely using radar. Because small variations in gravity caused by undersea ridges and valleys produce corresponding small variations in the ocean height, it is possible to infer the undersea topography from ocean height measurement. For example, the mass of a submerged seamount may attract enough sea water to raise sea levels above it by 1 to 2 m (3 to 6 ft), depending on its bulk. Similarly, the ocean surface level drops over a deep ocean trench, where there is less mass to attract sea water. The right page of the figure shows the North Atlantic Basin. The prominent central feature is the Mid-Atlantic Ridge. The left page shows the deep trenches of the western Pacific that mark the positions of subduction arcs—features of plate motion that we will discuss shortly.

CONTINENTAL MARGINS The *continental margin,* shown on the left and right sides of Figure 12.6, is the narrow zone in which oceanic lithosphere is in contact with continental lithosphere (see Figure 12.2b.) As the continental margin is approached from deep ocean, the ocean floor begins to slope gradually upward, forming the *continental rise*

> **Relief features of ocean basins include a midoceanic ridge with a central axial rift where crust is being pulled apart. Abyssal plains, with some abyssal hills, stretch to continental margins.**

(Figure 12.6). The floor then steepens greatly on the *continental slope.* At the top of this slope we arrive at the edge of the *continental shelf,* a gently sloping platform that is about 120 to 160 km (about 75 to 100 mi) wide along the eastern margin of North America. Water depth is about 150 m (about 500 ft) at the outer edge of the shelf. These three features—rise, slope, and shelf—form the continental margin.

Figure 12.6 illustrates a symmetrical ocean floor model—a midoceanic ridge with ocean basin floors on either side. This model nicely fits the North Atlantic and South Atlantic Ocean basins. It also applies rather well to the Indian Ocean and Arctic Ocean basins. The margins of these symmetrical basins are described as *passive continental margins* because they have not been subjected to strong tectonic and volcanic activity during the last 50 million years. This is because the continental and oceanic lithosphere that join at a passive continental margin are part of the same lithospheric plate and move together, away from the axial rift.

Many passive continental margins have accumulated great thicknesses of continental sediments. Near the continent, the strata of the shelf form a wedge-shaped deposit, thickening oceanward. A block diagram, Figure 12.8, shows details of this deposit. The river-borne sediments brought from the land are spread over the shallow seafloor by currents. Beneath the continental rise and its adjacent abyssal plain is another thick sediment deposit. It is formed from deep-sea sediments carried down the continental slope by swift muddy currents. The weight of these accumulated sediments causes them to sink, depressing the oceanic crust at its margin with the continental crust.

Deltas built by rivers contribute a great deal of the shelf sediment. Tongue-like turbid currents thickened

> **Continental lithosphere contacts oceanic lithosphere at a continental margin. Passive continental margins accumulate thick deposits of continental sediments. Active continental margins have oceanic trenches where oceanic crust is sliding beneath continental crust.**

12.7   Eye on the Landscape   Undersea topography
*Portions of a map of variations in the pull of gravity, which
indicate undersea topography. Deeper regions are shown in
tones of purple, blue, and green, while shallower regions are in
tones of yellow and reddish brown. On the left page, the ring of
subduction trenches of the western Pacific Basin is a prominent
feature. On the right page, the mid-Atlantic ridge and other
features of seafloor spreading are visible. Data were acquired by
the U.S. Navy Geosat satellite altimeter. (Copyright © 1995,
David T. Sandwell. Used by permission.)*   **What else would the
geographer see? ... Answers at the end of the chapter.**

12.8 **Structure of the continental shelf** *This block diagram shows an inner wedge of sediments beneath the continental shelf and an outer wedge of deep-sea sediments beneath the continental rise and abyssal plain.*

with fine silt and clay from these deltas carve submarine canyons into the continental shelf. Where these currents emerge from the canyons, they deposit their sediment and build *deep-sea cones* (Figure 12.9).

The margins of the Pacific Ocean Basin are very different from those of the Atlantic. Although the Pacific has a midoceanic ridge with ocean basin floors on either side, its margins are characterized by mountain arcs or island arcs with deep offshore *oceanic trenches* (see Figure 12.7). Geologists refer to these trenched ocean-basin edges as *active continental margins.* As we will describe in more detail in the next section, oceanic crust is here being bent downward and forced under continental crust, creating trenches and inducing volcanic activity. The locations of the major trenches are shown in Figure 12.4. Trench floors can reach depths of nearly 11,000 m (about 36,000 ft), although most range from 7000 to 10,000 m (about 23,000 to 33,000 ft).

12.9 **Deep-sea cone** *A deep-sea cone of sediment forms at the base of a submarine canyon cut into the continental shelf. (Copyright © A. N. Strahler.)*

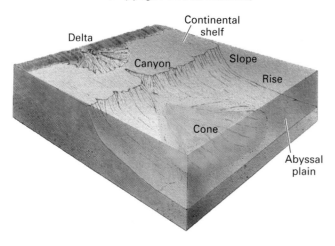

PLATE TECTONICS

With the structures and features of continents and ocean basins as a background, we can now turn to the motion of lithospheric plates and their interactions at boundaries—that is, to **plate tectonics. Tectonics** is a noun meaning "the study of tectonic activity." It refers here to all forms of breaking and bending of the entire lithosphere, including the crust. Breaking and bending occur on a very large scale when lithospheric plates collide. Sedimentary strata that lay flat and undisturbed for tens of millions of years on ocean floors and continental margins are crumpled and sheared into fragments by such collisions. The collision process itself may take millions of years to complete, but in the perspective of geologic time, it's only a brief episode.

Tectonic Processes

Generally speaking, prominent mountain masses and mountain chains (other than volcanic mountains) are elevated by one of two basic tectonic processes: *extension* and *compression* (Figure 12.10). Extensional tectonic activity—"pulling apart"—occurs where oceanic plates are separating or where a continental plate is undergoing breakup into fragments. As the crust thins, it is fractured and pushed upward, producing block mountains. We will return to a more detailed description of this process in a later section on continental rupture.

Compressional tectonic activity—"squeezing together" or "crushing"—acts at converging plate boundaries. The result is often an alpine mountain chain consisting of intensely deformed strata of marine origin. The strata are tightly compressed into wave-like structures called *folds.*

12.10 **Two basic forms of tectonic activity** *Flat-lying rock layers may be compressed to form folds or pulled apart to form faults by rifting.*

Overturned fold

Recumbent fold

Overthrusting

Thrust sheet

Overthrust fault

12.11 Folding in compressional tectonic activity
These schematic diagrams show the development of a recumbent fold, broken by a low-angle overthrust fault to produce a thrust sheet, or nappe, in alpine structure. (Based on diagrams by A. Heim, 1922, Geologie der Schweiz, vol. II-1, Tauschnitz, Leipzig.)

Typically, alpine folds can be traced through a history. First, they are overturned, and then they become *recumbent* as they are further overturned upon themselves (Figure 12.11). Accompanying the folding is a form of faulting in which slices of rock move over the underlying rock on fault surfaces with gentle inclination angles. These are called *overthrust faults.* The individual rock slices, called *thrust sheets,* are carried many tens of kilometers over the underlying rock. In the European Alps, thrust sheets of this kind were named *nappes* (from the French word meaning "cover sheet" or "tablecloth"). Nappes may be thrust one over the other to form a great pile. The entire deformed rock mass produced by such compressional mountain-making is called an *orogen,* and the event that produced it is an *orogeny.*

Plate Motions and Interactions

Figure 12.12 shows the major features of plate interactions. The vertical dimension of the block diagram (*a*) is greatly exaggerated, as are the landforms. A true-scale cross section (*b*) shows the correct relationships between crust and lithosphere, but surface relief features are too small to be shown. Diagram (*c*) is a sketch of plate motions on a spherical Earth.

Note that the figure shows two different kinds of lithospheric plates (see Figure 12.12*b*). Plates that lie beneath the ocean basins consist of *oceanic lithosphere.* This form of lithosphere is comparatively thin (about 50 km or 30 mi thick). The diagrams show two plates, X and Y, both made up of oceanic lithosphere. Plate Z bears continental crust and is made up of *continental lithosphere,* which is much thicker (about 150 km or 95 mi).

The lithosphere can be thought of as "floating" on the soft asthenosphere, but there is a difference in the relative surface heights of the two kinds of floating lithosphere. Consider two blocks of wood, one thicker than the other, floating in a pan of water. The surface of the thick block will ride higher above the water surface than that of the thin block. This principle explains why the continental surfaces rise high above the ocean floors.

As shown in Figure 12.12, plates X and Y are pulling apart along their common boundary, which lies along the axis of a midoceanic ridge. This pulling apart tends to create a gaping crack in the crust, but magma continually rises from

A simple experiment you can do will demonstrate how folding occurs in mountain-building. Take an ordinary towel of thick, limp cloth and lay it out on a smooth table top. Using both hands, palms down, bring the ends of the cloth slowly together. First, a simple up-fold will develop and grow, until it overturns to one side or the other. Then more folds will form, grow, and overturn, giving you a simple model of a new mountain range.

Tectonic processes include extension and compression. Extension causes fracturing and faulting of the crust, while compression produces folds and overthrust faults.

the mantle beneath to fill it. The magma appears as basaltic lava in the floor of the rift and quickly congeals. At greater depth under the rift, magma solidifies into *gabbro,* an intrusive rock of the same composition as basalt. Together, the basalt and gabbro continually form

12.12 Schematic cross sections showing some of the important elements of plate tectonics *Diagram (a) is greatly exaggerated in vertical scale, and emphasizes surface and crustal features. Only the uppermost 30 km (20 mi) is shown. Diagram (b) is drawn to true scale and shows conditions to a depth of 250 km (155 mi). Here the actual relationships between lithospheric plates can be examined, but surface features are too small to be shown. Diagram (c) is a pictorial rendition of plates on a spherical Earth and is not to scale. (Copyright © A. N. Strahler.)*

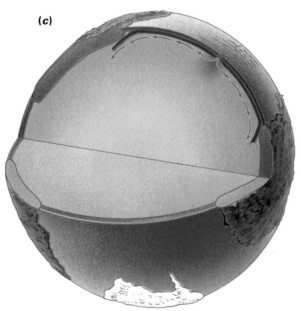

new oceanic crust. This type of boundary between plates is termed a *spreading boundary*.

At the right, the oceanic lithosphere of plate Y is moving toward the thick mass of continental lithosphere that comprises plate Z. Where these two plates collide, they form a *converging boundary*. Because the oceanic plate is comparatively thin and dense, in contrast to the thick, buoyant continental plate, the oceanic lithosphere bends down and plunges into the soft layer, or asthenosphere. The process in which one plate is carried beneath another is called **subduction.**

The leading edge of the descending plate is cooler and therefore denser than the surrounding hot, soft asthenos-

phere. As a result, the slab "sinks under its own weight," once subduction has begun. However, the slab is gradually heated by the surrounding hot rock and thus eventually softens. The underportion, which is mantle rock in composition, simply reverts to mantle rock as it softens.

The descending plate is covered by a thin upper layer of less dense mineral matter derived from oceanic and continental sediments. This material can melt and become magma. The magma tends to rise because it is less dense than the surrounding material. Figure 12.12b shows some magma pockets formed from the upper edge of the slab. They are pictured as rising like hot-air balloons through the overlying continental lithosphere. When they reach the

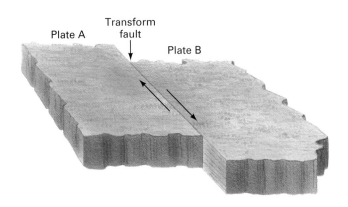

12.13   Transform fault *A transform fault involves the horizontal motion of two adjacent lithospheric plates, one sliding past the other. (Copyright © A. N. Strahler.)*

Earth's surface, they form a chain of volcanoes lying about parallel with the deep oceanic trench that marks the line of descent of the oceanic plate.

Viewing plate Y as a unit in Figure 12.12, we see that this single lithospheric plate is simultaneously undergoing both *accretion* (growth by addition) and *consumption* (loss by softening and melting in subduction zones). If rates of accretion and consumption are equal, the plate will maintain its overall size. If consumption is greater, the area of the plate will decrease. If accretion is greater, the area of the plate will expand.

We have yet to consider a third type of lithospheric plate boundary. Two lithospheric plates may be in contact along a common boundary on which one plate merely slides past the other with no motion that would cause the plates either to separate or to converge (Figure 12.13). This is a *transform boundary.* The plane along which motion occurs is a nearly vertical fracture extending down through the entire lithosphere, and it is called a *transform fault.* A *fault* is a rock plane along which there is motion of the rock mass on one side with respect to that on the other. (More about faults will appear in Chapter 13.) Transform boundaries are often associated with midoceanic ridges and are shown in Figures 12.7 and 12.12. GEODISCOVERIES

PLATE BOUNDARIES SUMMARIZED   In summary, there are three major kinds of active plate boundaries:

- *Spreading boundaries.* New lithosphere is being formed by accretion. Example: Seafloor spreading along the axial rift.
- *Converging boundaries.* Subduction is in progress, and lithosphere is being consumed. Example: Active continental margin.
- *Transform boundaries.* Plates are gliding past one another on a transform fault. Example: Transform boundary associated with midoceanic ridge.

> Boundaries between lithospheric plates may be of spreading, converging, or transform types. At a spreading boundary, crust is being pulled apart. At a converging boundary, plates are in collision and one plate may be subducted beneath another. At a transform boundary, two plates glide past each other.

Let us put these three boundaries into a pattern to include an entire lithospheric plate (Figure 12.14). Visualize the sunroof of an automobile, in which a portion of the roof slides to open. Just as the sunroof can move by sliding past the fixed portion of the roof at its sides, so a lithospheric plate can move by sliding past other plates on transform faults. Where the sunroof opens, the situation is similar to a spreading boundary. Where the sunroof slides under the rear part of the roof, the situation is similar to a converging boundary. Boundaries of lithospheric plates can be curved as well as straight, and individual plates can pivot as they move. There are many geometric variations in the shapes and motions of individual plates.

The motions of lithospheric plates follow a time cycle in which a single supercontinent is split apart, then joined back together, and then split apart again. Named the Wilson Cycle, it is described in more detail in *Eye on the Environment 12.1 • The Wilson Cycle and Supercontinents.*

The Global System of Lithospheric Plates

The global system of lithospheric plates consists of six great plates. These are listed in Table 12.2 and are shown on a world map in Figure 12.15. Several lesser plates and subplates are also recognized. They range in size from intermediate to comparatively small. Plate boundaries are shown by symbols, explained in the key accompanying the map.

The great Pacific plate (Figure 12.15*a*) occupies much of the Pacific Ocean Basin and consists almost entirely of oceanic lithosphere. Its relative motion is northwesterly, so that it has a subduction boundary

12.14   A simple lithospheric plate *A schematic diagram of a single rectangular lithospheric plate with two transform boundaries. (Copyright © A. N. Strahler.)*

The Wilson Cycle and Supercontinents

As you have learned in this chapter, the geography of the Earth and its varied environments has been changing slowly but continuously for hundreds of millions of years. The driving force for that slow change is the motion of the Earth's lithospheric plates. While it seems remarkable that the continents are actually in constant motion, what is more remarkable is the strong evidence suggesting that all continents were once joined in a single land mass that subsequently broke apart. Not only that, but over the billions of years of the Earth's geologic history, the union of the continents and their breakup is actually a repeating process—the *Wilson cycle*—that has occurred half a dozen times or more.

The Wilson cycle bears the name of the distinguished Canadian geophysicist Professor J. Tuzo Wilson of the University of Toronto. He made important new discoveries and interpretations that helped lead to a full-blown theory of plate tectonics. By the time he formulated his theory, all of the parts were known—continental rupture to start the opening of a new ocean basin, plate subduction to generate island arcs, and collisions to weld continents together. Using all of these separate lithospheric activities, Wilson forged a logical time chain of plate tectonic events, and it quickly gained wide acceptance as the *Wilson cycle*.

The Wilson cycle can best be appreciated by use of a cartoon-like presentation of the essential elements of plate tectonics. In our illustration at left, these elements are shown arranged throughout six stages and are listed with references to our text diagrams and examples found on the planet today. This ideal cycle requires some 300 to 500 million years to complete.

- *Stage 1.* Embryonic ocean basin. (See also Figure 12.19*b*.) The Red Sea separating the Arabian Peninsula from Africa is an active example.
- *Stage 2.* Young ocean basin. The Labrador Basin, a branch of the North Atlantic lying between Labrador and Greenland, is a fair example of this stage.
- *Stage 3.* Old ocean basin. (Figure 12.19*c*.) Includes all of the vast expanse of the North and South Atlantic oceans and the Antarctic Ocean. Passive margin sedimentary wedges have become wide and thick.
- *Stage 4a.* The ocean basin begins to close as continental plates collide with it. New subduction boundaries begin to form.
- *Stage 4b.* Island arcs have risen and grown into great volcanic island chains. (Figure 12.18*a*.) These are found surrounding the Pacific plate, with the Aleutian arc as a fine example.
- *Stage 5.* Closing continues. Formation of new subduction margins close to the continents is

The Wilson cycle *Schematic diagram depicting the six stages of the Wilson cycle. The diagrams are not to true scale. (From Plate Tectonics, copyright © 1998 by Arthur N. Strahler. Used by permission.)*

Convergence

Rotation

Rotation

Rotation

Dispersal

Supercontinent formation

Supercontinent

World ocean

| Plate | Arc–continent collision orogen |

| Plate / Plate | Continent–continent collision orogen |

Supercontinent cycle *Schematic diagram of the supercontinent cycle in three phases. A pattern of wiggly dashes shows collision orogens that accumulate around and between the original continents. (From Plate Tectonics, copyright © 1998 by Arthur N. Strahler. Used by permission.)*

followed by arc-continent collisions. The Japanese Islands represent this stage.

- *Stage 6.* The ocean basin has finally closed with a collision orogen forming a continental suture (Figure 12.18c). The Himalayan orogen is a fine recent example, with activity continuing today.

Note that the continent of the final stage is wider than the continent of the first stage because of the additional continental lithosphere formed during the intervening collisions. This is how continents can grow over geologic time.

Our stage diagrams show a single continent splitting into two continents, then rejoining to form one. Suppose that over the entire globe several continents are splitting apart and reclosing at about the same time. Expanding on this idea, imagine that these fragments

become detached from a single world continent that can be called a *supercontinent.*

Several powerful lines of evidence show that a supercontinent actually came into existence, starting about 200 million years ago. Called *Pangea,* it is described in our chapter text. Good evidence has now been found that an earlier supercontinent, dubbed *Rodinia,* was fully formed about 700 million years ago. It consisted of early representatives of the same continents that later made up Pangea. Rodinia broke apart, and its fragments were carried away in different directions. Then they reversed their motions and headed back toward a common center, where many continent–continent collisions bonded them together by sutures to comprise Pangea. Some interesting evidence has also pointed to the former existence of a supercontinent even older than Rodinia.

The illustration above presents the supercontinent cycle as a loop with three stages. Analogous to Stages 1–3 of the Wilson cycle is the dispersal phase, followed by a convergence phase that corresponds to Wilson stages 4 through 6. The cycle ends in a complete new supercontinent. Many new collision orogens are formed between the original continental fragments. Given, say, 3000 million years ago (Middle Archean time) as the first occurrence of a full-blown supercontinent cycle, there could have been some 6 to 10 such cycles. The hypothesis of a time cycle of supercontinents, repeating the Wilson cycle over and over again, now holds its place as the basic theme of the geologic evolution of our planet.

(a)

(b)

(c)

Polar stereographic projection

(d)

Table 12.2 | **The Lithospheric Plates**

Great plates	Lesser plates
Pacific	Nazca
American (North, South)	Cocos
Eurasian	Philippine
Persian subplate	Caribbean
African	Arabian
Somalian subplate	Juan de Fuca
Austal-Indian	Caroline
Antarctic	Bismark
	Scotia

along most of the western and northern edge. The eastern and southern edge is mostly a spreading boundary. A sliver of continental lithosphere is included and makes up the coastal portion of California and all of Baja California. The California portion is bounded by an active transform fault (the San Andreas fault).

The American plate (Figure 12.15*a*) includes most of the continental lithosphere of North and South America as well as the entire oceanic lithosphere lying west of the midoceanic ridge that divides the Atlantic Ocean Basin down the middle. For the most part, the western edge of the American plate is a subduction boundary, with oceanic lithosphere diving beneath the continental lithosphere. The eastern edge is a spreading boundary. (Some classifications recognize separate North American and South American plates.)

The Eurasian plate (Figure 12.15*b*) is mostly continental lithosphere, but it is fringed on the west and north by a belt of oceanic lithosphere. The African plate has a central core of continental lithosphere nearly surrounded by oceanic lithosphere.

The Austral-Indian plate takes the form of a long rectangle. It is mostly oceanic lithosphere but contains two cores of continental lithosphere—Australia and peninsular India. Recent evidence shows that these two continental masses are moving independently and may actually be considered to be parts of separate plates. The Antarctic plate (Figure 12.15*c*) has an elliptical shape and is almost completely enclosed by a spreading plate boundary. This means that the other plates are moving away from the pole. The continent of Antarctica forms a central core of continental lithosphere completely surrounded by oceanic lithosphere.

Of the nine lesser plates, the Nazca and Cocos plates of the eastern Pacific are rather simple fragments of oceanic lithosphere bounded by the Pacific midoceanic spreading boundary on the west and by a subduction boundary on the east. The Philippine plate is noteworthy as having subduction boundaries on both east and west edges. Two small but distinct lesser plates—Caroline and Bismark—lie to the southeast of the Philippine plate. The Arabian plate has two transform fault boundaries, and its relative motion is northeasterly. The

Caribbean plate also has important transform fault boundaries. The tiny Juan de Fuca plate is steadily diminishing in size and will eventually disappear by subduction beneath the American plate. Similarly, the Scotia plate is being consumed by the American and Antarctic plates.

Figure 12.16 is a schematic circular cross section of the lithosphere along a great circle in low latitudes. It shows several of the great plates and their boundaries. The great circle is tilted by about 30° with respect to the equator. The section crosses the African plate, heads northeast across the Eurasian plate through the Himalayas to Japan and Korea, then dips southeast across the Pacific plate, cutting across the South American plate, and finally returns to Africa. Three spreading boundaries at midoceanic ridges are encountered, with two subduction zones (the Japan and Peru trenches) and a continent-to-continent collision, where the Austral-Indian plate dives under the Eurasian plate.

Subduction Tectonics

Converging plate boundaries, with subduction in progress, are zones of intense tectonic and volcanic activity. The narrow zone of a continent that lies above a plate undergoing subduction is therefore an active continental margin. (See plates Y and Z in Figure 12.12.) Figure 12.17 shows some details of the geologic processes that are associated with plate subduction. Two diagrams are used. Part *a* is exaggerated to show crustal and surface details, and part *b* is drawn to true scale to show the lithospheric plates. **GEODISCOVERIES**

The Earth's lithospheric shell is divided into six great lithospheric plates and at least nine lesser plates.

12.16 Arrangement of lithospheric plates *Schematic circular cross section of the major plates on a great circle tilted about 30 degrees with respect to the equator. (A. N. Strahler.)*

12.17 Typical features of an active subduction zone *The upper diagram (a) uses vertical exaggeration to show surface and crustal details. Sediments scraped off the moving plate form tilted wedges that accumulate in a rising tectonic mass. Near the mainland is a shallow trough in which sediment brought from the land accumulates. Metamorphic rock forms above the descending plate. Magma rising from the top of the descending plate reaches the surface to build a chain of volcanoes. Diagram (b) is a true-scale cross section showing the entire thickness of the lithospheric plates. (Copyright © A. N. Strahler.)*

The oceanic trench receives sediment coming from two sources. Carried along on the moving oceanic plate is deep ocean sediment—fine clay and ooze—that has settled to the ocean floor. From the continent comes terrestrial sediment in the form of sand and mud brought by streams to the shore and then swept into deep water by currents. In the bottom of the trench, both types of sediment are intensely deformed and are dragged down with the moving plate. The

When oceanic crust collides with continental crust, deep ocean sediment accumulates in an accretionary prism. Subducted oceanic crust melts and rises as andesite magma, creating a mountain arc.

deformed sediment is then scraped off the plate and shaped into wedges that ride up, one over the other, on steep fault planes. The wedges accumulate at the plate boundary, forming an *accretionary prism* in which metamorphism takes place. In this way, new continental crust of metamorphic rock is formed, and the continental plate is built outward.

The accretionary prism is of relatively low density and tends to rise, forming a *tectonic crest*. The tectonic crest is shown to be submerged in the figure, but in some cases it forms an island chain paralleling the coast—that is, a *tectonic arc*. Between the tectonic crest and the mainland is a shallow trough, the *forearc trough* (Figure 12.17b). This trough traps a great deal of terrestrial sediment, which accumulates in a basin-like structure. The bottom of the forearc trough continually subsides under the load of the added sediment. In some cases, the seafloor of the trough is flat and shallow, forming a type of continental shelf. Sediment carried across the shelf moves down the steep outer slope of the accretionary prism in tongue-like flows of turbidity currents.

The lower diagram of Figure 12.17 shows the descending lithospheric plate entering the asthenosphere. Intense heating of the upper surface of the plate melts the oceanic crust, forming basaltic magma. As this magma rises, it is changed in chemical composition at the base of the crust and becomes andesite magma. The rising andesite magma reaches the surface to form volcanoes of andesite lava, such as those we see in the Andes of South America.

Orogens and Collisions

Visualize, now, a situation in which two continental litho-spheric plates converge along a subduction boundary. Ultimately, the two masses must collide because the impacting masses are too thick and too buoyant to allow either plate to slip under the other. The result is an orogeny in which various kinds of crustal rocks are crumpled into folds and sliced into nappes. This process has been called "telescoping" after

the behavior of a folding telescope that collapses from a long tube into a short cylinder. Collision permanently unites the two plates, terminating further tectonic activity along that collision zone. Appropriately, the collision zone is named a **continental suture.** Our world map of plates, Figure 12.15, uses a special symbol for these sutures, shown in northern Africa and near the Aral Sea.

Continent–continent collisions occurred in the Cenozoic Era along a great tectonic line that marks the southern boundary of the Eurasian plate. (See Figure 12.15, map *b*.) The line begins with the Atlas Mountains of North Africa, and it runs across the Aegean Sea region into western Turkey. Beyond a major gap in Turkey, the line takes up again in the Zagros Mountains of Iran. Jumping another gap in southeastern Iran and Pakistan, the collision line sets in again in the great Himalayan Range, where it is still active.

Each segment of this collision zone represents the collision of a different north-moving plate against the single and relatively immobile Eurasian plate. A European segment containing the Alps was formed when the African plate collided with the Eurasian plate in the Mediterranean region. A Persian segment resulted from the collision of the Arabian plate with the Eurasian plate. A Himalayan segment represents the collision of the Indian continental portion of the Austral-Indian plate with the Eurasian plate.

Figure 12.18 is a series of cross sections in which the tectonic events of a typical continent–continent collision are reconstructed. Diagram *a* shows a passive margin at the left and an active subduction margin at the right. As the ocean between the converging continents is eliminated, a succes-

12.18 Continent–continent collision *Schematic cross sections showing continent–continent collision and the formation of a suture zone with nappes. (Copyright © A. N. Strahler.)*

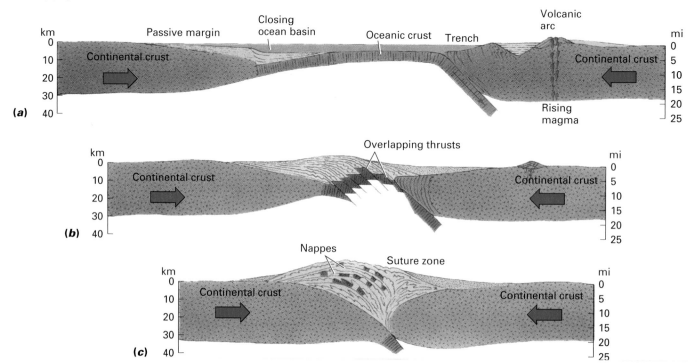

sion of overlapping thrust faults cuts through the oceanic crust in diagram *b.* The thrust slices ride up, one over the other, telescoping the oceanic crust and the sediments above it. As the slices become more and more tightly squeezed, they are forced upward. The upper part of each thrust sheet assumes a horizontal attitude to form a nappe in diagram *c,* which then glides forward under gravity on a low downgrade. A mass of metamorphic rock is formed between the joined continental plates, welding them together. This new rock mass is the continental suture. It is a distinctive type of orogen.

Continent–continent collisions have occurred many times since the late Precambrian time. Several ancient sutures have been identified in the continental shields. The Ural Mountains, which divide Europe from Asia, are one such suture, formed near the end of the Paleozoic Era.

Another type of collision is the arc–continent collision. Recall that island arcs often form in subduction zones as sediments of lesser density are carried into the mantle, then melt and rise as magma. In some situations, an ancient island arc can be carried toward a continent and collide with a passive continental margin. This creates a complex orogen that includes folded, faulted, and metamorphosed rocks derived from both the island arc and the sediments of the forearc trough (Figure 12.17*b*). Examples of orogens resulting from arc–continent collisions can be found on both eastern and western sides of North America. They include the ancient Appalachian and Ouachita Mountain belts as well as the younger Cordilleran Ranges of western North America.

Continental Rupture and New Ocean Basins

We have already noted that the continental margins bordering the Atlantic Ocean Basin on both its eastern and western sides are very different from the active margin of a subduction zone. At present, the Atlantic margins have no important tectonic activity and are passive continental margins. Even so, they represent the contact between continental lithosphere and oceanic lithosphere, with continental crust meeting oceanic crust, as shown in Figure 12.8. The formation of passive margins begins when a single plate of continental lithosphere is rifted apart. This process is called *continental rupture.*

Figure 12.19 uses three schematic block diagrams to show how continental rupture takes place and leads to the development of passive continental margins. At first, the crust is both lifted and stretched apart as the lithospheric plate is arched upward. The crust fractures and moves along faults—upthrown blocks form mountains, while down-dropped blocks form basins.

Eventually a long narrow valley, called a *rift valley* appears (block *a*). The widening crack in its center is continually filled in with magma rising from the mantle below. The magma solidifies to form new crust in the floor of the

> When two continental lithospheric plates collide in an orogeny, continental rocks are crumpled and overthrust. The plates eventually become joined in a continental suture.

rift valley. Crustal blocks on either side slip down along a succession of steep faults, creating a mountainous landscape. The Rift Valley system of East Africa—described in more detail in Chapter 13—is a notable example of this stage of continental rupture. As separation continues, the rift valley widens and opens to the ocean. The ocean enters the narrow valley with a spreading plate boundary running down its center (block *b*). Rising magma in the central rift produces new oceanic crust and lithosphere.

The Red Sea is a narrow ocean formed by a continental rupture. Its straight coasts mark the edges of the rupture. The widening of such an ocean basin can continue until a large ocean has formed and the continents are widely separated (Figure 12.19, block *c*).

During the process of opening of an ocean basin, the spreading boundary develops a series of offset breaks, one

12.19 Continental rupture and spreading *Schematic block diagrams showing stages in continental rupture and the opening of a new ocean basin. The vertical scale is greatly exaggerated to emphasize surface features. (a) The crust is uplifted and stretched apart, causing it to break into blocks that become tilted on faults. (b) A narrow ocean is formed, floored by new oceanic crust. (c) The ocean basin widens, while the passive continental margins subside and receive sediments from the continents. (Copyright © A. N. Strahler.)*

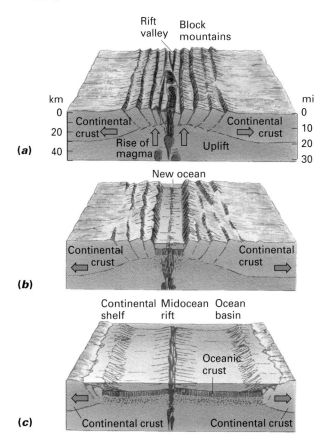

of which is shown in the upper left-hand part of diagram *a* of Figure 12.12. The offset ends of the axial rift are connected by an active transform fault. As spreading continues, the offsets slide past each other, and a scar-like feature is formed on the ocean floor as an extension of the transform fault. These *transform scars* take the form of narrow ridges or cliff-like features and may extend for hundreds of kilometers across the ocean floors. Before the true nature of these scars was understood, they were called "fracture zones." That name still persists and can be seen on maps of the ocean floor. The scars are not, for the most part, associated with active faults.

> **Continental rupture begins with the formation of a rift valley and tilted block mountains. Ocean soon invades the rift. As the continental crust recedes, oceanic crust fills the gap.**

The Power Source for Plate Movements

The system of lithospheric plates in motion represents a huge flow system of dense mineral matter. Its operation requires enormous power. It is generally agreed that the source of this power lies in the heat released by radioactivity. You may recall from basic chemistry that some elements are found in forms that differ with respect to their atomic mass. These forms are known as *isotopes*. For example, the most common isotope of uranium (U) is uranium-238, a form with an atomic mass of 238. Another form is uranium-235, which has an atomic mass number of 235. Chemists distinguish these two forms by writing them as ^{238}U and ^{235}U.

A key to understanding radioactivity is that certain isotopes are unstable, meaning that the composition of the atomic nucleus of the isotope can experience an irreversible change. This change process is known as *radioactive decay.* When a nucleus decays, it emits matter and energy. The energy is absorbed by the surrounding matter and thus ultimately takes the form of sensible heat. The flow of heat from radioactive decay is termed *radiogenic heat.* Nearly all of this heat is generated by the decay of isotopes of uranium (^{238}U, ^{235}U), thorium (^{232}Th), and potassium (^{40}K). Most of the radiogenic heat is liberated in the rock beneath the continents, within the uppermost 100 km or so (about 60 mi). Radiogenic heat is sufficient to keep Earth layers below the crust close to the melting point and thus provides the power source for the formation of magma.

How does radiogenic heating produce plate motions? The answer is not completely known, but one theory is that plate motions are produced by *convection currents* in hot mantle rock. Recall that in the case of the atmosphere, upward motion of warmer, less dense air takes place by convection. It is thought that, in a somewhat similar type of convection, unequal heating produces streams of upwelling mantle rock that rise steadily beneath spreading plate boundaries. Geologists do not yet have a full understanding of how this rise of heated rock causes plates to move. One hypothesis states that, as the rising mantle lifts the lithospheric plate to a higher elevation,

> **The motion of lithospheric plates is thought to result from convection currents of upwelling plastic mantle rock that lift the plates and cause them to split apart.**

the lithospheric plate tends to move horizontally away from the spreading axis under the influence of gravity.

At the opposite edge of the plate, subduction occurs because the oceanic plate is colder and denser than the asthenosphere through which it is sinking. The motion of the oceanic plate exerts a drag on the underlying asthenosphere, setting in motion flow currents in the upper mantle. Thus, slow convection currents probably exist in the asthenosphere beneath the moving plates, but their pathways and depths of operation are not well understood.

CONTINENTS OF THE PAST

Although modern plate tectonic theory is only a few decades old, the concept of a breakup of an early supercontinent into fragments that drifted apart dates back into the nineteenth century and beyond. Almost as soon as good navigational charts became available to show the continental outlines, geographers became intrigued with the close correspondence in outline between the eastern coast of South America and the western coastline of Africa. Credit for the first full-scale scientific hypothesis of the breakup of a single large continent belongs to Alfred Wegener, a German meteorologist and geophysicist, who offered geologic evidence as early as 1915 that the continents had once been united and had drifted apart (Figure 12.20). He reconstructed a supercontinent named *Pangea,* which existed intact as early as about 300 million years ago, in the Carboniferous Period.

A storm of controversy followed, and many American geologists denounced Wegener's "continental drift" hypothesis. However, he had some loyal supporters in Europe, South Africa, and Australia. Several lines of hard scientific evidence presented by Wegener strongly favored the former existence of Pangea. The prominent geographer and climatologist, Vladimir Köppen collaborated with Wegener in introducing favorable evidence from distribution patterns of fossil and present-day plant species. But Wegener's explanation of the physical process that separated the continents was weak and was strongly criticized on valid physical grounds. He had proposed that a continental layer of less dense rock had moved like a great floating "raft" through a "sea" of denser oceanic crustal rock. Geologists could show by use of established principles of physics that this proposed mechanism was impossible. In 1930, Wegener perished of cold and exhaustion while on an expedition to the Greenland ice sheet. There followed three decades in which only a small minority of scientists promoted his hypothesis.

In the 1960s, however, seismologists showed beyond doubt that thick lithospheric plates are in motion, both along

12.20 Wegener's Pangea *Alfred Wegener's 1915 map fits together the continents that today border the Atlantic Ocean Basin. The sets of dashed lines show the fit of Paleozoic tectonic structures between Europe and North America and between southernmost Africa and South America. (From A. Wegener, 1915, Die Entstehung der Kontinente und Ozeane, F. Vieweg, Braunschweig.)*

the midoceanic ridge and beneath the deep offshore trenches of the continents. Other geophysicists used magnetic data "frozen" into crustal rock to conclude that the continents had moved great distances apart. Within only a few years, Wegener's scenario was validated, but only by applying a mechanism for the process that was never dreamed of in his time.

The continents are moving today. Data from orbiting satellites have shown that rates of separation, or of convergence, between two plates are on the order of 5 to 10 cm (about 2 to 4 in.) per year, or 50 to 100 km (about 30 to 60 mi) per million years. At that rate, global geography must have been very different in past geologic eras than it is today. Many continental riftings and many plate collisions have taken place over the past two billion years. Single continents have fragmented into smaller ones, while at other times, small continents have merged to form large ones. (See *Eye on the Environment 12.1 • The Wilson Cycle and Supercontinents.*)

A brief look at the changing continental arrangements and locations over the past 250 m.y. is worthwhile in our

> A single great continent, Pangea, surrounded by a single great ocean, Panthalassa, broke apart over hundreds of millions of years to form the present-day pattern of continents, islands, and oceans.

study of physical geography because those changes brought with them changes in climate, soils, and vegetation on each continent as the continents moved across the parallels of latitude. Geography of the geologic past bears the name "paleogeography." If you decide to become a paleogeographer, you will need to carry with you into the past all you have learned about the physical geography of the present.

Wegener had made a crude map showing stages in the breakup of Pangea, but his timetable had been in part incorrect. Modern reconstructions of the global arrangements of past continents have been available since the mid-1960s. Global maps drawn for each geologic period have been repeatedly revised in the light of new evidence, but differences of interpretation affect only the minor details.

Figure 12.21 shows stages in the separations and travels of our continents, starting in the Permian Period of the Mesozoic Era, about 250 m.y. ago. In the first map, Pangea lies astride the equator and extends nearly to the two poles. Regions that are now North America and western Eurasia lie in the northern hemisphere. Jointly they are called *Laurasia*. Regions that are now South America, Africa, Antarctica, Australia, New Zealand, Madagascar, and peninsular India lie south of the equator. Jointly, they go by the name *Gondwana*. Subsequent maps (*b* through *d*) show the breaking apart and dispersal of the Laurasia and Gondwana plates to yield their modern components and locations, as seen in map *e*.

Note in particular that North America traveled from a low-latitude location into high latitudes, finally closing off the Arctic Ocean as a largely landlocked sea. This change may have been a major factor in bringing on the Ice Age in late Cenozoic time, about two million years ago. (See Chapter 19.) Another traveler was the Indian Peninsula, which started out from a near-subarctic southerly location in Permian time and streaked northeast across the Thethys Sea to collide with Asia in the northern tropical savanna zone. India's bedrock still bears the grooves and scratches of a great glaciation in Permian time, when it was nearer to the south pole. For the future, we may safely predict that the Atlantic Ocean basins will become much wider, and the westward motion of the Americas will cause a reduction in the width of the Pacific Basin.

A LOOK AHEAD

Over the long spans of geologic time, our planet's surface is shaped and reshaped endlessly. Oceans open and close, mountains rise and fall. Rocks are folded, faulted, and fractured. Arcs of volcanoes spew lava to build chains of lofty peaks. Deep trenches consume sediments that are carried beneath adjacent continents. These activities are byproducts of the motions of lithospheric plates and are described by plate tectonics.

For geographers, the great value of plate tectonics is that it provides a grand scheme for understanding the

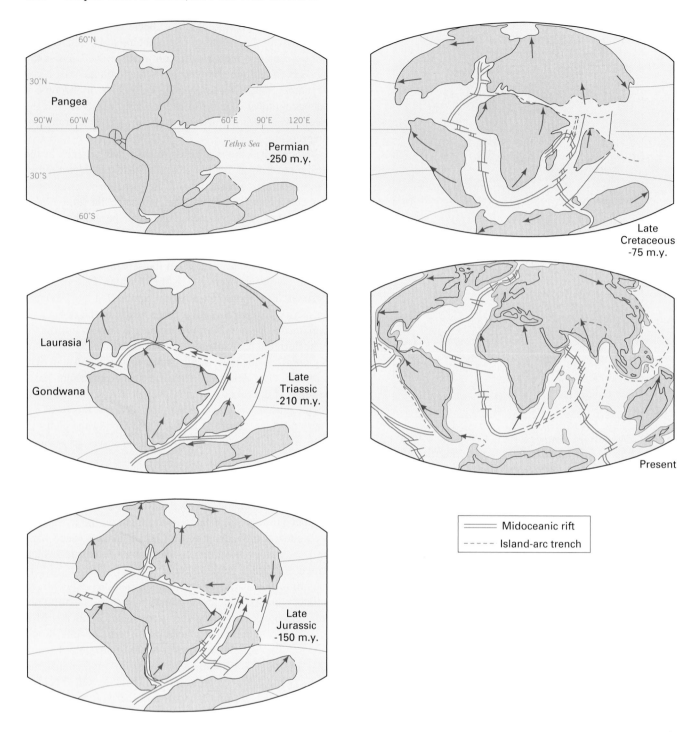

12.21   *Pangea's breakup* *The breakup of Pangea is shown in five stages. Inferred motion of lithospheric plates is indicated by arrows. (Redrawn and simplified from maps by R. S. Dietz and J. C. Holden, Jour. Geophysical Research, vol. 75, pp. 4943–4951, Figures 2 to 6. Copyrighted by the American Geophysical Union. Used by permission.*

nature and distribution of the largest and most obvious features of our planet's surface—its continents and ocean basins and their major relief features. However, large features, such as mountain ranges, are made up of many smaller features—individual peaks, for example. In the next chapter, we will look at the continental surface in more detail, examining the volcanic and tectonic landforms that result when volcanoes erupt and rock layers are folded and faulted. As we will see, the formation of these landforms is often marked by powerful earthquakes—phenomena that can produce major impacts, indeed, on modern technological society.

In Review | The Lithosphere and Plate Tectonics

- At the center of the Earth lies the **core**—a dense mass of liquid iron and nickel that is solid at the very center. Enclosing the metallic core is the **mantle,** composed of ultramafic rock. The outermost layer is the **crust. Continental crust** consists of two zones—a lighter zone of felsic rocks atop a denser zone of mafic rocks. **Oceanic crust** consists only of denser, mafic rocks.

- The **lithosphere,** the outermost shell of rigid, brittle rock, includes the crust and an upper layer of the mantle. Below the lithosphere is the **asthenosphere,** a region of the mantle in which mantle rock is soft or plastic.

- Geologists trace the history of the Earth through the geologic time scale. *Precambrian time* includes the Earth's earliest history. It is followed by three major divisions—the *Paleozoic, Mesozoic,* and *Cenozoic* eras.

- Continental masses consist of active belts of mountain-making and inactive regions of old, stable rock. Mountain-building occurs by *volcanism* and *tectonic activity. Alpine chains* include **mountain arcs** and **island arcs.** They occur in two principal mountain belts—the circum-Pacific and Eurasian-Indonesian belts.

- **Continental shields** are regions of low-lying igneous and metamorphic rocks. They may be *exposed* or *covered* by layers of sedimentary rocks. Ancient *mountain roots* lie within some shield regions.

- The ocean basins are marked by a *midoceanic ridge* with its central *axial rift.* This ridge occurs at the site of crustal spreading. Most of the *ocean basin floor* is *abyssal plain,* covered by fine sediment. As *passive continental margins* are approached, the *continental rise, slope,* and *shelf* are encountered. At *active continental margins,* deep *oceanic trenches* lie offshore.

- The two basic tectonic processes are extension and compression. Both processes can lead to the formation of mountains. *Extension* occurs in the splitting of plates, when the crust thins, is fractured, and then pushed upward to produce block mountains. When lithospheric plates collide, *compression* occurs, shaping rock layers into *folds* that then break and move atop one another along *overthrust faults.*

- *Continental lithosphere* includes the thicker, lighter continental crust and a rigid layer of mantle rock beneath. *Oceanic lithosphere* is comprised of the thinner, denser oceanic crust and rigid mantle below. The lithosphere is fractured and broken into a set of **lithospheric plates,** large and small, that move with respect to each other.

- Where plates move apart, a *spreading boundary* occurs. At *converging boundaries,* plates collide. At *transform boundaries,* plates move past one another on a *transform fault.* There are six major lithospheric plates.

- When oceanic lithosphere and continental lithosphere collide, the denser oceanic lithosphere plunges beneath the continental lithospheric plate, a process called **subduction.** A trench marks the site of downplunging. Some subducted oceanic crust melts and rises to the surface, producing volcanoes. Under the severe compression that occurs with continent–continent collision, the two continental plates are welded together in a zone of metamorphic rock named a **continental suture.**

- In *continental rupture,* extensional tectonic forces move a continental plate in opposite directions, creating a *rift valley.* Eventually, the rift valley widens and opens to the ocean, and new oceanic crust forms as spreading continues.

- Plate movements are thought to be powered by *convection currents* in the plastic mantle rock of the asthenosphere.

- During the Permian Period, the continents were joined in a single, large supercontinent—Pangea—that broke apart, leading eventually to the present arrangement of continents and ocean basins.

Key Terms

core, p. 397
mantle, p. 397
crust, p. 397
continental crust, p. 398
oceanic crust, p. 398

lithosphere, p. 398
asthenosphere, p. 398
lithospheric plates,
    p. 398
mountain arc, p. 401

island arcs, p. 401
continental shields,
    p. 401
plate tectonics, p. 406
tectonics, p. 406

subduction, p. 408
continental suture,
    p. 416

Review Questions

1. Describe the Earth's inner structure, from the center outward. What types of crust are present? How are they different?

2. How do geologists use the term *lithosphere?* What layer underlies the lithosphere, and what are its properties? Define the term *lithospheric plate.*

3. More recent geologic time is divided into three eras. Name them in order from oldest to youngest. How do geologists use the terms *period* and *epoch?* What age is applied to time before the earliest era?

4. What proportion of the Earth's surface is in ocean? in land? Do these proportions reflect the true proportions of continents and oceans? If not, why not?

5. What are the two basic subdivisions of continental masses?

6. What term is attached to belts of active mountain-making? What are the two basic processes by which mountain belts are constructed? Provide examples of mountain arcs and island arcs.

7. What is a continental shield? How old are continental shields? What two types of shields are recognized?

8. Describe how compressional mountain-building produces folds, faults, overthrust faults, and thrust sheets (nappes).

9. Name the six great lithospheric plates. Identify an example of a spreading boundary by general geographic location and the plates involved. Do the same for a converging boundary.

10. Describe the process of subduction as it occurs at a converging boundary of continental and oceanic lithospheric plates. How is the continental margin extended? How is subduction related to volcanic activity?

11. How does continental rupture produce passive continental margins? Describe the process of rupturing and its various stages.

12. What are transform faults? Where do they occur?

13. How is the principle of convection thought to be related to the power source for plate tectonic motions? What role does gravity play in the motion of lithospheric plates?

14. Provide a brief history of the idea of "drifting continents."

15. What was Wegener's theory about "continental drift?" Why was it opposed at the time?

16. Briefly summarize the history of our continents that geologists have reconstructed beginning with the Permian period.

Eye on the Environment 12.1 • The Wilson Cycle and Supercontinents

1. What is the Wilson cycle of plate tectonics? What is the net effect of a Wilson cycle on the continental mass?

2. Using a sketch, describe the cycle by which supercontinents are formed and reformed.

Visualizing Exercises

1. Sketch a cross section of an ocean basin with passive continental margins. Label the following features: midoceanic ridge, axial rift, abyssal plain, continental rise, continental slope, and continental shelf.

2. Identify and describe two types of lithospheric plates. Sketch a cross section showing a collision between the two types. Label the following features: oceanic crust, continental crust, mantle, oceanic trench, and rising magma. Indicate where subduction is occurring.

3. Sketch a continent–continent collision and describe the formation process of a continental suture. Provide a present-day example where a continental suture is being formed, and give an example of an ancient continental suture.

4. Figure 12.15 is a Mercator map of lithospheric plates, whereas Figure 12.16 presents a cross section of plates on a great circle. Construct a similar cross section of plates on the 30° S parallel of latitude. As in Figure 12.16, label the plates and major geographic features. Refer to Figure 12.4 for arcs, trenches, and midocean ridges.

Essay Question

1. Suppose astronomers discover a new planet that, like Earth, has continents and oceans. They dispatch a reconnaissance satellite to photograph the new planet. What features would you look for, and why, to detect past and present plate tectonic activity on the new planet?

Eye on the Landscape

Chapter Opener **Sinai Peninsula**
The Sinai Peninsula lies at a corner of the African lithospheric plate where the Arabian plate is pulling away. The Gulfs of Suez (left) and Aqaba (right) mark down-dropped fault zones formed as the crust pulls apart. Notice how the pattern of highlands (A), (B), and (C) connects across the gulfs, showing that the land masses on either sides were once continuous. Notice also that the Arabian plate (C) has moved significantly northward along a fault line (D). Lithospheric plates and their motions are the subject of this chapter.

12.7 **Undersea** **topography**
Notice the Hawaiian Islands (A) and the trail of seamounts (B) leading up to them. The islands are outpourings of lava from a "hotspot" in the mantle. The hotspot has remained fixed, however, as the Pacific lithospheric plate has moved across it. The seamounts thus mark the path of motion of the Pacific plate. Note that the direction of motion changed abruptly at one point in time. Some hotspots are relatively recent and don't show a trail of ancient eruptions. The Cape Verde Islands (C) are an example. Hotspot volcanoes are described in Chapter 13.

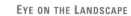

EYE ON THE LANDSCAPE

Diamond Head with Waikiki Beach (upper left), Honolulu, Hawaii. What else would the geographer see? … Answers at the end of the chapter.

13 | VOLCANIC AND TECTONIC LANDFORMS

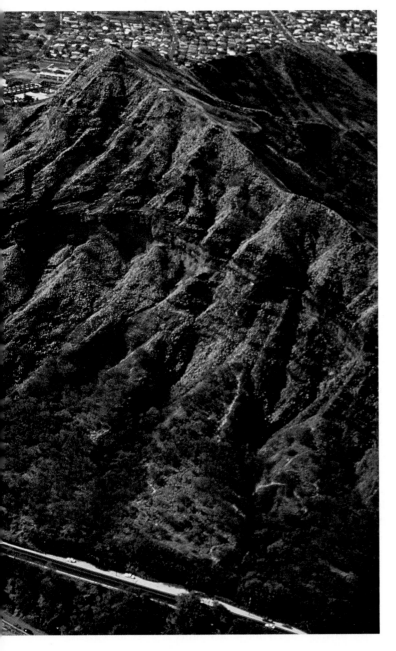

. . . .
PUTTING YOU IN THE PICTURE

Some of the most devastating natural phenomena are eruptions of volcanoes and great earthquakes, both of which are subjects of this chapter. You've probably seen videos or television news coverage of these spectacular events. But did you ever think what it would be like to experience a tsunami—a seismic sea wave generated by an earthquake or explosive volcanic eruption?

Imagine yourself on the balcony of an ocean-view room at an exclusive beach hotel at Waikiki, taking in a view that is postcard-perfect. The arc of hotels stretches to your right and left, fringed with palms and pools. Beyond is the beach, lined with Sun worshippers. Sailboats glide lazily back and forth in the lagoon, while outrigger canoes move more deliberately under paddle power.

    As you gaze out beyond the surfriders and sailors, you notice a white foaming of the surf out at the edge of the reef. Slowly the foaming intensifies, and within a minute or so, the reef is a white crescent of churning, seething water. Something very wrong is happening. The water level in the lagoon is rising, and very rapidly, as ocean water pours in over the reef. Accompanying the water are ocean waves of full force, no longer damped by the effects of the reef and lagoon. Waves are breaking higher and higher on the beach, and within a few moments, tongues, then torrents, of sea water begin to pour toward the hotels. Bathers running landward are overtaken and tumbled among the tables and chairs swept from beachside terraces. Palm trees bend and fall under the onslaught.

    As the first waves hit your hotel, you feel a shaking motion, like an earthquake. The beach has completely disappeared, and the ocean waves are crashing full-force against the hotel, shattering glass windows and doors and breaking interior partitions. You can feel the thumping and tearing of steel and concrete as the building shudders with the blows.

Just as you are beginning to think about what to do, the water begins to retreat, first slowly, then as a great outgoing flood. Splintered woodwork, uprooted plants, and even floating cars flash past as the ebbing waters carry them seaward. The beach is visible once again, now gullied by the ocean-bound water. The surf withdraws far out in the lagoon, leaving a vast sandy plain littered with debris and shallow water pools, glistening in the Sun.

Awestruck by the devastation spread in front of you, you thank your lucky stars that you were not a casualty of the great tsunami. But then you glance seaward, and stare with horror. The reef is awash with foam once more. Another great sea wave is headed for the beach!

Seem far-fetched? Not a bit. On April 1, 1946, a major earthquake close to the Aleutian Islands generated a tsunami that struck the Hawaiian Islands five hours later, killing 170 persons. An observer on the north shore of Oahu documented the wave as a rise and fall of sea level, repeated several times at 12-minute intervals. The first wave inundated the coast to a height of 3.9 m (13 ft) above tide level. The water then retreated, but it was fol-

lowed minutes later by an even higher landward surge, this time reaching 5.1 m (17 ft) above tide level. A third, even higher surge, followed the second. On the island of Hawaii, the city of Hilo was ravaged by breaking waves at a height of 9 m (30 ft) above sea level.

A tsunami is a very dramatic example of Earth forces at work—in this case, a secondary effect of a huge, submarine earthquake. This chapter focuses on how Earth forces act to shape the landscape, from the building of steep-sided volcanoes by volcanic eruptions to the lifting of huge masses of rock by motions along earthquake faults. As we will see, the power of the Earth in building these landforms can be awesome indeed. ■

VOLCANIC AND TECTONIC LANDFORMS

Have you ever seen an active volcano? If you live on the west coast, perhaps you've visited Mount St. Helens to view the devastation of its huge explosion in 1980. If you've been to the island of Hawaii, you may have seen flowing lava at first-hand. Even if you've never visited a volcano, you've probably seen pictures or films of erupting volcanoes. These lofty conical forms are unmistakable and are one example of the landforms we will study in this chapter.

Even if you've never seen a real volcano, you've certainly seen mountains. Some mountains are simply masses of rock that are resistant to erosion and so stand out against a surrounding area of weaker rocks. Others, however, are the direct result of tectonic forces that have recently uplifted them. The mountains of the western United States contain many ranges that are examples of uplifted mountain blocks bounded by faults—planes of slippage along which rock masses move in opposite directions. Tectonic activity can also produce mountains by bending rock layers into long, upfolded ridges, with downfolded valleys in between.

By looking in detail at landforms created by volcanic and tectonic activity, we are moving down from the global perspective of enormous lithospheric plates that collide to form vast mountain arcs or separate to form yawning ocean basins. We are now "zooming in" to examine landscape features that are actually small enough to view from a single vantage point. As we will see,

the global perspective of plate tectonics helps explain how these landscape features are created and where they occur.

LANDFORMS

Landforms are the surface features of the land—for example, mountain peaks, cliffs, canyons, plains, beaches, and sand dunes. Landforms are created by many processes, and much of the remainder of this book describes landforms and how they are produced. **Geomorphology** is the scientific study of the processes that shape landforms. In this chapter, we will examine landforms produced directly by volcanic and tectonic processes.

The shapes of continental surfaces reflect the "balance of power," so to speak, between two sets of forces. Internal Earth forces act to move crustal materials upward through volcanic and tectonic processes, thus bringing fresh rock to the Earth's surface. In opposition are processes and forces that act to lower continental surfaces by removing and transporting mineral matter through the action of running water, waves and currents, glacial ice, and wind. We can refer to these processes of land sculpture collectively as **denudation**.

Seen in this perspective, landforms in general fall into two basic groups—initial landforms and sequential land-

---

Landforms reflect a "balance of power" between internal Earth forces that bring fresh rock to the surface and denudation processes that remove and transport mineral matter from fresh rock masses.

(*a*)

Mountain block created:
an initial landform

Earth's crust fractured,
dislocated by internal
earth forces

Mountain block carved into
sequential landforms

(a) Erosional:
Canyon
Divide

(*b*)

Earth forces dormant

(b) Depositional (fan)

13.1 Initial and sequential landforms *An initial landform is created, here by tectonic activity, then carved into sequential landforms. (Drawn by A. N. Strahler.)*

forms. *Initial landforms* are produced directly by volcanic and tectonic activity. They include volcanoes and lava flows, as well as downdropped rift valleys and elevated mountain blocks in zones of recent crustal deformation. Figure 13.1*a* shows a mountain block uplifted by crustal activity. The energy for lifting molten rock and rigid crustal masses to produce the initial landforms comes from an internal heat source. As we explained in Chapter 12, this heat energy is generated largely by natural radioactivity in rock of the Earth's crust and mantle. It is the fundamental energy source for the motions of lithospheric plates.

Landforms shaped by processes and agents of denudation belong to the group of *sequential landforms*. The word "sequential" means that they follow in sequence after the initial landforms have been created, and a crustal mass has been raised to an elevated position. Figure 13.1*b* shows an uplifted crustal block (an initial landform) that has been attacked by agents of denudation and carved up into a large number of sequential landforms.

You can think of any landscape as representing the existing stage in a great contest of opposing forces. As lithospheric plates collide or pull apart, internal Earth forces periodically elevate parts of the crust to create initial landforms. The external agents of denudation—running water, waves, wind, and glacial ice—persistently wear these masses down and carve them into vast numbers of smaller sequential landforms.

Today we can observe the many stages of this endless struggle between internal and external forces by traveling to different parts of the globe. Where we find high alpine mountains and volcanic chains, internal Earth forces have recently dominated the contest. In the rolling low plains of the continental interiors, the agents of denudation have won a temporary victory. At other locations, we can find many intermediate stages. Because the internal Earth forces act repeatedly and violently, new initial landforms keep coming into existence as old ones are subdued.

This chapter focuses on initial landforms that are created directly by volcanic activity and tectonic activity. After we have described the most important agents of denudation in Chapters 14 and 16, we will return in Chapter 17 to the sequential landforms that develop on volcanic and tectonic landforms.

VOLCANIC ACTIVITY

We have already identified *volcanism,* or volcanic activity, as one of the forms of mountain-building. The extrusion of magma builds landforms, and these landforms can accumulate in a single area both as volcanoes and as thick lava flows. Through these volcanic activities, imposing mountain ranges are constructed.

A **volcano** is a conical or dome-shaped initial landform built by the emission of lava and its contained gases from a constricted vent in the Earth's surface (Figure 13.2). The magma rises in a narrow, pipe-like conduit from a magma reservoir lying beneath. Upon reaching the surface, igneous material may pour out in sluggish, tongue-like lava flows. Magma may also be violently ejected in the form of solid fragments driven skyward under pressure of confined gases. Ejected fragments, ranging in size from boulders to fine dust, are collectively called *tephra.* Forms and dimensions of

13.2 Anatomy of a stratovolcano *Idealized cross section of a stratovolcano with feeders from magma chamber beneath. The steep-sided cone is built up from layers of lava and tephra. (Copyright © A. N. Strahler.)*

Lava

Tephra

Magma chamber

a...***Mount St. Helens*** *This stratovolcano of the Cascade Range in southwestern Washington erupted violently on the morning of May 18, 1980, emitting a great cloud of condensed steam, heated gases, and ash from the summit crater. Within a few minutes, the plume had risen to a height of 20 km (12 mi), and its contents were being carried eastward by stratospheric winds. The eruption was initiated when an explosion demolished the northern portion of the cone, which is concealed from this viewpoint.*

b...**Mount Mayon** Located in southeastern Luzon, the Philippines, Mount Mayon is often considered the world's most nearly perfect stratovolcanic cone. Its summit rises to an altitude of nearly 2400 m (about 8000 ft). Mount Mayon has erupted at least 40 times since 1616. Its most recent eruption was in March 2000, causing the evacuation of a local population of about 55,000 people.

c...**Crater Lake** Crater Lake, Oregon, is a water-filled caldera marking the remains of the summit of Mount Mazama, which exploded about 6600 years ago. Wizard Island (center foreground) was built on the floor of the caldera after the major explosive activity had ceased. It is an almost perfectly shaped cone of cinders capping small lava flows.

a volcano are quite varied, depending on the type of lava and the presence or absence of tephra.

Stratovolcanoes

The nature of volcanic eruption, whether explosive or quiet, depends on the type of magma. Recall from Chapter 11 that there are two main types of igneous rocks: felsic and mafic. The felsic lavas (rhyolite and andesite) have a high degree of viscosity—that is, they are thick and gummy, and resist flow. So, volcanoes of felsic composition typically have steep slopes, and lava usually does not flow long distances from the volcano's vent. When the volcano erupts, tephra falls on the area surrounding the crater and contributes to the structure of the cone. Volcanic bombs are also included in the tephra. These solidified masses of lava range up to the size of large boulders and fall close to the crater. **GEODISCOVERIES**

The interlayering of sluggish streams of felsic lava and falls of tephra produces a **stratovolcano.** This tall, steep-sided cone usually steepens toward the summit, where a bowl-shaped depression—the *crater*—is located. The crater is the principal vent of the volcano. Felsic lavas usually hold large amounts of gas under high pressure. As a result, these lavas can produce explosive eruptions. The eruption of Mount St. Helens in 1980 is an example of an explosive eruption of felsic lavas (Figure 13.3). Very fine volcanic dust from such eruptions can rise high into the troposphere and stratosphere, traveling hundreds or thousands of kilometers before settling to the Earth's surface.

Lofty, conical stratovolcanoes are well known for their scenic beauty. Fine examples are Mount Hood and Mount St. Helens in the Cascade Range, Mount Fuji in Japan, Mount Mayon in the Philippines (Figure 13.3), and Mount Shishaldin in the Aleutian Islands. Stratovolcanoes are sometimes referred to as *composite volcanoes,* or composite cones, since they are composed of composites of ash and lava.

Another important form of emission from explosive stratovolcanoes is a cloud of white-hot gases and fine ash. This intensely hot cloud, or "glowing avalanche," travels rapidly down the flank of the volcanic cone, searing everything in its path. On the Caribbean island of Martinique, in 1902, a glowing cloud issued without warning from Mount Pelée. It swept down on the city of St. Pierre, destroying the city and killing all but two of its 30,000 inhabitants. More recently, glowing avalanches spawned by a 1997 eruption of the Soufrière Hills volcano on Montserrat, another island of the Lesser Antilles, killed about 20 people in small villages. Plymouth, the island's capital, was flooded with hot ash and debris, causing extensive fires that devastated the evacuated city (Figure 13.4). The southern two-thirds of the small island was left uninhabitable.

> Stratovolcanoes are built of layers of felsic lava and volcanic ash. Because felsic lava is viscous, it builds a tall, steep conical form. Felsic magma can contain gases under high pressure, so felsic eruptions are often explosive.

CALDERAS   One of the most catastrophic of natural phenomena is a volcanic explosion so violent that it destroys the entire central portion of the volcano. Vast quantities of ash and dust are emitted and fill the atmosphere for many hundreds of square kilometers around the volcano. Only a great central depression, named a *caldera,* remains after the explosion. Although some of the upper part of the volcano is blown outward in fragments, most of it settles back into the cavity formed beneath the former volcano by the explosion.

Krakatoa, a volcanic island in Indonesia, exploded in 1883, leaving a huge caldera. Great seismic sea waves generated by the explosion killed many thousands of persons living in low coastal areas of Sumatra and Java. About 75 km^3 (18 mi^3) of rock was blown out of the crater during the explosion. Vast quantities of gas and fine particles of dust were carried into the stratosphere, where they contributed to a rosy glow of sunrises and sunsets that was seen around the world for several years afterward.

A classic example of a caldera produced in prehistoric times is Crater Lake, Oregon (Figure 13.3). The former volcano, named Mount Mazama, is estimated to have risen 1200 m (about 4000 ft) higher than the present caldera rim. The great explosion and collapse occurred about 6600 years ago.

STRATOVOLCANOES AND SUBDUCTION ARCS   Most of the world's active stratovolcanoes lie within the circum-Pacific mountain belt. Here, subduction of the Pacific, Nazca, Cocos, and Juan de Fuca plates is active. In Chapter 12, we explained how andesitic magmas rise beneath volcanic arcs of active continental margins and island arcs (see Figure 12.12). One good example is the volcanic arc of Sumatra and Java, lying over the subduction zone between the Australian plate and the Eurasian plate. Another is the Aleutian volcanic arc, located where the Pacific plate dives beneath the North American plate. The Cascade Mountains of northern California, Oregon, and Washington form a similar chain. Important segments of the Andes Mountains in South America consist of stratovolcanoes.

Shield Volcanoes

In contrast to thick, gassy felsic lava, mafic lava (basalt) is often highly fluid. It typically has a low viscosity and holds little gas. As a result, eruptions of basaltic lava are usually quiet, and the lava can travel long distances to spread out in thin layers. Typically, then, large basaltic volcanoes are broadly rounded domes with gentle slopes. They are referred to as **shield volcanoes.** Hawaiian volcanoes are of this type.

The shield volcanoes of the Hawaiian Islands are characterized by gently rising, smooth slopes that flatten near the top, producing a broad-topped volcano (Figure 13.5).

13.4 Ash cloud *A cloud of hot, dense volcanic ash, emitted by the Soufrière Hills volcano, courses down this narrow valley on the island of Monserrat in the Lesser Antilles.*

13.5 Basaltic shield volcanoes of Hawaii *At lower left is the now-cold Halemaumau pit crater, formed in the floor of the central depression of Kilauea volcano. On the distant skyline is the snow-capped summit of Mauna Kea volcano, its elevation over 4000 m (about 13,000 ft).*

Domes on the island of Hawaii rise to summit elevations of about 4000 m (about 13,000 ft) above sea level. Including the basal portion lying below sea level, they are more than twice that high. In width they range from 16 to 80 km (10 to 50 mi) at sea level and up to 160 km (about 100 mi) at the submerged base. The basalt lava of the Hawaiian volcanoes is highly fluid and travels far down the gentle slopes. Most of the lava flows issue from fissures (long, gaping cracks) on the flanks of the volcano.

Hawaiian lava domes have a wide, steep-sided central depression that may be 3 km (2 mi) or more wide and several hundred meters deep. These large depressions are a type of collapsed caldera. Molten basalt is sometimes seen in the floors of deep pit craters that occur on the floor of the central depression or elsewhere over the surface of the lava dome.

HOTSPOTS, SEAFLOOR SPREADING, AND SHIELD VOLCANOES

The chain of Hawaiian volcanoes was created by the motion of the Pacific plate over a *hotspot*—a plume of upwelling basaltic magma deep within the mantle, arising far down in the asthenosphere. Figure 13.6 diagrams this process. As the hot mantle rock rises, magma forms in bodies that melt their way through the lithosphere and reach the seafloor. Each major pulse of the plume sets off a cycle of volcano formation. However, the motion of the oceanic lithosphere eventually carries the volcano away from the location of the deep plume, and so it becomes extinct. Erosion processes wear the volcano away, and ultimately it

Hotspots are plumes of rising basaltic magma that generate volcanic activity. When they occur beneath moving plates of oceanic crust, they create chains of islands and seamounts. On land, they cover large regions with flood basalts.

becomes a low island. Continued attack by waves and slow settling of the island reduce it to a coral-covered platform. Eventually only a sunken island, or *guyot,* exists.

The process of crustal motion over a hotspot thus produces a long trail of islands and guyots. The Hawaiian trail, shown in Figure 13.7, trends northwestward. It is 2400 km (about 1500 mi) long and includes a sharp bend to the north caused by a sudden change of direction of the Pacific plate. This distant leg consists of the Emperor Seamounts. Several other long trails of volcanic seamounts cross the Pacific Ocean basin. They, too, follow parallel paths that reveal the plate motion.

A few basaltic volcanoes also occur along the midoceanic ridge, where seafloor spreading is in progress. Perhaps the outstanding example is Iceland, in the North Atlantic Ocean. Iceland is constructed entirely of basalt. Basaltic flows are superimposed on older basaltic rocks as dikes and sills formed by magma emerging from deep within the spreading rift. Mount Hekla, an active volcano on Iceland, is a shield volcano somewhat similar to those of Hawaii. In Chapter 11, *Eye on the Environment 11.1 ● Battling Iceland's Heimaey Volcano* documents how valiant Icelanders coped with a recent eruption. Other islands consisting of basaltic volcanoes located along or close to the axis of the Mid-Atlantic Ridge are the Azores, Ascension, and Tristan da Cunha.

Where a mantle plume lies beneath a continental lithospheric plate, the hotspot may generate enormous volumes of basaltic lava that emerge from numerous vents and fissures and accumulate

13.6 Hotspot volcano chain
A chain of volcanic islands is formed as oceanic crust moves across a hotspot of rising magma. (Copyright © A. N. Strahler.)

Guyot   Beveled island   Extinct volcano   Active volcano   Sea-floor spreading

Oceanic crust

Hotspot

13.7 Hawaiian seamount chain *This sketch map locates the Hawaiian seamount chain in the northwest Pacific Ocean basin. Dots are summits. The enclosing colored area marks the base of the volcano at the ocean floor. (Copyright © A. N. Strahler.)*

layer upon layer. The basalt may ultimately attain a thickness of thousands of meters and cover thousands of square kilometers. These accumulations are called *flood basalts.*

An important American example is found in the Columbia Plateau region of southeastern Washington, northeastern Oregon, and westernmost Idaho. Here, basalts of Cenozoic age cover an area of about 130,000 km^2 (about 50,000 mi^2)—nearly the same area as the state of New York. Individual basalt flows are exposed along the walls of river gorges as cliffs in which vertical joint columns are conspicuous (Figure 13.8).

Associated with flood basalts, shield volcanoes, and scattered occurrences of basaltic lava flows is a small volcano known as a *cinder cone* (Figure 13.9). Cinder cones form when frothy basalt magma is ejected under high pressure from a narrow vent, producing tephra. The rain of tephra accumulates around the vent to form a roughly circular hill with a central crater. Cinder cones rarely grow to heights of more than a few hundred meters. An exceptionally fine example of a cinder cone is Wizard Island, built on the floor of Crater Lake long after the caldera was formed (Figure 13.3).

HOT SPRINGS AND GEYSERS Where hot rock material is near the Earth's surface, it can heat nearby ground water to high temperatures. When the ground water reaches the surface, it provides *hot springs* at temperatures not far below the boiling point of water (Figure 13.10). At some places, jet-like emissions of steam and hot water occur at intervals from small vents—producing *geysers* (Figure 13.11). Since the water that emerges from hot springs

and geysers is largely ground water that has been heated in contract with hot rock, this water is recycled surface water. Little, if any, is water that was originally held in rising bodies of magma.

The heat from masses of lava close to the surface in areas of hot springs and geysers provides a source of energy for electric power generation. *Eye on the Environment 13.1 • Geothermal Energy Sources* provides more information about this application.

Volcanic Activity over the Globe

Figure 13.12 shows the locations of volcanoes that have been active within the last 12,000 years. By comparing this map with that of Figure 12.4, it is easy to see that many volcanoes are located along subduction boundaries. In fact, the "ring of fire" around the Pacific Rim is the most obvious feature on the map. Other volcanoes are located on or near oceanic spreading centers. Iceland is an example. In continental regions, spreading in East Africa has also produced volcanoes. (We'll return to the East African rift later in this chapter.) Hotspot activity, as in the location of the Hawaiian Islands, is also present.

> Volcanic activity is frequent along subduction boundaries, which accounts for the "ring of fire" around the Pacific Rim. Midocean spreading centers and continental rifts are also locations of volcanic activity.

Volcanic Eruptions as Environmental Hazards

The eruptions of volcanoes and lava flows are environmental hazards of the severest sort, often taking a heavy toll of plant and animal life and devastating human habitations. What natural phenomenon can compare with the Mount

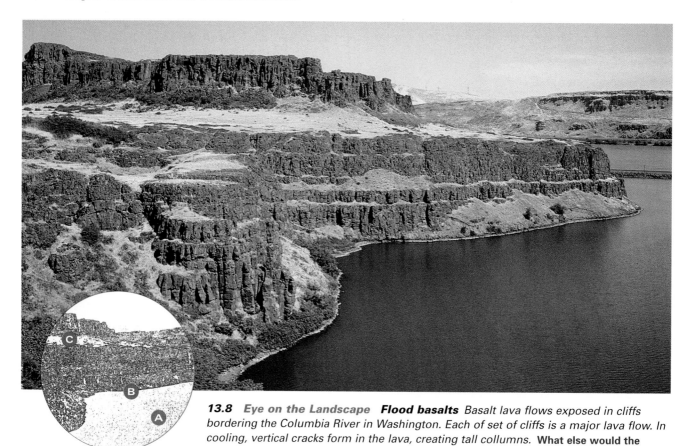

13.8 *Eye on the Landscape* **Flood basalts** *Basalt lava flows exposed in cliffs bordering the Columbia River in Washington. Each of set of cliffs is a major lava flow. In cooling, vertical cracks form in the lava, creating tall collumns.* **What else would the geographer see? ... Answers at the end of the chapter.**

Pelée disaster in which thousands of lives were snuffed out in seconds? Perhaps only an earthquake or storm surge of a tropical cyclone is equally disastrous.

Wholesale loss of life and destruction of towns and cities are frequent in the history of peoples who live near active volcanoes. Loss occurs principally from sweeping clouds of incandescent gases that descend the volcano

---

Volcanic eruptions can have extreme environmental impacts. Flows of hot gas, showers of ash, cinders, and bombs, violent earthquakes, and accompanying seismic sea waves can cause great loss of life.

---

slopes like great avalanches, from lava flows whose relentless advance engulfs whole cities, from showers of ash, cinders, and bombs, and from violent earthquakes associated with volcanic activity. For habitations along low-lying coasts, there is the additional peril of great seismic sea waves, generated elsewhere by explosive destruction of undersea or island volcanoes.

13.9 *Cinder cone* *This young cinder cone is built atop rough-surfaced basalt lava flows. Lava Beds National Monument, northern California.*

13.11 Old Faithful Geyser *An eruption of Old Faithful Geyser in Yellowstone National Park, Wyoming.*

13.10 Mammoth Hot Springs *Small terraces ringed by mineral deposits hold steaming pools of hot water as the spring cascades down the slope. This example of geothermal activity is from Yellowstone National Park, Wyoming.*

In 1985, an explosive eruption of Ruiz Volcano in the Colombian Andes caused the rapid melting of ice and snow in the summit area. Mixing with volcanic ash, the water formed a variety of mudflow known as a *lahar.* Rushing downslope at speeds up to 145 km (90 mi) per hour, the lahar became channeled into a valley on the lower slopes, where it engulfed a town and killed more than 20,000 persons.

Scientific monitoring techniques are reducing the toll of death and destruction from volcanoes. By analyzing the gases emitted from the vent of an active volcano, as well as the minor earthquakes and local land tilting that precede a major quake, scientists have successfully predicted periods of volcanic activity. Extensive monitoring of Mount Mayon and the Mexican volcano Popocatepetl to predict recent eruptions has led to evacuations that saved hundreds or thousands of lives. However, not every volcano is well monitored or predictable.

Despite their potential for destructive activity, volcanoes are a valuable natural

There are two basic forms of tectonic activity: compression and extension. Compression occurs at converging lithospheric plate margins, while extension occurs along continental and oceanic rifting.

resource in terms of recreation and tourism. Few landscapes can rival in beauty the mountainous landscapes of volcanic origin. National parks have been made of Mount Rainier, Mount Lassen, and Crater Lake in the Cascade Range, a mountain mass largely of volcanic construction. Hawaii Volcanoes National Park recognizes the natural beauty of Mauna Loa and Kilauea—their breathtaking displays of molten lava are a living textbook of igneous processes. *Focus on Remote Sensing 13.2 • Remote Sensing of Volcanoes* shows Mount Vesuvius, Mount Fuji, and Popocatepetl as imaged by several different remote sensing systems.

LANDFORMS OF TECTONIC ACTIVITY

Recall from Chapter 12, our introduction to global plate tectonics, that there are two basic forms of tectonic activity: compression and extension (see Figure 12.10). Along converging lithospheric plate boundaries, tectonic activity is primarily compression. In subduction zones, sedimentary layers of the ocean floor are compressed within a trench as the descending plate forces them against the overlying plate. In continental collision, compression is of the severest kind. In zones of rifting of continental plates, however, the brittle continental crust is pulled apart and yields

Eye on the Environment | **13.1**

Geothermal Energy Sources

Geothermal energy is energy in the form of sensible heat that originates within the Earth's crust and makes its way to the surface by conduction. Heat may be conducted upward through solid rock or carried up by circulating ground water that is heated at depth and returns to the surface. Concentrated geothermal heat sources are usually associated with igneous activity, but there also exist deep zones of heated rock and ground water that are not directly related to igneous activity.

Observations made in deep mines and bore holes show that the temperature of rock increases steadily with depth. Although the rate of increase falls off quite rapidly with increasing depth, temperatures attain very high values in the upper mantle, where rock is close to its melting point. Heat within the Earth's crust and mantle is produced largely by radioactive decay. Slow as this internal heat production is, it scarcely diminishes with time, and the basic energy resource it provides can be regarded as limitless on a human scale.

It might seem simple enough to obtain all our energy needs by drilling deep holes at any desired location into the crust and letting the hot rock turn injected fresh water into steam, which we could use to generate electricity as our primary energy resource. Unfortunately, at the depths usually required to furnish the needed heat intensity, crustal rock tends to close any cavity or opening by rupture and slow flowage. This phenomenon would either prevent the holes from being drilled or would close them in short order. Generally, then, we must look for geothermal localities, where special conditions have caused hot rock and hot ground water to lie within striking distance of conventional drilling methods. Areas of hot springs and geysers are primary candidates.

Natural hot-water and steam localities were the first type of geothermal energy source to be developed and at present account for nearly all production of geothermal electrical power. Wells are drilled to tap the hot water. When it reaches the surface, the water flashes into steam under the reduced pressure of the atmosphere. The steam is fed into generating turbines to produce electricity, then condensed in large cooling towers (see photo below). The resulting hot water is usually released into surface stream flow, where it may create a thermal pollution problem. The larger steam fields have sufficient energy to generate at least 15 megawatts of electric power, and a few can generate 200 megawatts or more.

In certain areas, the intrusion of magma has been sufficiently

recent that solid igneous rock of a batholith is still very hot in a depth range of perhaps 2 to 5 km (about 1 to 3 mi). At this depth, the rock is strongly compressed and contains little, if any, ground water. Rock in this zone may be as hot as 300°C (about 575°F) and could supply an enormous quantity of heat energy. The planned development of this resource includes drilling into the hot zone and then shattering the surrounding rock by hydrofracture—a method using water under pressure that is widely used in petroleum development. Surface water would then be pumped down one well into the fracture zone and heated water pumped up another well. Although some experiments have been conducted, this heat source has not yet been exploited in any practical way.

Geothermal power plant This electricity-generating power plant at the Geysers, California, runs on steam produced by superheated ground water. Steam pipes in the foreground lead to the plant. After use in generating turbines, the steam is condensed in large cylindrical towers.

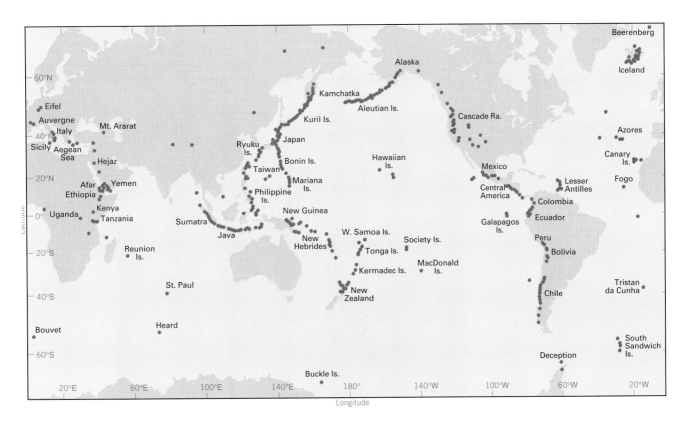

13.12   Volcanic activity of the Earth *Dots show the locations of volcanoes known or believed to have erupted within the past 12,000 years. Each dot represents a single volcano or cluster of volcanoes. (After data of NOAA. Copyright © A.N. Strahler.*

by faulting. This motion is extensional and produces faults. We begin our examination of landforms of tectonic activity with folding produced by compression.

Fold Belts

In Chapter 12, we saw that two plates of continental lithosphere can collide, causing severe compressional stress. When flat-lying strata from a continental shelf or margin are caught in the collision, they can experience **folding** (see Figure 12.10). The wave-like shapes imposed on the strata consist of alternating arch-like upfolds, called *anticlines,* and through-like downfolds, called *synclines.* Thus, the initial landform associated with an anticline is a broadly rounded mountain ridge, and the landform corresponding to a syncline is an elongate, open valley.

An example of open folds of comparatively young geologic age that has long attracted the interest of geographers is the Jura Mountains of France and Switzerland. Figure 13.13 is a block diagram of a small portion of that fold belt. The rock strata are mostly limestone layers and were capable of being deformed by bending with little brittle fracturing. Folding occurred in late Cenozoic (Miocene) time. Notice that each mountain crest is associated with the axis of an anticline, while each valley lies over the axis of a syncline. Some of the anticlinal arches have been partially removed by erosion processes. The rock structure can be seen clearly in the walls

of the winding gorge of a major river that crosses the area. The Jura folds lie just to the north of the main collision orogen of the Alps. Because of their location near a mountain mass, they are called *foreland folds.* The ridge and valley region of Pennsylvania, Maryland, and Virginia is another example but of much greater age.

Faults and Fault Landforms

A **fault** in the brittle rocks of the Earth's crust occurs when rocks suddenly yield to unequal stresses by fracturing. Faulting is accompanied by a displacement—a slipping motion—along the plane of breakage, or *fault plane.* Faults are often of great horizontal extent, so that the surface trace, or fault line, can sometimes be followed along the ground for many kilometers. Most major faults extend down into the crust for at least several kilometers.

Faulting occurs in sudden slippage movements that generate earthquakes. A single fault movement may result in slippage of as little as a centimeter or as much as 15 m (about 50 ft). Successive movements may occur many years or decades apart, even several centuries apart. Over long time spans, the accumulated displacements can amount to tens or hundreds of kilometers. In some places, clearly recognizable sedimentary rock layers are offset on opposite sides of a fault, allowing the total amount of displacement to be measured accurately.

Focus on Remote Sensing | 13.2

Remote Sensing of Volcanoes

Volcanoes are always attractive subjects for remote sensing. As monumental landforms, they have distinctive shapes and appearances that are easy to recognize in satellite imagery. They are also very dynamic subjects with the capacity to erupt in spectacular fashion, providing smoke plumes, ash falls, and lava flows. Here's a gallery of images of volcanoes from a suite of remote imagers with different characteristics.

Mount Vesuvius. This famous volcano erupted in A.D. 79, ejecting a huge cauliflower-shaped cloud of ash and debris that rapidly settled to the Earth. The nearby Roman city of Pompeii was buried in as much as 30 m (100 ft) of ash and dust, leaving an archeological treasure that was not discovered until 1748. In recent history, major eruptions were recorded in 1631, 1794, 1872, 1906, and 1944.

The image of Mount Vesuvius was acquired by the Advanced Spaceborne Thermal Emission and Reflection Radiometer (ASTER) (see *Focus on Remote Sensing 11.2 • Geologic Mapping with ASTER*) on September 26, 2000. Spatial resolution is 15 by 15 m (49 by 49 ft). Red, green, and blue colors in the image have been assigned to near-infrared, red, and green spectral bands, respectively. Vegetation appears bright red, with urban areas in blue and green tones. The magnitude of development around the volcano shows that the impact of a major eruption would be particularly catastrophic. If sudden, thousands of deaths would be expected.

Mount Fuji. Like Mount Vesuvius, Mount Fuji lies

Mount Vesuvius imaged by ASTER *(Image courtesy NASA/GSFC/MITI/ERSDAC/JAROS and U. S./Japan ASTER Science Team.)*

within striking distance of a large population center—Tokyo, Japan, located about 100 km (about 60 mi) to the northeast. Although it has the symmetry of a simple cone, it is actually a complex structure with two former volcanic cones buried within its outer

form. Mount Fuji is considered an active volcano, with 16 eruptions since A.D. 781. Its last eruption, in 1707, darkened the noontime sky and shed dust on the present-day Tokyo region.

This striking image of Mount Fuji, showing the

Mount Fuji imaged by the Shuttle Radar Topography Mission *(Image courtesy NASA/JPL/NIMA.)*

Tokyo region in the foreground, was acquired by the Shuttle Radar Topography Mission interferometric synthetic aperture radar on February 21, 2000. This type of radar sends simultaneous pulses of radio waves toward the ground from two antennas spaced 60 m (about 200 ft) apart. Very slight differences in the return signals can be related to the ground height. In this way, the radar can map elevations very precisely. The image is constructed by computer, simulating a viewpoint above and to the east of Tokyo. A color scale is assigned to elevation, ranging from white to green to brown. Vertical scale is doubled for visualization, so Mount Fuji and surrounding peaks appear twice as steep as they actually are.

Popocatepetl. On December 18, 2000, Popocatepetl came to life, spewing molten lava in plumes and rivers from its summit of 5470 m (17,946 ft). Located only 65 km (40 mi) from Mexico City, the "smoking mountain" of the Nahuatl language is located within view of nearly 30 million people. Following its last eruption in 1994, the volcano was under constant monitoring, and as a result, scientists were able to anticipate the eruption. About half the population in the valleys directly below the volcano were evacuated, and no injuries were reported.

The close-up image of Popocatepetl was acquired by Landsat-7 on January 4, 1999, at a spatial resolution of 30 by 30 m (98 by 98 ft). Snow and ice flank the summit crater. Canyons carved into the volcano lead away from the summit. The lower slopes are thickly covered with vegetation, which appears green in this image. The surrounding plain shows intensive agricultural development.

The eruption of Popocatepetl was captured by the Sea-viewing Wide Field-of-View Sensor (SeaWiFS) on December 19, 2000, at a spatial resolution of 1 by 1 km (0.62 by 0.62 mi). This global-scale imager pictured the width of

Popocatepetl imaged by Landsat-7 *(Image courtesy Ron Beck, EROS Data Center.)*

southern Mexico from the Pacific to the Caribbean in true color. Popocatepetl (P) is shown very near the center of the image, emitting a smoke plume moving south and east. A large cloud bank lies to the east and obscures much of the right-hand side of the image.

High, thin cirrus clouds overlie part of the plume. Mexico City (M) is visible north and west of the volcano as a gray-brown patch flanked by north-south mountain ranges.

Eruption of Popocatepetl imaged by SeaWiFS *(Image courtesy SeaWiFS Project, NASA/Goddard Space Flight Center, and ORBIMAGE.)*

13.13 Anticlines and synclines *Structural diagrams for the Jura Mountains, France and Switzerland. A cross section shows the folds and ground surface (a). The landscape developed on the folds is shown in the block diagram (b). (After E. Raisz.)*

NORMAL FAULTS   One common type of fault associated with crustal rifting is the **normal fault** (Figure 13.14*a*). The plane of slippage, or fault plane, is steeply inclined. The crust on one side is raised, or upthrown, relative to the other, which is downthrown. A normal fault results in a steep, straight, cliff-like feature called a *fault scarp* (Figure 13.14). Fault scarps range in height from a few meters to a few hundred meters (Figure 13.15). Their length is usually measurable in kilometers. In some cases, they attain lengths as great as 300 km (about 200 mi).

Normal faults are not usually isolated features. Commonly, they occur in multiple arrangements, often as a set of parallel faults. These arrangements give rise to a grain or pattern of rock structure and topography. A narrow block dropped down between two normal faults is a *graben* (Figure 13.16). A narrow block elevated between two normal faults is a *horst*. Grabens make conspicuous topographic

> **In a fault, rocks break apart and move along a fault plane. Normal faults are caused by extension and produce downdropped blocks (grabens) and upthrown blocks (horsts).**

trenches, with straight, parallel walls. Horsts make block-like plateaus or mountains, often with a flat top but steep, straight sides.

In rifted zones of the continents, regions where normal faulting occurs on a grand scale, mountain masses called *block mountains* are produced. The up-faulted mountain blocks can be described as either tilted or lifted (Figure 13.17). A tilted block has one steep face—the fault scarp—and one gently sloping side. A lifted block, which is a type of horst, is bounded by steep fault scarps on both sides.

TRANSCURRENT FAULTS   Recall that lithospheric plates also slide past one another horizontally along major transform faults and that these features comprise one type of lithospheric plate boundary. Long before the principles of plate tectonics became known, geologists referred to such faults as **transcurrent faults** (Figure 13.14*c*), or sometimes as

(a) Normal fault

(b) Reverse fault

(c) Transcurrent fault

(d) Overthrust fault

13.14 Four types of faults

strike-slip faults. In a transcurrent fault, the movement is predominantly horizontal. Where the land surface is nearly flat, no scarp, or a very low one at most, results. Only a thin fault line is traceable across the surface. In some places a narrow trench, or rift, marks the fault.

The best known of the active transcurrent faults is the great San Andreas fault, which can be

Transcurrent faults occur where lithospheric plates slide past each other. Movement is predominantly horizontal. The San Andreas fault of California is a famous transcurrent fault.

followed for a distance of about 1000 km (about 600 mi) from the Gulf of California to Cape Mendocino, a location well north of the San Francisco area, where it heads out to sea. It is a transform fault that marks the active boundary between the Pacific plate and the North American plate (see Figure 12.4). The Pacific plate is moving toward the northwest, which means that a great portion of the state of California and all of Lower (Baja) California is moving bodily northwest with respect to the North American mainland.

Throughout many kilometers of its length, the San Andreas fault appears as a straight, narrow scar. In some

13.15 Fault scarp *This fault scarp was formed during the Hebgen Lake, Montana, earthquake of 1959. In a few moments, a displacement of 6 m (20 ft) took place on a normal fault.*

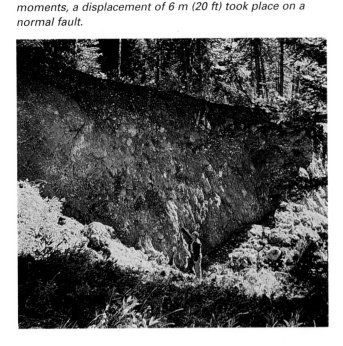

13.16 Initial landforms of normal faulting *A graben is a downdropped block, often forming a long, narrow valley. A horst is an upthrown block, forming a plateau, mesa, or mountain. (A. N. Strahler)*

13.17 Fault block mountains *Fault block mountains may be of tilted type (left) or lifted type (right). (After W. M. Davis.)*

Tilted block

Lifted block

places this scar is a trench-like feature, and elsewhere it is a low scarp (Figure 13.18). Frequently, a stream valley takes an abrupt jog—for example, first right, then left—when crossing the fault line. This offset in the stream's course shows that many meters of movement have occurred in fairly recent time.

Reverse and overthrust faults are caused by compression. In a reverse fault, an overhanging scarp that slumps downward is formed. In an overthrust fault, one rock mass slides up and over another.

REVERSE AND OVERTHRUST FAULTS In a *reverse fault*, the inclination of the fault plane is such that one side rides up over the other and a crustal shortening occurs (Figure 13.14*b*). Reverse faults produce fault scarps similar to those of normal faults, but the possibility of landsliding is greater because an overhanging scarp tends to be formed. The San Fernando, California, earthquake of 1971 was generated by slippage on a reverse fault.

The *low-angle overthrust fault* (Figure 13.14*d*) involves predominantly horizontal movement. One slice of rock rides over the adjacent ground surface. A thrust slice may be up to 50 km (30 mi) wide. The evolution of low-angle thrust faults was explained in Chapter 12 and illustrated in Figure 12.11.

The Rift Valley System of East Africa

Rifting of continental lithosphere is the very first stage in the splitting apart of a continent to form a new ocean basin. The process is beautifully illustrated by the East African Rift Valley system. This region has attracted the attention of geologists since the early 1900s. They gave the name *rift valley* to what is basically a graben but with a more complex history that includes the building of volcanoes on the graben floor.

Figure 13.19 is a sketch map of the East African Rift Valley system. It is about 3000 km (1900 mi) long and extends from the Red Sea southward to the Zambezi River. Along this axis, the Earth's crust is being lifted and spread apart in a long, ridge-like swell. The Rift Valley system consists of a number of graben-like troughs. Each is a separate rift valley ranging in width from about 30 to 60 km (20 to 40 mi). As geologists noted in early field surveys of this system, the rift valleys are like keystone blocks of a masonry arch that have slipped down between neighboring blocks because the arch has spread apart somewhat. Thus, the floors of the rift valleys are above the elevation of most of the African continental surface. Major rivers and several long, deep lakes—Lake Nyasa and Lake Rudolph, for example—occupy some of the valley floors.

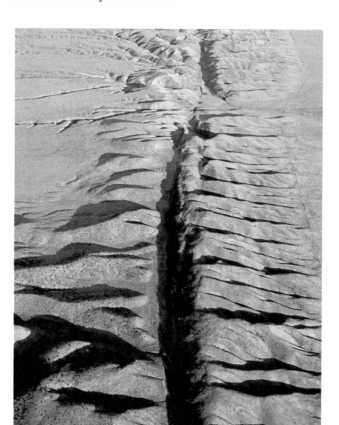

13.18 The San Andreas fault in southern California *The fault is marked by a narrow trough. The fault is slightly offset in the middle distance.*

13.19 East African Rift Valley *This sketch map shows the East African Rift Valley system and the Red Sea to the north.*

The sides of the rift valley typically consist of multiple fault steps (Figure 13.20). Sediments, derived from the high plateaus that form the flanks of troughs, make thick fills in the floors of the valleys. Two great stratovolcanoes have been built close to the Rift Valley east of Lake Victoria. One is Mount Kilimanjaro, whose summit rises to over 6000 m (about 19,000 ft). The other, Mount Kenya, is only a little lower and lies right on the equator.

EARTHQUAKES

You have probably seen television news accounts of disastrous earthquakes and their destructive effects (Figure 13.21). Californians know about severe earthquakes from first-hand experience, but several other areas in North America have also experienced strong earthquakes, and a few of these have been very severe. An **earthquake** is a motion of the ground surface, ranging from a faint tremor to a wild motion capable of shaking buildings apart.

The earthquake is a form of energy of wave motion transmitted through the surface layer of the Earth. Waves move outward in widening circles from a point of sudden energy release, called the *focus.* Like ripples produced when a pebble is thrown into a quiet pond, these seismic waves gradually lose energy as they travel outward in all directions. (The term *seismic* means "pertaining to earthquakes.")

13.20 Eye on the Landscape **The Rift Valley wall in Ethiopia** *Multiple fault scarps give the landscape a stepped appearance.* **What else would the geographer see?** **... Answers at the end of the chapter.**

Most earthquakes are produced by sudden slip movements along faults. They occur when rock on both sides of the fault is slowly bent over many years by tectonic forces. Energy accumulates in the bent rock, just as it does in a bent archer's bow. When a critical point is reached, the strain is relieved by slippage on the fault, and the rocks on opposite sides of the fault move in different directions. A large quantity of energy is instantaneously released in the form of seismic waves, which shake the ground. In the case of a transcurrent fault, on which movement is in a horizontal direction, slow bending of the rock that precedes the shock takes place over many decades. Sometimes a slow, steady displacement known as *fault creep* occurs, which tends to reduce the accumulation of stored energy.

The devastating San Francisco earthquake of 1906 resulted from slippage along the San Andreas fault, which is dominantly a transcurrent fault. This fault also passes about 60 km (about 40 mi) inland of Los Angeles, placing the densely populated metropolitan Los Angeles region in great jeopardy. Associated with the San Andreas fault are several important parallel and branching transcurrent faults, all of which are capable of generating severe earthquakes.

Earthquakes can also be produced by volcanic activity, as when magma rises or recedes within a volcanic chamber.

A scale of earthquake magnitudes was devised in 1935 by the distinguished seismologist Charles F. Richter. Now called the *Richter scale* it describes the quantity of energy released by a single earthquake. Scale numbers range from 0 to 9, but there is really no upper limit other than nature's own energy release limit. For each whole unit of increase (say, from 5.0 to 6.0), the quantity of energy release increases by a factor of 32. A value of 9.5 is the largest observed to date—the Chilean earthquake of 1960. The great San Francisco earthquake of 1906 is now rated as magnitude 7.9. Table 13.1 and Figure 13.22 show how the Richter scale relates to the energy released by earthquakes.

> **An earthquake is a wave motion transmitted through the surface layer of the Earth. It is usually produced by sudden slippage on a fault.**

> **The Richter scale is used to measure the energy released by earthquakes. It ranges from 0 to 9 or more. Earthquakes in populous regions measuring 7 or greater on the Richter scale are usually associated with major damage and loss of life.**

13.21 Earthquake devastation *This building did not survive the Mexico City earthquake of September 1985.*

Table 13.1 | Richter Magnitude and Energy Release

Magnitude, Richter Scale*	Energy Release (joules)	Comment
2.0	6×10^7	Smallest quake normally detected by humans.
2.5–3.0	10^8–10^9	Quake can be felt if it is nearby. About 100,000 shallow quakes of this magnitude per year.
4.5	4×10^{11}	Can cause local damage.
5.7	2×10^{13}	Energy released by Hiroshima atom bomb.
6.0	6×10^{13}	Destructive in a limited area. About 100 shallow quakes per year of this magnitude.
6.7	7×10^{14}	Northridge earthquake of 1994.
7.0	2×10^{15}	Rated a major earthquake above this magnitude. Quake can be recorded over whole Earth. About 14 per year this great or greater.
7.1	3×10^{15}	Loma Prieta earthquake of 1989.
8.25 (7.9)	4.5×10^{16}	San Francisco earthquake of 1906.
8.1	10^{17}	Mexican earthquake of 1986.
8.4 (9.2)	4×10^{18}	Alaskan earthquake of 1964.
8.3 (9.5)	10^{19}	Chilean earthquake of 1960, near the border of Equador.

* ( ) indicates magnitude as adjusted by Kanamori.

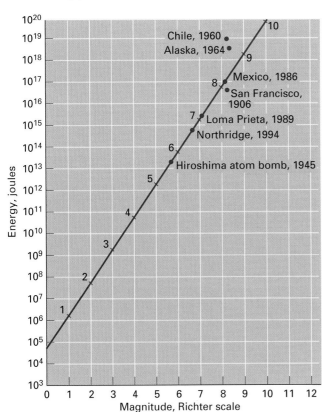

13.22 Great earthquakes on the Richter scale *This graph plots the energy released and the Richter scale rating for some great earthquakes mentioned in this chapter. Note that the energy scale is logarithmic and increases by powers of ten.*

Earthquakes and Plate Tectonics

Seismic activity—the repeated occurrence of earthquakes—occurs primarily near lithospheric plate boundaries. Figure 13.23, which shows the location of all earthquake centers during a typical seven-year period, clearly reveals this pattern. The greatest intensity of seismic activity is found along converging plate boundaries where oceanic plates are undergoing subduction. Strong pressures build up at the downward-slanting contact of the two plates, and these are relieved by sudden fault slippages that generate earthquakes of large magnitude. This mechanism explains the great earthquakes experienced in Japan, Alaska, Central America, Chile, and other narrow zones close to trenches and volcanic arcs of the Pacific Ocean Basin.

Good examples can be cited from the Pacific coast of Mexico and Central America, where the subduction boundary of the Cocos plate lies close to the shoreline. The great earthquake that devastated Mexico City in 1986 was centered in the deep trench offshore. Two great shocks in close succession, the first of magnitude 8.1 and the second of 7.5, damaged cities along the coasts of the Mexican states of Msauichocoan and Guerrero. Although Mexico City lies inland about 300 km (about 185 mi) distant from the earthquake epicenters, it experienced-intense ground shaking of underlying saturated clay formations, with the resulting death toll of some 10,000 persons (Figure 13.21).

Transcurrent faults on transform boundaries that cut through the continental lithosphere are also sites of intense seismic activity, with moderate to strong earthquakes. The most familiar example is the San Andreas fault, discussed further below.

13.23 Earthquake locations *This world map plots earthquake center locations and centers of great earthquakes. Center locations of all earthquakes originating at depths of 0 to 100 km (62 mi) during a six-year period are shown by red dots. Each dot represents a single location or a cluster of centers. Black circles identify centers of earthquakes of Richter magnitude 8.0 or greater during an 80-year period. The map clearly shows the pattern of earthquakes occurring at subduction boundaries. (Compiled by A. N. Strahler from data of U.S. government. Copyright @ A. N. Strahler.)*

Spreading boundaries are the locations of a third class of narrow zones of seismic activity related to lithospheric plates. Most of these boundaries are identified with the midoceanic ridge and its branches. For the most part, earthquakes in this class are limited to moderate intensities.

Earthquakes also occur at scattered locations over the continental plates, far from active plate boundaries. In many cases, no active fault is visible, and the geologic cause of the earthquake is uncertain. For example, the great New Madrid earthquake of 1811 was centered in the Mississippi River floodplain in Missouri. It produced three great shocks in close succession, rated from 8.1 to 8.3 on the Richter scale.

> **Earthquakes are frequent at spreading and converging boundaries of lithospheric plates. Transcurrent faults on transform boundaries are also sites of earthquake occurrence.**

Seismic Sea Waves

An important environmental hazard often associated with a major earthquake centered on a subduction plate boundary is the *seismic sea wave,* or **tsunami,** as it is known to the

Japanese. A train of these water waves is often generated in the ocean by a sudden movement of the seafloor at a point near the earthquake source. The waves travel over the ocean in ever-widening circles, but they are not perceptible at sea in deep water. Seismic sea waves are sometimes referred to as "tidal waves," but since they have nothing to do with tides, the name is quite misleading.

When a tsunami arrives at a distant coastline, it causes a rise in water level. Normal wind-driven waves, superimposed on the heightened water level, attack places inland that are normally above their reach. For example, in this century several destructive seismic sea waves in the Pacific Ocean attacked ground as high as 10 m (30 ft) above normal high-tide level, causing widespread destruction and many deaths by drowning in low-lying coastal areas. It is thought that the coastal flooding that occurred in Japan in 1703, with an estimated loss of life of 100,000 persons, may have been caused by seismic sea waves.

13.24 San Andreas fault system in California *A sketch map of the San Andreas fault and Transverse Ranges showing the locked sections of the fault alternating with active sections. (Based on data of the U.S. Geological Survey.)*

Earthquakes along the San Andreas Fault

Almost a hundred years have passed since the great San Francisco earthquake of 1906 was generated by movement on the San Andreas fault. The maximum horizontal displacement of the ground was about 6 m (20 ft). Since then, this sector of the fault has been locked—that is, the rocks on the two sides of the fault have been held together without sudden slippage (Figure 13.24). In the meantime, the two lithospheric plates that meet along the fault have been moving steadily with respect to one another. This means that a huge amount of unrelieved strain energy has already accumulated in the crustal rock on either side of the fault.

The San Andreas fault and related faults in southern and central California are potential sources of great earthquakes occurring in densely populated regions. The Loma Prieta, Landers, San Fernando, Whittier Narrows, and Northridge earthquakes are recent examples.

On October 17, 1989, the San Francisco Bay area was severely jolted by an earthquake with a Richter magnitude of 7.1. The earthquake's epicenter was located near Loma Prieta peak, about 80 km (50 mi) southeast of San Francisco, at a point only 12 km (7 mi) from the city of Santa Cruz, on Monterey Bay. The city of Santa Cruz suffered severe structural damage to older buildings. In the distant San Francisco Bay area, destructive ground shaking proved surprisingly severe. Buildings, bridges, and viaducts on landfills were particularly hard hit (Figure 13.25). Altogether, 62 lives were lost in this earthquake, and the damage was estimated to be about $6 billion. In comparison, the 1906 earthquake took a toll of 700 lives and property damage equivalent to about 30 billion present-day dollars.

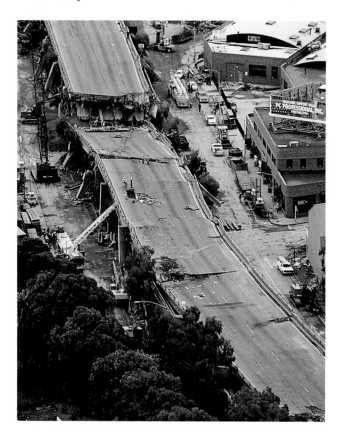

13.25   *Earthquake damage in Oakland, California*
*This section of the double-decked Nimitz Freeway
(Interstate 880) in Oakland, California, collapsed during
the Loma Prieta earthquake, crushing at least 39 people in
their cars.*

The displacement that caused the Loma Prieta earthquake occurred deep beneath the surface not far from the
San Andreas fault, which has not slipped since the great San
Francisco earthquake of 1906. The slippage on the Loma
Prieta fault amounted to about 1.8 m (6 ft) horizontally and
1.2 m (4 ft) vertically, but did not break the ground surface
above it. Geologists state that the Loma Prieta slippage,
though near the San Andreas fault, probably has not relieved
more than a small portion of the strain on the San Andreas.
While the occurrence of another major earthquake in the
San Francisco region cannot be predicted with precision, it
is inevitable. As each decade passes, the probability of that
event becomes greater.

We may need only to look to Japan for a scenario of
what the citizens of the San Francisco Bay area could experience when a branch of the San Andreas fault lets go. The
Hyogo-ken Nanbu earthquake that devastated the city of
Kobe in January of 1995 (Figure 13.26) occurred on a

short side branch of a major transcurrent fault quite similar tectonically to the San Andreas. A similar side branch
of the main San Andreas fault, called the Hayward fault,
runs through the East Bay region of San Francisco. Seismologists estimate that a slip on the Hayward fault would
generate a quake of magnitude 7.0—about the same as the
Kobe quake—and has a 28 percent probability of occurring by the year 2018. Although the nation of Japan prides
itself on its earthquake preparedness, the Kobe catastrophe
left 5000 dead and 300,000 homeless, and did property
damage estimated at 10 times that of the Northridge earthquake of 1994, which was the most costly earthquake in
American history.

Along the southern California portion of the San Andreas
fault, a recent estimate placed at about 50 percent the likelihood that a very large earthquake will occur within the next
30 years. In 1992 three severe earthquakes occurred in close
succession along active local faults a short distance north of
the San Andreas fault in the southern Mojave Desert (Figure
13.24). The second of these, the Landers earthquake, a powerful 7.5 on the Richter scale, occurred on a transcurrent
fault trending north-northwest. It caused a 80-km (50-mi)
rupture across the desert landscape. These three events have
led to speculation that the likelihood of a major slip on the
nearby San Andreas fault in the near future has substantially
increased.

For residents of the Los Angeles area, and additional
serious threat lies in the large number of active faults close
at hand. Movements on these local faults have produced
more than 40 damaging earthquakes since 1800, including
the Long Beach earthquakes of the 1930s and the San Fernando earthquake of 1971. The San Fernando earthquake
measured 6.6 on the Richter scale and produced severe
structural damage near the earthquake center. The brief but
intense primary shock and aftershocks that followed damaged beyond repair many older structures built of unrein-
forced brick masonary. In 1987 an earthquake of magnitude
6.1 struck the vicinity of Pasadena and Whittier, located
within about 20 km (12 mi) of downtown Lost Angeles.
Known as the Whittier Narrows earthquake, it was generated along a local fault system that had not previously shown
significant seismic activity. The Northridge earthquake of
1994, described in *Eye on the Environment 13.3 • The
Northridge Earthquake,* measured 6.7 on the Richter scale
and produced severe structural damage near the earthquake
center.

A slip along the San Andreas fault, some 50 km (31 mi)
to the north of the densely populated region of Los Angeles,
will release an enormously larger quantity of energy than
local earthquakes, such as the Northridge or San Fernando
earthquakes. On the other hand, the desructive effects of a
San Andreas earthquake in downtown Los Angeles will be

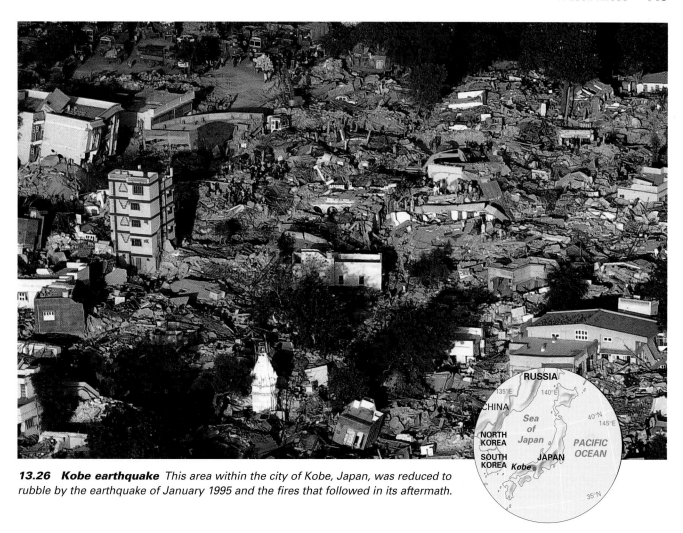

13.26   Kobe earthquake *This area within the city of Kobe, Japan, was reduced to rubble by the earthquake of January 1995 and the fires that followed in its aftermath.*

somewhat moderated by the greater travel distance. Although the intensity of ground shaking might not be much different from that of the San Fernando earthquake, for example, it will last much longer and cover a much wider area of the Los Angeles region. The potential for damage and loss of life is enormous.

A LOOK AHEAD
In the last three chapters, we have surveyed the composition, structure, geologic activity, and initial landforms of the Earth's crust. We began with a study of the rocks and minerals that make up the Earth's crust and core. We saw how the cycle of rock transformation involves a continuous cycling and recycling of rocks and minerals that has occurred over some 3 billion years or more of geologic time. In the second chapter, we developed plate tectonics as the mechanism powering the rock cycle. The global pattern of plate tectonics also explains the geo-

graphical distribution of mountain ranges, ocean basins, and continental shields occurring on the Earth's surface. In this chapter, we described the initial landforms that result directly from volcanic and tectonic activity, occurring primarily at the boundaries of spreading or colliding lithospheric plates.

With this survey of the Earth's crust and the geologic processes that shape it now completed, we can turn to other landform-creating processes in the following chapters. First will be the processes of weathering, which breaks rock into small particles, and mass wasting, which moves them downhill as large and small masses under the influence of gravity. Then we will turn to running water in three chapters that describe the behavior of rivers and streams, their work in shaping landforms, and how running water dissects rock layers to reveal their inner structures. In our two concluding chapters, we will examine landforms created by waves, wind, and glacial ice.

The Northridge Earthquake

The January 17, 1994, Northridge earthquake struck a heavily populated suburb of Los Angeles, California, at 4:30 A.M., PST. Assigned a Richter scale rating of 6.7, it produced the strongest ground motions ever recorded in an urban setting in North America and the greatest financial losses from a natural disaster in the United States since the great San Francisco earthquake of 1906.

Northridge lies in an east-west trending fault-rich valley close to the base of the lofty San Gabriel Mountains. This range is an uplifted fault-block that has produced earthquakes at many locations along the base of its steep southern flank. One of these was the equally severe San Fernando earthquake of 1971, centered not far to the north in the same valley. These active earthquake faults belong to a tectonic belt known as the Transverse Ranges, which runs more or less east-west through the Los Angeles Basin and heads out across the Pacific coastline beyond Santa Barbara (see Figure 13.24). Although the San Andreas Fault lies not far distant to the north, these two fault systems have different trends and operate quite independently.

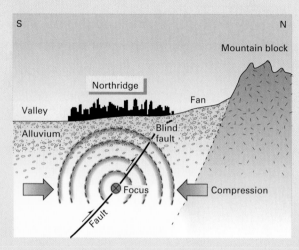

Geologic setting of the Northridge earthquake

The Northridge and San Fernando earthquake faults are of the reverse type (see Figure 13.14*b*). Our sketch shows how it works. Large arrows pointing toward each other indicate that along the Transverse Ranges the Earth's crust is under a strong, north-south, horizontal compressional force. The fault plane slants steeply upward. When the fault suddenly starts to slip at the focus, located about 20 km below the valley surface, the upper block, on the right, delivers a powerful upward punch that sharply lifts the ground surface. As with a boxer's "uppercut," what counts is the acceleration of that first upward slip motion. Seismic waves also spread out radially from the focus, generating strong ground shaking with horizontal motions. Notice that the displacement along the earthquake fault did not extend upward to the ground surface; that is, it is a "blind fault." Nevertheless, disturbances of the ground surface occurred over a wide area.

In Review | Volcanic and Tectonic Landforms

- **Landforms** are the surface features of the land, and **geomorphology** is the scientific study of landforms. *Initial landforms* are shaped by volcanic and tectonic activity, while *sequential landforms* are sculpted by agents of denudation, including running water, waves, wind, and glacial ice.

- **Volcanoes** are landforms marking the eruption of lava at the Earth's surface. **Stratovolcanoes,** formed by the emission of thick, gassy, felsic lavas, have steep slopes and tend toward explosive eruptions that can form *calderas.* Most active stratovolcanoes lie along the Pacific Rim, where subduction of oceanic lithospheric plates is occurring.

- At *hotspots,* rising mantle material provides mafic magma that erupts as basaltic lavas. because these lavas are more fluid and contain little gas, they form broadly rounded **shield volcanoes.** Hotspots occurring beneath continental crust can also provide vast areas of *flood basalts.* Some basaltic volcanoes occur along the midoceanic ridge.

- The two forms of tectonic activity are compression and extension. Compression occurs at lithospheric plate colli-

Northridge earthquake damage
This building suffered severe structural damage in the Northridge earthquake.

Northridge had more severe structural damage than San Fernando, although both were Richter 6.7 energy releases. In the Northridge case, the focus lay directly beneath the city, whereas the focus of the San Fernando earthquake was located some distance back under the mountain block. Northridge casualties included 33 persons dead as a direct result of the earthquake, more than 7000 injuries treated at local hospitals, and over 20,000 people left homeless. Total financial losses were estimated as high as $20 billion. Sections of three freeways were closed, including the busiest highway in the country, I-5.

In the wake of the severe structural damages of the 1971 San Fernando earthquake, a program of strengthening (retrofitting) of buildings, freeways, and dams had been in progress for some two decades throughout the Los Angeles region by the time of the Northridge quake. While the retrofitting may have reduced some forms of damage, it was soon discovered that the long established practice of making welded connections of the steel frames of buildings and freeway bridges failed badly during the quake. The welded joints proved brittle and easy to fracture. In contrast, the few modern buildings designed to move sidewise easily on a fixed base performed well and without damage. The implications of these revelations are obviously of enormous consequence to the residents of the California "Southland." Is it now too late to correct a monstrous design error? There will be many more earthquakes like those of Northridge and San Fernando, occurring unexpectedly at widely separated points within the Transverse Ranges.

sions. At first, the compression produces **folding**—anticlines (upfolds) and synclines (downfolds). If compression continues, folds may be overturned and eventually overthrust faulting can occur.

- Extension occurs where lithospheric plates are spreading apart, generating **normal faults.** These can produce upthrown and downdropped blocks that are sometimes as large as mountain ranges or *rift valleys.* **Transcurrent faults** occur where two rock masses move horizontally past each other.

- **Earthquakes** occur when rock layers, bent by tectonic activity, suddenly fracture and move. The sudden motion at the fault produces earthquake waves that shake and move the ground surface in the adjacent region. The energy released by an earthquake is measured by the *Richter scale.* Large earthquakes occurring near developed areas can cause great damage. Most severe earthquakes occur near plate collision boundaries.

- The San Andreas fault is a major transcurrent fault located near two great urban areas—Los Angeles and San Francisco. The potential for a severe earthquake on this fault is high, and the probability of a major Earth movement increases every year.

Key Terms

landforms, p. 426
geomorphology, p. 426
denudation, p. 426

volcano, p. 427
stratovolcano, p. 430
shield volcano, p. 430

folding, p. 437
fault, p. 437
normal fault, p. 440

transcurrent fault, p. 440
earthquake, p. 443
tsunami, p. 446

Review Questions

1. Distinguish between initial and sequential landforms. How do they represent the balance of power between internal Earth forces and external forces of denudation agents?

2. What is a stratovolcano? What is its characteristic shape, and why does that shape occur? Where do stratovolcanoes generally occur and why?

3. What is a shield volcano? How is it distinguished from a stratovolcano? Where are shield volcanoes found, and why? Give an example of a shield volcano. How are flood basalts related to shield volcanoes?

4. Describe the stages in the life cycle of a basaltic shield volcano of the Hawaiian type.

5. What is a hotspot? What produces it? How is it related to the life cycle?

6. How can volcanic eruptions become natural disasters? Be specific about the types of volcanic events that can devastate habitations and extinguish nearby populations.

7. Briefly describe the Rift Valley System of East Africa as an example of normal faulting.

8. How does a transcurrent fault differ from a normal fault? What landforms are expected along a transcurrent fault? How are transcurrent faults related to plate tectonic movements?

9. What is an earthquake, and how does it arise? How are the locations of earthquakes related to plate tectonics?

10. Describe the tsunami, including its origin and effects.

11. Briefly summarize the geography and recent history of the San Andreas fault system in California. What are the prospects for future earthquakes along the San Andreas fault?

Eye on the Environment 13.1 ● Geothermal Energy Sources

1. What is the ultimate source of geothermal power? Where would you go, and why, to find a geothermal power source?

2. How is geothermal energy extracted? What environmental concerns arise in this process?

Eye on the Environment 13.3 ● The Northridge Earthquake

1. Explain the geologic setting of the Northridge quake of 1994. What type of fault was was involved? Where was it located? What type of motion occured?

2. What discovery was made concerning construction practices as a result of the Northridge quake?

Visualizing Exercises

1. Sketch a cross section through a normal fault, labeling the fault plane, upthrown side, downthrown side, and fault scarp.

2. Sketch a cross section through a foreland fold belt showing rock layers in different colors or patterns. Label anticlines and synclines.

Essay Questions

1. Write a fictional news account of a volcanic eruption. Select a type of volcano—composite or shield—and a plausible location. Describe the eruption and its effects as it was witnessed by observers. Make up any details you need, but be sure they are scientifically correct.

2. How are mountains formed? Provide the plate tectonics setting for mountain formation, and then describe now specific types of mountain landforms arise.

Eye on the Landscape

Chapter Opener Diamond Head and Waikiki Beach Diamond Head (**A**) is a peak on the rim of a crater (**B**) of a now-extinct volcano. Erosion of the volcano's slopes has revealed the traces of gently sloping individual lava flows that once built the volcano (**C**). Coral reef (**D**) is visible in near-shore waters as a collection of dark masses marked by breaking waves. Volcanoes are one of the main subjects of this chapter, while coral reef coasts are examined in Chapter 18.

13.8 Flood basalts The Columbia River and its tributaries drain a large mountainous area reaching as far east as the Bitterroot Mountains of Idaho and as far north as Jasper National Park in Alberta. The river has been extensively developed for hydropower generation with the building of many large dams. Here we see the still waters (**A**) of a reservoir and a narrow lake shore (**B**) at the water's edge. Note also the sparse vegetation (**C**) characteristic of this location in the rain shadow of the Cascades. Lakes and dams are discussed in Chapters 15 and 16. Rain shadow effects on precipitation were covered in Chapter 4.

13.20 Ethopian Rift Valley The local agriculture exploits the stepped landscape at the edge of the Rift Valley with cultivated fields (**A**) laid out on the flat tops of terraces. Given the mix of fallow and still-green fields, this photo is probably from the end of the growing season. Note also the flat faces of the slopes on the fault scarps between gullies (**B**). These are likely to be remnants of the original planes of motion, given the recent geologic age of the Rift Valley. Chapter 9 shows the natural vegetation of the Ethiopean highlands as rainforest; precipitation is over 100 cm (40 in.) per year (Chapter 7). Fault planes and scarps are covered in this chapter.

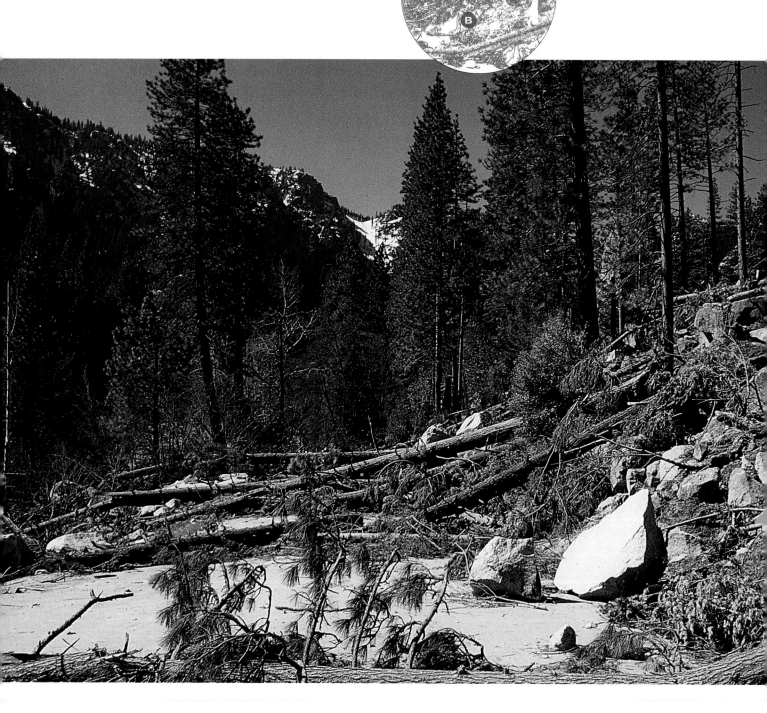

14 | WEATHERING AND MASS WASTING

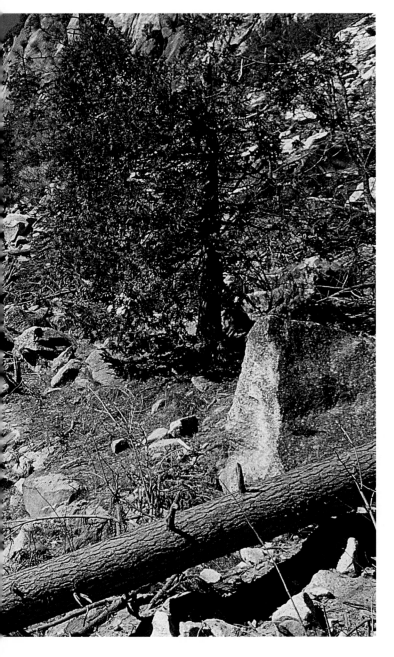

• • • • • • • • • • • • • • • • • • •
PUTTING YOU IN THE PICTURE

The pull of gravity is relentless and unforgiving. It acts to move all things closer to the Earth's center, and for objects on the Earth's surface, that generally means downhill. If you are a skier, you may be grateful for that downhill pull (providing the ski lift is working to handle the uphill part of skiing). As a hiker, the downhill force makes the trek home from the summit a lot easier (if your knees and leg muscles are in good shape). But if you are an inhabitant of a peaceful valley below a steep mountain slope that suddenly gives way, you may not be so grateful for the force that has buried your home under tons of rocky debris.

Landslide, earthflow, mudflow, soil creep—these words describe the different ways that Earth materials move downhill under the force of gravity. Sometimes the materials move very quickly, other times very slowly. Perhaps you've walked the sidewalks of a town or city built on a steep slope and seen retaining walls leaning over or even toppled by the slow downhill creep of soil. While annoying and expensive for a homeowner, a failing retaining wall is not likely to result in death or destruction. But a landslide, earthflow, or mudflow can easily take lives and destroy property.

In 1903, an enormous mass of limestone rock slid from the face of Turtle Mountain, near Frank, Alberta, into the valley below (see Figure 14.22). The volume of the landslide was estimated at about 27 million m³ (35 million yd³). It moved down through 900 m (3000 ft) of elevation at great speed and slid uphill on the opposite side of the valley to regain over 100 m (300 ft) of elevation. The rocky debris buried part of the town of Frank, Alberta, taking 70 lives.

A similar landslide in the Madison Canyon of Montana was triggered by the Hebgen Lake earthquake of 1959. About 28 million m³ (about 36 million yd³) of rock broke away from the south wall of the canyon of the Madison River and descended more than half a kilometer (0.3 mi) at a speed later estimated to have exceeded 45 m/s (about 100 mi/hr). The slide debris blocked the Madison River and a new lake quickly began to form. Within three weeks' time, it was

about 60 m (about 200 ft) deep. (*Eye on the Environment 14.1* provides more information about this extreme event.)

Earthflows are slower but no less destructive. For residents of the Palos Verdes Hills, a coastal peninsula in Los Angeles County, the Portuguese Bend "landslide" was a nightmare that wouldn't go away. About 160 hectares (400 acres) of residential land settled and moved downslope some 20 m (70 ft) over a three-year period. (see Figure 14.24). Houses were slowly tilted, cracked, and broken apart as the land underneath them crept along a shale layer softened by water from the septic systems and abundant lawn-watering of the residential area.

The Nicolet Earthflow of 1955 occurred when a clay layer beneath the city of Nicolet, Quebec, unexpectedly became liquid. A large chunk of the town quickly flowed into the Nicolet River. (see Figure 14.19). Only three

lives were lost, but damage was in the millions.

Some large mudflows can follow stream and river courses for long distances. Immediately following the eruption of Mount St. Helens in 1980, a mudflow of fresh ash and melted snow hurtled down the Toutle River, destroying fields and forests. It traveled a total distance of more than 120 km (75 mi), eventually reaching the Columbia River.

Mass movements of earth under the influence of gravity are one of the two main subjects of this chapter, and we'll return to some of the examples cited above in our text. However, the subject we will present first is weathering—a set of physical and chemical processes that break rocks apart to create large and small particles and to alter their mineral composition. Weathering prepares the way for mass movements, and it can also create its own distinctive landforms, as we will see. ■

WEATHERING AND MASS WASTING

Now that we have completed our study of the Earth's crust—including its mineral composition, its moving lithospheric plates, and its tectonic and volcanic landforms—we can focus on the shallow life layer itself. At this sensitive interface, the external processes of denudation carve sequential landforms from the rocks uplifted by the Earth's internal processes. Our investigation of what happens to rock upon exposure at the surface began in Chapter 11 with a description of mineral alteration of rock and the production of sediment, which is an essential part of the cycle of rock change. In this chapter, we look further at the softening and breakup of rock, termed *weathering,* and how the resulting particles move downhill under the force of gravity, a process termed *mass wasting.*

Weathering is the general term applied to the combined action of all processes that cause rock to disintegrate physically and decompose chemically because of exposure near the Earth's surface. We can recognize two types of weathering. In *physical weathering,* rocks are fractured and broken apart, primarily by the growth of ice or salt crystals along rock planes and mineral contacts that are penetrated by watery solutions.

Weathering is the combined action of physical weathering, in which rocks are fractured and broken, and chemical weathering, in which rock minerals are transformed to softer or more soluble forms.

In *chemical weathering,* rock minerals are transformed from types that were stable when the rocks were formed to types that are now stable at surface temperatures and pressures. As we saw in Chapter 11, these chemical processes include oxidation, hydrolysis, and acid solution. Weathering acts to produce **regolith**—a surface layer of weathered rock particles that lies above solid, unaltered rock. Weathering also leads to a number of distinctive landforms, as we will see in this chapter.

This chapter also examines how gravity acts on rock fragments, created by weathering, to produce landforms. In this process, gravity induces the spontaneous downhill movement of soil, regolith, and rock fragments, but without the action of moving water, air, or ice. The downhill movement is referred to as **mass wasting.** Movement of a mass of soil or weathered rock to lower levels takes place when the internal strength of the soil or weathered rock declines to a critical point below which the force of gravity cannot be resisted. This failure of strength under the ever-present force of gravity takes many forms and scales, and we will see that human activity causes or aggravates several forms of mass wasting.

We conclude our chapter with a look at a suite of special landforms and geomorphic processes that are found in the arctic lands. They are created primarily by the freezing and thawing of water, acting in concert with gravity.

PHYSICAL WEATHERING

Physical weathering produces regolith from massive rock by the action of forces strong enough to fracture the rock. The physical weathering processes discussed in this chapter include frost action, salt-crystal growth, unloading, and wedging by plant roots. The term *mechanical weathering* is also used for physical weathering.

Frost Action

One of the most important physical weathering processes in cold climates is *frost action,* the repeated growth and melting of ice crystals in the pore spaces of soil and in rock fractures. In contrast to most other liquids, water expands when it freezes. Perhaps you've left a bottle of water chilling in the freezer overnight and observed this phenomenon first hand, removing a mass of ice surrounded by broken glass the next morning. The expansion of water in freezing can fragment even extremely hard rocks, given many cycles of freeze and thaw. Frost action and ice crystal growth produce a number of conspicuous effects and forms in all climates that have cold winters. Features caused by frost action and the buildup of ice below the surface are particularly visible in the tundra climate of arctic coasts and islands, and above the timberline in high mountains.

Almost everywhere, bedrock is cut through by systems of fractures called *joints.* These fractures are thought to occur as rocks exposed to heat and pressure cool and contract. Joints typically occur in parallel and intersecting planes, creating surfaces of weakness along which weathering can act to break rock into individual blocks. Since there is no relative movement of rock along joints, they cannot be considered to be faults. Joints are important to the physical weathering of rocks because they admit water to the rock. As we will see shortly, this allows ice and salt-crystal growth to further fracture the rock, creating rock fragments and regolith.

In sedimentary rocks, the planes of stratification, or bedding planes, comprise another natural set of planes along which water can penetrate. Often these are cut at right angles by sets of joints. Comparatively weak stresses can separate joint blocks, while strong stresses are required to make fresh fractures through solid rock. The process of separating rock along joints and bedding planes can be called *block separation* (Figure 14.1).

As coarse-grained igneous rock becomes weakened by chemical decomposition, water is able to penetrate the contact surfaces between mineral grains. In this location, the water can freeze and exert forces strong enough to separate the grains. This form of breakup is termed *granular disintegration* (Figure 14.1). The end product is a fine gravel or coarse sand in which each grain consists of a single mineral

14.1 Bedrock disintegration *Joint-block separation and granular disintegration are two common forms of bedrock disintegration. (Drawn by A. N. Strahler.)*

particle separated from the others along its original crystal or grain boundaries.

The effects of frost action can be seen in all climates that have a winter season with many alternations of freeze and thaw. Where bedrock is exposed on knolls and mountain summits, joint blocks are pried apart by water that freezes in joint cracks (Figure 14.2). Under the most favorable conditions, seen on high mountain summits and in the arctic tun-

14.2 Frost-shattered blocks *Ice crystal growth within the joint planes of rock can cause the rock to split apart. This split boulder, on the high country of the Sierra Nevada of California, is an example.*

14.3   Quartzite blocks *These frost-shattered blocks of quartzite are on the summit of the Snowy Range, Wyoming, at an elevation of about 3700 m (about 12,000 ft). (A. N. Strahler.)*

dra, large angular rock fragments accumulate in a layer that completely blankets the bedrock beneath. The German name *felsenmeer* ("rock sea") has been given to such expanses of broken rock (Figure 14.3).

In high mountains, frost action on cliffs of bare rocks detaches rock fragments that fall to the cliff base. These loose fragments are known as *talus.* Where production of fragments is rapid, they accumulate to form *talus slopes.* Most cliffs are notched by narrow ravines that funnel the rock fragments into separate tracks. Each track, or chute, feeds a growing, cone-like talus body. Talus cones are arranged side by side along the base of the cliff (Figure 14.4). Fresh talus slopes are unstable, so that the disturbance created by walking across the slope or dropping a large rock fragment from the cliff above will easily set off a sliding or rolling motion within the surface layer of fragments.

In fine-textured soils and sediments, composed largely of silt and clay, soil water freezes in horizontal layers or lens-shaped bodies. As these ice layers thicken, the overlying soil layer is heaved upward. Prolonged frost heaving can produce minor irregularities and small mounds on the soil surface. A rock fragment at the surface can sometimes conduct heat from the soil to the cold night air and sky in such a way as to cause perpendicular ice needles to grow beneath the fragment and lift it above the surface (Figure 14.5). The same process acting on a rock fragment below the soil surface can eventually push the fragment to the surface.

When water freezes in bedrock joints and bedding planes, it expands and can split rocks apart. Talus slopes of angular rock fragments accumulate below cliffs and outcrops exposed to such frost action.

Freezing of soil water produces lenses of ice that heave soil upward. Ice needle growth can lift individual soil and rock particles as well.

Frost action is a dominant process in arctic and high-mountain tundra environments, where it is a factor in the formation of a wide variety of unique landforms. We investigate these landforms in the last section of this chapter.

Salt-Crystal Growth

Closely related to the growth of ice crystals is the weathering process of rock disintegration by growth of salt crystals in rock pores. This process, *salt-crystal growth,* operates extensively in dry climates and is responsible for many of the niches, shallow caves, rock arches, and pits seen in sandstone formations. During long drought periods, ground water is moved to the surface of the rock by *capillary action.* In this process, water is drawn into fine openings and passages in the rock by the same surface tension that gives a water droplet its rounded shape. As water evaporates from this porous outer zone of the sandstone, tiny crystals of salts such as halite (sodium chloride), calcite (calcium carbonate), or gypsum (calcium sulfate) are left behind. Over time, the growth force of these crystals produces grain-by-grain breakup of the sandstone, which crumbles into a sand and is swept away by wind and rain.

Zones of rock lying close to the base of a cliff are especially susceptible to breakup by salt-crystal growth because at that location the ground water seeps downward and outward to reach the rock surface (Figure 14.6). In the southwestern United States, many of the deep niches or cave-like recesses formed in this

14.4 *Eye on the Landscape* **Talus cones** *Frost action has caused these cliffs to shed angular blocks of rock that accumulate in talus cones. In the foreground is the shore of Lake Louise in the Canadian Rockies.* **What else would the geographer see? ... Answers at the end of the chapter.**

14.5 **Needle ice growth** *At night, water in the soil freezes at the surface, creating ice needles that can lift particles of soil or move larger stones.*

14.6 Niche formation *In dry climates there is a slow seepage of water from the cliff base. Salt-crystal growth separates the grains of permeable sandstone, breaking them loose and creating a niche.*

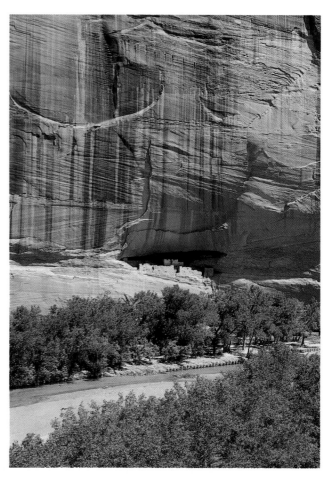

14.7 The White House Ruin *A former habitation of Native Americans occupies a large niche in sandstone in the lower wall of Canyon de Chelly, Arizona.*

way were occupied by Native Americans. Their cliff dwellings gave them protection from the elements and safety from armed attack (Figure 14.7).

Salt crystallization also damages masonry buildings, as well as concrete sidewalks and streets. Brick and concrete in contact with moist soil are highly susceptible to grain-by-grain disintegration from salt crystallization. On damp basement floors and walls, the salt crystals can be seen as a soft, white, fibrous layer. The deicing salts spread on streets and highways can be quite destructive. Sodium chloride (rock salt), widely used for this purpose, is particularly damaging to concrete pavements and walks, curb-stones, and other exposed masonry structures.

Salt-crystal growth occurs naturally in arid and semiarid regions. In humid climates, abundant rainfall dissolves salts and carries them downward to ground water. (Chapter 15 describes how rainwater infiltrates soils and moves to ground water below).

In arid climates, slow evaporation of ground water from outcropping sandstone surfaces causes the growth of salt crystals. Crystal growth breaks the rock apart grain by grain, producing niches, shallow caves, and rock arches.

Unloading

A widespread process of rock disruption related to physical weathering results from *unloading,* a process that relieves the confining pressure on underlying rock. *Exfoliation* is another term used for unloading. Unloading occurs as rock is brought near the surface by erosion of overlying layers. Rock formed at great depth beneath the Earth's surface (particularly igneous and metamorphic rock) is in a slightly compressed state because of the confining pressure of overlying rock. As the rock above is slowly worn away, the pressure is reduced, and the rock expands slightly in volume.

This causes the rock to crack in layers that are more or less parallel to the surface, creating a type of jointing called *sheeting structure.* In massive rocks like granite or marble, thick curved layers or shells of rock break free in succession from the parent mass below, much as you might peel the layers of an onion.

Where sheeting structure has formed over the top of a single large knob or hill of massive rock, an *exfoliation dome* is produced (Figure 14.8). Domes are among the largest of the landforms shaped primarily by weathering. In Yosemite Valley, California, where domes are spectacularly displayed, the individual rock sheets may be as thick as 15 m (50 ft).

Other Physical Weathering Processes

Most rock-forming minerals expand when heated and contract when cooled. Where rock surfaces are exposed daily to the intense heating of the Sun alternating with nightly cooling, the resulting expansion and contraction exert powerful disruptive forces on the rock. Although first-hand evidence

14.8 Exfoliation domes *The sheeting structure of these exfoliation domes is visible in the lower part of the photo, where successive shells of rock have fallen away. North Dome and Royal Arches, Yosemite National Park, California.*

is lacking, it seems likely that daily temperature changes can cause the breakup of a surface layer of rock already weakened by other agents of weathering.

Another mechanism of rock breakup is the growth of plant roots, which can wedge joint blocks apart. You have probably observed concrete sidewalk blocks uplifted and fractured by the growth of tree roots. This process is also active when roots grow between rock layers or joint blocks. Even fine rootlets in joint fractures can cause the loosening of small rock fragments and grains.

CHEMICAL WEATHERING AND ITS LANDFORMS

We investigated **chemical weathering** processes in Chapter 11 under the heading of mineral alteration. Recall that the dominant processes of chemical change affecting silicate minerals are oxidation, hydrolysis, and carbonic acid action. *Oxidation* and *hydrolysis* change the chemical structure of minerals, turning them into new minerals that are typically softer and bulkier and therefore more susceptible to erosion and mass movement. *Carbonic acid action* dissolves minerals, washing them away in runoff. Note also that chemical reactions proceed more rapidly at warmer temperatures. Thus, chemical weathering is most effective in warm, moist climates.

Hydrolysis and Oxidation

Decomposition by hydrolysis and oxidation changes the minerals of strong rock into weaker forms that are rich in clay minerals and oxides. In warm, humid climates of the equatorial, tropical, and subtropical zones, hydrolysis and oxidation over thousands of years have resulted in the decay of igneous and metamorphic rocks to depths as great as 100 m (about 300 ft). The decayed rock material is soft, clay-rich, and easily eroded. To the construction engineer, deeply weathered rock is of major concern in the building of highways, dams, or other heavy structures. Although the weathered rock is soft and easy to move, its high clay content reduces its strength, and foundations built on the weathered rock can fail under heavy loads.

In dry climates, exposed granite weathers by hydrolysis to produce many interesting boulder and pinnacle forms (Figure 14.9). Although rainfall is infrequent, water penetrates the granite along planes between crystals of quartz and feldspar. Chemical weathering of these surfaces then breaks individual crystal grains away from the main mass of rock, leaving the rounded forms shown in the photo. The grain-by-grain breakup forms a fine desert gravel consisting largely of quartz and partially decomposed feldspar crystals.

Acid Action

Chemical weathering is also produced by *acid action*, largely that of *carbonic acid*. This weak acid is formed when carbon dioxide dissolves in water. Rainwater, soil water, and stream water all normally contain dissolved carbon dioxide. Carbonic acid slowly dissolves some types of minerals. Carbonate sedimentary rocks, such as limestone and marble, are particularly susceptible to the acid action. In this process, the mineral calcium carbonate is dissolved and carried away in solution in stream water.

Carbonic acid reaction with limestone produces many interesting surface forms, mostly of small dimensions. Outcrops of limestone typically show cupping (formation of rounded cavities), rilling (formation of surface valleys), grooving, and fluting in intricate designs (Figure 14.10). In a few places, the scale of deep grooves and high wall-like rock fins reaches proportions that keep people and animals

> In chemical weathering, the minerals that make up rocks are chemically altered or dissolved. The end products are often softer and bulkier forms that are more susceptible to erosion and mass movement.

14.9 Granular disintegration
Large joint blocks of granite are gradually rounded into smooth forms by grain-by-grain disintegration in a desert environment. Alabama Hills, Owens Valley, California.

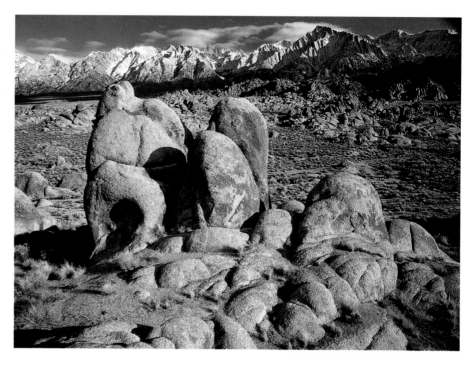

from passing through. Carbonic acid in ground water can dissolve limestone to produce underground caverns as well as distinctive landscapes that form when underground caverns collapse. These landforms and landscapes will be described in Chapter 15.

14.10 Solution features in limestone This outcrop of pure limestone shows grooves and cavities formed by carbonic acid action. County Clare, Ireland.

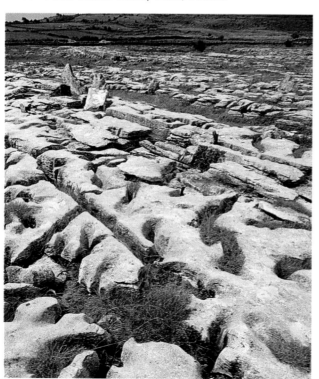

In urban areas, air is commonly polluted by sulfur and nitrogen oxides. When these gases dissolve in rainwater, the result is acid precipitation (see Chapter 4). The acids rapidly dissolve limestone and chemically weather other types of building stones. The result can be very damaging to stone sculptures, building decorations, and tombstones (Figure 14.11).

In the wet low-latitude climates, mafic rock, particularly basaltic lava, dissolves rapidly under attack by soil acids. The effects of solution removal of basaltic lava are displayed in spectacular grooves, fins, and spires on the walls of deep alcoves in part of the Hawaiian Islands (Figure 14.12). The landforms produced are quite similar to those formed by carbonic acid action on massive limestones in the moist climates of the midlatitudes.

MASS WASTING

With our discussion of weathering, we have described an array of processes that act to alter rock chemically and break up rock into fragments. These rock fragments are subjected to gravity, running water, waves, wind, and the flow of glacial ice in landform-making processes. In the remainder of this chapter, we consider the first of these landform agents—gravity. We will return to the others in the following chapters.

Everywhere on the Earth's surface, gravity pulls continuously downward on all materials. Bedrock is usually so strong and well supported that it remains fixed in place. However, when a mountain slope becomes too steep, bedrock masses can break free and fall or slide to new positions of rest. In cases where huge masses of bedrock are involved, the result can be catastrophic to towns and villages

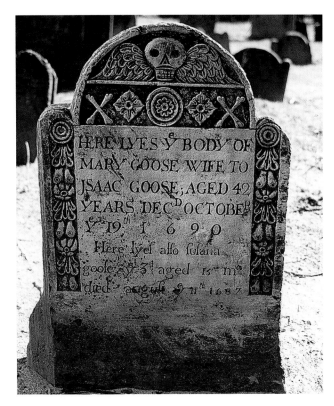

14.11 Chemical weathering of tombstones *These weathered tombstones are from a burying ground in Boston, Massachusetts. The marker on the left, carved in marble, has been strongly weathered, weakening the lettering. The marker on the right, made of slate, is much more resistant to erosion.*

14.12 Eye on the Landscape Solution weathering of basalt *The steep walls of many narrow coastal ravines are deeply grooved by chemical weathering of the basaltic lava on the Napali Coast, Kauai, Hawaii.* **What else would the geographer see? ... Answers at the end of the chapter.**

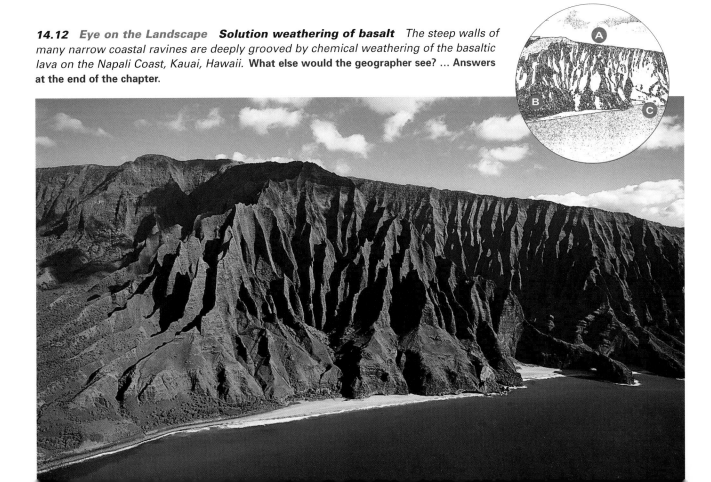

Kinds of earth materials:	Rock (dry)	Regolith, soil, alluvium, clays + water	Water + sediment
Physical properties:	Hard, brittle, solid	Plastic substance	Fluid
Kinds of motion:	Falling, rolling, sliding	Flowage within the mass	Fluid flow

Forms of mass movement: Very slow → Very fast

ROCK CREEP TALUS CREEP | SOIL CREEP
SOLIFLUCTION
LANDSLIDES: | EARTHFLOW (slump or flowage)
BEDROCK SLUMP | ROCKSLIDE
MUDFLOW
STREAM FLOW
ROCKFALL | ALPINE DEBRIS AVALANCHE | DEBRIS FLOOD

14.13   *Processes and forms of mass wasting*

in the path of the slide. Such slides are a major environmental hazard in mountainous regions. They are but one form of *mass wasting,* which we defined earlier as spontaneous downhill movement of soil, regolith, and rock under the influence of gravity.

Because soil, regolith, and many forms of sediment are held together poorly, they are much more susceptible to movement under the force of gravity than hard, massive bedrock. On most slopes, at least a small amount of downhill movement is going on constantly. Although much of this motion is imperceptible, the regolith sometimes slides or flows rapidly.

> Mass wasting is spontaneous movement of soil, regolith, and rock under the influence of gravity. There are many forms of mass wasting, depending on the speed of the motion and the amount of water involved.

The processes of mass wasting and the landforms they produce are extremely varied and tend to grade one into another. We have selected only a few of the most important forms of mass wasting to present in this chapter—they are shown in Figure 14.13. At the top of the diagram, three categories of information are provided, including the Earth material involved, the properties of the material, and the kind of motion that occurs. Boxes in the body of the diagram contain the names of several forms of mass movements, arranged according to composition (left to right) and speed (top to bottom). As we describe each of these forms, this diagram will help you understand how it compares to the others.

Slopes

Mass wasting occurs on slopes. As used in physical geography, the term *slope* designates a small strip or patch of the land surface that is inclined from the horizontal. Thus, we speak of "mountain slopes," "hill slopes," or "valley-side slopes" to describe some of the inclined ground surfaces that we might encounter in traveling across a landscape. Slopes guide the downhill flow of surface water and fit together to form drainage systems within which surface-water flow converges into stream channels (Chapter 16). Nearly all natural surfaces slope to some degree. Very few are perfectly horizontal or vertical.

Most slopes are mantled with regolith, which grades downward into solid, unaltered rock, known simply as **bedrock.** Regolith provides the source for **sediment,** which consists of rock and mineral particles that are transported and deposited in a fluid medium. This fluid may be water, air, or even glacial ice. Both regolith and sediment comprise parent materials for the formation of *soil,* as we saw in Chapter 10.

Figure 14.14 shows a typical hill slope that forms one wall of the valley of a small stream. Soil and regolith blanket the bedrock, except in a few places where the bedrock is particularly hard and projects in the form of *out-*

14.14   *Soil, regolith, and outcrops on a hillslope*
Colluvium accumulates at the foot of the slope, while alluvium, or transported regolith, lies in the floor of an adjacent stream valley. (Drawn by A. N. Strahler.)

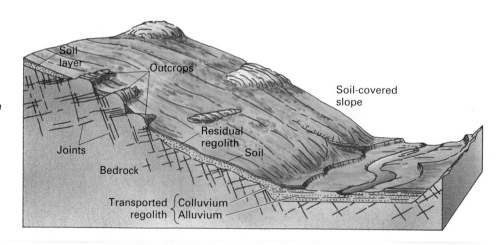

Soil layer · Outcrops · Soil-covered slope · Residual regolith · Soil · Joints · Bedrock · Transported regolith { Colluvium / Alluvium }

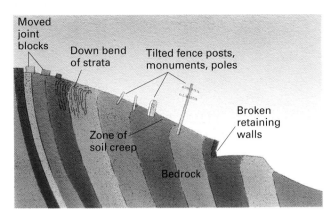

14.15 Indicators of soil creep *The slow, downhill creep of soil and regolith shows up in many ways on a hillside. (After C. F. S. Sharpe.)*

crops. Residual regolith is derived directly from the rock beneath and moves very slowly down the slope toward the stream. Accumulations of regolith at the foot of a slope are called *colluvium.* Beneath the valley bottom are layers of transported regolith, called **alluvium,** which is sediment transported and deposited by the stream. This sediment had its source in regolith prepared on hill slopes many kilometers or miles upstream. All accumulations of sediment on the land surface, whether deposited by streams, waves and currents, wind, or glacial ice, can be designated *transported regolith,* in contrast to residual regolith.

The thickness of soil and regolith is quite variable. Although the soil is rarely more than 1 or 2 m thick (about 3 to 6 ft), residual regolith on decayed and fragmented rock may extend down to depths of 5 to 100 m (about 15 to 300 ft), or more at some locations. On the other hand, soil or regolith, or both, may be missing. In some places, everything is stripped off down to the bedrock, which then appears at the surface as an outcrop. In other places, following cultivation or forest fires, the fertile soil is partly or entirely eroded away, and severe erosion exposes the regolith.

Soil Creep

On almost any soil-covered slope, there is extremely slow downhill movement of soil and regolith, a process called **soil creep.** Figure 14.15 shows some of the signs indicating soil creep. Joint blocks of distinctive rock types are found moved far downslope from the outcrop. In some layered rocks such as shales or slates, edges of the strata seem to "bend" in the downhill direction (Figure 14.16). This is not true plastic bending but is the result of downhill creep of many rock pieces on small joint cracks. Creep causes fence posts and utility poles to lean downslope and even shift measurably out of line. Roadside retaining walls can buckle and break under the pressure of soil creep.

Soil creep is caused by disturbance of the soil under the influence of gravity. Alternate drying and wetting of the soil, growth of ice needles and lenses, heating and cooling of the soil, trampling and burrowing by animals, and shaking by earthquakes all produce some disturbance of the soil and regolith. Because gravity exerts a downhill pull on every such rearrangement, the particles very gradually work their way downslope.

> Slopes are mantled with residual regolith, which accumulates at the foot of slopes as colluvium. Regolith that is transported by moving water is termed *alluvium.*

14.16 Soil creep *Slow, downhill creep of regolith on this mountainside near Downieville, California, has caused vertical rock layers to seem to bend rightward.*

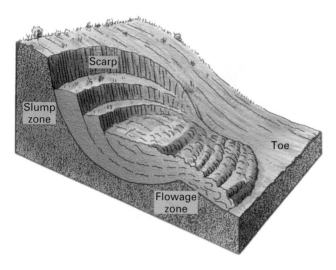

14.17 Sketch of an earthflow *An earthflow with slump terraces well developed in the upper part. Flowage has produced a bulging toe. (After A. N. Strahler.)*

Earthflow

In regions of humid climate, a mass of water-saturated soil, regolith, or weak shale may move down a steep slope during a period of a few hours in the form of an **earthflow,** sketched in Figure 14.17. At the top, the material slumps away, leaving a curved, wall-like scarp. Often the soil farther from the scarp moves more rapidly, leaving a series of steps as the soil mass slips downward. At the bottom, the sodden soil flows in a sluggish mass that piles up in ridges or lobes to form a bulging toe.

Shallow earthflows, affecting only the soil and regolith, are common on sod-covered and forested slopes that have been saturated by heavy rains. An earthflow may affect a few square meters, or it may cover an area of a few hectares (several acres). If the bedrock of a mountainous region is rich in clay (derived from shale or deeply weathered volcanic rocks), earthflows sometimes involve millions of metric tons of bedrock moving by plastic flowage like a great mass of thick mud.

Earthflows are a common cause of blockage of highways and railroad lines, usually during periods of heavy rains. Generally, the rate of flowage is slow, so that the flows are not a threat to life. However, property damage to buildings, pavements, and utility lines is often severe where construction has taken place on unstable soil slopes.

GEODISCOVERIES

Environmental Impact of Earthflows

One special form of earthflow has proved to be a major environmental hazard in parts of Norway and Sweden and along the St. Lawrence River and its tributaries in Quebec Province of Canada. This type of flowage involves horizon-

> *An earthflow is a mass of water-saturated soil that moves slowly downhill. Some earthflows occur when quick clays liquify after a mechanical shock.*

tally layered clays, sands, and silts accumulated during the Ice Age that form low, flat-topped terraces adjacent to rivers and lakes. Over a large area, which may be 600 to 900 m (about 2000 to 3000 ft) across, a layer of silt and sand 6 to 12 m (about 20 to 40 ft) thick begins to move toward the river, sliding on a layer of soft clay that has spontaneously turned into a near-liquid state. The moving mass also settles downward and breaks into step-like masses. Carrying along houses or farms, the layer ultimately reaches the river, into which it pours as a great disordered mass of mud. Figure 14.18 provides a block diagram of a classic example that occurred in 1898 near St. Thuribe, Quebec.

This type of earthflow is caused by *quick clays*—clays that spontaneously change from a solid condition to a near-liquid condition when subjected to a shock or disturbance. Quick clays are thought to have formed in the shallow waters of saltwater bays near the end of the Ice Age. When deposited, the thin plates of clay in these layers have a "house of cards" structure with a large proportion of water-filled void space between plates. The saltwater provides positively and negatively charged atoms that help to bind the structure and give it strength. But with the regional uplift that often occurs after ice sheets melt away, the quick clay is elevated above sea level, and fresh ground water replaces the saltwater. A mechanical shock, such as an Earth tremor, then causes the house of cards structure to collapse. Because such a large proportion of water (from 45 to 80 percent by volume) is present, the clay–water mixture behaves as a liquid, with almost no strength remaining.

A particularly spectacular example of this type of earthflow occurred in Nicolet, Quebec, in 1955. A clay layer beneath the town liquefied, carrying much of the town into the Nicolet River (Figure 14.19). Fortunately, only three lives were lost, but the damage to buildings and a bridge ran into millions of dollars.

14.18 Earthflow at St. Thuribe *This block diagram depicts the earthflow of 1898 near St. Thuribe, Quebec. (After C. F. S. Sharpe, Landslides and Related Phenomena, Columbia Univ. Press, New York.)*

14.19 The Nicolet earthflow *Much of Nicolet, Quebec, flowed into the Nicolet River when an underlying layer of quick clay liquefied, causing an earthflow.*

Mudflow and Debris Flood

One of the most spectacular forms of mass wasting and a potentially serious environmental hazard is the **mudflow.** This mud stream of fluid consistency pours swiftly down canyons in mountainous regions (Figure 14.20). In deserts, where vegetation does not protect the mountain soils, local thunderstorms produce rain much faster than it can be absorbed by the soil. As the water runs down the slopes, it forms a thin mud that flows down to the canyon floors and then follows the stream courses. As it flows, it picks up additional sediment, becoming thicker and thicker until it is too thick to flow further. Great boulders are carried along, buoyed up in the mud. Roads, bridges, and houses in the canyon floor are engulfed and destroyed. Where the mudflow emerges from a canyon and spreads across an alluvial fan, severe property damage and even loss of life may be the result (Figure 14.21).

Note that mudflows are rapid events in which water, sediment, and debris follow slopes and river valleys to lower elevations. In contrast, earthflows are slower and are confined to collapsing slopes. This difference is shown by the relative positions of these two types of mass wasting in Figure 14.13.

As explained in Chapter 13, mudflows that occur on the slopes of erupting volcanoes are called *lahars.* Freshly fallen volcanic ash and dust are turned into mud by heavy rains or melting snows and flow down the slopes of the volcano. Herculaneum, a city at the base of Mount Vesuvius, was destroyed by a mudflow during the eruption of A.D. 79. At the same time, the neighboring city of Pompeii was buried under volcanic ash.

Mudflows are rapid events in which water, sediment, and debris cascade down slopes and valleys to lower elevations. They are produced by very heavy rainfall or snowmelt caused by volcanic activity.

Mudflows vary in consistency, from a mixture like concrete emerging from a mixing truck to consistencies similar to that of turbid river floodwaters. The watery type of mudflow is called a *debris flood* or *debris flow* in the western United States, and particularly in southern California, where it occurs commonly and with disastrous effects. The material carried in a debris flood ranges from fine particles to boulders to tree trunks and limbs. In mountainous regions on steep slopes, these flows are termed *alpine debris ava-*

14.20 Mudflows in an arid environment *Thin, stream-like mudflows issue occasionally from canyon mouths in arid regions. The mud spreads out on the fan slopes below in long, narrow tongues. (Drawn by A. N. Strahler.)*

14.21 Mudflow deposit
This mudflow, carrying numerous large boulders, issued from a steep mountain canyon in the Wasatch Mountains, Utah.

lanches. The intense rainfall of hurricanes striking the eastern United States often causes debris avalanches in the hollows and valleys of the Blue Ridge and Smoky Mountains.

Landslide

A **landslide** is the rapid sliding of large masses of bedrock or regolith. Wherever mountain slopes are steep, there is a possibility of large, disastrous landslides (Figure 14.22). In Switzerland, Norway, or the Canadian Rockies, for example, villages built on the floors of steep-sided valleys have been destroyed by the sliding of millions of cubic meters of rock set loose without warning.

Landslides are triggered by earthquakes or sudden rock failures rather than by heavy rains or sheet floods. Thus, they are different from earthflows and mudflows, which are typically induced by heavy rains. Landslides can also result when the base of a slope is oversteepened by excavation or river erosion. As shown in Figure 14.13, landslides can range from *rockslides* of jumbled bedrock fragments to *bedrock slumps* in which most of the bedrock remains more or less intact as it moves. The amazing speed with which rockslide rubble travels down a mountainside is thought to be due to the presence of a layer of compressed air trapped between the slide and the ground surface. The air layer reduces frictional resistance and may keep the rubble from making contact with the surface beneath.

Severe earthquakes in mountainous regions are a major cause of landslides. An example is the landslide of Santa Tecla, El Salvador, that killed hundreds of people on Janu-

> In a landslide, a large mass of rock suddenly moves from a steep mountain slope to the valley below. Landslides are triggered by earthquakes or rock failures rather than heavy rains.

ary 13, 2001 (Figure 14.23). The earthquake was centered off the southern coast of El Salvador and measured 7.6 on the Richter scale. It was felt as far away as Mexico City. In the shaking, a steep slope above the neighborhood of Las Colinas collapsed, creating a wave of earth that swept across the ordered grid of houses and streets, burying hundreds of homes and their inhabitants. The same earthquake also triggered other landslides in the region, with additional loss of life and property. *Eye on the Environment 14.1 • The Great Hebgen Lake Disaster* presents the history of the Madison Slide, another earthquake-triggered landslide that dammed a river to create a new, natural lake. *GEODISCOVERIES*

Aside from occasional local catastrophes, landslides have rather limited environmental influence because they occur only sporadically and usually in thinly populated mountainous regions. Small slides can, however, repeatedly block or break important mountain highways or railway lines. *GEODISCOVERIES*

INDUCED MASS WASTING

Human activities can induce mass wasting in forms ranging from mudflow and earthflow to landslide. These activities include (1) piling up of waste soil and rock into unstable accumulations that can move spontaneously, and (2) removing the underlying support of natural masses of soil, regolith, and bedrock. We can refer to mass movements produced by human activities as *induced mass wasting*.

14.22 The Turtle Mountain slide *A classic example of an enormous, disastrous landslide is the Turtle Mountain slide, which took place near Frank, Alberta. A huge mass of limestone slid from the face of Turtle Mountain between South and North peaks, descended to the valley, then continued up the low slope of the opposite valley side until it came to rest as a great sheet of bouldery rock debris. (Data from Canadian Geological Survey, Department of Mines.) (Drawn by A. N. Strahler. Copyright © A. N. Strahler.)*

In Los Angeles County, California, real estate development has been carried out on very steep hillsides and mountainsides. Roads and home sites have been bulldozed out of the deep regolith. The excavated regolith is piled up to form nearby embankments. When saturated by heavy winter rains, these embankments can give way. This produces earthflows, as well as mudflows and debris floods that travel far down the canyon floors and spread out on the alluvial fan surfaces below, burying streets and houses in mud and boulders.

Many debris floods of this area are also produced by heavy rains falling on mountain slopes denuded of vegetation by fire in the preceding dry season. Some of these fires are set carelessly or deliberately by humans.

Human activities can induce mass wasting by piling up unstable materials or undercutting natural supports for slopes or rock masses. Cutting and filling to build highways and runways and extract mineral resources is termed scarification.

14.24). These movements occur in shales that tend to become plastic when water is added. The upper part of the earthflow subsides and slumps backward, as illustrated in Figure 14.17. The interior and lower parts of the mass move forward by slow flowage, producing a toe of flowage material.

The largest of the earthflows in the Palos Verdes Hill area was the Portuguese Bend "landslide" that affected an area of about 160 hectares (about 400 acres). It was caused by the slippage of sedimentary rock layers on an underlying layer of clay. The total movement over a three-year period was about 20 m (about 70 ft). Damage to residential and other structures totaled some $10 million. Geologists attributed the earthflow to infiltration of water from septic tanks and from irrigation water applied to lawns and gardens. A discharge of over 115,000 liters (about 30,000 gallons) of water per day from some 150 homes is believed to have sufficiently weakened the clay layer to start and sustain the flowage.

Induced Earthflows

Examples of both large and small earthflows induced or aggravated by human activities are found in the Palos Verdes Hills of Los Angeles County, California (Figure

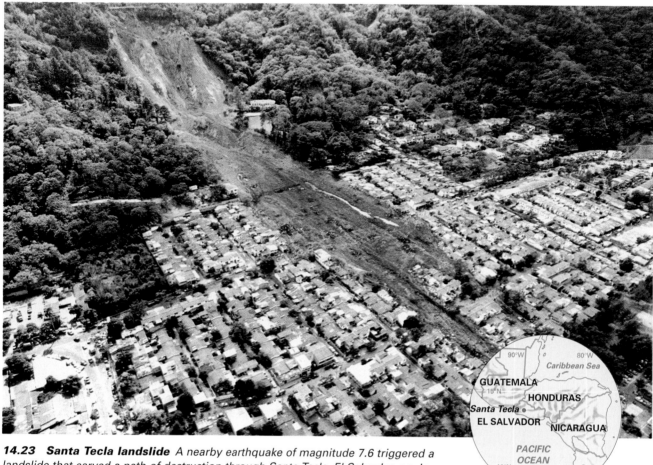

14.23 Santa Tecla landslide *A nearby earthquake of magnitude 7.6 triggered a landslide that carved a path of destruction through Santa Tecla, El Salvador, on January 13, 2001.*

Scarification of the Land

Industrial societies now possess enormous machine power and explosives capable of moving great masses of regolith and bedrock from one place to another. This technology is used to extract mineral resources. Another use is to move earth in the construction of highway grades, airfields, building foundations, dams, canals, and various other large structures. Both activities involve removal of Earth materials, a process that destroys the preexisting ecosystems and habitats of plants and animals. When the materials are used to build up new land on adjacent surfaces, ecosystems and habitats are also destroyed—by burial. What distinguishes artificial forms of mass wasting from the natural forms is that machinery is used to raise Earth materials against the force of gravity. Explosives used in blasting can produce disruptive forces many times more powerful than the natural forces of physical weathering.

Scarification is a general term for excavations and other land disturbances produced to extract mineral resources. The rock waste, known as *spoil* or *tailings,* accumulates at the site. Scarification takes the form of open-pit mines, strip mines, quarries for structural materials, sand and gravel pits, clay pits, phosphate pits, scars from hydraulic mining, and stream gravel deposits reworked by dredging.

Strip mining is a particularly destructive scarification activity. Strip mining of coal is practiced where coal seams lie close to the surface or actually appear as outcroppings along hillsides (Figure 14.25). Earth-moving equipment removes the covering strata, or *overburden,* to bare the coal, which is lifted out by power shovels. The piled-up overburden remains as spoil. When saturated by heavy rains and melting snows, the spoil can generate earthflows and mudflows that descend on houses, roads, and forest. The spoil also supplies sediment that clogs stream channels far down the valleys.

As a result of these problems, strip mining is under strict control in most locations. Mine operators are not permitted to create hazardous spoil slopes, and they must restore spoil banks and ridges to a natural condition. In this process, termed *reclamation,* topsoil is removed and saved as the mine is opened and worked. When spoil is returned to the mine cavity after extraction of the coal, it is covered with the topsoil and suitable plants are introduced. With good management and proper planning, the resulting reclaimed land becomes usable again. But there are also many reclamation failures in which the land has been damaged beyond repair.

Scarification is on the increase. Driven by an ever increasing human population, the demand for coal and

The Great Hebgen Lake Disaster

For some 200 vacationers camping near the Madison River in a deep canyon just downstream from Hebgen Lake, not far west of Yellowstone National Park, the night of August 17, 1959, began quietly, with almost everyone safely bedded down in their tents or camping trailers. Up to a certain point, it was everything a great vacation should be—that point in time was 11:37 P.M., Mountain Standard Time. At precisely that instant, not one but four terrifying forms of disaster were set loose on the sleeping vacationers—earthquake, landslide, hurricane-force wind, and raging flood. The earthquake, which measured 7.1 on the Richter scale, was the cause of it all. The first shocks, lasting several minutes, rocked the campers violently in their trailers and tents. Those who struggled to go outside could scarcely stand up, let alone run for safety.

Then came the landslide. A dentist and his wife watched through the window of their trailer as a mountain seemed to move across the canyon in front of them, trees flying from its surface like toothpicks in a gale. Then, as rocks began to bang against the sides and top of their trailer, they got out and raced for safer ground. Later, they found that the slide had stopped only about 25 m (80 ft) from the trailer. Pushed by the moving mountain was a vicious blast of wind. It swept upriver, tumbling trailers end over end.

Then came the flood. Two women schoolteachers, sleeping in their car only about 5 m (15 ft) from the river bank, awoke to the violent shaking of the earthquake. Like other campers, they first thought they had a marauding bear on their hands. Puzzled and frightened, they started the engine and headed the car for higher ground. As they did so, they were greeted by a great roar coming from the mountainside above and behind them. An instant later, the car was completely engulfed by a wall of water that surged up the river bank, then quickly drained back. With the

Madison slide *Seen from the air, the Madison slide forms a great dam of rubble across the Madison River Canyon. The slide was triggered by a nearby earthquake.*

screams of drowning campers in their ears, the two women managed to drive the car to safe ground, high above the river.

After the first surge of water, generated as the landslide mass hit the river, the river began a rapid rise. This rise was aided by great surges of water topping the Hebgen Dam, located upstream, as earthquake aftershocks rocked the water of Hebgen Lake back and forth along its length of about 50 km (30 mi). In the darkness of night, the terrified victims of the flood had no idea what was happening. In the panic that ensued, a 71-year-old man performed an almost unbelievable act of heroism to save his wife and himself from drowning. As the water rose inside their house trailer, he forced open the door, pulled his wife with him to the trailer roof, then carried her up on the branches of a nearby pine tree. Here they were finally able to reach safety. The water had risen 10 m (about 30 ft)

above ground level in just minutes.

The Madison Slide, as the huge earth movement was later named, had a bulk of 28 million m^3 (37 million yd^3) of rock (see photo). It consisted of a chunk of the south wall of the canyon, measuring over 600 m (about 2000 ft) in length and 300 m (about 1000 ft) in thickness. The mass descended over a half of a kilometer (about a third of a mile) to the Madison River, its speed estimated at 160 km (100 mi) per hour. Pulverized into bouldery debris, the slide crossed the canyon floor, its momentum carrying it over 120 m (about 400 ft) in vertical distance up the opposite canyon wall. At least 26 persons died beneath the slide, and their bodies have never been recovered. Acting as a huge dam, the slide caused the Madison River to back up, forming a new lake. In three weeks' time, the lake was nearly 100 m (330 ft) deep. Today it is a permanent feature, named Earthquake Lake.

14.24 Induced earthflow *This aerial view shows houses on Point Fermun, in Palos Verdes, California, disintegrating as they slide downward toward the sea.*

14.25 Contour strip mining *This strip mine near Lynch, Kentucky, follows an outcrop of coal along a hillside contour. A highway makes use of the winding bench at the base of the high rock wall.*

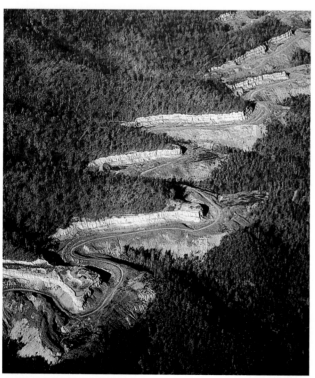

industrial minerals used in manufacturing and construction is on the rise. At the same time, as the richer and more readily available mineral deposits are consumed, industry turns to poorer and less easily accessible grades of ore. As a result, the rate of scarification is further increased, making the success of reclamation efforts increasingly important.

PROCESSES AND LANDFORMS OF ARCTIC AND ALPINE TUNDRA

The landscape of the treeless arctic and alpine tundra environment shows many distinctive effects of weathering and mass wasting under a regime of severe winter cold. In the tundra, soil water is solidly frozen throughout a winter of many months duration. During the short summer season, however, surface thawing occurs, leaving the soil saturated and vulnerable to mass wasting and water erosion. With the return of cold temperatures, the freezing of soil water exerts a strong mechanical influence on the surface layer, creating a number of distinctive landforms.

The tundra environment can be thought of as a unique natural system. Today it is often found in close proximity to glacial ice—near tongue-like alpine glaciers and along the fringes of ice sheets. This unique location has long been described by the adjective **periglacial** (*peri,* meaning "near"). Here, intense frost action creates a distinctive set of landforms and land-forming processes that we may term the *periglacial system.* This system is driven by a very large annual range in temperature, which creates a strong annual cycle in the flow of energy into and out of the surface layer of the ground. These cycles provide the mechanism for the movement of soil particles individually and collectively by changing water from liquid to solid and back to liquid again, thus creating the distinctive processes that operate in the periglacial environment. We preface our discussion of periglacial processes with a look at permafrost.

Permafrost includes ground, water, and bedrock that is permanently below the freezing temperature of water. The active layer of permafrost is the shallow surface layer that freezes and thaws each year.

Arctic Permafrost

Ground and bedrock that are perennially (year-round) below the freezing point of fresh water (0°C; 32°F) are called **permafrost.** By "ground" we mean to include mineral matter falling within the full range of particle sizes described in Chapter 11—that is, clay, silt, sand, pebbles, and boulders. The term *permafrost* strictly refers only to ground temperature. Thus, solid bedrock perennially below the freezing point is also included in permafrost. Water is commonly present in pore spaces in the ground, and in the frozen state it is known as **ground ice.**

Permafrost prevails over most of the tundra climate region and in a wide area of boreal forest climate bordering the tundra. A distinctive feature of permafrost terrains is a shallow surface layer subject to seasonal thawing called the *active layer.* It ranges in thickness from about 15 cm (6 in.) to about 4 m (13 ft), depending on the latitude and nature of the ground. Below the active layer is the upper surface of the perennially frozen zone, called the *permafrost table.*

The distribution of permafrost in the northern hemisphere is shown in Figure 14.26. Four zones are shown on the map. *Continuous permafrost,* which extends without gaps or interruptions under all surface features, coincides largely with the tundra climate, but it also includes a large part of the boreal forest climate in Siberia. *Discontinuous permafrost,* which occurs in patches separated by frost-free zones under lakes and rivers, occupies much of the boreal forest climate zone of North America and Eurasia. A third zone—*sub-sea permafrost*—lies beneath sea level in a shallow offshore zone. On our map you will find it beneath the Arctic sea offshore of the Asian coast and the coasts of Alaska, the Yukon, and the Northwest Territories in North America. Last is *alpine permafrost,* which is found at high elevations where temperatures are always below freezing.

Cross sections of the permafrost zone in Alaska and Asia are shown in Figure 14.27. In this diagram, the active layer includes the layer of annual freezing and thawing over all terrain, not just permafrost. Figure 14.27 also

shows *sporadic permafrost,* a zone to the south of discontinuous permafrost in which occasional patches of permafrost occur.

Permafrost reaches to a depth of 300 to 450 m (about 1000 to 1500 ft) in the continuous zone near lat. 70° N (Figure 14.27). Much of this permanently frozen zone is an inheritance from the more severe conditions of the last Ice Age, but some permafrost bodies may be growing under existing climate conditions.

Even in the areas of deepest continuous permafrost, there exist isolated pockets that never fall below the freezing point. The Siberian word *talik* has been applied to these pockets. One kind of talik lies beneath a deep lake that freezes over in winter and thaws in summer. The talik may extend to a depth of 100 m in the continuous permafrost zone. The heat necessary for a talik to be formed and maintained may come from circulating surface water that penetrates to these depths.

14.26 Permafrost map *Distribution of permafrost in the northern hemisphere (Adapted from Troy L. Péwé, Geotimes, vol. 29, no. 2, p. 11, copyright, 1984, by the American Geologic Institute. Used by permission.)*

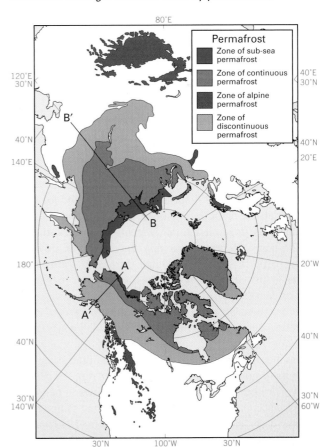

14.27 Permafrost map cross sections *These north-south cross sections of permafrost in Alaska and Asia are located on Figure 14.26. (From Robert F. Black, "Permafrost," Chapter 14 of P. D. Trask's Applied Sedimentation. Copyright © by John Wiley & Sons. Reprinted by permission of John Wiley & Sons, Inc.)*

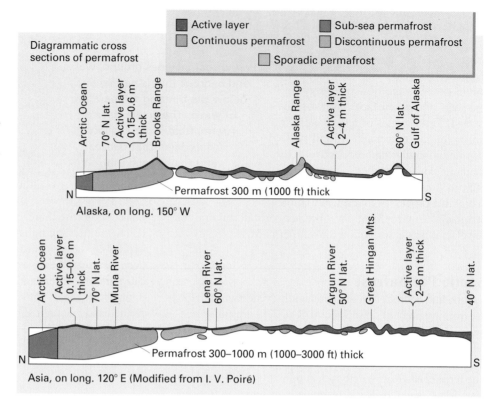

Diagrammatic cross sections of permafrost

Active layer
Continuous permafrost
Sub-sea permafrost
Discontinuous permafrost
Sporadic permafrost

Arctic Ocean — 70° N lat. — Active layer 0.15–0.6 m thick — Brooks Range — Alaska Range — Active layer 2–4 m thick — 60° N lat. — Gulf of Alaska

Permafrost 300 m (1000 ft) thick

Alaska, on long. 150° W

Arctic Ocean — Active layer 0.15–0.6 m thick — 70° N lat. — Muna River — Lena River 60° N lat. — Argun River 50° N lat. — Great Hingan Mts. — Active layer 2–6 m thick — 40° N lat.

Permafrost 300–1000 m (1000–3000 ft) thick

Asia, on long. 120° E (Modified from I. V. Poiré)

Forms of Ground Ice

The amount of ground ice present below the permafrost table varies greatly from place to place. Near the surface, it can take the form of a body of almost 100 percent ice. At increasing depth, the percentage of ice becomes lower and approaches zero at great depth.

A common form of ground ice is the **ice wedge.** In silty alluvium, such as that formed on river floodplains and delta plains in the arctic environment, ice accumulates in vertical wedge-forms in deep cracks in the sediment (Figure 14.28). Ice wedges are thought to originate in cracks that form when permafrost shrinks during extreme winter cold. As shown in Figure 14.29, surface melt-water enters the crack during the spring and becomes frozen in the fall, widening the crack. Repeated cracking and filling with new ice cause the ice wedge to thicken until it becomes as wide as 3 m (about 10 ft) and as deep as 30 m (about 50 ft). Ice wedges are typically interconnected into a system of polygons, called *ice-wedge polygons* (Figure 14.30).

Another remarkable ice-formed feature of the arctic tundra is a conspicuous conical mound, called a *pingo* (Figure 14.31). The pingo has a core of ice and grows in height as more ice accumulates, forcing up the overlying sediment. In extreme cases, pingos reach a height of 50 m (164 ft) and a basal diameter as great as 600 m (about 2000 ft).

Annual cycles of freeze and thaw build ice wedges on arctic river plains and deltas. On slopes, freeze–thaw cycles in the active layer move coarser particles into ring-like arrangements called stone polygons.

Patterned Ground and Solifluction

In areas of coarse-textured regolith consisting of rock particles in a wide range of sizes, the annual cycle of thawing and freezing causes the coarsest fragments—pebbles and cobbles—to move horizontally as well as vertically and thus become sorted out from the finer particles. This type of sorting produces ring-like arrangements of coarse fragments. The linking of adjacent rings produces a net-like pattern to form a system of *stone polygons* (also called "stone rings" or "stone nets") (Figure 14.32). On steep slopes, the process of soil creep causes the polygons to be elongated in the downslope direction and to become transformed into parallel stone stripes (Figure 14.33). Both stone polygons and ice-wedge polygons are examples of a class of features called **patterned ground.**

A special variety of earthflow characteristic of arctic permafrost tundra regions is **solifluction** (from Latin words meaning "soil" and "to flow"). It is active in early summer, when thawing has penetrated the upper few tenths of meters of the surface. At this time, the soil is fully saturated with water that cannot escape downward because of the permafrost below. Moving almost imperceptibly, this saturated soil is deformed into *solifluction terraces* and *solifluction lobes* that give the tundra slope a stepped appearance (Figure 14.34).

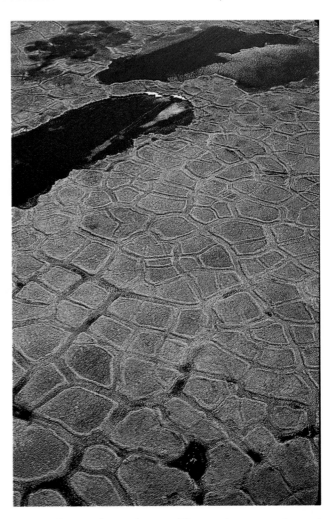

14.28 Ice wedge *Exposed by river erosion, this great wedge of solid ice fills a vertical crack in organic-rich floodplain silt along the Yukon River, near Galena in western Alaska.*

14.30 Ice-wedge polygons *These polygons are formed by the growth of ice wedges. Alaskan north slope, near the border of Alaska and Yukon Territory, Canada.*

14.29 Formation of an ice wedge *(a) In the beginning stage, an open crack appears during the winter and is filled with meltwater, which freezes into ice during the following autumn. (b) After several hundred winters, the ice wedge has grown, and continues to grow as the same seasonal sequence is repeated. (Adapted by permission from A. H. Lachenbruch in Rhodes W. Fairbridge, Ed., The Encyclopedia of Geomorphology, New York, Reinhold Publishing Corp.)*

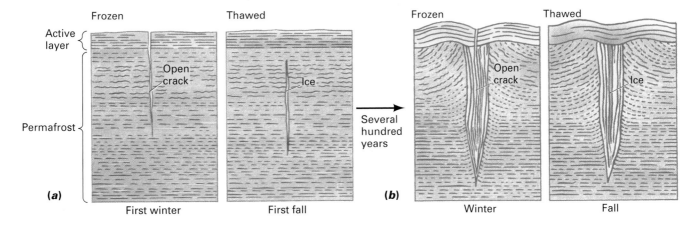

14.31 *Pingo* This large pingo, formed by the growth of ice beneath its surface, rises above the flat plain of the McKenzie River delta, N.W.T., Canada.

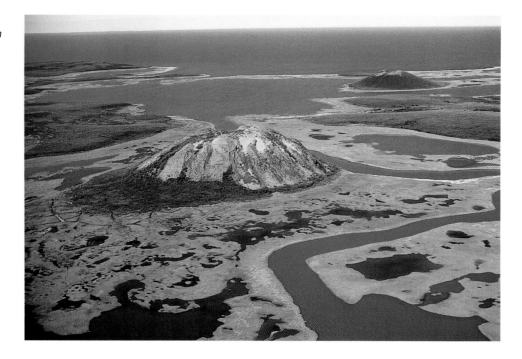

14.32 *Stone polygons* Sorted circles of gravel form a network of stone rings on this nearly flat land surface where water drainage is poor because of underlying permafrost. The circles in the foreground are 3 to 4 m (10 to 13 ft) across; the gravel ridges are 20 to 30 cm (8 to 12 in.) high. Broggerhalvoya, western Spitsbergen, latitude 78° N.

14.33 Stone stripes *This diagram shows how stone rings are often drawn out downslope into stone stripes. (Adapted from C. F. S. Sharpe, Landslides and Related Phenomena, p. 37, Figure 5. New York, Columbia Univ. Press. Used by permission of the publisher.)*

Alpine Tundra

Most of the periglacial processes and forms of the low arctic tundra are also found in the high alpine tundra throughout high mountains of the middle and high latitudes. Major regions of *alpine permafrost* in the northern hemisphere are shown on our map, Figure 14.26. In the alpine tundra, there is a dominance of steep mountainsides with large exposures of hard bedrock that has been strongly abraded by glacial ice. Postglacial talus slopes and cones formed of large angular blocks are conspicuously developed (Figure 14.4). Patterned ground and solifluction terraces occupy relatively small valley floors where slopes are low and finer sediment tends to accumulate.

Alpine permafrost is restricted to elevations at which the mean annual temperature is below freezing. The general elevation at which mean annual temperatures fall to

14.34 Solifluction *Solifluction has created this landscape of soil mounds in the Richardson Mountains, Northwest Territories, Canada. Bulging masses of water-saturated regolith have slowly moved downslope, overriding the permafrost below, while carrying their covers of plants and soil.*

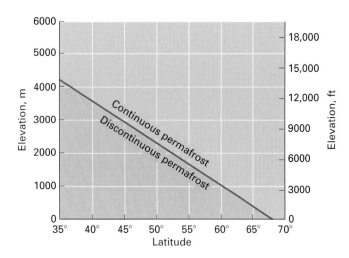

14.35   Boundary between continuous and discontinuous permafrost *This graph shows how the elevation of the boundary between continuous and discontinuous alpine permafrost changes with latitude.*

0°C (32°F) will, of course, be related to latitude. In subtropical latitudes, elevations of 4000 m (about 13,000 ft) or greater are required, while at latitudes above 70° permafrost can be encountered at sea level. Figure 14.35 is a schematic graph showing the boundary between continuous and discontinuous permafrost as it varies with elevation and latitude. While the boundary at lat. 40° lies at about 3500 m (about 11,500 ft), it has fallen to 2200 m (about 6600 ft) at lat. 50°, and by lat. 68° it has descended to sea level.

Environmental Problems of Permafrost

Environmental degradation of permafrost regions arises from surface changes produced by human activity. One such activity is the destruction or removal of an insulating surface cover that typically consists of a layer of decaying organic matter in combination with living plants of the tundra or arctic forest. When this layer is scraped off or burned away, the summer thaw is extended to a greater depth, with the result that ice wedges and other ice bodies melt in the summer and waste downward. Meltwater mixes with silt and clay to form mud, which is then eroded and transported by water streams, leaving trench-like morasses. This total activity is called *thermal erosion.*

Over large expanses of nearly flat arctic tundra, removal of the natural surface cover is followed by widespread subsidence of the ground and the formation of depressions. Water erosion tends to enlarge these depressions, and thus numerous new lakes come into existence. The resulting terrain is described as *thermokarst.* (The word "karst," to be explained in Chapter 15, means a hummocky terrain developed by uneven solution removal of limestone bedrock.)

The consequences of disturbance of permafrost terrain became evident in World War II, when military bases, airfields, and highways were constructed hurriedly without regard for the maintenance of the natural protective surface insulation. In extreme cases, scraped areas turned into mud-filled depressions and even into small lakes that expanded in area with successive seasons of thaw, engulfing nearby buildings. Engineering practices now call for placing buildings on piles with an insulating air space below or for the deposition of a thick insulating pad of coarse gravel over the surface prior to construction. Steam and hot-water lines are placed above ground to prevent thaw of the permafrost layer.

Another serious engineering problem of arctic regions is the behavior of streams in winter. As the surfaces of streams and springs freeze over, the water beneath bursts out from place to place, freezing into huge accumulations of ice. If this phenomenon occurs at a highway bridge or culvert, the roadway may become impassable.

The lessons of superimposing our technology on a highly sensitive natural environment were learned the hard way—by encountering undesirable and costly effects that were not anticipated. Continued economic development of arctic regions is likely to teach additional hard lessons.

A LOOK AHEAD   In this chapter, we have examined two related topics—weathering and mass wasting. In the weathering process, rock near the surface is broken up into smaller fragments and often altered in chemical composition. In the mass wasting process, weathered rock and soil move downhill in slow to sudden mass movements. The landforms of mass wasting are produced by gravity acting directly on soil and regolith. Gravity also powers another landform-producing agent—running water—which we take up in the next three chapters. The first deals with water in the hydrologic cycle, in soil, and in streams. The second deals specifically with how streams and rivers erode regolith and deposit sediment to create landforms. The third describes how stream erosion strips away rock layers of different resistance, providing large landforms that reveal underlying rock structures.

Clearing of natural surface layers over permafrost can induce rapid thawing, leading to erosion, settling, and the growth of shallow lakes.

In Review | Weathering and Mass Wasting

- **Weathering** is the action of processes that cause rock near the surface to disintegrate and decompose into **regolith**. **Mass wasting** is the spontaneous downhill motion of soil, regolith, or rock under gravity.

- **Physical weathering** produces regolith from solid rock by breaking bedrock into pieces. *Frost action* breaks rock apart by the repeated growth and melting of ice crystals in rock fractures and *joints,* as well as between individual mineral grains. In mountainous regions of vigorous frost, fields of angular blocks accumulate as *felsenmeers.* Slopes of rock fragments form *talus cones.* In soils and sediments, needle ice and ice lenses push rock and soil fragments upward. *Salt-crystal growth* in dry climates breaks individual grains of rock free, and can damage brick and concrete. *Unloading* of the weight of overlying rock layers can cause some types of rock to expand and break loose into thick shells, producing *exfoliation domes.* Daily temperature cycles in arid environments are thought to cause rock breakup. Wedging by plant roots also forces rock masses apart.

- **Chemical weathering** results from mineral alteration. Igneous and metamorphic rocks can decay to great depths through *hydrolysis* and *oxidation,* producing a regolith that is often rich in clay minerals. *Carbonic acid* action dissolves limestone. In warm, humid environments, basaltic lavas can also show features of solution weathering produced by acid action.

- Mass wasting occurs on slopes that are mantled with regolith. **Soil creep** is a process of mass wasting in which regolith moves down slopes almost imperceptibly under the influence of gravity. In an **earthflow,** water-saturated soil or regolith slowly flows downhill. Quick clays, which are unstable and can liquefy when exposed to a shock, have produced earthflows in previously glaciated regions. A **mudflow** is much swifter than an earthflow. It follows stream courses, becoming thicker as it descends and picks up sediment. A watery mudflow with debris ranging from fine particles to boulders to tree trunks and limbs is called a *debris flow.* A **landslide** is a rapid sliding of large masses of bedrock, sometimes triggered by an earthquake.

- The **scarification** of land by human activities, such as mining of coal or ores, can heap up soil and regolith into unstable masses that produce earthflows or mudflows. Mass wasting can also be caused by removal of supporting layers, undermining the natural support of soil and regolith. These actions are termed *induced mass wasting.*

- The tundra environment is dominated by the **periglacial** *system* of distinctive landforms and processes related to freezing and thawing of water in the active layer of **permafrost.** *Continuous permafrost* and *sub-sea permafrost* occur at the highest latitudes, flanked by a band of *discontinuous permafrost* that is transitional to warmer regions.

- *Pingos* and **ice wedges** are forms of **ground ice** found in permafrost terrains. **Patterned ground** occurs when ice growth and melting in freeze–thaw cycles create *stone polygons* and *ice-wedge polygons.* During the brief summer thaw, saturated soils flow to form **solifluction** *terraces* and *lobes.* Alpine tundra and permafrost extend to lower latitudes in high-mountain environments. Removal of the surface layer for construction in permafrost terrains can cause semipermanent thawing of the upper permafrost, creating *thermal erosion* and *thermokarst.*

Key Terms

weathering, p. 456	bedrock, p. 464	landslide, p. 468	ground ice, p. 473
regolith, p. 456	sediment, p. 464	scarification, p. 470	ice wedge, p. 474
mass wasting, p. 456	alluvium, p. 465	periglacial, p. 472	patterned ground, p. 474
physical weathering, p. 457	soil creep, p. 465	permafrost, p. 473	solifluction, p. 474
chemical weathering, p. 461	earthflow, p. 466		
	mudflow, p. 467		

Review Questions

1. What is meant by the term *weathering?* What types of weathering are recognized?
2. Define the terms *regolith, bedrock, sediment,* and *alluvium.*
3. How does frost action break up rock? Describe some landforms created by frost action and how they are formed.
4. How does salt-crystal growth break up rock? Give an example of a landform that arises from salt-crystal growth.
5. What is an exfoliation dome, and how does it arise? Provide an example.
6. Name three types of chemical weathering. Describe how limestone is often altered by a chemical weathering process.

7. Define mass wasting and identify the processes it includes.
8. What is soil creep, and how does it arise?
9. What is an earthflow? What features distinguish it as a landform?
10. Contrast earthflows and mudflows, providing an example of each.
11. Define the term *landslide.* How does a landslide differ from an earthflow?
12. Define and describe induced mass wasting. Provide some examples.
13. Explain the term *scarification.* Provide an example of an activity that produces scarification.
14. What is meant by the term *periglacial?*

15. Define and describe permafrost and some of its features, including ground ice, active layer, and permafrost table.
16. Identify and describe two forms of ground ice.
17. What is patterned ground? Identify two types of patterned ground.
18. What environmental problems are associated with human activity in regions of permafrost?

Eye on the Environment 14.1 • The Great Hebgen Lake Earthquake

1. Describe the formation of Earthquake Lake. What four forms of natural environmental disaster were involved?

Visualizing Exercises

1. Define the terms *regolith, bedrock, sediment,* and *alluvium.* Sketch a cross section through a part of the landscape showing these features and label them on the sketch.
2. Copy or trace Figure 14.13; then identify and plot on the diagram the mass movement associated with each of the following locations: Turtle Mountain, Palos Verdes Hills, Nicolet, Mount St. Helens, Madison River, and Herculaneum.

Essay Questions

1. A landscape includes a range of lofty mountains elevated above a dry desert plain. Describe the processes of weathering and mass wasting that might be found on this landscape and identify their location.
2. Imagine yourself as the newly appointed director of public safety and disaster planning for your state or province. One of your first jobs is to identify locations where human populations are threatened by potential disasters, including those of mass wasting. Where would you look for mass wasting hazards and why? In preparing your answer, you may want to consult maps of your state or province.

Eye on the Landscape

Chapter Opener Rockslide in Yosemite National Park Rockslides and landslides are thought to travel on a cushion of air, accounting for their rapid velocities. Notice how the tree at **(A)** is broken off 2 to 3 meters from the ground. The lower part is still erect and untouched. Perhaps the lower part of this tree was within that cushion of air underneath the rock mass and so survived destruction. Note also that the rock fragments are very angular, with sharp edges and faces **(B)**. This indicates that travel has not been far. As rocks are tumbled and rolled in transport, they rapidly lose sharp edges and become more rounded. Rockslides and landslides are subjects of this chapter.

14.4 Talus cones Individual rock strata **(A)** are readily visible in this photo. They are ancient sedimentary rocks that were thrust over and above younger rocks of the plains during an arc-continent collision. Strong and resistant to erosion, they now form the magnificent peaks of the northern Rockies in Alberta. At **(B)** are the remains of a small glacier, coated with gray blocks of talus. Chapter 12, on plate tectonics, describes overthrust faults and arc-continent collisions. Alpine glaciers are covered in Chapter 19.

14.12 Solution weathering of basalt Kauai is the oldest of the Hawaiian Islands. Note the rounded dome shape of its outline **(A)**, which is characteristic of shield volcanoes (Chapter 13). The red-brown colors of soil, exposed on lower slopes **(B)**, are those of iron oxides and indicate Oxisols (Chapter 10). Pocket beaches and an arch at **(C)** are products of wave action (Chapter 18).

EYE ON THE LANDSCAPE

This pair of Landsat images shows northern St. Louis and vicinity during a normal year (left) and in the midst of the Mississippi flood of 1993. What else would the geographer see? … Answers at the end of the chapter.

15 | FRESH WATER OF THE CONTINENTS

· · · · · · · · · · · · · · · · · ·
PUTTING YOU IN THE PICTURE

The violent blast cut the still night air with a force that broke windows and set off car alarms on the far side of the mile-wide river. The muddy waters of the mighty Mississippi, thrown back at first by the force of the dynamite, surged forward through a new gap in the levee just upstream from Prairie du Rocher, Illinois. Fifteen minutes later, a second thundering explosion resounded across the width of the seething river, and floodwaters spilled through another new gash.

At 3:30 that morning, August 4, 1993, the commissioners of the Fort Chartres–Ivy Landing Drainage and Levee District had approved a risky plan to dynamite two new crevasses in the levee upstream from the historic village of Prairie du Rocher. They hoped that the new breaches would skim the top off the crest of high waters approaching the main levee that had so far held the town secure from flooding. The waters from the breached levee would take hostage the fields and farms of a large area of upriver floodplain in return for sparing the village. It wasn't long before their vote was affirmed by the two blasts that echoed for miles up and down the river.

In the eyes of the commissioners, Prairie du Rocher was a prize worth saving. The town traces its roots back to its founding as a French trading post in 1722 and considers itself to be the birthplace of Illinois. Its quaint, turn-of-the-century architecture and its reputation for bed-and-breakfast inns and fine French food made the small town famous in the region as an upscale tourist mecca.

The vote of the commissioners to dynamite the levee was against the advice of the U.S. Army Corps of Engineers, the agency charged with building and maintaining the levees, dams, reservoirs, and other flood control structures that normally keep the powerful Mississippi River shackled and controlled. The Corps's engineers were afraid that the shock wave from the blast, transmitted through the river, would damage the levee near the town, flooding the village instead of protecting it. The pressure wave might also cause the levee at the site to lose its strength, turning rapidly into a mixture of sediment and water in a process called liquefaction. The resulting break would be vast enough to send a huge

volume of water through the levee that could swamp other levees and dikes and eventually attack the town from the rear. Even without liquefaction, the flooding could be severe enough to engulf areas that would have otherwise remained dry.

On August 3, one day earlier, the Corps had created its own breach in the upstream levee, dispatching a crane shovel on a barge to remove the top 1.2 m (4 ft) of a 120-m (400 ft) section of the levee. According to plan, an area of about 50 km^2 (20 mi^2) was flooded sooner rather than later, thus helping to blunt the rapidly approaching flood crest. But the water continued to rise, relentlessly threatening the historic village and leading to the late-night meeting and vote by the district commissioners.

Fortunately, none of the catastrophes foretold by the Corps' engineers came about. Prairie du Rocher was saved from flooding, but whether the extreme measures adopted by the desperate commissioners made the difference in keeping

the floodwaters from overtopping the levees is not really known.

The summer of 1993 was a time of extreme measures all along the upper reaches of the Mississippi. In June and July, unprecedented rainfall deluged the upper Mississippi Basin, including southern Minnesota, western Wisconsin, Iowa, Illinois, and Missouri. The result was a flood of a magnitude estimated to recur only once in 500 years. Devastation was widespread, and costs of the flood control, damage, and cleanup exceeded $12 billion.

Rivers are but one part of the hydrologic cycle—the set of complex flow pathways by which water moves as vapor over the land, condenses as precipitation and falls to Earth, then runs off as surface water and sinks in as ground water. This chapter focuses on these latter parts of the hydrologic cycle because they provide the fresh water that supports not only human activities, but nearly all biological activity on land. ■

FRESH WATER OF THE CONTINENTS

Water is essential to life. Nearly all organisms require a constant flux of water or at least a water-rich environment for survival. Humans are no exception. Our activity depends on a constant supply of fresh water that is provided by precipitation over the lands. Some of this water infiltrates the land surface and is stored in soils, regolith, and pores in bedrock. An even smaller portion is in motion as flowing fresh water in streams and rivers. In this chapter, we focus on these two parts of the hydrologic cycle—water at the land surface and water that lies within the ground.

Recall from Chapter 4 that fresh water on the continents in surface and subsurface water is only a very small part—about 3 percent—of the hydrosphere's total water. Most of this fresh water is locked into ice sheets and mountain glaciers. Ground water accounts for a little more than half of 1 percent. Although this is a very small fraction of global water, it is still many times larger than the amount of fresh water in lakes, streams, and rivers, which account for only three-hundredths of 1 percent of the total water. Note also that ground water can be found at almost every location on

---

When precipitation reaches the ground, it either runs off or infiltrates into the soil. As runoff, it flows into streams. As infiltration, it either returns to the air through evapotranspiration or seeps downward to become ground water.

---

land that receives rainfall. In contrast, fresh surface water varies widely in abundance. In many arid regions, streams and rivers are nonexistent for most or all of the year.

Let's review the hydrologic cycle, focusing on surface and ground water. In Chapter 4, we discussed atmospheric moisture and precipitation, describing the part of the hydrologic cycle in which water evaporates from ocean and land surfaces and then precipitates, falling to Earth as rain or snow. What happens then to this precipitation? Figure 15.1 provides the answer. As the diagram shows, a portion of the precipitation returns directly to the atmosphere through evaporation from the soil. Another portion travels downward, moving through the soil under the force of gravity to become part of the underlying ground water. Following underground flow paths, this subsurface water eventually emerges to become surface water, or it may emerge directly in the shore zone of the ocean. A third portion flows over the ground surface as runoff to lower levels. As it travels, the water flow becomes collected into streams, which eventually conduct the running water to the ocean.

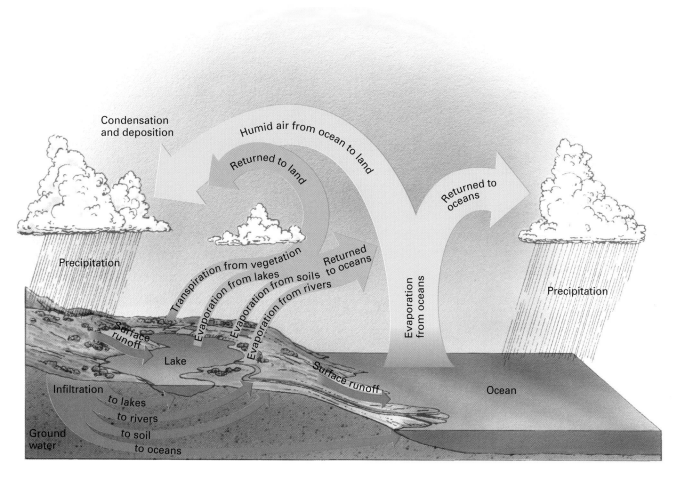

15.1 The hydrologic cycle *The hydrologic cycle traces the various paths of water from oceans, through the atmosphere, to land, and its return to oceans.*

In this chapter, we will trace the parts of the hydrologic cycle that include both the subsurface and surface pathways of water flow. The study of these flows is part of the science of *hydrology,* which is the study of water as a complex but unified system on the Earth.

Figure 15.2 shows what happens to water from precipitation as it first reaches the land surface. Most soil surfaces in their undisturbed, natural states are capable of absorbing the water from light or moderate rains by **infiltration.** In this process, water enters the small natural passageways between irregularly shaped soil particles, as well as the larger openings in the soil surface. These openings are formed by the borings of worms and animals, earth cracks produced by soil drying, cavities left from decay of plant roots, or spaces made by the growth and melting of frost crystals. A mat of decaying leaves and stems breaks the force of falling water drops and helps to keep these openings clear.

The precipitation that infiltrates the soil is temporarily held in the soil layer as soil water, occupying the *soil water belt* (Figure 15.2). Water within this belt can be returned to the surface and then to the atmosphere through a process

that combines two components—direct evaporation from soil and transpiration by vegetation. As we explained in Chapter 3, these two forms of water vapor transport are combined in the term *evapotranspiration.*

When rain falls too rapidly to be passed downward through soil openings, **runoff** occurs and a surface water layer runs over the surface and down the direction of ground slope. This surface runoff is called **overland flow.** In periods of heavy, prolonged rain or rapid snowmelt, overland flow feeds directly to streams.

Overland flow also occurs when soil that is already saturated receives rainfall or snowmelt. Since soil openings and pores are already filled, water cannot infiltrate the soil and drain down to deeper layers. Under these conditions, nearly all of the precipitation or snowmelt will run off.

Runoff as overland flow moves surface particles from hills to valleys, and so it is an agent that shapes landforms. Because runoff supplies water to streams and rivers, it also allows rivers to cut canyons and gorges, and carry sediment to the ocean—but we are getting ahead of our story. We'll return to landforms carved by running water in Chapter 16.

15.2 Paths of precipitation Precipitation falling on the land follows three primary paths. Some precipitation is returned to the atmosphere through evapotranspiration, while some runs off the soil surface. The remainder sinks into the soil water belt, where it is accessible to plants. A portion of the infiltrating water passes through the soil water belt and percolates down to the ground water zone. (A. H. Strahler.)

GROUND WATER

As shown in Figure 15.2, water derived from precipitation can continue to flow downward beyond the soil water belt. This slow downward flow under the influence of gravity is termed *percolation.* Eventually, the percolating water reaches ground water. **Ground water** is the part of the subsurface water that fully saturates the pore spaces in bedrock, regolith, or soil, and so occupies the *saturated zone* (Figure 15.3). The **water table** marks the top of this zone. Above it is the *unsaturated zone* in which water does not fully saturate the pores. Here, water is held in thin films adhering to mineral surfaces. This zone also includes the soil water belt.

Ground water moves slowly along deep flow paths, eventually emerging by seepage into streams, ponds, lakes, and marshes. In these places the land surface dips below the water table. Streams that flow throughout the year—perennial streams—derive much of their water from ground water seepage. GEODISCOVERIES

The Water Table Surface

Where there are many wells in an area, the position of the water table can be mapped in detail (Figure 15.4). This is done by plotting the water heights in the wells and noting the trend of change in elevation from one well to the other. The

15.3 Zones of subsurface water Water in the soil water belt is available to plants. Water in the unsaturated zone percolates downward to the saturated zone of ground water, where all pores and spaces are filled with water.

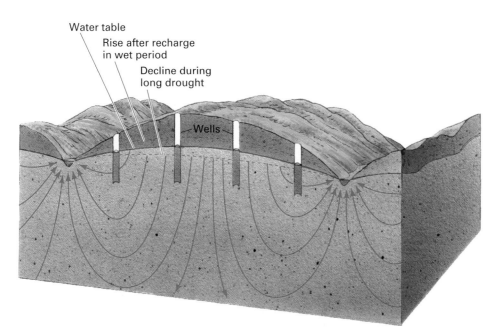

Water table
Rise after recharge
in wet period
Decline during
long drought
Wells

15.4 Water table surface
The configuration of the water table surface conforms broadly with the land surface above it. It varies in response to prolonged wet and dry periods. Ground water flow paths circulate water to deep levels in a very slow motion and eventually feed streams by seepage.

water table is highest under the highest areas of land surface—hilltops and divides. The water table declines in elevation toward the valleys, where it appears at the surface close to streams, lakes, or marshes.

The reason for this water table configuration is that water percolating down through the unsaturated zone tends to raise the water table, while seepage into lakes, streams, and marshes tends to draw off ground water and to lower its level. These differences in water table level are built up and maintained because ground water only moves extremely slowly through the fine chinks and pores of bedrock and regolith. In periods of high precipitation, the water table rises under hilltops or divide areas. In periods of water deficit, or during a drought, the water table falls (Figure 15.4).

Figure 15.4 also shows paths of ground water flow. The direction of flow at any point depends on the direction of pressure at that point. Gravity always exerts a downward pressure, while the difference in the height of the water table between hilltop and streambed produces a sideward pressure force. In the presence of resistance to flow, the result of these downward and sideward pressures causes ground water to flow in curving paths. Water that enters the hillside midway between the hilltop and the stream flows rather directly toward the stream. Water reaching the water table midway between the two streams, however, flows almost straight down to great depths before recurving and rising upward again. Progress along these deep paths is incredibly slow, while flow near the surface is much faster. The most rapid flow is close to the stream, where the arrows converge. Over time, the level of the water table tends to remain stable, and the flow of water released to streams and lakes must balance the flow of water percolating down into the water table.

> The water table marks the top of the saturated zone of ground water. It is highest under hilltops and divides, and it slopes to intersect the surface at lakes, marshes, and streams.

Aquifers

Sedimentary layers often exert a strong control over the storage and movement of ground water. For example, clean, well-sorted sand—such as that found in beaches, dunes, or stream alluvium—can hold an amount of ground water about equal to about one-third of its bulk volume. A bed of sand or sandstone is thus often a good *aquifer*—that is, a layer of rock or sediment that contains abundant, freely flowing ground water. In contrast, beds of clay and shale are relatively impermeable and hold little free water. They are known as *aquicludes*. Figure 15.5 shows a situation in which a shale bed is overlain by a bed

15.5 Ground water in horizontal strata *Water flows freely within the sandstone aquifer but only very slowly in the shale aquiclude. A lens of shale creates a perched water table above the main water table. (Copyright © A.N. Strahler.)*

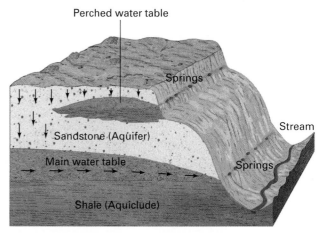

Perched water table
Springs
Stream
Sandstone (Aquifer)
Main water table
Springs
Shale (Aquiclude)

15.6 Artesian well *A dipping sandstone layer interleaved between impervious layers provides a source of water under pressure that flows naturally to the surface. (Drawn by Erwin Raisz. Copyright © A.N. Strahler.)*

of sandstone. Ground water moves freely through the sandstone aquifer and across the top of the shale aquiclude, emerging in springs along the canyon shown at the right. Also shown in a small lens of shale, which accumulates a shallow *perched water table* above it.

When an aquifer is situated between two aquicludes, ground water in the aquifer may be under pressure and thus flow freely from a well. This type of self-flowing well is an **artesian well.** Figure 15.6 illustrates this situation, in which a porous sandstone bed (aquifer) is sandwiched between two impervious rock layers (aquicludes). Precipitation on the hills where the sandstone outcrops provides water that saturates the sandstone layer, filling it to that elevation. Since the elevation of the well that taps the aquifer is below that of the range of hills feeding the aquifer, hydrostatic pressure causes water to rise in the well.

LIMESTONE SOLUTION BY GROUND WATER

We saw in Chapter 14 that in moist climates limestone at the surface is slowly dissolved by carbonic acid action, produc-

ing lowland areas. The slow flow of ground water in the saturated zone can also dissolve limestone below the surface, producing deep underground caverns. These features can then collapse, producing a sinking of the ground above and the development of a unique type of landscape.

Limestone Caverns

You are probably familiar with such famous caverns as Mammoth Cave or Carlsbad Caverns. *Limestone caverns* are interconnected subterranean cavities in bedrock formed by the corrosive action of circulating ground water on limestone. Figure 15.7 shows how caverns develop. As shown in the left diagram, the action of carbonic acid is particularly concentrated in the saturated zone just below the water table. This removal process forms many kinds of underground "landforms," such as tortuous tubes and tunnels, great open chambers, and tall chimneys. Subterranean streams can be found flowing in the lowermost tunnels, and these carry the products of solution to emerge along the banks of surface streams and rivers.

In a later stage, shown in the right diagram, the stream has deepened its valley, and the water table has been corre-

15.7 Cavern development *Limestone dissolves at the top of the ground water zone. When rapid erosion of streams lowers the water table, caverns result in the unsaturated zone. Water flow through the caverns results in travertine deposition. (Copyright © A.N. Strahler.)*

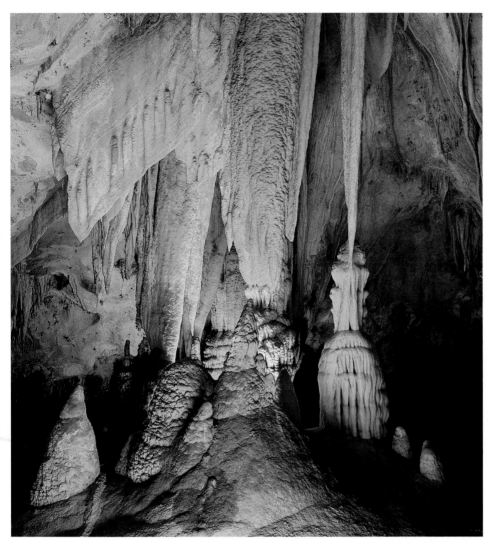

15.8 Inside a cavern
Travertine deposits in the Papoose Room of Carlsbad Caverns in New Mexico. Deposits include stalactites (slender rods hanging from the ceiling), stalagmites (upward pointing rods), and sturdy columns, all formed when dripping ground water evaporates, leaving travertine deposits of calcium carbonate behind.

spondingly lowered to a new position. The cavern system previously formed is now in the unsaturated zone. Deposition of carbonate matter, known as *travertine,* takes place on exposed rock surfaces in the caverns. Encrustations of travertine take many beautiful forms—stalactites (hanging rods), stalagmites (upward pointing rods), columns, and drip curtains (Figure 15.8).

Karst Landscapes

Where limestone solution is very active, we find a landscape with many unique landforms. This is especially true along the Dalmatian coastal area of Croatia, where the landscape is called *karst.* Geographers apply that term to the topography of any limestone area where sinkholes are numerous and small surface streams are nonexistent. A *sinkhole* is a surface depression in a region of cavernous limestone (Figure 15.9). Some

---

Carbonic acid action dissolves limestone, producing caverns. Cavern collapse creates sinkholes and a karst landscape.

---

sinkholes are filled with soil washed from nearby hillsides, while others are steep-sided, deep holes. They develop where the limestone is more susceptible to solution weathering, or where an underground cavern near the surface has collapsed.

Development of a karst landscape is shown in Figure 15.10. In an early stage, funnel-like sinkholes are numerous. Later, the caverns collapse, leaving open, flat-floored valleys. Examples of some important regions of karst or karst-like topography are the Mammoth Cave region of Kentucky, the Yucatan Peninsula, and parts of Cuba and Puerto Rico. In regions such as southern China and west Malaysia, the karst landscape is dominated by steep-sided, conical limestone hills or towers, 100 to 500 m (about 300 to 1500 ft) high (Figure 15.11). The towers are sometimes capped by beds of more resistant rock or by impure limestones that dissolve more slowly. They are often riddled with caverns and passageways.

15.9 *Eye on the Landscape* **Sinkholes** *Sinkholes in limestone are created by solution. These sinkholes are near Roswell, New Mexico.* **What else would the geographer see? ... Answers at the end of the chapter.**

15.10 **Evolution of a karst landscape** *(a) Rainfall enters the cavern system through sinkholes in the limestone. (b) Extensive collapse of caverns reveals surface streams flowing on shale beds beneath the limestone. Some parts of the flat-floored valleys can be cultivated. (Drawn by Erwin Raisz. Copyright © A. N. Strahler.)*

PROBLEMS OF GROUND WATER MANAGEMENT

Rapid withdrawal of ground water has seriously impacted the environment in many places. Increased urban populations and industrial developments require larger water supplies—needs that cannot always be met by constructing new surface water reservoirs. To fill these needs, vast numbers of wells using powerful pumps draw huge volumes of ground water to the surface, greatly altering nature's balance of ground water discharge and recharge.

In dry climates, agriculture is often heavily dependent on irrigation water from pumped wells—especially since major river systems are likely to be already fully utilized for irrigation. Wells are also convenient water sources. They can be drilled within the limits of a given agricultural or industrial property and can provide immediate supplies of water without any need to construct expensive canals or aqueducts.

In earlier times, the small well that supplied the domestic and livestock needs of a home or farmstead was actually dug by hand and sometimes lined with masonry. By contrast, a modern well supplying irrigation and industrial water is drilled by powerful machinery that can bore a hole 40 cm (16 in.) or more in diameter to depths of 300 m (about 1000 ft) or more. Drilled wells are lined with metal casings that exclude impure near-surface water and prevent the walls from caving in and clogging the tube. Near the lower end of

15.11 *Eye on the Landscape* **Tower karst** *White limestone can be seen exposed in the nearly vertical sides of these towers. Near Guilin (Kweilin), Guanxi Province, southern China at lat. 25° N.* **What else would the geographer see? ... Answers at the end of the chapter.**

the hole, in the ground water zone, the casing is perforated to admit the water. The yield of a single drilled well ranges from as low as a few hundred liters or gallons per day in a domestic well to many millions of liters or gallons per day for a large industrial or irrigation well.

Water Table Depletion

As water is pumped from a well, the level of water in the well drops. At the same time, the surrounding water table is lowered in the shape of a downward-pointing cone, termed the *cone of depression* (Figure 15.12). The difference in height between the cone tip and the original water table is the *drawdown*. The cone of depression may extend out as far

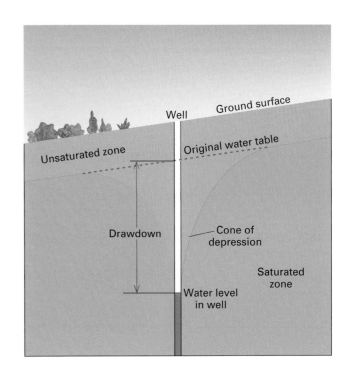

15.12 *Drawdown and cone of depression in a pumped well* *As the well draws water, the water table is depressed in a cone shape centered on the well. (A. H. Strahler.)*

as 16 km (10 mi) or more from a well where heavy pumping is continued. Where many well are in operation, their intersecting cones will produce a general lowering of the water table.

Water table depletion often greatly exceeds recharge—the rate at which infiltrating water moves downward to the saturated zone. In arid regions, much of the ground water for irrigation is drawn from wells driven into thick aquifers of sand and gravel. These deposits are often recharged by the seasonal flow of streams that originate high in adjacent mountains. Fanning out across the dry lowlands, the streams lose water, which sinks into the sands and gravels and eventually percolates to the water table below. The extraction of ground water by pumping can greatly exceed this recharge by stream flow, lowering the water table. Deeper wells and more powerful pumps are then required. The result is exhaustion of a natural resource that is not renewable except over very long periods of time.

Another byproduct of water table depletion is subsidence—a sinking of the land in response to the removal of water from underlying sediments. This problem has plagued a number of major cities that rely heavily on ground water wells for their water supplies. *Eye on Global Change 15.1* • *Sinking Cities* provides more information on this topic.

Contamination of Ground Water

Another major environmental problem related to ground water withdrawal is contamination of wells by pollutants that infiltrate the ground and reach the water table. Both solid and liquid wastes are responsible. Disposal of solid wastes poses a major environmental problem in the United States because our advanced industrial economy provides an endless source of garbage and trash. Traditionally, these waste products were trucked to the town dump and burned there in continually smoldering fires that emitted foul smoke and gases. The partially consumed residual waste was then buried under earth.

Wells draw down the water table at a point, creating a cone of depression. As many wells exploit an aquifer, their cones of depression merge to create a general lowering of the water table.

Sanitary landfills can release pollutants and toxic compounds that infiltrate to the water table, causing ground water contamination that renders adjacent well waters unfit for consumption.

In recent decades, a major effort has been made to improve solid-waste disposal methods. One method is high-temperature incineration, but it often leads to air pollution. Another is the sanitary landfill method in which waste is not allowed to burn. Instead, layers of waste are continually buried, usually by sand or clay available on the landfill site. The waste is thus situated in the unsaturated zone. Here it can react with rainwater that infiltrates the ground surface. This water picks up a wide variety of chemical compounds from the waste body and carries them down to the water table (Figure 15.13).

Once in the water table, the pollutants follow the flow paths of the ground water. As the arrows in the figure indicate, the polluted water may flow toward a supply well that draws in ground water from a large radius. Once the polluted water has reached the well, the water becomes unfit for human consumption. Polluted water may also move toward a nearby valley, causing pollution of the stream flowing there (left side of Figure 15.13).

Another source of contamination in coastal wells is *saltwater intrustion*. Since fresh water is less dense than saltwater, a coastal aquifer can be underlain by a layer of saltwater from the ocean. When the aquifer is depleted, the level of saltwater rises and eventually reaches the well from below, rendering the well unusable.

SURFACE WATER

So far, we have examined how water moves below the land surface. Now, we turn to tracing the flow paths of surplus water that runs off the land surface and ultimately reaches the sea. Here, we will be concerned primarily with rivers and streams. (In general usage, we speak of "rivers" as large watercourses and "streams" as smaller ones. However, the word "stream" is also used as a scientific term designating the channeled flow of surface water of any amount.)

GEODISCOVERIES

15.13 *Movement of polluted ground water*
Polluted water, leached from a waste disposal site, moves toward a supply well (right) and a stream (left). (Copyright © A. N. Strahler.)

15.14 Overland flow
Overland flow, in the form of a thin sheet of water, covers the nearly flat plain in the middle distance. The water converges into stream flow in a narrow, steep-sided gully (left). This photograph was taken shortly after a summer thunderstorm had deluged the area. The locality, near Raton, New Mexico, shows steppe grassland vegetation.

Overland Flow and Stream Flow

As we saw earlier in this chapter, runoff that flows down the slopes of the land in broadly distributed sheets is overland flow. We can distinguish overland flow from *stream flow,* in which the water occupies a narrow channel confined by lateral banks. Overland flow can take several forms. Where the soil or rock surface is smooth, the flow may be a continuous thin film, called *sheet flow* (Figure 15.14). Where the ground is rough or pitted, flow may take the form of a series of tiny rivulets connecting one water-filled hollow with another. On a grass-covered slope, overland flow is subdivided into countless tiny threads of water, passing around the stems. Even in a heavy and prolonged rain, you might not notice overland flow in progress on a sloping lawn. On heavily forested slopes, overland flow may pass entirely concealed beneath a thick mat of decaying leaves.

Overland flow eventually contributes to a stream, which is a much deeper, more concentrated form of runoff. We can define a **stream** as a long, narrow body of flowing water occupying a trenchlike depression, or channel, and moving to lower levels under the force of gravity. The **channel** of a stream is a narrow trough. The forces of flowing water shape the trough to its most effective form for moving the quantities of water and sediment supplied to the stream (Figure 15.15). A channel may be so narrow that you can easily jump across it, or, in the case of the Mississippi River, as wide as 1.5 km (about 1 mi) or more.

As a stream flows under the influence of gravity, the water encounters resistance—a form of friction—with the channel walls. As a result, water close to the bed and banks moves slowly, while water in the central part of the flow moves faster. If the channel is straight and symmetrical, the single line of maximum velocity is located in midstream. If the stream curves, the maximum velocity shifts toward the bank on the outside of the curve.

Actually, the arrows in Figure 15.15 only show average velocity. In all but the most sluggish streams, the water is

15.15 Characteristics of stream flow *The velocity of flow is greatest in the middle (a) and at the top (b) of the stream. (c) Mean velocity, cross-sectional area and slope change in the pools and rapids of a stream section of uniform discharge. (b, Copyright © A. N. Strahler.)*

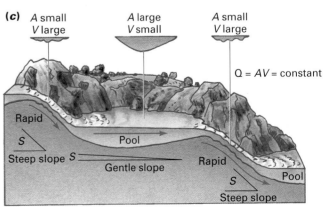

Eye on Global Change | 15.1

Sinking Cities

Ground water is a resource on which human civilization now depends to meet much of its demand for fresh water. However, it may take thousands of years to accumulate large reservoirs of underground water. When this water is removed rapidly by pumping, the rate of removal far exceeds the rate of recharge. The volume of water in the reservoir is reduced, and the water table falls. What are the effects of this change?

One important environmental effect of excessive ground water withdrawal is subsidence of the ground surface. Venice, Italy, provides a dramatic example of this side effect. Venice was built in the eleventh century A.D. on low-lying islands in a coastal lagoon, sheltered from the ocean by a barrier beach. Underlying the area are some 1000 m (about 3300 ft) of layers of sand, gravel, clay, and silt, with some layers of peat. Compaction of these soft layers has been going on gradually for centuries under the heavy load of city buildings. However, ground water withdrawal, which has been greatly accelerated in recent decades, has aggravated the condition.

Many ancient buildings in Venice now rest at lower levels and have suffered severe damage as a result of flooding during winter storms on the adjacent Adriatic Sea. Coastal storms normally raise sea level, and when high tides occur at the same time, water rises even higher. The problem of flooding during storms is aggravated by the fact that many of the canals of Venice receive raw sewage, so that the floodwater is contaminated.

Most of the subsidence in recent decades has been attributed to withdrawals of large amounts of ground water from industrial wells at Porto Marghere, the modern port of Venice, located a few kilometers distant on the mainland shore. This pumping has now been greatly curtailed, reducing the rate of subsidence to a very small natural rate (about 1 mm per year). However, the threat of flooding and damage to churches and other buildings of great historical value remains. Flood control now depends on the construction of seawalls and floodgates on the barrier beach that lies between Venice and the open ocean.

By the early 1980s, the port city of Bangkok, located on the Chao Phraya River a short distance from the Gulf of Thailand, had become the world's most rapidly sinking city as a result of massive ground water withdrawals from soft marine sediments beneath the city. During the 1980s the rate of subsidence reached 14 cm (6 in.) per year, and major floods were frequent. Some reduction in the rate of ground water withdrawal has produced a modest decrease in the rate of subsidence, but as in the case of Venice, the flood danger remains.

Ground water withdrawal has affected several regions in California, where ground water for irrigation has been pumped from basins filled with alluvial sediments. Water table levels in these basins have dropped over 30 m (98 ft), with a maximum drop of 120 to 150 m (394 to 492 ft) being recorded in one locality of California's San Joaquin Valley. In the Los Bannos–Kettleman City area, ground subsidence of as much as 4.2 to 4.8 m (14 to 16 ft) was measured at some locations in a 35-year period.

Another important area of ground subsidence accompanying water withdrawal is beneath Hous-

affected by *turbulence,* a system of countless eddies that are continually forming and dissolving. If we follow a particular water molecule, it will travel a highly irregular, corkscrew path as it is swept downstream. Motions include upward, downward, and sideward directions. Only if we measure the water velocity at a certain fixed point for a long period of time, say, several minutes, will the average motion at that point be downstream and in a line parallel with the surface and bed.

Stream Discharge

Stream flow at a given location is measured by its **discharge,** which is defined as the volume of water per unit time passing through a cross section of the stream at that location. It is measured in cubic meters (cubic feet) per second. The cross-sectional area and average velocity of a stream can change within a short distance, even though the stream discharge does not change. These changes occur because of changes in the *gradient* of the stream channel. The gradient is the rate of fall in elevation of the stream surface in the downstream direction (Figure 15.15).

When the gradient is steep, the force of gravity will act more strongly and flow velocity will be greater. When the stream channel has a gentle gradient, the velocity will be slower. However, in a short stretch of stream, the discharge will remain constant. As shown in Figure 15.15, this means that in stretches of rapids, where the stream flows swiftly, the stream channel will be shallow and narrow. In pools,

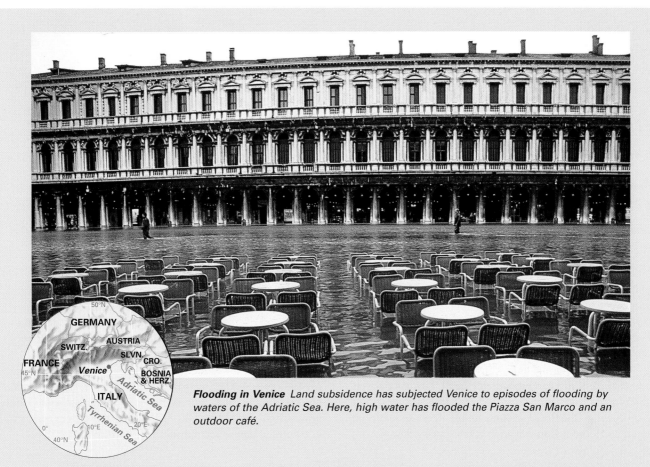

Flooding in Venice *Land subsidence has subjected Venice to episodes of flooding by waters of the Adriatic Sea. Here, high water has flooded the Piazza San Marco and an outdoor café.*

ton, Texas, where the ground surface has subsided from 0.3 to 1 m (1 to 3 ft) in a metropolitan area 50 km (31 mi) across. Damage to buildings, pavements, airport runways, and other structures has resulted.

Perhaps the most celebrated case of ground subsidence is that affecting Mexico City. Carefully measured ground subsidence has ranged from about 4 to 7 m (13 to 23 ft). The subsidence has resulted from the withdrawal of ground water from an aquifer system beneath the city and has caused many serious engineering problems. As water has been drained out, the volume of clay beds overlying the aquifer has contracted greatly. To combat the ground subsidence, recharge wells were drilled to inject water into the aquifer. In addition, new water supplies from sources outside the city area were developed to replace local ground water use.

---

where the stream flows more slowly, the stream channel will be wider and deeper to maintain the same discharge. Sequences of pools and rapids can be found along streams of all sizes.

Stream discharge at a location on a stream is determined by noting the height, or *stage*, of the surface of the stream with respect to a fixed level nearby, such as a marker on a bridge abutment. Stage is measured by a *stream gauge,* which uses a float inside a stilling well to record the height of the water surface (Figure 15.16). Stage is then converted to discharge by consulting a table made from flow measurements acquired at that location at various stages in the past.

The discharge of a stream measures its volume rate of flow. Discharge is measured by recording the height of the water surface, or stage, at a gauging station where the relation between height and discharge is known.

The discharges of streams and rivers change from day to day, and records of daily and flood discharges of major streams and rivers are important information. They are used in planning the development and distribution of surface waters, as well as in designing flood-protection structures and generating models of how floods progress down a particular river system. An important activity of the U.S. Geological Survey is the measurement, or gauging, of stream discharge in the United States. In cooperation with states and municipalities, this organization maintains over 6000 gauging stations on principal streams and their tributaries.

15.16 *Current meter* *A hydrologist sits in a cable car, lowering a current meter into a large river. At the bottom of the tether is a streamlined weight with a vertical fin. The current meter, located just above the weight, has a set of three cups that are propelled in a circular motion by the force of the flowing water. Each revolution of the cup set closes and opens an electrical connection, providing a series of "clicks" that can be counted manually or automatically. The rate of clicking determines the velocity of the water flow.*

Figure 15.17 is a map showing the relative discharge of major rivers of the United States. The mighty Mississippi with its tributaries dwarfs all other North American rivers. Two other discharges are of major proportions—the Columbia River, draining a large segment of the Rocky Mountains in southwestern Canada and the northwestern United States, and the Great Lakes, discharging through the St. Lawrence River. The Colorado River, a much smaller stream, has its origin in the snowmelt of the Rocky Mountains and then crosses a vast semiarid and arid region that adds little tributary flow.

As the figure shows, the discharge of major rivers increases downstream as a natural consequence of the way streams and rivers combine to deliver runoff and sediment to the oceans. The gradient also changes in a downstream direction. The general rule is the larger the cross-sectional area of the stream, the lower the gradient. Great rivers, such as the Mississippi and Amazon, have gradients so low that they can be described as "flat." For example, the water surface of the lower Mississippi River falls in elevation about 3 cm for each kilometer of downstream distance (1.9 in. per mi).

15.17 *River discharge* *This schematic map shows the relative magnitude of the discharge of U.S. rivers. Width of the river as drawn is proportional to mean annual discharge. (After U.S. Geological Survey.)*

Rivers with headwaters in high mountains have characteristics that are especially desirable for utilizing river flow for irrigation and for preventing floods. The higher ranges serve as snow storage areas, slowly releasing winter and spring precipitation through early or midsummer. As summer progresses, melting proceeds to successively higher levels. In this way, a continuous, sustained river flow is maintained. Among the snow-fed rivers of the western United States and Cananda are the Columbia, Snake, Missouri, Arkansas, and Colorado.

Drainage Systems

As runoff moves to lower and lower levels and eventually to the sea, it becomes organized into a **drainage system.** The system consists of a branched network of stream channels, as well as the adjacent sloping ground surfaces that contribute overland flow to those channels. Between the channels on the crests of ridges are *drainage divides,* which mark the boundary between slopes that contribute water to different streams or drainage systems. The entire system is bounded by an outer drainage divide that outlines a more-or-less pear-shaped **drainage basin** or **watershed** (Figure 15.18).

A typical stream network within a drainage basin is shown in Figure 15.18. Each fingertip tributary receives runoff from a small area of land surface surrounding the channel. This runoff is carried downstream and merged with runoff from other small tributaries as they join the main stream. The drainage system thus provides a converging mechanism that funnels overland flow into streams and smaller streams into larger ones.

> A drainage basin consists of a branched network of stream channels and adjacent slopes that feed the channels. It is bounded by a drainage divide.

STREAM FLOW

The discharge of a stream will increase in response to a period of heavy rainfall or snowmelt. However, the response is delayed, as the movement of water into stream channels takes time. The length of delay depends on a number of factors. The most important factor is the size of the drainage basin feeding the stream. Larger drainage basins show a longer delay.

The relationship between stream discharge and precipitation is best studied by means of a simple graph, called a *hydrograph,* which plots the discharge of a stream with time at a particular stream gauge. Figure 15.19 is a hydrograph for a drainage basin of about 800 km^2 (300 mi^2) in area located in Ohio within the moist continental climate ⑩. The graph shows the discharge of Sugar Creek (smooth line), the main stream of the drainage basin, during a four-day period that included a rainstorm. Rainfall for the 12-hour storm is shown by a bar graph giving the number of centimeters of precipitation in each two-hour period. The average total rainfall over the watershed of Sugar Creek was about 15 cm (6 in.). About half of this amount passed down the stream within three days' time. Some rainfall was held in the soil as soil water, some evaporated, and some infiltrated to the water table to be held in long-term storage as ground water.

Studying the rainfall and runoff graphs in Figure 15.19, we see that prior to the onset of the storm, Sugar Creek was carrying a small discharge. This flow, which is supplied by the seepage of ground water into the channel, is termed *base flow.* After the heavy rainfall began, several hours elapsed before the stream gauge began to show a rise in discharge. This inter-

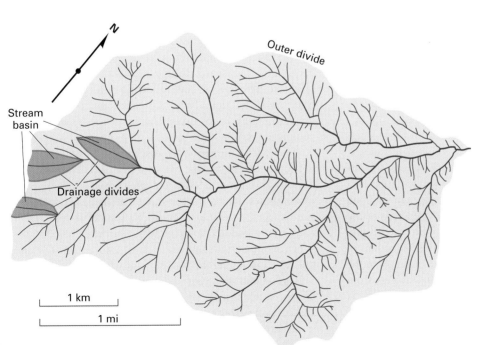

15.18 Channel network of a stream *Smaller and larger streams merge in a network that carries runoff downstream. Each small tributary has its own small drainage basin, bounded by drainage divides. An outer drainage divide delineates the stream's watershed at any point on the stream. (Data of U.S. Geological Survey and Mark A. Melton.)*

15.19 Sugar Creek hydrograph *Four days of precipitation and stream flow at Sugar Creek, Ohio, following a heavy rainstorm in August. (After Hoyt and Langbein, Floods, copyright © Princeton University Press. Used by permission.)*

val, called the *lag time,* indicates that the branching system of channels was acting as a temporary reservoir. The channels were at first receiving inflow more rapidly than it could be passed down the channel system to the stream gauge.

Lag time is measured as the difference between the time at which half of the precipitation has occurred and that at which half the runoff has passed downstream. In the Sugar Creek example, the lag time was about 18 hours, with the peak flow reached almost 24 hours after the rainfall began. Note also that the stream's discharge rose much more abruptly than it fell. In general, the larger a watershed, the longer is the lag time between peak rainfall and peak discharge, and the more gradual is the rate of decline of discharge after the peak has passed. Another typical feature of a flood hydrograph is the slow but distinct rise in the amount of discharge contributed by base flow.

> A hydrograph plots stream flow with time. Peaks in the hydrograph occur after rainfall events. Between rains, stream flow falls to base flow, which is fed by ground water seepage into the stream's channels.

How Urbanization Affects Stream Flow

The growth of cities and suburbs affects the flow of small streams in two ways. First, an increasing percentage of the surface becomes impervious to infiltration as it is covered by buildings, driveways, walks, pavements, and parking lots. In a closely built-up residential area with small lot sizes, the percentage of impervious surface may run as high as 80 percent.

As the proportion of impervious surface increases, overland flow from the urbanized area generally increases. This change acts to increase the frequency and height of flood peaks during heavy storms for small watersheds lying largely within the urbanized area. Recharge to the ground water body beneath is also reduced, and this reduction, in turn, decreases the base flow contribution to channels in the same area. Thus, the full range of stream discharges, from low stages in dry periods to flood stages in wet periods, is made greater by urbanization.

A second change caused by urbanization is the introduction of storm sewers that quickly carry storm runoff from

paved areas directly to stream channels for discharge. Thus, runoff travel time to channels is shortened, while the proportion of runoff is increased by the expansion in impervious surfaces. The two changes together act to reduce the lag time of urban streams and increase their peak discharge levels. Many rapidly expanding suburban communities are finding that low-lying, formerly flood-free, residential areas now experience periodic flooding as a result of urbanization.

The Annual Flow Cycle of a Large River

In regions of humid climates, where the water table is high and normally intersects the important stream channels, the hydrographs of larger streams will show clearly the effects of two sources of water—base flow and overland flow. Figure 15.20 is a hydrograph of the Chattahoochee River in Georgia, a fairly large river draining a watershed of 8700 km^2 (3350 mi^2), much of it in the southern Appalachian Mountains. The sharp, abrupt fluctuations in discharge are produced by overland flow following rain periods of one to three days' duration. These are each similar to the hydrograph of Figure 15.19, except that here they are shown much compressed by the time scale.

After each rain period the discharge falls off rapidly, but if another storm occurs within a few days, the discharge rises to another peak. The enlarged inset graph in Figure 15.20 shows details for the month of January. Here, three storms occur in rapid succession. Runoff drops between each storm but not completely to the level of base flow. When a long period intervenes between storms, the discharge falls to a low value, the base flow, at which it levels off.

Throughout the year the base flow, which represents ground water inflow into the stream, undergoes a marked annual cycle. During winter and early spring, water table levels are raised, and the rate of inflow into streams is increased. For the Chattahoochee River, the rate of base flow

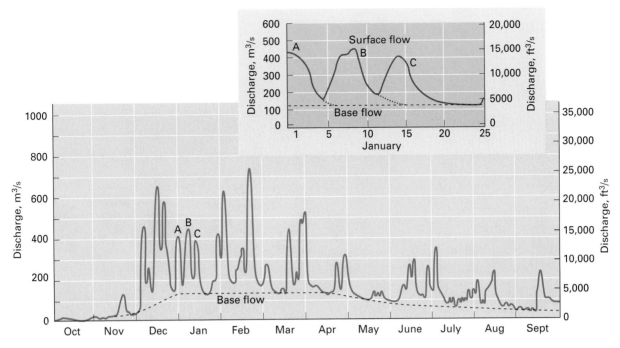

15.20 *Chattahoochee River hydrograph* This hydrograph shows the fluctuating discharge of the Chattahoochee River, Georgia, throughout a typical year. The high peaks are caused by runoff from streams that produced heavy overland flow. The inset graph expands the graph for the month of January. (Data of U.S. Geological Survey.)

during January, February, March, and April holds uniform at about 100 m³/s (about 3500 ft³/s). The base flow begins to decline in spring, as heavy evapotranspiration losses reduce soil water and therefore cut off the recharge of ground water. The decline continues through the summer, reaching a low of about 30 m³/s (about 1000 ft³/s) by the end of October.

River Floods

You've probably seen enough media coverage of river floods to have a good idea of the appearance of floodwaters and the damage caused by flood erosion and deposition of silt and clay. We can define the term **flood** as the condition that exists when the discharge of a river cannot be accommodated within its normal channel. As a result, the water spreads over the adjoining ground, which is normally cropland or forest. Sometimes, however, the ground is occupied by houses, factories, or transportation corridors.

Most rivers of humid climates have a *floodplain* a broad belt of low, flat ground bordering the channel on one or both sides that is flooded by stream waters about once a year (Figure 15.21). This flood usually occurs in the season when abundant surface runoff combines with the effects of a high water table to supply more runoff than can be carried in the channel. Annual inundation is considered a flood, even though its occurrence is expected and does not prevent the cultivation of crops after the flood has subsided. Annual flooding does

> A flood occurs when a river rises to leave its bed and cover adjacent lands. Flood forecasts rely on precipitation patterns, past history, and present river levels to predict when and how high a flood will crest.

not interfere with the growth of dense forests, which are widely distributed over low, marshy floodplains in all humid regions of the world. The National Weather Service, which provides a flood-warning service, designates a particular river stage at a given place as the *flood stage*. Above this critical level, inundation of the floodplain will occur. Still higher discharges of water, the rare and disastrous floods that may occur as seldom as once in 30 or 50 years, inundate ground lying well above the floodplain (Figure 15.22).

Flash floods are characteristic of streams draining small watersheds with steep slopes. These streams have short lag times—perhaps only an hour or two—so when an intense rainfall event occurs, the stream rises very quickly to a high level. The flood arrives as a swiftly moving wall of turbulent water, sweeping away buildings and vehicles in its path. In arid western watersheds, great quantities of coarse rock debris are swept into the main channel and travel with the floodwater, producing debris floods (Chapter 14). In forested landscapes, tree limbs and trunks, soil, rocks, and boulders are swept downstream in the floodwaters. Because flash floods often occur too quickly to warn affected populations, they can cause significant loss of life.

Flood Prediction

The magnitude of a flood is usually measured by the peak discharge or highest stage of a river during the period of flooding. As the flood history of any river shows, large

15.21 Floodplain *This aerial view shows the floodplain of the vast Amazon River in the foreground and far distance.*

15.22 Red River in flood *This photo shows the Red River, near Fargo, North Dakota, in flood in April 1997. The normal channel is marked by the sinuous bands of trees running across the lower part of the photo. Dikes have kept the water from flooding the town in the center of the image.*

floods occur less frequently than smaller ones—that is, the greater the discharge or higher the stage, the less likely is the flood. A tool used by hydrologists to present the flood history of a river is the flood expectancy graph. Figure 15.23 shows expectancy graphs for two rivers. The meaning of the bar symbols is explained in the key. The Mississippi River at Vicksburg illustrates a great river responding largely to spring floods, with high waters in April and May and low flows in September and October. The Sacramento River shows the effects of the Mediterranean climate, with its winter wet season and long severe summer drought. Winter floods in January and February are caused by torrential rainstorms and snowmelt in the mountain watersheds of the Sierra Nevada and southern Cascades. By midsummer, river flow has shrunk to a very low stage.

The National Weather Service operates a River and Flood Forecasting Service through 85 offices located at strategic points along major river systems of the United States. When a flood threatens, forecasters analyze precipitation patterns and the progress of high waters moving downstream. Examining the flood history of the rivers and streams concerned,

they develop specific flood forecasts. These are delivered to communities within the associated district, which usually covers one or more large watersheds. Flood warnings are publicized by every possible means, and various agencies cooperate closely to plan the evacuation of threatened areas and the removal or protection of property.

The Mississippi Flood of 1993

The Mississippi River Flood of 1993 provides an example of one of the most serious and damaging floods to strike the United States in recent years. (See chapter opener photo and *Putting You in the Picture.*) Although the usual time for the flooding of the Mississippi is in spring (Figure 15.23), this extreme event occurred in midsummer. The 1993 flood was triggered by a succession of unprecedented rainfalls in the upper Mississippi Basin. In June of that year, rainfall totaled over 30 cm (12 in.) for large areas of the region. Southern Minnesota and western Wisconsin received even larger amounts. Monthly rainfall totals were the highest on many records dating back more than 100 years. The heavy rainfall

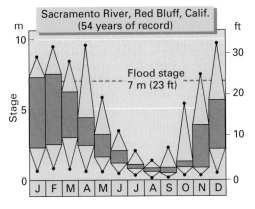

15.23 *Flood expectancy graphs* Maximum and minimum monthly stages of the Mississippi River at Vicksburg, Mississippi, and the Sacramento River at Red Bluff, California.

continued into July, concentrated in Iowa, Illinois, and Missouri. Inundated by huge volumes of rain, the vast wet landscape of the upper Midwest came to resemble a sixth Great Lake when viewed from the air.

As the water drained from the landscape, so the Mississippi and its tributaries rose, creating a flood crest of swiftly moving turbid water that worked its way slowly downstream through July and early August. At nearly every community along the length of the Mississippi and Missouri rivers, citizens fought to raise and reinforce the levees that protected their lands, homes, and businesses from the ravages of the flood. Where levees failed to keep them in check, the rivers filled their broad floodplains from bluff to bluff, as shown in the two Landsat images of our chapter opener photo. By the time it was over, the waters had crested at 16.8 m (49.4 ft) at St. Louis, nearly 6 m (20 ft) above flood stage.

The Mississippi River Flood of 1993 was an environmental disaster of the first magnitude. An area twice the size of New Jersey was flooded, taking 50 lives and driving nearly 70,000 people from their homes. Property damage, including loss of agricultural crops, was estimated at $12 billion. Of the region's 1400 levees, at least 800 were breached or overtopped. Some flood engineers classed the event as a 500-year flood, meaning that only once in a 500-year interval is a flood of this magnitude likely to occur. However, this does not mean that a similar flood, or an even larger one, could not occur in the near future. For the people living on its banks and bottomlands, the mighty Mississippi is a sleeping giant that awakens with devastating consequences.

> Lakes serve as vital reservoirs of fresh water on the land. They are formed in many different ways but are generally short-lived over geologic time.

LAKES

A **lake** is a water body that has an upper surface exposed to the atmosphere and no appreciable gradient. The term *lake* includes a wide range of water bodies. Ponds (which are small, usually shallow water bodies), marshes, and swamps with standing water can all be included under the definition of a lake. Lakes receive water input from streams, overland flow, and ground water, and so are included as parts of drainage systems. Many lakes lose water at an outlet, where water drains over a dam (natural or constructed) to become an outflowing stream. Lakes also lose water by evaporation. Lakes, like streams, are landscape features but are not usually considered to be landforms.

Lakes are quite important from the human viewpoint. They are frequently used as sources of fresh water and food, such as fish. Where dammed to a high level above the outlet stream, they can provide hydroelectric power as well. Lakes and ponds are also important recreation sites and sources of natural beauty.

Where lakes are not naturally present in the valley bottoms of drainage systems, we create lakes as needed by placing dams across the stream channels. Many regions that formerly had almost no natural lakes are now abundantly supplied. Some are small ponds built to serve ranches and farms, while others cover hundreds of square kilometers. In some areas, the number of artificial lakes is large enough to have significant effects on the region's hydrologic cycle.

Basins occupied by lakes show a wide range of origins as well as a vast range in dimensions. Lake basins, like stream channels, are true landforms. Basins are created by a number of geologic processes. For example, the tectonic process of crustal faulting creates many large, deep lakes. Lava flows often form a dam in a river valley, causing water to back up as a lake. Landslides suddenly create lakes, as we saw in the case of the Madison Slide. (See *Eye on the Environment 14.1 ● The Great Hebgen Lake Disaster.*)

An important point about lakes in general is that they are short-lived features on the geologic time scale. Lakes disappear by one of two processes, or a combination of both. First, lakes that have stream outlets will be gradually drained as the outlets are eroded to lower levels. Where a strong bedrock threshold underlies the outlet, erosion will be slow but nevertheless certain. Second, lakes accumulate inorganic sediment carried by streams entering the lake and organic matter produced by plants and animals within the lake. Eventually, they fill up, forming a boggy wetland with little or no free water surface (Figure 15.24).

Lakes can also disappear when climate changes. If precipitation is reduced within a region, or temperatures and net radiation increase, evaporation can exceed input and the lake will dry up. Many former lakes of the southwestern United States flourished in moister periods of glacial advance during the Ice Age. Today, they have shrunk greatly or have disappeared entirely under the present arid regime.

In moist climates, the water level of lakes and ponds coincides closely with the water table in the surrounding area. Seepage of ground water into the lake, as well as direct runoff of precipitation, maintains these water surfaces permanently throughout the year. Examples of such freshwater ponds are found widely distributed in glaciated regions of North America and Europe. Here, plains of glacial sand and gravel contain natural pits and hollows left by the melting of stagnant ice masses that were buried in the sand and gravel deposits (see Chapter 19). Figure 15.25 is a block diagram showing small freshwater ponds on Cape Cod. The surface elevation of these ponds coincides closely with the level of the surrounding water table.

Many former freshwater water table ponds have become partially or entirely filled by organic matter from the growth and decay of water-loving plants (Figure 15.24). The ultimate result is a bog with its surface close to the water table. Freshwater marshes and swamps, in which water stands at or close to the ground surface over a broad area, also represent the appearance of the water table at the surface. Such areas of poor surface drainage are included under the name of *wetlands*.

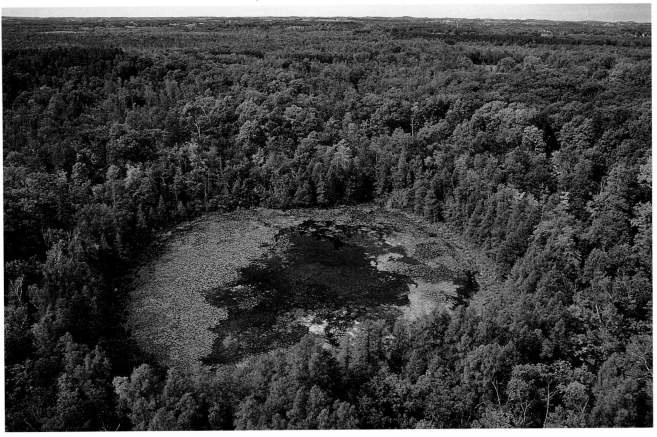

15.24 A freshwater pond in Wisconsin *Vegetation is slowly growing inward at the edges, and eventually the pond will become a bog supporting wet forest.*

Saline Lakes and Salt Flats

Lakes with no surface outlet are characteristic of arid regions. Here, the average rate of water loss by evaporation balances the average rate of stream inflow. If the rate of inflow increases, the lake level will rise. At the same time, the lake surface will increase in area, allowing a greater rate of evaporation. A new balance can then be achieved. Similarly,

if the region becomes more arid, reducing input and increasing evaporation, the water level will fall to a lower level.

Lakes without outlets often show salt buildup. Dissolved solids are brought into the lake by streams—usually streams that head in distant highlands where a water surplus exists. Since evaporation removes only pure water, the salts remain behind and the salinity of the water slowly

15.25 Freshwater ponds *A sketch of freshwater ponds in sandy glacial deposits on Cape Cod, Massachusetts. The water level of the ponds corresponds closely to the water table. (Redrawn from A Geologist's View of Cape Cod, copyright © A. N. Strahler, 1966. Used by permission of Doubleday, a division of Bantam Doubleday Dell Publishing Group, Inc.)*

Glacial sand and gravel

Water table

Pond

Well

Ground water zone

15.26 Salt encrustations
These salt encrustations at the edge of Great Salt Lake, Utah, were formed when the lake level dropped during a dry period.

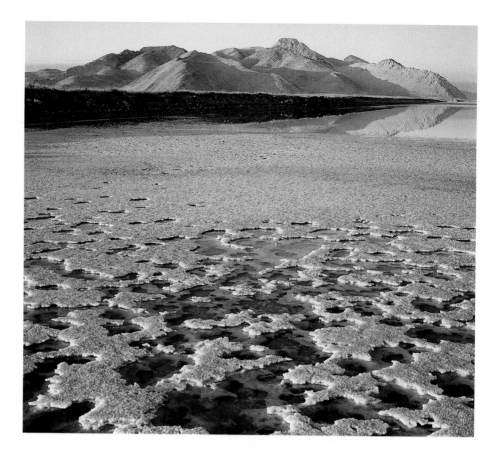

increases. Salinity, or degree of "saltiness," refers to the abundance of certain common ions in the water. Eventually, salinity level reach a point where salts are precipitated as solids (Figure 15.26).

Sometimes the surfaces of such lakes lie below sea level. An example is the Dead Sea, with a surface elevation of –396 m (–1299 ft). The largest of all lakes, the Caspian Sea, has a surface elevation of –25 m (82 ft). Both of these large lakes are saline.

In regions where climatic conditions consistently favor evaporation over input, the lake may be absent. Instead, a shallow basin covered with salt deposits (see Figure 15.32), otherwise known as a *salt flat* or dry lake, will occur. In Chapter 16 we will describe dry lakebeds as landforms. On rare occasions, these flats are covered by a shallow layer of water, brought by flooding streams heading in adjacent highlands.

A number of desert salts are of economic value and have been profitably extracted from salt flats. In shallow coastal estuaries in the desert climate, sea salt for human consumption is commercially harvested by allowing it to evaporate in shallow basins. One well-known source of this salt is the Rann of Kutch, a coastal lowland in the tropical desert of westernmost India, close to Pakistan. Here the evaporation of shallow water of the Arabian Sea has long provided a major source of salt for inhabitants of the interior.

The Aral Sea is an inland lake that is rapidly becoming saline through human activity. The two major rivers that feed this vast central Asian water body have been largely diverted into irrigation of agricultural lands, and with its inflow greatly reduced, the lake has shrunk and become increasingly salty. *Eye on the Environment 15.2 • The Aral Sea—A Dying Saline Lake* tells the story of this environmental disaster.

Desert Irrigation

Human interaction with the tropical desert environment is as old as civilization itself. Two of the earliest sites of civilization—Egypt and Mesopotamia—lie in the tropical deserts. Successful occupation of these deserts requires irrigation with large supplies of water from nondesert sources. For Egypt and Mesopotamia, the water sources of ancient times were the rivers that cross the desert but derive their flow from regions that have a water surplus. These are referred to as *exotic rivers* because their flows are derived from an outside region.

Irrigation systems in arid lands divert the discharge of an exotic river such as the Nile, Indus, Jordan, or Colorado into a distribution system that allows the water to infiltrate the soil of areas under crop cultivation. Ultimately, such irrigation projects can suffer from two undersirable side effects: salinization and waterlogging of the soil.

Eye on the Environment | 15.2

The Aral Sea—A Dying Saline Lake

East of the Caspian Sea, astride the former Soviet republics of Kazakhstan and Uzbekistan, lies an immense saline lake—the Aral Sea. Fed by meltwaters of high glaciers and snowfields in the lofty Hindu Kush, Pamir, and Tien Shan Ranges, the lake endured through thousands of years as an oasis for terrestrial and aquatic wildlife deep in the heart of the central Asian desert.

But in the last 30 years, the Aral Sea, once larger than Lake Huron, has shrunk to a shadow of its former extent. The volume of its waters has decreased by 66 percent, and its salinity has increased from 1 percent to nearly 3 percent, making it almost as salty as sea water. Twenty of the 24 fish species native to the lake have disappeared. Its catch of commercial fish, which once supplied 10 percent of the total for the Soviet Union, has dwindled to zero. The deltas of the Amu Darya and Syr Darya rivers, which enter the south and east sides of the lake, were islands of great ecological diversity, teeming with fingerling fishes, birds, and their predators. Now only about half of the species of nesting birds remain. Many species of aquatic plants, shrubs, and grasses have vanished. Commercial hunting and trapping have almost ceased.

What caused this ecological catastrophe? The answer is simple—the lake's water supply was cut off. As an inland lake with no outlet, the Aral Sea receives water from the Amu Darya and Syr Darya, as well as a small amount from direct precipitation, but it loses water by evaporation. Its gains balanced its losses, and, although these gains and losses varied from year to year, the area, depth, and volume of the lake remained nearly constant until about 1960.

In the late 1950s the Soviet government embarked on the first phases of a vast irrigation program, using water from the Amu Darya and Syr Darya for cotton cropping on the region's desert plain. The diversion of water soon became significant as more and more land came under irrigation. As a result, the influx fell to nearly zero by the early 1980s. The surface level of the Aral was sharply lowered and its area reduced. The sea became divided into two separate parts.

As the lake's shoreline receded, the exposed lakebed became encrusted with salts. The once-flourishing fishing port of Muynak became a ghost town, 50 km (30 mi) from the new lake shoreline. Strong winds now blow salt particles and mineral dusts in great clouds southwestward over the irrigated cotton fields and westward over grazing pastures. These salts—particularly the sodium chloride and sodium sulfate components—are toxic to plants. The salt dust permanently poisons the soil and can only be flushed away with more irrigation water.

The future of the lake appears grim indeed. With the present fresh water inflow nearly zero, the salinity of the remaining lake will soon reach the level of the ocean. Without the sacrifice of agricultural production for water to fill the lake, or the importation of more water from vast distances to the north, there is little that can be done to save the lake.

The Aral Sea shrinks This pair of satellite images shows the Aral Sea in 1976 and 1997. About two-thirds of the sea's volume has been lost in the last 30 years.

Salinization occurs when salts build up in the soil to levels that inhibit plant growth. This happens because an irrigated area within a desert loses large amounts of soil water through evapotranspiration. Salts contained in the irrigation water remain in the soil and increase to high concentrations. Salinization may be prevented or cured by flushing the soil salts downward to lower levels by the use of more water. This remedy requires greater water use than for crop growth alone. In addition, new drainage systems must be installed to dispose of the excess saltwater.

Waterlogging occurs when irrigation with large volumes of water causes a rise in the water table, bringing the zone of saturation close to the surface. Most food crops cannot grow in perpetually saturated soils. When the water table rises to the point at which upward movement under capillary action can bring water to the surface, evaporation is increased and salinization is intensified.

Agricultural areas of major salinization include the Indus River Valley in Pakistan, the Euphrates Valley in Syria, the Nile Delta of Egypt, and the wheat belt of western Australia. In the United States, extensive regions of heavily salinized agriculture are found in the San Joaquin and Imperial valleys of California. Other areas of salinization occur throughout the entire semiarid and arid regions of the western United States. *Eye on the Environment 10.1 • Death of a Civilization* documented the effects of salinization and waterlogging on the civilizations of ancient Mesopotamia.

Pollution of Surface Water

Streams, lakes, bogs, and marshes are specialized habitats of plants and animals. Their ecosystems are particularly sensitive to changes induced by human activity in the water balance and in water chemistry. Our industrial society not only makes radical physical changes in water flow by construction of engineering works (dams, irrigation systems, canals, dredged channels), but also pollutes and contaminates our surface waters with a large variety of wastes.

The sources of water pollutants are many and varied. Some industrial plants dispose of toxic metals and organic compounds by discharging them directly into streams and lakes. Many communities still discharge untreated or partly treated sewage wastes into surface waters. In urban and suburban areas, pollutant matter entering streams and lakes includes deicing salt and lawn conditioners (lime and fertilizers), which can also contaminate ground water. In agricultural regions, important sources of pollutants are fertilizers and livestock wastes. Mining and processing of mineral deposits are also major sources of water pollution. Even

Salinization and waterlogging are undesirable side effects of long-continued irrigation. Arid regions watered by exotic rivers are most affected.

Water pollutants include various types of common ions and salts, as well as heavy metals, organic compounds, and acids. Excessive plant nutrients in runoff feeding lakes can lead to eutrophication— the artificial "aging" of a lake.

contamination by radioactive substances released from nuclear power and processing plants can occur.

Among the common chemical pollutants of both surface water and ground water are sulfate, chloride, sodium, nitrate, phosphate, and calcium ions. (Recall from Chapter 10 that an ion is the charged form of a molecule or an atom and that many chemical compounds dissolve in water by forming ions. Sulfate ions enter runoff both by fallout from polluted urban air and as sewage effluent. Chloride and sodium ions are contributed both by fallout from polluted air and by deicing salts used on highways. In some locations with snowy winters, community water supplies located close to highways have become polluted from deicing salts. Important sources of nitrate ions are fertilizers and sewage effluent. Excessive concentrations of nitrate in freshwater supplies are highly toxic, and, at the same time, their removal is difficult and expensive. Phosphate ions are contributed in part by fertilizers and by detergents in sewage effluent.

Phosphate and nitrate are plant nutrients and can lead to excessive growth of algae and other aquatic plants in streams and lakes. Applied to lakes, this process is known as *eutrophication,* which is often described as the "aging" of a lake. In eutrophication, the accumulation of nutrients stimulates plant growth, producing a large supply of dead organic matter in the lake. Microorganisms break down this organic matter but require oxygen in the process. However, oxygen dissolves only slightly in water, and so it is normally present only in low concentrations. The added burden of oxygen use by the decomposers reduces the oxygen level to the point where other organisms, such as desirable types of fish, cannot survive. After a few years of nutrient pollution, the lake can take on the characteristics of a shallow pond that results when a lake is slowly filled with sediment and organic matter over thousands of years by natural "aging" processes.

A particular form of chemical pollution of surface water goes under the name of *acid mine drainage.* It is an important form of environmental degradation in parts of Appalachia where abandoned coal mines and strip-mine workings are concentrated (Figure 15.27). Ground water emerges from abandoned mines and as soil water percolating through strip-mine waste banks. This water contains sulfuric acid and various salts of metals, particularly of iron. Acid of this origin in stream waters can have adverse effects on animal life. In sufficient concentrations, it is lethal to certain species of fish and has at times caused massive fish kills.

Toxic metals, among them mercury, along with pesticides and a host of other industrial chemicals, are introduced into

15.27 Strip-mine water pollution *This small stream is badly polluted by acid waters that have percolated through strip-mine waste.*

streams and lakes in quantities that are locally damaging or lethal to plant and animal communities. In addition, sewage introduces live bacteria and viruses that are classed as biological pollutants. These pose a threat to the health of humans and animals alike.

Another form of pollution is *thermal pollution*, which refers to the discharge of heat into the environment from combustion of fuels and from the conversion of nuclear energy into electric power. Thermal pollution of water takes the form of discharges of heated water into streams, estuaries, and lakes, which can have drastic effects on local aquatic life. The impact may be quite large in a small area.

> **Human society is heavily dependent on fresh surface water for irrigation, potable water supplies, and industrial consumption. Artificial dams can increase freshwater supplies and generate power, but may have environmental costs.**

SURFACE WATER AS A NATURAL RESOURCE

Fresh surface water is a basic natural resource essential to human agricultural and industrial activities. Runoff held in reservoirs behind dams provides water supplies for great urban centers, such as New York City and Los Angeles. When diverted from large rivers, it provides irrigation water for highly productive lowlands in arid lands, such as the Sacramento and San Joaquin valleys of California and the Nile Valley of Egypt. To these uses of runoff are added hydroelectric power, where the gradient of a river is steep, or routes of inland navigation, where the gradient is gentle.

Our heavily industrialized society requires enormous supplies of fresh water for its sustained operation. Urban dwellers consume water in their homes at rates of 150 to 400 liters (50 to 100 gallons) per person per day (Figure 15.28). Large quantities of water are used for cooling in air conditioning units and power plants. Much of this water is obtained from surface water supplies, and the demand increases daily.

Unlike ground water, which represents a large water storage body, fresh surface water in the liquid state is stored only in small quantities. (An exception is the Great Lakes system.) Recall from Chapter 4 that the global quantity of available ground water is about 20 times as large as that stored in freshwater lakes and that the water held in streams is only about one one-hundredth of that in lakes. Because of small natural storage capacities, surface water can be drawn only at a rate comparable with its annual renewal through precipitation. Dams are built to develop useful storage

15.28 Domestic water use *This pie chart shows how water is used in an average home in Akron, Ohio. (U.S. Geological Survey.)*

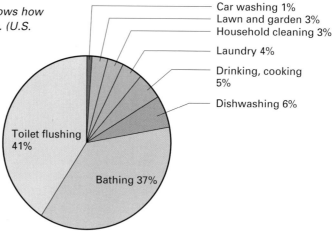

Car washing 1%
Lawn and garden 3%
Household cleaning 3%
Laundry 4%
Drinking, cooking 5%
Dishwashing 6%
Toilet flushing 41%
Bathing 37%

capacity for runoff that would otherwise escape to the sea, but once the reservoir has been filled, water use must be scaled to match the natural supply rate averaged over the year. Development of surface water supplies brings on many environmental changes, both physical and biological, and these must be taken into account in planning for future water developments.

A LOOK AHEAD This chapter has focused on water, including the precipitation that runs off the land and flows into the sea. Think for a moment about the drainage system of a stream that conducts this flow of runoff. The smallest streams catch runoff from slopes, carrying the runoff into larger streams. The larger streams, in turn, receive runoff from their side slopes and also pass their flow on to still larger streams. However, this cannot happen unless the gradients of slopes and streams are adjusted so that water keeps flowing downhill. This means that the landscape is shaped and organized into landforms that are an essential part of the drainage system. The shaping of landforms within the drainage system occurs as running water erodes the landscape, which is the subject of the next chapter.

In Review | Fresh Water of the Continents

- The fresh water of the lands accounts for only a small fraction of the Earth's water. Since it is produced by precipitation over land, it depends on the continued operation of the hydrologic cycle for its existence. The soil layer plays a key role in determining the fate of precipitation by diverting it in three ways: to the atmosphere as *evapotranspiration,* to ground water through *percolation,* and to streams and rivers as **runoff.**

- **Ground water** occupies the pore spaces in rock and regolith. The water table marks the upper surface of the *saturated zone* of ground water, where pores are completely full of water. Ground water moves in slow paths deep underground, recharging rivers, streams, ponds, and lakes by upward seepage and thus contributing to runoff. Solution of limestone by ground water can produce *limestone caverns* and generate *karst* landscapes.

- Wells draw down the water table and, in some regions, lower the water table more quickly than it can be recharged. Ground water contamination can occur when precipitation percolates through contaminated soils or waste materials. Landfills and dumps are common sources of ground water contaminants.

- Runoff includes *overland flow,* moving as a sheet across the land surface, and *stream flow* in **streams** and rivers, which is confined to a **channel.** Rivers and streams are organized into a **drainage system** that moves runoff from slopes into channels and from smaller channels into larger ones. The **discharge** of a stream measures the flow rate of water moving past a given location. Discharge increases downstream as tributary streams add more runoff.

- The *hydrograph* plots the discharge of a stream at a location through time. Since it takes time for water to move down slopes and into progressively larger stream channels, peak discharge differs from peak precipitation by a *lag time.* The larger the stream, the longer the lag time and the more gradual the rate of decline of discharge after the peak. Because urbanization typically involves covering ground surfaces with impervious materials, urban streams exhibit shorter lag times and higher peak discharges. Annual hydrographs of streams from humid regions show an annual cycle of *base flow* on which are superimposed discharge peaks related to individual rainfall episodes.

- **Floods** occur when river discharge increases and the flow can no longer be contained within the river's usual channel. Water spreads over the *floodplain,* inundating low fields and forests adjacent to the channel. When discharge is high, floodwaters can rise beyond normal levels to inundate nearby areas of development, causing damage and sometimes taking lives. *Flash floods* occur in small, steep watersheds and can be highly destructive. The Mississippi Flood of 1993, occurring at an unusual time—during the summer rather than the spring—was an extreme flood event that caused widespread destruction and damage.

- **Lakes** are especially important parts of the drainage system because they are sources of fresh water. They are also used for recreation, and, in many cases, they can supply hydroelectric power. Where lakes occur in inland basins, they are often saline. When climate changes, such lakes can dry up, creating *salt flats.*

- Irrigation is the diversion of fresh water from streams and rivers to supply the water needs of crops. In desert regions, where irrigation is most needed, problems of *salinization* and *waterlogging* can occur, reducing productivity and eventually creating unusable land.

- Water pollution arises from many sources, including industrial sites, sewage treatment plants, agricultural activities, mining, and processing of mineral deposits. Sulfate, nitrate, phosphate, chloride, sodium, and calcium ions are frequent contaminants. Toxic metals, pesticides, and industrial chemicals are also hazards.

- Ground water and surface water resources are essential for human activities. Human civilization is dependent on abundant supplies of fresh water for many uses. But because the fresh water of the continents is such a small part of the global water pool, utilization of water resources takes careful planning and management.

Key Terms

infiltration, p. 485
runoff, p. 485
overland flow, p. 485
ground water, p. 486

water table, p. 486
artesian well, p. 488
stream, p. 493
channel, p. 493

discharge, p. 494
drainage system, p. 497
drainage basin, p. 497
watershed, p. 497

flood, p. 499
lake, p. 502

Review Questions

1. What happens to precipitation falling on soil? What processes are involved?
2. How and under what conditions does precipitation reach ground water?
3. How do caverns come to be formed in limestone? Describe the key features of a karst landscape.
4. How do wells affect the water table? What happens when pumping exceeds recharge?
5. How is ground water contaminated? Describe how a well might become contaminated by a nearby landfill dump.
6. Define discharge (of a stream) and the two quantities that determine it. How does discharge vary in a downstream direction? How does gradient vary in a downstream direction?
7. What is a drainage system? How are slopes and streams arranged in a drainage basin?
8. Define the term *flood.* What is the floodplain? What factors are used in forecasting floods?
9. Why was the Mississippi Flood of 1993 so unusual?
10. How are lakes defined? What are some of their characteristics? What factors influence the size of lakes?

11. Describe some of the problems that can arise in long-continued irrigation of desert areas.

12. Identify common surface water pollutants and their sources.

13. How is surface water utilized as a natural resource?

Eye on Global Change 15.1 • Sinking Cities

1. Why has land subsidence occurred in Venice? What are the effects?

2. Identify four other locations that have suffered land subsidence caused by ground water withdrawal.

Eye on the Environment 15.2 • The Aral Sea—A Dying Saline Lake

1. What changes have occurred in the Aral Sea in the past 30 years?

2. Why have these changes occurred? What would be required to reverse them?

Visualizing Exercises

1. Sketch a cross section through the land surface showing the position of the water table and indicating flow directions of subsurface water motion with arrows. Include the flow paths of ground water. Be sure to provide a stream in your diagram. Label the saturated and unsaturated zones.

2. Why does water rise in an artesian well? Illustrate with a sketched cross-sectional diagram showing the aquifer, aquicludes, and the well.

Essay Questions

1. A thundershower causes heavy rain to fall in a small region near the headwaters of a major river system. Describe the flow paths of that water as it returns to the atmosphere and ocean. What human activities influence the flows? in what ways?

2. Imagine yourself a recently elected mayor of a small city located on the banks of a large river. What issues might you be concerned with that involve the river? In developing your answer, choose and specify some characteristics for this city—such as its population, its industries, its sewage systems, and the present uses of the river for water supply or recreation.

Eye on the Landscape

Chapter Opener Mississippi Flood of 1993 Note the former river channels that have been cut off by migration of the river channel **(A)**. These are called ox-bow lakes and are described in Chapter 16. These images also demonstrate a type of false-color image in which the red color shows shortwave infrared reflectance, green shows near-infrared reflectance, and blue shows red reflectance. Recall from Chapter 3 that vegetation reflects very strongly in near-infrared wavelengths, and thus vegetation appears bright green **(B)**. Urban surfaces are bright in the shortwave infrared and thus appear red **(C)**. Sediment-laden river water is brown and so provides a strong red signal that translates as blue in the image **(D)**.

15.9 Sinkholes Note the amount of bare rock visible in many large patches on this arid plain **(A)**. The lack of soil suggests that the limestone is so pure that its dissolution leaves little or no residual material behind. Also, look at the vegetation ringing the sinkholes **(B)**. The plants are probably drawing on ground water. From this observation, we might conclude that ground water is not very far below the vegetated region between the two roads **(C)**. Solution weathering of limestone was covered in Chapter 14. Phreatophytes, which are plants that survive in an arid environment by tapping the water table, are discussed in Chapter 8.

15.11 **Tower karst** As we saw in Chapter 14, solution weathering of certain types of bedrock in warm and wet environments can produce a landscape of steep, vertical slopes. Compare these towers **(A)**, formed by solution of limestone, with the fins and grooves of Kauai in Figure 14.12, formed by solution of basaltic lava. Note also the flooded fields **(B)**. They are probably rice paddies in the spring, just before planting with young rice stalks.

EYE ON THE LANDSCAPE

White-water rafting on the Colorado River in Grand Canyon National Park, Arizona.
What else would the geographer see? … Answers at the end of the chapter.

16 | LANDFORMS MADE BY RUNNING WATER

• • • • • • • • • • • • • • • •

PUTTING YOU IN THE PICTURE

Have you ever experienced the power of a river first-hand? One of the best ways to learn about rivers is to ride one for a while. Picture yourself on a rafting trip down the Colorado River as you follow the narrative below.

The rickety school bus turns off the pavement and bumps down the dirt road toward the riverbank. With a lurch, it stops on a wide, gravel-covered spot next to the river, and you clamber off. Behind you is the ancient truck carrying the rafts, paddles, and gear that you and your group will need for your 12-day rafting trip through the majestic Grand Canyon of the Colorado River.

Soon, you are afloat, seven to a raft, slowly drifting with the current down the wide river. "Left side, forward!" "Right side, back!" The river guide barks the orders as you practice paddling together as a team. Sitting on the outer tube of the raft, you twist sideways, dipping your canoe paddle deeply into the blue-green water and pushing hard. This is real work! The river guide explains that safe passage through the hundred or more rapids you'll encounter on the trip will depend on how well you and your raftmates can move your craft across the swift currents of the mighty river.

Then, almost too quickly, you are at the first riffle. The raft glides easily through the rough water, and you think that perhaps this won't be so difficult after all. Then your river guide tells you that before the day is done, you'll be shooting the Badger Creek and Soap Creek Rapids, which have claimed several lives over the years. Back to paddle exercises, this time in earnest!

As the days pass, you come to feel quite at home with the river. The thrill of wild white-water rides has alternated with the calm of slow drifting in the bright summer sun. Although you've taken foot trips to explore the side canyons and historic sites along the way, you've always returned to the river to continue your journey, and you feel a close bond with it. But this comfortable feeling about the river is due to change. Lava Falls—the last major obstacle on your river trip—lies just downriver. With its 11 m (37 ft) drop, it is the highest in the canyon.

A booming noise begins to build, telling you that the rapids are approaching. The smooth, swift tongue of water that leads into the rapids accelerates the raft. You and your crew paddle furiously across the current, aiming for the side of a huge drop that lies ahead. Suddenly, you're falling over the edge and down into the froth below. Gliding and twisting, the raft splashes into the white water and is thrown violently to the side. Flying spray hits you, hard. The shock of the cold water on your face and arms makes you gasp for breath. The bow rises as the raft hits the huge standing wave on the downstream side of the hole.

"Pull! Pull!" The guide's urgent shout rises above the roar. Your paddle splits the water, backed by every ounce of strength you can muster. Perched on top of the standing wave, the raft seems for a sickening moment about to slip back into the pitching waters of the hole behind. But suddenly, the raft's bow drops and you're yanked forward as the raft falls into the trough beyond. Then it's a wild roller coaster ride through the haystacks of waves below. The raft rebounds to the right off a huge pillow of foam, then bounces into a side eddy, a patch of calmer water well below the roaring fall. You catch your breath, your heart pounding with the excitement, surrounded by white water on every side. What an amazing trip!

Rivers are powerful, indeed, and especially so in flood. From the physical geographer's perspective, running water is the most important of all the agents that create landforms. The processes of erosion, transportation, and deposition by fluvial systems, and the landforms they create, are the subjects of this chapter. ∎

LANDFORMS MADE BY RUNNING WATER

Most of the world's land surface has been sculpted by running water. Flowing as a sheet across a land surface, running water picks up particles and moves them downslope into a stream channel. When rainfall is heavy, streams and rivers swell, lifting large volumes of sediment and carrying them downstream. In this way, running water erodes mountains and hills, carves valleys, and deposits sediment. This chapter describes the work of running water and the landforms it creates.

Running water is one of four flowing substances that erode, transport, and deposit mineral and organic matter. The other three are waves, glacial ice, and wind. These four fluid agents carry out the processes of *denudation,* which we discussed in Chapter 13. Recall that denudation is the total action of all processes by which the exposed rocks of the continents are worn away and the resulting sediments are transported to the sea or closed inland basins.

Denudation is an overall lowering of the land surface. If left unchecked to operate over geologic time, denudation will reduce a continent to a nearly featureless, sea-level surface. However, recall that plate tectonics has always kept continental crust elevated well above the ocean basins. The result is that running water, waves, glacial ice, and wind have always had plenty of raw material available to create the many landforms that we see around us.

Thanks to denudation processes and plate tectonics, the land environments of life are always in constant change, even as plants and animals have undergone evolutionary development. Wind, water, waves, and ice produce and maintain a wide variety of landforms that are the habitats for evolving life forms. In turn, the life forms become adapted to those habitats and diversify to a degree that matches the diversity of the landforms themselves.

FLUVIAL PROCESSES AND LANDFORMS

Landforms shaped by running water are described as **fluvial landforms.** They are shaped by the **fluvial processes** of overland flow and stream flow, which we described in Chapter 15. Weathering and the slower forms of mass wasting (Chapter 14), such as soil creep, operate hand in hand with overland flow, providing the rock and mineral fragments that are carried into stream systems.

Fluvial landforms and fluvial processes dominate the continental land surfaces the world over. Throughout geologic history, glacial ice has been present only in continental areas located in mid- and high latitudes and in high mountains. Landforms made by wind

---

Fluvial landforms are shaped by the fluvial processes of overland flow and stream flow. Wherever rain falls, these processes act to create erosional and depositional landforms.

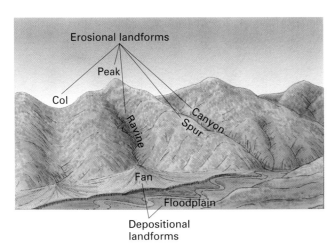

16.1 Erosional and depositional landforms (A. N. Strahler.)

action are found only in very small parts of the continental surfaces. And landforms made by waves and currents are restricted to the narrow contact zone between oceans and continents. That is why, in terms of area, the fluvial landforms dominate the environment of terrestrial life.

Areas of fluvial landforms are also exploited for agriculture. Except for areas in the northern hemisphere that were formerly occupied by glacial ice, most land areas used in crop cultivation or for grazing have been shaped by fluvial processes. For this reason, fluvial landforms are very important to the support of the human species.

Erosional and Depositional Landforms

All agents of denudation perform the geological activities of erosion, transportation, and deposition. Consequently, there are two major groups of landforms—erosional landforms and depositional landforms. When an initial landform, such as an uplifted crustal block, is created, it is attacked by the processes of denudation and especially by fluvial action. Valleys are formed where rock is eroded away by fluvial agents. Between the valleys are ridges, hills, or mountain summits, representing the remaining parts of the crustal block that are as yet uncarved by running water. These sequential landforms, shaped by progressive removal of the bedrock mass, are *erosional landforms*. Fragments of soil, regolith, and bedrock that are removed from the parent rock mass are transported and deposited elsewhere to make an entirely different set of surface features—the *depositional landforms*.

Figure 16.1 illustrates the two groups of landforms as produced by fluvial processes. The ravine, canyon, peak,

Soil erosion occurs when overland flow moves soil particles downslope. Erosion is greatest on bare slopes of fine particles, carving rills and gullies. A vegetation cover greatly reduces soil erosion.

spur, and col are erosional landforms. The fan, built of rock fragments below the mouth of the ravine, is a depositional landform. The floodplain, built of material transported by a stream, is also a depositional landform.

SLOPE EROSION

Fluvial action starts on the uplands as *soil erosion*. By exerting a dragging force over the soil surface, overland flow picks up particles of mineral matter ranging in size from fine colloidal clay to coarse sand or even gravel. The size grade selected depends on the speed of the current and the degree to which the particles are bound by plant rootlets or held down by a mat of leaves. Added to this solid matter is dissolved mineral matter in the form of ions produced by acid reactions or direct solution.

This ongoing removal of soil is part of the natural geological process of denudation. It occurs everywhere that precipitation falls on land. Under stable natural conditions in a humid climate, the erosion rate is slow enough that a soil with distinct horizons is formed and maintained. Each year a small amount of soil is washed away, while a small amount of solid rock material becomes altered to new regolith and soil. These conditions also enable plant communities to maintain themselves in a stable equilibrium. Soil scientists refer to this state of activity as the *geologic norm*.

Accelerated Erosion

In contrast, the rate of soil erosion may be enormously speeded up by human activities or by rare natural erosional events to produce a state of *accelerated erosion*. What happens then is that the soil is removed much faster than it can be formed, and the uppermost soil horizons are progressively exposed. Accelerated erosion arises most commonly when the plant cover and the physical state of the ground surface change. Destruction of vegetation by the clearing of land for cultivation or by forest fires sets the stage for a series of drastic changes. No foliage remains to intercept rain, and the protection of a ground cover of fallen leaves and stems is removed. Consequently, the raindrops fall directly on the mineral soil.

The direct force of falling drops on bare soil causes a geyser-like splashing in which soil particles are lifted and then dropped into new positions. This process is termed *splash erosion* (Figure 16.2). Soil scientists estimate that a torrential rainstorm has the ability to disturb as much as 225 metric tons of soil per hectare (about 100 U.S. tons per acre). On a sloping ground surface, splash erosion shifts the soil slowly downhill. An even more important effect is to cause the soil surface to become much less able to absorb water. This occurs because the natural soil openings become sealed by particles shifted by raindrop splash. Reduced infil-

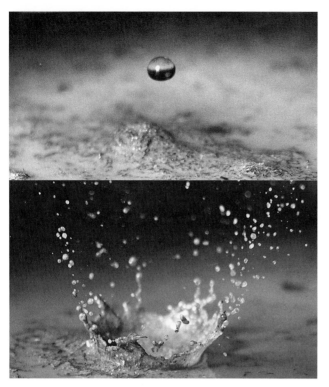

16.2 Soil erosion by rain splash *A large raindrop (above) lands on a wet soil surface, producing a miniature crater (below). Grains of clay and silt are thrown into the air, and the soil surface is disturbed.*

tration, in turn, permits a much greater depth of overland flow to occur from a given amount of rain. So the rate of soil erosion is intensified.

Another effect of the destruction of vegetation is to reduce greatly the resistance of the ground surface to the force of erosion under overland flow. On a slope covered by grass sod, even a deep layer of overland flow causes little soil erosion. This protective action is present because the energy of the moving water is dissipated in friction with the grass stems, which are tough and elastic. On a heavily forested slope, the surface layer of leaves, twigs, roots, and even fallen tree trunks take up the force of overland flow.

Without such a cover, the eroding force is applied directly to the bare soil surface, easily dislodging the grains and sweeping them downslope.

We can get a good appreciation of the contrast between normal and accelerated erosion rates by comparing the quantity of sediment derived from cultivated surfaces with that derived from naturally forested or reforested surfaces. The comparison is made within a single region in which climate, soil, and topography are fairly uniform. *Sediment yield* is a technical term for the quantity of sediment removed by overland flow from a unit area of ground surface in a given unit of time. Yearly sediment yield is stated in metric tons per hectare, or tons per acre.

Figure 16.3 gives data of annual average sediment yield and runoff by overland flow from several types of upland surfaces in northern Mississippi. Notice that both surface runoff and sediment yield decrease greatly with the increased effectiveness of the protective vegetative cover. Sediment yield from cultivated land undergoing accelerated erosion is over 10 times greater than that from pasture and about one thousand times greater than that from pine plantation land.

Sheet Erosion and Rilling

Accelerated soil erosion is a constant problem in cultivated regions with a substantial water surplus. When the natural cover of forest or prairie grasslands is first removed and the soil is plowed for cultivation, little erosion will occur until the action of rain splash has broken down the soil aggregates and sealed the larger openings. Then, however, overland flow begins to remove the soil in rather uniform thin layers, a process termed *sheet erosion*. Because of seasonal cultivation, the effects of sheet erosion are often little noticed until the upper horizons of the soil are removed or greatly thinned.

Where land slopes are steep, runoff from torrential rains produces a more destructive activity, *rill erosion*, in which many closely spaced channels are scored into the soil and regolith. If the rills are not destroyed by soil tillage, they may soon begin to join together into still larger channels. These deepen rapidly and soon become *gullies*—steep-

16.3 Runoff and sediment yield *This bar graph shows that both runoff and sediment yield are much greater for open land than for land covered by shrubs and forest. (Data of S.J. Ursic U.S.D.A.)*

	Land use or cover type	Average annual runoff: cm/yr (in./yr)	Average annual sediment yield: metric tons/hectare (tons/acre)
Open land	Cultivated	40 (16)	50 (22)
	Pasture	38 (15)	3.6 (1.6)
Forest land	Abandoned fields	18 (7)	0.3 (0.13)
	Depleted hardwoods	13 (5)	0.2 (0.1)
	Pine plantations	2.5 (1)	0.05 (0.02)

16.4 Gullies *Deep branching gullies have carved up an overgrazed pasture near Shawnee, Oklahoma. Contour terracing and check dams have halted the headward growth of the gullies.*

walled, canyon-like trenches whose upper ends grow progressively upslope (Figure 16.4). Ultimately, a rugged, barren topography results from accelerated soil erosion that is allowed to proceed unchecked.

Colluvium and Alluvium

With an increased depth of overland flow and no vegetation cover to absorb the eroding force, soil particles are easily picked up and moved downslope. Eventually, they reach the base of the slope, where the surface slope becomes more gentle and meets the valley bottom. There the particles come to rest and accumulate in a thickening layer termed *colluvium* (see Figure 14.14). Because this deposit is built by overland flow, it has a sheet-like distribution and may be little noticed, except where it eventually buries fence posts or tree trunks.

If not deposited as colluvium, sediment carried by overland flow eventually reaches a stream in the adjacent valley floor. Once in the stream, it is carried farther downvalley and may accumulate as alluvium in layers on the valley floor. As we saw in Chapter 15, the term **alluvium** is used to describe any stream-laid sediment deposit. Deposition of alluvium can bury fertile floodplain soil under infertile, sandy layers.

Coarse alluvium chokes the channels of small streams and can cause the water to flood broadly over the valley bottoms.

Slope Erosion in Semiarid and Arid Environments

Thus far, we have discussed slope erosion in moist climates with a natural vegetation of forest or a dense prairie grassland. Conditions are quite different in a midlatitude semiarid climate with summer drought. Here, the natural plant cover consists of short-grass prairie (steppe). Although it is sparse and provides a rather poor ground cover of plant litter, the grass cover is normally strong enough that a slow pace of erosion can be sustained.

Much the same conditions are also found in the tropical savanna grasslands. In these semiarid environments, however, the natural equilibrium is highly sensitive and easily upset. Depletion of the plant cover by fires or the grazing of herds of domesticated animals can easily set off rapid erosion. These sensitive, marginal environments require cautious use because they lack the potential to recover rapidly from accelerated erosion once it has begun.

Erosion at a very high rate by overland flow is actually a natural process in certain favorable locations in semiarid and

16.5 Badlands *Clay beds form badlands at Zabriskie Point, Death Valley National Monument, California. (A. H. Strahler.)*

arid lands. Here, the erosion produces *badlands*. Badlands are underlain by clay formations, which are easily eroded by overland flow. Erosion rates are too fast to permit plants to take hold, and no soil can develop. A maze of small stream channels is developed, and ground slopes are very steep (Figure 16.5).

One well-known area of badlands in the semiarid short-grass prairie is the Big Badlands of South Dakota, along the White River. Badlands such as these are self-sustaining and have been in existence on continents throughout much of geologic time. Badlands can also result from poor agricultural processes, especially when the vegetation cover of clay formations is disturbed by plowing or overgrazing.

THE WORK OF STREAMS

The work of streams consists of three closely related activities—erosion, transportation, and deposition. **Stream erosion** is the progressive removal of mineral material from the floor and sides of the channel, whether bedrock or regolith. **Stream transportation** consists of movement of the eroded particles dragged over the stream bed, suspended in the body of the stream, or held in solution as ions. **Stream deposition** is the accumulation of transported particles on the stream bed and floodplain, or on the floor of a standing body of water into which the stream empties. Erosion cannot occur

without some transportation taking place, and the transported particles must eventually come to rest. Thus, erosion, transportation, and deposition are simply three phases of a single activity.

Stream Erosion

Streams erode in various ways, depending on the nature of the channel materials and the tools with which the current is armed. The force of the flowing water not only sets up a dragging action on the bed and banks, but also causes particles to impact the bed and banks. Dragging and impact can easily erode alluvial materials, such as gravel, sand, silt, and clay. This form of erosion, called *hydraulic action,* can excavate enormous quantities in a short time. The undermining of the banks causes large masses of alluvium to slump into the river, where the particles are quickly separated and become part of the stream's load. This process of bank caving is an important source of sediment during high river stages and floods.

Where rock particles carried by the swift current strike against bedrock channel walls, chips of rock are detached. The large, strong fragments become rounded as they travel. The rolling of cobbles and boulders over the stream bed further crushes and grinds the smaller grains to produce a wide assortment of grain sizes. This process of mechanical wear is called *abrasion*. It is the principal means of erosion in

> Streams erode their beds and banks by hydraulic action, abrasion, and corrosion. Abrasion by stones on a bedrock riverbed can create deep depressions known as potholes.

16.6 Potholes *These potholes in lava bedrock attest to the abrasion that takes place on the bed of a swift mountain stream. McCloud River, California. (A. H. Strahler.)*

of corrosion are conspicuous in limestone, which develops cupped and fluted surfaces.

Stream Transportation

The solid matter carried by a stream is the **stream load.** It is carried in three forms (Figure 16.7). *Dissolved matter* is transported invisibly in the form of chemical ions. All streams carry some dissolved ions resulting from mineral alteration. Sand, gravel, and larger particles move as *bed load* close to the channel floor by rolling or sliding. Clay and silt are carried in *suspension*—that is, they are held within the water by the upward elements of flow in turbulent eddies in the stream. This fraction of the transported matter is the *suspended load* (Figure 16.8). Of the three forms, suspended load is generally the largest.

A large river such as the Mississippi carries as much as 90 percent of its load in suspension. Most of the suspended load comes from its great western tributary, the Missouri River, which is fed from semiarid lands, including the Dakota Badlands. (Figure 15.17 shows the Mississippi River system and its major branches.)

bedrock too strong to be affected by simple hydraulic action. A striking example of abrasion is the erosion of a *pothole.* This occurs when a shallow depression in the bedrock of a stream bed acquires one or several grinding stones, which are spun around and around by the flowing water and carve a deep depression (Figure 16.6).

Finally, the chemical processes of rock weathering—acid reactions and solution—are effective in removing rock from the stream channel. The process is called *corrosion.* Effects

---

Streams carry dissolved matter, sediment in suspension, and a bed load of larger particles bumped and rolled along the bottom. A stream's capacity to carry sediment increases sharply with its velocity.

The Yellow River (Huanghe) of China heads the world list in annual suspended sediment load, and its watershed sediment yield is one of the highest known for a large river basin. This is because much of its basin consists of cultivated upland surfaces of wind-deposited silt that is very easily eroded. (See Chapter 18 and Figure 18.32.) In addition, much of the river's upper watershed is in a semiarid climate with dry winters. Vegetation is sparse, and the runoff from heavy summer rains sweeps up a large amount of sediment.

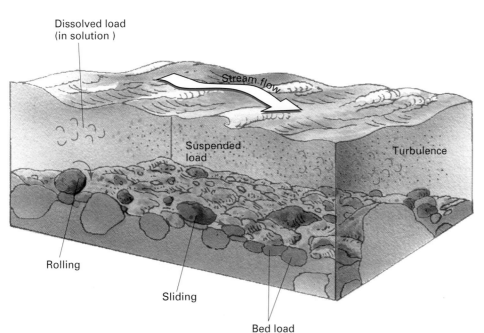

Dissolved load (in solution)

Stream flow

Suspended load

Turbulence

Rolling

Sliding

Bed load

16.7 Sediment load
Streams carry their load as dissolved, suspended, and bed load. Suspended load is kept in suspension by turbulence. Bed load moves by sliding or rolling.

16.8 Suspended sediment *This turbulent stream in Madagascar carries a heavy load of suspended sediment. Gullying erodes the denuded hills in the foreground.*

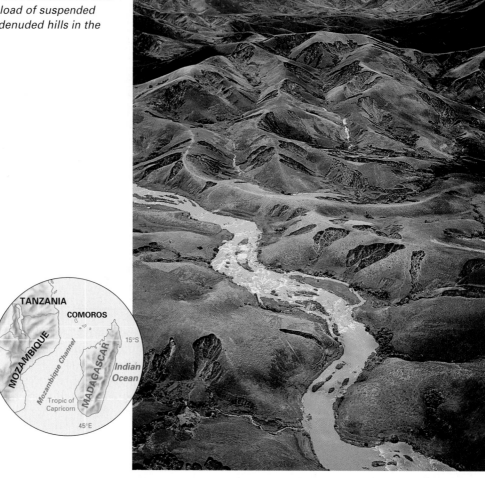

Capacity of a Stream to Transport Load

The maximum solid load of debris that can be carried by a stream at a given discharge is a measure of the *stream capacity*. This load is usually measured in units of metric tons per day passing downstream at a given location. Total solid load includes both bed load and suspended load.

A stream's capacity to carry suspended load increases sharply with an increase in the stream's velocity because the swifter the current, the more intense the turbulence. The capacity to move bed load also increases with velocity—the faster water motion produces a stronger dragging force against the bed. In fact, the capacity to move bed load increases according to the third to fourth power of the velocity. In other words, when a stream's velocity is doubled in times of flood, its ability to transport bed load is increased from eight to sixteen times. Thus, most of the conspicuous changes in the channel of a stream occur in a flood stage.

When water flow increases, a stream flowing in a channel that is cut into thick layers of silt, sand, and gravel will easily widen and deepen its channel. When

---

Over time, a stream develops a graded profile in which the gradient is just sufficient to carry the average annual load of water and sediment produced by its drainage basin.

---

the flow slackens, the stream will deposit material in the bed, filling the channel again. Where a stream flows in a channel of hard bedrock, the channel cannot be quickly deepened in response to rising waters and may not change much during a single flood. Such conditions exist in streams that occupy deep canyons and have steep gradients.

STREAM GRADATION

Most major stream systems have gone through thousands of years of runoff, erosion, and deposition. Over time, the gradients of stream segments tend to become adjusted so that they just carry the average load of sediment that they receive from slopes and inflowing channels. How does this come about? *GEODISCOVERIES*

Visualize a small stream basin in which runoff and overland flow carry sediment to a stream channel. If more sediment accumulates each year in the stream channel than can be carried away, the surface of the channel will be built channel will be built up and the slope of the stream will increase. But with increased slope comes increased stream velocity and increased ability to carry sediment. Eventually, the slope will reach a point at which the stream just carries away the sediment that it receives. If sediment flow to the stream is reduced, the stream will gradually erode its channel downward. This will reduce its slope and also reduce its ability to carry sediment until it can only carry the reduced amount it receives from the hill slopes. Since every stream channel experiences this process, eventually the whole stream will tend toward a state in which the slopes of all its segments form a coordinated network that just carries the sediment load contributed by the drainage basin. A stream in this equilibrium condition is referred to as a **graded stream.**

Figure 16.9 shows how a graded stream might develop on a landscape that is rapidly uplifted, perhaps by a series of fault steps or blocks, or possibly uncovered after erosion by continental glaciers. The side of the block shows a series of *stream profiles*—plots of elevation of the stream with distance from the sea. At first, the stream is ungraded, with large fluctuations in the slope profile. Water accumulates in shallow depressions to form lakes, which overflow from higher to lower levels. Rapids and falls are abundant. As time passes, the landscape is slowly eroded by fluvial action. Each stream segment seeks its own equilibrium slope, and the stream profile is smoothed out into a uniform curve, shown as profile line 3. The profile has now been graded. From that time forward, this *graded profile* is steadily lowered in elevation as the landscape is further eroded (curves 4 through 6).

Landscape Evolution of a Graded Stream

Figure 16.10 illustrates the stream gradation process over a landscape in a series of block diagrams. In (*a*), we see *waterfalls* and *rapids,* which are simply portions of the channel with steep gradients. Flow velocity at these points is greatly increased, and abrasion of bedrock is therefore most intense. As a result, the falls are cut back and the rapids are trenched.

At the same time, the ponded stretches of the stream are first filled by sediment and are later lowered in level as the lake outlets are cut down. In time, the lakes disappear and the falls are transformed into rapids.

In the early stages of gradation, the capacity of the stream exceeds the load supplied to it, so that little or no alluvium accumulates in the channels. Abrasion continues to deepen the major channels, with the result that they come to occupy steep-walled *gorges* or *canyons* (Figure 16.11). Weathering and mass wasting of these rock walls contribute an increasing supply of rock debris to the channels. Also on the increase is debris shed from land surfaces that contribute overland flow to the newly developed branches.

The erosion of rapids reduces the gradient to a slope angle that more closely approximates the average gradient of that section of the stream (Figure 16.10*b*). At the same time, branches of the main stream are being extended into higher parts of the original land mass. These carve out many new small drainage basins. Thus, the original tectonic landscape is transformed into a complete fluvial landform system.

As landscape change continues, a gradual decrease in a stream's capacity to move bed load results from the gradual reduction in the channel gradient. Simultaneously, the load supplied to the stream from the entire upstream area is on the increase. So, the time comes when the supply of load exactly matches the stream's capacity to transport it. Figure 16.10*c* shows the landscape at this point in time. Now, all the major streams have achieved the graded condition and possess graded profiles that descend smoothly and uniformly.

The first indication that a stream has attained a graded condition is the beginning of floodplain development. For reasons that are not completely understood, the river begins to wander sidewards, cutting into the side slopes flanking its channel. A sinuous path of arcs and curving bends develops. On the inside of each bend, alluvium accumulates as a long, curving deposit of sediment—termed a point bar. Widening of the bar deposit produces a crescent-shaped area of low ground, which is the first stage in floodplain development.

16.9 Stream gradation
Schematic diagram of gradation of a stream. Originally, the channel consists of a succession of lakes, falls, and rapids. (Copyright © A. N. Strahler.)

16.10 *Evolution of a graded stream and its valley* *(Drawn by E. Raisz. Copyright © A. N. Strahler.) (a) Stream established on a land surface dominated by landforms of recent tectonic activity. (b) Gradation in progress. The lakes and marshes drained. The gorge is deepening, and the tributary valleys are extending. (c) Graded profile attained. Floodplain development is beginning, and the widening of the valley is in progress. (d) Floodplain widened to accommodate meanders. Floodplains now extend up tributary valleys.*

plain development. This stage is illustrated in Figure 16.10c. As lateral cutting continues, the floodplain strips are widened, and the channel develops sweeping bends (Figure 16.12). These winding river bends are called **alluvial meanders** (*d*). In this way, the floodplain is widened into a continuous belt of flat land between steep valley walls. GEODISCOVERIES

Floodplain development reduces the frequency with which the river attacks and undermines the adjacent valley wall. Weathering, mass wasting, and overland flow can then act to reduce the steepness of the valley-side slopes (Figure 16.13). As a result, in a humid climate, the gorge-like

aspect of the valley gradually disappears and eventually gives way to an open valley with soil-covered slopes protected by a dense plant cover. The time required for the stream to reach a graded condition and erode a broad valley will be a few tens of millions of years.

As we have noted, the gradient and channel form of each portion of the stream network is adjusted to carry the sediment load it receives. The gradient and channel form varies, however, from the small streams at the headwaters of the network to the broad river channels found farther downstream.

One way to study the properties of streams is to organize them by stream order, in which stream segments are ordered

16.11 **River gorge** *The Grand Canyon of the Yellowstone River has been carved in volcanic rock (rhyolite) of the Yellowstone Plateau. Yellowstone National Park, Wyoming.*

16.12 **Meander bend** *Idealized map of a meander bend of a large alluvial river, such as the lower Mississippi. Arrows show the position of the swiftest current. (Copyright © A. N. Strahler.)*

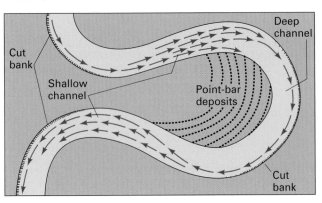

from those draining the smallest of slopes to those formed far downstream from the progressive joining of many upstream segments. Figure 16.14 illustrates this concept. The smallest tributaries are first-order streams, which receive overland flow from within their first-order basin. When two first-order streams join, the result is a second-order stream and so forth. As you might well imagine, many stream properties are related to stream order, including drainage area and discharge.

Great Waterfalls

As we see from Figure 16.9, the stream gradation process acts to smooth the profile of a stream by draining lakes and removing falls and rapids. Thus, large waterfalls on major rivers are comparatively rare the world over. Faulting and dislocation of large crustal blocks have caused

16.13 Evolution of side slopes *Following stream gradation, the valley walls become gentler in slope, and the bedrock is covered by soil and weathered rock. (After W. M. Davis. Copyright © A. N. Strahler.)*

spectacular waterfalls on several east African rivers. Victoria Falls, shown in Figure 16.15, is a prime example. As we explained in Chapter 13, this is the Rift Valley region, where block faulting has been taking place.

Another class of large waterfalls involves new river channels resulting from glacial activity in the Ice Age. Erosion and deposition by large moving ice sheets greatly disrupted drainage patterns in northern continental regions, creating lakes and causing river courses to be shifted to new locations. Niagara Falls is a good example (Figure 16.16). The outlet from Lake Erie into Lake Ontario, the Niagara River, is situated over a gently inclined layer of limestone, beneath which lies easily eroded shale. The river has gradually eroded the edge of the limestone layer, producing a steep gorge marked by Niagara Falls at its head (Figure 16.17). The height of the falls is now 52 m (171 ft), and its discharge is about 5700 m^3 per second (about 200,000 ft^3/s). The drop of Niagara Falls is utilized for the production of hydroelectric power by the Niagara Power Project. Water is withdrawn upstream from the falls

16.14 Stream order *Schematic diagram of a third-order drainage basin showing both the stream channel system and the drainage divide network.*

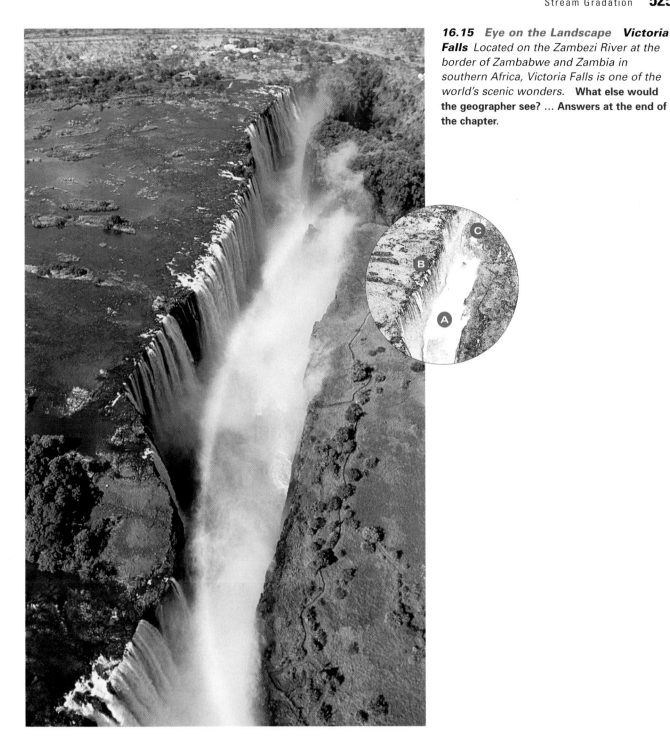

16.15 *Eye on the Landscape* **Victoria Falls** *Located on the Zambezi River at the border of Zambabwe and Zambia in southern Africa, Victoria Falls is one of the world's scenic wonders.* **What else would the geographer see? ... Answers at the end of the chapter.**

and carried in tunnels to generating plants located about 6 km (4 mi) downstream from the falls. *GEODISCOVERIES*

Dams and Resources

Because most large rivers of steep gradient do not have falls, dams are necessary to create the vertical drop required to spin the turbines of electric power generators. Hydroelectric power is cheap, nonpolluting, and renewable, and large dams provide fresh water for urban use and irrigation. However, lakes behind dams drown river valleys and can inundate the gorges, rapids, and waterfalls of major rivers.

Scenic and recreational resources such as these can be lost. White-water boating and rafting, for example, are now popular sports on our wild rivers. The Grand Canyon of the Colorado River, probably more than any single product of fluvial processes, demonstrates the scenic and recreational value of a great river gorge. *Focus on Remote Sensing 16.1* • *A Canyon Gallery* shows how some of the world's more spectacular canyons and gorges are imaged by sensors on orbiting spacecraft.

Dam construction also destroys ecosystems adapted to the river environment. In addition, deposition of sediment behind the dam rapidly reduces the holding capacity of the

16.16 Niagara River topographic setting *A bird's-eye view of the Niagara River with its falls and gorge carved in strata of the Niagara Escarpment. View is toward the southwest from a point over Lake Ontario. (After a sketch by G. K. Gilbert. Redrawn from A. N. Strahler. Copyright © A. N. Strahler.)*

lake and within a century or so may fill the lake basin. This reduces the ability of the lake to provide consistent supplies of water and hydroelectric power. It is small wonder, then, that new dam projects can meet with stiff opposition from concerned local citizens' groups and national environmental organizations.

Aggradation and Alluvial Terraces

A graded stream, delicately adjusted to its supply of water and rock waste from upstream sources, is highly sensitive to changes in those inputs. Changes in climate or vegetation cover bring changes in discharge and load at downstream points, and these changes in turn require channel readjustments. One kind of change is the buildup of alluvium in the valley floors.

Consider first what happens when bed load increases, exceeding the transporting capacity of the stream. Along a section of channel where the excess load is introduced, the coarse sediment accumulates on the stream bed in the form of bars of sand, gravel, and pebbles. These deposits raise the elevation of the stream bed, a process called *aggradation*. As more bed materials accumulate, the stream channel gradient is steepened and flow velocity increases. This increase enables bed materials to be dragged downstream and spread over the channel floor at more and more distant downstream sections. In this way, sediment introduced at the head of a stream will be gradually spread along the whole length of the stream.

16.17 Niagara Falls *Niagara Falls is formed where the river passes over the eroded edge of a massive limestone layer. Pictured here is the Canadian side of the falls.*

16.19   Alluvial terrace formation *Alluvial terraces form when a graded stream slowly cuts away the alluvial fill in its valley. (Drawn by A. N. Strahler.)*

16.18   Braided stream *The braided channel of the Chitina River, Wrangell Mountains, Alaska, shows many separate channels separating and converging on a floodplain filled with glacial debris.*

Aggradation typically changes the channel cross section from a narrow and deep form to a wide and shallow one. Because bars are continually being formed, the flow is divided into multiple threads. These rejoin and subdivide repeatedly to give a typical *braided stream* (Figure 16.18). The coarse channel deposits spread across the former floodplain, burying fine-textured alluvium under the coarse material.

What processes cause alluvium to build up on valley floors and induce stream aggradation? One way is for alluvium to accumulate as a result of accelerated soil erosion. Other causes of aggradation are related to major changes in global climate, such as the onset of an ice age. Figure 16.18 illustrates one natural cause of aggradation—glacial activity—that has been of major importance in stream systems of North America and Eurasia during the recent Ice Age. In our photo example, a modern valley glacier has provided a large quantity of coarse rock debris at the head of the valley. Valley aggradation of a similar kind was widespread in a broad zone near the edges of the great ice sheets of the

> **Accelerated soil erosion and glacial activity can increase the amount of sediment brought to a river system, causing aggradation and the development of braided streams and river segments.**

Ice Age. The accumulated alluvium filled most valleys to depths of several tens of meters. Figure 16.19a shows a valley filled in this manner by an aggrading stream. The case could represent any one of a large number of valleys in New England or the Middle West.

Suppose, next, that the source of bed load is cut off or greatly diminished. In the case illustrated in Figure 16.19, the ice sheets have disappeared from the distant headwater areas and, with them, the supplies of coarse rock debris. Reforestation of the landscape restores a protective cover to valley-side and hill slopes of the region, keeping coarse mineral particles from entering the stream through overland flow. Now the streams have abundant water discharges but little bed load. In other words, they are operating below their transporting capacity. The result is channel scour. The channel form becomes both deeper and narrower, and it also begins to develop meanders.

Focus on Remote Sensing | **16.1**

A Canyon Gallery

Deep canyons, carved by powerful rivers crossing high terrain, are among the most dramatic features of the landscape. They provide a stunning testament to the power of fluvial systems to shape the Earth. Who can forget the awesome spectacle of a yawning canyon, seen from a precarious perch on a high rim? Here's a gallery of satellite images of canyons as acquired by the MISR and ASTER instruments.

The Grand Canyon of the Colorado River is among the most famous in the world. Spanning a length of about 450 km (about 280 mi) with vertical drops up to about 1500 m (about 5000 ft), it is indeed spectacular. Image (*a*), in true color, was acquired by MISR on December 31, 2000. Trace the path of the Colorado from Lake Powell, at the upper right, through narrow Marble Canyon, until the canyon broadens to the southeast of the snow-covered Kaibab Plateau. Here the Grand Canyon begins, revealing a dissected landscape between the two canyon rims as the river curves around the plateau. Note the sharp black shadow at the south rim, indicating that the rim is steep indeed. The canyon continues as the river flows westward to the edge of the frame. To the south of the canyon are cloud streaks that have formed from condensation trails of jet aircraft. Looking closely, you can see the shadow of each cloud as a dark streak to the north of the cloud.

Image (*b*), acquired by ASTER on May 12, 2000, is a computer-generated perspective view of the Grand Canyon looking north up Bright Angel Canyon from the South Rim. In this false color image, vegetation appears green and water appears blue, but rocks are not shown in their true colors. The blue and black patches on the green Kaibab Plateau are burned areas from a fire that is still smoldering. A pall of smoke stretches to

Grand Canyon, Arizona, imaged by MISR *(Courtesy NASA/GSFC/LARC/JPL, MISR team.)*

the east. Compare this with Figure 17.5, which is a ground-level photo of Bright Angel Canyon.

Image (*c*), another MISR image, shows the coast of southern Peru in the Ariquipa region. Here the Pacific coast runs nearly east-west. The coastline is obscured by low clouds and fog at the bottom of the picture. Note the two vast canyons reaching

Grand Canyon perspective view from ASTER *(Courtesy NASA/GSFC/MITI/ERSDAC/JAROS and U.S. Japan ASTER Science Team.)*

(c)

Canyons of the Andes as seen by MISR *(Courtesy NASA/GSFC/LARC/JPL, MISR team.)*

from the sea deep into the Andean Plateau. They are the canyons of the Rio Ocoña to the west and Rio Camaná to the east. Dwarfing the American Grand Canyon, they are far wider and deeper. In fact, the canyon of the middle branch of the Rio Ocoña (Rio Cotahuasi) reaches a depth of 3354 m (11,001 ft) below the plateau, which is more than twice as deep as the Grand Canyon. The white patch between the two canyon systems is the snow-covered Nevado Coropuna, at 6425 m (21,074 ft) elevation. Peaks to the west and east are Nevado Solimana (6117 m, 20,064 ft) and Nevado Ampato (5795 m, 18,909 ft).

Our last image (*d*) is another ASTER scene, a color infrared image of the Yangtze River canyon in the Three Gorges region of the provinces of Hubei and Sichuan, China. Although not a candidate for the deepest canyon, it is among the most scenic of the world's canyons, with steep limestone cliffs and forest-covered slopes separating clusters of quaint villages where tributaries meet the main channel. The inset image shows the construction site for the Three Gorges Dam, which will create a reservoir 175 m (574 ft) deep and about 600 km (about 375 mi) long. The dam will provide vast amounts of hydroelectric power and sharply reduce extreme flooding of the Yangtze. However, two of the three gorges will be flooded, and many local residents will be displaced. The Chinese government is making a major effort to forecast and mitigate environmental problems that will be created by the huge project. Our book's cover provides a view of the Qutang Gorge as well as more information about the dam.

Three Gorges region of the Yangtze River imaged by ASTER *(Courtesy NASA/GSFC/MITI/ERSDAC/JAROS and U.S. Japan ASTER Science Team.)*

(d)

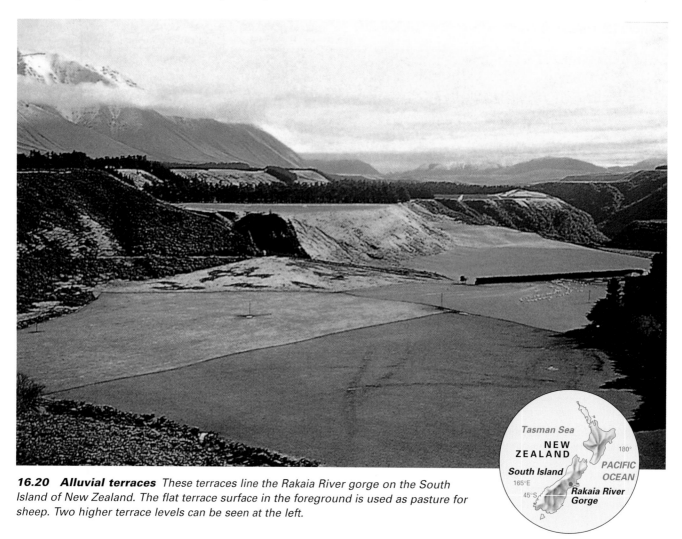

16.20 Alluvial terraces *These terraces line the Rakaia River gorge on the South Island of New Zealand. The flat terrace surface in the foreground is used as pasture for sheep. Two higher terrace levels can be seen at the left.*

Gradually, the stream profile level is lowered, in a process called *degradation*. Because the stream is very close to being in the graded condition at all times, its dominant activity is lateral (sidewise) cutting by growth of meander bends, as shown in Figure 16.19*b*. By this process, the valley alluvium is gradually excavated and carried downstream, but not all of it can be removed because the channel encounters hard bedrock masses in many places. These obstructions prevent the cutting away of more alluvium. Consequently, as shown in (*c*) step-like alluvial surfaces remain on both sides of the valley. The treads of these steps are called *alluvial terraces.*

Alluvial terraces have always attracted human settlement because of the advantages over both the valley-bottom floodplain—which is subject to annual flooding—and the hill slopes beyond—which may be too steep and rocky to cultivate. Besides, terraces are easily tilled and make prime agricultural land (Figure 16.20). Towns are also easily laid out on the flat ground of a terrace.

> Alluvial terraces form when an aggrading river loses its sediment input and begins degrading its bed, leaving terraces behind as it cuts deeper into its sediment-filled valley.

> An alluvial river, with its low gradient and broad floodplain, creates characteristic landforms including bluffs, meanders, cutoffs, ox-bow lakes, and natural levees.

Roads and railroads can be constructed along the terrace surfaces with little difficulty.

Alluvial Rivers and Their Floodplains

We turn now to the graded river with its floodplain. As time passes, the floodplain is widened, so that broad areas of floodplain lie on both sides of the river channel. Civil engineers have given the name **alluvial river** to a large river of very low channel gradient. It flows on a thick floodplain accumulation of alluvium constructed by the river itself in earlier stages of its activity. Characteristically, an alluvial river experiences overbank floods each year or two. These floods occur during the season of large water surplus over the watershed. Overbank flooding of an alluvial river normally inundates part or all of a floodplain that is bounded on either side by rising steep slopes, called *bluffs.*

Typical landforms of an alluvial river and its floodplain are illustrated in

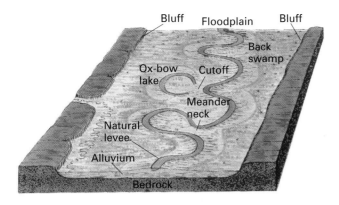

16.21   Floodplain landforms of an alluvial river
(Drawn by A. N. Strahler.)

Figure 16.21. Dominating the floodplain is the meandering river channel itself and abandoned stretches of former channels. Meanders develop narrow necks, which are cut through, thus shortening the river course and leaving a meander loop abandoned. This event is called a *cutoff.* It is quickly followed by deposition of silt and sand across the ends of the abandoned channel, producing an *ox-bow lake.* The ox-bow lake is gradually filled in with fine sediment brought in during high floods and with organic matter produced by aquatic plants. Eventually, the ox-bows are converted into swamps, but their identity is retained indefinitely (Figure 16.22).

During periods of overbank flooding, when the entire floodplain is inundated, water spreads from the main channel over adjacent floodplain deposits (Figure 16.23). As the current slackens, sand and silt are deposited in a zone adjacent to the channel. The result is an accumulation of higher land on either side of the channel known as a **natural levee.** Because deposition is heavier closest to the channel and decreases away from the channel, the levee surface slopes away from the channel (Figure 16.21). Between the levees and the bluffs is lower ground, called the *backswamp.* In Figure 16.23, the small settlements are located on the higher ground of the levee, next to the river, while the agricultural fields occupy the backswamp area.

Overbank flooding not only results in the deposition of a thin layer of silt on the floodplain, but also brings an infusion of dissolved mineral substances that enter the soil. As a result of the resupply of nutrients, floodplain soils retain their remarkable fertility, even though they are located in regions of rainfall surplus from which these nutrients are normally leached away.

> Where rapid uplift causes meandering rivers to cut deeply into bedrock, entrenched meanders are formed.

Entrenched Meanders

What happens when a broadly meandering river is uplifted by rapid tectonic activity? The uplift increases the river's gradient, so that it cuts downward into the bedrock below. This forms a steep-walled inner gorge. On either side lies the former floodplain, now a flat terrace

16.22   Ox-bow lakes *This aerial photo of the Mississippi River floodplain shows two ox-bow lakes— former river bends that were cut off as the river shifted its course. At the bottom is a thin former meander channel still receiving river water. If flood deposits seal it off from the river, it will become an ox-bow lake as well. Vegetation appears bright red in this color-infrared photo.*

high above river level. Any river deposits left on the terrace are rapidly stripped off by runoff because floods no longer reach the terraces to restore eroded sediment.

Uplift may cause the meanders to become impressed into the bedrock and give the inner gorge a meandering pattern (Figure 16.24). These sinuous bends are termed *entrenched meanders* to distinguish them from the floodplain meanders of an alluvial river (Figure 16.25). Although entrenched meanders are not free to shift about as floodplain meanders do, they can enlarge slowly so as to produce cutoffs. Cutoff of an entrenched meander leaves a high, round hill separated from the valley wall by the deep abandoned river channel and the shortened river course (Figure 16.24). As you might guess, such hills formed ideal

16.23 Inundated floodplain *This river floodplain in Bangladesh is largely under water during a 1973 flood. Villages occupy the higher ground of the natural levees bordering the river channel.*

16.24 Block diagram of entrenched meanders *Uplift of a meandering stream has produced entrenched meanders. One meander neck has been cut through, forming a natural bridge. (Drawn by E. Raisz. Copyright © A.N. Strahler.)*

16.25 Entrenched meanders *The Goosenecks of the San Juan River in Utah are deeply entrenched river meanders in horizontal sedimentary rock layers. The canyon, carved from sandstones and limestones, is about 370 m (1399 ft) deep.*

natural fortifications. Many European fortresses of the Middle Ages were built on such cutoff meander spurs. Under unusual circumstances, where the bedrock includes a strong, massive sandstone formation, meander cutoff leaves a *natural bridge* formed by the narrow meander neck.

The Geographic Cycle

The Earth's fluvial landscapes are quite diverse. They range from mountain regions of steep slopes and rugged peaks to regions of gentle hills and valleys to nearly flat plains that stretch from horizon to horizon. One way to view these landscapes is to consider them as stages of evolution in a cycle that begins with rapid uplift and follows with long erosion by streams in a graded condition. This cycle, called the *geographic cycle,* was first described by William Morris Davis, a prominent geographer and geomorphologist of the late nineteenth and early twentieth centuries. Let's look at this idea in more detail.

Consider a landscape made up of many drainage basins and their branching stream networks that has been rapidly uplifted by tectonic forces. The region is rugged, with steep mountainsides and high, narrow crests (Figure 16.26*a*). We refer to this landscape as being in a *youthful stage.*

After initial uplift, the main streams draining the region establish a graded condition. Rock debris is transported out of each drainage basin at the same average rate as the debris is being contributed from the land surfaces within the basin. Eventually, the export of debris lowers the land surface generally, and the average altitude of the land surface steadily declines. This decline must be accompanied by a reduction in the average gradients of all streams (*b*). As the sharp mountain peaks and gorges of the youthful stage give way to rounded hills and broad valleys, we reach the *mature stage* of the geographic cycle.

As time passes, the streams and valley-side slopes of the drainage basins undergo gradual change to lower gradients. In theory, the ultimate goal of the denudation process is to

> The geographic cycle traces the fate of rivers and fluvial landforms from an initial uplift creating steep slopes and canyons to a final low, gently rolling surface called a peneplain.

reduce the land mass to a featureless plain at sea level. In this process, a sea-level surface imagined to lie beneath the entire land mass represents the lower limiting level, or *base level,* of the fluvial denudation (labeled in Figures 16.9 and 16.26). But because the rate of denudation becomes progressively slower, the land surface approaches the base-level surface of zero elevation at a slower and slower pace. Under this scenario, the ultimate goal can never be reached. Instead, after the passage of some millions of years, the land surface is reduced to a gently rolling surface of low elevation, called a *peneplain* (*c*). You can think of this strange term as meaning an "almost-plain." With the evolution of the peneplain, the landscape has reached *old age.*

Production of a peneplain requires a high degree of crustal and sea-level stability for a period of many millions of years. One region that has been cited as a possible example of a contemporary peneplain is the Amazon-Orinoco Basin of South America. This vast region is a stable continental shield of ancient rock with very low relief.

What happens if a peneplain is uplifted? Figure 16.26*d* shows the peneplain of (*a*) uplifted to an elevation of several hundred meters. The base level is now far below the land surface. Soon streams begin to trench the land mass and to carve deep, steep-walled valleys, shown in (*e*). This process is called *rejuvenation.* With the passage of many millions of years, the landscape will be carved into the rugged stage shown in (*a*), and the later stages of (*b*) and (*c*) will follow.

Equilibrium Approach to Landforms

While Davis's idealized geographic cycle is useful for understanding landscape evolution over very long periods of time, it does little to explain the diversity of the features observed in real landscapes. Most geomorphologists now approach landforms and landscapes from the viewpoint of *equilibrium.* This approach explains a fluvial landform as the product of forces acting upon it, including both forces of

(a)

(b)

(c)

(d)

(e)

16.26   The geographic cycle *(a) In the youthful stage, relief is great, slopes are steep, and the rate of erosion is rapid. (b) In the mature stage, relief is greatly reduced, slopes are gentle, and rate of erosion is slow. Soils are thick over the broadly rounded hill summits. (c) In old age, after many millions of years of fluvial denudation, a peneplain is formed. Slopes are very gentle, and the landscape is an undulating plain. Floodplains are broad, and the stream gradients are extremely low. All of the land surface lies close to base level. (d) The peneplain is uplifted. (e) Streams trench a new system of deep valleys in the phase of land-mass rejuvenation. (Drawn by A.N. Strahler.)*

in which hill slopes and stream gradients remain steep in order to maintain a graded condition while eroding a strong rock like massive granite. In the next chapter, we will provide many examples of landforms that result from erosive processes acting on both strong and weak rock within the same region.

Another problem with Davis's geographic cycle is that it only applies where the land surface is stable over long periods of time. As we know from our study of plate tectonics in Chapter 12, crustal movements are frequent on the geologic time scale, and few regions of the land surface remain untouched by tectonic forces in the long run. Recall also that continental lithosphere floats on a soft asthenosphere. As layer upon layer of rock is stripped from a land mass by erosion, the land mass becomes lighter and is buoyed upward. The process of crustal rise in response to unloading is known as *isostatic compensation*. The proper model, then, is one of uplift as an ongoing process to which erosional processes are constantly adjusting rather than as a sudden event followed by denudation.

FLUVIAL PROCESSES IN AN ARID CLIMATE

Desert regions look strikingly different from humid regions in both vegetation and landforms. Obviously, the lower precipitation makes the difference. Vegetation is sparse or absent, and land surfaces are mantled with mineral material—sand, gravel, rock fragments, or bedrock itself.

Although deserts have low precipitation, rain falls in dry climates as well as in moist, and most landforms of desert regions are formed by running water. A particular locality in a dry desert may experience heavy rain only once in several

uplift and denudation, with the characteristics of the rock material playing an important role. Thus, we find steep slopes and high relief where the underlying rock is strong and highly resistant to erosion. Even a "youthful" landscape may be in a long-lived equilibrium state

The equilibrium approach sees fluvial landforms as reflecting a balance between processes of uplift and denudation acting on rocks of varying resistance to erosion.

years. But when rain does fall, stream channels carry water and perform important work as agents of erosion, transportation, and deposition. Fluvial processes are especially effective in shaping desert landforms because of the sparse vegetation cover. The few small

16.27 Flash flood *A flash flood has filled this desert channel in the Tucson Mountains of Arizona with raging, turbid waters. A distant thunderstorm produced the runoff.*

plants that survive offer little or no protection to soil or bedrock. Without a thick vegetative cover to protect the ground and hold back the swift downslope flow of water, large quantities of coarse rock debris are swept into the streams. A dry channel is transformed in a few minutes into a raging flood of muddy water heavily charged with rock fragments (Figure 16.27).

An important contrast between regions of arid and humid climates lies in the way in which the water enters and leaves a stream channel (Figure 16.28). In a humid region (*a*) with a high water table sloping toward a stream channel, ground water moves steadily toward the channel and seeps into the stream bed, producing permanent (perennial) streams. In arid regions (*b*), the water table normally lies far below the channel floor. Where a stream flows across a plain of gravel and sand, water is lost from the channel by seepage. Loss of discharge by seepage and evaporation strongly depletes the flow of streams in alluvium-filled valleys of arid regions. As a result, aggradation occurs and braided channels are common. Streams of desert regions are often short and end in alluvial deposits or on the floors of shallow, dry lakes.

16.28 Stream water flow to ground water *In humid regions (a), a stream channel receives ground water through seepage. In arid regions (b), stream water seeps out of the channel and into the water table below. (Copyright © A.N. Strahler.)*

(a)

(b)

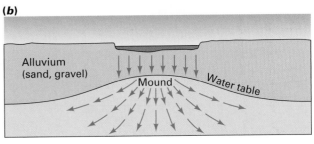

16.29 Features of an alluvial fan *A cross section shows mudflow layers interbedded with sand layers, providing water (arrows) for a well in the fan. (Copyright © A.N. Strahler.)*

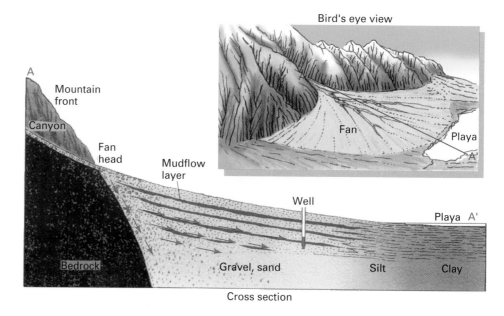

Alluvial Fans

One very common landform built by braided, aggrading streams is the **alluvial fan,** a low cone of alluvial sands and gravels resembling in outline an open Japanese fan (Figure 16.29). The apex, or central point of the fan, lies at the mouth of a canyon or ravine. The fan is built out on an adjacent plain. Alluvial fans are of many sizes. In fact, some desert fans are many kilometers across (Figure 16.30).

> Although rain is infrequent in desert environments, running water shapes desert landforms with great effectiveness because of the lack of vegetation cover.

Fans are built by streams carrying heavy loads of coarse rock waste from a mountain or an upland region. The braided channel shifts constantly, but its position is firmly fixed at the canyon mouth. The lower part of the channel, below the apex, sweeps back and forth. This activity accounts for the semicircular fan form and the downward slope in all radial directions away from the apex.

16.30 Alluvial fans *Great alluvial fans extend out upon the floor of Death Valley. The canyons from which they originate have carved deeply into a great uplifted fault block.*

Large, complex alluvial fans also include mudflows (Figure 16.29). Mud layers are interbedded with sand and gravel layers. Water infiltrates the fan at its head, making its way to lower levels along sand layers. The mudflow layers serve as barriers to ground water movement. The trapped ground water is under pressure from water higher in the fan apex. When a well is drilled into the lower slopes of the fan, water rises spontaneously as artesian flow. (See Chapter 15 and Figure 15.6.)

> **Alluvial fans are common features of arid landscapes. They occur where streams discharge water and sediment from a narrow canyon or gorge onto an adjacent plain.**

Alluvial fans are the primary sites of ground water reservoirs in the southwestern United States. In many fan areas, sustained heavy pumping of these reserves for irrigation has lowered the water table severely. The rate of recharge is extremely slow in comparison. At some locations, recharge is increased by building water-spreading structures and infiltrating basins on the fan surfaces. A serious side effect of excessive ground water withdrawal is subsidence (sinking) of the land surface. (See *Eye on Global Change 15.1 ● Sinking Cities.*)

The Landscape of Mountainous Deserts

Where tectonic activity has recently produced block faulting in an area of continental desert, the assemblage of fluvial landforms is particularly diverse. The basin-and-range region of the western United States is such an area. It includes large parts of Nevada and Utah, southeastern California, southern Arizona and New Mexico, and adjacent parts of Mexico. The uplifted and tilted blocks, separated by downdropped tectonic basins, provide an environment for the development of spectacular erosional and depositional fluvial landforms.

Figure 16.31 demonstrates some landscape features of mountainous deserts. Figure 16.31*a* shows two uplifted fault blocks with a down dropped block between them. Although denudation acts on the uplifted blocks as they are being raised, we have shown them as very little modified at the time tectonic activity has ceased. At first, the faces of the fault block are extremely steep. They are scored with deep ravines, and talus blocks form cones at the bases of the blocks.

Figure 16.31*b* shows a later stage of denudation in which streams have carved up the mountain blocks into a rugged landscape of deep canyons and high divides. Rock waste furnished by these steep mountain slopes is carried from the mouths of canyons to form numerous large alluvial fans. The fan deposits form a continuous apron extending far out into the basins.

In the centers of desert basins lie the saline lakes and dry lake basins mentioned in Chapter 15. Accumulation of fine sediment and precipitated salts produces an extremely flat basin floor, called a **playa** in the southwestern United States and in Mexico. Salt flats are found where an evaporite layer forms the surface. In some playas, shallow water forms a salt lake.

Figure 16.32 is an air photograph of a mountainous desert landscape in the Death Valley region of eastern California. Three environmental zones can be seen in the photo. First is a region of rugged mountain masses dissected into canyons with steep rocky walls. At lower elevations, a zone of coalescing alluvial fans lines the mountain front. Last is the white playa occupying the central part of the basin.

In this type of mountainous desert, fluvial processes are limited to local transport of rock particles from a mountain range to the nearest adjacent basin, which receives all the sediment. The basin gradually fills as the mountains diminish in elevation. Because there is no outflow to the sea, the concept of a base level of denudation has no meaning. Each arid basin

16.31 Mountainous desert landforms *Idealized diagrams of landforms of the mountainous deserts of the southwestern United States. (a) Initial uplift of two blocks, with a downdropped valley between them. (b) Stage of rapid filling of tectonic basins with debris from high, rugged mountain blocks. (c) Advanced stage with small mountain remnants and broad playa and fan slopes. (Drawn by A. N. Strahler.)*

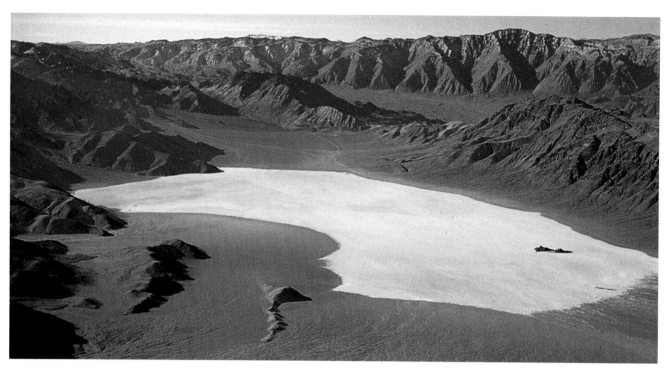

16.32 **Desert playa** *Racetrack Playa, a flat, white plain, is surrounded by alluvial fans and rugged mountains. This desert valley lies in the northern part of the Panamint Range, not far west of Death Valley, California. In the distance rises the steep eastern face of the Inyo Mountains, a great fault block.*

becomes a closed system as far as transport of sediment is involved. Only the hydrologic system is open, with water entering as precipitation and leaving as evaporation.

As the desert mountain masses are lowered in height and reduced in extent, a gently sloping rock floor, called a *pediment* develops close to the receding mountain front (Figure 16.33). As the remaining mountains

Landforms of mountainous deserts include alluvial fans, dry lakes or playas, and pediments—rock platforms veneered with alluvium.

shrink further in size, the pediment surfaces expand to form wide rock platforms thinly veneered with alluvium. This advanced stage is shown in Figure 16.31*c*. The desert land surface produced in an advanced stage of fluvial activity is an undulating plain consisting largely of areas of pediment surrounded by areas of alluvial fan and playa surfaces.

16.33 **Pediment** *A pediment formerly carved in granite at the foot of a desert mountain slope is now seen in profile as a straight, sloping line. Later erosion has exposed the bedrock, and weathering has formed the joint blocks into boulders (intermediate distance). Little Dragoon Mountains, near Adams Peak, about 20 km (about 12 mi) northeast of Benson, Arizona.*

A LOOK AHEAD   Running water, the primary agent of denudation, does not act equally on all types of rocks. Some rocks are more resistant to erosion, while others are less so. As a result, fluvial erosion can create unique and interesting landscapes that reveal rock structures of various kinds. This will be the subject of our next chapter.

In Review | Landforms Made by Running Water

- This chapter has covered the landforms and land-forming processes of running water, one of the four active agents of *denudation*. Like the other agents, running water erodes, transports, and deposits rock material, forming both *erosional* and *depositional landforms.*

- The work of running water begins on slopes, producing *colluvium* where overland flow moves soil particles downslope, and producing **alluvium** when the particles enter stream channels and are later deposited. In most natural landscapes, *soil erosion* and soil formation rates are more or less equal, a condition known as the *geologic norm. Badlands* are an exception in which natural erosion rates are very high.

- The work of streams includes **stream erosion** and **stream transportation.** Where stream channels are carved into soft materials, large amounts of sediment can be obtained by *hydraulic action.* Where stream channels flow on bedrock, channels are deepened only by the abrasion of bed and banks by mineral particles, large and small. Both the *suspended load* and *bed load* of rivers increase greatly as velocity increases. Velocity, in turn, depends on gradient.

- Over time, streams tend to a **graded** condition, in which their gradients are adjusted to move the average amount of water and sediment supplied to them by slopes. Lakes and *waterfalls,* created by tectonic, volcanic, or glacial activity, are short-lived events, geologically speaking, that give way to a smooth, *graded* stream *profile.* Grade is maintained as landscapes are eroded toward base level.

- When provided with a sudden inflow of rock material, as, for example, by glacial action, streams build up their beds by *aggradation.* When that inflow ceases, streams resume downcutting, leaving behind *alluvial terraces.*

- Large rivers with low gradients that move large quantities of sediment are termed **alluvial rivers.** The **meandering** of these rivers forms *cutoff* meanders, *ox-bow lakes,* and other typical landforms. Alluvial rivers are sites of intense human activity. Their fertile floodplains yield agricultural crops and provide easy transportation paths. When a region containing a meandering alluvial river is uplifted, *entrenched meanders* can result.

- The *geographic cycle* organizes fluvial landscapes according to their age in a cycle of uplift that forms mountains and subsequent erosion to nearly flat surfaces called *peneplains.* In the *equilibrium* approach, landforms are viewed as products of uplift and erosion as continuous processes acting on rocks of varying resistance to erosion.

- Although rainfall is scarce in deserts, running water is very effective there in producing **fluvial landforms.** Desert streams, subject to flash flooding, build **alluvial fans** at the mouths of canyons. Water sinks into the fan deposits, creating local ground water reservoirs. Eventually, desert mountains are worn down into gently sloping *pediments.* Fine sediments and salts, carried by streams, accumulate in **playas,** from which water evaporates, leaving sediment and salt behind.

Key Terms

fluvial landforms,
  p. 514
fluvial processes, p. 514
alluvium, p. 517
stream erosion, p. 518

stream transportation,
  p. 518
stream deposition, p.
  518
stream load, p. 519

graded stream, p. 521
alluvial meanders,
  p. 522
alluvial river, p. 530
natural levee, p. 531

alluvial fan, p. 536
playa, p. 537

Review Questions

1. List and briefly identify the four flowing substances that serve as agents of denudation.
2. Describe the process of slope erosion. What is meant by the geologic norm?
3. Contrast the two terms *colluvium* and *alluvium*. Where on a landscape would you look to find each one?
4. What special conditions are required for badlands to form?
5. When and how does sheet erosion occur? How does it lead to rill erosion and gullying?
6. In what ways do streams erode their bed and banks?
7. What is stream load? Identify its three components. In what form do large rivers carry most of their load?
8. How is velocity related to the ability of a stream to move sediment downstream?
9. How does stream degradation produce alluvial terraces?
10. Define the term *alluvial river.* Identify some characteristic landforms of alluvial rivers. Why are alluvial rivers important to human civilization?
11. Describe the evolution of a fluvial landscape according to the geographic cycle. What is meant by rejuvenation?
12. What is the equilibrium approach to landforms? How does it differ from interpretation using the geographic cycle?
13. Why is fluvial action so effective in arid climates, considering that rainfall is scarce? How do streams in arid climates differ from streams in moist climates?
14. Describe the evolution of the landscape in a mountainous desert. Use the terms *alluvial fan, playa,* and *pediment* in your answer.

Visualizing Exercises

1. Compare erosional and depositional landforms. Sketch an example of each type.
2. What is a graded stream? Sketch the profile of a graded stream and compare it with the profile of a stream draining a recently uplifted set of landforms.
3. Sketch the floodplain of a graded, meandering river. Identify key landforms on the sketch. How do they form?

Essay Questions

1. A river originates high in the Rocky Mountains, crosses the high plains, flows through the agricultural regions of the Midwest, and finally reaches the sea. Describe the fluvial processes and landforms you might expect to find on a journey along the river from its headwaters to the ocean.
2. What would be the effects of climate change on a fluvial system? Choose either the effects of cooler temperatures and higher precipitation in a mountainous desert, or warmer temperatures and lower precipitation in a humid agricultural region.

Eye on the Landscape

Chapter Opener **White-water rafting on the Colorado** Grand Canyon is an entrenched river gorge cutting through layers of flat-lying sandstone and shale **(A)**. Note the large accumulations of talus at the base of the cliffs **(B)**. Most of the canyon shows only sparse desert vegetation, but deep-rooted plants appear near the river and on sedimentary deposits of side streams **(C)**. Surprisingly, the river water in the canyon **(D)** is quite cold. Nearly all of it is released from the cold, deep waters of Lake Powell behind the Glen Canyon Dam, which is upstream. We will see more of landforms of the Grand Canyon in Chapter 17. Plants and water were covered in Chapter 8, and dams were discussed in the preceding chapter (15).

16.15 **Victoria Falls** A fault has shattered and broken rock layers, creating a zone of weakness that has been eroded by the Zambezi River to form the gorge of Victoria Falls **(A)**. Note the resistant rock layer that keeps the waterfall vertical **(B)**. Native vegetation here is thorntree-tall grass savanna, which is visible in the near distance **(C)**. Faults were covered in Chapter 13, while the vegetation of Africa is mapped and described in Chapter 9.

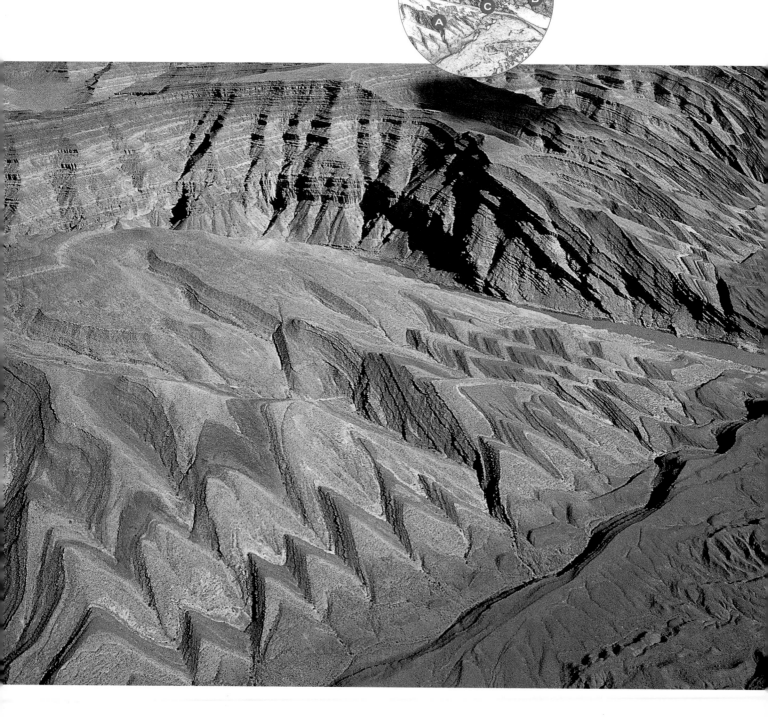

17 | LANDFORMS AND ROCK STRUCTURE

• • • • • • • • • • • • • • • • •

PUTTING YOU IN THE PICTURE

An aerial view of the Earth provides a unique perspective on the landscape. At the scale of a high-flying passenger aircraft, an overview emerges that brings out the broader patterns of landscape. In regions of diverse rock types, this big picture emphasizes the contrast between easily eroded rock formations and those that are more resistant. The central Rocky Mountains and Colorado Plateau are such regions. Imagine yourself on a flight from Denver to Las Vegas on a clear day in June. Here's what you might experience:

The force of the powerful jet engines pushes you deeply into your window seat as you roll down the runway. Quickly, the plane is airborne and heads westward toward the magnificent Rockies. The slope of the Front Range looms large, and as you head across the mountain barrier, the ground seems to rise up to meet you. Below is I-70, twisting and turning its way up the narrow valleys toward the summit. There's the entrance to the Eisenhower Tunnel! The tiny cars below disappear into the mountain slope for the long traverse beneath the mountain crest.

Looking south, you see the peaks of the Front Range spread out in a long line. The overlying sedimentary rocks have long been eroded away here in the center of this vast elongated dome, exposing the hard granite at the core. You easily identify Pikes Peak in the far distance, its summit silhouetted against the dark blue sky. Looking straight down, you see the gray of the ancient granite rock, mantling the slopes between the patches of snow still lying in the low places. Snowpacks in high valley heads below you hint at how the mountains must have looked when their glaciers slowly and majestically marched valleyward, carving broad, U-shaped valleys and leaving the ridges and piles of sediment now resting on the valley floors.

As the flight progresses, the landscape becomes drier, and the vast high desert of the Colorado Plateau replaces the forested lower slopes of the ranges. Below is a landscape of nearly flat-lying rocks. Long lines of cliffs, marking the edges of resistant rock beds, advance and recede as the aircraft moves steadily southwestward.

"Well, folks, you're in luck this afternoon." Suddenly, the voice of the captain cuts the low rushing sound that fills the airplane's cabin. "Air traffic control has vectored us to the south, so we'll be approaching Las Vegas from just over the North Rim of the Grand Canyon. The views should be spectacular!"

And spectacular they are. The gaping chasm cuts open the surrounding plateaus with a broad red gash. Within the canyon, the flat-lying rock formations form a succession of giant steps leading down to a narrow inner gorge that contains a blue-gray ribbon of water. The canyons of tributary creeks cut deeply into the rock sequence, creating wedding cakes and turrets of multileveled rock.

But it is the coloring of the rocks that is the most breathtaking. Buff and gray layers of limestone alternate with reds and purples of siltstones and shales. The green tones of vegetation overshadow the earth colors on the wider benches and platforms. The play of light and shade on the landscape etches the shapes in relief, enhancing

their coloration and making the canyon's landforms look somehow even more grand and monumental.

Descending now, the airplane approaches Las Vegas. This flight will remain etched in your memory for years to come, not only because of its sheer scenic beauty, but also for the way in which the landscape revealed its inner structure to your educated eye.

This chapter focuses on the adjustment of landforms to the nature of the rock beneath them. As we will see, weak rocks tend to form valleys or lowlands, while strong rocks tend to stand out as highlands, ridges, or cliffs. Keep this principle in mind as we examine the relation between landforms and rock structure. ■

LANDFORMS AND ROCK STRUCTURE

Denudation acts to wear down all rock masses exposed at the land surface. However, different types of rocks are worn down at different rates. Some are easily eroded, while others are highly resistant to erosion. Generally, weak rocks will underlie valleys, and strong rocks will underlie hills, ridges, and uplands. This means that the pattern of landforms on a landscape can reveal the sequence of rock layers underneath it. The sequence, in turn, is determined by the type of rock structure present—for example, a series of folds, or perhaps a dome, in a set of rock layers. Thus, there is often a direct relationship between landforms and rock structure.

Over the world's vast land area, many types of rock and rock structures appear. Repeated episodes of uplift, followed by periods of denudation, can bring to the surface rocks and rock structures formed deep within the crust. In this way, even ancient mountain roots, formed during continental collisions, may eventually appear at the surface and be subjected to denudation. Batholiths and other plutons—igneous rock structures produced by magmas that cool at depth—can similarly be exposed.

As we will see in this chapter, rock structure controls not only the locations of uplands and lowlands, but also the placement of streams and the shapes and heights of the intervening divides. A distinctive assemblage of landforms and stream patterns develops on each of the major types of crustal structures. Recall that in Chapter 13 we classified

all landforms as either initial or sequential in origin. Initial landforms are produced directly by volcanic activity, folding, and faulting. Denudation soon converts these initial landforms into sequential landforms, which are controlled in shape, size, and arrangement by the underlying rocks and their structure. The subject of this chapter is therefore the sequential landforms that arise through erosion of rock structures.

ROCK STRUCTURE AS A LANDFORM CONTROL

As denudation takes place, landscape features develop according to patterns of bedrock composition and structure. Figure 17.1 shows an arrangement of five layers of sedimentary rock, together with a mass of much older igneous rock. The sediments were originally deposited in horizontal layers atop the igneous rock, but the block of igneous and sedimentary rock has been bent and tilted in an ancient orogeny. The diagram shows the usual landform shape of each rock type and whether it forms valleys or mountains. The cross section on the front of the block diagram shows conventional symbols used by geologists for each rock type—for example, fine dots for sandstone or "bricks" for limestone.

As indicated by the diagram, shale is a weak rock that is easily eroded and

> As denudation wears down landscapes, landforms emerge that reflect bedrock composition and structure. Shale is a weak rock that forms valleys, while sandstone and conglomerate are strong rocks that form ridges.

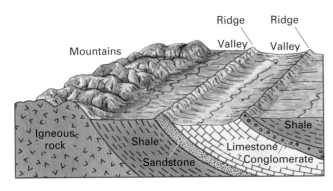

17.1 Landforms and rock types *Landforms evolve through the slow erosional removal of weaker rock, leaving the more resistant rock standing as ridges or mountains. (Drawn by A. N. Strahler.)*

forms the low valley floors of the region. Limestone, subjected to solution by carbonic acid in rainwater and surface water, also forms valleys in humid climates. In arid climates, on the other hand, limestone is a resistant rock and usually stands high to form ridges and cliffs. Sandstone and conglomerate are typically resistant to denudation, and they form ridges or uplands. As a group, the igneous rocks are also resistant—they typically form uplands or mountains rising above adjacent areas of shale and limestone. Although metamorphic rocks in general are more resistant to denudation, individual metamorphic rock types—marble or schist, for example—vary in their resistance.

Figure 17.2 provides a preview of the important rock structures and their accompanying landforms that we will discuss in this chapter. In these sketches, resistant beds

17.2 Landforms and rock structure *This sequence of block diagrams illustrates the principal rock structures and associated landforms covered in this chapter. (Drawn by A. N. Strahler.)*

17.3 Strike and dip *The orientation of a natural rock plane, such as the top of a sedimentary bed, is described by its strike and dip. (Drawn by A. N. Strahler.)*

as joints—occur within rock layers and are likely to be tilted. We need a system of geometry to enable us to measure and describe the position in which these natural planes are held and to indicate them on maps. Examples of such planes include the bedding layers of sedimentary strata, the sides of a dike, and the joints in granite.

The tilt and orientation of a natural rock plane are measured with reference to a horizontal plane. Figure 17.3 shows a water surface resting against tilted sedimentary strata. Since it is horizontal, we can use the water surface as the horizontal plane. The acute angle formed between the rock plane and the horizontal plane is termed the *dip.* The amount of dip is stated in degrees, ranging from 0 for a horizontal rock plane to 90 for a vertical rock plane. Next, we examine the line of intersection between the inclined rock plane and the horizontal water plane. This horizontal line will have an orientation with respect to the compass. This orientation is termed the *strike.* In Figure 17.3, the strike is along a north-south line.

form ridges or belts of hills, while weaker rocks form lowlands or valleys. The upper four sketches show landforms and rock structures of sedimentary rock layers. The structures include horizontal layers, gently sloping layers, and layers that are folded or warped into domes. Other structures shown are fault blocks, plutons, metamorphic belts, and eroded volcanoes.

The broad features of landscapes controlled by the rock structures shown in Figure 17.2 are particularly visible from the perspective of an orbiting spacecraft. *Focus on Remote Sensing 17.1 • Landsat Views Rock Structures* provides some satellite images in which rock structures are especially striking.

Strike and Dip

The surfaces of most rock layers are not flat but are tipped away from the horizontal. In addition, internal planes—such

LANDFORMS OF HORIZONTAL STRATA AND COASTAL PLAINS

Extensive areas of the ancient continental shields are covered by thick sequences of horizontal sedimentary rock layers. At various times in the 600 million years following the end of Precambrian time, these strata were deposited in shallow inland seas. Following crustal uplift, with little disturbance other than minor crustal warping or faulting, these areas became continental surfaces undergoing denudation.

Arid Regions

In arid climates, where vegetation is sparse and the action of overland flow is especially effective, sharply defined landforms develop on horizontal sedimentary strata (Figure 17.4). The normal sequence of landforms is a sheer rock wall, called a *cliff,* which develops at the edge of a resistant

17.4 Arid climate landforms *In arid climates, distinctive erosional landforms develop in horizontal strata. (Drawn by A. N. Strahler.)*

rock layer. At the base of the cliff is an inclined slope, which flattens out into a plain beyond.

In these arid regions, erosion strips away successive rock layers, leaving behind a broad platform capped by hard rock layers. This platform is usually called a **plateau.** Cliffs retreat as near-vertical surfaces because the weak clay or shale formations exposed at the cliff base are rapidly washed away by storm runoff and channel erosion. When undermined, the rock in the upper cliff face repeatedly breaks away along vertical fractures.

Cliff retreat produces a **mesa,** which is a table-topped plateau bordered on all sides by cliffs. Mesas represent the remnants of a formerly extensive layer of resistant rock. As a mesa is reduced in area by retreat of the rimming cliffs, it maintains its flat top. Eventually, it becomes a small steep-sided hill known as a **butte.** Further erosion may produce a single, tall column before the landform is totally consumed.

In the walls of the great canyons of the Colorado Plateau region, these landforms are wonderfully displayed. Figure 17.5 shows a view of the Grand Canyon. The flat surface of the canyon rim marks the edge of the vast plateau surrounding the canyon. Within the canyon, resistant rock layers crop out at different levels, producing a series of steps. Cliffs of resistant rock form the risers, while weaker rocks form the sloping treads below. Benches or platforms develop atop

> Plateaus, mesas, and buttes are landforms of flat-lying erosion-resistant strata in arid regions. Slopes in these regions form stairstep sequences of vertical risers of resistant rock with sloping treads of weaker rock.

resistant strata, extending the treads and producing a series of levels that break the decent into the canyon. Small mesas and buttes remain within the canyon, isolated by erosion of surrounding rock strata.

Drainage Patterns on Horizontal Strata

Imagine making a map of a region that showed only its stream channels. This map would show the stream pattern, or **drainage pattern,** of the region. The drainage pattern that develops on each land-mass type is often related to the underlying rock type and structure.

Regions of horizontal strata show broadly branching stream networks formed into a *dendritic drainage pattern.* Figure 17.6 is a map of the stream channels in a small area within the Appalachian Plateau region of Kentucky. The smaller streams in this pattern take a great variety of compass directions. Because the rock layers lie flat, no one direction is favored.

Coastal Plains

Coastal plains are found along passive continental margins that are largely free of tectonic activity. They are underlain by nearly horizontal strata that slope gently toward the ocean. Figure 17.7*a* shows a coastal zone that has recently

17.5 *Stairstep landscape* *Bright Angel Canyon, Grand Canyon National Park, Arizona. Note the "stairstep" landscape in which resistant sandstones and limestones form flat-topped benches and mesas, while weaker shales form slopes connecting them.*

Focus on Remote Sensing | 17.1

Landsat Views Rock Structures

With its 30 by 30 m (98 by 98 ft) spatial resolution and its 175 by 175 km (109 by 109 mi) image size, Landsat provides a perspective from space that captures a large region in a single view. This scale is ideal for revealing the relationship between landforms and rock structure that is the subject of this chapter.

Image (*a*) shows the ridge-and-valley country of south-central Pennsylvania in a color-infrared presentation. Like an old wooden plank deeply etched by drifting sand to reveal the grain, the land surface shows zigzag ridges formed by bands of hard quartzite. The strata were crumpled into folds during a continental collision that took place over 200 million years ago to produce the Appalachians. In this September image, the red color depicts vegetation cover, while the blue colors identify mainly agricultural fields. The cloud bands are formed by condensation from passing aircraft; note their shadows above and to the left.

Image (*b*) is a color-infrared view of Canyonlands National Park, Utah, and the surrounding area of the Colorado Plateau. The Colorado River makes its way across the scene from upper right to lower left, joined by the Green River near the center of the image. Notice the dendritic drainage pattern that is typical of flat-lying sedimentary rocks. As the river and its tributaries have eroded the landscape, individual rock layers have been stripped from large areas, producing regions dominated by different colors. The river system has cut valleys through beds of different color and thickness, creating a series of colored outlines around the drainage channels. Vegetation is nearly absent in this arid region, but deep red tones indicate a sparse tree cover on Abajo Peak, in the southeast corner of the image. Agriculture near the town of Moab

(a)

Ridge-and-valley landscape *(Courtesy Earth Satellite Corporation.)*

(b)

Canyonlands National Park and vicinity, near Moab, Utah *(NASA/JPL.)*

(c)

mass. Near the head of the valley and to the west is snow-capped Mount Whitney. At 4418 M (14,491 ft), it is the highest peak in the conterminous United States.

Kauai (image *d*) is the oldest and most heavily dissected of the shield volcanoes that make up the island chain of Hawaii. The radial drainage pattern of streams and ridge crests leading away from the central summit is a primary feature of this image. The intense red colors of the northern and western slopes identify lush vegetation, watered by orographic rainfall from the prevailing northeast trade winds. Note the tiny popcorn cumulus clouds that have appeared on the lower slopes and ridge crests where the orographic effect is strong. On the far side of the island is an arid region of rain shadow, with bright green tones identifying the famous Waimea Canyon. The intense red color of the rocks here (see Figure 17.26), produced by weathered iron compounds from the basaltic lava, appears bright green in this color infrared image.

also appears in red tones in the northeast corner of the image.

The southern portion of California's Sierra Nevada Mountains is the region imaged in (*c*). Bounded by faults on the east and west, the range is a huge uplifted block eroded to reveal a core of resistant granite. To the east is the down-dropped Owens Valley, with Owens Lake shrunken by diversion of surface and ground water to meet the needs of Los Angeles. To the west is the fertile Central Valley, with its lush irrigated croplands appearing bright red in this color infrared image. Note the valley of the Kern River, which has carved a straight, north-south gorge bisecting the southern end of the mountain

(d)

17.6 Dendritic drainage *This dendritic drainage pattern is from an area of horizontal strata in the Appalachian Plateau of northern Kentucky. (Derived from data of the U.S. Geological Survey.)*

emerged from beneath the sea. Formerly, this zone was a shallow continental shelf that accumulated successive layers of sediment brought from the land distributed by currents. On the newly formed land surface, streams flow directly seaward, down the gentle slope. A stream of this origin is a *con-*

sequent stream. It is a stream whose course is controlled by the initial slope of a new land surface. Consequent streams form on many kinds of initial landforms, such as volcanoes, fault blocks, or beds of drained lakes.

In an advanced stage of coastal-plain development, a new series of streams and topographic features has developed (Figure 17.7b). Where more easily eroded strata (usually clay or shale) are exposed, denudation is rapid, making *lowlands.* Between them rise broad belts of hills called **cuestas.** Cuestas are commonly underlain by layers of sand, sandstone, limestone, or chalk that dip away from the cuesta at a low angle. The lowland lying between the area of older rock—the oldland—and the first cuesta is called the *inner lowland.*

Streams that develop along the trend of the lowlands, parallel with the shoreline, are of a class known as *subsequent streams.* They take their position along any belt or zone of weak rock and therefore follow closely the pattern of rock exposure. Subsequent streams occur in many regions; we will mention them again in the discussion of folds, domes, and faults.

The drainage lines on a fully dissected coastal plain combine to form a *trellis drainage pattern.* In this type of pattern, a main stream has tributaries that are arranged at right angles. You can see this pattern in the streams of Figure 17.7b. The subsequent streams trend at about right angles to the consequent streams.

The coastal plain of the United States is a major geographical region, ranging in width from 160 to 500 km (about 100 to 300 mi) and extending for 3000 km (about 2000 mi) along the Atlantic and Gulf coasts. In Alabama and

17.7 Development of a broad coastal plain *(a) Early stage—plain recently emerged. (b) Advanced stage—cuestas and lowlands developed. (Drawn by A. N. Strahler.)*

17.8 The Alabama–Mississippi coastal plain *This coastal plain is belted by a series of sandy cuestas and shale lowlands. (After A. K. Lobeck.)*

are underlain by sandy formations and support pine forests. Limestone forms lowlands, such as the Black Belt in Alabama. This belt is named for its dark, fertile soils.

LANDFORMS OF WARPED ROCK LAYERS

Although sedimentary rocks are horizontal or gently sloping in some regions, in other areas rock layers may be upwarped or downwarped. In some cases, these form domes and basins. In others, they form anticlines and synclines—wave-like forms of crests and troughs—as we noted in Chapter 13.

Sedimentary Domes

A distinctive land-mass type is the **sedimentary dome,** a circular or oval structure in which strata have been forced upward into a domed shape. Sedimentary domes occur in various places within the covered shield areas of the continents. Igneous intrusions at great depth may have been responsible for some of these uplifts. For others, upthrusting on deep faults may have been the cause.

Erosion features of a sedimentary dome are illustrated in the two block diagrams of Figure 17.9. Strata are first removed from the summit region of the dome, exposing older strata beneath. Eroded edges of steeply dipping strata form sharp-crested sawtooth ridges called *hogbacks* (Figure 17.10). When the last of the strata have been removed, the ancient shield rock is exposed in the central core of the dome, which then develops a mountainous terrain.

Mississippi (Figure 17.8), the coastal plain shows many of the features of Figure 17.7. In this region, the coastal plain sweeps far inland, as the oldland boundary shows. Cuestas and lowlands run in belts curving to follow the ancient coast. The cuestas, named "hills" or "ridges,"

> **Coastal plains exhibit alternating belts of cuestas and lowlands. Consequent streams flow to the sea across the belts, fed by subsequent streams that drain the cuestas and lowlands.**

17.9 Dome erosion *Erosion of sedimentary strata from the summit of a dome structure. (a) The strata are partially eroded, forming an encircling hogback ridge. (b) The strata are eroded from the center of dome, revealing a core of older igneous or metamorphic rock. (Drawn by A. N. Strahler.)*

(a)

Sandstone

Shale

Igneous and metamorphic rock

(b)

Hogback ridge

Central core

17.10 Hogbacks *Sandstone hogbacks lie along the eastern base of the Colorado Front Range, which is a large domed structure. The view is southward toward the city of Boulder. The rugged forested terrain in the distance is developed on igneous and metamorphic rock exposed in the core of the dome.*

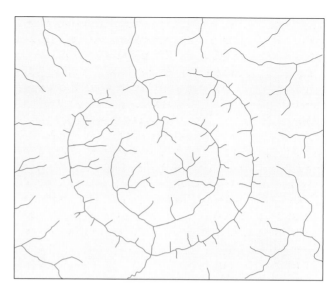

17.11 Dome drainage *The drainage pattern on an eroded dome combines annular and radial elements.*

17.12 The Black Hills dome *The Black Hills consist of a broad, flat-topped dome deeply eroded to expose a core of igneous and metamorphic rocks. (Drawn by A. N. Strahler.)*

The stream network on a deeply eroded dome shows dominant subsequent streams forming a circular system, called an *annular drainage pattern* (Figure 17.11). ("Annular" means "ring-shaped.") The shorter tributaries make a radial arrangement. The total pattern resembles a trellis pattern bent into a circular form.

THE BLACK HILLS DOME A classic example of a large and rather complex sedimentary dome is the Black Hills dome of western South Dakota and eastern Wyoming (Figure 17.12). The Red Valley is continuously developed around the entire dome and is underlain by a weak shale, which is easily washed away. On the outer side of the Red Valley is a high, sharp hogback of Dakota sandstone, known as Hogback Ridge, rising some 150 m (about 500 ft) above the level of the Red Valley. Farther out toward the margins of the dome, the strata are less steeply inclined and form a series of cuestas.

The eastern central part of the Black Hills consists of a mountainous core of intrusive and metamorphic rocks. These mountains are richly forested, while the surrounding valleys are beautiful open parks. Thus, the region is very attractive as a summer resort area. In the northern part of the central core, there are valuable ore deposits. At Lead is the fabulous Homestake Mine, one of the world's richest gold-producing mines. The west-central part of the Black Hills consists of a limestone plateau deeply carved by streams. The plateau represents one of the last remaining sedimentary rock layers to be stripped from the core of the dome.

> The Black Hills dome of South Dakota is a sedimentary dome that is eroded to produce hogback ridges, a circular valley, and a mountainous core of igneous and metamorphic rocks.

Fold Belts

According to the model of mountain-making developed in Chapter 12, strata of the continental margins are deformed into folds along narrow belts during continental collision. As we explained in Chapter 13, a trough-like downbend of strata is a *syncline,* and the arch-like upbend next to it is an *anticline.* In a belt of folded strata, synclines alternate with anticlines, like a succession of wave troughs and wave crests on the ocean surface. GEODISCOVERIES

The two block diagrams of Figure 17.13 show some of the distinctive landforms resulting from fluvial denudation of a belt of folded strata. Deep erosion of these simple, open folds in this land-mass type produces a **ridge-and-valley landscape.** Weaker formations such as shale and limestone are eroded away, leaving hard strata, such as sandstone or quartzite, to stand in bold relief as long,

> Fold belts create a ridge-and-valley landscape of alternating resistant-rock ridges and weak-rock valleys. These linear features lead to the formation of a trellis drainage pattern.

narrow ridges. Note that anticlines, or upwarps, are not always ridges. If a resistant rock type at the center of the anticline is eroded through to reveal softer rocks underneath, an *anticlinal valley* may form. An *anticlinal valley* is shown in Figure 17.13*a.* A *synclinal mountain* is also possible. It occurs when a resistant rock type is exposed at the center of a syncline, and the rock stands up as a ridge (Figure 17.13*b*).

On eroded folds, the stream network is distorted into a trellis drainage pattern (Figure 17.14). The principal elements of this pattern are long, parallel subsequent streams occupying the narrow valleys of weak rock. In a few places, a larger stream cuts across a ridge in a deep, steep-walled *water gap.*

The folds illustrated in Figure 17.13 are continuous and even-crested. They produce ridges that are approximately parallel in trend and continue for great distances. In some fold regions, however, the folds are not continuous and level-crested. Instead, the fold crests rise or descend from place to place. A descending fold crest is said to "plunge." Plunging folds give rise to a zigzag line of ridges (Figure 17.15). Excellent examples of plunging folds can be seen in *Focus on Remote Sensing 17.1 • Landsat Views Rock Structures.* GEODISCOVERIES

LANDFORMS DEVELOPED ON OTHER LAND-MASS TYPES

We now turn to the landforms developed on a few remaining land-mass types. First are erosional forms that develop on a land mass exposed to faulting. Then we discuss the landforms that develop on metamorphic rocks and

17.13   Stages in the erosional development of folded strata (a) Erosion exposes a highly resistant layer of sandstone or quartzite, which controls much of the ridge-and-valley landscape. (b) Continued erosion partly removes the resistant formation but reveals another below it. (Drawn by A. N. Strahler.)

17.14 Trellis drainage *A trellis drainage pattern on deeply eroded folds in the Central Appalachians.*

on exposed igneous batholiths. Last are the landforms of eroded volcanoes.

Erosion Forms on Fault Structures

In Chapter 13, we found that active normal faulting produces a sharp surface break called a *fault scarp* (see Figure 13.15). Repeated faulting may produce a great rock cliff hundreds of meters high. Erosion quickly modifies a fault scarp, but because the fault plane extends hundreds of meters down into the bedrock, its effects on erosional landforms persist for long spans of geologic time. Figure 17.16 shows both the original fault scarp (above) and its later landform expression (below), known as a *fault-line scarp.* Even though the cover of sedimentary strata has been completely removed, exposing the ancient shield rock, the fault continues to produce a landform. Because the fault plane is a zone of weak rock that has been crushed during faulting, it is occupied by a subsequent stream. A scarp persists along the upthrown side. Such scarps on ancient fault

Slate and marble are weak metamorphic rocks that underlie valleys. Schist, gneiss, and quartzite are more resistant and underlie uplands and ridges.

lines are numerous in the exposed and covered continental shields.

Figure 17.17 shows the eastern slope of the Sierra Nevada Mountains east of Yosemite National Park. This steep face is a fault scarp along which an enormous mountain block has been uplifted by many small movements within a few million years. As the uplift has proceeded, erosion has cut deeply into the rock face. The sheer slope has been further modified by glacial activity during the Ice Age, but it still retains its form as created by fault motion.

Metamorphic Belts

Where strata have been tightly folded and altered into metamorphic rocks during continental collision, denudation eventually develops a landscape with a strong grain of ridges and valleys. These features lack the sharpness and straightness of ridges and valleys in belts of open folds. Even so, the principle that governs the development of ridges and valleys in open folds applies to metamorphic landscapes as well—namely, that rocks resistant to denudation form highlands and ridges, while rocks that are more susceptible to denudation form lowlands and valleys.

Figure 17.18 shows typical erosional forms associated with parallel belts of metamorphic rocks, such as schist, slate, quartzite, and marble. Marble forms valleys, while slate and schist make hill belts. Quartzite stands out boldly and may produce conspicuous narrow hogbacks. Areas of gneiss form highlands.

Parts of New England, particularly the Taconic and Green Mountains of New Hampshire and Vermont, illustrate the landforms eroded on an ancient

17.16 Fault scarp evolution *(a) A recently formed fault scarp. (b) Despite continental denudation over several million years, the fault plane causes a long narrow valley bounded by a scarp. (Drawn by A. N. Strahler. Copyright © A. N. Strahler.)*

(a)

Fault scarp

(b)

Fault-line scarp

Subsequent stream

17.15 Plunging folds *Folds with crests that plunge downward give rise to zigzag ridges following erosion. (Drawn by Erwin Raisz. Copyright © by A. N. Strahler.)*

Syncline   Anticline

17.17 Eye on the Landscape Erosion of an uplifted fault block *The near-vertical faces of the foreground peaks in this photo mark a fault scarp along which the rocks have been uplifted by a series of successive motions in a short span of geologic time. Talus slopes of rock debris line the base of the scarp. Eastern Sierra Nevada, California.* **What else would the geographer see? ... Answers at the end of the chapter.**

17.18 Metamorphic belts *Metamorphic rocks typically form long, narrow parallel belts of valleys and mountains. (Drawn by A. N. Strahler.)*

metamorphic belt. The larger valleys trend north and south and are underlain by marble. These are flanked by ridges of gneiss, schist, slate, or quartzite.

Exposed Batholiths and Monadnocks

Recall from Chapter 13 that batholiths—huge plutons of intrusive igneous rock—are formed deep below the Earth's surface. Some are eventually uncovered by erosion and appear at the surface. Because batholiths are typically composed of resistant rock, they are eroded into hilly or mountainous uplands.

Batholiths of granitic composition are a major ingredient in the mosaic of ancient rocks comprising the continental shields. A good example is the Idaho batholith, a granite

17.19 Dendritic drainage *This dendritic drainage pattern is developed on the dissected Idaho batholith.*

mass exposed over an area of about 40,000 sq km (about 16,000 sq mi)—a region almost as large as New Hampshire and Vermont combined. Another example is the Sierra Nevada batholith of California, which makes up most of the high central part of that great mountain block.

A dendritic drainage pattern is often well developed on an eroded batholith (Figure 17.19). Note how similar this pattern is to that on horizontal strata (Figure 17.6).

Small bodies of granite, representing dome-like projections of batholiths that lie below, are often found surrounded by ancient metamorphic rocks into which the granite was intruded (Figure 17.20). A good example is Stone Mountain, which rises above the Piedmont Upland of Georgia. Stone Mountain and other monolithic domes are unique landforms that are discussed further in *Eye on the Environment 17.2 • Marvelous, Majestic, Monolithic Domes.*

The name **monadnock** has been given to an isolated mountain or hill, such as Stone Mountain, that rises conspicuously above a peneplain. A monadnock develops because the rock within the monadnock is much more resistant to denudation processes than the bedrock of the surrounding region. The name is taken from Mount Monadnock in southern New Hampshire.

Deeply Eroded Volcanoes

Recall from Chapter 13 that volcanoes can be divided into two types—stratovolcanoes and shield volcanoes. Stratovolcanoes are typically constructed by explosive eruptions accumulating layers of thick, gummy andesite lava flows and airborne tephra, while shield volcanoes are built primarily of thin, runny flows of basalt. Thus, these two types of volca-

noes produce different types of landforms as they erode away.

Figure 17.21 shows successive stages in the erosion of stratovolcanoes, lava flows, and a caldera. Shown in block (*a*) are active volcanoes in the process of building. These are initial landforms. Lava flows issuing from the volcanoes have spread down into a stream valley, following the downward grade of the valley and forming a lake behind the lava dam.

In block (*b*) some changes have taken place. The most conspicuous change is the destruction of the largest volcano to produce a caldera. A lake occupies the caldera, and a small cone has been built inside. One of the other volcanoes, formed earlier, has become extinct. It has been dissected by streams and has lost its smooth conical form. Smaller, neighboring volcanoes are still active, and the contrast in form is marked.

The system of streams on a dissected volcano cone is a *radial drainage pattern.* Because these streams take their position on a slope of an initial land surface, they are of the consequent variety. It is often possible to recognize volcanoes from a drainage map alone because of the perfection of the radial pattern (Figure 17.22). A good example of a partly dissected volcano is Mount Shasta in California (Figure 17.23).

In Figure 17.21*c* all volcanoes are extinct and have been deeply eroded. The caldera lake has been drained, and the rim has been worn to a low, circular ridge. The lava flows have been able to resist erosion far better than the rock of the surrounding area and have come to stand as mesas high above the general level of the region.

Block (*d*) of Figure 17.21 shows an advanced stage of erosion of stratovolcanoes. All that remains now is a small, sharp peak, called a *volcanic neck*—which is the remains of

> Eroded volcanoes develop a radial drainage pattern of streams leading away from the summit. At the last stage of erosion, all that remains is a volcanic neck with radial dikes.

17.20 Exposed batholiths *Batholiths appear at the land surface only after long-continued erosion has removed thousands of meters of overlying rocks. Small projections of the granite intrusion appear first and are surrounded by older rock. (Drawn by A. N. Strahler.)*

17.21 Erosion in stratovolcanoes These block diagrams illustrate stages in the erosional development of stratovolcanoes and lava flows. (Drawn by Erwin Raisz. Copyright © by A. N. Strahler.)

(a) Lava flow / Stratovolcanoes

(b) Extinct volcano / Caldera

(c) Lava mesas / Remains of caldera

(d) Lava mesa / Necks, dikes

17.22 Radial drainage Radial drainage patterns of volcanoes in the East Indies. The letter "C" marks locations of craters.

Eye on the Environment | 17.2

Marvelous Majestic, Monolithic Domes

There is something awesome about a granite dome. Perhaps it's the way it projects above its surroundings, emerging abruptly from the ground and rising so steeply to its rounded summit. Perhaps it's the smooth, barren surface, hardly marred by a crack or crevice. Or maybe it's just the size—a huge, single, uniform object that dwarfs an observer like no other. In any event, these geological curiosities have aroused the interest of observers for centuries—perhaps even millennia.

Probably the most famous of all granite domes is Rio de Janeiro's Sugar Loaf, a monument that serves as a symbol for that cosmopolitan Brazilian city (*a*). Of course, Australia has its Ayres Rock, a gigantic rock mass in the desert of Northern Territory. But, alas, Ayres Rock is mere sandstone, not solid granite. Australia actually has some

fine granite domes, located on the Eyre Peninsula of South Australia, but they have not achieved the international fame of Ayres Rock. Splendid granite domes also occur in the Nubian Desert of North Africa. Even Pasadena, California, has its own Little League contender in Eagle Rock, a small but prominent dome of massive conglomerate. These monolithic domes of uniform rock owe their rounded shapes to grain-by-grain disintegration of their rock surfaces. Because the domes lack the joints found in most other rocks, oxidation and hydrolysis act uniformly to weather the rock surface, producing a smooth, rounded form.

And what about the spectacular domes of Yosemite National Park in California's Sierra Nevada Range? Yosemite Valley has several splendid domes high up on either side of the deep canyon (*b*). As we

mentioned in Chapter 14, these are exfoliation domes, bearing thick rock shells. As the dome weathers, the outermost shells break apart and slide off as new ones are generated within. Beneath their shells, however, the Yosemite domes seem to be genuine monoliths.

Another fine granite dome is Stone Mountain, near Atlanta (*c*). Like Sugar Loaf, it is a uniform mass of granite seemingly without joints or fissures. It is the surface remains of an ancient, knob-like granite pluton intruded into older metamorphic rocks of the region. For millions of years, this region has been undergoing denudation, and most of it has been reduced to a rolling upland. Numerous monadnocks of quartzite with elongate, ridge-like forms rise above the upland, but Stone Mountain is unique in terms of its compact outline and smoothly rounded form.

(*a*)

Sugar Loaf Mountain *The monolithic rock dome of Sugar Loaf graces the harbor of Rio de Janiero, Brazil.*

(b)

Yosemite Domes *These fine examples of monolithic granite domes are found in the Upper Yosemite Valley, Yosemite National Park, California.*

(c)

Stone Mountain *This dome is a striking erosional remnant about 2.4 km (1.5 mi) long that rises 193 m (650 ft) above the surrounding Piedmont plain.*

17.23 Partly dissected volcano *Mount Shasta in the Cascade Range, northern California, is a partly dissected composite volcano. It has been eroded by streams and small alpine glaciers. Partway up the right-hand side is a more recent subsidiary volcanic cone, called Shastina, with a sharp crater rim.*

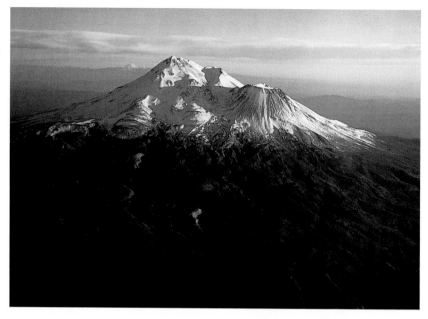

lava that solidified in the pipe of the volcano. Radiating from it are wall-like dikes, formed of magma, which filled radial fractures around the base of the volcano. Perhaps the finest illustration of a volcanic neck with radial dikes is Ship Rock, New Mexico (Figure 17.24).

Shield volcanoes show erosion features that are quite different from those of stratovolcanoes. Figure 17.25*a* shows the first stage of erosion of Hawaiian shield volcanoes, beginning in the upper diagram with the active volcano and its central depression. These are initial landforms. Radial

17.24 Volcanic neck *Ship Rock, New Mexico, is a volcanic neck enclosed by a week shale formation. The peak rises about 520 m (about 1700 ft) above the surrounding plain. In the foreground and to the left, wall-like dikes extend far out from the central peak.*

(a)

(b)

(c)

17.25 Erosion of shield volcanoes Shield volcanoes in various stages of erosion make up the Hawaiian Islands. (a) Newly formed dome with central depression. (b) Early stage of erosion with deeply eroded valley heads. (c) Advanced erosion stage with steep slopes and mountainous relief. (Drawn by A. N. Strahler.)

consequent streams cut deep canyons into the flanks of the extinct shield volcano *(b)*, and these canyons are opened out into deep, steep-walled amphitheaters such as Waimea Canyon, on the island of Kauai (Figure 17.26). In the last stages *(c)*, the original surface of the shield volcano is entirely obliterated, leaving a rugged mountain mass made up of sharp-crested divides and deep canyons.

A LOOK AHEAD

With this chapter, we conclude our series of three chapters on running water as a landforming-making agent. Because the landscapes of most regions on the Earth's land surface are produced by fluvial processes acting on differing rock types, running water is by far the most important agent in creating the variety of landforms that we see around us. The three remaining agents of denudation are waves, wind, and glacial ice. These are the subjects of the final two chapters of our book. In the first, we describe how waves shape the landforms of beaches and coasts, and how wind shapes the dunes of coasts and deserts. In the second, we show how glacial ice scrapes, pushes, and deposits rock materials to form the distinctive landforms of glaciated regions.

17.26 Advanced erosion of a shield volcano Waimea Canyon, over 760 m (about 2500 ft) deep, has been eroded into the flank of an extinct shield volcano on Kauai, Hawaii. Gently sloping layers of basaltic lava flows are exposed in the canyon walls.

In Review | Landforms and Rock Structure

- This chapter has added a new dimension to the realm of landforms produced by fluvial denudation—the variety and complexity introduced by differences in rock composition and crustal structure. Because the various kinds of rocks offer different degrees of resistance to the forces of denudation, they exert a controlling influence on the shapes of landforms. We have seen, too, that streams carving up a land mass are controlled to a high degree by the structure of the rock on which they act. In this way, distinctive drainage patterns evolve.

- In arid regions of horizontal strata, resistant rock layers produce vertical *cliffs* separated by gentler slopes on less resistant rocks. The **drainage pattern** is *dendritic.* Where strata are gently dipping, as on a **coastal plain,** the more resistant rock layers stand out as **cuestas,** interspersed with lowland valleys on weaker rocks. *Consequent streams* cut across the cuestas toward the ocean, and *subsequent streams* follow the valleys. This forms a *trellis drainage pattern.*

- Where rock layers are uparched into a **dome,** erosion produces a circular arrangement of rock layers outward from the center of the dome. Resistant strata form *hogbacks,* and weaker rocks form lowlands. Igneous rocks are often revealed in the center. The drainage pattern is *annular.*

- In fold belts, the sequence of *synclines* and *anticlines* brings a linear pattern of rock layers to the surface. Resistant strata form ridges, and weaker strata form valleys. Like the coastal plain, the drainage pattern will be trellised.

- Faulting provides an initial surface along which rock layers are moved—the *fault scarp.* This feature can persist as a fault-line scarp long after the initial scarp is gone. In metamorphic belts, weak layers of shale, slate, and marble underlie valleys, while gneiss and schist form uplands. Quartzite stands out as a ridge-former.

- Exposed batholiths are often composed of uniform, resistant igneous rock. They erode to form a dendritic drainage pattern. **Monadnocks** of intrusive igneous rock stand up above a plain of weaker rocks.

- Stratovolcanoes produce lava flows that initially follow valleys but are highly resistant to erosion. After erosion of the surrounding area, they can remain as highlands, lava ridges, or mesas. At the last stages of erosion, all that remains of stratovolcanoes are necks and dikes. Hawaiian shield volcanoes are eroded by streams that form deeply carved valleys with steeply sloping heads. Eventually, these merge to produce a landscape of steep slopes and knife-like ridges.

Key Terms

plateau, p. 547	drainage pattern, p. 547	sedimentary dome, p. 551	monadnock, p. 556
mesa, p. 547	coastal plain, p. 547	ridge-and-valley	
butte, p. 547	cuestas, p. 550	landscape, p. 553	

Review Questions

1. Why is there often a direct relationship between landforms and rock structure? How are geologic structures formed deep within the Earth exposed at the surface?

2. Which of the following types of rocks—shale, limestone, sandstone, conglomerate, igneous rocks—tends to form lowlands? uplands?

3. How are the tilt and orientation of a natural rock plane measured?

4. How are coastal plains formed? Identify and describe two landforms found on coastal plains. What drainage pattern is typical of coastal plains?

5. What types of landforms are associated with sedimentary domes? How are they formed? What type of drainage pattern would you expect for a dome and why? Provide an example of an eroded sedimentary dome and describe it briefly.

6. How does a ridge-and-valley landscape arise? Explain the formation of the ridges and valleys. What type of drainage pattern is found on this landscape? Where in the United States can you find a ridge-and-valley landscape?

7. What type of landform(s) might you expect to find in the presence of a fault? Why?

8. Which of the following types of rocks—schist, slate, quartzite, marble, gneiss—tend to form lowlands? uplands?

9. Do shield volcanoes erode differently from stratovolcanoes? How so?

Eye on the Environment 17.2 ● Marvelous, Majestic, Monolithic Domes

1. What are the characteristics of domes? What process gives the dome its unique shape?

2. Briefly describe and locate two examples of granite domes.

Visualizing Exercises

1. Identify and sketch a typical landform of flat-lying rock layers found in an arid region.
2. What types of landforms and drainage patterns develop on batholiths? Provide a sketch of a batholith. What is the difference between a batholith and a monadnock?
3. Describe and sketch the stages through which a landscape of stratovolcanoes is eroded. What distinctive landforms remain as the last remnants?

Essay Questions

1. Imagine the following sequence of sedimentary strata—sandstone, shale, limestone, and shale. What landforms would you expect to develop in this structure if the sequence of beds is (*a*) flat-lying in an arid landscape; (*b*) slightly tilted as in a coastal plain; (*c*) folded into a syncline and an anticline in a fold belt; (*d*) fractured and displaced by a normal fault? Use sketches in your answer.
2. A region of ancient mountain roots, now exposed at the surface, includes a central core of plutonic rocks surrounded on either side by belts of metamorphic rocks. What landforms would you expect for this landscape and why?

Eye on the Landscape

Chapter Opener Rapley Monocline A monocline is a rock structure in which flat layers bend down (or up), then come to lie flat again. Here we see such a structure eroded by fluvial processes to expose the beds. As streams cut into the monocline, they create triangular facets, or flatirons **(A)**. Long, linear ridges, or hogbacks, are also formed **(B)**. Note the entrenched meander of the sediment-laden San Juan River **(C)** that cuts through the structure in the foreground. Fluvial erosion and entrenched meanders were covered in Chapter 16.

17.17 Uplifted fault block This photo of the dramatic fault scarp at the eastern edge of the Sierra Nevada shows a landscape with abundant evidence of glacial action. Note the knife-edge ridges and peaks that crown the top of the range **(A)**. These were formed as ice capping the range flowed down and out of the region. The entire basin in the foreground, from the near peaks **(B)** to the lakes **(C)**, was filled with ice that scoured loose rock from the basin, leaving an irregular topography of hills and depressions that are now filled with water. Chapter 19 describes landforms created by moving ice.

Coastal erosion   *An eroding bluff of Ice Age sediment. Sankaty Head Lighthouse, Nantucket Island, Massachusetts.*

18 | LANDFORMS MADE BY WAVES AND WIND

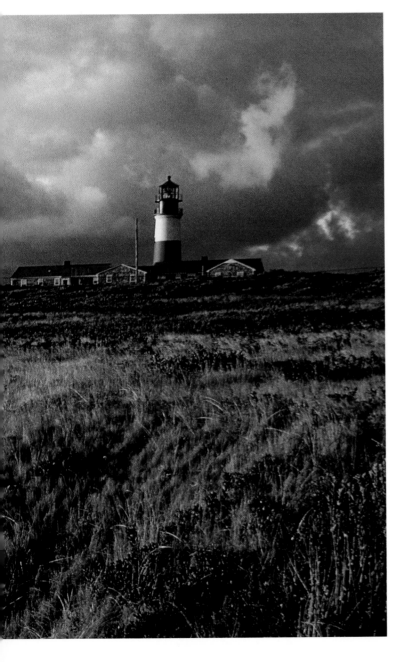

· · · · · · · · · · · · · · · · · ·
PUTTING YOU IN THE PICTURE

A coastal storm, occurring at the interface between land, ocean, and atmosphere, can be an awesome event. Lashed by unrelenting wind and rain, the surf attacks the beach, moving large quantities of sand offshore. At high tide, waves cut new paths through coastal dunes to inland bays. Shoreline bluffs are undercut and collapse, releasing new sediment to the shoreline. Picture yourself witnessing the power of a coastal storm as you read the account below.

You slow the car and carefully turn into a narrow, black-topped road that leads to the lighthouse. The drive wasn't easy, with all those downed tree limbs and flooded spots along the way. What a way to end your Thanksgiving vacation! Before arriving at the cottage on Cape Cod, you had visions of long treks along the beach, inhaling the salt air and watching the sea birds rush to and fro with the swash of the breaking waves.

Soon after you arrived, however, the skies turned gray, and the wind and rain began—a "nor'easter" was in store. Last night the storm reached its peak, as rain fell in sheets and the wind howled without letup. This morning the rain stopped, and so you decided to venture out to the beach.

It's not long before the lighthouse comes into view, a wide, glass-topped column set against the low clouds of the dark sky. You enter the parking lot, which is perched on top of the coastal bluff overlooking the beach. The far portion of the lot is blocked off with red- and white-striped sawhorses anchored by sandbags to keep them erect in the strong wind.

Stepping out of the car, you find that the force of the wind is strong, though not overwhelming. Beyond the sawhorses, you see that a large section of pavement has broken away and slid down the bluff to the beach below. The storm waves, striking the bluff at high tide, must have undermined the pavement, allowing it to fall into the breaker zone beneath. The wooden stairs down the bluff have fared somewhat better. The sand underneath the bottom part of the lowest flight has been undercut, but access to the beach is not difficult, and you descend. Fortunately, the tide is now out.

The beach looks totally different from the way it looked when you visited last summer. Then, the beach had a wide, flat upper portion, ideal for sunbathing, and sloped steeply in the breaker zone. Now the beach is just a long, gentle slope stretching from the bluffs to the breakers. The waves are breaking much farther out, where the gray and white of the surf and foam converges with the gray and white of the sky. Rolls of frothing water move landward, eventually collapsing and running far up the low slope.

The roar of the surf is intense and continuous. The beach feels dense and hard. A few inches above the wet, smooth surface, sheets of wind-blown sand move downwind, braiding and separating into flowing, stream-like currents. You turn to look back at the bluff face. The freshly eroded layers of sand and silt are clear and distinct. Large, wave-washed lumps of sediment lie at the base of the bluff. It won't be long before the ocean's erosive power claims the lighthouse itself.

You linger for a time, enjoying the dark beauty of the scene, before heading for home. Although this trip to the beach wasn't exactly what you had in mind for the holiday, you are glad to have experienced first-hand the power of waves and wind.

Wind and waves are the primary agents that shape the shoreline. Unrelenting, they often work against human efforts to develop the coastline and preserve it in an unchanging configuration. Thus, human and natural forces often conflict in the coastal zone, as we will see in this chapter. ∎

LANDFORMS MADE BY WAVES AND WIND

In Chapter 5, we saw how the rotation of the Earth on its axis, combined with unequal solar heating of the Earth's surface, produces a system of global winds in which air moves as a fluid across the Earth's surface. This motion of air produces a frictional drag that can move surface materials. When winds blow over broad expanses of water, waves are generated. These waves then expend their energy at coastlines, eroding rock, moving sediment, and creating landforms. When winds blow over expanses of soil or alluvium that are not protected by vegetation, fine particles can be picked up and carried long distances before coming to rest. Coarser particles, such as sand, can also be moved to build landforms, such as sand dunes. Because wave action on beaches provides abundant sand, we often find landforms of both wind and waves near coastlines.

This chapter describes the processes of erosion and deposition, as well as the landforms that result when wind power moves surface materials either directly as wind or indirectly as waves. Unlike fluvial action or mass wasting, wind and breaking waves can move materials uphill against the force of gravity. For this reason, wind power is a unique land-forming agent. Another agent that acts in coastal regions is the tide—the rhythmic rise and fall in sea level that is generated by the gravitational attraction of the Sun and Moon. This subject is also covered in our chapter.

Wave action is the most important agent in shaping shoreline landforms. Waves attack weak materials to make marine scarps and resistant materials to make marine cliffs.

Throughout this chapter we will use the term **shoreline** to mean the shifting line of contact between water and land. The broader term **coastline,** or simply **coast,** refers to a zone in which coastal processes operate or have a strong influence. The coastline includes the shallow water zone in which waves perform their work, as well as beaches and cliffs shaped by waves, and coastal dunes. Also often present are **bays**—bodies of water that are sheltered by the configuration of the coast from strong wave action. Where a river empties into an ocean bay, the bay is termed an **estuary.** In an estuary, fresh and ocean water mix, creating a unique habitat for many plants and animals that is neither fresh water nor ocean.

THE WORK OF WAVES

The most important agent shaping coastal landforms is wave action. The energy of waves is expended primarily in the constant churning of mineral particles and water as waves break at the shore. This churning erodes shoreline materials, moving the shoreline landward. But, as we will see shortly, the action of waves and currents can also move sediment along the shoreline for long distances. This activity can build beaches outward as well as form barrier islands just offshore.

18.1 A breaking wave *As the wave approaches the beach (1–3), it steepens (4–5), and finally falls forward (6–7), rushing up the beach slope (8). (After W. M. Davis.)*

Waves travel across the deep ocean with little loss of energy. When waves reach shallow water, the drag of the bottom slows and steepens the wave. However, the wave top maintains its forward velocity and eventually falls down onto the face of the wave, creating a *breaker* (Figure 18.1). Many tons of water surge forward, riding up the beach slope.

Where weak or soft materials—various kinds of regolith, such as alluvium—make up the coastline, the force of the forward-moving water alone easily cuts into the coastline. Here, erosion is rapid, and the shoreline may recede rapidly. Under these conditions, a steep bank—a *marine scarp*—is the typical coastal landform (Figure 18.2). It retreats steadily under attack of storm waves.

GEODISCOVERIES

Marine Cliffs

Where a **marine cliff** lies within reach of the moving water, it is impacted with enormous force. Rock fragments of all sizes, from sand to cobbles, are carried by the surging water and thrust against the bedrock of the cliff. The impact breaks away new rock fragments, and the cliff is undercut at the base. In this way, the cliff erodes shoreward, maintaining its form as it retreats. The retreat of a marine cliff formed of hard bedrock is exceedingly slow, when judged in terms of a human life span.

18.2 Eye on the Landscape
Retreating shoreline *Mohegan Bluffs, Block Island, Rhode Island. This marine scarp of Ice Age sediments is being rapidly eroded, threatening historic Southeast Lighthouse. In 1993 the lighthouse was moved to a safer spot 73.5 m (245 ft) away.* **What else would the geographer see? ... Answers at the end of the chapter.**

18.3 *Landforms of sea cliffs* Wave action undercuts the marine cliff, maintaining its form. (Drawn by E. Raisz.)

Figure 18.3 illustrates some details of a typical marine cliff. A deep basal indentation, the *wave-cut notch,* marks the line of most intense wave erosion. The waves find points of weakness in the bedrock and penetrate deeply to form crevices and *sea caves.* Where a more resistant rock mass projects seaward, it may be cut through to form a picturesque *sea arch.* After an arch collapses, a rock column, known as a *stack,* remains. Eventually, the stack is toppled by wave action and is leveled. As the sea cliff retreats landward, continued wave abrasion forms an *abrasion platform* (Figure 18.4). This sloping rock floor continues to be eroded and widened by abrasion beneath the breakers. If a beach is present, it is little more than a thin layer of gravel and cobblestones atop the abrasion platform.

Beaches

Where sand is in abundant supply, it accumulates as a thick, wedge-shaped deposit, or **beach.** Beaches absorb the energy of breaking waves. During short periods of storm activity, the beach is cut back, and sand is carried offshore a short distance by the heavy wave action. However, the sand is slowly returned to the beach during long periods when waves are weak. In this way, a beach may retain a fairly stable but alternating configuration over many years' time.

Although beaches are most often formed from particles of fine to coarse quartz sand, some beaches are built from rounded pebbles or cobbles. Still others are formed from fragments of volcanic rock (Hawaii) or even shells (Florida).

18.4 A wave-cut abrasion platform Continued wave abrasion formed this rock platform on the central coast of California, Montana d'Oro State Park.

18.5 How waves move sediment by littoral drift (a) Swash and backwash move particles along the beach in beach drift. (b) Waves set up a longshore current that move particles by longshore drift. (c) Littoral drift, produced by these two processes, creates a sandspit.

Beaches are shaped by alternate landward and seaward currents of water generated by breaking waves. After a breaker has collapsed, a foamy, turbulent sheet of water rides up the beach slope. This *swash* is a powerful surge that causes a landward movement of sand and gravel on the beach. When the force of the swash has been spent against the slope of the beach, a return flow, or *backwash*, pours down the beach (Figure 18.5a). This "undercurrent" or "undertow," as it is popularly called, can be strong enough to sweep unwary bathers off their feet and carry them seaward beneath the next oncoming breaker.

> Breakers attacking the shore at an angle produce littoral drift, which includes beach drift—a movement of sediment along the shoreline—as well as longshore drift—a sediment movement just offshore.

Littoral Drift

The unceasing shifting of beach materials with swash and backwash of breaking waves also results in a sidewise movement known as *beach drift* (Figure 18.5a). Wave fronts usually approach the shore at less than a right angle, so that the swash and its burden of sand ride obliquely up the beach. After the wave has spent its energy, the backwash flows down the slope of the beach in the most direct downhill direction. The particles are dragged directly seaward and come to rest at a position to one side of the starting place. This movement is repeated many times, and individual rock particles travel long distances along the shore. Multiplied many thousands of times to include the numberless particles of the beach, beach drift becomes a very significant form of sediment transport.

Sediment in the shore zone is also moved along the beach in a related, but different, process. When waves approach a shoreline at an angle to the beach, a current is set up parallel to the shore in a direction away from the wind. This is known as a *longshore current* (Figure 18.5b). When wave and wind conditions are favorable, this current is capable of carrying sand along the sea bottom. The process is called *longshore drift*. Beach drift and longshore drift, acting together, move particles in the same direction for a given set of onshore winds. The total process is called **littoral drift** (Figure 18.5c. ("Littoral" means "pertaining to a coast or shore.")

Littoral drift operates to shape shorelines in two quite different situations. Where the shoreline is straight or broadly

18.6 **Eye on the Landscape** **Sandspit** *This white sandspit is growing in a direction toward the observer. Monomoy Point, Cape Cod. Massachusetts.* **What else would the geographer see? ... Answers at the end of the chapter.**

curved for many kilometers at a stretch, littoral drift moves the sand along the beach in one direction for a given set of prevailing winds. This situation is shown in Figure 18.5*c*. Where a bay exists, the sand is carried out into open water as a long finger, or *sandspit* (Figure 18.6). As the sandspit grows, it forms a barrier, called a *bar,* across the mouth of the bay.

A second situation is shown in Figure 18.7. Here the coastline consists of prominent headlands, projecting seaward, and deep bays. Approaching wave fronts slow when the water becomes shallow, and this slowing effect causes the wave front to wrap around the headland. High, wave-cut cliffs develop shoreward of an abrasion platform. Sediment from the eroding cliffs is carried by littoral drift along the sides of the bay, converging on the head of the bay. The result is a crescent-shaped beach, often called a *pocket beach.* GEODISCOVERIES

Littoral Drift and Shore Protection

When sand arrives at a particular section of the beach more rapidly than it is carried away, the beach is widened and built oceanward. This change is called **progradation** (building out). When sand leaves a section of beach more rapidly than it is brought in, the beach is narrowed and the shoreline

moves landward. This change is called **retrogradation** (cutting back).

Along stretches of shoreline affected by retrogradation, the beach may be seriously depleted or even entirely disappear, destroying valuable shore property. In some circum-

18.7 **Pocket beaches** *On an embayed coast, sediment is carried from eroding headlands to the bayheads, where pocket beaches are formed. (Copyright © A. N. Strahler.)*

18.8   Breaching of a barrier beach *Erosion by severe storms during the winter of 1993 carved out an inlet in this barrier beach on the south shore of Long Island, New York. A system of groins has trapped sand, protecting the far stretch of beach. In the foreground, the beach has receded well inland of the houses that were once located on its edge. (© Vie De Lucia/NYT Pictures.)*

stances, structures can be installed that will cause progradation, and so build a broad, protective beach. This is done by installing groins at close intervals along the beach. A *groin* is simply a wall or embankment built at right angles to the shoreline. It may be constructed of large rock masses, of concrete, or of wooden pilings. The groins act to trap sediment moving along the shore as littoral drift (Figure 18.8).

In some cases, the source of beach sand is sediment delivered to the coast by a river. Construction of dams far upstream on the river may drastically reduce the sediment load of the river, cutting off the source of sand for littoral drift. Retrogradation can then occur on a long stretch of shoreline.

Groins are walls or embankments built at right angles to the shoreline. They trap littoral drift and help prevent beach retrogradation.

TIDAL CURRENTS

Most marine coastlines are influenced by the *ocean tide,* a rhythmic rise and fall of sea level under the influence of changing attractive forces of Moon and Sun on the rotating Earth. Where tides are great, the effects of changing water level and the tidal currents thus set in motion are of major importance in shaping coastal landforms.

The tidal rise and fall of water level is graphically represented by the *tide*

Ocean tides produce tidal currents at the shoreline. These currents scour inlets and distribute fine sediment in bays and estuaries.

curve. Figure 18.9 is a tide curve for Boston Harbor covering a day's time. The water reached its maximum height, or high water, at 3.6 m (11.8 ft) on the tide staff and then fell to its minimum height, or low water, at 0.8 m (2.6 ft), occurring about $6\frac{1}{4}$ hours later. A second high water occurred about $12\frac{1}{2}$ hours after the previous high water, completing a single tidal cycle. In this example, the tidal range, or difference between heights of successive high and low waters, is 2.8 m (9.2 ft).

In bays and estuaries, the changing tide sets in motion currents of water known as *tidal currents.* The relationships between tidal currents and the tide curve are shown in Figure 18.10. When the tide begins to fall, an *ebb current* sets in. This flow ceases about the time when the tide is at its lowest point. As the tide begins to rise, a landward current, the *flood current,* begins to flow. GEODISCOVERIES

Tidal Current Deposits

Ebb and flood currents generated by tides perform several important functions along a shoreline. First, the currents that flow in and out of bays through narrow inlets are very swift

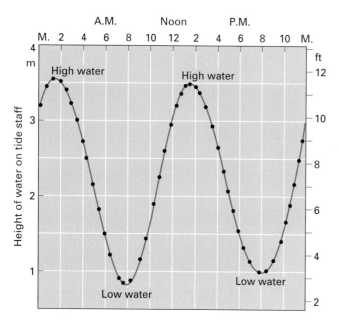

18.9 Tide curve *Height of water at Boston Harbor measured every half hour.*

and can scour the inlet strongly. This keeps the inlet open, despite the tendency of shore-drifting processes to close the inlet with sand.

Second, tidal currents carry large amounts of fine silt and clay in suspension. This fine sediment is derived from streams that enter the bays or from bottom muds agitated by storm wave action. It settles to the floors of the bays and

18.10 Tidal currents *The ebb current flows seaward as the tide level falls. The flood current flows landward as the tide level rises. Tidal currents are strongest in the middle of the cycle when water level is changing most rapidly.*

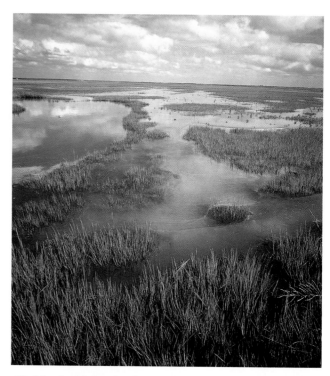

18.11 A tideland salt marsh *At high tide, most of this salt marsh is flooded. Chincoteague National Wildlife Refuge, Virginia.*

estuaries, where it accumulates in layers and gradually fills the bays. Much organic matter is present in this sediment.

In time, tidal sediments fill the bays and produce mud flats, which are barren expanses of silt and clay. They are exposed at low tide but covered at high tide. Next, a growth of salt-tolerant plants takes hold on the mud flat. The plant stems trap more sediment, and the flat is built up to approximately the level of high tide, becoming a *salt marsh* (Figure 18.11). A thick layer of peat is eventually formed at the surface. Tidal currents maintain their flow through the salt marsh by means of a highly complex network of winding tidal streams.

TYPES OF COASTLINES

The world's coastlines present a number of different coastline types. Each type is unique because of the distinctive land mass against which the ocean water has come to rest. One group of coastline types derives its qualities from **submergence,** the partial drowning of a coast by a rise of sea level or a sinking of the crust. Another group derives its qualities from **emergence,** the exposure of submarine landforms by a falling of sea level or a rising of the crust. Another group of coastline types results when new land is built out into the ocean by volcanoes and lava flows, by the growth of river deltas, or by the growth of coral reefs.

Several important types of coastlines are illustrated in Figure 18.12. The first two are the result of submergence. The *ria coast* is a deeply embayed coast resulting from submergence of a land mass dissected by streams. The *fiord coast* is deeply

Ria coast

Fiord coast

Barrier-island coast

Delta coast

Volcano coast (*left*)   Coral-reef coast (*right*)

Fault coast

18.12 Seven types of coastlines *These sketches illustrate the important features of seven types of coastlines. (Drawn by A. N. Strahler.)*

indented by steep-walled fiords, which are submerged glacial troughs (discussed further in Chapter 19). The *barrier-island coast* is associated with a recently emerged coastal plain. Here, the offshore slope is very gentle, and a barrier island of sand, lying a short distance from the coast, is created by wave action. large rivers build elaborate deltas, producing *delta coasts*. The *volcano coast* is formed by the eruption of volcanoes and lava flows, partly constructed below water level. Reef-building corals create new land and make a *coral-reef coast*. Down-faulting of the coastal margin of a continent can allow the shoreline to come to rest against a fault scarp, producing a *fault coast*.

> Ria coasts and fiord coasts result from submergence of a land mass. Islands, bars, estuaries, and deep bays are characteristic of these "drowned" coastlines.

Coastlines of Submergence

Coastlines of submergence include ria coasts and fiord coasts. The ria coast, which takes its name from the Spanish

word for estuary, *ria*, has many offshore islands. A ria coast is formed when a rise of sea level or a crustal sinking (or both) brings the shoreline to rest against the sides of valleys previously carved by streams (Figure 18.13*a*). Wave attack forms cliffs on the exposed seaward sides of islands and headlands (*b*). Sediment produced by wave action accumulates in the form of beaches along the cliffed headlands and at the heads of bays. This sediment is carried by littoral drift and is often built into sandspits across bay mouths and as connecting links between islands and mainland (*c*). Eventually, the sandspits seal off the bays, forming estuaries (*d*). If sea level remains at the same height with respect to the land for a long time, the coast may evolve into a cliffed shoreline of narrow beaches (*e*).

The fiord coast is similar to the ria coast. However, the submerged valleys were carved by flowing glaciers instead of streams. As a result, the valleys are deep, with straight,

18.13 Stages in the evolution of a ria coastline *Wave action on a coastline of submergence creates many interesting coastal landforms. (Drawn by A. N. Strahler.)*

steep sides. Because sediment rapidly sinks into the deep water, beaches are rare. Fiords are described further in Chapter 19 (see Figures 19.7 and 19.8).

Many glaciated coastlines, such as those of New England and the Canadian Maritime Provinces, are coastlines of submergence. During the Ice Age, the weight of ice sheets depressed the crust of these regions, and they are still slowly rising now that the ice sheets are gone. Meanwhile, ocean waves and currents are rapidly eroding their rocky shorelines, creating bays, bars, and estuaries, as shown in Figure 18.13.

> Barrier island coasts are found where a coastal plain of gently sloping sediments emerges from the sea. Barrier island beaches protect a shallow lagoon that is serviced by tidal inlets in the barrier beach.

Barrier-Island Coasts

In contrast to the bold relief and deeply embayed outlines of coastlines of submergence are low-lying coasts from which the land slopes gently beneath the sea. The coastal plain of the Atlantic and Gulf coasts of the United States presents a

particularly fine example of such a gently sloping surface. As we explained in Chapter 17, this coastal plain is a belt of relatively young sedimentary strata, formerly accumulated beneath the sea as deposits on the continental shelf. During the latter part of the Cenozoic Era and into recent time, the coastal plain emerged from the ocean as a result of repeated crustal uplifts.

Along much of the Atlantic Gulf coast there exist *barrier islands,* low ridges of sand built by waves and further increased in height by the growth of sand dunes (Figure 18.14). Behind the barrier island lies a *lagoon.* It is a broad expanse of shallow water in places largely filled with tidal deposits.

A characteristic feature of barrier islands is the presence of gaps, known as *tidal inlets.* Strong currents flow alternately landward and seaward through these gaps as the tide rises and falls. In heavy storms, the barrier island may be breached by new inlets (Figure 18.8). After that occurs, tidal current scour will tend to keep a new inlet open. In some cases, the inlet is closed later by shore drifting of beach sand. Perhaps the finest example of a barrier-island coast is the Gulf coast of Texas (Figure 18.15).

Delta and Volcano Coasts

The deposit of clay, silt, and sand made by a stream or river where it flows into a body of standing water is known as a **delta.** Deposition is caused by rapid reduction in velocity of the current as it pushes out into the standing water. Typically, the river channel divides and subdivides into lesser channels called *distributaries.* The coarser sand and silt particles settle out first, while the fine clays continue out farthest and eventually come to rest in fairly deep water. Contact of fresh with saltwater causes the finest clay particles to clot together and form larger particles that settle to the seafloor.

Deltas show a wide variety of outlines. The Nile Delta has the basic triangular shape of the Greek letter delta. In outline, it resembles an alluvial fan. The Mississippi Delta has a different shape. Long, branching fingers grow far out into the Gulf of Mexico at the ends of the distributaries, giving the impression of a bird's foot. A satellite image of the

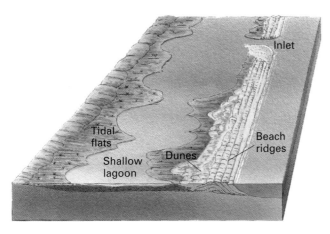

18.14 Barrier island *A barrier island is separated from the mainland by a wide lagoon. Sediments fill the lagoon, while dune ridges advance over the tidal flats. An inlet allows tidal flows to pass in and out of the lagoon. (Drawn by A. N. Strahler.)*

18.15 Barrier-island coast *(a) A Landsat image of the Texas barrier-island coast. Red colors on the Landsat image indicate vegetation. Bright white areas are dunes or beach sand. Lighter blue tones mark the presence of sediment in water inside of the barrier beach. (b) Index map of the location.*

(a)

(b)

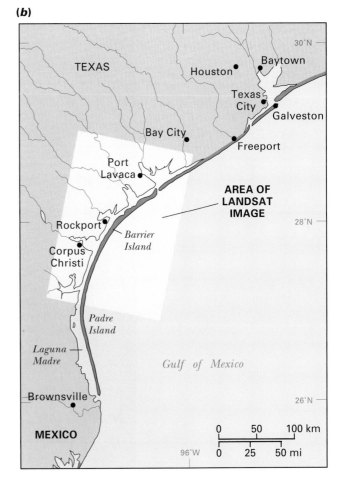

delta, Figure 18.16, shows the great quantity of suspended sediment—clay and fine silt—being discharged by the river into the Gulf. It amounts to about 1 million metric tons (about 1.1 million English tons) per day.

Delta growth is often rapid, ranging from 3 m (about 10 ft) per year for the Nile to 60 m (about 200 ft) per year for the Mississippi Delta. Some cities and towns that were at river mouths several hundred years ago are today several kilometers inland.

Volcano coasts arise where volcanic deposits—lava and ash—flow from active volcanoes into the ocean. Low cliffs occur when wave action erodes the fresh deposits. Beaches are typically narrow, steep, and composed of fine particles of the extrusive rock.

Coral-Reef Coasts

Coral-reef coasts are unique in that the addition of new land is made by organisms—corals and algae. Growing together, these organisms secrete rock-like deposits of mineral carbonate, called **coral reefs.** As coral colonies

> **A delta results when a river empties into the ocean. The Mississippi Delta has a bird's foot plan, with branching distributaries carrying river water and sediment to the ocean.**

> **Coral reef coasts occur in warm oceans where corals build reefs at the land-sea margin. Volcano coasts are found where fresh volcanic deposits of lava and ash reach the ocean. Raised shorelines are marked by marine terraces.**

die, new ones are built on them, accumulating as limestone. Coral fragments are torn free by wave attack, and the pulverized fragments accumulate as sand beaches.

Coral-reef coasts occur in warm tropical and equatorial waters between the limits of lat. 30° N and 25° S. Water temperatures above 20°C (68°F) are necessary for dense growth of coral reefs. Reef corals live near the water surface. The sea water must be free of suspended sediment and well aerated for vigorous coral growth. For this reason, corals thrive in positions exposed to wave attack from the open sea. Because muddy water prevents coral growth, reefs are missing opposite the mouths of muddy streams. Coral reefs are remarkably flat on top. They are exposed at low tide and covered at high tide.

There are three distinctive types of coral reefs—fringing reefs, barrier reefs, and atolls. *Fringing reefs* are built as platforms attached to shore (Figure 18.17). They are widest in front of headlands where wave attack is strongest. *Barrier reefs* lie out from shore and are separated from the mainland by a

18.16 The Mississippi Delta viewed by Landsat *The natural levees of the bird-foot delta appear as lace-like filaments in a great pool of turbid river water. New Orleans can be seen at the upper left, occupying the region between the natural levees of the Mississippi River and the southern shore of Lake Pontchartrain.*

18.17 Fringing reefs *Coral reefs fringe the Island of Moorea, Society Islands, South Pacific Ocean. The island is a deeply dissected volcano with a history of submergence. Tahiti lies in the background.*

lagoon (Figure 18.17). Narrow gaps occur at intervals in barrier reefs. Through these openings, excess water from breaking waves is returned from the lagoon to the open sea.

Atolls are more or less circular coral reefs enclosing a lagoon but have no land inside. On large atolls, parts of the reef have been built up by wave action and wind to form low island chains connected by the reef. Most atolls are built on a foundation of volcanic rock that has subsided below sea level. (See Figure 13.6 and accompanying text.)

Raised Shorelines and Marine Terraces

The active life of a shoreline is sometimes cut short by a sudden rise of the coast. When this tectonic event occurs, a *raised shoreline* is formed. If present, the marine cliff and abrasion platform are abruptly raised above the level of wave action. The former abrasion platform has now become a *marine terrace* (Figure 18.18). Of course, fluvial denudation acts to erode the terrace as soon as it is formed. The terrace may also undergo partial burial under alluvial fan deposits.

Raised shorelines are common along the continental and island coasts of the Pacific Ocean because here tectonic processes are active along the mountain and island arcs.

Repeated uplifts result in a series of raised shorelines in a step-like arrangement. Fine examples of these multiple marine terraces are seen on the western slope of San Clemente Island, off the California coast (Figure 18.19).

18.18 A marine terrace *A raised shoreline becomes a cliff parallel with the newer, lower shoreline. The former abrasion platform is now a marine terrace. (Drawn by A. N. Strahler.)*

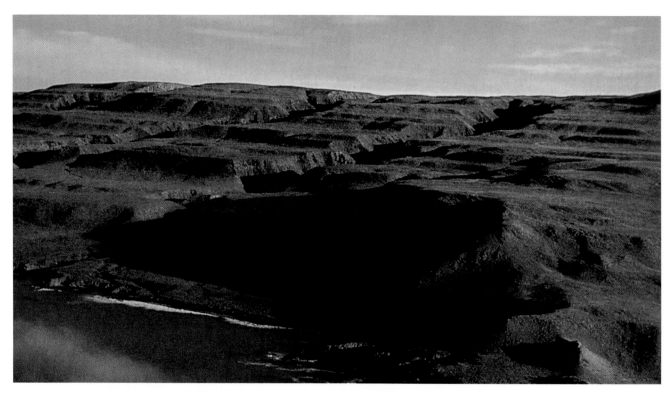

18.19 *A raised shoreline* *Marine terraces on the western slope of San Clemente Island, off the southern California coast. More than 20 terraces have been identified in this series. The highest has an elevation of about 400 m (about 1300 ft).*

WIND ACTION

Transportation and deposition of sand by wind is an important process in shaping certain coastal landforms. We have already mentioned coastal sand dunes, which are derived from beach sand. In the remainder of this chapter, we investigate the transport of mineral particles by wind and the shaping of dune forms. Our discussion also provides information about dune forms far from the coast—in desert environments, where the lack of vegetation cover allows dunes to develop if a source of abundant sand particles is present.

Wind blowing over the land surface is one of the active agents of landform development. Ordinarily, wind is not strong enough to dislodge mineral matter from the surfaces of unweathered rock, or from moist, clay-rich soils, or from soils bound by a dense plant cover. Instead, the action of wind in eroding and transporting sediment is limited to land surfaces where small mineral and organic particles are in a loose, dry state. These areas are typically deserts and semiarid lands (steppes). An exception is the coastal environment, where beaches provide abundant supplies of loose sand. In this environment, wind action shapes coastal dunes, even where the climate is humid and the land surface inland from the coast is well protected by a plant cover.

> Wind action moves mineral particles when they are in a dry state and unprotected by a vegetation cover. These conditions are found in deserts and semiarid regions of the world, as well as on sandy shorelines.

Erosion by Wind

Wind performs two kinds of erosional work: abrasion and deflation. Loose particles lying on the ground surface may be lifted into the air or rolled along the ground by wind action. In the process of *wind abrasion,* wind drives sand and dust particles against an exposed rock or soil surface. This causes the surface to be worn away by the impact of the particles. Abrasion requires cutting tools—mineral particles—carried by the wind, while deflation is accomplished by air currents alone.

The sandblasting action of wind abrasion against exposed rock surfaces is limited to the basal meter or two of a rock mass that rises above a flat plain. This height is the limit to which sand grains can rise high into the air. Wind abrasion produces pits, grooves, and hollows in the rock. Wooden utility poles on windswept sandy plains are quickly cut through at the base unless a protective metal sheathing or heap of large stones is placed around the base.

The removal of loose particles from the ground is termed **deflation.** Deflation acts on loose soil or sediment. Dry river courses, beaches, and areas of recently formed glacial deposits are susceptible to deflation. In dry climates, much of the ground surface is subject to deflation because the soil or rock is largely bare of vegetation.

Wind is selective in its deflational action. The finest particles, those of clay and silt sizes, are lifted and raised into the air—sometimes to a height of a thousand meters (about 3300 ft) or more. Sand grains are moved only when winds are at least moderately strong and usually travel within a meter or two (about 3 to 6 ft) of the ground. Gravel fragments and rounded pebbles can be rolled or pushed over flat ground by strong winds, but they do not travel far. They become easily lodged in hollows or between other large grains. Consequently, where a mixture of size of particles is present on the ground, the finer sized particles are removed and the coarser particles remain behind.

A landform produced by deflation is a shallow depression called a **blowout.** The size of the depression may range from a few meters (10 to 20 ft) to a kilometer (0.6 mi) or more in diameter, although it is usually only a few meters deep. Blowouts form in plains regions of dry climate. Any small depression in the surface of the plain, especially where the grass cover has been broken or disturbed, can form a blowout. Rains fill the depression and create a shallow pond or lake. As the water evaporates, the mud bottom dries out and cracks, leaving small scales or pellets of dried mud. These particles are lifted out by the wind.

> Wind deflation can produce blowouts and help form desert pavement—a surface armor of coarse particles that reduces further deflation.

Deflation is also active in semidesert and desert regions. In the southwestern United States, playas often occupy large areas on the flat floors of tectonic basins (see Figure 16.32). Deflation has reduced many playas several meters in elevation.

Rainbeat, overland flow, and deflation may be active for a long period on the gently sloping surface of a desert alluvial fan or alluvial terrace. These processes remove fine particles, leaving coarser, heavier materials behind. As a result, rock fragments ranging in size from pebbles to small boulders become concentrated into a surface layer known as a **desert pavement** (Figure 18.20). The large fragments become closely fitted together, concealing the smaller particles—grains of sand, silt, and clay—that remain beneath. The pavement acts as an armor that effectively protects the finer particles from rapid removal by deflation. However, the pavement is easily disturbed by the wheels of trucks and motorcycles, exposing the finer particles and allowing severe deflation and water erosion to follow.

Dust Storms

Strong, turbulent winds blowing over barren surfaces lift great quantities of fine dust into the air, forming a dense, high cloud called a **dust storm.** In semiarid grasslands, a dust storm is generated where ground surfaces have been stripped of protective vegetation cover by cultivation or grazing. Strong winds cause soil particles and coarse sand grains to hop along the ground. This motion breaks down the soil particles and disturbs more soil. With each impact, fine dust is released that can be carried upward by turbulent winds.

A dust storm approaches as a dark cloud extending from the ground surface to heights of several thousand meters (Figure 18.21). Typically, the advancing cloud wall represents a rapidly moving cold front. Within the dust cloud there is deep gloom or even total darkness. Visibility is cut to a few meters, and a fine choking dust penetrates everywhere.

SAND DUNES

A **sand dune** is any hill of loose sand shaped by the wind. Active dunes constantly change form under wind currents.

18.20 Desert pavement *A desert pavement is formed of closely fitted rock fragments. Lying on the surface are fine examples of wind-faceted rocks, which attain their unusual shapes by long-continued sandblasting.*

Eye on Global Change | 18.1

The Threat of Rising Sea Level

The Earth's climate seems to be warming, probably in response to human activity. One of the possible effects of climatic warming is the rise of sea level. Imagine, for a moment, the effect of an increase of several meters. Low-lying areas—such as tidal estuaries and deltas—would be flooded. Coastal cities would have to be defended from wave attack. Barrier islands would be swept landward or over-topped. Atolls and other low islands would disappear. The threat of a rise in sea level is real, but like other possible impacts of global warming, it is far from certain how long it will take sea level to rise and how great the effect will be.

A review of the scientific principles involved here may be helpful. First, these changes in sea level are eustatic—that is, they refer to worldwide sea-level change. Local relative changes in sea level resulting from the uplift or downsinking of the Earth's crust are tectonic effects and not included in eustatic change. Eustatic change can result from a change either in water volume of the world ocean (as by

melting of glaciers) or in the water-holding capacity of the ocean basins. However, any change in the capacity of the ocean basins would be a result of plate tectonics and thus occur on a longer time scale than that involved in global warming.

Climate affects the water volume of the oceans in two ways. First, a volume increase results from water temperature increase. Global warming will raise the temperature of the uppermost layer of the oceans—at most a few hundred meters thick—and the resulting thermal expansion of that water will cause a small but significant eustatic rise.

Second, global warming may increase the rate of glacial melting, causing mountain glaciers and ice sheets to shrink in volume. Melt-water would then increase the volume of the world ocean. Keep in mind, however, that, by increasing precipitation, global warming might instead promote the growth of glaciers by increasing precipitation. Although it is not clear which process will dominate in the short

term, we do know that large swings in world sea level have taken place throughout the past 3 million years as ice sheets have cyclically expanded and contracted. (In Chapter 19 we will return to a discussion of the mechanism of possible sea-level rise by ice melting.)

Has a slow eustatic rise of sea level accompanied the overall global temperature rise that started in the mid-1880s? Apparently so. Scientists examining tide gauge records have concluded that global average sea level rose between 10 and 20 cm (3.9 and 7.9 in.) during the twentieth century. About half of this increase is attributed to thermal expansion, with the remainder from melting ice in glaciers and ice sheets.

Projections of the rate of sea-level rise during the twenty-first century are varied. The Intergovernmental Panel on Climate Change, a United Nations-sponsored group of scientists, predicted in 2001 a global mean sea-level rise ranging from 9 to 88 cm (3.5 to 34.6 in.) between 1990 and 2100.

18.21 **Dust storm** *A cloud of fine dust sweeps across this savanna plain in eastern Kenya.*

Coastal erosion *Wave action has undermined the bluff beneath these two buildings, depositing them on the beach below.*

population of about a quarter of a million people, is an example.

Recent studies by the Environmental Defense Fund and the National Environmental Trust project how a rise of 1 m (3.3 ft) might affect the New York area. Flooding from severe storms, occurring more frequently with global warming, would shut down the city's subways, riverside highways, and La Guardia Airport, bringing the city's transportation system to a virtual halt. On Long Island, category 2 hurricanes would demolish beach communities and submerge Jones Beach, Fire Island, and the Hampton beaches. The south fork of Long Island would be split into smaller sections. Water tables would rise, creating more frequent local flooding at locations far from the beach front. Although these impact studies are based on the high end of sea-level rise projections, they demonstrate some of the impacts that could be felt by low-lying coastal plains even with lesser rises.

It seems certain that some degree of sea-level rise, brought on by global warming, is inevitable. The question now is, when and how much?

The impacts of such a rise, especially if toward the high end of the estimate, could be very costly for major coastal cities of the world. Large river deltas, including both the cities built on them and their reclaimed tidal estuaries and marshes, would face catastrophic damage through storm surf and storm surges in concert with high tides. Along cliffed coasts under direct attack by waves, retrogradation would be intensified, resulting in massive losses of shorefront properties (see photo). Some island nations would lose most or all of their land. The Maldives, a chain of atolls in the Indian Ocean with a

---

Dunes form where there is a source of sand—for example, a sandstone formation that weathers easily to release individual grains, or perhaps a beach supplied with abundant sand from a nearby river mouth. Dunes must be free of a vegetation cover in order to form and move. They become inactive when stabilized by a vegetation cover, or when patterns of wind or sand sources change.

Dune sand is most commonly composed of the mineral quartz, which is extremely hard and largely immune to chemical decay. The grains are beautifully rounded by abrasion (see Figure 11.12). Figure 18.22 shows how sand grains are moved by strong winds—in long, low leaps, bouncing after impact with other grains. Rebounding grains rarely rise more than half a centimeter above the dune surface. Grains struck by bouncing grains are pushed forward, and, in this way, the surface sand layer creeps downwind. This type of hopping, bouncing movement is termed *saltation.* GEODISCOVERIES

The barchan dune is a crescent-shaped heap of sand that moves across a flat, pebble-covered plain. The points of the crescent are directed downwind.

Types of Sand Dunes

One common type of sand dune is an isolated heap of free sand called a *barchan,* or *crescentic dune.* This type of dune has the outline of a crescent, and the points of the crescent are directed downwind (Figure 18.23). On the upwind side of the crest, the sand slope is gentle and smoothly rounded. On the downwind side of the dune, within the crescent, is a steep dune slope, the *slip face.* This face maintains a more or less

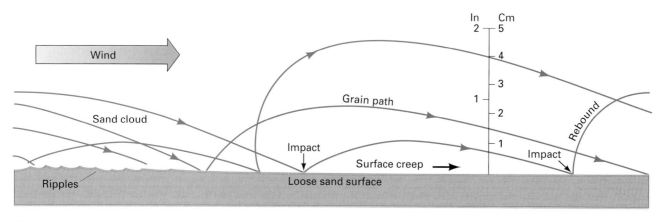

18.22 **Saltation** *Sand particles travel in a series of long leaps. (After R. A. Bagnold.)*

18.23 **Barchan dunes** *The arrow indicates wind direction. (Drawn by A. N. Strahler.)*

18.24 **Migrating barchan dunes** *This aerial view shows a large barchan dune moving from right to left. At its apex is a smaller barchan dune that is overtaking it.*

18.25 A sand sea *Transverse dunes of a sand sea, near Yuma, Arizona. The view is eastward; prevailing winds are northerly (from the left side of the photo).*

constant angle from the horizontal (Figure 18.24), which is known as the *angle of repose.* The slip face is oversteepened slightly by the accumulation of individual wind-carried sand grains until it becomes unstable and the outermost layer of sand on the face slips down the dune slope, restoring the angle of repose. For loose sand, this angle is about 35°.

GEODISCOVERIES

Barchan dunes usually rest on a flat, pebble-covered ground surface. The life of a barchan dune may begin as a sand drift in the lee of some obstacle, such as a small hill, rock, or clump of brush. Once a sufficient mass of sand has formed, it begins to move downwind, taking on the crescent form. For this reason, the dunes are usually arranged in chains extending downwind from the sand source.

Where sand is so abundant that it completely covers the solid ground, dunes take the form of wave-like ridges separated by trough-like furrows. These dunes are called *transverse dunes* because, like ocean waves, their crests trend at right angles to the direction of the dominant wind (Figure 18.25). The entire area may be called a *sand sea* because it resembles a storm-tossed sea suddenly frozen to immobility. The sand ridges have sharp crests and are asymmetrical, the gentle slope being on the windward and the steep slip face on the lee side. Deep depressions lie between the dune ridges. Sand seas require enormous quantities of sand, supplied by material weathered from sandstone forma-

Parabolic dunes are bowed outward in the downwind direction—the opposite of barchan dunes. They include coastal blowout dunes and hairpin dunes.

tions or from sands in nearby alluvial plains. Transverse dune belts also form adjacent to beaches that supply abundant sand and have strong onshore winds.

Wind is a major agent of landscape development in the Sahara Desert. Enormous quantities of reddish dune sand have been derived from weathering of sandstone formations. The sand is formed into a great sand sea, called an *erg.* Elsewhere, there are vast flat-surfaced sheets of sand that are armored by a layer of pebbles that forms a desert pavement. A surface of this kind in the Sahara is called a *reg.*

Some of the Saharan dunes are elaborate in shape. For example, the *star dune* (heaped dune) is a large hill of sand whose base resembles a many-pointed star in plan. The Arabian star dunes remain fixed in position and have served for centuries as reliable landmarks for desert travelers. The star dune also occurs in the deserts of the border region between the United States and Mexico.

Another group of dunes belongs to a family in which the curve of the dune crest is bowed outward in the downwind direction. (This curvature is the opposite of the barchan dune.) These are termed *parabolic dunes.* A common type of parabolic dune is the *coastal blowout dune,* formed adjacent to beaches. Here, large supplies of sand are available, and the sand is blown landward by prevailing winds (Figure 18.26a). A saucer-shaped depression is formed by deflation,

(a)

(b)

(c)

18.26 Three types of parabolic dunes *The prevailing wind direction is the same for all three types. (a) Coastal blowout dunes. (b) Parabolic dunes on a semiarid plain. (c) Parabolic dunes drawn out into hairpin forms. (Drawn by A. N. Strahler.)*

18.27 Coastal blowout dune *This coastal blowout dune is advancing over a coniferous forest, with the slip face gradually burying the tree trunks. Pacific coast, near Florence, Oregon.*

and the sand is heaped in a curving ridge resembling a horseshoe in plan. On the landward side is a steep slip face that advances over the lower ground and buries forests, killing the trees (Figure 18.27). Coastal blowout dunes are well displayed along the southern and eastern shore of Lake Michigan. Dunes of the southern shore have been protected for public use as the Indiana Dunes State Park.

On semiarid plains, where vegetation is sparse and winds are strong, groups of parabolic blowout dunes develop to the lee of shallow deflation hollows (Figure 18.26*b*). Sand is caught by low bushes and accumulates on a broad, low ridge. These dunes have no steep slip faces and may remain relatively immobile. In some cases, the dune ridge migrates downwind, drawing the dune into a long, narrow form with parallel sides resembling a hairpin in outline (Figure 18.26*c*).

Another class of dunes, described as *longitudinal dunes*, consists of long, narrow ridges oriented parallel with the direction of the prevailing wind (Figure 18.28). These dune ridges may be many kilometers long and cover vast areas of tropical and subtropical deserts in Africa and Australia.

Coastal Foredunes

Landward of sand beaches, we usually find a narrow belt of dunes in the form of irregularly shaped hills and depressions. These are the *foredunes*. They normally bear a cover of beachgrass and a few other species of plants capable of survival in the severe environment (Figure 18.29). On coastal foredunes, the sparse cover of beachgrass and other small plants acts as a baffle to trap sand moving landward from the adjacent beach. As a result, the foredune ridge is built upward to become a barrier standing several meters above high-tide level.

Foredunes form a protective barrier for tidal lands that often lie on the landward side of a beach ridge or barrier island. In a severe storm, the swash of storm waves cuts

18.28 *Longitudinal dunes* Longitudinal dunes run parallel to the direction of the wind. (Drawn by A. N. Strahler.)

onset of a storm that brings high water levels and intense wave action, swash is funneled through the gap and spreads out on the tidal marsh or tidal lagoon behind the ridge. Sand swept through the gap is spread over the tidal deposits. If eroded, the gap can become a new tidal inlet for ocean water to reach the bay beyond the beach. For many coastal communities of the eastern United States seaboard, the breaching of a dune ridge with its accompanying overwash may bring unwanted change to the tidal marsh or estuary.

LOESS

In several large midlatitude areas of the world, the surface is covered by deposits of wind-transported silt that has settled out from dust storms over many thousands of years. This material is known as **loess.** (The pronunciation of this German word is somewhere between "lerse" and "luss.") It generally has a uniform yellowish to buff color and lacks any visible layering. Loess tends to break away along vertical cliffs wherever it is exposed by the cutting of a stream or grading of a roadway (Figure 18.30). It is also very easily eroded by running water and is subject to rapid gullying when the vegetation cover that protects it is broken. Because it is easily excavated, loess has been widely used for cave dwellings both in China and in Central Europe.

away the upper part of the beach. Although the foredune barrier may then be attacked by wave action and partly cut away, it will not usually yield. Between storms, the beach is rebuilt, and, in due time, wind action restores the dune ridge, if plants are maintained.

If the plant cover of the dune ridge is reduced by vehicular and foot traffic, a blowout will rapidly develop. The new cavity can extend as a trench across the dune ridge. With the

> Coastal foredunes form a protective barrier against storm wave action that keeps waves from overwashing a beach ridge or barrier island.

The thickest deposits of loess are in northern China, where a layer over 30 m (about 100 ft) thick is common and a maximum thickness of 100 m (about 300 ft) has been measured. This layer covers many hundreds of

18.29 *Coastal foredunes* Beachgrass thriving on coastal foredunes has trapped drifting sand to produce a dune ridge. Queen's County, Prince Edward Island, Canada.

18.30 Wind-transported silt *This thick layer of loess in New Zealand was deposited during the Ice Age. Loess has excellent cohesion and often forms vertical faces as it wastes away.*

square kilometers and appears to have been brought as dust from the interior of Asia. Loess deposits are also of major importance in the United States, Central Europe, Central Asia, and Argentina.

In the United States, thick loess deposits lie in the Missouri-Mississippi Valley (Figure 18.31). Large areas of the prairie plains region of Indiana, Illinois, Iowa, Missouri, Nebraska, and Kansas are underlain by loess ranging in thickness from 1 to 30 m (about 3 to 100 ft). Extensive deposits also occur in Tennessee and Mississippi in areas bordering the lower Mississippi River floodplain. Still other loess deposits are in the Palouse region of northeast Washington and western Idaho.

The American and European loess deposits are directly related to the continental glaciers of the Pleistocene Epoch. At the time when the ice covered much of North America and Europe, a generally dry winter climate prevailed in the land bordering the ice sheets. Strong winds blew southward and eastward over the bare ground, picking up silt from the floodplains of braided streams that discharged the meltwater from the ice. This dust settled on the ground between streams, gradually building up a smooth, level ground surface. The loess is particularly thick along the eastern sides of the valleys because of prevailing westerly winds. It is well exposed along the bluffs of most streams flowing through these regions today.

Loess is of major importance in world agricultural resources. Loess forms the parent matter of rich black soils (Mollisols, Chapter 10) especially suited to cultivation of

> **Loess is a deposit of wind-blown silt that may be as thick as 30 m (about 100 ft) in some regions of North America. It forms highly productive but easily eroded soils.**

grains. The highly productive plains of southern Russia, the Argentine pampa, and the rich grain region of north China are underlain by loess. In the United States, corn is extensively cultivated on the loess plains in Kansas, Iowa, and Illinois, where rainfall is sufficient. Wheat is grown farther west on loess plains of Kansas and Nebraska and in the Palouse region of eastern Washington.

The thick loess deposit covering a large area of north-central China in the province of Shanxi and adjacent

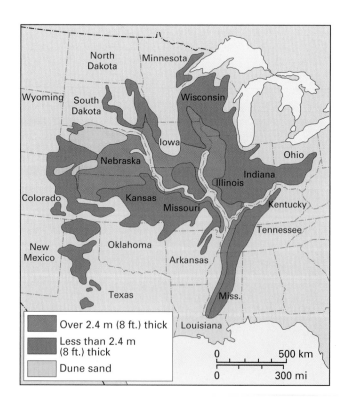

18.31 Map of loess distribution in the central United States *(Data from Map of Pleistocene Eolian Deposits of the United States, Geological Society of America.)*

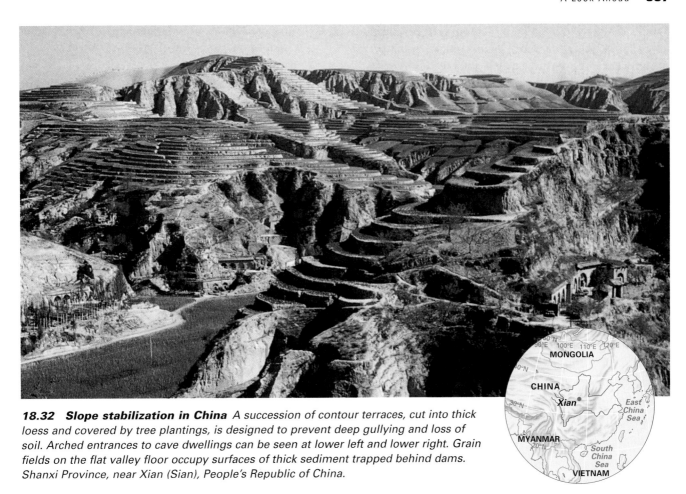

18.32 Slope stabilization in China *A succession of contour terraces, cut into thick loess and covered by tree plantings, is designed to prevent deep gullying and loss of soil. Arched entrances to cave dwellings can be seen at lower left and lower right. Grain fields on the flat valley floor occupy surfaces of thick sediment trapped behind dams. Shanxi Province, near Xian (Sian), People's Republic of China.*

provinces poses a difficult problem of severe soil erosion. Although the loess is capable of standing in vertical walls, it also succumbs to deep gullying during the period of torrential summer rains. From the steep walls of these great scars, fine sediment is swept into streams and carried into tributaries of the Huang He (Yellow River). The Chinese government has implemented an intensive program of slope stabilization by using artificial contour terraces (seen in Figure 18.32) in combination with tree planting. Valley bottoms have been dammed so as to trap the silt to form flat patches of land suitable for cultivation.

Induced Deflation

Induced deflation, is a frequent occurrence when short-grass prairie in a semiarid region is cultivated without irrigation. Plowing disturbs the natural soil surface and grass cover, and in drought years, when vegetation dies out, the unprotected soil is easily eroded by wind action. Much of the Great Plains region of the United States has suffered such deflation, experiencing dust storms generated by turbulent winds. Strong cold fronts frequently sweep over this area and lift dust high into the troposphere at times when soil moisture is low. The "Dust Bowl" of the 1930s is an example.

Human activities in very dry, hot deserts contribute measurably to the raising of high dust clouds. In the desert of northwest India and Pakistan (the Thar Desert bordering the Indus River), the continued trampling of fine-textured soils by hooves of grazing animals and by human feet produces a dust cloud that hangs over the region for long periods. It extends to a height of 9 km (about 30,000 ft).

A LOOK AHEAD This chapter has described the processes and landforms associated with wind action, either directly or through the medium of wind-driven ocean waves. The power source for wind action is, of course, the Sun. By heating the Earth's surface in a nonuniform pattern, the flow of solar energy produces pressure gradients that cause wind. The Sun also powers the last active landform-making agent of erosion, transportation, and deposition on our list—glacial ice. By evaporating water from the oceans and returning that water to the lands as snow, solar power creates the bodies of solid ice that we distinguish as mountain glaciers and ice sheets.

Compared to wind and water, glacial ice moves much more slowly but is far steadier in its motion. Like a vast conveyor belt, glacial ice moves sediment forward relentlessly, depositing the sediment at the ice margin, where the ice melts. By plowing its way over the landscape, glacial ice also shapes the local terrain—bulldozing loose rock from hillsides and plastering sediments underneath its vast bulk. This slow but steady action is very different from that of water, wind, and waves, and produces a set of landforms that is the subject of our last text chapter.

In Review | Landforms Made by Waves and Wind

- This chapter has described the landforms of waves and wind, both of which are indirectly powered by the Earth's rotation and the unequal heating of its surface by the Sun. Waves act at the **shoreline**—the boundary between water and land. Waves expend their energy as *breakers,* which erode hard rock into **marine cliffs** and create *marine scarps* in softer materials.

- **Beaches,** usually formed of sand, are shaped by the swash and backwash of waves, which continually work and rework beach sediment. Wave action produces **littoral drift,** which moves sediment parallel to the beach. This sediment accumulates in bars and sandspits, which further ther extend the beach. Depending on the nature of *longshore currents* and the availability of sediment, shorelines can experience **progradation** or **retrogradation.**

- Tidal forces cause sea level to rise and fall rhythmically, and this change of level produces *tidal currents* in bays and estuaries. Tidal flows redistribute fine sediments within bays and estuaries, which can accumulate with the help of vegetation to form *salt marshes.*

- Coastlines of **submergence** result when coastal lands sink below sea level or sea level rises rapidly. Scenic *ria* and *fiord coasts* are examples. Coastlines of **emergence** include *barrier-island coasts* and *delta coasts.* **Coral-reef** coasts occur in regions of warm tropical and equatorial waters. Along some coasts, rapid uplift has occurred, creating *raised shorelines* and *marine terraces.*

- Global sea level has varied widely in the past and is presently increasing slowly. A possible explanation is that global warming is increasing the volume of sea water by thermal expansion and by causing glacial ice on lands to melt. Future rises may be very costly to human society as coasts erode, islands are submerged, and coastal cites are subjected to frequent flooding.

- Wind is a landform-creating agent that acts by moving sediment. **Deflation** occurs when wind removes mineral particles—especially clay and silt, which can be carried long distances. Deflation creates **blowouts** in semidesert regions and lowers playa surfaces in deserts. In arid regions, deflation produces **dust storms.**

- **Sand dunes** form when a source, such as a sandstone outcrop or a beach, provides abundant sand that can be moved by wind action. *Barchan dunes* are arranged individually or in chains leading away from the sand source. *Transverse dunes* form a sand sea of frozen "wave" forms arranged perpendicular to the wind direction. *Parabolic dunes* are arc-shaped—*coastal blowout dunes* are an example. *Longitudinal dunes* parallel the wind direction and cover vast desert areas. Coastal *foredunes* are stabilized by dune grass and help protect the coast against storm wave action.

- **Loess** is a surface deposit of fine, wind-transported silt. It can be quite thick, and it typically forms vertical banks. Loess is very easily eroded by water and wind. In eastern Asia, the silt forming the loess was transported by winds from extensive interior deserts located to the north and west. In Europe and North America, the silt was derived from fresh glacial deposits during the Pleistocene Epoch. Human activities can hasten the action of deflation by breaking protective surface covers of vegetation and desert pavement.

Key Terms

shoreline, p. 566	beach, p. 568	delta, p. 575	dust storm, p. 579
coastline, p. 566	littoral drift, p. 569	coral reef, p. 576	sand dune, p. 579
coast, p. 566	progradation, p. 570	atolls, p. 577	loess, p. 585
bay, p. 566	retrogradation, p. 572	deflation, p. 578	
estuary, p. 566	submergence, p. 572	blowout, p. 579	
marine cliff, p. 567	emergence, p. 572	desert pavement, p. 579	

Review Questions

1. What is the energy source for wind and wave action?
2. What landforms can be found in areas where bedrock meets the sea?
3. What is littoral drift, and how is it produced by wave action?
4. Identify progradation and retrogradation. How can human activity influence retrogradation?
5. How are salt marshes formed? How can they be reclaimed for agricultural use?
6. What key features identify a coastline of submergence? Identify and compare the two types of coastlines of submergence.
7. Under what conditions do barrier-island coasts form? What are the typical features of this type of coastline? Provide and sketch an example of a barrier-island coast.
8. What conditions are necessary for the development of coral reefs? Identify three types of coral-reef coastlines.
9. How are marine terraces formed?

10. What is deflation, and what landforms does it produce? What role does the dust storm play in deflation?

11. How do sand dunes form? Describe and compare barchan dunes, transverse dunes, star dunes, coastal blowout dunes, parabolic dunes, and longitudinal dunes.

12. What is the role of coastal dunes in beach preservation? How are coastal dunes influenced by human activity? What problems can result?

13. Define the term *loess*. What is the source of loess, and how are loess deposits formed?

Eye on Global Change 18.1 • The Threat of Rising Sea Level

1. How might global climate change bring about a rise in sea level?

2. Is a sea-level rise occurring at present? What are the projections for the future, and what are the implications of the projections?

Visualizing Exercises

1. Describe the features of delta coasts and their formation. Sketch and compare the shapes of the Mississippi and Nile deltas.

2. Take a piece of paper and let it represent a map with winds coming from the north, at the top of the page. Then sketch the shapes of the following types of dunes: barchan, transverse, parabolic, and longitudinal.

Essay Questions

1. Consult an atlas to identify a good example of each of the following types of coastlines: ria coast, fiord coast, barrier-island coast, delta coast, coral-reef coast, and fault coast. For each example, provide a brief description of the key features you used to identify the coastline type.

2. Wind action moves sand close to the ground in a bouncing motion, whereas silt and clay are lifted and carried longer distances. Compare landforms and deposits that result from wind transportation of sand with those that result from wind transportation of silt and finer particles.

Eye on the Landscape

18.2 Retreating shoreline The sediments exposed in the bluff were laid down by streams fed by melting stagnant continental ice sheets at the end of the Ice Age. Because the ice melted irregularly, the deposits are not very uniform, and the exposed bedding **(A)** gives that impression. Note also the many large blocks of rock present within the deposits. They were let down from melting ice and were too large to be moved by the streams laying the sediments. A large lag deposit **(B)** of these stones remains at the water's edge, where it is being worked by waves. Deposits of continental ice sheets are covered in Chapter 19.

18.6 Sandspit The freshness of the sand here **(A)** indicates that Monomoy Point is growing very rapidly. The sand is supplied by the erosion of marine cliffs at Nauset Beach to the north. Note the overwash channels **(B)** that are cut through the vegetated sand dunes by the high waters and high surf of severe storms. To the upper left is the town of Chatham **(C)**, its low hills formed from rock debris shed by melting ice sheets during the Ice Age. The coastline has the "drowned" look of a coastline of submergence, with its many lakes and bays. See this chapter for more details.

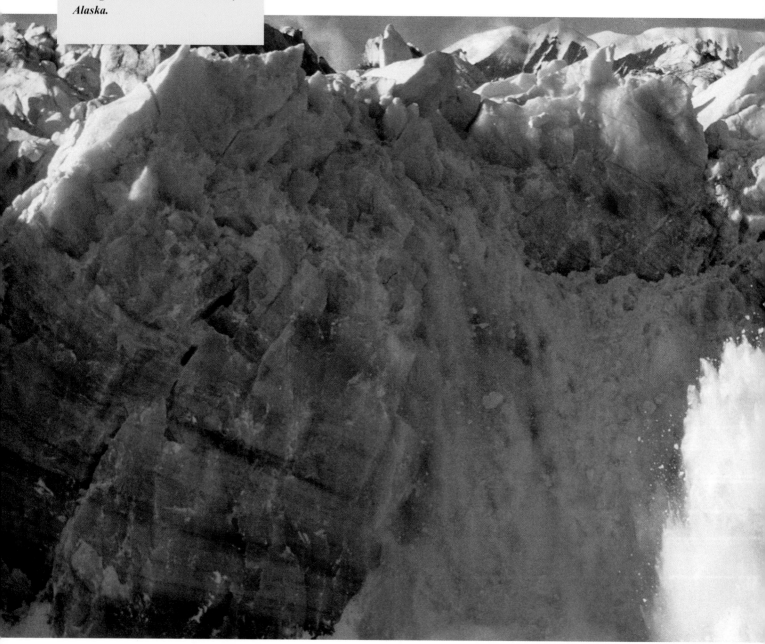

A huge block of ice falls from the front of a tidewater glacier, hurling tons of seawater high in the air. Wrangell–St. Elias National Park, Alaska.

19 | GLACIAL LANDFORMS AND THE ICE AGE

. . . .
PUTTING YOU IN THE PICTURE

Glacial ice, flowing slowly and majestically under the influence of gravity, is a powerful geomorphic agent that creates many types of distinctive landforms. Perhaps the best way to appreciate the power of glaciers is to visit one. Imagine yourself on an ocean cruise along the inland passage from Vancouver to Juneau with the ultimate destination of an Alaskan tidewater glacier. Here's what you might experience:

The huge cruise ship moves silently and effortlessly between the two steep, forest-covered slopes. The green walls of spruce and hemlock glide by slowly, the spires of the forest giants fading into the low ceiling of gray clouds overhead. Tiny droplets of drizzle cool your face as you lean against the rail, observing the somber scene. The trees seem almost close enough to reach out and touch. Below you, the ship's bow parts the gray-green water with little resistance, providing only a low rushing noise that barely breaks the stillness. Ahead, the sea and forest merge, fade, and then blend into the clouds and fog.

So this is Alaska! The travel brochure showed your ship cruising in a crystal-blue channel past magnificent vistas of snow-capped peaks, but so far the trip has provided only low clouds, fog, and conifer-clad slopes. The weather may soon change, however. The forecast calls for slow clearing later in the day, about the time that your ship will reach Glacier Bay. This steep-walled fiord is a drowned valley that was once filled with a huge tongue of glacial ice. Now, however, it is filled only with sea water and small drifting ice fragments released by the many streams of glacial ice that enter the bay along its 100 km (60 mi) length.

Soon you notice a change in the ship's rhythm. A graceful curving turn, and you leave Icy Strait to enter the mouth of Glacier Bay. Miraculously, the overcast skies have yielded to a blue and white patchwork of sky and cloud, and the sun has appeared. Your ship moves slowly down the narrow passage leading to the Grand Pacific Glacier. To either side are low, rocky peaks, sculpted by the ice into stark angular shapes composed of points and edges. Ahead are the high peaks of the St. Elias Mountains, entirely capped with white.

As the glacier comes into view, the ship slows further and approaches the steep ice face carefully. It is fractured into tall, vertical columns with step-like horizontal facets at the top. The color of the glacial ice is what strikes you most. Beneath a top white layer of decaying ice and snow, the ice is a bright greenish-blue. The color is so intense that the ice seems to glow from within. Above the clear aqua of the ice are the grays and browns of the rocks, the white of the snow-capped peaks, and the blue and white patchwork of sky. Below is the gray-green of the water, dotted with white ice fragments of many shapes and sizes.

The ice is also noisy. It groans, screeches, clicks, and pops. A loud crack startles you, then reverberates against the walls of the fiord. A monumental column of ice breaks free and falls, as if in slow motion. A great splash wells upward as it breaks the surface of the water. Next to the huge ice front, the bulk of the cruise ship seems very small. Suddenly, you connect the force of the moving ice with the shaping of the vast fiords that have contained your small ship on its passage. The power that carved these huge, water-filled canyons is awesome indeed.

If you live in a region that was covered by ice during the last Ice Age, the evidence of glaciation is all around. From ponds formed by the melting of buried ice blocks to the great hills of sediments dumped by glacial ice sheets at their margins, the action of glaciers has drastically modified much of the landscape of northern regions. In this chapter, you will learn about how glaciers create unique landforms like these. ■

GLACIAL LANDFORMS AND THE ICE AGE

In this chapter we turn to the last of the active agents that create landforms—glacial ice. Not long ago, during the Ice Age, much of northern North America and Eurasia was covered by massive sheets of glacial ice. As a result, glacial ice has played a dominant role in shaping landforms of large areas in midlatitude and subarctic zones. Glacial ice still exists today in two great accumulations of continental dimensions—the Greenland and Antarctic Ice Sheets—and in many smaller masses in high mountains.

The glacial ice sheets of Greenland and Antarctica strongly influence the radiation and heat balance of the globe. Because of their intense whiteness, they reflect much of the solar radiation they receive. Their intensely cold surface air temperatures contrast with temperatures at more equatorial latitudes. This temperature difference helps drive the system of meridional heat transport that we described in Chapters 3 and 5. In addition, these enormous ice accumulations represent water in storage in the solid state. They figure as a major component of the global water balance. When the volume of glacial ice increases, as during an ice age, sea levels must fall. When ice sheets melt away, sea level rises. Today's coastal environments evolved during the rising sea level that followed the melting of the last ice sheets of the Ice Age.

GLACIERS

Most of us know ice only as a brittle, crystalline solid because we are accustomed to seeing it in small quantities. Where a great thickness of ice exists, the pressure on the ice at the bottom makes the ice lose its rigidity. That is, the ice becomes plastic. This allows the ice mass to flow in response to gravity, slowly spreading out over a larger area or moving downhill. On steep mountain slopes, the ice can also move by sliding. Movement is the key characteristic of a *glacier,* defined as any large natural accumulation of land ice affected by present or past motion.

Glacial ice accumulates when the average snowfall of the winter exceeds the amount of snow that is lost in summer by ablation. The term **ablation** means the loss of snow and ice by evaporation and melting. When winter snowfall exceeds summer ablation, a layer of snow is added each year to what has already accumulated. As the snow compacts by surface melting and refreezing, it turns into a granular ice and is then compressed by overlying layers into hard crystalline ice. When the ice mass is so thick that the lower layers become plastic, outward or downhill flow starts, and the ice mass is now an active glacier.

Glacial ice forms where temperatures are low and snowfall is high. These conditions can occur both at high eleva-

> When snow accumulates to a great thickness, it can turn into flowing glacial ice. Alpine glaciers form in high mountains, while ice sheets form on continental interiors at high latitudes.

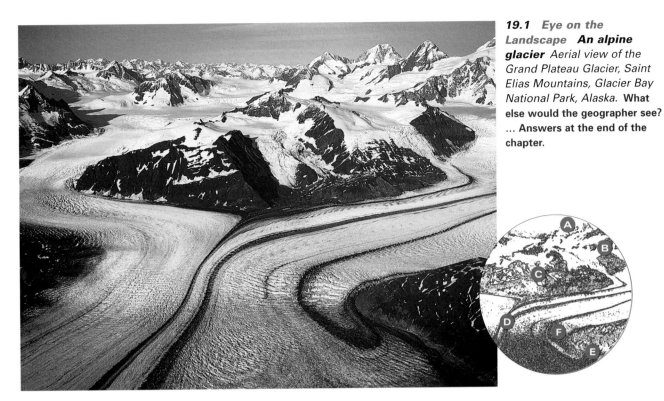

tions and at high latitudes. In mountains, glacial ice can form even in tropical and equatorial zones if the elevation is high enough to keep average annual temperatures below freezing. Orographic precipitation encourages the growth of glacial ice. In high mountains, glaciers flow from small high-elevation collecting grounds down to lower elevations, where temperatures are warmer. Here the ice disappears by ablation. Typically, mountain glaciers are long and narrow because they occupy former stream valleys. These **alpine glaciers** are a distinctive type (Figure 19.1).

In arctic and polar regions, prevailing temperatures are low enough that snow can accumulate over broad areas, eventually forming a vast layer of glacial ice. Accumulation starts on uplands that intercept heavy snowfall. The uplands become buried under enormous volumes of ice, which can reach a thickness of several thousand meters. The ice then spreads outward, over surrounding lowlands, and covers all landforms it encounters. This extensive type of ice mass is called an **ice sheet.** As already noted, ice sheets exist today in Greenland and Antarctica.

Glacial ice normally contains abundant rock fragments ranging from huge angular boulders to pulverized rock flour. Some of this material is eroded from the rock floor on which the ice moves. In alpine glaciers, rock debris is also derived from material that slides or falls from valley walls onto the ice.

Glaciers can erode and deposit great quantities of sediment. *Glacial abrasion* is a glacial erosion process caused by rock fragments that are held within the ice and scrape and grind against bedrock (Figure 19.2). Erosion also occurs by *plucking,* as moving ice lifts out blocks of bedrock that have been loosened by the freezing and expansion of water in

19.2 ***Glacial abrasion*** *This grooved and polished surface, now partly eroded, marks the former path of glacial ice. Cathedral Lakes, Yosemite National Park, California.*

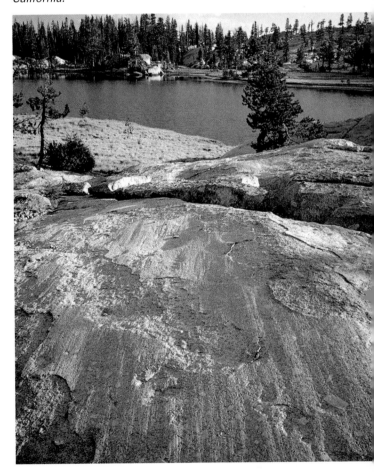

joint fractures. Abrasion and plucking act to smooth the bed of a glacier as glacial flow continues through time. Rock debris brought into a glacier is eventually deposited at the lower end of a glacier, where the ice melts. Both erosion and deposition result in distinctive glacial landforms.

ALPINE GLACIERS

Figure 19.3 illustrates a number of features of alpine glaciers. The illustration shows a simple glacier occupying a sloping valley between steep rock walls. Snow collects at the upper end in a bowl-shaped depression, the **cirque.** The upper end lies in a zone of accumulation. Layers of snow in the process of compaction and recrystallization are called *firn.*

The smooth firn field is slightly bowl-shaped in profile. Flowage in the glacial ice beneath the firn carries the ice downvalley out of the cirque. The rate of ice flow is accelerated at a steep rock step, where deep crevasses (gaping fractures) mark an ice fall. The lower part of the glacier lies in the zone of ablation. In this area, the rate of ice wastage is rapid, and old ice is exposed at the glacier surface. As the ice thins by ablation, it loses its plasticity and may develop deep crevasses. At its lower end, or terminus, the glacier carries abundant rock debris. As the downward-flowing ice melts, the debris accumulates.

> Cirques are rounded rock basins that contain the heads of alpine glaciers. They form high on glaciated peaks.

Although the uppermost layer of a glacier is brittle, the ice beneath behaves as a plastic substance and moves by slow flowage (Figure 19.4). Like stream flow, glacier flow is most rapid far from the glacier's bed—near the midline and toward the top of the glacier's surface. Alpine glaciers also move by basal sliding. In this process, the ice slides downhill, lubricated by meltwater and mud at its base.

A glacier establishes a dynamic balance in which the rate of accumulation at the upper end balances the rate of ablation at the lower end. This balance is easily upset by changes in the average annual rates of accumulation or ablation, causing the glacier's terminus to move forward or melt.

Glacial flow is usually very slow. It amounts to a few centimeters per day for large ice sheets and the more sluggish alpine glaciers, but as fast as several meters per day for an active alpine glacier. However, some alpine glaciers experience episodes of very rapid movement, termed *surges*. A surging glacier may travel downvalley at speeds of more than 60 m (about 200 ft) per day for several months. The reasons for surging are not well understood but probably involve mechanisms that increase the amount of meltwater beneath the ice, enhancing basal sliding. Most glaciers do not experience surging. **GEO*DISCOVERIES***

19.3 Cross section of an alpine glacier *Ice accumulates in the glacial cirque, then flows downhill, abrading and plucking the bedrock. Glacial debris accumulates at the glacier terminus. (After A. N. Strahler.)*

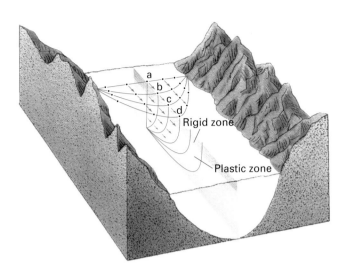

19.4 *Motion of glacial ice* *Ice moves most rapidly on the glacier's surface at its midline. Movement is slowest near the bed, where the ice contacts bedrock or sediment.*

Landforms Made by Alpine Glaciers

Landforms made by alpine glaciers are shown in a series of diagrams in Figure 19.5. Mountains are eroded and shaped by glaciers, and after the glaciers melt, the remaining landforms are exposed to view. Diagram *a* shows a region sculptured entirely by weathering, mass wasting, and streams. The mountains have a smooth, rounded appearance. Soil and regolith are thick.

Imagine now that a climatic change results in the accumulation of snow in the heads of the higher valleys. An early stage of glaciation is shown at the right side of diagram *b*, where snow is collecting and cirques are being carved by the grinding motion of the ice. Deepening of the cirques is aided by intensive frost shattering of the bedrock near the masses of compacted snow. At a later stage (center), glaciers have filled the valleys and are integrated into a system of tributaries that feed a trunk glacier. Tributary glaciers join the main glacier smoothly. The cirques grow steadily larger. Their rough, steep walls soon replace the smooth, rounded slopes of the original mountain mass. Where two cirque walls intersect from opposite sides, a jagged, knife-like ridge, called an *arête*, is formed. Where three or more cirques grow together, a sharp-pointed peak is formed. Such peaks are called *horns*. Where opposed cirques have intersected deeply, a pass or notch, called a *col*, is formed. On the left side of the diagram, a portion of the landscape is drawn as unglaciated. This portion retains its rounded forms and V-shaped valleys.

A ridge or pile of rock debris left by glacial action that marks the edge of a glacier is termed a **moraine.** A *lateral moraine* is a debris ridge formed along the edge of the ice adjacent to the trough wall (Figure 19.3). Where two ice streams join, this marginal debris is dragged along to form a narrow band riding on the ice in midstream (Figures

> Alpine glaciers, descending through preexisting valleys, excavate wide, U-shaped glacial troughs. When the glaciers retreat, meltwater streams can fill trough floors with sediment.

19.1, 19.5*b*), called a *medial moraine.* At the terminus of a glacier, rock debris accumulates in a *terminal moraine,* an embankment curving across the valley floor and bending upvalley along each wall of the trough (Figure 19.6).

Glacial Troughs and Fiords

Glacier flow constantly deepens and widens its rock channel, so that after the ice has finally melted, a deep, steep-walled **glacial trough** remains (Figure 19.5*c*). The trough typically has a U-shape in cross-profile (Figure 19.7). Tributary glaciers also carve U-shaped troughs, but they are smaller in cross section and less deeply eroded by their smaller glaciers. Because the floors of these troughs lie high above the level of the main trough, they are called *hanging valleys.* Streams later occupy the abandoned valleys, providing scenic waterfalls that cascade over the lip of the hanging valley to the main trough below. High up in the smaller troughs, the bedrock is unevenly excavated, so that the floors of troughs and cirques contain rock basins and rock steps. The rock basins are occupied by small lakes, called *tarns* (Figure 19.5*c*). Major troughs sometimes hold large, elongated trough lakes.

GEODISCOVERIES

Many large glacial troughs now are filled with alluvium and have flat floors. Aggrading streams that issued from the receding ice front were heavily laden with rock fragments so that the deposit of alluvium extended far downvalley. Figure 19.7 shows a comparison between a trough with little or no fill (*b*) and another with an alluvial-filled bottom (*c*).

When the floor of a trough open to the sea lies below sea level, the sea water enters as the ice front recedes. The result is a deep, narrow estuary known as a **fiord** (Figure 19.7*d*). Fiords are opening up today along the Alaskan coast, where some glaciers are melting back rapidly and ocean waters are filling their troughs. Fiords are found largely along mountainous coasts between lat. 50° and 70° N and S (Figure 19.8). On these coasts, glaciers were nourished by heavy orographic snowfall, associated with the marine west-coast climate ⑧.

As large, distinctive features in areas of rugged terrain, glaciers are readily viewed from the perspective of space. *Focus on Remote Sensing 19.1 • Remote Sensing of Glaciers* shows some spectacular images of glaciers acquired from Earth orbit. One of the images uses a special radar imaging technique to portray the motion of an ice stream.

GEODISCOVERIES

ICE SHEETS OF THE PRESENT

In contrast to alpine glaciers are the enormous ice sheets of Antarctica and Greenland. These are huge plates of ice, thousands of meters thick in the central areas, resting on land masses of subcontinental size. The Greenland Ice Sheet

19.5 Landforms produced by alpine glaciers *(a) Before glaciation sets in, the region has smoothly rounded divides and narrow, V-shaped stream valleys. (b) After glaciation has been in progress for thousands of years, new erosional forms are developed. (c) With the disappearance of the ice, a system of glacial troughs is exposed. (Drawn by A. N. Strahler.)*

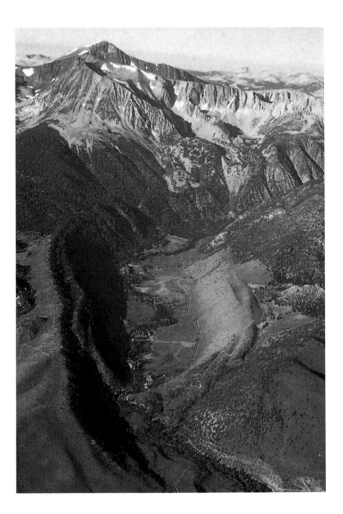

19.6 Alpine terminal moraine *A terminal moraine, shaped like the bow of a great canoe, lies at the mouth of a deep glacial trough on the east face of the Sierra Nevada. Cirques can be seen in the distance. Near Lee Vining, California.*

19.7 Development of a glacial trough *(a) During maximum glaciation, the U-shaped trough is filled by ice to the level of the small tributaries. (b) After glaciation, the trough floor may be occupied by a stream and lakes. (c) If the main stream is heavily loaded, it may fill the trough with alluvium. (d) Should the glacial trough have been deepened below sea level, it will be occupied by an arm of the sea, or fiord. (Drawn by E. Raisz.)*

(a)

(b)

(c)

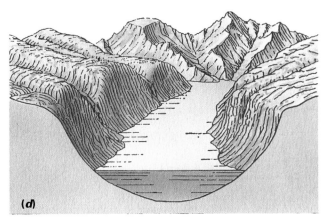

(d)

Remote Sensing of Glaciers

Because glaciers are often found in inaccessible terrain or in extremely cold environments, they are difficult to survey and monitor. Satellite remote sensing provides an invaluable tool for studying both continental and alpine glaciers in spite of these difficulties.

Some of the world's more spectacular alpine glaciers are found in South America, along the crest of the Andes in Chile and Argentina, where mountain peaks reach as high as 3700 m (about 12,000 ft). Image (*a*) is a true-color image, acquired by astronauts aboard the International Space Station in December 2000 of Cerro San Lorenzo (San Lorenzo Peak). The peak itself is a glacial horn. Leading away from the horn to the south is a long, sharp ridge, or arête. To the left of the peak is a cirque, now only partly filled with glacial ice. These features were carved during the Ice Age, when the alpine glaciers were larger and filled the now-empty glacial

troughs behind the ridge to the right.

Northwest of Cerro San Lorenzo lies the San Quentin glacier, shown in ASTER image (*b*), acquired on May 2, 2000. This color-infrared image was acquired at 15-m spatial resolution and shows the glacier in fine detail. The snout of the glacier ends in a shallow lake of sediment-laden water that drains by fine streams into the Golfo de Penas at lower left. Note the low, semicircular ridge a short distance from the lake—this is a terminal moraine, marking a stand of the ice tongue in the recent past. A high cloud partly obscures the southern end of the snout and nearby coastline. The intense red color indicates a thick vegetation cover.

The world's largest glacier is, of course, the ice cap that covers nearly all of Antarctica. At the edges of the ice cap are outlet glaciers, where glacial flow into the ocean is quite rapid. Image (*c*) shows the Lambert Glacier, one of

the largest and longest of Antarctica's outlet glaciers. The image was acquired by Canada's Radarsat radar imager. Because this type of radar system can see through the clouds, it is ideal for mapping a large area in a short period of time. It also allows the measurement of the velocity of glacier flow by comparing paired images acquired at different times (in this case, 24 days apart) using a technique called radar interferometry. The image uses color to show the velocity. Brown tones indicate little or no motion, and show both exposed mountains and stationary ice. Green, blue, and red tones indicate increasing velocity. Velocity and direction are also shown by the arrows superimposed on the image. Glacial flow is most rapid at the left, where the flow is channeled through a narrow valley, and at the right, where the glacier spreads out and thins to feed the Amery Ice Shelf.

Andean alpine glacial features *H, horn (San Lorenzo Peak); A, arête; C, cirque; T, glacial trough. (Courtesy NASA.)*

(a)

(b)

San Quentin Glacier, Chile   *The San Quentin Glacier as imaged by ASTER. (Image courtesy NASA/GSFC/MITI/ERSDAC/JAROS and U.S./Japan ASTER Science Team.*

Lambert Glacier, Antarctica   *The Lambert Glacier as imaged by Radarsat during the 2000 Antarctic Mapping Mission. (Image courtesy Canadian Space Agency/NASA/Ohio State University, Jet Propulsion Laboratory, Alaska SAR Facility.)*

(c)

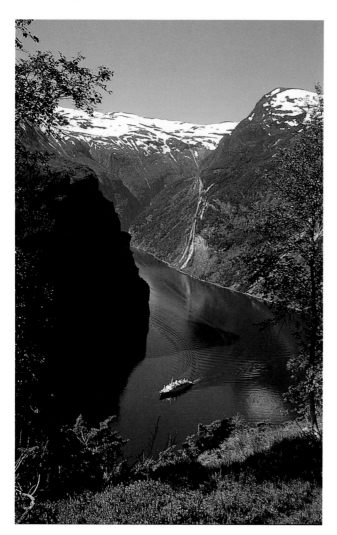

19.8 *A glacial trough Geirangerfjord, Norway, is a deeply carved glacial trough occupied by an arm of the sea.*

has an area of 1.7 million sq km (about 670,000 sq mi) and occupies about seven-eights of the entire island of Greenland (Figure 19.9). Only a narrow, mountainous coastal strip of land is exposed. The Antarctic Ice Sheet covers 13 million sq km (about 5 million sq mi) (Figure 19.10). Both ice sheets are developed on large, elevated land masses in high latitudes. No ice sheet exists near the north pole, which is positioned in the vast Arctic Ocean. Ice there occurs only as floating sea ice.

The surface of the Greenland Ice Sheet has the form of a very broad, smooth dome. Underneath the ice sheet, the rock floor lies near or slightly below sea level under the central region but is higher near the edges. The Antarctic Ice Sheet is thicker than the Greenland Ice Sheet—as much as 4000 m (about 13,000 ft) at maximum. At some locations, ice sheets extend long tongues, called outlet glaciers, to reach the sea at the heads of fiords. From the floating edge of the glacier, huge masses of ice break off and drift out to open sea with tidal currents to become icebergs. An important glacial feature of Antarctica is the presence of great plates of floating glacial ice, called *ice shelves* (Figure 19.10). Ice shelves

are fed by the ice sheet, but they also accumulate new ice through the compaction of snow. **GEODISCOVERIES**

SEA ICE AND ICEBERGS

Free-floating ice on the sea surface is of two types—sea ice and icebergs. *Sea ice* (Figure 19.11) is formed by direct freezing of ocean water. In contrast, *icebergs* are bodies of land ice that have broken free from glaciers that terminate in the ocean. Aside from differences in origin, a major difference between sea ice and icebergs is thickness. Sea ice does

19.9 *The Greenland Ice Sheet Contours show elevations of the ice sheet surface. (Based on data of R. F. Flint, Glacial and Pleistocene Geology, John Wiley & Sons, New York.)*

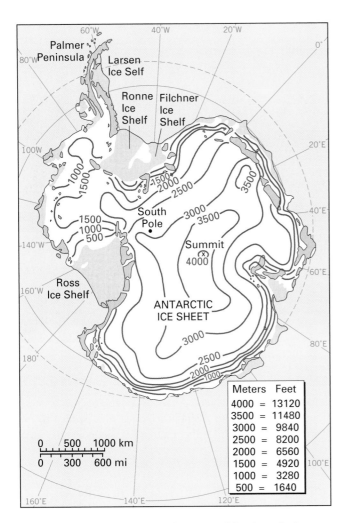

19.10 **The Antarctic Ice Sheet and its ice shelves**
Contours show elevations of the ice sheet surface. (Based on data of American Geophysical Union.)

19.11 **Sea ice** A Landsat image of a portion of the Canadian arctic archipelago. An ice cap on a land mass is visible in the center of the photo, with exposed mountainous ridges on the uncovered portions of the land mass. A branching glacial trough, now a water-filled fiord, is seen in the lower part of the image. In the upper left are huge chunks of free-floating sea ice.

not exceed 5 m (15 ft) in thickness, while icebergs may be hundreds of meters thick.

Pack ice is sea ice that completely covers the sea surface. Under the forces of wind and currents, pack ice breaks up into individual patches called ice floes. The narrow strips of open water between such floes are known as *leads*. Where ice floes are forcibly brought together by winds, the ice margins buckle and turn upward into pressure ridges that often resemble walls of ice. Travel on foot across the polar sea ice is extremely difficult because of such obstacles. The surface zone of sea ice is composed of fresh water, while the deeper ice is salty.

When a valley glacier or tongue of an ice sheet terminates in sea water, blocks of ice break off to form icebergs (Figure 19.12). Because they are only slightly less dense than sea water, icebergs float very low in the water. About five-sixths of the bulk of an iceberg is submerged. The ice is composed of

---

Sea ice is formed by freezing of ocean surface water and is rarely thicker than 5 m (16.4 ft). Icebergs are floating chunks of glacial ice and can be hundreds of meters thick.

An ice age includes cycles of glaciation, deglaciation, and interglaciation. The Earth is presently in an interglaciation within the Late-Cenozoic Ice Age.

---

fresh water since it is formed from compacted and recrystallized snow. GEODISCOVERIES

THE ICE AGE

The period during which continental ice sheets grow and spread outward over vast areas is known as a **glaciation.** Glaciation is associated with a general cooling of average air temperatures over the regions where the ice sheets originate. At the same time, ample snowfall must persist over the growth areas to allow the ice masses to build in volume.

When the climate warms or snowfall decreases, ice sheets become thinner and cover less area. Eventually, the ice sheets may melt completely. This period is called a *deglaciation.* Following a deglaciation, but preceding the next glaciation, is a period in which a mild climate prevails—an **interglaciation.** The last interglaciation began about 140,000 years ago and ended between

19.12 **Iceberg** *Penguins enjoy an outing on an iceberg off the coast of Antarctica near Elephant Island.*

120,000 and 110,000 years ago. A succession of alternating glaciations and interglaciations, spanning a period of 1 to 10 million years or more, constitutes an *ice age.*

Throughout the past 3 million years or so, the Earth has been experiencing the **Late-Cenozoic Ice Age** (or, simply, the **Ice Age**). As you may recall from Chapter 12, the Cenozoic Era has seven epochs (see Table 12.1). The Ice Age falls within the last three epochs: Pliocene, Pleistocene, and Holocene. These three epochs comprise only a small fraction—about one-twelfth—of the total duration of the Cenozoic Era.

About 50 years ago, most geologists associated the Ice Age with the Pleistocene Epoch, which began about 1.6 million years before present. However, new evidence obtained from deep-sea sediments shows that the glaciations of the Ice Age began in late Pliocene time, perhaps 2.5 to 3.0 million years ago.

At present, we are within an interglaciation of the Late-Cenozoic Ice Age, following a deglaciation that set in quite rapidly about 15,000 years ago. In the preceding glaciation, called the *Wisconsinan Glaciation,* ice sheets covered much of North America and Europe, as well as parts of northern Asia and southern South America. The maximum ice advance of the Wisconsinan Glaciation was reached about 18,000 years ago.

Glaciation During the Ice Age

Figures 19.13 and 19.14 show the maximum extent to which North America and Europe were covered during the last advance of the ice. Most of Canada was engulfed by the vast Laurentide Ice Sheet. It spread south into the United States, covering most of the land lying north of the Missouri and Ohio rivers, as well as northern Pennsylvania and all of New

York and New England. Alpine glaciers of the western ranges coalesced into a single ice sheet that spread to the Pacific shores and met the Laurentide sheet on the east.

19.13 **Maximum glaciation in North America** *Continental glaciers of the Ice Age in North America at their maximum extent reached as far south as the present Ohio and Missouri rivers. Note that during glaciations sea level was much lower. The present coastline is shown for reference only. (Based on data of R. F. Flint, Glacial and Pleistocene Geology, John Wiley & Sons, New York.)*

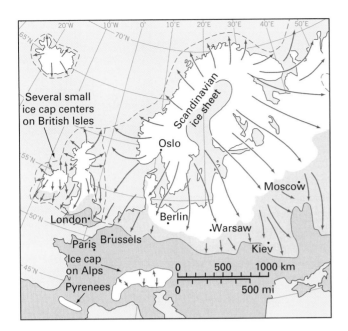

19.14 *Maximum glaciation in Europe* The *Scandinavian Ice Sheet dominated Northern Europe during the Ice Age glaciations. As noted in Figure 19.13, the present coastline is far inland from the coastline that prevailed during glaciations. (Based on data of R. F. Flint, Glacial and Pleistocene Geology, John Wiley & Sons, New York.)*

Notice that an area in southwestern Wisconsin escaped inundation. Known as the Driftless Area, it was apparently bypassed by glacial lobes moving on either side.

In Europe, the Scandinavian Ice Sheet centered on the Baltic Sea, covering the Scandinavian countries. It spread south into central Germany and far eastward to cover much of Russia. In north-central Siberia, large ice caps formed over the northern Ural Mountains and highland areas farther east. Ice from these centers grew into a large sheet covering much of central Siberia. The European Alps were capped by enlarged alpine glaciers. The British Isles were mostly covered by a small ice sheet that had several centers on highland areas and spread outward to coalesce with the Scandinavian Ice Sheet.

At the maximum spread of these ice sheets, sea level was as much as 125 m (410 ft) lower than today, exposing large areas of the continental shelf on both sides of the Atlantic Basin. The shelf supported a vegetated landscape populated with animal life, including Pleistocene elephants (mastodons and mammoths). The drawdown of sea level explains why the ice sheets shown on our maps extend far out into what is now the open ocean.

South America, too, had an ice sheet. It grew from ice caps on the southern Andes Range south of about latitude 40°S and spread westward to the Pacific shore, as well as eastward, to cover a broad belt of Patagonia. It covered all

of Tierra del Fuego, the southern tip of the continent. The South Island of New Zealand, which today has a high spine of alpine mountains with small relict glaciers, developed a massive ice cap in late Pleistocene time. All high mountain areas of the world underwent greatly intensified alpine glaciation at the time of maximum ice sheet advance. Today, most remaining alpine glaciers are small ones. In less favorable locations, the Ice Age alpine glaciers are entirely gone.

The weights of the continental ice sheets, covering vast areas with ice masses several kilometers thick, exerted downward forces on the crust, causing depressions of the crust of hundreds of meters at some locations. With the melting of the ice of the last glaciation, the crust began an *isostatic rebound* that is still going on today at some locations. Uplifted marine terraces provide the primary evidence for this recent upward motion. **GEODISCOVERIES**

LANDFORMS MADE BY ICE SHEETS

Landforms made by the last ice advance and recession are very fresh in appearance and show little modification by erosion processes. It is to these landforms that we now turn our attention.

Erosion by Ice Sheets

Like alpine glaciers, ice sheets are highly effective eroding agents. The slowly moving ice scraped and ground away much solid bedrock, leaving behind smoothly rounded rock masses. These bear countless grooves and scratches trending in the general direction of ice movement (see Figure 19.2). Sometimes the ice polishes the rock to a smooth, shining surface. The evidence of ice abrasion is common throughout glaciated regions of North America and may be seen on almost any hard rock surface that is freshly exposed. Conspicuous knobs of solid bedrock shaped by the moving ice are also common features (Figure 19.15). The side from which the ice approached is usually smoothly rounded. The lee side, where the ice plucked out angular joint blocks, is irregular and blocky.

> Although the largest ice sheets spread from northern hemisphere centers in Canada, Scandinavia, and Siberia, South America and New Zealand also developed ice sheets.

The ice sheets also excavated enormous amounts of rock at locations where the bedrock was weak and the flow of ice was channeled by the presence of a valley trending in the direction of ice flow. Under these conditions, the ice sheet behaved like a valley glacier, scooping out a deep, U-shaped trough. The Finger Lakes of western New York State are fine examples (Figure 19.16). Here, a set of former stream valleys lay largely parallel to the southward spread of the ice, and a set of long, deep basins was eroded. Blocked at their north ends by glacial debris, the basins now hold lakes. Many hundreds of lake basins were created by glacial erosion and deposition over the glaciated portion of North America.

19.15 A glacially abraded rock knob *Glacial action abrades the rock into a smooth form as it rides over the rock summit, then plucks bedrock blocks from the lee side, producing a steep, rocky slope. (Copyright © A. N. Strahler.)*

Abrasion · Ice movement · Lee side · Bedrock

Deposits Left by Ice Sheets

The term **glacial drift** includes all varieties of rock debris deposited in close association with glaciers. Drift is of two major types. *Stratified drift* consists of layers of sorted and stratified clays, silts, sands, or gravels. These materials were deposited by meltwater streams or in bodies of water adjacent to the ice. **Till** is an unstratified mixture of rock fragments, ranging in size from clay to boulders, that is deposited directly from the ice without water transport. As the glacial ice melts in a stagnant marginal zone, the rock particles it holds are lowered to the solid surface beneath, where they form a layer of debris (Figure 19.17). This *ablational till* shows no sorting and often consists of a mixture of sand and silt, with many angular pebbles and boulders. Beneath this residual layer there may be a basal layer of dense *lodgment till,* consisting of clay-rich debris previously dragged forward beneath the moving ice. Where till forms a thin, more or less even cover, it is referred to as *ground moraine.*

Over those parts of North America formerly covered by Late-Cenozoic ice sheets, glacial drift thickness averages

> The term *glacial drift* describes all types of rock debris deposited in close association with glaciers. *Stratified drift* includes sediment laid by water, while *till* is deposited directly by ice.

from 6 m (about 20 ft) over mountainous terrain, such as New England, to 15 m (about 50 ft) and more over the lowlands of the north-central United States. Over Iowa, drift thickness is from 45 to 60 m (about 150 to 200 ft), and over Illinois it averages more than 30 m (about 100 ft). In some places where deep stream valleys existed prior to glacial advance, as in parts of Ohio, drift is much thicker.

To understand the form and composition of deposits left by ice sheets, it will help to examine the conditions prevailing at the time of the ice sheet's existence. Figure 19.18*a* shows a region partly covered by an ice sheet with a stationary front edge. This condition occurs when the rate of ice ablation balances the amount of ice brought forward by spreading of the ice sheet. Although the ice fronts of the Ice Age advanced and receded in many minor and major fluctuations, there were long periods when the front was essentially stable and thick deposits of drift accumulated.

MORAINES   The transportational work of an ice sheet resembles that of a huge conveyor belt. Anything carried on the belt is dumped off at the end and, if not constantly

19.16 Eye on the Landscape Finger Lakes *The Finger Lakes region of New York is shown in an astronaut photo from the Space Shuttle. The view is looking toward the northeast. The lakes are oriented roughly in a north-south direction.* **What else would the geographer see? ... Answers at the end of the chapter.**

(a)

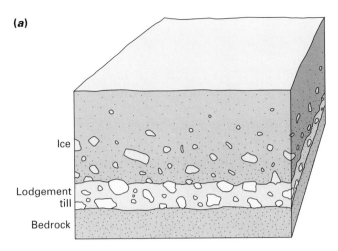

Ice

Lodgement
till

Bedrock

(b)

Ablation till

Lodgement
till

Bedrock

19.17   Glacial till *(a) As ice passes over the ground, sediment and coarse rock fragments are pressed into a layer of lodgment till. (b) When the overlying ice stagnates and melts, it leaves a residual deposit of ablation till. (Copyright © A. N. Strahler.)*

removed, will pile up in increasing quantity. Rock fragments brought within the ice are deposited at its outer edge as the ice evaporates or melts. Glacial till that accumulates at the immediate ice edge forms an irregular, rubbly heap—the *terminal moraine.* After the ice has disappeared (Figure 19.18*b*), the moraine appears as a belt of knobby hills interspersed with basin-like hollows, or kettles, some of which hold small lakes. The name *knob-and-kettle* is often applied to morainal belts.

Terminal moraines form great curving patterns. The outward curvature is southward and indicates that the ice advanced as a series of great *ice lobes,* each with a curved front (Figure 19.19). Where two lobes come together, the moraines curve back and fuse together into a single *interlobate moraine* pointed northward. In its general recession accompanying disappearance, the ice front paused for some time along a number of positions, causing morainal belts similar to the terminal moraine belt to be formed. These belts are known as *recessional moraines* (Figure 19.18*b*). They run roughly parallel with the terminal moraine but are often thin and discontinuous.

> Moraines are built of rock debris deposited at the edges of a melting glacier or ice sheet. Terminal moraines mark the limits of glaciation.

OUTWASH AND ESKERS   Figure 19.18 shows a smooth, sloping plain lying in front of the ice margin. This is the *outwash plain,* formed of stratified drift left by braided streams issuing from the ice. The plain is built of layer upon layer of sands and gravels.

Large streams carrying meltwater issue from tunnels in the ice. These form when the ice front stops moving for many kilometers back from the front. After the ice has gone, the position of a former ice tunnel is marked by a long, sinuous ridge of sediment known as an *esker* (Figure 19.20). The esker is the deposit of sand and gravel laid on the floor of the former ice tunnel. After the ice has melted away, only the stream-bed deposit remains, forming a ridge. Many eskers are several kilometers long.

DRUMLINS AND TILL PLAINS   Another common glacial form is the *drumlin,* a smoothly rounded, oval hill resembling the bowl of an inverted teaspoon. It consists in most cases of glacial till (Figure 19.21). Drumlins invariably lie in a zone behind the terminal moraine. They commonly occur in groups or swarms and may number in the hundreds. The long axis of each drumlin parallels the direction of ice movement. Drumlins are typically steeper at the broad end, which faces oncoming ice. The origin of drumlins is not well understood. They seem to have been formed under moving ice by a plastering action in which layer upon layer of bouldery clay was spread on the drumlin.

Between moraines, the surface overridden by the ice is covered by glacial till. This cover is often inconspicuous since it forms no prominent landscape feature. The till layer may be thick and may obscure, or entirely bury, the hills and valleys that existed before glaciation. Where thick and smoothly spread, the layer forms a level *till plain.* Plains of this origin are widespread throughout the central lowlands of the United States and southern Canada.

MARGINAL LAKES AND THEIR DEPOSITS   When the ice advanced toward higher ground, valleys that may have opened out northward were blocked by ice. Under such conditions, marginal glacial lakes formed along the ice front (see Figure 19.18*a*). These lakes overflowed along the lowest available channel between the ice and the rising ground slope, or over some low pass along a divide. Streams of meltwater from the ice built *glacial deltas* into these marginal lakes.

When the ice withered away, the lakes drained, leaving a flat floor exposed. Here, layers of fine clay and silt had accumulated. Glacial lake plains often contain extensive areas of marshland. The deltas are now curiously isolated, flat-topped landforms known as *delta kames,* composed of well-washed and well-sorted sands and gravels (Figure 19.18*b*).

PLUVIAL LAKES   During the Ice Age, some regions experienced a cooler, moister climate. In the western United States, closed basins filled with water, forming pluvial lakes.

19.18 Marginal landforms of continental glaciers *(a) With the ice front stabilized and the ice in a wasting, stagnant condition, various depositional features are built by meltwater. (b) The ice has wasted completely away, exposing a variety of new landforms made under the ice. (Drawn by A. N. Strahler.)*

The largest of these, glacial Lake Bonneville, was about the size of Lake Michigan and occupied a vast area of western Utah. With the warmer and drier climate of the present interglacial period, these lakes shrank greatly in volume. Lake Bonneville became the present-day Great Salt Lake. Many other lakes dried up completely, forming desert playas. The history of these pluvial lakes is known from their ancient shorelines, some as high as 300 m (about 1000 ft) above present levels.

Environmental Aspects of Glacial Deposits

Because much of Europe and North America was glaciated by the Pleistocene ice sheets, landforms associated with the ice are of major environmental importance. Agricultural influences of glaciation are both favorable and unfavorable, depending on preglacial topography and the degree and nature of ice erosion and deposition.

In hilly or mountainous regions, such as New England, the glacial till is thinly distributed and extremely stony. Till cultivation is difficult because of countless boulders and cobbles in the clay soil. Till accumulations on steep mountain or roadside slopes are subject to mass movement as earthflows when clay in the till becomes weakened after absorbing water from melting snows and spring rains. Along moraine belts, the steep slopes, the irregularity of knob-and-kettle topography, and the abundance of boulders hinder crop cultivation but are suitable for use as pasture.

19.19 Map of Midwest moraines *Moraine belts of the north-central United States have a curving pattern left by ice lobes. Some regions of interlobate moraines are shown by the color overlay. (Based on data of R. F. Flint and others, Glacial Map of North America, Geological Society of America.)*

Flat till plains, outwash plains, and lake plains, on the other hand, can sometimes provide very productive agricultural land. Bordering the Great Lakes, fertile soils have formed on till plains and on exposed lakebeds. This fertility is enhanced by a blanket of wind-deposited silt (loess) that covers these plains (Chapter 18).

Stratified drift deposits are of great commercial value. The sands and gravels of outwash plains, deltas, and eskers provide necessary materials for both concrete manufacture and highway construction. Where it is thick, stratified drift forms an excellent aquifer and is a major source of ground water supplies (see Chapter 15).

INVESTIGATING THE ICE AGE

A great scientific breakthrough in the study of Ice Age glacial history came in the 1960s. First, scientists learned how to measure the absolute age of certain types of water-laid sediments by means of ancient magnetism. The Earth's magnetic field experienced many sudden reversals of polarity in Cenozoic time, and the absolute ages of these reversals have been firmly established. Second, techniques were developed to take long sample cores of undisturbed fine-textured sediments of the deep ocean floor. Within each core, scientists could determine the age of sediment layers at various control points by identifying magnetic polarity reversals. By further studying the composition and chemistry of the layers within the core, a record of ancient temperature cycles in the air and ocean could be established.

Deep-sea cores reveal a long history of alternating glaciations and interglaciations going back at least as far as 2 million years and possibly 3 million years before present. The

19.20 Esker *The curving ridge of sand and gravel in this photo is an esker, marking the bed of a river of meltwater flowing underneath a continental ice sheet near its margin. Kettle-Moraine State Park, Wisconsin.*

19.21 Drumlin *This small drumlin, located south of Sodus, New York, shows a tapered form from upper right to lower left, indicating that the ice moved in that direction (north to south).*

cores show that in Late-Cenozoic time more than 30 glaciations occurred, spaced at time intervals of about 90,000 years. How much longer this sequence will continue into the future is not known, but perhaps for 1 or 2 million years, or even longer.

Possible Causes of the Late-Cenozoic Ice Age

What caused the Earth to enter into an Ice Age with its numerous cycles of glaciation and interglaciation? Three causes seem possible. First is a change in the placement of continents on the Earth's surface through plate tectonic activity. Second is an increase in the number and severity of volcanic eruptions. Third is a reduction in the Sun's energy output.

Perhaps the answer lies in plate tectonics, through the motions of lithospheric plates following the breakup of Pangea. Recall from Chapter 12 that in Permian time, only the northern tip of the Eurasian continent projected into the polar zone. But as the Atlantic Basin opened up, North America moved westward and poleward to a position opposite Eurasia, while Greenland took up a position between North America and Europe.

Three factors have been proposed to explain why the Earth has entered the Late-Cenozoic Ice Age: a change in the placement of the continents, increased volcanic activity, and a reduction in solar energy output.

The effect of these plate motions was to bring an enormous land-mass area to a high latitude and to surround a polar ocean with land. Because the flow of warm ocean currents into the polar ocean was greatly reduced, or at times totally cut off, this arrangement was favorable to the growth of ice sheets. The polar ocean was ice-covered much of the time, and average air temperatures in high latitudes were at times lowered enough to allow ice sheets to grow on the encircling continents. Furthermore, Antarctica moved southward during the breakup of Pangea and took up a position over the south pole. In that location, it was ideally situated to develop a large ice sheet. Some scientists have also proposed that the uplift of the Himalayan Plateau, a result of the collision of the Austral-Indian and Eurasian plates, could have modified weather patterns sufficiently to trigger the Ice Age.

The second geological mechanism suggested as a basic cause of the Ice Age is increased volcanic activity on a global scope in late-Cenozoic time. Volcanic eruptions produce dust veils that linger in the stratosphere and reduce the intensity of solar radiation reaching the ground (Chapter 3). Temporary cooling of near-surface air temperatures follows such eruptions. Although the geologic record shows periods of high levels of volcanic activity in the Miocene and Pliocene epochs, their role in initiating the Ice Age has not been convincingly demonstrated on the basis of present evidence.

Another possible cause of the Ice Age is a slow decrease in the Sun's energy output over the last several million years,

---

perhaps as part of a cycle of slow increase and decrease over many million years' duration. As yet, data are insufficient to identify this mechanism as a possible basic cause. However, research on this topic is being stepped up as new knowledge of the Sun is acquired from satellites that probe the Sun's atmosphere and monitor its changing surface.

Possible Causes of Glaciation Cycles

What timing and triggering mechanisms are responsible for the many cycles of glaciation and interglaciation that the Earth is experiencing during the present Ice Age? Although many causes for glacial cycles have been proposed, we will limit our discussion here to one major contender called the **astronomical hypothesis.** It has been under consideration for about 40 years and is now widely accepted.

The astronomical hypothesis is based on well-established motions of the Earth in its orbit around the Sun. Recall from Chapter 1 that the Earth's orbit around the Sun is an ellipse, not a circle, and that we refer to the point in the orbit nearest the Sun as perihelion, and the point farthest as aphelion. Presently, perihelion occurs around December 5 and aphelion around July 5. Astronomers have observed that the orbit slowly rotates on a 108,000-year cycle, thus shifting the absolute time of perihelion and aphelion by a very small amount each year. In addition, the orbit's shape varies on a cycle of 92,000 years, becoming more and less elliptical. This changes the Earth–Sun distance and therefore the amount of solar energy the Earth receives at each point of the annual cycle.

The Earth's axis of rotation also experiences cyclic motions. The tilt angle of the axis varies from about 22 to 24 degrees on a 40,000-year cycle. The axis also "wobbles" on a 26,000-year cycle, moving in a slow circular motion much like that of a spinning top or toy gyroscope.

According to the astronomical hypothesis, the timing of glaciations and interglaciations is determined by variations in insolation produced by minor cycles in the Earth's orbit and the Earth's axial rotation.

As a result of these cycles in axial rotation and solar revolution, the annual insolation experienced at each latitude changes from year to year. Figure 19.22 shows a graph of summer insolation received at 65° N latitude for the last 500,000 years as calculated from these cycles. The graph is known as the Milankovitch curve, named for Milutin Milankovich, the astronomer who first calculated it in 1938. The dominant cycle of the curve has a period of about 40,000 years, but note that every second or third peak seems to be higher. Peaks at about 12,000, 130,000, 220,000, 285,000, and 380,000 years ago have been associated with the rapid melting of ice sheets and the onset of deglaciations, as revealed by other dating methods involving ancient ice cores and deep lake sediment cores. Most scientists studying climate change during the Ice Age now agree that cyclic insolation changes are the primary factor explaining the cycles of glaciation of the Ice Age.

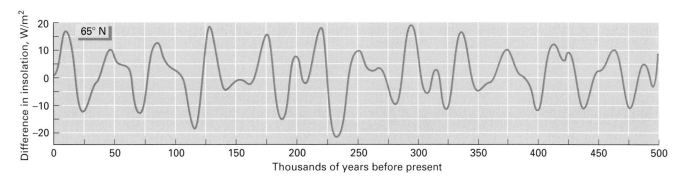

19.22   *The Milankovitch curve* *The vertical axis shows fluctuations in summer daily insolation at lat. 65° N for the last 500,000 years. These are calculated from mathematical models of the change in Earth–Sun distance and change in axial tilt with time. The zero value represents the present value. (Based on calculations by A. D. Vernekar, 1968. Copyright © A. N. Strahler.)*

The actual mechanisms by which insolation changes cause ice sheets to grow or to disappear are unclear. The entire subject is so complex that it is difficult for even a research scientist to grasp fully. Interactions between the atmosphere, the oceans, and the continental surfaces (including the ice sheets) are numerous and closely interrelated with many threads of cause and effect. Changes that occur in one Earth realm are fed back to the other realms in a most complex manner.

Holocene Environments

The elapsed time span of about 10,000 years since the Wisconsinan Glaciation ended is called the *Holocene Epoch*. It began with a rapid warming of ocean surface temperatures. Continental climate zones then quickly shifted poleward, and plants soon became reestablished in glaciated areas.

Three major climatic periods occurred during the Holocene Epoch leading up to the last 2000 years. These periods are inferred from studies of changes in vegetation cover types as observed in fossil pollen and spores preserved in glacial bogs. The earliest of the three is known as the Boreal stage and was characterized by boreal forest vegetation in midlatitude regions. There followed a general warming until the Atlantic stage, with temperatures somewhat warmer than today, was reached about 8000 years ago (–8000 years). Next came a period of temperatures that were below average, the Subboreal stage. This stage spanned the age range –5000 to –2000 years.

Through the availability of historical records and of more detailed evidence, the climate of the past 2000 years can be described on a finer scale. A secondary warm period occurred in the period A.D. 1000 to 1200 (–1000 to –800 years). This warm episode was followed by the Little Ice Age, A.D. 1450–1850 (–550 to –50 years). During the latter, valley glaciers made new advances and extended to lower elevations. More recent temperature cycles are discussed in Chapter 3.

Within the last century, global temperatures have been slowly warming, as we noted in Chapter 3. What will be the effect of global warming on the existing continental ice sheets of Greenland and Antarctica? *Eye on Global Change 19.2 • Ice Sheets and Global Warming* discusses the possibilities.

Cycles of glaciation and interglaciation, as well as the lesser climatic cycles of the Holocene Epoch, have proceeded without human influence for millions of years. They demonstrate the power of natural forces to make drastic swings from cold to warm climates, and they confound our efforts to understand human impact on climate.

A LOOK AHEAD   The group of chapters we have now completed has reviewed landform-making processes that operate on the surface of the continents. These have ranged from mass wasting to the erosion and deposition brought about by glacial ice. The great variety and complexity of landforms we have described are not difficult to understand when each agent of denudation is examined in turn.

Human influence on landforms is felt most strongly on surfaces of fluvial denudation because of the severity of surface changes caused by agriculture and urbanization. Landforms shaped by wind and by waves and currents are also highly sensitive to changes induced by human activity. Only glaciers maintain their integrity and are thus far largely undisturbed by human activity. Perhaps even this last realm of nature's superiority will eventually fall prey to human interference through climate changes induced by industrial activity.

Our closing Epilogue dramatizes four areas of ongoing conflict between the human species and its natural environment—sea-level rise from climatic warming, population pressures, rainforest clearing, and development in earthquake-prone areas. As the human presence continues to impact our planet's natural systems, such conflicts will become increasingly important to resolve.

Eye on Global Change | 19.2

Ice Sheets and Global Warming

What effects will global warming have on the Earth's ice sheets? The Antarctic Ice Sheet holds 91 percent of the Earth's ice. If it were to melt entirely, the Earth's mean sea level would rise by about 40 m (about 200 ft). Additional water could be released by the melting of the Greenland Ice Sheet, which holds most of the remaining volume of land ice. Although global warming will accelerate glacial melting and thinning of the ice at the edges of ice sheets, increased precipitation, held by warmer air over the ice sheets, will likely produce a net growth in ice sheet thickness. Thus, it seems most unlikely that global warming will melt these ice sheets completely, unless the warming is much greater than anticipated. However, there may be other factors at work than snow buildup and marginal melting in the case of individual ice sheets.

In 1998 NASA researchers repeated measurements of the height of the southern half of the Greenland Ice Sheet that were originally made in 1993 using a laser altimeter mounted on an aircraft. They discovered that marginal thinning had reduced the volume of the ice sheet by about 42 km^3 (about 10 mi^3) in that five-year period. The thinning was primarily on the eastern side of the Ice Sheet. On the western side, melting was more in balance with new snow accumulation. The loss of this ice volume is not large enough to have much effect on global sea level. But a substantial flow of fresh meltwater into the ocean could influence the thermohaline circulation of the northern Atlantic (see Chapter 5), triggering an ocean cycle of several centuries' duration that would bring very cold winter weather to Europe.

At the other pole, recent attention has focused on the Western Antarctic Ice Sheet, shown in the map to the right. Much of this vast expanse of Antarctic ice is "grounded" on a bedrock base that is well below sea level—that is, a large portion of the ice sheet rests directly on deep bedrock without sea water underneath. Attached to the grounded ice sheet are the Ross, Ronne, and Filchner ice shelves, which are underlain by sea water. They are also grounded at some points where the shelves enclose islands and overlay higher parts of undersea topography. At present, the grounded ice shelves act to hold back the flow of the main part of the ice sheet.

Geophysicists regard this as an unstable situation. A rapid melting or deterioriation of the ice shelves, perhaps in response to global warming, would release the back pressure on the main part of the ice sheet, which would then move forward and thin rapidly. With a reduced thickness, reduced pressure at the bottom of the ice would allow sea water to enter, and soon most of the sheet would be floating in ocean water. The added bulk would then raise sea level by as much as 6 m (about 20 ft).

A key feature of the Western Antarctic Ice Sheet is the presence of a number of ice streams within the sheet. Ice in these streams flows at a much more rapid rate than in the surrounding ice mass, probably because geothermal activity under the ice streams provides enough heat to melt the ice at the base. This creates a liquid bottom layer that lubricates the ice motion. The ice streams blend gradually into the ice shelves, providing a main source for their supply of ice.

The flow of the ice streams is essential to maintaining the ice shelves. If these streams slowed, the ice shelf might retreat, which could release the unstable western ice sheet, producing catastrophic flooding. In January, 2002, Drs. Ian Joughin and Slawek Tulacyzk reported in *Science* magazine that the ice streams feeding the Ross

West Antarctic Ice Sheet *A map of Antarctica showing the Western Antarctic Ice Sheet. (Adapted with permission from C.R. Bentley, Science Vol. 275, p. 1077. Copyright © American Association for the Advancement of Science.)*

Ice Shelf are getting thicker and flowing more slowly. They attribute the change to gradual thinning of the ice sheet since the end of the last glaciation, a situation that allows cold Antarctic temperatures to penetrate to the base of the ice and freeze the liquid bottom layer. If the slowing of the ice streams continues at the present rate for another 70 to 80 years, it would stop altogether, denying the West Antarctic Ice Sheet an influx of ice about equal to the average annual flow of the Missouri River. However, ice flow into other parts of the West Antarctic Ice Sheet seems to be speeding up, even as the ice streams feeding the Ross Ice Shelf are slowing. Whether this increasing flow will offset the effects of slowing Ross Shelf ice streams is not known.

Either way, new evidence suggests that there has been at least one collapse of the West Antarctic

Ice Sheet during an interglacial period. Sediments extracted from below the ice sheet show the presence of marine organisms that indicate freely circulating ocean water at some time within the last two million years. Moreover, temperatures at the time of the collapse are inferred to be not much greater than those of today.

Meanwhile, the climatic warming of the past few decades seems to be causing the ice on the shelves to thin and fracture more easily, freeing huge icebergs to drift near the shelf front. The photo below shows six satellite images of Ross Island, at the western edge of the Ross Ice Shelf, acquired in 2000 and 2001 by the MISR instrument. The island is about 75 km (47 mi) long and includes Mount Erebus (leftmost peak) with an elevation of 3743 m (12, 277 ft). The first three show a vast iceberg, designated C-16, as it rotates a quarter-turn

counterclockwise in a month's time and then stops. Slowly moving in from the left is another huge berg, designated B-15A. In the last image, sea ice from the previous winter has expanded to surround C-16, and B-15A presses snugly against it.

The motion of these huge chunks of floating glacial ice has affected local penguin rookeries. The extension of the sea ice coupled with the icebergs has trapped large numbers of penguins in their feeding grounds on the open sea, making it difficult for them to get to their breeding areas on land. One colony is in danger of extinction.

The possible effects of climate change on global ice sheets are far from certain. At best, we can hope that human-induced global warming will result in slow, progressive changes to which human civilization can readily adapt.

E19.5   Icebergs and Ross Island seen by MISR *This sequence of images, acquired by NASA's MISR instrument, shows the motion of huge icebergs near Ross Island, Antarctica, at the edge of the Ross Ice Shelf.*

In Review | Glacial land Forms and the Ice Age

- Glaciers form when snow accumulates to a great depth, creating a mass of ice that is plastic in lower layers and flows outward or downhill from a center in response to gravity. As they move, glaciers can deeply erode bedrock by abrasion and plucking. The eroded fragments, incorporated into the flowing ice, leave depositional landforms when the ice melts.

- **Alpine glaciers** develop in **cirques** in high mountain locations. Alpine glaciers flow downvalley on steep slopes, picking up rock debris and depositing it in *lateral* and *terminal* **moraines.** Through erosion, glaciers carve U-shaped **glacial troughs** that are distinctive features of glaciated mountain regions. They become **fiords** if later submerged by rising sea level.

- **Ice sheets** are huge plates of ice that cover vast areas. They are present today in Greenland and Antarctica. The Antarctic Ice Sheet includes *ice shelves*—great plates of floating glacial ice. Icebergs form when glacial ice flowing into an ocean breaks into great chunks and floats free. *Sea ice,* which is much thinner and more continuous, is formed by direct freezing of ocean water and accumulation of snow.

- An ice age includes alternating periods of **glaciation,** *deglaciation,* and **interglaciation.** During the past 2 to 3 million years, the Earth has experienced the **Late-Cenozoic Ice Age.** During this ice age, continental ice sheets have grown and melted as many as 30 times. The most recent glaciation is the *Wisconsinan Glaciation,* in which ice sheets covered much of North America and Europe, as well as parts of northern Asia and southern South America.

- Moving ice sheets create many types of landforms. Bedrock is grooved and scratched. Where rocks are weak, long valleys can be excavated to depths of hundreds of meters. The melting of glacial ice deposits **glacial drift,** which may be stratified by water flow or deposited directly as **till.** Moraines accumulate at ice edges. *Outwash plains* are built up by meltwater streams. Tunnels within the ice leave streambed deposits as *eskers.* Till may be spread smooth and thick under an ice sheet, leaving a *till plain.* This may be studded with elongated till mounds, termed *drumlins.* Meltwater streams build *glacial deltas* into lakes formed at the ice margin and line lake bottoms with clay and silt. When the lakes drain, these features remain.

- Several factors have been proposed to explain the cause of present ice age glaciations and interglaciations. These factors include ongoing change in the global position of continents, an increase in volcanism, and a reduction in the Sun's energy output. Individual cycles of glaciation seem strongly related to cyclic changes in Earth–Sun distance and axial tilt.

- The effects of global warming on the Earth's ice sheets are not well understood at present. Some scientists fear that warming could release vast volumes of Antarctic ice that would quickly slide into the ocean and increase sea level by 6 m (20 ft) or more. Others have noted that global warming enhances precipitation, which causes snow to accumulate atop ice sheets and may cancel some of the potential rise in sea level due to slow melting.

Key Terms

ablation, p. 592	moraine, p. 595	interglaciation, p. 601	till, p. 604
alpine glacier, p. 593	glacial trough, p. 595	Late-Cenozoic Ice Age,	astronomical hypothesis,
ice sheet, p. 593	fiord, p. 595	p. 602	p. 608
cirque, p. 594	glaciation, p. 601	glacial drift, p. 604	

Review Questions

1. How does a glacier form? What factors are important? Why does a glacier move?
2. Distinguish between alpine glaciers and ice sheets.
3. What is a glacial trough and how is it formed? What is its basic shape? In what ways can a glacial trough appear after glaciation is over?
4. Where are ice sheets present today? How thick are they?
5. Contrast sea ice and icebergs, including the processes by which they form.
6. Identify the Late-Cenozoic Ice Age. When did it begin? What was the last glaciation in this cycle? When did it end?
7. What areas were covered with ice sheets by the last glaciation? How was sea level affected?

8. What are moraines? How are they formed? What types of moraines are there?
9. Identify the landforms and deposits associated with stream action at or near the front of an ice sheet.
10. Identify the landforms and deposits associated with deposition underneath a moving ice sheet.
11. Identify the landforms and deposits associated with lakes that form at ice sheet margins.
12. What cycles are known to affect the amount of solar radiation received by polar regions of the Earth?
13. What is the Milankovitch curve? What does it show about warm and cold periods during the last 500,000 years?

14. How have environments changed during the Holocene Epoch? What periods are recognized, and what are their characteristics?

Eye on Global Change 19.2 ● Ice Sheets and Global Warming

1. What process seems to be offsetting the rate of melting of ice sheets as climate warms? Where has it been observed?

2. Why is the Western Antarctic Ice Sheet regarded as unstable? What is the role of ice streams in maintaining the present size of the ice sheet?

3. What change has recently been observed in the flow of ice streams into the Ross Ice Shelf? What are the implications?

4. How can changes in icebergs and sea ice at the ice front affect penguin populations?

Visualizing Exercises

1. What are some typical features of an alpine glacier? Sketch a cross section along the length of an alpine glacier and label it.

2. Refer to Figure 19.10, which shows the Antarctic continent and its ice cap. Identify the Ross, Filchner, and Larsen ice shelves. Use the scale to measure the approximate area of each in square kilometers or miles. Consulting an atlas or almanac, identify the state or province of the United States or Canada that is nearest in area to each.

Essay Questions

1. Imagine that you are planning a car trip to the Canadian Rockies. What glacial landforms would you expect to find there? Where would you look for them?

2. At some time during the latter part of the Pliocene Epoch, the Earth entered an ice age. Describe the nature of this ice age and the cycles that occur within it. What explanations are proposed for causing an ice age and its cycles? What cycles have been observed since the last ice sheets retreated?

Eye on the Landscape

19.1 Alpine glacier This alpine glacial landscape shows many of the glacial landforms and features described in this chapter. In the near distance are sharp peaks, or horns **(A)**, and knife-edge ridges, or arêtes **(B)**. In the middle foreground is a bowl-shaped cirque **(C)** that no longer bears glacial ice, although the glacier from the adjacent cirque spills into the empty cirque across a low pass (col). Glacial flow features are especially evident, including medial moraines **(D)**, lateral moraines **(E)**, and crevasses **(F)** that open as the ice falls across a rock step.

19.16 Finger Lakes The Finger Lakes of western New York State have been called "inland fiords" for their resemblance to the fiords of glaciated regions. (Fiords are discussed in this chapter.) As explained in our main text, the lakes were eroded from preexisting stream valleys by ice action during the multiple continental glaciations of the Ice Age. The two largest lakes in the center of the image **(A)** are Cayuga Lake (upper) and Seneca Lake (lower). The pattern of agricultural fields in the center and upper left part of the image **(B)** marks the intensive agricultural development of the Lake Erie lowlands and foothills of the Appalachian Plateau, which are mantled with productive soils of glacial origin (Udalfs; see Chapter 10). To the south, at the bottom of the image **(C),** the terrain becomes more dissected, with more pronounced valleys and ridges, as it slopes upward to the higher elevations of the plateau. Stream dissection of the landscape was covered in Chapter 16.

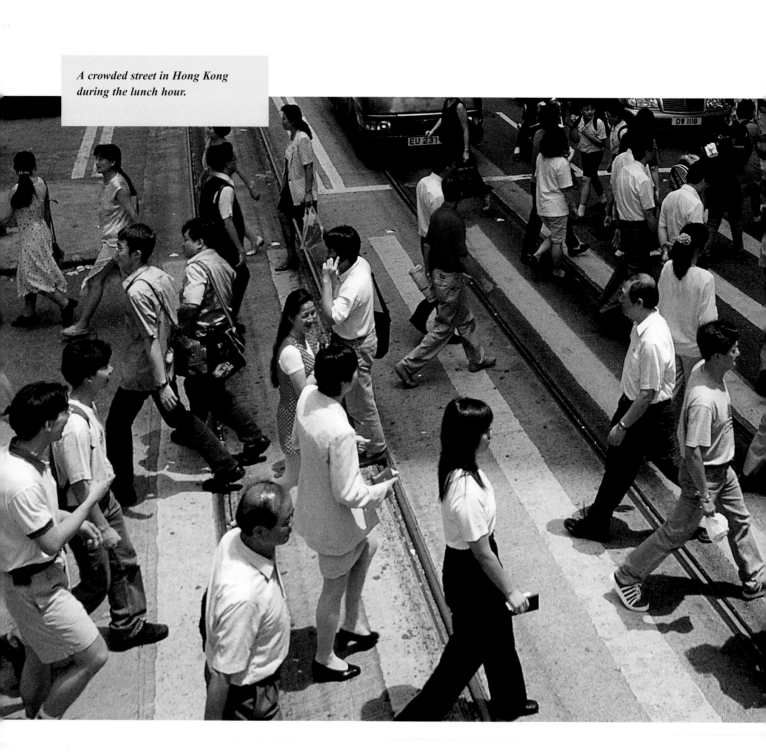

A crowded street in Hong Kong during the lunch hour.

NEWS FROM THE FUTURE

Our text began with a feature describing a trip you might take as a lunar astronaut, traveling home from the Moon to Earth in 2050. What will life be like in the middle of the century? Imagine turning on a futuristic television set and watching the news for June 26, 2050. It might go something like this.

"Good evening, America, and welcome to the Nightly News. This is Dan Rogers in Washington. Tonight's big stories are from the environment. First, we'll go to Galveston, where the story is the flooding from the first storm of the season, Hurricane Alfred. Here's our correspondent, Lorraine Jackson, on the scene. Lorraine, what's it like there now?"

"Thanks, Dan. The storm has slackened off in the last few hours, but the system of dams and pumps they have here has worked, at least for now, to keep the floodwaters out of the center of the city. As you can see, the rain is still falling and the winds are still strong. But the real problem with this storm has not been the wind and rain. Alfred is one of the weaker storms to reach the Texas coast in the past several hurricane seasons. The problem is with the ocean flooding that this coastal outpost endures when a storm like Alfred passes close by.

"The ocean dikes have been taking a terrible beating from the wave action. Here's some shots of the waves breaking on the dikes that we got earlier. All that wave water spilling over the dikes has to be pumped out, and fast. We understand from the local authorities that several smaller dikes west of the city were breached this morning, adding to the floodwaters. Apparently, several subdivisions out there are well under water now. There was a scare when the power went out earlier in the day, and one of the huge backup generators supplying power to the pumps refused to start. But the other generators are keeping up with the load of the pumps."

"When the citizens of this Texas city decided 30 years ago to fight against the rise in sea level rather than move inland like most other beach communities, they knew that the battle against the ocean would be costly and could eventually be a losing one. But for now, the valiant city of Galveston seems to be holding its own, Dan."

"Lorraine, what about the death toll?"

"So far, Dan, the authorities have not reported any deaths from the storm, although there were some injuries in the evacuation. We had a report that one of the huge buses they use for evacuations hit the rear end of another bus, but only the driver and two passengers were injured. Apparently, the accident didn't slow down the river of huge vehicles streaming across the viaduct to the mainland. As I reported this morning, last night's evacuation went pretty smoothly. The residents here have been through so many evacuations in the last few years that they are all familiar with the emergency routine."

"Thanks, Lorraine. That is relatively good news."

Climatic change may raise sea level, enhancing the ability of hurricanes to destroy coastal settlements. Here, Hurricane Alicia lashes the Texas coast at Galveston.

"Our next story marks a momentous event for the human race—the birth of the 7 billionth citizen of Earth. For the story, we go to Beijing, to our correspondent Li Feng Sun. Where are you, Feng Sun?"

"Dan, I'm here in the corridor outside the maternity ward in this small birthing center in a suburb west of Beijing, waiting word about the birth of Zhu Peng, destined to be the world's 7 billionth citizen. Of course, we're not pre-

cisely certain that baby Zhu Peng will be the world's 7 billionth citizen, but scientists have predicted that in precisely four minutes and fourteen seconds, the Earth's population will reach 7 billion. They also tell us that the 7 billionth citizen is most likely to be Chinese.

"Economists and population planners have been marking the event as a real milestone. Before the turn of the century, they projected that the world's population would be

As progressively larger areas of the Earth's surface are developed, the potential for damage from extreme natural events, such as earthquakes, increases. This interchange on the Golden State Freeway, Los Angeles, was destroyed in the Whittier Narrows Earthquake of 1987.

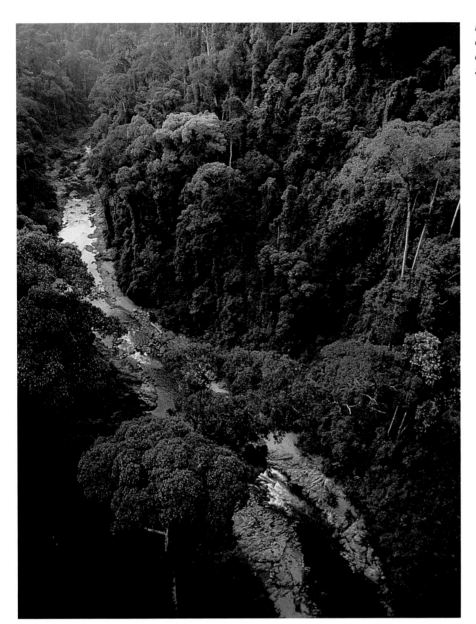

High biodiversity is a feature of equatorial rainforests, such as this example in the Datum River Gorge, Sabah, Borneo.

over 9 billion by 2050. You may remember some years ago the many predictions that the plans for sustainable world development would crash along with populations of Africa and Southeast Asia, which were decimated by the AIDS epidemic. China was only little impacted because AIDS spread quite slowly here, and because of China's crash program to inoculate its entire population with the AIDS vaccine shortly after it was developed. That left this huge country to continue its slow population growth coupled with rapid economic growth fueled by this country's vast natural and human resources. The result is the success story we're all familiar with—China's economy surpassing the economies of Europe, the Americas, and finally the East Asian trading block of Japan, Taiwan, Korea, and Singapore."

"Feng Sun, I hear some chanting in the background. Are those demonstrators?"

"That's right, Dan. The militant arm of the Chinese Green party has mounted a demonstration here. In fact, we

have one of the leaders here, Huang Chung, ready to talk with us. Huang Chung, what are the demonstrators here for?"

"Well, Feng Sun, we're protesting this event because we don't think the world needs any more people. Every consumer requires more food and uses more resources. We've already cut down almost half of the world's forests. Global warming has made the deserts spread and our cropland become less productive. We all know that global warming was generated by consumers demanding products and lifestyles that depend on fossil fuel..."

"Feng Sun, I hate to break in, but we've got an interrupt here. Apparently, there's been a breakthrough on the biodiversity talks at the U.N. For the story, let's go to Maria Remarquez in New York. Maria?"

"Hi, Dan. We've just gotten word here that the rainforest nations will back down from their threat to withdraw from the biodiversity treaty. Apparently, the U.S., the European

Union, and Chinese-East Asian bloc have agreed to make some of the payments demanded for the use of the rainforest genetic materials that the rainforest nations contend were stolen from them by biotechnology companies.

"Here to tell us about it is Professor Kate Carver of New York's City University. Professor Carver, why is this agreement important?"

"Well, Maria, because with this agreement in place, we can go ahead with the plan to set up rainforest preservation zones, perhaps as early as next year. The rainforest ecosystems are our most complex ecological communities, and they hold the greatest diversity of life of all habitats on Earth. They are a reservoir of genetic material that is irreplaceable, and they need to be protected as vast areas forever. Without these genetic reserves, we'd be missing many of our cures for diseases ranging from cancer to AIDS.

"You know, Maria, I might also point out that this agreement is good for the environment in another way. The money will be used by the rainforest nations to stimulate the use of nonpolluting technologies for sustainable development. That is, the funds will be helping these nations promote more use of solar power, wind power, and methane gas production. In this way, they will reduce their reliance on old-fashioned fossil fuels and with it the release of carbon dioxide that produces climate change."

"Thanks for that perspective, Professor. It looks like a win for the environment here at the United Nations this evening, Dan."

"And thanks for that fine story, Maria. For our next story, remember that it was exactly one month ago when the disastrous El Cajon earthquake struck Los Angeles. We go now to Pasadena for a report on the cleanup and rebuilding. Our reporter with the story is Keith Mancini. Keith?"

"It was early on a Sunday morning just one month ago when Angelinos were thrown from their beds by the violent shaking that devastated so much of Los Angeles. It was the largest earthquake to hit Southern California in more than a century, measuring magnitude 8.1 on the Richter Scale, flattening thousands of buildings, and rendering thousands more unusable. The death toll, estimated at over 16,000 with more than 100,000 injured, was also by far the largest in this century for an American natural disaster.

"Today Los Angeles is still digging out. The magnitude of the task ahead is mind-boggling, Dan. The electric freeway system is virtually closed for all but a few segments. Nearly every viaduct and bridge experienced some damage, and engineers are struggling to inspect them all for safety and set the priorities for repair. Most of the city's skyscrapers are still unoccupied, since structural repairs are expected to take many more months to complete. In the hardest-hit areas containing old masonry buildings, almost nothing was left standing. Now that the searches for casualties are over, bulldozers and dump trucks are removing the debris.

"In the Long Beach area, the cleanup is beginning from the fires that started in the oil processing plants and spread to the chemical factories. The toxic compounds carried by the smoke and ash contaminated a huge area. It has taken several weeks for the National Guard to replace their cordon

of troops and vehicles with the fencing and barbed wire that will keep looters out while the area is being detoxified, which is expected to take six to nine months.

"Meanwhile, homeless residents are flocking to the tent cities constructed of portable geodomes now being set up by the National Guard. The demand increases daily as troops and local authorities evict inhabitants still trying to live in the dangerous, crumbling buildings. The geodome cities at least provide shelter, hot food, and sanitation for the thousands of displaced residents, luxuries that are not easy to come by here.

"But the amazing thing, Dan, is the dedication of these residents to rebuilding their city. Everywhere you go in this vast metropolis, there is activity. People are pulling together to rebuild their homes, their businesses, and their lives. There's a real 'can-do' spirit here, and L.A. should emerge from this tragedy all the stronger for it."

"Thanks, Keith, for that fine report. It won't be easy for the folks in Los Angeles in the weeks and months ahead, but at least they know that the rest of the nation is pulling for them.

"And, that's it for the top news stories today. Stay with us for the local report, coming up after these messages."

PERSPECTIVE ON THE NEWS FROM THE FUTURE

This news report dramatizes some of the present concerns about the environment that have arisen in our survey of physical geography. The first news story calls attention to the sea-level rise that is predicted to accompany global warming. The latest predictions are that by the year 2100 global temperature will increase by 1.4°C to 5.8°C (2.5 °F to 10.4°F). Sea level is expected to rise by 9 to 88 cm (3.5 to 34.6 in.).

The problem with rising sea level is not so much with day-to-day levels, but rather the levels that occur when tides combine with storms to produce intense wave action on low-lying coasts. The Republic of Maldives, which is a nation of 1190 small islands, would be entirely submerged by a rise of 2 m (6.6 ft). Even with a rise of only 1 m (3.3ft), the country could be wiped out by a severe storm surge.

Sea-level rise is just one of the possible effects of climatic warming. Even though the extent of warming is uncertain, scientists have concluded that warming would bring not only warmer temperatures, but also more global precipitation. However, this precipitation increase will not be evenly distributed. Computer models predict that high latitudes will experience more precipitation, while midlatitude continental interiors will experience drier summers. These types of changes would have important effects on crop growth, shifting the varieties and types of crops grown within large regions. Increased irrigation could also be needed.

Because the effects of climatic warming could be so far-reaching, scientists have managed to alert the public about the danger of continued increases of CO_2 and other gases that contribute to global warming. At the 1992 Earth Summit in Rio de Janeiro, most of the world's nations agreed in principle on a treaty to reduce human impact on climate. The industrialized countries pledged to reduce the levels of greenhouse gases they release, thus slowing the rapid increases in atmospheric CO_2, methane, chlorofluorocarbons, and nitrous oxide (N_2O) that have been measured. At subsequent meetings in Kyoto, Buenos Aires, and Marrakesh, specific reduction targets were hammered out and agreed upon. Hailed as the "strongest environmental treaty that's ever been drafted," it calls for a group of industrialized nations, including Russia, Canada, Japan, Australia, and the European Union, to reduce emissions of greenhouse gases to levels that are 5 percent lower than those of 1990. United States President George W. Bush rejected the treaty, however, leaving the United States on the sidelines of this pioneering effort to fight human-induced global change.

Another action of the Rio Earth Summit was the adoption of a treaty on biodiversity, which is the subject of the third news story. In the treaty, nations pledged to protect the genetic resources of their plant and animal populations by shielding the populations from extinction. Forest preservation was addressed as a need as well. The low-latitude rainforests cover only about 7 percent of the Earth's surface, but they contain over half of the world's species. They are especially rich in insect species and flowering plants. Plants of these regions are a particularly valuable resource. They have provided most of the species on which humans rely for food. Many highly effective medical drugs are produced from these plants.

Not only are low-latitude rainforests valuable resources for human exploitation, but they are also very complex ecosystems that have evolved over millions of years. Many environmentalists believe that humans have a responsibility to preserve our planet's natural systems and maintain their diversity. It is encouraging that the world's nations have recognized the importance of ecological resources such as the low-latitude rainforests and are working toward their preservation.

The last future news story, about an earthquake in Los Angeles, demonstrates the impact of natural disasters on human activities. The story is only slightly exaggerated—an earthquake of magnitude 8.1 on the San Andreas fault would most certainly have devastating effects on a nearby city the size of Los Angeles.

By nature, the human species has a short-term outlook, preferring to prepare for dangers that are likely to occur within a time span of a few years to a few decades. Many natural processes work on far longer time scales, however. Although a major earthquake on the San Andreas fault may not happen more than once every few generations, there were probably tens of thousands of severe earthquakes on that fault in the Los Angeles region during the Ice Age, with hundreds of these occurring since the last advance of glacial ice. A major flood on the Mississippi River, like that of 1993, is a still more likely event. Although infrequent in human terms, these rare events have shaped much of the Earth's landscape. For the geologist or physical geographer

trained in Earth processes, the evidence of these past events is everywhere.

Earthquakes, floods, and tropical storms are not new phenomena. What is new is that human civilization has come to dominate most areas of the planet, so that these natural occurrences now have major impacts on human societies in the regions in which they occur. Taken in the aggregate, they become a factor of global change.

The event of the second future news story, which dramatizes the birth of the 7 billionth citizen of Earth, probably has the most important implications. It is actually quite an optimistic scenario, for present estimates of the world population in 2050 are closer to 9.5 billion. The story refers to future population crashes in Africa and Southeast Asia produced by the AIDS epidemic that now rages throughout the world, and advances the idea that China will be impacted less severely than other developing nations. Whether this future will come to pass is uncertain. But in any event, the human population will continue to increase until either it regulates itself or destroys itself.

Why is population growth so important? Humans are the most powerful force transforming the surface of the Earth today. About 15 to 20 percent of the world's forest area has been cleared since human civilization began to exploit forest resources. In the last 300 years, humans have increased the area of cropland by 450 percent. Populations of the largest urban areas have increased 25 times in that same time span. These changes have directly influenced the global energy balance, the hydrologic cycle, and the dynamics of the atmosphere, with effects that are unknown in the long term.

Recognizing these trends, the nations of the world have come together in an effort to promote sustainable development. Under this policy, the needs and aspirations of present and future generations are met without impacting the ability of future generations to satisfy their needs. The idea is to focus on social and economic progress that is compatible with the environment. The concept draws on the premise that human knowledge, sophisticated technology, and access to resources are now so well developed that they can be used to allow developing countries to increase the well-being of their populations in a way that does not damage the environment. In this effort, much help and cooperation is required from developed nations, to provide not only the funds to stimulate sustainable development, but also the technological base that will be required.

There are now some hopeful signs that the long-continued increase of human population will slow. In 1996, reports revealed for the first time that fertility rates had declined in all regions of the world. A survey of population experts predicted that the Earth's population will peak at around 11 billion by 2075 and remain level or decrease slightly from 2075 to 2100. Other studies of the economic impact of population growth have noted that technological innovation seems to naturally lead to increasing use of technologies that have a smaller environmental impact. Although the citizens of underdeveloped countries will want and need to advance their standard of living, their consumption will perhaps be less damaging to the environment. These recent trends in population and resource utilization can be expected to ease the transition from environmental exploitation to sustainable development in many world regions.

The social and economic aspects of continued population growth and policies such as sustainable development are far beyond the scope of physical geography, but they are concerns of which every responsible citizen of our planet must be aware. This book has provided a background and context on the workings of the Earth and the environment that we, as authors, hope will be valuable to you, its readers, as you face the challenges of world citizenship in the twenty-first century.

Appendix

1

THE CANADIAN SYSTEM OF SOIL CLASSIFICATION*

The formation of the National Soil Survey Committee of Canada in 1940 was a milestone in the development of soil classification and of pedology generally in Canada. Prior to that time, Canada used the 1938 U.S. Department of Agriculture (USDA) system. Although the Canadian experience showed that the concept of zonal soils was useful in the western plains, it proved less applicable in eastern Canada where parent materials and relief factors had a dominant influence on soil properties and development in many areas.

Canadian pedologists closely observed the evolution of the U.S. Comprehensive Soil Classification System (CSCS) during the 1950s and 1960s and ultimately adopted several important features of that system. Nevertheless, the special needs of a workable Canadian national system required that a completely independent classification system be established. Because Canada lies entirely in a latitude zone poleward of the 40th parallel, there was no need to incorporate those soil orders found only in lower latitudes. Furthermore, the vast expanse of Canadian territory lying within the boreal forest and tundra climates necessitated the recognition at the highest taxonomic level (the order) of soils of cold regions that appear only as suborders and even as great groups in the CSCS—for example, the Cryaquents and Cryorthents that occupy much of

the soils map of northern Canada above the 50th parallel (Figure 10.13).

The overall philosophy of the Canadian system is pragmatic: the aim is to organize knowledge of soils in a reasonable and usable way. The system is a natural or taxonomic one in which the classes (taxa) are based on properties of the soils themselves and not on interpretations of the soils for various uses. Thus, the taxa are concepts based on generalization of properties of real, not idealized, bodies of soils. Although taxa in the Canadian system are defined on the basis of actual soil properties that can be observed and measured, the system has a genetic bias in that properties or combinations of properties that reflect genesis are favored in distinguishing among the higher taxa. Thus, the soils brought together under a single soil order are seen as the product of a similar set of dominant soil-forming processes resulting from broadly similar climatic conditions.

The Canadian system recognizes the *pedon* as the basic unit of soils; it is defined as in the CSCS. Major mineral horizons of the soil (A, B, C) are defined in much the same way as in the U.S. system. Thus, the Canadian system of soil taxonomy is more closely related to the U.S. system than to any other. Both are hierarchical, and the taxa are defined on the basis of measurable soil properties. However, they differ in several respects. The Canadian system is designed to classify only the soils that occur in Canada and is not a comprehensive system. The U.S. system includes the suborder, a taxon not recognized in the Canadian system. Because 90 percent of the area of Canada is not likely to be cultivated, the Canadian system does not recognize as diagnostic those horizons strongly affected by plowing and application of soil conditioners and fertilizers.

---

* Throughout this section, numerous sentences and phrases are taken verbatim or paraphrased from the following work: Canada Soil Survey Committee, *The Canadian System of Soil Classification,* Research Branch, Canada Department of Agriculture, Publication 1646, 1978. Table A1.1 and Figure A1.2 are also compiled from this source.

Table A1.1 | Subhorizons and Organic Horizons of the Canadian System of Soil Classification

Subhorizons; Lowercase Suffixes

b   Buried soil horizon.

c   Cemented (irreversible) pedogenic horizon.

ca Horizon of secondary carbonate enrichment in which the concentration of lime exceeds that in the unenriched parent material.

e   Horizon characterized by the eluviation of clay, Fe, Al, or organic matter alone or in combination.

f   Horizon enriched with amorphous material, principally Al and Fe combined with organic matter; reddish near upper boundary, becoming yellower at depth.

g   Horizon characterized by gray colors, or prominent mottling, or both, indicative of permanent or intense reduction.

h   Horizon enriched with organic matter.

j   Used as a modifier of suffixes e, f, g, n, and t to denote an expression of, but failure to meet, the specified limits of a suffix it modifies.

k   Denotes presence of carbonate as indicated by visible effervescence when dilute HCl is added.

m   Horizon slightly altered by hydrolysis, oxidation, or solution, or all three to give a change in color or structure, or both.

n   Horizon in which the ratio of exchangeable Ca to exchangeable Na is 10 or less. It must also have the following distinctive morphological characteristics: prismatic or columnar structure, dark coatings on ped surfaces, and hard to very hard consistence when dry.

p   Horizon disturbed by human activities such as cultivation, logging, and habitation.

s   Horizon of salts, including gypsum, which may be detected as crystals or veins or as surface crusts of salt crystals.

sa Horizon with secondary enrichment of salts more soluble than Ca and Mg carbonates; the concentration of salts exceeds that in the unenriched parent material.

t   Illuvial horizon enriched with silicate clay.

u   Horizon that is markedly disrupted by physical or faunal processes other than cryoturbation.

x   Horizon of fragipan character. (A fragipan is a loamy subsurface horizon of high bulk density and very low organic matter content. When dry it is hard and seems to be cemented.)

y   Horizon affected by cryoturbation as manifested by disrupted or broken horizons, incorporation of materials from other horizons, and mechanical sorting.

z   A frozen layer.

Organic Horizons

O   Organic horizon developed mainly from mosses, rushes, and woody materials.

L   Organic horizon characterized by an accumulation of organic matter derived mainly from leaves, twigs, and woody materials in which the organic structures are easily discernible.

F   Same as L, above, except that original structures are difficult to recognize.

H   Organic horizon characterized by decomposed organic matter in which the original structures are indiscernible.

SOIL HORIZONS AND OTHER LAYERS

The definitions of classes in the Canadian system are based mainly on kinds, degrees of development, and sequence of soil horizons and other layers in pedons. The major mineral horizons are A, B, and C. The major organic horizons are L, F, and H, which are mainly forest litter at various stages of decomposition, and O, which is derived mainly from bog, marsh, or swamp vegetation. Subdivisions of horizons are labeled by adding lowercase suffixes to the major horizon symbols—for example, Ah or Ae.

Besides the horizons, nonsoil layers are recognized. Two such layers are R, rock, and W, water. Lower mineral layers not affected by pedogenic processes are also identified. In organic soils, layers are described as tiers.

The principal mineral horizons, A, B, and C, are defined as follows:

A   Mineral horizon found at or near the surface in the zone of leaching or eluviation of materials in solution or suspension, or of maximum *in situ* accumulation of organic matter, or both.

B   Mineral horizon characterized by enrichment in organic matter, sesquioxides, or clay; or by the development of soil structure; or by change of color denoting hydrolysis, reduction, or oxidation.

C   Mineral horizon comparatively unaffected by the pedogenic processes operative in A and B horizons. The processes of gleying and the accumulation of calcium and magnesium and more soluble salts can occur in this horizon.

Lowercase suffixes, used to designate subdivisions of horizons, are shown in Table A1.1.

SOIL ORDERS OF THE CANADIAN SYSTEM

Nine soil orders make up the highest taxon of the Canadian System of Soil Classification. Listed in alphabetical order, they are as follows.

Brunisolic	Gleysolic	Podzolic
Chernozemic	Luvisolic	Regosolic
Cryosolic	Organic	Solonetzic

Table A1.2 lists the great groups within each order.

Table A1.2 | **Great Groups of the Canadian Soil Classification System**

Order	Great Group
Brunisolic	Melanic Brunisol
	Eutric Brunisol
	Sombric Brunisol
	Dystric Brunisol
Chernozemic	Brown
	Dark Brown
	Black
	Dark Gray
Cryosolic	Turbic Cryosol
	Static Cryosol
	Organic Cryosol
Gleysolic	Humic Gleysol
	Gleysol
	Luvic Gleysol
Luvisolic	Gray Brown Luvisol
	Gray Luvisol
Organic	Fibrisol
	Mesisol
	Humisol
	Folisol
Podzolic	Humic Podzol
	Ferro-Humic Podzol
	Humo-Ferric Podzol
Regosolic	Regosol
	Humic Regosol
Solonetzic	Solentz
	Solodized Solonetz
	Solod

Brunisolic Order

The central concept of the *Brunisolic order* is that of soils under forest having brownish-colored Bm horizons. Most Brunisolic soils are well to imperfectly drained. They occur in a wide range of climatic and vegetative environments, including boreal forest; mixed forest, shrubs, and grass; and heath and tundra. As compared with the Chernozemic soils, the Brunisolic soils show a weak B horizon of accumulation attributable to their moister environment. Brunisolic soils lack the diagnostic podzolic B horizon of the Podzolic soils, in which accumulation in the B horizon is strongly developed. The Melanic Brunisol shown in profile in Figure A1.1 can be found in the St. Lawrence Lowlands, surrounded by Podzolic soils (Figure A1.2).

Chernozemic Order

The general concept of the *Chernozemic order* is that of well to imperfectly drained soils having surface horizons darkened by the accumulation of organic matter from the decomposition of xerophytic or mesophytic grasses and forms representative of grassland communities or of grassland-forest communities with associated shrubs and forbs.

The major area of Chernozemic soils is the cool, subarid to subhumid interior plains of western Canada. Most Chernozemic soils are frozen during some period each winter, and the soil is dry at some period each summer. The mean annual temperature is higher than O°C and usually less than 5.5°C. The associated climate is typically the semiarid (steppe) variety of the dry midlatitude climate ⑨ (Figure A1.2).

Essential to the definition of soils of the Chernozemic order is that they must have an A horizon (typically, Ah) in which organic matter has accumulated, and they must meet several other requirements. The A horizon is at least 10 cm thick; its color is dark brown to black. It usually has sufficiently good structure that it is neither massive and hard nor single-grained when dry. The profile shown in Figure A1.1 is that of the Orthic subgroup of the Brown great group; it shows a Bm horizon that is typically of prismatic structure. The C horizon is one of lime accumulation (Cca). Clearly, the Chernozemic soils can be closely correlated with the Mollisols of the CSCS.

Cryosolic Order

Soils of the *Cryosolic order* occupy much of the northern third of Canada where permafrost remains close to the surface of both mineral and organic deposits. Cryosolic soils predominate north of the tree line, are common in the subarctic forest area in fine-textured soils, and extend into the boreal forest in some organic materials and into some alpine areas of mountainous regions. Cryoturbation (intense disturbance by freeze–thaw activity) of these soils is common, and it may be indicated by patterned ground features such as sorted and nonsorted nets, circles, polygons, stripes, and Earth hummocks.

Cryosolic soils are found in either mineral or organic materials that have permafrost either within 1 m of the surface or within 2 m if more than one-third of the pedon has been strongly cryoturbated, as indicated by disrupted, mixed, or broken horizons. The profile shown in Figure A1.1 is that of the Orthic subgroup of the Static Cryosol great group. Note the presence of organic L, H, and O surface horizons and the thin Ah horizon. The Cryosolic soils are closely correlated with the Cryaquepts of the CSCS.

Gleysolic Order

Soils of the *Gleysolic order* have features indicating periodic or prolonged saturation with water and reducing conditions. They commonly occur in patchy association with other soils in the landscape. Gleysolic soils are usually associated with either a high ground water table at some period of the year or temporary saturation above a relatively impermeable layer. Some Gleysolic soils may be submerged under shallow water throughout the year. The profile shown in Figure A1.1 is that of the Gleysol great group. It has a thick Ah horizon. The underlying Bg horizon is grayish and shows mottling typical of reducing conditions.

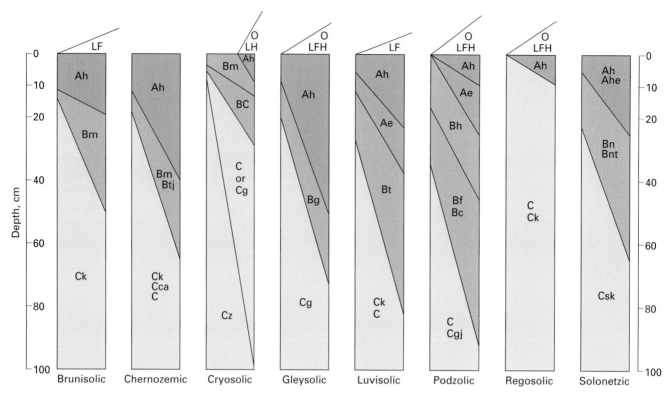

A1.1 Soil profiles *Representative schematic profiles of eight of the nine orders of the Canadian system of soil classification. Slanting lines show the range in depth and thickness of each horizon. (The horizon planes are actually approximately horizontal within the pedon.) See Table A1.1 for an explanation of symbols. (From Canada Soil Survey Committee, Research Branch, Canada Department of Agriculture, 1978.)*

Luvisolic Order

Soils of the *Luvisolic order* generally have light-colored, eluvial horizons (Ae), and they have illuvial B horizons (Bt) in which silicate clay has accumulated. These soils develop characteristically in well to imperfectly drained sites, in sandy loam to clay base-saturated parent materials under frost vegetation in subhumid to humid, mild to very cold climates. The genesis of Luvisolic soils is thought to involve the suspension of clay in the soil solution near the soil surface, downward movement of the suspended clay with the soil solution, and deposition of the translocated clay at a depth where downward motion of the soil solution ceases or becomes very slow. The representative profile shown in Figure A1.1 is that of the Orthic subgroup of the Gray Brown Luvisol great group.

Luvisolic soils occur from the southern extremity of Ontario to the zone of permafrost and from Newfoundland to British Columbia. The largest area of these soils are Gray Luvisols occurring in the central to northern interior plains under deciduous, mixed, and coniferous forest. In this location they appear to correlate with the Boralfs of the Alfisol order in the CSCS. Gray-Brown Luvisolic soils of southern Ontario would correlate with the suborder of Udalfs.

Organic Order

Soils of the *Organic order* are composed largely of organic materials. They include most of the soils commonly known as peat, muck, or bog soils. Organic soils contain 17 percent or more organic carbon (30 percent organic matter) by weight. Most Organic soils are saturated with water for prolonged periods. They occur widely in poorly and very poorly drained depressions and level areas in regions of subhumid to humid climate and are derived from vegetation that grows in such sites. However, one group of Organic soils consists of leaf litter overlying rock or fragmental material; soils of this group may occur on steep slopes and may rarely be saturated with water. (No profile of the Organic soils is shown in Figure A1.1.) Organic soils can be correlated with the Histosols of the CSCS.

Podzolic Order

Soils of the *Podzolic order* have B horizons in which the dominant accumulation product is amorphous material composed mainly of humified organic matter in varying degrees with Al and Fe. Typically, Podzolic soils occur in coarse- to medium-textured, acid parent materials, under forest and heath vegetation in cool to very cold humid to

1. Podzols, Brunisols
2. Luvisols, Brunisols, rock
3. Brunisols, rock
4. Luvisols, Organics
5. Chernozemic soils
6. Luvisols
7. Brunisols, Gleysols
8. Organics, Cryosols
9. Podzols, Brunisols, Organics
10. Brunisols, Cryosols, rock
11. Cryosols, rock, ice
12. Mainly rock

Approximate Equivalents
in Soil Taxonomy

Brunisol–Inceptisol	Gleysol–Inceptisol, wet
Chernozemic–Mollisol	Luvisol–Alfisol
Cryosol–Inceptisols with permafrost	Organic–Histosol
	Podzol–Spodosol

A1.2 Soil regions *Generalized map of soil regions of Canada. (Courtesy of Land Resources Research Institute, Agriculture Canada.) (Illustration is taken from Fundamentals of Soil Science, 7th ed., by Henry D. Foth, John Wiley & Sons.)*

very humid climates. Podzolic soils can usually be readily recognized in the field. Generally, they have organic surface horizons that are commonly L, F, and H. Most Podzolic soils have a reddish brown to black B horizon (Bh) with an abrupt upper boundary. The profile shown in Figure A1.1 is that of the Orthic subgroup of the Humic Podzol great group.

The Podzolic soils correspond closely to the Spodosols (Orthods) of the CSCS.

Regosolic Order

Regosolic soils have weakly developed horizons. The lack of development of genetic horizons may be due to any number of factors: youthfulness of the parent material, for example, recent alluvium; instability of the material, for example, colluvium on slopes subject to mass wasting; nature of the material, for example, nearly pure quartz sand; climate, for example, dry cold conditions. Regosolic soils are generally rapidly to imperfectly drained. They occur in a wide range of vegetation and climates. The profile shown in Figure A1.1 is that of the Orthic subgroup of the Regosol great group. It has only a thin humic A horizon (Ah) and a surface horizon of organic materials.

Regosolic soils correspond with the Entisols of the CSCS.

Solonetzic Order

Soils of the *Solonetzic order* have B horizons that are very hard when dry and swell to a sticky mass of very low permeability when wet. Typically, the Solonetzic B horizon has prismatic or columnar macrostructure that breaks into hard to extremely hard, blocky peds with dark coatings. Solonetzic soils occur on saline parent materials in some areas of the semiarid to subhumid interior plains in association with Chernozemic soils and to a lesser extent with Luvisolic and Gleysolic soils. Most Solonetzic soils are associated with a vegetative cover of grasses and forbs. The profile shown in Figure A1.1 is that of the Brown subgroup of the Solonetz great group.

Solonetzic soils are thought to have developed from parent materials that were more or less uniformly salinized with salts high in sodium. Leaching of salts by descending rainwater presumably mobilizes the sodium-saturated colloids. The colloids are apparently carried downward and deposited in the B horizon. Further leaching results in depletion of alkali cations in the A horizon, which becomes acidic, and a platy Ahe horizon usually develops. The underlying Solonetzic B horizon (Bn, Bnt) usually consists of darkly stained, fused, intact columnar peds. This stage is followed by the structural breakdown of the upper part of the B horizon and eventually its complete destruction in the most advanced stage, known as solodization. Solonetzic soils are correlated with the suborder of Argids in the order of Aridisols under the CSCS.

CLIMATE DEFINITIONS AND BOUNDARIES

The following table summarizes the definitions and boundaries of climates and climate subtypes based on the soil-water balance, as described in Chapter 15 and shown on the world climate map, Figure 7.6. All definitions and boundaries are provisional.

Ep   Water need (potential evapotranspiration)
D   Soil water shortage (deficit)
R   Water surplus (runoff)
S   Storage (limited to 30 cm)

Group I: Low-Latitude Climates

1. **Wet equatorial climate** ①
   Ep ≥ 10 cm in every month, and
   S ≥ 20 cm in 10 or more months.

2. **Monsoon and trade-wind coastal climate** ②
   Ep ≥ 4 cm in every month, or
   Ep > 130 cm annual total, or both, and
   S ≥ 20 cm in 6, 7, 8, or 9 consecutive months, or, if
   S > 20 cm in 10 or more months, then Ep ≤ 10 cm in 5 or more consecutive months.

3. **Wet-dry tropical climate** ③
   D ≥ 20cm, and
   R ≥ 10 cm, and
   Ep ≥ 130 cm annual total, or Ep ≥ 4 cm in every month, or both, and
   S ≥ 20 cm in 5 months or fewer, or minimum monthly S < 3 cm.

4. **Dry tropical climate** ④
   D ≥15 cm, and
   R = 0, and
   Ep ≥ 130 cm annual total, or Ep ≥ 4 cm in every month, or both.

Subtypes of dry climates (④, ⑤, ⑦, and ⑨)
   s Semiarid subtype (Steppe subtype) At least 1 month with S > 2 cm.
   a Desert subtype
     No month with S > 2 cm.

Group II: Midlatitude Climates

5. **Dry subtropical climate** ⑤
   D ≥ 15 cm, and
   R = 0, and Ep < 130 cm annual total, and
   Ep ≥ 0.8 cm in every month, and
   Ep < 4 cm in 1 month.
   (Subtypes ⑤a and ⑤s as defined under ④.)

6. **Moist subtropical climate** ⑥
   D < 15 cm when R = 0, and
   Ep < 4 cm in at least 1 month, and
   Ep ≥ 0.8 cm in every month.

7. **Mediterranean climate** ⑦
   D ≥ 15 cm, and
   R ≥ 0, and
   Ep ≥ 0.8 cm in every month, and storage index > 75%, or P/Ea × 100 < 40%. (Subtypes ⑦a and ⑦s as defined under ④.)

8. **Marine west-coast climate** ⑧
   D < 15 cm, and
   Ep < 80 cm annual total, and
   Ep ≥ 0.8 cm in every month.

9. **Dry midlatitude climate** ⑨
   D ≥ 15 cm, and
   R = 0, and
   Ep ≤ 0.7 cm in at least 1 month, and

Ep > 52.5 cm annual total.
(Subtypes ⑨a and ⑨s as defined under ④.)

10. **Moist continental climate** ⑩
D < 15 cm when R = 0, and
Ep ≤ 0.7 cm in at least 1 month, and
Ep > 52.5 cm annual total.

Group III: High-Latitude Climates

11. **Boreal forest climate** ⑪

52.5 cm > Ep > 35 cm annual total, and
Ep = 0 in fewer than 8 consecutive months.

12. **Tundra climate** ⑫
Ep < 35 cm annual total, and
Ep = 0 in 8 or more consecutive months.

13. **Ice sheet climate** ⑬
Ep = 0 in all months.

Appendix

3 | CONVERSION FACTORS

Metric to English

Metric Measure	Multiply by*	English Measure
LENGTH		
Millimeters (mm)	0.0394	Inches (in.)
Centimeters (cm)	0.394	Inches (in.)
Meters (m)	3.28	Feet (ft)
Kilometers (km)	0.621	Miles (mi)
AREA		
Square centimeters (cm^2)	0.155	Square inches (in^2)
Square meters (m^2)	10.8	Square feet (ft^2)
Square meters (m^2)	1.12	Square yards (yd^2)
Square kilometers (km^2)	0.386	Square miles (mi^2)
Hectares (ha)	2.47	Acres
VOLUME		
Cubic centimeters (cm^3)	0.0610	Cubic inches (in^3)
Cubic meters (m^3)	35.3	Cubic feet (ft^3)
Cubic meters (m^3)	1.31	Cubic yards (yd^3)
Milliliters (ml)	0.0338	Fluid ounces (fl oz)
Liters (l)	1.06	Quarts (qt)
Liters (l)	0.264	Gallons (gal)
MASS		
Grams (g)	0.0353	Ounces (oz)
Kilograms (kg)	2.20	Pounds (lb)
Kilograms (kg)	0.00110	Tons (2000 lb)
Tonnes (t)	1.10	Tons (2000 lb)

English to Metric

English Measure	Multiply by*	Metric Measure
LENGTH		
Inches (in.)	2.54	Centimeters (cm)
Feet (ft)	0.305	Meters (m)
Yards (yd)	0.914	Meters (m)
Miles (mi)	1.61	Kilometers (km)
AREA		
Square inches (in^2)	6.45	Square centimeters (cm^2)
Square feet (ft^2)	0.0929	Square meters (m^2)
Square yards (yd^2)	0.836	Square meters (m^2)
Square miles (mi^2)	2.59	Square kilometers (km^2)
Acres	0.405	Hectares (ha)
VOLUME		
Cubic inches (in^3)	16.4	Cubic centimeters (cm^3)
Cubic feet (ft^3)	0.0283	Cubic meters (m^3)
Cubic yards (yd^3)	0.765	Cubic meters (m^3)
Fluid ounces (fl oz)	29.6	Milliliters (ml)
Pints (pt)	0.473	Liters (l)
Quarts (qt)	0.946	Liters (l)
Gallons (gal)	3.79	Liters (l)
MASS		
Ounces (oz)	28.4	Grams (g)
Pounds (lb)	0.454	Kilograms (kg)
Tons (2000 lb)	907	Kilograms (kg)
Tons (2000 lb)	0.907	Tonnes (t)

* Conversion factors shown to 3 decimal-digit precision.

Appendix

4

TOPOGRAPHIC MAP SYMBOLS

VARIATIONS WILL BE FOUND ON OLDER MAPS

Hard surface, heavy duty road, four or more lanes . . . . . .

Hard surface, heavy duty road, two or three lanes . . . . . .

Hard surface, medium duty road, four or more lanes . . . .

Hard surface, medium duty road, two or three lanes . . . . .

Improved light duty road . . . .

Unimproved dirt road and trail . . . .

Dual highway, dividing strip 25 feet or less . . . .

Dual highway, dividing strip exceeding 25 feet . . . .

Road under construction . . . .

Railroad, single track and multiple track . . . .

Railroads in juxtaposition . . . .

Narrow gage, single track and multiple track . . . .

Railroad in street and carline . . . .

Bridge road and railroad . . . .

Drawbridge, road and railroad . . . .

Footbridge . . . .

Tunnel, road and railroad . . . .

Overpass and underpass . . . .

Important small masonry or earth dam . . . .

Dam with lock . . . .

Dam with road . . . .

Canal with lock . . . .

Buildings (dwelling, place of employment, etc.) . . . .

School, church, and cemetery . . . . Cem

Buildings (barn, warehouse, etc.) . . . .

Power transmission line . . . .

Telephone line, pipeline, etc. (labeled as to type) . . . . . .

Wells other than water (labeled as to type) . . . . o Oil   o Gas

Tanks; oil, water, etc. (labeled as to type) . . . . • ● ⊘ Water

Located or landmark object; windmill . . . . o   ⏚

Open pit, mine, or quarry; prospect . . . . ⤬   x

Shaft and tunnel entrance . . . . ▪   Y

Horizontal and vertical control station:

Tablet, spirit level elevation . . . . BM△ 5653

Other recoverable mark, spirit level elevation . . . . △ 5455

Horizontal control station: tablet, vertical angle elevation VABM△ 9519

Any recoverable mark, vertical angle or checked elevation △ 3775

Vertical control station: tablet, spirit level elevation . . . . . BM✕ 957

Other recoverable mark, spirit level elevation . . . . ✕ 954

Checked spot elevation . . . . ✕ 4675

Unchecked spot elevation and water elevation . . . . ✕ 5657 870

Boundary, national . . . .

State . . . .

County, parish, municipio . . . .

Civil township, precinct, town, barrio . . . .

Incorporated city, village, town, hamlet . . . .

Reservation, national or state . . . .

Small park, cemetery, airport, etc. . . . .

Land grant . . . .

Township or range line, United States land survey . . . . .

Township or range line, approximate location . . . .

Section line, United States land survey . . . .

Section line, approximate location . . . .

Township line, not United States land survey . . . .

Section line, not United States land survey . . . .

Section corner, found and indicated . . . . +   +

Boundary monument: land grant and other . . . . □   □

United States mineral or location monument . . . . ▲

Index contour	Intermediate contour . .
Supplementary contour	Depression contours . . .
Fill	Cut . . . .
Levee	Levee with road . . . . . .
Mine dump	Wash . . . .
Tailings	Tailings pond . . . .
Strip mine	Distorted surface . . . . .
Sand area	Gravel beach . . . .

Perennial streams	Intermittent streams . . .
Elevated aqueduct	Aqueduct tunnel . . . .
Water well and spring . o . . . . . o	Disappearing stream . . .
Small rapids	Small falls . . . .
Large rapids	Large falls . . . .
Intermittent lake	Dry lake . . . .
Foreshore flat	Rock or coral reef . . . .
Sounding, depth curve . .	Piling or dolphin . . . . .
Exposed wreck	Sunken wreck . . . .
Rock, bare or awash; dangerous to navigation	*	⊛

Marsh (swamp)	Submerged marsh . . . . .
Wooded marsh	Mangrove . . . .
Woods or brushwood . . .	Orchard . . . .
Vineyard	Scrub . . . .
Inundation area	Urban area . . . .

SCALE 1:24 000

CONTOUR INTERVAL 40 FEET
DATUM IS MEAN SEA LEVEL

SAN RAFAEL, CALIF.

NE/4 MT. TAMALPAIS 15' QUADRANGLE

N 3752.5—W 12230/7.5

QUADRANGLE LOCATION

GLOSSARY

This glossary contains definitions of terms shown in the text in italics or boldface. Terms that are *italicized* within the definitions will be found as individual entries elsewhere in the glossary.

A horizon   *mineral* horizon of the *soil,* overlying the *E* and *B horizons.*

ablation   a wastage of glacial ice by both *melting* and *evaporation.*

abrasion   erosion of *bedrock* of a *stream channel* by impact of particles carried in a stream and by rolling of larger *rock* fragments over the stream bed; abrasion is also an activity of glacial ice, waves, and *wind.*

abrasion platform   sloping, nearly flat *bedrock* surface extending out from the foot of a *marine cliff* under the shallow water of breaker zone.

absorption of radiation   transfer of *electromagnetic energy* into heat *energy* within a *gas* or *liquid* through which the radiation is passing or at the surface of a *solid* struck by the radiation.

abyssal hills   small hills rising to heights from a few tens of meters to a few hundred meters above the deep ocean floor.

abyssal plain   large expanse of very smooth, flat ocean floor found at depths of 4600 to 5500 m (15,000 to 18,000 ft).

accelerated erosion   *soil erosion* occurring at a rate much faster than *soil horizons* can be formed from the parent *regolith.*

accretion of lithosphere   production of new *oceanic lithosphere* at an active *spreading plate boundary* by the rise and solidification of *magma* of basaltic composition.

accretionary prism   mass of deformed trench *sediments* and ocean floor *sediments* accumulated in wedge-like slices on the underside of the overlying plate above a plate undergoing *subduction.*

acid action   solution of *minerals* by acids occurring in *soil* and *ground water.*

acid deposition   the *deposition* of acid raindrops and/or dry acidic dust particles on vegetation and ground surfaces.

acid mine drainage   sulfuric acid effluent from *coal* mines, mine tailings, or spoil ridges made by *strip mining.*

acid rain   rainwater having an abnormally low *pH,* between 2 and 5, as a result of air pollution by sulfur oxides and nitrogen oxides.

active continental margins   continental margins that coincide with tectonically active plate boundaries. (See also *continental margins, passive continental margins.*)

active layer   shallow surface layer subject to seasonal thawing in *permafrost* regions.

active pool   type of pool in the *biogeochemical cycle* in which the materials are in forms and places easily accessible to life processes. (See also *storage pool.*)

active systems   *remote sensing* systems that emit a beam of wave *energy* at a source and measure the intensity of that *energy* reflected back to the source.

actual evapotranspiration (water use)   Actual rate of *evapotranspiration* at a given time and place.

adiabatic lapse rate   (See *dry adiabatic lapse rate, wet adiabatic lapse rate.*)

adiabatic principle   the physical principle that a *gas* cools as it expands and warms as it is compressed, provided that no heat flows into or out of the gas during the process.

adiabatic process   change of temperature within a *gas* because of compression or expansion, without gain or loss of heat from the outside.

advection fog   *fog* produced by *condensation* within a moist basal air layer moving over a cold land or water surface.

aerosols   tiny particles present in the *atmosphere,* so small and light that the slightest movements of air keep them aloft.

aggradation   raising of *stream channel* altitude by continued *deposition* of *bed load.*

air   a mixture of *gases* that surrounds the Earth.

air mass extensive body of air within which upward gradients of temperature and moisture are fairly uniform over a large area.

air pollutant an unwanted substance injected into the *atmosphere* from the Earth's surface by either natural or human activities; includes *aerosols, gases,* and *particulates.*

air temperature temperature of air, normally observed by a *thermometer* under standard conditions of shelter and height above the ground.

albedo percentage of *electromagnetic energy* reflected from a surface.

albic horizon pale, often sandy *soil horizon* from which *clay* and free iron oxides have been removed. Found in the profile of the *Spodosols.*

Alfisols *soil order* consisting of *soils* of humid and subhumid climates, with high *base status* and an *argillic horizon.*

allele specific version of a particular gene.

allelopathy interaction among *species* in which a plant secretes substances into the soil that are toxic to other organisms.

allopatric speciation type of *speciation* in which populations are geographically isolated and gene flow between the populations does not take place.

alluvial fan gently sloping, conical accumulation of coarse *alluvium* deposited by a *braided stream* undergoing *aggradation* below the point of emergence of the channel from a narrow *gorge* or *canyon.*

alluvial meanders sinuous bends of a *graded stream* flowing in the alluvial deposit of a *floodplain.*

alluvial river stream of low *gradient* flowing upon thick deposits of *alluvium* and experiencing approximately annual overbank flooding of the adjacent *floodplain.*

alluvial terrace bench-like landform carved in *alluvium* by a *stream* during *degradation.*

alluvium any stream-laid *sediment* deposit found in a *stream channel* and in low parts of a stream valley subject to flooding.

alpine chains high mountain ranges that are narrow belts of *tectonic activity* severely deformed by *folding* and thrusting in comparatively recent geologic time.

alpine debris avalanche *debris flood* of steep mountain *slopes,* often laden with tree trunks, limbs, and large boulders.

alpine glacier long, narrow, mountain *glacier* on a steep downgrade, occupying the floor of a trough-like valley.

alpine permafrost *permafrost* occurring at high altitudes equatorward of the normal limit of *permafrost.*

alpine tundra a plant *formation class* within the *tundra biome,* found at high altitudes above the limit of *tree* growth.

amphibole group silicate minerals rich in calcium, magnesium, and iron, dark in color, high in *density,* and classed as *mafic minerals.*

amplitude for a smooth wave-like curve, the difference in height between a crest and the adjacent trough.

andesite *extrusive igneous rock* of diorite composition, dominated by *plagioclase feldspar;* the extrusive equivalent of *diorite.*

Andisols a *soil order* that includes *soils* formed on volcanic ash; often enriched by organic matter, yielding a dark soil color.

anemometer weather instrument used to indicate *wind* speed.

aneroid barometer *barometer* using a mechanism consisting of a partially evacuated air chamber and a flexible diaphragm.

angle of repose natural surface inclination *(dip)* of a *slope* consisting of loose, coarse, well-sorted *rock* or *mineral fragments;* for example, the *slip face* of a *sand dune,* a *talus slope,* or the sides of a *cinder cone.*

annuals plants that live only a single growing season, passing the unfavorable season as a seed or spore.

annular drainage pattern a stream network dominated by concentric (ring-like) major *subsequent streams.*

antarctic circle *parallel of latitude* at 66 $\frac{1}{2}$°S.

antarctic front zone frontal zone of interaction between antarctic *air masses* and polar air masses.

antarctic zone *latitude* zone in the latitude range 60° to 75°S (more or less), centered on the *antarctic circle,* and lying between the *subantarctic zone* and the *polar zone.*

anticlinal valley valley eroded in weak *strata* along the central line or axis of an eroded *anticline.*

anticline upfold of *strata* or other layered *rock* in an arch-like structure; a class of *folds.* (See also *syncline.*)

anticyclone center of high *atmospheric pressure.*

aphelion point on the Earth's elliptical orbit at which the Earth is farthest from the Sun.

aquatic ecosystem *ecosystem* of a *lake, bog,* pond, river, *estuary,* or other body of water.

Aquepts *suborder* of the *soil order Inceptisols;* includes Inceptisols of wet places, seasonally saturated with water.

aquiclude *rock* mass or layer that impedes or prevents the movement of *ground water.*

aquifer *rock* mass or layer that readily transmits and holds *ground water.*

arc curved line that forms a portion of a circle.

arc-continent collision collision of a volcanic arc with *continental lithosphere* along a *subduction* boundary.

arctic circle *parallel of latitude* at 66 $\frac{1}{2}$°N.

arctic front zone frontal zone of interaction between arctic *air masses* and polar air masses.

arctic tundra a plant *formation class* within the *tundra biome,* consisting of low, mostly herbaceous plants, but with some very small stunted *trees,* associated with the *tundra climate* ⑫.

arctic zone *latitude* zone in the latitude range 60° to 75°N (more or less), centered about on the *arctic circle,* and lying between the *subarctic zone* and the *polar zone.*

arête sharp, knife-like divide or crest formed between two *cirques* by alpine glaciation.

argillic horizon *soil horizon,* usually the B *horizon,* in which *clay minerals* have accumulated by *illuviation.*

arid (dry climate subtype) subtype of the dry climates that is extremely dry and supports little or no vegetation cover.

Aridisols *soil order* consisting of soils of dry climates, with or without *argillic horizons,* and with accumulations of *carbonates* or soluble salts.

artesian well   drilled well in which water rises under hydraulic pressure above the level of the surrounding *water table* and may reach the surface.

aspect   compass orientation of a *slope* as an inclined element of the ground surface.

association   plant-animal community type identified by the typical organisms that are likely to be found together.

asthenosphere   soft layer of the upper *mantle*, beneath the rigid *lithosphere*.

astronomical hypothesis   explanation for glaciations and interglaciations making use of cyclic variations in the form of solar *energy* received at the Earth's surface.

atmosphere   envelope of gases surrounding the Earth, held by *gravity*.

atmospheric pressure   pressure exerted by the atmosphere because of the force of *gravity* acting upon the overlying column of air.

atoll   circular or closed-loop *coral reef* enclosing an open *lagoon* with no island inside.

atomic mass number   total number of *protons* and *neutrons* within the nucleus of an atom.

atomic number   number of *protons* within the nucleus of an atom; determines element name and chemical properties of the atom.

autogenic succession   form of *ecological succession* that is self-producing—that is, results from the actions of plants and animals themselves.

autumnal equinox   *equinox* occurring on September 22 or 23.

average deviation   difference between a single value and the mean of all values, taken without respect to sign.

axial rift   narrow, trench-like depression situated along the center line of the *mid-oceanic ridge* and identified with active seafloor spreading.

axis of rotation   center line around which a body revolves, as the Earth's axis of rotation.

B horizon   mineral *soil horizon* located beneath the A *horizon*, and usually characterized by a gain of *mineral matter* (such as *clay minerals* and oxides of aluminum and iron) and organic matter *(humus)*.

backswamp   area of low, swampy ground on the *floodplain* of an *alluvial river* between the *natural levee* and the *bluffs*.

backwash   return flow of *swash* water under influence of gravity.

badlands   rugged land surface of steep *slopes*, resembling miniature mountains, developed on weak *clay* formations or clay-rich *regolith* by fluvial erosion too rapid to permit plant growth and soil formation.

bar   low ridge of *sand* built above water level across the mouth of a *bay* or in shallow water paralleling the shoreline. May also refer to embankment of sand or gravel on floor of a *stream channel*.

bar (pressure)   unit of pressure equal to 10^5 Pa *(pascals)*; approximately equal to the pressure of the Earth's *atmosphere* at sea level.

barchan dune   *sand dune* of crescentic base outline with a sharp crest and a steep lee *slip face*, with crescent points (horns) pointing downwind.

barometer   instrument for measurement of *atmospheric pressure*.

barrier (to dispersal)   a zone or region that a *species* is unable to colonize or perhaps even occupy for a short time, thus halting *diffusion*.

barrier island   long narrow island, built largely of beach *sand* and dune sand, parallel with the mainland and separated from it by a *lagoon*.

barrier reef   *coral reef* separated from mainland *shoreline* by a *lagoon*.

barrier-island coast   *coastline* with broad zone of shallow water offshore (a *lagoon*) shut off from the ocean by a *barrier island*.

basalt   *extrusive igneous rock* of *gabbro* composition; occurs as *lava*.

base flow   that portion of the *discharge* of a *stream* contributed by *ground water* seepage.

base level   lower limiting surface or level that can ultimately be attained by a *stream* under conditions of stability of the Earth's crust and sea level; an imaginary surface equivalent to sea level projected inland.

base status of soils   quality of a *soil* as measured by the presence or absence of *clay minerals* capable of holding large numbers of *bases*. Soils of high *base status* are rich in base-holding *clay*

minerals; soils of low *base status* are deficient in such minerals.

bases   certain positively charged *ions* in the *soil* that are also plant nutrients; the most important are calcium, magnesium, potassium, and sodium.

batholith   large, deep-seated body of *intrusive igneous rock*, usually with an area of surface exposure greater than 100 km² (40 mi²).

bauxite   mixture of several *clay minerals*, consisting largely of aluminum oxide and water with impurities; a principal ore of aluminum.

bay   a body of water sheltered from strong wave action by the configuration of the *coast*.

beach   thick, wedge-shaped accumulation of *sand*, gravel, or cobbles in the zone of breaking waves.

beach drift   transport of *sand* on a beach parallel with a *shoreline* by a succession of landward and seaward water movements at times when *swash* approaches obliquely.

bearing   direction angle between a line of interest and a reference line, which is usually a line pointing north.

bed load   that portion of the *stream load* moving close to the stream bed by rolling and sliding.

bedrock   solid *rock* in place with respect to the surrounding and underlying *rock* and relatively unchanged by *weathering* processes.

bedrock slump   landslide of *bedrock* in which most of the *bedrock* remains more or less intact as it moves.

bioclimatic frontier   geographic boundary corresponding with a critical limiting level of climate stress beyond which a *species* cannot survive.

biodiversity   the variety of biological life on Earth or within a region.

biogas   mixture of methane and *carbon dioxide* generated by action of anaerobic bacteria in animal and human wastes enclosed in a digesting chamber.

biogeochemical cycle   total system of *pathways* by which a particular type of *matter* (a given element, compound, or ion, for example) moves through the Earth's *ecosystem* or *biosphere;* also called a *material cycle* or *nutrient cycle*.

biogeographic region   region in which the same or closely related plants and animals tend to be found together.

biogeography the study of the distributions of organisms at varying spatial and temporal *scales,* as well as the processes that produce these distribution patterns.

biomass dry weight of living organic matter in an *ecosystem* within a designated surface area; units are kilograms of organic matter per square meter.

biome largest recognizable subdivision of *terrestrial ecosystems,* including the total assemblage of plant and animal life interacting within the *life layer.*

biosphere all living organisms of the Earth and the environments with which they interact.

biota plants and animals, referred to collectively; a list of plants and animals found at a location or in a region.

bitumen combustible mixture of hydrocarbons that is highly viscous and will flow only when heated; considered a form of petroleum.

bituminous sand (See *bitumen.*)

blackbody ideal object or surface that is a perfect radiator and absorber of *energy;* absorbs all radiation it intercepts and emits radiation perfectly according to physical theory.

block mountains class of mountains produced by block faulting and usually bounded by *normal faults.*

block separation separation of individual joint blocks during the process of *physical weathering.*

blowout shallow depression produced by continued *deflation.*

bluffs steeply rising ground slopes marking the outer limits of a *floodplain.*

bog a shallow depression filled with organic matter, for example, a glacial lake or pond basin filled with *peat.*

Boralfs *suborder* of the *soil order Alfisols;* includes Alfisols of *boreal forests* or high mountains.

boreal forest variety of *needleleaf forest* found in the *boreal forest climate* ⑪ regions of North America and Eurasia.

boreal forest climate ⑪ cold climate of the *subarctic zone* in the northern *hemisphere* with long, extremely severe winters and several consecutive months of zero *potential evapotranspiration (water need).*

Borolls *suborder* of the *soil order Mollisols;* includes Mollisols of cold-winter semiarid plants *(steppes)* or high mountains.

braided stream *stream* with shallow channel in coarse *alluvium* carrying multiple threads of fast flow that subdivide and rejoin repeatedly and continually shift in position.

breaker sudden collapse of a steepened water wave as it approaches the shoreline.

broadleaf deciduous forest *forest* type consisting of broadleaf *deciduous trees* and found in the *moist subtropical climate* ⑥ in parts of the *marine westcoast climate* ⑧. (See also *midlatitude deciduous forest.*)

broadleaf evergreen forest *forest* type consisting of broadleaf *evergreen trees* and found in the wet equatorial and tropical climates. (See also *low-latitude rainforest.*)

Brunisolic order a class of *forest soils* in the Canadian soil classification system with brownish *B horizon.*

budget in flow systems, an accounting of *energy* and *matter* flows that enter, move within, and leave a system.

bush-fallow farming agricultural system practiced in the African *savanna woodland* in which *trees* are cut and burned to provide cultivation plots.

butte prominent, steep-sided hill or peak, often representing the final remnant of a resistant layer in a region of flat-lying *strata.*

C horizon *soil horizon* lying beneath the *B horizon,* consisting of *sediment* or *regolith* that is the *parent material* of the soil.

calcification accumulation of *calcium carbonate* in a soil, usually occurring in the *B* or *C horizons.*

calcite mineral having the composition *calcium carbonate.*

calcium carbonate compound consisting of calcium (Ca) and carbonate (CO_3) *ions,* formula $CaCo_3$, occurring naturally as the mineral *calcite.*

caldera large, steep-sided circular depression resulting from the explosion and subsidence of a *stratovolcano.*

canyon (See *gorge.*)

capillary action process by which *capillary tension* draws water into a small opening, such as a *soil* pore or a *rock joint.*

capillary tension a cohesive force among surface molecules of a *liquid* that gives a droplet its rounded shape.

carbohydrate class of organic compounds consisting of the elements carbon, hydrogen, and oxygen.

carbon cycle *biogeochemical cycle* in which carbon moves through the *biosphere;* includes both *gaseous cycles* and *sedimentary cycles.*

carbon dioxide the chemical compound CO_2, formed by the union of two atoms of oxygen and one atom of carbon; normally a gas present in low concentration in the *atmosphere.*

carbon fixation (See *photosynthesis.*)

carbonates (carbonate minerals, carbonate rocks) *minerals* that are carbonate compounds of calcium or magnesium or both, i.e., *calcium carbonate* or magnesium carbonate. (See also *calcite.*)

carbonic acid a weak acid created when CO_2 gas dissolves in water.

carbonic acid action chemical reaction of *carbonic acid* in rainwater, *soil water,* and *ground water* with *minerals;* most strongly affects carbonate minerals and *rocks,* such as limestone and marble; an activity of *chemical weathering.*

cartography the science and art of making maps.

Celsius scale temperature scale in which the *freezing* point of water is 0° and the boiling point is 100°.

Cenozoic Era last (youngest) of the *eras* of geologic time.

channel (See *stream channel.*)

chaparral sclerophyll scrub and dwarf *forest* plant *formation class* found throughout the coastal mountain ranges and hills of central and southern California.

chemical energy *energy* stored within an organic molecule and capable of being transformed into *heat* during metabolism.

chemical weathering chemical change in *rock-forming* minerals through exposure to atmospheric conditions in the presence of water; mainly involving *oxidation, hydrolysis, carbonic acid action,* or direct solution.

chemically precipitated sediment *sediment* consisting of *mineral matter* precipitated from a water solution in which the matter has been transported in the dissolved state as *ions.*

chernozem type of *soil order* closely equivalent to *Mollisol;* an order of the Canadian Soil Classification System.

Chernozemic order a class of grassland *soils* in the Canadian soil classification system with a thick *A horizon* rich in organic matter.

chert *sedimentary rock* composed largely of silicon dioxide and various impurities, in the form of nodules and layers, often occurring with *limestone* layers.

chinook wind a *local wind* occurring at certain times to the lee of the Rocky Mountains; a very dry *wind* with a high capacity to evaporate *snow.*

chlorofluorocarbons (CFCs) synthetic chemical compounds containing chlorine, fluorine, and carbon atoms that are widely used as coolant fluids in refrigeration systems.

cinder cone conical hill built of coarse *tephra* ejected from a narrow volcanic vent; a type of *volcano.*

circle of illumination great circle that divides the globe at all times into a sunlit *hemisphere* and a shadowed hemisphere.

circum-Pacific belt chains of andesite *volcanoes* making up mountain belts and *island arcs* surrounding the Pacific Ocean basin.

cirque bowl-shaped depression carved in *rock* by glacial processes and holding the *firn* of the upper end of an *alpine glacier.*

clast rock or mineral fragment broken from a parent *rock* source.

clastic sediment *sediment* consisting of particles broken away physically from a parent *rock* source.

clay *sediment* particles smaller than 0.004 mm in diameter.

clay minerals class of *minerals* produced by alteration of *silicate minerals,* having plastic properties when moist.

claystone *sedimentary rock* formed by lithification of *clay* and lacking *fissile* structure.

cliff sheer, near-vertical *rock* wall formed from flat-lying resistant layered *rocks,* usually *sandstone, limestone,* or *lava* flows; may refer to any near-vertical rock wall. (See also *marine cliff.*)

climate generalized statement of the prevailing weather conditions at a given place, based on statistics of a long period of record and including mean values, departures from those means,

and the probabilities associated with those departures.

climatic frontier a geographical boundary that marks the limit of survival of a plant *species* subjected to climatic stress.

climatology the science that describes and explains the variability in space and time of the heat and moisture states of the Earth's surface, especially its land surfaces.

climax stable community of plants and animals reached at the end point of *ecological succession.*

climograph a graph on which two or more climatic variables, such as monthly mean temperature and monthly mean precipitation, are plotted for each month of the year.

closed flow system flow system that is completely self-contained within a boundary through which no *matter* or *energy* is exchanged with the external environment. (See also *open flow system.*)

cloud forest a type of low evergreen rainforest that occurs high on mountain slopes, where *clouds* and *fog* are frequent.

clouds dense concentrations of suspended water or ice particles in the diameter range 20 to 50 μm. (See also *cumuliform clouds, stratiform clouds.*)

coal *rock* consisting of hydrocarbon compounds, formed of compacted, lithified, and altered accumulations of plant remains *(peat).*

coarse textured (rock) having *mineral* crystals sufficiently large that they are at least visible to the naked eye or with low magnification.

coast (See *coastline.*)

coastal and marine geography the study of the geomorphic processes that shape shores and coastlines and their application to coastal development and marine resource utilization.

coastal blowout dune high *sand dune* of the *parabolic dunes* class formed adjacent to a beach, usually with a deep *deflation* hollow (blowout) enclosed within the dune ridge.

coastal forest subtype of *needleleaf evergreen forest* found in the humid coastal zone of the northwestern United States and western Canada.

coastal plain coastal belt, emerged from beneath the sea as a former *conti-*

nental shelf, underlain by *strata* with gentle *dip* seaward.

coastline (coast) zone in which coastal processes operate or have a strong influence.

coefficient of variation in statistics, the ratio of the standard deviation to the mean.

col natural pass or low notch in an *arête* between opposed *cirques.*

cold front moving weather *front* along which a cold *air mass* moves underneath a warm air mass, causing the latter to be lifted.

cold-blooded animal animal whose body temperature passively follows the temperature of the environment.

cold-core ring circular eddy of cold water, surrounded by warm water and lying adjacent to a warm, polewardmoving *ocean current,* such as the Gulf Stream. (See also *warm-core ring.*)

colloids particles of extremely small size, capable of remaining indefinitely in suspension in water. May be mineral or organic in nature.

colluvium deposit of *sediment* or *rock* particles accumulating from overland flow at the base of a *slope* and originating from higher slopes where *sheet erosion* is in progress. (See also *alluvium.*)

community an assemblage of organisms that live in a particular *habitat* and interact with one another.

competition form of interaction among plant or animal *species* in which both draw resources from the same *pool.*

component in flow systems, a part of the system, such as a *pathway,* connection, or flow of *matter* or *energy.*

composite volcano volcano composed of layers of ash and lava. (See also *stratovolcano.*)

compression (tectonic) squeezing together, as horizontal compression of crustal layers by *tectonic* processes.

condensation process of change of *matter* in the gaseous state *(water vapor)* to the liquid state (liquid water) or solid state (ice).

condensation nucleus a tiny bit of solid *matter (aerosol)* in the *atmosphere* on which *water vapor* condenses to form a tiny water droplet.

conduction of heat transmission of *sensible heat* through *matter* by transfer of *energy* from one atom or molecule to the next in the direction of decreasing temperature.

cone of depression   conical configuration of the lowered *water table* around a well from which water is being rapidly withdrawn.

conformal projection   *map* projection that preserves without shearing the true shape or outline of any small surface feature of the Earth.

conglomerate   a *sedimentary rock* composed of pebbles in a matrix of finer *rock* particles.

conic projections   a group of *map projections* in which the *geographic grid* is transformed to lie on the surface of a developed cone.

consequent stream   *stream* that takes its course down the slope of an *initial landform,* such as a newly emerged *coastal plain* or a *volcano.*

consumers   animals in the *food chain* that live on organic matter formed by *primary producers* or by other *consumers.* (See also *primary consumers, secondary consumers.*)

consumption (of a lithospheric plate)   destruction or disappearance of a subducting *lithospheric plate* in the *asthenosphere,* in part by *melting* of the upper surface, but largely by softening because of heating to the temperature of the surrounding *mantle rock.*

continental collision   event in *plate tectonics* in which subduction brings two segments of the *continental lithosphere* into contact, leading to formation of a *continental suture.*

continental crust   crust of the continents, of felsic composition in the upper part; thicker and less dense than *oceanic crust.*

continental drift   hypothesis, introduced by Alfred Wegener and others early in the 1900s, of the breakup of a parent continent, *Pangea,* starting near the close of the *Mesozoic Era,* and resulting in the present arrangement of *continental shields* and intervening *ocean-basin floors.*

continental lithosphere   *lithosphere* bearing *continental crust* of *felsic igneous rock.*

continental margins   (1) Topographic: one of three major divisions of the ocean basins, being the zones directly adjacent to the continent and including the *continental shelf, continental slope,* and *continental rise.* (2) Tectonic: marginal belt of continental crust and lithosphere that is in contact with *oceanic crust* and *lithosphere,* with or without an active plate boundary being present at the contact. (See also *active continental margins, passive continental margins.*)

continental rise   gently sloping seafloor lying at the foot of the *continental slope* and leading gradually into the *abyssal plain.*

continental rupture   crustal spreading apart affecting the *continental lithosphere,* so as to cause a *rift valley* to appear and to widen, eventually creating a new belt of *oceanic lithosphere.*

continental scale   scale of observation at which we recognize continents and other large Earth surface features, such as ocean currents.

continental shelf   shallow, gently sloping belt of seafloor adjacent to the continental shoreline and terminating at its outer edge in the *continental slope.*

continental shields   ancient crustal *rock* masses of the continents, largely *igneous rock* and *metamorphic rock,* and mostly of *Precambrian age.*

continental slope   steeply descending belt of seafloor between the *continental shelf* and the *continental rise.*

continental suture   long, narrow zone of crustal deformation, including underthrusting and intense *folding,* produced by a *continental collision.* Examples: Himalayan Range, European Alps.

continuous permafrost   *permafrost* that extends without gaps or interruptions under all surface features.

convection (atmospheric)   air motion consisting of strong updrafts taking place within a *convection cell.*

convection cell   individual column of strong updrafts produced by atmospheric *convection.*

convection current   in *plate tectonics,* a stream of upwelling *mantle* rock that rises steadily beneath a spreading plate boundary.

convection loop   circuit of moving *fluid,* such as *air* or water, created by unequal heating of the *fluid.*

convectional precipitation   a form of *precipitation* induced when warm, moist air is heated at the ground surface, rises, cools, and condenses to form water droplets, raindrops, and eventually, rainfall.

converging boundary   boundary between two crustal plates along which *subduction* is occurring and *lithosphere* is being consumed.

coral reef   rock-like accumulation of *carbonates* secreted by corals and algae in shallow water along a marine shoreline.

coral-reef coast   *coast* built out by accumulations of *limestone* in *coral reefs.*

core of Earth   spherical central mass of the Earth composed largely of iron and consisting of an outer liquid zone and an interior solid zone.

Coriolis effect   effect of the Earth's rotation tending to turn the direction of motion of any object or *fluid* toward the right in the northern *hemisphere* and to the left in the southern *hemisphere.*

corrosion   erosion of *bedrock* of a *stream channel* (or other *rock* surface) by chemical reactions between solutions in stream water and *mineral* surfaces.

cosmopolitan species   *species* that are found very widely.

counterradiation   *longwave radiation* of atmosphere directed downward to the Earth's surface.

covered shields   areas of *continental shields* in which the ancient *rocks* are covered beneath a thin layer of sedimentary *strata.*

crater   central summit depression associated with the principal vent of a *volcano.*

crescentic dune   (See *barchan dunes.*)

crevasse   gaping crack in the brittle surface ice of a *glacier.*

crude oil   liquid fraction of *petroleum.*

crust of Earth   outermost solid shell or layer of the Earth, composed largely of *silicate minerals.*

Cryaquepts   great group within the soil *suborder* of *Aquepts;* includes Aquepts of cold climate regions and particularly the *tundra climate* ⑫.

Cryosolic order   a class of *soils* in the Canadian soil classification system associated with strong frost action and underlying *permafrost.*

cuesta   *erosional landform* developed on resistant *strata* having low to moderate *dip* and taking the form of an asymmetrical low ridge or hill belt with one side a steep slope and the other a gentle slope; usually associated with a *coastal plain.*

cultural energy *energy* in forms exclusive of solar *energy* of *photosynthesis* that is expended on the production of raw food or feed crops in agricultural *ecosystems.*

cumuliform clouds *clouds* of globular shape, often with extended vertical development.

cumulonimbus cloud large, dense *cumuliform cloud* yielding *precipitation.*

cumulus cloud type consisting of low-lying, white cloud masses of globular shape well separated from one another.

cutoff cutting-through of a narrow neck of land, so as to bypass the *stream flow* in an *alluvial meander* and cause it to be abandoned.

cycle in flow systems, a closed flow system of *matter.* Example: *biogeochemical cycle.* (See also *closed flow system.*)

cycle of rock change total cycle of changes in which *rock* of any one of the three major *rock* classes—*igneous rock, sedimentary rock, metamorphic rock*—is transformed into *rock* of one of the other classes.

cyclone center of low *atmospheric pressure.* (See also *tropical cyclone, wave cyclone.*)

cyclonic precipitation a form of *precipitation* that occurs as warm moist air is lifted by air motion occurring in a *cyclone.*

cyclonic storm intense weather disturbance within a moving *cyclone* generating strong winds, cloudiness, and *precipitation.*

cylindric projections group of *map projections* in which the *geographic grid* is transformed to lie on the surface of a developed cylinder.

data acquisition component component of a *geographic information system* in which data are gathered together for input to the system.

data management component component of a *geographic information system* that creates, stores, retrieves, and modifies data layers and *spatial objects.*

daughter product new *isotope* created by decay of an *unstable isotope.*

daylight saving time time system under which time is advanced by one hour with respect to the *standard time* of the prevailing *standard meridian.*

debris flood (debris flow) stream-like flow of muddy water heavily charged with *sediment* of a wide range of size grades, including boulders, generated by sporadic torrential rains upon steep mountain watersheds.

decalcification removal of *calcium carbonate* from a *soil horizon* as *carbonic acid* reacts with *carbonate mineral matter.*

December solstice (See *winter solstice.*)

deciduous plant *tree* or *shrub* that sheds its leaves seasonally.

declination of Sun latitude at which the Sun is directly overhead; varies from $-23^1/_2°$ ($23^1/_2°$ S lat.) to $+23^1/_2°$ N lat.

décollement detachment and extensive sliding of a *rock* layer, usually *sedimentary,* over a near-horizontal basal *rock* surface; a special form of low-angle thrust *faulting.*

decomposers organisms that feed on dead organisms from all levels of the *food chain;* most are microorganisms and bacteria that feed on decaying organic matter.

deep sea cone a fan-shaped accumulation of undersea *sediment* on the *continental rise* produced by sediment-rich currents flowing down the *continental slope.*

deficit (soil water shortage) in the soil water budget, the difference between *water use* and *water need;* the quantity of irrigation water required to achieve maximum growth of agricultural crops.

deflation lifting and transport in *turbulent suspension* by *wind* of loose particles of *soil* or *regolith* from dry ground surfaces.

deglaciation widespread recession of *ice sheets* during a period of warming global climate, leading to an interglaciation. (See also *glaciation, interglaciation.*)

degradation lowering or downcutting of a *stream channel* by *stream erosion* in *alluvium* or *bedrock.*

degree of arc measurement of the angle associated with an *arc,* in degrees.

delta *sediment* deposit built by a stream entering a body of standing water and formed of the *stream load.*

delta coast *coast* bordered by a *delta.*

delta kame flat-topped hill of *stratified drift* representing a glacial *delta* constructed adjacent to an *ice sheet* in a marginal glacial lake.

dendritic drainage pattern *drainage pattern* of tree-like branched form, in which the smaller streams take a wide variety of directions and show no parallelism or dominant trend.

denitrification biochemical process in which nitrogen in forms usable to plants is converted into molecular nitrogen in the gaseous form and returned to the atmosphere—a process that is part of the *nitrogen cycle.*

density of matter quantity of mass per unit of volume, stated in kg m^{-2}.

denudation total action of all processes whereby the exposed *rocks* of the continents are worn down and the resulting *sediments* are transported to the sea by the *fluid agents;* includes also *weathering* and *mass wasting.*

deposition (atmosphere) the change of state of a substance from a *gas (water vapor)* to a *solid* (ice); in the science of *meteorology,* the term *sublimation* is used to describe both this process and the change of state from solid to vapor. (See also *sublimation.*)

deposition (of sediment) (See *stream deposition.*)

depositional landform *landform* made by *deposition* of *sediment.*

desert biome *biome* of the dry climates consisting of thinly dispersed plants that may be *shrubs,* grasses, or perennial *herbs,* but lacking in *trees.*

desert pavement surface layer of closely fitted pebbles or coarse *sand* from which finer particles have been removed.

desertification (See *land degradation.*)

detritus decaying organic matter on which *decomposers* feed.

dew-point lapse rate rate at which the dew point of an air mass decreases with elevation; typical value is 1.8°C/1000 m (1.0°F/1000 ft).

dew-point temperature temperature of an *air mass* at which the air holds its full capacity of water vapor.

diagnostic horizons *soil horizons,* rigorously defined, that are used as diagnostic criteria in classifying *soils.*

diffuse radiation solar radiation that has been *scattered* (deflected or reflected) by minute dust particles or cloud particles in the *atmosphere.*

diffuse reflection solar *radiation* scattered back to space by the Earth's *atmosphere.*

diffusion the slow extension of the range of a *species* by normal processes of dispersal.

digital image numeric representation of a picture consisting of a collection of numeric brightness values (pixels) arrayed in a fine grid pattern.

dike thin layer of *intrusive igneous rock,* often near-vertical or with steep *dip,* occupying a widened fracture in the surrounding *rock* and typically cutting across older *rock* planes.

diorite *intrusive igneous rock* consisting dominantly of plagioclase feldspar and pyroxene; a *felsic igneous rock.*

dip acute angle between an inclined natural *rock* plane or surface and an imaginary horizontal plane of reference; always measured perpendicular to the *strike.* Also a verb, meaning to incline toward.

diploid having two sets of chromosomes, one from each parent organism.

discharge volume of flow moving through a given cross section of a stream·in a given unit of time; commonly given in cubic meters (feet) per second.

discontinuous permafrost *permafrost* that occurs in patches separated by frost-free zones under lakes and rivers.

disjunction geographic distribution pattern of *species* in which one or more closely related species are found in widely separated regions.

dispersal the capacity of a *species* to move from a location of birth or origin to new sites.

distributary branching *stream channel* that crosses a *delta* to discharge into open water.

diurnal adjective meaning "daily."

doldrums belt of calms and variable winds occurring at times along the *equatorial trough.*

dolomite carbonate mineral or *sedimentary rock* having the composition calcium magnesium carbonate.

dome (See *sedimentary dome.*)

drainage basin total land surface occupied by a *drainage system,* bounded by a *drainage divide* or watershed.

drainage divide imaginary line following a crest of high land such that overland flow on opposite sides of the line enters different *streams.*

drainage pattern the plan of a network of interconnected *stream channels.*

drainage system a branched network of *stream channels* and adjacent land *slopes,* bounded by a *drainage divide* and converging to a single channel at the outlet.

drainage winds *winds,* usually cold, that flow from higher to lower regions under the direct influence of *gravity.*

drawdown (of a well) difference in height between base of cone of depression and original water table surface.

drought occurrence of substantially lower-than-average *precipitation* in a season that normally has ample precipitation for the support of food-producing plants.

drumlin hill of glacial *till,* oval or elliptical in basal outline and with smoothly rounded summit, formed by plastering of till beneath moving, debris-laden glacial ice.

dry adiabatic lapse rate rate at which rising air is cooled by expansion when no *condensation* is occurring; 10°C per 1000 m (5.5°F per 1000 ft).

dry desert plant *formation class* in the *desert biome* consisting of widely dispersed xerophytic plants that may be small, hard-leaved or spiny *shrubs,* succulent plants (cacti), or hard grasses.

dry midlatitude climate ⑨ dry climate of the *midlatitude zone* with a strong annual cycle of *potential evapotranspiration (water need)* and cold winters.

dry subtropical climate ⑤ dry climate of the *subtropical zone,* transitional between the *dry tropical climate* ④ and the *dry midlatitude climate* ⑨.

dry tropical climate ④ climate of the *tropical zone* with large total annual *potential evapotranspiration (water need).*

dune (See *sand dune.*)

dust bowl western Great Plains of the United States, which suffered severe wind deflation and soil drifting during the drought years of the middle 1930s.

dust storm heavy concentration of dust in a turbulent *air mass,* often associated with a *cold front.*

E horizon soil mineral horizon lying below the *A horizon* and characterized by the loss of *clay minerals* and oxides of iron and aluminum; it may show a concentration of *quartz* grains and is often pale in color.

Earth's crust (See *crust of Earth.*)

earthflow moderately rapid downhill flowage of masses of water-saturated *soil, regolith,* or weak *shale,* typically forming a step-like terrace at the top and a bulging toe at the base.

earthquake a trembling or shaking of the ground produced by the passage of *seismic waves.*

earthquake focus point within the Earth at which the *energy* of an *earthquake* is first released by rupture and from which *seismic waves* emanate.

easterly wave weak, slowly moving trough of low pressure within the belt of *tropical easterlies;* causes a weather disturbance with rain showers.

ebb current oceanward flow of *tidal current* in a *bay* or tidal stream.

ecological succession time-succession (sequence) of distinctive plant and animal communities occurring within a given area of newly formed land or land cleared of plant cover by burning, clear cutting, or other agents.

ecology science of interactions between life forms and their environment; the science of *ecosystems.*

ecosystem group of organisms and the environment with which the organisms interact.

edaphic factors factors relating to soil that influence a terrestrial ecosystem.

El Niño episodic cessation of the typical *upwelling* of cold deep water off the coast of Peru; literally, "The Christ Child," for its occurrence in the Christmas season once every few years.

electromagnetic radiation (electromagnetic energy) wave-like form of *energy* radiated by any substance possessing heat; it travels through space at the speed of light.

electromagnetic spectrum the total *wavelength* range of *electromagnetic energy.*

eluviation soil-forming process consisting of the downward transport of fine particles, particularly the *soil colloids* (both mineral and organic), carrying them out of an upper *soil horizon.*

emergence exposure of submarine landforms by a lowering of sea level or a rise of the crust, or both.

endemic species a *species* found only in one region or location.

energy the capacity to do work, that is, to bring about a change in the state or motion of *matter*.

energy balance (global) balance between *shortwave* solar *radiation* received by the Earth-atmosphere system and *radiation* lost to space by *shortwave* reflection and *longwave radiation* from the Earth-atmosphere system.

energy balance (of a surface) balance between the flows of *energy* reaching a surface and the flows of energy leaving it.

energy flow system *open system* that receives an input of *energy*, undergoes internal energy flow, energy transformation, and energy storage, and has an energy output.

Entisols *soil order* consisting of mineral soils lacking *soil horizons* that would persist after normal plowing.

entrenched meanders winding, sinuous valley produced by *degradation* of a *stream* with trenching into the *bedrock* by downcutting.

environmental temperature lapse rate rate of temperature decrease upward through the *troposphere;* standard value is 6.4 C°/km (3 F°/1000 ft).

epipedon *soil horizon* that forms at the surface.

epiphytes plants that live above ground level out of contact with the soil, usually growing on the limbs of *trees* or *shrubs;* also called "air plants."

epoch a subdivision of geologic time.

equal-area projections class of *map projections* on which any given area of the Earth's surface is shown to correct relative areal extent, regardless of position on the globe.

equator *parallel of latitude* occupying a position midway between the Earth's poles of *rotation;* the largest of the parallels, designated as *latitude* 0°.

equatorial current westward-flowing *ocean current* in the belt of the *trade winds*.

equatorial easterlies upper-level easterly air flow over the *equatorial zone*.

equatorial rainforest plant *formation class* within the *forest biome,* consisting of tall, closely set broadleaf *trees* of evergreen or semideciduous habit.

equatorial trough atmospheric low-pressure trough centered more or less over the *equator* and situated between the two belts of *trade winds*.

equatorial zone *latitude* zone lying between lat. 10° S and 10° N (more or less) and centered upon the *equator*.

equilibrium in flow systems, a state of balance in which flow rates remain unchanged.

equinox instant in time when the *subsolar point* falls on the Earth's equator and the *circle of illumination* passes through both poles. *Vernal equinox* occurs on March 20 or 21; *autumnal equinox* on September 22 or 23.

era major subdivision of geologic time consisting of a number of geologic periods. The three *eras* following *Precambrian time* are *Paleozoic, Mesozoic,* and *Cenozoic*.

erg large expanse of active *sand dunes* in the Sahara Desert of North Africa.

erosional landforms class of the *sequential landforms* shaped by the removal of *regolith* or *bedrock* by agents of erosion. Examples: *gorge,* glacial *cirque, marine cliff*.

esker narrow, often sinuous embankment of coarse gravel and boulders deposited in the bed of a meltwater *stream* enclosed in a tunnel within stagnant ice of an *ice sheet*.

estuary *bay* that receives fresh water from a river mouth and saltwater from the ocean.

Eurasian-Indonesian belt mountain arc system extending from southern Europe across southern Asia and Indonesia.

eustatic referring to a true change in sea level, as opposed to a local change created by upward or downward *tectonic* motion of land.

eutrophication excessive growth of algae and other related organisms in a *stream* or *lake* as a result of the input of large amounts of nutrient *ions,* especially phosphate and nitrate.

evaporation process in which water in liquid state or solid state passes into the vapor state.

evaporites class of *chemically precipitated sediment* and *sedimentary rock* composed of soluble salts deposited from saltwater bodies.

evapotranspiration combined water loss to the atmosphere by *evaporation* from the *soil* and *transpiration* from plants.

evergreen plant *tree* or *shrub* that holds most of its green leaves throughout the year.

evolution the creation of the diversity of life forms through the process of natural selection.

exfoliation (See *unloading.*)

exfoliation dome smoothly rounded *rock* knob or hilltop bearing *rock* sheets or shells produced by spontaneous expansion accompanying *unloading*.

exotic river *stream* that flows across a region of dry climate and derives its *discharge* from adjacent uplands where a *water surplus* exists.

exponential growth increase in number or value over time in which the increase is a constant proportion or percentage within each time unit.

exposed shields areas of *continental shields* in which the ancient basement *rock,* usually of *Precambrian* age, is exposed to the surface.

extension (tectonic) drawing apart of crustal layers by *tectonic activity* resulting in *faulting*.

extinction the event that the number of organisms of a *species* shrinks to zero so that the species no longer exists.

extrusion release of molten *rock magma* at the surface, as in a flow of *lava* or shower of volcanic ash.

extrusive igneous rock *rock* produced by the solidification of *lava* or ejected fragments of *igneous rock (tephra)*.

Fahrenheit scale temperature scale in which the *freezing* point of water is 32° and the boiling point 212°.

fair weather system a *traveling anticyclone,* in which the descent of air suppresses clouds and precipitation and the weather is typically fair.

fallout *gravity* fall of atmospheric particles of *particulates* reaching the ground.

fault sharp break in *rock* with a displacement (slippage) of the block on one side with respect to an adjacent block. (See also *normal fault, overthrust fault, strike-slip fault, transform fault.*)

fault coast   *coast* formed when a *shoreline* comes to rest against a *fault scarp.*

fault creep   more or less continuous slippage on a *fault plane,* relieving some of the accumulated strain.

fault plane   surface of slippage between two Earth blocks moving relative to each other during faulting.

fault scarp   cliff-like surface feature produced by faulting and exposing the *fault plane;* commonly associated with a *normal fault.*

fault-line scarp   erosion scarp developed upon an inactive *fault* line.

feedback   in flow systems, a linkage between flow paths such that the flow in one *pathway* acts either to reduce or increase the flow in another *pathway.*

feldspar   group of *silicate minerals* consisting of silicate of aluminum and one or more of the metals potassium, sodium, or calcium. (See also *plagioclase feldspar, potash feldspar.*)

felsenmeer   expanse of large blocks of *rock* produced by *joint* block separation and shattering by *frost action* at high altitudes or in high latitudes; from the German for "rock sea."

felsic igneous rock   *igneous rock* dominantly composed of *felsic minerals.*

felsic minerals (felsic mineral group)   *quartz* and *feldspars* treated as a mineral group of light color and relatively low *density.* (See also *mafic minerals.*)

fine textured (rock)   having *mineral* crystals too small to be seen by eye or with low magnification.

fiord   narrow, deep ocean embayment partially filling a *glacial trough.*

fiord coast   deeply embayed, rugged coast formed by partial *submergence* of *glacial troughs.*

firn   granular old *snow* forming a surface layer in the zone of accumulation of a *glacier.*

fissile   adjective describing a *rock,* usually *shale,* that readily splits up into small flakes or scales.

flash flood   flood in which heavy rainfall causes a stream or river to rise very rapidly.

flocculation   the clotting together of colloidal mineral particles to form larger particles; it occurs when a colloidal suspension in fresh water mixes with sea water.

flood   *stream flow* at a stream *stage* so high that it cannot be accommodated within the *stream channel* and must spread over the banks to inundate the adjacent *floodplain.*

flood basalts   large-scale outpourings of basalt *lava* to produce thick accumulations of *basalt* over large areas.

flood current   landward flow of a *tidal current.*

flood stage   designated stream-surface level for a particular point on a *stream,* higher than which overbank flooding may be expected.

floodplain   belt of low, flat ground, present on one or both sides of a *stream channel,* subject to inundation by a *flood* about once annually and underlain by alluvium.

flow system   a physical *system* in which *matter, energy,* or both move through time from one location to another.

fluid   substance that flows readily when subjected to unbalanced stresses; may exist as a *gas* or a *liquid.*

fluid agents   *fluids* that erode, transport, and deposit *mineral matter* and organic matter; they are running water, waves and currents, glacial ice, and *wind.*

fluvial landforms   *landforms* shaped by running water.

fluvial processes   geomorphic processes in which running water is the dominant *fluid* agent, acting as *overland flow* and *stream flow.*

focus   (See *earthquake focus.*)

fog   cloud layer in contact with land or sea surface, or very close to that surface. (See also *advection fog, radiation fog.*)

folding   process by which *folds* are produced; a form of *tectonic activity.*

folds   wave-like corrugations of *strata* (or other layered *rock* masses) as a result of crustal *compression.*

food chain (food web)   organization of an *ecosystem* into steps or levels through which *energy* flows as the organisms at each level consume *energy* stored in the bodies of organisms of the next lower level.

forb   broad-leaved *herb,* as distinguished from the grasses.

forearc trough   in plate tectonics, a shallow trough between a *tectonic arc* and a continent; accumulates *sediment* in a basin-like structure.

foredunes   ridge of irregular *sand dunes* typically found adjacent to *beaches* on low-lying *coasts* and bearing a partial cover of plants.

foreland folds   *folds* produced by *continental collision* in *strata* of a *passive continental margin.*

forest   assemblage of *trees* growing close together, their crowns forming a layer of foliage that largely shades the ground.

forest biome   *biome* that includes all regions of *forest* over the lands of the Earth.

formation classes   subdivisions within a *biome* based on the size, shape, and structure of the plants that dominate the vegetation.

fossil fuels   naturally occurring hydrocarbon compounds that represent the altered remains of organic materials enclosed in *rock;* examples are *coal, petroleum (crude oil),* and *natural* gas.

fractional scale   (See *scale fraction.*)

freezing   change from liquid state to solid state accompanied by release of *latent heat,* becoming *sensible heat.*

fringing reef   *coral reef* directly attached to land with no intervening *lagoon* of open water.

front   surface of contact between two unlike *air masses.* (See also *cold front, occluded front, polar front, warm front.*)

frost action   *rock* breakup by forces accompanying the *freezing* of water.

gabbro   *intrusive igneous rock* consisting largely of pyroxene and *plagioclase feldspar,* with variable amounts of *olivine;* a *mafic igneous rock.*

gas (gaseous state)   *fluid* of very low density (as compared with a liquid of the same chemical composition) that expands to fill uniformly any small container and is readily compressed.

gaseous cycle   type of *biogeochemical cycle* in which an element or compound is converted into *gaseous* form, diffuses through the *atmosphere,* and passes rapidly over land or sea where it is reused in the *biosphere.*

gene flow   speciation process in which evolving populations exchange *alleles* as individuals move among populations.

genetic drift *speciation* process in which chance mutations change the genetic composition of a breeding population until it diverges from other populations.

genotype the gene set of an individual organism or *species*.

genus a collection of closely related *species* that share a similar genetic evolutionary history.

geographic grid complete network of *parallels* and *meridians* on the surface of the globe, used to fix the locations of surface points.

geographic information system (GIS) a system for acquiring, processing, storing, querying, creating, and displaying *spatial data;* normally computer-based.

geographic isolation *speciation* process in which a breeding population is split into parts by an emerging geographic barrier, such as an uplifting mountain range or a changing *climate*.

geography the study of the evolving character and organization of the Earth's surface.

geography of soils the study of the distribution of soil types and properties and the processes of soil formation.

geologic norm stable natural condition in a moist climate in which slow *soil erosion* is paced by maintenance of *soil horizons* bearing a plant community in an equilibrium state.

geology science of the solid Earth, including the Earth's origin and history, materials comprising the Earth, and the processes acting within the Earth and upon its surface.

geomorphology science of Earth surface processes and *landforms,* including their history and processes of origin.

geostrophic wind *wind* at high levels above the Earth's surface blowing parallel with a system of straight, parallel *isobars.*

geyser periodic jet-like emission of hot water and steam from a narrow vent at a geothermal locality.

glacial abrasion *abrasion* by a moving *glacier* of the *bedrock* floor beneath it.

glacial delta *delta* built by meltwater streams of a *glacier* into standing water of a marginal glacial lake.

glacial drift general term for all varieties and forms of *rock* debris deposited

in close association with *ice sheets* of the *Pleistocene Epoch.*

glacial plucking removal of masses of *bedrock* from beneath an *alpine glacier* or *ice sheets* as ice moves forward suddenly.

glacial trough deep, steep-sided *rock* trench of U-shaped cross section formed by *alpine glacier* erosion.

glaciation (1) general term for the total process of glacier growth and *landform* modification by *glaciers.* (2) single episode or time period in which *ice sheets* formed, spread, and disappeared.

glacier large natural accumulation of land ice affected by present or past flowage. (See also *alpine glacier.*)

Gleysolic order a class of *soils* in the Canadian soil classification system characterized by indicators of periodic or prolonged water saturation.

global radiation balance the energy flow process by which the Earth absorbs shortwave solar radiation and emits longwave radiation. In the long run, the two flows must balance.

global scale scale at which we are concerned with the Earth as a whole, for example in considering Earth–Sun relationships.

gneiss variety of *metamorphic rock* showing banding and commonly rich in *quartz* and *feldspar.*

Gondwana a *supercontinent* of the Permian *Period* including much of the regions that are now South America, Africa, Antarctica, Australia, New Zealand, Madagascar, and peninsular India.

Goode projection an equal-area *map projection,* often used to display areal thematic information, such as *climate* or *soil* type.

gorge (canyon) steep-sided *bedrock* valley with a narrow floor limited to the width of a *stream channel.*

graben trench-like depression representing the surface of a crustal block dropped down between two opposed, infacing *normal faults.* (See also *rift valley.*)

graded profile smoothly descending profile displayed by a *graded stream.*

graded stream *stream* (or *stream channel*) with *stream gradient* so adjusted as to achieve a balanced state in which average *bed load* transport is matched to average bed load input; an

average condition over periods of many years' duration.

gradient degree of *slope,* as the gradient of a river or a flowing glacier.

granite *intrusive igneous rock* consisting largely of *quartz, potash feldspar,* and *plagioclase feldspar,* with minor amounts of biotite and hornblende; a *felsic igneous rock.*

granitic rock general term for *rock* of the upper layer of the *continental crust,* composed largely of *felsic igneous* and *metamorphic rock; rock* of composition similar to that of *granite.*

granular disintegration grain-by-grain breakup of the outer surface of coarse-grained *rock,* yielding *sand* and gravel and leaving behind rounded boulders.

graphic scale *map* scale as shown by a line divided into equal parts.

grassland biome *biome* consisting largely or entirely of *herbs,* which may include grasses, grass-like plants, and *forbs.*

gravitation mutual attraction between any two masses.

gravity gravitational attraction of the Earth upon any small mass near the Earth's surface. (See also *gravitation.*)

gravity gliding the sliding of a *thrust sheet* away from the center of an *orogen* under the force of *gravity.*

great circle circle formed by passing a plane through the exact center of a perfect sphere; the largest circle that can be drawn on the surface of a sphere.

greenhouse effect accumulation of heat in the lower *atmosphere* through the absorption of *longwave radiation* from the Earth's surface.

greenhouse gases atmospheric gases such as CO_2 and *chlorofluorocarbons (CFCs)* that absorb outgoing *longwave radiation,* contributing to the *greenhouse effect.*

groin wall or embankment built out into the water at right angles to the *shoreline.*

gross photosynthesis total amount of *carbohydrate* produced by *photosynthesis* by a given organism or group of organisms in a given unit of time.

ground ice frozen water within the pore spaces of *soils* and *regolith.*

ground moraine moraine formed of till distributed beneath a large expanse of land surface covered at one time by an ice sheet.

ground water *subsurface water* occupying the *saturated zone* and moving under the force of *gravity.*

growth rate (of a population) rate at which a population grows or shrinks with time; usually expressed as a percent or proportion of increase or decrease in a given unit of time.

gullies deep, V-shaped trenches carved by newly formed *streams* in rapid headward growth during advanced stages of *accelerated soil erosion.*

guyot sunken remnant of a volcanic island.

gyres large circular *ocean current* systems centered upon the oceanic subtropical *high-pressure cells.*

habitat subdivision of the environment according to the needs and preferences of organisms or groups of organisms.

Hadley cell atmospheric circulation cell in low latitudes involving rising air over the *equatorial trough* and sinking air over the *subtropical high-pressure belts.*

hail form of *precipitation* consisting of pellets or spheres of ice with a concentric layered structure.

half-life time required for an initial quantity at time-zero to be reduced by one-half in an exponential decay system.

hanging valley stream valley that has been truncated by marine erosion so as to appear in cross section in a *marine cliff,* or truncated by glacial erosion so as to appear in cross section in the upper wall of a *glacial trough.*

hazards assessment a field of study blending *physical* and *human geography* to focus on the perception of risk of natural hazards and on developing public policy to mitigate that risk.

haze minor concentration of *pollutants* or natural forms of *aerosols* in the atmosphere causing a reduction in visibility.

heat (See *latent heat, sensible heat.*)

heat island persistent region of higher air temperatures centered over a city.

hemisphere half of a sphere; that portion of the Earth's surface found between the *equator* and a pole.

herb tender plant, lacking woody stems, usually small or low; may be *annual* or *perennial.*

herbivory form of interaction among *species* in which an animal (herbivore) grazes on herbaceous plants.

heterosphere region of the *atmosphere* above about 100 km in which *gas* molecules tend to become increasingly sorted into layers by molecular weight and electric charge.

hibernation dormant state of some vertebrate animals during the winter season.

high base status (See *base status of soils.*)

high-latitude climates group of climates in the *subarctic zone, arctic zone,* and *polar zone,* dominated by arctic *air masses* and polar air masses.

high-level temperature inversion condition in which a high-level layer of warm air overlies a layer of cooler air, reversing the normal trend of cooling with altitude.

high-pressure cell center of high barometric pressure; an *anticyclone.*

Histosols *soil order* consisting of *soils* with a thick upper layer of organic matter.

hogbacks sharp-crested, often sawtooth ridges formed of the upturned edge of a resistant *rock* layer of *sandstone, limestone,* or *lava.*

Holocene Epoch last *epoch* of geologic time, commencing about 10,000 years ago; it followed the *Pleistocene Epoch* and includes the present.

homosphere the lower portion of the *atmosphere,* below about 100 km altitude, in which atmospheric *gases* are uniformly mixed.

horse latitudes *subtropical high-pressure belt* of the North Atlantic Ocean, coincident with the central region of the Azores high; a belt of weak, variable winds and frequent calms.

horst crustal block uplifted between two *normal faults.*

hot springs springs discharging heated *ground water* at a temperature close to the boiling point; found in geothermal areas and thought to be related to a *magma* body at depth.

hotspot (biogeography) geographic region of high biodiversity.

hotspot (plate tectonics) center of intrusive *igneous* and *volcanic* activity thought to be located over a rising *mantle plume.*

human habitat the lands of the Earth that support human life.

human geography the part of *systematic geography* that deals with social, economic and behavioral processes that differentiate *places.*

human-influenced vegetation vegetation that has been influenced in some way by human activity, for example, through cultivation, grazing, timber cutting, or urbanization.

humidity general term for the amount of *water vapor* present in the air. (See also *relative humidity, specific humidity.*)

humification *pedogenic process* of transformation of plant tissues into *humus.*

humus dark brown to black organic matter on or in the *soil,* consisting of fragmented plant tissues partly digested by organisms.

hurricane *tropical cyclone* of the western North Atlantic and Caribbean Sea.

hydraulic action *stream erosion* by impact force of the flowing water upon the bed and banks of the *stream channel.*

hydrograph graphic presentation of the variation in *stream discharge* with elapsed time, based on data of stream gauging at a given station on a stream.

hydrologic cycle total plan of movement, exchange, and storage of the Earth's free water in gaseous state, liquid state, and solid state.

hydrology science of the Earth's water and its motions through the *hydrologic cycle.*

hydrolysis chemical union of water molecules with *minerals* to form different, more stable mineral compounds.

hydrosphere total water realm of the Earth's surface zone, including the oceans, surface waters of the lands, *ground water,* and water held in the *atmosphere.*

hygrometer instrument that measures the *water vapor* content of the *atmosphere;* some types measure *relative humidity* directly.

ice age span of geologic time, usually on the order of one to three million years, or longer, in which glaciations alternate with interglaciations repeatedly in rhythm with cyclic global climate changes. (See also *glaciation, interglaciation.*)

Ice Age (Late-Cenozoic Ice Age) the present ice age, which began in late Pliocene time, perhaps 2.5 to 3 million years ago.

ice lobes (glacial lobes) broad tongue-like extensions of an *ice sheet* resulting from more rapid ice motion where terrain was more favorable.

ice sheet large thick plate of glacial ice moving outward in all directions from a central region of accumulation.

ice sheet climate ⑬ severely cold climate, found on the Greenland and Antarctic *ice sheets*, with *potential evapotranspiration (water need)* effectively zero throughout the year.

ice shelf thick plate of floating glacial ice attached to an *ice sheet* and fed by the ice sheet and by *snow* accumulation.

ice storm occurrence of heavy glaze of ice on solid surfaces.

ice wedge vertical, wall-like body of ground ice, often tapering downward, occupying a shrinkage crack in *silt* of *permafrost* areas.

ice-wedge polygons polygonal networks of *ice wedges.*

iceberg mass of glacial ice floating in the ocean, derived from a glacier that extends into tidal water.

igneous rock *rock* solidified from a high-temperature molten state; *rock* formed by cooling of *magma.* (See also *extrusive igneous rock, felsic igneous rock, intrusive igneous rock, mafic igneous rock, ultramafic igenous rock.*)

illuviation accumulation in a lower *soil horizon* (typically, the *B horizon*) of materials brought down from a higher horizon; a soil-forming process.

image processing mathematical manipulation of digital images, for example, to enhance contrast or edges.

Inceptisols *soil order* consisting of soils having weakly developed *soil horizons* and containing weatherable *minerals.*

induced deflation loss of *soil* by wind erosion that is triggered by human activity such as cultivation or overgrazing.

induced mass wasting *mass wasting* that is induced by human activity, such

as creation of waste *soil* and *rock* piles or undercutting of *slopes* in construction.

infiltration absorption and downward movement of *precipitation* into the *soil* and *regolith.*

infrared imagery images formed by *infrared radiation* emanating from the ground surface as recorded by a remote sensor.

infrared radiation *electromagnetic energy* in the *wavelength* range of 0.7 to about 200 μm.

initial landforms *landforms* produced directly by internal Earth processes of *volcanism* and *tectonic activity.* Examples: *volcano, fault scarp.*

inner lowland on a *coastal plain*, a shallow valley lying between the first *cuesta* and the area of older *rock* (oldland).

input flow of *matter* or *energy* into a system.

insolation interception of solar *energy (shortwave radiation)* by an exposed surface.

inspiral horizontal inward spiral or motion, such as that found in a *cyclone.*

interglaciation within an *ice age*, a time interval of mild global climate in which continental *ice sheets* were largely absent or were limited to the Greenland and Antarctic ice sheets; the interval between two glaciations. (See also *deglaciation, glaciation.*)

interlobate moraine *moraine* formed between two adjacent lobes of an *ice sheet.*

International Date Line the 180° *meridian of longitude*, together with deviations east and west of that meridian, forming the time boundary between adjacent *standard time zones* that are 12 hours fast and 12 hours slow with respect to Greenwich standard time.

interrupted projection projection subdivided into a number of sectors (gores), each of which is centered on a different central meridian.

intertropical convergence zone (ITCZ) zone of convergence of *air masses* of *tropical easterlies (trade winds)* along the axis of the *equatorial trough.*

intrusion body of *igneous rock* injected as *magma* into preexisting crustal *rock;* example: *dike* or *sill.*

intrusive igneous rock *igneous rock* body produced by solidification of *magma* beneath the surface, surrounded by preexisting *rock.*

inversion (See *temperature inversion.*)

ion atom or group of atoms bearing an electrical charge as the result of a gain or loss of one or more electrons.

island arcs curved lines of volcanic islands associated with active *subduction* zones along the boundaries of *lithospheric plates.*

isobars lines on *map* passing through all points having the same *atmospheric pressure.*

isohyet line on a *map* drawn through all points having the same numerical value of *precipitation.*

isopleth line on a *map* or globe drawn through all points having the same value of a selected property or entity.

isostasy principle describing the flotation of the *lithosphere*, which is less dense, on the plastic *asthenosphere*, which is more dense.

isostatic compensation crustal rise or sinking in response to unloading by *denudation* or loading by sediment deposition, following the principle of *isostasy.*

isostatic rebound local crustal rise after the melting of *ice sheets*, following the principle of *isostasy.*

isotherm line on a *map* drawn through all points having the same air temperature.

isotope form of an element with a unique *atomic mass number.*

jet stream high-speed air flow in narrow bands within the *upper-air westerlies* and along certain other global *latitude* zones at high levels.

joints fractures within *bedrock*, usually occurring in parallel and intersecting sets of planes.

Joule unit of work or energy in the metric system; symbol, J.

June solstice (See *summer solstice.*)

karst landscape or topography dominated by surface features of *limestone* solution and underlain by a *limestone cavern* system.

Kelvin scale (K) temperature scale on which the starting point is absolute zero, equivalent to –273°C.

kinetic energy form of *energy* represented by *matter* (mass) in motion.

knob and kettle terrain of numerous small knobs of *glacial drift* and deep depressions usually situated along the *moraine* belt of a former *ice sheet.*

lag time interval of time between occurrence of precipitation and peak discharge of a *stream.*

lagoon shallow body of open water lying between a *barrier island* or a *barrier reef* and the mainland.

lahar rapid downslope or downvalley movement of a tongue-like mass of water-saturated *tephra* (volcanic ash) originating high up on a steep-sided volcanic cone; a variety of *mudflow.*

lake terrestrial body of standing water surrounded by land or glacial ice.

laminar flow smooth, even flow of a *fluid* shearing in thin layers without *turbulence.*

land breeze local wind blowing from land to water during the night.

land degradation *degradation* of the quality of plant cover and *soil* as a result of overuse by humans and their domesticated animals, especially during periods of *drought.*

land mass large area of *continental crust* lying above sea level (base level) and thus available for removal by *denudation.*

land-mass rejuvenation episode of rapid fluvial *denudation* set off by a rapid crustal rise, increasing the available *land mass.*

landforms configurations of the land surface taking distinctive forms and produced by natural processes. Examples: hill, valley, plateau. (See also *depositional landform, erosional landforms, initial landforms, sequential landforms.*)

landslide rapid sliding of large masses of *bedrock* on steep mountain slopes or from high *cliffs.*

lapse rate rate at which temperature decreases with increasing altitude. (See also *dry adiabatic lapse rate, environmental temperature lapse rate, wet adiabatic lapse rate.*)

large-scale map *map* with *fractional scale* greater than 1:100,000; usually shows a small area.

Late-Cenozoic Ice Age the series of *glaciations, deglaciations,* and *interglaciations* experienced during the late *Cenozoic Era.*

latent heat heat absorbed and held in storage in a *gas* or *liquid* during the processes of *evaporation,* or *melting,* or *sublimation;* distinguished from *sensible heat.*

latent heat transfer flow of *latent heat* that results when water absorbs heat to change from a *liquid* or *solid* to a *gas* and then later releases that heat to new surroundings by *condensation* or *deposition.*

lateral moraine *moraine* forming an embankment between the ice of an *alpine glacier* and the adjacent valley wall.

laterite rock-like layer rich in *sequioxides* and iron, including the minerals *bauxite* and *limonite,* found in low latitudes in association with *Ultisols* and *Oxisols.*

latitude arc of a *meridian* between the *equator* and a given point on the globe.

Laurasia a *supercontinent* of the Permian *Period,* including much of the region that is now North America and western Eurasia.

lava *magma* emerging on the Earth's solid surface, exposed to air or water.

leaching *pedogenic process* in which material is lost from the *soil* by downward washing out and removal by percolating surplus soil water.

leads narrow strips of open ocean water between ice floes.

level of condensation elevation at which an upward-moving parcel of moist air cools to the *dew point* and *condensation* begins to occur.

liana woody vine supported on the trunk or branches of a *tree.*

lichens plant forms in which algae and fungi live together (in a symbiotic relationship) to create a single structure; they typically form tough, leathery coatings or crusts attached to *rocks* and *tree* trunks.

life cycle continuous progression of stages in a growth or development process, such as that of a living organism.

life layer shallow surface zone containing the *biosphere;* a zone of interaction between *atmosphere* and land surface, and between atmosphere and ocean surface.

life zones series of vegetation zones describing vegetation types that are

encountered with increasing elevation, especially in the southwestern U.S.

life form characteristic physical structure, size, and shape of a plant or of an assemblage of plants.

lightning an electric *arc* passing between differently charged parts of a *cloud* mass or between the cloud and the ground.

limestone nonclastic *sedimentary rock* in which *calcite* is the predominant *mineral,* and with varying minor amounts of other minerals and *clay.*

limestone caverns interconnected subterranean cavities formed in *limestone* by *carbonic acid action* occurring in slowly moving *ground water.*

limonite mineral or group of *minerals* consisting largely of iron oxide and water, produced by *chemical weathering* of other iron-bearing minerals.

line type of *spatial object* in a *geographic information system* that has starting and ending *nodes;* may be directional.

liquid *fluid* that maintains a free upper surface and is only very slightly compressible, as compared with a *gas.*

lithosphere strong, brittle outermost *rock* layer of the Earth, lying above the *asthenosphere.*

lithospheric plate segment of *lithosphere* moving as a unit, in contact with adjacent lithospheric plates along plate boundaries.

littoral drift transport of *sediment* parallel with the *shoreline* by the combined action of *beach drift* and *longshore current* transport.

loam soil-texture class in which no one of the three size grades *(sand, silt, clay)* dominates over the other two.

local scale scale of observation of the Earth in which local processes and phenomena are observed.

local winds general term for *winds* generated as direct or immediate effects of the local terrain.

loess accumulation of yellowish to buff-colored, fine-grained *sediment,* largely of *silt* grade, upon upland surfaces after transport in the air in *turbulent suspension* (i.e., carried in a *dust storm*).

logistic growth growth according to a mathematical model in which the *growth rate* eventually decreases to near zero.

longitude *arc* of a *parallel* between the *prime meridian* and a given point on the globe.

longitudinal dunes class of *sand dunes* in which the dune ridges are oriented parallel with the prevailing wind.

longshore current current in the breaker zone, running parallel with the *shoreline* and set up by the oblique approach of waves.

longshore drift *littoral drift* caused by action of a *longshore current*.

longwave radiation *electromagnetic energy* emitted by the Earth, largely in the range from 3 to 50 μm.

low base status (See *base status of soils*.)

low-angle overthrust fault *overthrust fault* in which the *fault plane* or fault surface has a low angle of *dip* or may be horizontal.

low-latitude climates group of climates of the *equatorial zone* and *tropical zone* dominated by the subtropical high-pressure belt and the *equatorial trough*.

low-latitude rainforest evergreen broadleaf forest of the wet equatorial and tropical climate zones.

low-latitude rainforest environment low-latitude environment of warm temperatures and abundant *precipitation* that characterizes rainforest in the *wet equatorial* ① and *monsoon and trade-wind coastal* ② climates.

low-level temperature inversion atmospheric condition in which temperature near the ground increases, rather than decreases, with elevation.

low-pressure trough zone of low pressure between two *anticyclones*.

lowlands broad, open valleys between two *cuestas* of a *coastal plain*. (The term may refer to any low areas of land surface.)

Luvisolic order a class of *forest soils* in the Canadian soil classification system in which the *B horizon* accumulates *clay*.

mafic igneous rock *igneous rock* dominantly composed of *mafic minerals*.

mafic minerals (mafic mineral group) *minerals*, largely *silicate minerals*, rich in magnesium and iron, dark in color, and of relatively great density.

magma mobile, high-temperature molten state of *rock*, usually of *silicate mineral* composition and with dissolved *gases*.

manipulation and analysis component component of a *geographic information system* that responds to spatial queries and creates new data layers.

mantle *rock* layer or shell of the Earth beneath the *crust* and surrounding the *core*, composed of *ultramafic igneous rock* of *silicate mineral* composition.

mantle plume a column-like rising of heated *mantle rock*, thought to be the cause of a *hotspot* in the overlying *lithospheric plate*.

map a paper representation of space showing point, line, or area data.

map projection any orderly system of *parallels* and *meridians* drawn on a flat surface to represent the Earth's curved surface.

marble variety of *metamorphic rock* derived from *limestone* or dolomite by recrystallization under pressure.

marine cliff *rock* cliff shaped and maintained by the undermining action of breaking waves.

marine scarp steep seaward *slope* in poorly consolidated *alluvium, glacial drift,* or other forms of *regolith,* produced along a coastline by the undermining action of waves.

marine terrace former *abrasion platform* elevated to become a step-like coastal *landform*.

marine west-coast climate ⑧ cool moist climate of west coasts in the *midlatitude zone,* usually with a substantial annual *water surplus* and a distinct winter *precipitation* maximum.

mass number (See *atomic mass number*.)

mass wasting spontaneous downhill movement of *soil, regolith,* and *bedrock* under the influence of *gravity,* rather than by the action of *fluid* agents.

material cycle a closed matter flow system in which matter flows endlessly, powered by energy inputs (See also *biogeochemical cycle*.)

mathematical modeling using variables and equations to represent real processes and systems.

matter physical substance that has mass and density.

matter flow system total system of *pathways* by which a particular type of *matter* (a given element, compound, or ion, for example) moves through the Earth's *ecosystem* or *biosphere*.

mean annual temperature mean of daily air temperature means for a given year or succession of years.

mean daily temperature sum of daily maximum and minimum air temperature readings divided by two.

mean monthly temperature mean of daily air temperature means for a given calendar month.

mean velocity mean, or average, speed of flow of water through an entire stream cross section.

meanders (See *alluvial meanders*.)

mechanical energy *energy* of motion or position; includes *kinetic energy* and *potential energy*.

mechanical weathering (See *physical weathering*.)

medial moraine long, narrow deposit of fragments on the surface of a *glacier*; created by the merging of *lateral moraines* when two glaciers join into a single stream of ice flow.

Mediterranean climate ⑦ climate type of the *subtropical zone*, characterized by the alternation of a very dry summer and a mild, rainy winter.

melting change from solid state to liquid state, accompanied by absorption of *sensible heat* to become *latent heat*.

Mercator projection conformal *map projection* with horizontal parallels and vertical meridians and with *map* scale rapidly increasing with increase in *latitude*.

mercury barometer *barometer* using the Torricelli principle, in which *atmospheric pressure* counterbalances a column of mercury in a tube.

meridian of longitude north–south line on the surface of the global *oblate ellipsoid*, connecting the *north pole* and *south pole*.

meridional transport flow of *energy* (heat) or *matter* (water) across the *parallels of latitude*, either poleward or equatorward.

mesa table-topped *plateau* of comparatively small extent bounded by *cliffs* and occurring in a region of flat-lying *strata*.

mesopause upper limit of the *mesosphere*.

mesosphere atmospheric layer of upwardly diminishing temperature, situated above the *stratopause* and below the *mesopause.*

Mesozoic Era second of three geologic *eras* following *Precambrian time.*

metamorphic rock *rock* altered in physical structure and/or chemical *(mineral)* composition by action of heat, pressure, *shearing* stress, or infusion of elements, all taking place at substantial depth beneath the surface.

meteorology science of the *atmosphere;* particularly the physics of the lower or inner atmosphere.

mica group aluminum-silicate *mineral* group of complex chemical formula having perfect cleavage into thin sheets.

microburst brief onset of intense *winds* close to the ground beneath the downdraft zone of a *thunderstorm* cell.

microcontinent fragment of *continental crust* and its *lithosphere* of subcontinental dimensions that is embedded in an expanse of *oceanic lithosphere.*

micrometer metric unit of length equal to one-millionth of a meter (0. 000001 m); abbreviated μm.

microwaves waves of the *electromagnetic radiation* spectrum in the *wavelength* band from about 0.03 cm to about 1 cm.

midlatitude climates group of climates of the *midlatitude zone* and *subtropical zone,* located in the *polar front zone* and dominated by both tropical *air masses* and polar air masses.

midlatitude deciduous forest plant *formation class* within the *forest biome* dominated by tall, broadleaf deciduous *trees,* found mostly in the *moist continental climate* ⑩ and *marine west-coast climate* ⑧.

midlatitude zones latitude zones occupying the *latitude* range 35° to 55° N and S (more or less) and lying between the *subtropical zones* and the *subarctic (subantarctic) zones.*

mid-oceanic ridge one of three major divisions of the ocean basins, being the central belt of submarine mountain topography with a characteristic *axial rift.*

millibar unit of *atmospheric pressure;* one-thousandth of a bar. *Bar* is a force of one million dynes per square centimeter.

mineral naturally occurring inorganic substance, usually having a definite chemical composition and a characteristic atomic structure. (See also *felsic minerals, mafic minerals, silicate minerals.*)

mineral alteration chemical change of *minerals* to more stable compounds upon exposure to atmospheric conditions; same as *chemical weathering.*

mineral matter (soils) component of *soil* consisting of weathered or unweathered mineral grains.

mineral oxides (soils) secondary *minerals* found in *soils* in which original minerals have been altered by chemical combination with oxygen.

minute (of arc) 1/60 of a degree.

mistral local drainage wind of cold air affecting the Rhone Valley of southern France.

Moho contact surface between the Earth's *crust* and *mantle;* a contraction of Mohorovic, the name of the seismologist who discovered this feature.

moist continental climate ⑩ moist climate of the *midlatitude zone* with strongly defined winter and summer seasons, adequate *precipitation* throughout the year, and a substantial annual *water surplus.*

moist subtropical climate ⑥ moist climate of the *subtropical zone,* characterized by a moderate to large annual *water surplus* and a strongly seasonal cycle of *potential evapotranspiration (water need).*

mollic epipedon relatively thick, dark-colored surface *soil horizon,* containing substantial amounts of organic matter *(humus)* and usually rich in *bases.*

Mollisols *soil order* consisting of *soils* with a *mollic horizon* and high *base status.*

monadnock prominent, isolated mountain or large hill rising conspicuously above a surrounding *peneplain* and composed of a *rock* more resistant than that underlying the peneplain; a *landform* of *denudation* in moist climates.

monsoon and trade-wind coastal climate ② moist climate of low latitudes showing a strong rainfall peak in the season of high Sun and a short period of reduced rainfall.

monsoon forest *formation class* within the *forest biome* consisting in part of deciduous *trees* adapted to a long dry season in the *wet–dry tropical climate* ③.

monsoon system system of low-level *winds* blowing into a continent in summer and out of it in winter, controlled by *atmospheric pressure* systems developed seasonally over the continent.

montane forest plant *formation class* of the *forest biome* found in cool upland environments of the *tropical zone* and *equatorial zone.*

moraine accumulation of *rock* debris carried by an *alpine glacier* or an *ice sheet* and deposited by the ice to become a *depositional landform.* (See also *lateral moraine, terminal moraine.*)

morphology the outward form and appearance of individual organisms or *species.*

mountain arc curving section of an *alpine chain* occurring on a *converging boundary* between two crustal plates.

mountain roots erosional remnants of deep portions of ancient *continental sutures* that were once *alpine chains.*

mountain winds daytime movements of air up the *gradient* of valleys and mountain slopes; alternating with nocturnal *valley winds.*

mucks organic *soils* largely composed of fine, black, sticky organic matter.

mud *sediment* consisting of a mixture of *clay* and *silt* with water, often with minor amounts of *sand* and sometimes with organic matter.

mudflow a form of *mass wasting* consisting of the downslope flowage of a mixture of water and *mineral* fragments *(soil, regolith,* disintegrated *bedrock),* usually following a natural drainage line or *stream channel.*

mudstone *sedimentary rock* formed by the lithification of *mud.*

multipurpose map *map* containing several different types of information.

multispectral image image consisting of two or more images, each of which is taken from a different portion of the spectrum (e.g., blue, green, red, infrared).

multispectral scanner *remote sensing* instrument, flown on an aircraft or spacecraft, that simultaneously collects multiple *digital images (multispectral images)* of the ground. Typically, images are collected in four or more spectral bands.

mutation change in genetic material of a reproductive cell.

nappe overturned recumbent *fold* of *strata,* usually associated with *thrust sheets* in a collision *orogen.*

natural bridge natural *rock* arch spanning a *stream channel,* formed by cutoff of an *entrenched meander* bend.

natural flow systems flow systems of *energy* or naturally occurring substances that are powered largely or completely by natural power sources.

natural gas naturally occurring mixture of hydrocarbon compounds (principally methane) in the gaseous state held within certain porous *rocks.*

natural levee belt of higher ground paralleling a meandering *alluvial river* on both sides of the *stream channel* and built up by *deposition* of fine *sediment* during periods of overbank flooding.

natural selection selection of organisms by environment in a process similar to selection of plants or animals for breeding by agriculturalists.

natural vegetation stable, mature plant cover characteristic of a given area of land surface largely free from the influences and impacts of human activities.

needleleaf evergreen forest *needleleaf forest* composed of evergreen tree species, such as spruce, fir, and pine.

needleleaf forest plant *formation class* within the *forest biome,* consisting largely of needleleaf *trees.* (See also *boreal forest.*)

needleleaf tree *tree* with long, thin or flat leaves, such as pine, fir, larch, or spruce.

negative exponential mathematical form of a curve that smoothly decreases to approach a steady value, usually zero.

negative feedback in flow systems, a linkage between flow paths such that the flow in one *pathway* acts to reduce the flow in another pathway. (See also *feedback, positive feedback.*)

net photosynthesis *carbohydrate* production remaining in an organism after *respiration* has broken down sufficient carbohydrate to power the metabolism of the organism.

net primary production rate at which *carbohydrate* is accumulated in the tissues of plants within a given *ecosystem;* units are kilograms of dry organic matter per year per square meter of surface area.

net radiation difference in intensity between all incoming *energy* (positive quantity) and all outgoing energy (negative quantity) carried by both *shortwave radiation* and *longwave radiation.*

neutron atomic particle contained within the nucleus of an atom; similar in mass to a *proton,* but without a magnetic charge.

nitrogen cycle *biogeochemical cycle* in which nitrogen moves through the *biosphere* by the processes of *nitrogen fixation* and *denitrification.*

nitrogen fixation chemical process of conversion of *gaseous* molecular nitrogen of the *atmosphere* into compounds or *ions* that can be directly utilized by plants; a process carried out within the *nitrogen cycle* by certain microorganisms.

node point marking the end of a *line* or the intersection of lines as *spatial objects* in a *geographic information system*

noon (See *solar noon.*)

noon angle (of the Sun) angle of the Sun above the horizon at its highest point during the day.

normal fault variety of *fault* in which the *fault plane* inclines *(dips)* toward the downthrown block and a major component of the motion is vertical.

north pole point at which the northern end of the Earth's *axis of rotation* intersects the Earth's surface.

northeast trade winds surface *winds* of low *latitudes* that blow steadily from the northeast. (See also *trade winds.*)

nuclei (atmospheric) minute particles of solid *matter* suspended in the *atmosphere* and serving as cores for *condensation* of water or ice.

nutrient cycle (See *biogeochemical cycle.*)

O_1 horizon surface *soil horizon* containing decaying *organic matter* that is recognizable as leaves, twigs, or other organic structures.

O_a horizon *soil horizon* below the O_1 horizon containing decaying *organic matter* that is too decomposed to recognize as specific plant parts, such as leaves or twigs.

oasis desert area where *ground water* is tapped for crop irrigation and human needs.

oblate ellipsoid geometric *solid* resembling a flattened sphere, with polar axis shorter than the equatorial diameter.

occluded front weather *front* along which a moving *cold front* has overtaken a *warm front,* forcing the warm *air mass* aloft.

ocean basin floors one of the major divisions of the ocean basins, comprising the deep portions consisting of *abyssal plains* and low hills.

ocean current persistent, dominantly horizontal flow of ocean water.

ocean tide periodic rise and fall of the ocean level induced by gravitational attraction between the Earth and Moon in combination with Earth *rotation.*

oceanic crust crust of basaltic composition beneath the ocean floors, capping *oceanic lithosphere.* (See also *continental crust.*)

oceanic lithosphere *lithosphere* bearing *oceanic crust.*

oceanic trench narrow, deep depression in the seafloor representing the line of *subduction* of an oceanic *lithospheric plate* beneath the margin of a continental lithospheric plate; often associated with an *island arc.*

oil sand (See *bituminous sand.*)

old-field succession form of *secondary succession* typical of an abandoned field, such as might be found in eastern or central North America.

olivine *silicate mineral* with magnesium and iron but no aluminum, usually olive-green or grayish-green; a *mafic mineral.*

open flow system system of interconnected flow paths of *energy* and/or *matter* with a boundary through which that *energy* and/or *matter* can enter and leave the system.

organic matter (soils) material in *soil* that was originally produced by plants or animals and has been subjected to decay.

Organic order a class of *soils* in the Canadian soil classification system that is composed largely of organic materials.

organic sediment *sediment* consisting of the organic remains of plants or animals.

orogen the mass of tectonically deformed *rocks* and related *igneous rocks* produced during an *orogeny.*

orogeny major episode of *tectonic activity* resulting in *strata* being deformed by *folding* and faulting.

orographic pertaining to mountains.

orographic precipitation *precipitation* induced by the forced rise of moist air over a mountain barrier.

outcrop surface exposure of *bedrock.*

output the flow of *matter* or *energy* out of a system.

outspiral horizontal outward spiral or motion, such as that found in an *anticyclone.*

outwash glacial deposit of stratified drift left by *braided streams* issuing from the front of a *glacier.*

outwash plain flat, gently sloping plain built up of *sand* and gravel by the *aggradation* of meltwater *streams* in front of the margin of an *ice sheet.*

overburden *strata* overlying a layer or *stratum* of interest, as overburden above a *coal* seam.

overland flow motion of a surface layer of water over a sloping ground surface at times when the *infiltration* rate is exceeded by the *precipitation* rate; a form of *runoff.*

overthrust fault *fault* characterized by the overriding of one crustal block (or *thrust sheet*) over another along a gently inclined *fault plane;* associated with crustal *compression.*

ox-bow lake crescent-shaped lake representing the abandoned channel left by the *cutoff* of an *alluvial meander.*

oxidation chemical union of free oxygen with metallic elements in *minerals.*

oxide chemical compound containing oxygen; in *soils,* iron oxides and aluminum oxides are examples.

Oxisols *soil order* consisting of very old, highly weathered *soils* of low *latitudes,* with an oxic horizon and low *base status.*

oxygen cycle *biogeochemical cycle* in which oxygen moves through the *biosphere* in both *gaseous* and *sedimentary* forms.

ozone a form of oxygen with a molecule consisting of three atoms of oxygen, O_3.

ozone layer layer in the *stratosphere,* mostly in the altitude range 20 to 35 km (12 to 31 mi), in which a concentration of *ozone* is produced by the action of solar *ultraviolet radiation.*

pack ice floating *sea ice* that completely covers the sea surface.

Paleozoic Era first of three geologic *eras* comprising all geologic time younger than *Precambrian time.*

Pangea hypothetical parent continent, enduring until near the close of the *Mesozoic Era,* consisting of the *continental shields* of *Laurasia* and *Gondwana* joined into a single unit.

parabolic dunes isolated low *sand dunes* of parabolic outline, with points directed into the prevailing *wind.*

parallel of latitude east-west circle on the Earth's surface, lying in a plane parallel with the *equator* and at right angles to the *axis of rotation.*

parasitism form of negative interaction between *species* in which a small species (parasite) feeds on a larger one (host) without necessarily killing it.

parent material inorganic, *mineral* base from which the *soil* is formed; usually consists of *regolith.*

particulates *solid* and *liquid* particles capable of being suspended for long periods in the *atmosphere.*

pascal metric unit of pressure, defined as a force of one newton per square meter (1 N/m^2); symbol, Pa; 100 Pa = 1 mb, 10^5 Pa = 1 bar.

passive continental margins continental margins lacking active plate boundaries at the contact of *continental crust* with *oceanic crust.* A passive margin thus lies within a single *lithospheric plate.* Example: Atlantic continental margin of North America. (See also *active continental margins, continental margins.*)

passive systems electromagnetic remote sensing systems that measure radiant *energy* reflected or emitted by an object or surface.

pathway in an *energy flow system,* a mechanism by which *matter* or *energy* flows from one part of the system to another.

patterned ground general term for a ground surface that bears polygonal or ring-like features, including stone polygons, and ice wedge polygons; typically produced by *frost action* in severe climates.

peat partially decomposed, compacted accumulation of plant remains occurring in a *bog* environment.

ped individual natural *soil* aggregate.

pediment gently sloping, rock-floored land surface found at the base of a mountain mass or *cliff* in an arid region.

pedogenic processes group of recognized basic soil-forming processes, mostly involving the gain, loss, *translocation,* or transformation of materials within the *soil* body.

pedology science of the *soil* as a natural surface layer capable of supporting living plants; synonymous with *soil science.*

peneplain land surface of low elevation and slight relief produced in the late stages of *denudation* of a *land mass.*

perched water table surface of a lens of *ground water* held above the main body of ground water by a discontinuous impervious layer.

percolation slow, downward flow of water by *gravity* through *soil* and subsurface layers toward the *water table.*

perennials plants that live for more than one growing season.

peridotite *igneous rock* consisting largely of olivine and pyroxene; an *ultramafic igneous rock* occurring as a *pluton,* also thought to compose much of the upper *mantle.*

periglacial located near the margin of *alpine glaciers* or large *ice sheets,* in an environment of intense *frost action.*

periglacial system a distinctive set of landforms and land-forming processes that are created by intense *frost action.*

perihelion point on the Earth's elliptical orbit at which the Earth is nearest to the Sun.

period of geologic time time subdivision of the *era,* each ranging in duration between about 35 and 70 million years.

permafrost condition of permanently frozen water in the *soil, regolith,* and *bedrock* in cold climates of subarctic and arctic regions.

permafrost table in *permafrost,* the upper surface of perennially frozen ground; lower surface of the *active layer.*

petroleum (crude oil) natural *liquid* mixture of many complex hydrocarbon compounds of organic origin, found in accumulations (oil pools) within certain *sedimentary rocks.*

pH measure of the concentration of hydrogen *ions* in a solution. (The number represents the logarithm to the base

10 of the reciprocal of the weight in grams of hydrogen *ions* per liter of water.) Acid solutions have pH values less than 6, and basic solutions have pH values greater than 6.

phenotype   the morphological expression of the *genotype* of an individual. It includes all the physical aspects of its structure that are readily perceivable.

photoperiod   duration of daylight on a given day of the year at a given *latitude*.

photosynthesis   production of *carbohydrate* by the union of water with *carbon dioxide* while absorbing light *energy*.

phreatophytes   plants that draw water from the *ground water table* beneath *alluvium* of dry *stream channels* and valley floors in desert regions.

phylum   highest division of higher plant and animal life.

physical geography   the part of *systematic geography* that deals with the natural processes occurring at the Earth's surface that provide the physical setting for human activities; includes the broad fields of *climatology, geomorphology, coastal and marine geography, geography of soils,* and *biogeography.*

physical weathering   breakup of massive *rock (bedrock)* into small particles through the action of physical forces acting at or near the Earth's surface. (See also *weathering.*)

phytoplankton   microscopic floating plants found largely in the uppermost layer of ocean water.

pingo   conspicuous conical mound or circular hill, having a core of ice, found on plains of the *arctic tundra* where *permafrost* is present.

pioneer plants   plants that first invade an environment of new land or a *soil* that has been cleared of vegetation cover; often these are annual *herbs.*

pioneer stage   first stage of an *ecological succession.*

place   in geography, a location on the Earth's surface, typically a settlement or small region with unique characteristics.

plagioclase feldspar   aluminum-silicate *mineral* with sodium or calcium or both.

plane of the ecliptic   imaginary plane in which the Earth's orbit lies.

plant ecology   the study of the relationships between plants and their environment.

plant nutrients   *ions* or chemical compounds that are needed for plant growth.

plate tectonics   theory of *tectonic activity* dealing with *lithospheric plates* and their activity.

plateau   upland surface, more or less flat and horizontal, upheld by resistant beds of *sedimentary rock* or lava flows and bounded by a steep *cliff.*

playa   flat land surface underlain by fine *sediment* or evaporite minerals deposited from shallow lake waters in a dry climate in the floor of a closed topographic depression.

Pleistocene Epoch   *epoch* of the *Cenozoic Era,* often identified as the *Ice Age;* it preceded the *Holocene Epoch.*

plinthite   iron-rich concentrations present in some kinds of *soils* in deeper *soil horizons* and capable of hardening into rock-like material with repeated wetting and drying.

plucking   (See *glacial plucking.*)

pluton   any body of *intrusive igneous rock* that has solidified below the surface, enclosed in preexisting *rock.*

pocket beach   *beach* of crescentic outline located at a *bay* head.

podzol   type of *soil order* closely equivalent to *Spodosol;* an order of the Canadian Soil Classification System.

Podzolic order   a class of *forest* and heath *soils* in the Canadian soil classification system in which an amorphous material of humified organic matter with Al and Fe accumulates.

point   *spatial object* in a *geographic information system* with no area.

point bar   deposit of coarse bed-load *alluvium* accumulated on the inside of a growing *alluvial meander.*

polar easterlies   system of easterly surface winds at high *latitude,* best developed in the southern *hemisphere,* over Antarctica.

polar front   *front* lying between cold polar *air masses* and warm tropical *air masses,* often situated along a *jet stream* within the *upper-air westerlies.*

polar front jet stream   *jet stream* found along the *polar front,* where cold polar air and warm tropical air are in contact.

polar front zone   broad zone in midlatitudes and higher latitudes, occupied by the shifting *polar front.*

polar high   persistent low-level center of high *atmospheric pressure* located over the *polar zone* of Antarctica.

polar outbreak   tongue of cold polar air, preceded by a *cold front,* penetrating far into the *tropical zone* and often reaching the *equatorial zone;* it brings rain squalls and unusual cold.

polar projection   *map projection* centered on Earth's *north pole* or *south pole.*

polar zones   *latitude* zones lying between 75° and 90° N and S.

poleward heat transport   movement of heat from equatorial and tropical regions toward the poles, occurring as *latent* and *sensible heat transfer.*

pollutants   in air pollution studies, foreign matter injected into the lower *atmosphere* as *particulates* or as chemical pollutant *gases.*

pollution dome   broad, low dome-shaped layer of polluted air, formed over an urban area at times when winds are weak or calm prevails.

pollution plume   (1) the trace or path of pollutant substances, moving along the flow paths of *ground water.* (2) trail of polluted air carried downwind from a pollution source by strong winds.

polygon   type of *spatial object* in a *geographic information system* with a closed chain of connected *lines* surrounding an area.

polypedon   smallest distinctive geographic unit of the *soil* of a given area.

polyploidy   mechanism of *speciation* in which entire chromosome sets of organisms are doubled, tripled, quadrupled, etc.

pool   in flow systems, an area or location of concentration of *matter.* (See also *active pool, storage pool.*)

positive feedback   in flow systems, a linkage between flow paths such that the flow in one *pathway* acts to increase the flow in another *pathway.* (See also *feedback, negative feedback.*)

potash feldspar   aluminum-silicate *mineral* with potassium the dominant metal.

potential energy   *energy* of position; produced by *gravitational* attraction of the Earth's mass for a smaller mass on or near the Earth's surface.

potential evapotranspiration (water need)   ideal or hypothetical rate of *evapotranspiration* estimated to occur from a complete canopy of green foliage of growing plants continuously supplied with all the *soil water* they can use; a real condition reached in those situations where *precipitation* is sufficiently great or irrigation water is supplied in sufficient amounts.

pothole cylindrical cavity in hard *bedrock* of a *stream channel* produced by *abrasion* of a rounded *rock* fragment rotating within the cavity.

power source flow of *energy* into a *flow system* that causes *matter* to move.

prairie plant f*ormation class* of the *grassland biome,* consisting of dominant tall grasses and subdominant *forbs,* widespread in subhumid continental climate regions of the *subtropical zone* and *midlatitude zone.* (See also *short-grass prairie, tall-grass prairie.*)

Precambrian time all of geologic time older than the beginning of the Cambrian Period, i.e., older than 600 million years.

precipitation particles of *liquid* water or ice that fall from the *atmosphere* and may reach the ground. (See also *convectional precipitation, cyclonic precipitation, orographic precipitation.*)

predation form of negative interaction among animal *species* in which one species (predator) kills and consumes the other (prey).

preprocessing component component of a *geographic information system* that prepares data for entry to the system.

pressure gradient change of *atmospheric pressure* measured along a line at right angles to the *isobars.*

pressure gradient force force acting horizontally, tending to move air in the direction of lower *atmospheric pressure.*

prevailing westerly winds (westerlies) surface winds blowing from a generally westerly direction in the *midlatitude zone,* but varying greatly in direction and intensity.

primary consumers organisms at the lowest level of the *food chain* that ingest *primary producers* or *decomposers* as their *energy* source.

primary minerals in *pedology (soil science),* the original, unaltered *silicate minerals* of *igneous rocks* and *metamorphic rocks.*

primary producers organisms that use light *energy* to convert *carbon dioxide*

and water to *carbohydrates* through the process of *photosynthesis.*

primary succession *ecological succession* that begins on a newly constructed substrate.

prime meridian reference *meridian* of zero *longitude;* universally accepted as the Greenwich meridian.

product generation component component of a *geographic information system* that provides output products such as maps, images, or tabular reports.

progradation shoreward building of a *beach, bar,* or *sandspit* by addition of coarse *sediment* carried by *littoral drift* or brought from deeper water offshore.

proton positively charged particle within the nucleus of a atom.

pyroxene group complex aluminum-silicate *minerals* rich in calcium, magnesium, and iron, dark in color, high in density, classed as *mafic minerals.*

quartz mineral of silicon dioxide composition.

quartzite *metamorphic rock* consisting largely of the mineral *quartz.*

quick clays *clay* layers that spontaneously change from a solid condition to a near-liquid condition when disturbed.

radar an active *remote sensing* system in which a pulse of radiation is emitted by an instrument, and the strength of the echo of the pulse is recorded.

radial drainage pattern stream pattern consisting of *streams* radiating outward from a central peak or highland, such as a *sedimentary dome* or a *volcano.*

radiant energy transfer net flow of radiant *energy* between an object and its surroundings.

radiation (See *electromagnetic radiation.*)

radiation balance condition of balance between incoming *energy* of solar *shortwave radiation* and outgoing *longwave radiation* emitted by the Earth into space.

radiation fog *fog* produced by radiation cooling of the basal air layer.

radioactive decay spontaneous change in the nucleus of an atom that leads to the emission of *matter* and *energy.*

radiogenic heat heat from the Earth's interior that is slowly released by the *radioactive decay* of *unstable isotopes.*

radiometric dating a method of determining the geologic age of a *rock* or *mineral* by measuring the proportions of certain of its elements in their different isotopic forms.

rain form of *precipitation* consisting of falling water drops, usually 0.5 mm or larger in diameter.

rain gauge instrument used to measure the amount of *rain* that has fallen.

rain shadow belt of arid climate to lee of a mountain barrier, produced as a result of adiabatic warming of descending air.

rain-green vegetation vegetation that puts out green foliage in the wet season but becomes largely dormant in the dry season; found in the *tropical zone,* it includes the *savanna biome* and *monsoon forest.*

raised shoreline former *shoreline* lifted above the limit of wave action; also called an elevated *shoreline.*

rapids steep-*gradient* reaches of a *stream channel* in which *stream* velocity is high.

recessional moraine *moraine* produced at the ice margin during a temporary halt in the recessional phase of *deglaciation.*

reclamation in *strip mining,* the process of restoring *spoil* banks and ridges to a natural condition.

recombination source of variation in organisms arising from the free interchange of *alleles* of genes during the reproduction process.

recumbent overturned, as a folded sequence of *rock* layers in which the *folds* are doubled back upon themselves.

reflection outward scattering of *radiation* toward space by the *atmosphere* and/or Earth's surface.

reg desert surface armored with a pebble layer, resulting from long-continued *deflation;* found in the Sahara Desert of North Africa.

regional geography that branch of *geography* concerned with how the Earth's surface is differentiated into unique *places.*

regional scale the scale of observation at which subcontinental regions are discernible.

regolith   layer of *mineral* particles overlying the *bedrock;* may be derived by *weathering* of underlying bedrock or be transported from other locations by *fluid* agents. (See also *residual regolith, transported regolith.*)

Regosolic order   a class of *soils* in the Canadian soil classification system that exhibits weakly developed *horizons.*

relative humidity   ratio of *water vapor* present in the air to the maximum quantity possible for *saturated air* at the same temperature.

relative variability   ratio of a variability measure, such as the average deviation, to the mean of all observations.

remote sensing   measurement of some property of an object or surface by means other than direct contact; usually refers to the gathering of scientific information about the Earth's surface from great heights and over broad areas, using instruments mounted on aircraft or orbiting space vehicles.

remote sensor   instrument or device measuring *electromagnetic radiation* reflected or emitted from a target body.

removal   in soil science, the set of processes that result in the removal of material from a *soil horizon,* such as surface erosion or *leaching.*

representative fraction (R.F.)   (See *scale fraction.*)

residual regolith   *regolith* formed in place by alteration of the *bedrock* directly beneath it.

resolution   on a *map,* power to resolve small objects present on the ground.

respiration   the oxidation of organic compounds by organisms that powers bodily functions.

retrogradation   cutting back (retreat) of a *shoreline, beach, marine cliff,* or *marine scarp* by wave action.

reverse fault   type of *fault* in which one fault block rides up over the other on a steep *fault plane.*

revolution   motion of a planet in its orbit around the Sun, or of a planetary satellite around a planet.

rhyolite   *extrusive igneous rock* of *granite* composition; it occurs as *lava* or *tephra.*

ria   coastal embayment or *estuary.*

ria coast   deeply embayed *coast* formed by partial *submergence* of a *land mass* previously shaped by fluvial *denudation.*

Richter scale   scale of magnitude numbers describing the quantity of *energy* released by an *earthquake.*

ridge-and-valley landscape   assemblage of *landforms* developed by *denudation* of a system of open *folds* of *strata* and consisting of long, narrow ridges and valleys arranged in parallel or zigzag patterns.

rift valley   trench-like valley with steep, parallel sides; essentially a *graben* between two *normal faults;* associated with crustal spreading.

rill erosion   form of *accelerated erosion* in which numerous, closely spaced miniature channels (rills) are scored into the surface of exposed *soil* or *regolith.*

rock   natural aggregate of *minerals* in the solid state; usually hard and consisting of one, two, or more *mineral* varieties.

rock terrace   terrace carved in *bedrock* during the *degradation* of a *stream channel* induced by the crustal rise or a fall of the sea level. (See also *alluvial terrace, marine terrace.*)

Rodinia   early *supercontinent,* predating *Pangea,* that was fully formed about 700 million years ago.

rockslide   *landslide* of jumbled *bedrock* fragments.

Rossby waves   horizontal undulations in the flow path of the *upper-air westerlies;* also known as upper-air waves.

rotation   spinning of an object around an axis.

runoff   flow of water from continents to oceans by way of *stream flow* and *ground water* flow; a term in the water balance of the *hydrologic cycle.* In a more restricted sense, runoff refers to surface flow by *overland flow* and channel flow.

Sahel (Sahelian zones)   belt of *wet–dry tropical* ③ and *semiarid dry tropical* ④ climate in Africa in which *precipitation* is highly variable from year to year.

salic horizon   *soil horizon* enriched by soluble salts.

salinization   precipitation of soluble salts within the *soil.*

salt flat   shallow basin covered with salt deposits formed when stream input to the basin is subjected to severe *evaporation;* may also form by *evaporation* of a saline lake when *climate* changes.

salt marsh   *peat*-covered expanse of *sediment* built up to the level of high tide over a previously formed tidal mud flat.

salt water intrusion   occurs in a coastal well when an upper layer of fresh water is pumped out, leaving a saltwater layer below to feed the well.

salt-crystal growth   a form of *weathering* in which *rock* is disintegrated by the expansive pressure of growing salt crystals during dry weather periods when *evaporation* is rapid.

saltation   leaping, impacting, and rebounding of sand grains transported over a *sand* or pebble surface by *wind.*

sample standard deviation   in statistics, the square root of the average squared deviation from the mean for a sample.

sample statistics   numerical values that give basic information about a sample and its variability.

sand   *sediment* particles between 0.06 and 2 mm in diameter.

sand dune   hill or ridge of loose, well-sorted *sand* shaped by *wind* and usually capable of downwind motion.

sand sea   field of *transverse dunes.*

sandspit   narrow, finger-like embankment of *sand* constructed by *littoral drift* into the open water of a *bay.*

sandstone   variety of *sedimentary rock* consisting largely of mineral particles of sand grade size.

Santa Ana   easterly *wind,* often hot and dry, that blows from the interior desert region of southern California and passes over the coastal mountain ranges to reach the Pacific Ocean.

saturated air   air holding the maximum possible quantity of *water vapor* at a given temperature and pressure.

saturated zone   zone beneath the land surface in which all pores of the *bedrock* or *regolith* are filled with *ground water.*

savanna   a vegetation cover of widely-spaced *trees* with a grassland beneath.

savanna biome   *biome* that consists of a combination of *trees* and grassland in various proportions.

savanna woodland   plant *formation class* of the *savanna biome* consisting of a *woodland* of widely spaced *trees* and a grass layer, found throughout the *wet–dry tropical climate* ③ regions in a belt adjacent to the *monsoon forest* and *low-latitude rainforest.*

scale the magnitude of a phenomenon or system, as for example, *global scale* or *local scale.*

scale fraction ratio that relates distance on the Earth's surface to distance on a *map* or surface of a globe.

scale of globe ratio of size of a globe to size of the Earth, where size is expressed by a measure of length or distance.

scale of map ratio of distance between two points on a *map* and the same two points on the ground.

scanning systems *remote sensing* systems that make use of a scanning beam to generate images over the frame of surveillance.

scarification general term for artificial excavations and other land disturbances produced for purposes of extracting or processing *mineral* resources.

scattering turning aside of radiation by an atmospheric molecule or particle so that the direction of the scattered ray is changed.

schist foliated *metamorphic rock* in which mica flakes are typically found oriented parallel with foliation surfaces.

sclerophyll forest plant *formation class* of the *forest biome,* consisting of low sclerophyll *trees* and often including sclerophyll woodland or *scrub,* associated with regions of *Mediterranean climate* ⑦.

sclerophyll woodland plant *formation class* of the *forest biome* composed of widely spaced sclerophyll *trees* and *shrubs.*

sclerophylls hard-leaved evergreen *trees* and *shrubs* capable of enduring a long, dry summer.

scoria *lava* or *tephra* containing numerous cavities produced by expanding gases during cooling.

scrub plant *formation class* or subclass consisting of *shrubs* and having a canopy coverage of about 50 percent.

sea arch arch-like *landform* of a rocky, cliffed coast created when waves erode through a narrow headland from both sides.

sea breeze local wind blowing from sea to land during the day.

sea cave cave near the base of a *marine cliff,* eroded by breaking waves.

sea fog *fog* layer formed at sea when warm moist air passes over a cool ocean current and is chilled to the *condensation* point.

sea ice floating ice of the oceans formed by direct *freezing* of ocean water.

second of arc 1/60 of a minute, or 1/3600 of a degree.

secondary consumers animals that feed on *primary consumers.*

secondary minerals in *soil science, minerals* that are stable in the surface environment, derived by *mineral alteration* of the *primary minerals.*

secondary succession *ecological succession* beginning on a previously vegetated area that has been recently disturbed by such agents as fire, flood, windstorm, or humans.

sediment finely divided *mineral matter* and *organic matter* derived directly or indirectly from preexisting *rock* and from life processes. (See also *chemically precipitated sediment, organic sediment.*)

sediment yield quantity of sediment removed by *overland flow* from a land surface of given unit area in a given unit of time.

sedimentary cycle type of *biogeochemical cycle* in which the compound or element is released from *rock* by *weathering,* follows the movement of running water either in solution or as *sediment* to reach the sea, and is eventually converted into *rock.*

sedimentary dome up-arched *strata* forming a circular structure with domed summit and flanks with moderate to steep outward *dip.*

sedimentary rock *rock* formed from accumulation of *sediment.*

seismic sea wave (tsunami) train of sea waves set off by an *earthquake* (or other seafloor disturbance) traveling over the ocean surface.

seismic waves waves sent out during an *earthquake* by faulting or other crustal disturbance from an *earthquake focus* and propagated through the solid Earth.

semiarid (steppe) dry climate subtype subtype of the dry climates exhibiting a short wet season supporting the growth of grasses and *annual* plants.

semidesert plant *formation class* of the *desert biome,* consisting of xerophytic *shrub* vegetation with a poorly developed herbaceous lower layer; subtypes are semidesert scrub and *woodland.*

sensible heat heat measurable by a *thermometer;* an indication of the intensity of *kinetic energy* of molecular motion within a substance.

sensible heat transfer flow of heat from one substance to another by direct contact.

sequential landforms *landforms* produced by external Earth processes in the total activity of *denudation.* Examples: *gorge, alluvial fan, floodplain.*

seral stage stage in a *sere.*

sere in an *ecological succession,* the series of biotic communities that follow one another on the way to the stable stage, or *climax.*

sesquioxides oxides of aluminum or iron with a ratio of two atoms of aluminum or iron to three atoms of oxygen.

shale fissile, *sedimentary rock* of *mud* or *clay* composition, showing lamination.

shearing (of rock) slipping motion between very thin *rock* layers, like a deck of cards fanned with the sweep of a palm.

sheet erosion type of *accelerated soil erosion* in which thin layers of *soil* are removed without formation of rills or *gullies.*

sheet flow overland flow taking the form of a continuous thin film of water over a smooth surface of *soil, regolith,* or *rock.*

sheeting structure thick, subparallel layers of massive *bedrock* formed by spontaneous expansion accompanying *unloading.*

shield volcano low, often large, dome-like accumulation of basalt lava flows emerging from long radial fissures on flanks.

shoreline shifting line of contact between water and land.

short-grass praire plant *formation class* in the *grassland biome* consisting of short grasses sparsely distributed in clumps and bunches and some *shrubs,* widespread in areas of semiarid climate in continental interiors of North America and Eurasia; also called *steppe.*

shortwave infrared *infrared radiation* with wavelengths shorter than 3 μm.

shortwave radiation *electromagnetic energy* in the range from 0.2 to 3 μm, including most of the *energy* spectrum of solar radiation.

shrubs woody *perennial* plants, usually small or low, with several low-branching stems and a foliage mass close to the ground.

silica silicon dioxide in any of several mineral forms.

silicate minerals (silicates) *minerals* containing silicon and oxygen atoms, linked in the crystal space lattice in units of four oxygen atoms to each silicon atom.

sill *intrusive igneous rock* in the form of a plate where *magma* was forced into a natural parting in the *bedrock,* such as a bedding surface in a sequence of *sedimentary rocks.*

silt *sediment* particles between 0.004 and 0.06 mm in diameter.

sinkhole surface depression in *limestone,* leading down into *limestone caverns.*

slash-and-burn agricultural system, practiced in the *low-latitude rainforest,* in which small areas are cleared and the *trees* burned, forming plots that can be cultivated for brief periods.

slate compact, fine-grained variety of *metamorphic rock,* derived from *shale,* showing well-developed cleavage.

sleet form of *precipitation* consisting of ice pellets, which may be frozen raindrops.

sling psychrometer form of *hygrometer* consisting of a wet-bulb thermometer and a dry-bulb thermometer.

slip face steep face of an active *sand dune,* receiving sand by *saltation* over the dune crest and repeatedly sliding because of oversteepening.

slope (1) degree of inclination from the horizontal of an element of ground surface, analogous to *dip* in the geologic sense. (2) any portion or element of the Earth's solid surface. (3) verb meaning "to incline."

small circle circle formed by passing a plane through a sphere without passing through the exact center.

small-scale map *map* with *fractional scale* of less than 1:100,000; usually shows a large area.

smog mixture of *aerosols* and chemical *pollutants* in the lower atmosphere, usually found over urban areas.

snow form of *precipitation* consisting of ice particles.

soil natural terrestrial surface layer containing living matter and supporting or capable of supporting plants.

soil colloids mineral particles of extremely small size, capable of remaining suspended indefinitely in water; typically, they have the form of thin plates or scales.

soil creep extremely slow downhill movement of *soil* and *regolith* as a result of continued agitation and disturbance of the particles by such activities as *frost action,* temperature changes, or wetting and drying of the soil.

soil enrichment additions of materials to the *soil* body; one of the *pedogenic processes.*

soil erosion erosional removal of material from the *soil* surface.

soil horizon distinctive layer of the *soil,* more or less horizontal, set apart from other soil zones or layers by differences in physical and chemical composition, organic content, structure, or a combination of those properties, produced by soil-forming processes.

soil orders those eleven *soil* classes forming the highest category in the classification of soils.

soil profile display of *soil horizons* on the face of a freshly cut vertical exposure through the *soil.*

soil science (See *pedology.*)

soil solum that part of the *soil* made up of the A, E, and B *soil horizons;* the soil zone in which living plant roots can influence the development of soil horizons.

soil structure presence, size, and form of aggregations (lumps or clusters) of *soil* particles.

soil texture descriptive property of the *mineral* portion of the *soil* based on varying proportions of *sand, silt,* and *clay.*

soil water water held in the *soil* and available to plants through their root systems; a form of *subsurface water.*

soil water balance balance among the component terms of the *soil water budget;* namely, *precipitation, evapotranspiration,* change in soil water storage, and water surplus.

soil water belt *soil* layer from which plants draw *soil water.*

soil water budget accounting system evaluating the daily, monthly, or yearly amounts of *precipitation, evapotranspi-*

ration, soil water storage, water deficit, and water surplus.

soil water recharge restoring of depleted *soil water* by *infiltration* of *precipitation.*

soil water shortage (See *deficit.*)

soil water storage actual quantity of water held in the *soil water belt* at any given instant; usually applied to a soil layer of given depth, such as 300 cm (about 12 in.).

solar constant intensity of solar radiation falling upon a unit area of surface held at right angles to the Sun's rays at a point outside the Earth's *atmosphere;* equal to an *energy* flow of about 1400 W/m^2.

solar day average time required for the Earth to complete one *rotation* with respect to the Sun; time elapsed between one *solar noon* and the next, averaged over the period of one year.

solar noon instant at which the *subsolar point crosses* the *meridian of longitude* of a given point on the Earth; instant at which the Sun's shadow points exactly due north or due south at a given location.

solids substances in the solid state; they resist changes in shape and volume, are usually capable of withstanding large unbalanced forces without yielding, but will ultimately yield by sudden breakage.

solifluction tundra (arctic) variety of *earthflow* in which the saturated thawed layer over *permafrost* flows slowly downhill to produce multiple terraces and *solifluction* lobes.

solifluction lobe bulging mass of saturated *regolith* with steep curved front moved downhill by *solifluction.*

solifluction terrace mass of saturated *regolith* formed by *solifluction* into a flat-topped terrace.

Solonetzic order a class of *soils* in the Canadian soil classification system with a B horizon of sticky *clay* that dries to a vary hard condition.

solution a *weathering* process in which minerals dissolve in water; may be enhanced by the action of *carbonic acid* or weak organic acids.

sorting separation of one grade size of *sediment* particles from another by the action of currents of air or water.

source region extensive land or ocean surface over which an *air mass* derives its temperature and moisture characteristics.

south pole   point at which the southern end of the Earth's *axis of rotation* intersects the Earth's surface.

southeast trade winds   surface *winds* of low latitudes that blow steadily from the southeast. (See also *trade winds.*)

Southern Oscillation   episodic reversal of prevailing barometric pressure differences between two regions, one centered on Darwin, Australia, in the eastern Indian Ocean, and the other on Tahiti in the western Pacific Ocean; a precursor to the occurrence of an *El Niño* event. (See also *El Ñino.*)

southern pine forest   pine forest that is typically found on sandy soils of the Atlantic and Gulf Coast coastal plains.

southern pine forest   subtype of *needleleaf forest* dominated by pines and occurring in the *moist subtropical climate* ⑥.

spatial data   information associated with a specific location or area of the Earth's surface.

spatial object   a geographic area, *line* or *point* to which information is attached.

speciation   the process by which *species* are differentiated and maintained.

species   a collection of individual organisms that are capable of interbreeding to produce fertile offspring.

specific heat   property of a substance that governs the temperature change of the substance with a given input of heat *energy.*

specific humidity   mass of *water vapor* contained in a unit mass of air.

spit   (See *sandspit.*)

splash erosion   *soil erosion* caused by direct impact of falling raindrops on a wet surface of *soil* or *regolith.*

spodic horizon   *soil horizon* containing precipitated amorphous materials composed of organic matter and *sesquioxides* of aluminum, with or without iron.

Spodosols   *soil order* consisting of *soils* with a *spodic horizon,* an *albic horizon,* with low *base status,* and lacking in *carbonate* materials.

spoil   *rock* waste removed in a mining operation.

spreading plate boundary   *lithospheric plate* boundary along which two plates of *oceanic lithosphere* are undergoing separation, while at the same time new lithosphere is being formed by

accretion. (See also *transform plate boundary.*)

stable air mass   *air mass* in which the *environmental temperature lapse rate* is less than the *dry adiabatic lapse rate,* inhibiting convectional uplift and mixing.

stack (marine)   isolated columnar mass of *bedrock* left standing in front of a retreating *marine cliff.*

stage   height of the surface of a river above its bed or a fixed level near the bed.

standard meridians   *standard time* meridians separated by 15° of *longitude* and having values that are multiples of 15°. (In some cases meridians are used that are multiples of $7^1/_2°$.)

standard time system   time system based on the local time of a *standard meridian* and applied to belts of *longitude* extending $7^1/_2°$ (more or less) on either side of that meridian.

standard time zone   zone of the Earth in which all inhabitants keep the same time, which is that of a *standard meridian* within the zone.

star dune   large, isolated *sand dune* with radial ridges culminating in a peaked summit; found in the deserts of North Africa and the Arabian Peninsula.

statistics   a branch of mathematical sciences that deals with the analysis of numerical data.

steppe   semiarid grassland occurring largely in dry continental interiors. (See also *short-grass prairie.*)

steppe climate   (See *semiarid (steppe) dry climate subtype.*)

stone polygons   linked ring-like ridges of cobbles or boulders lying at the surface of the ground in *arctic* and *alpine tundra* regions.

storage capacity   maximum capacity of *soil* to hold water against the pull of *gravity.*

storage pool   type of pool in a *biogeochemical cycle* in which materials are largely inaccessible to life. (See also *active pool.*)

storage recharge   restoration of stored *soil water* during periods when *precipitation* exceeds *potential evapotranspiration (water need).*

storage withdrawal   depletion of stored *soil water* during periods when *evapotranspiration* exceeds *precipitation,* calculated as the difference

between *actual evapotranspiration (water use)* and *precipitation.*

storm surge   rapid rise of coastal water level accompanying the onshore arrival of a *tropical cyclone.*

strata   layers of *sediment* or *sedimentary rock* in which individual beds are separated from one another along bedding planes.

stratified drift   *glacial drift* made up of sorted and layered *clay, silt, sand,* or gravel deposited from meltwater in *stream channels,* or in marginal lakes close to the ice front.

stratiform clouds   clouds of layered, blanket-like form.

stratopause   upper limit of the *stratosphere.*

stratosphere   layer of *atmosphere* lying directly above the *troposphere.*

stratovolcano   *volcano* constructed of multiple layers of *lava* and *tephra* (volcanic ash).

stratus   *cloud* type of the low-height family formed into a dense, dark gray layer.

stream   long, narrow body of flowing water occupying a *stream channel* and moving to lower levels under the force of *gravity.* (See also *consequent stream, graded stream, subsequent stream.*)

stream capacity   maximum *stream load* of solid matter that can be carried by a *stream* for a given *discharge.*

stream channel   long, narrow, trough-like depression occupied and shaped by a *stream* moving to progressively lower levels.

stream deposition   accumulation of transported particles on a *stream* bed, upon the adjacent *floodplain,* or in a body of standing water.

stream erosion   progressive removal of mineral particles from the floor or sides of a *stream channel* by drag force of the moving water, or by *abrasion,* or by *corrosion.*

stream flow   water flow in a *stream channel;* same as channel flow.

stream gradient   rate of descent to lower elevations along the length of a *stream channel,* stated in m/km, ft/mi, degrees, or percent.

stream load   solid matter carried by a *stream* in dissolved form (as *ions*), in turbulent *suspension,* and as *bed load.*

stream profile a graph of the elevation of a *stream* plotted against its distance downstream.

stream transportation downvalley movement of eroded particles in a *stream channel* in solution, in *turbulent suspension,* or as *bed load.*

strike compass direction of the line of intersection of an inclined *rock* plane and a horizontal plane of reference. (See also *dip.*)

strike-slip fault variety of *fault* on which the motion is dominantly horizontal along a near-vertical *fault plane.*

strip mining mining method in which *overburden* is first removed from a seam of *coal,* or a sedimentary ore, allowing the coal or ore to be extracted.

structure (of a system) the pattern of the pathways and their interconnections within a flow system.

subantarctic low-pressure belt persistent belt of low *atmospheric pressure* centered about at lat. 65°S over the Southern Ocean.

subantarctic zone *latitude* zone lying between lat. 55° and 60°S (more or less) and occupying a region between the *midlatitude zone* and the *antarctic zone.*

subarctic zone *latitude* zone between lat. 55° and 60° N (more or less), occupying a region between the *midlatitude zone* and the *arctic zone.*

subduction descent of the downbent edge of a *lithospheric plate* into the *asthenosphere* so as to pass beneath the edge of the adjoining plate.

sublimation process of change of ice (solid state) to *water vapor* (*gaseous state*); in *meteorology,* sublimation also refers to the change of state from water vapor (liquid) to ice (solid), which is referred to as *deposition* in this text.

submergence inundation or partial drowning of a former land surface by a rise of sea level or a sinking of the *crust* or both.

suborder a unit of *soil* classification representing a subdivision of the *soil order.*

subsea permafrost *permafrost* lying below sea level, found in a shallow offshore zone fringing the arctic seacoast.

subsequent stream *stream* that develops its course by *stream erosion* along a band or belt of weaker *rock.*

subsolar point point on the Earth's surface at which solar rays are perpendicular to the surface.

subsurface water water of the lands held in *soil, regolith,* or *bedrock* below the surface.

subtropical broadleaf evergreen forest a formation class of the *forest biome* composed of broadleaf evergreen *trees;* occurs primarily in the regions of the *moist subtropical climate* ⑥.

subtropical evergreen forest a subdivision of the *forest biome* composed of both broadleaf and needleleaf evergreen *trees.*

subtropical high-pressure belts belts of persistent high *atmospheric pressure* trending east–west and centered about on lat. 30° N and S.

subtropical jet stream *jet stream* of westerly winds forming at the *tropopause,* just above the *Hadley cell.*

subtropical needleleaf evergreen forest a *formation class* of the *forest biome* composed of needleleaf evergreen *trees* occurring in the *moist subtropical climate* ⑥ of the southeastern U.S.; also referred to as the *southern pine forest.*

subtropical zones *latitude* zones occupying the region of lat. 25° to 35° N and S (more or less) and lying between the *tropical zones* and the *midlatitude zones.*

succulents plants adapted to resist water losses by means of thickened spongy tissue in which water is stored.

summer monsoon inflow of maritime air at low levels from the Indian Ocean toward the Asiatic low pressure center in the season of high Sun; associated with the rainy season of the *wet–dry tropical climate* ③ and the Asiatic monsoon climate.

summer solstice solstice occurring on June 21 or 22, when the *subsolar point* is located at 23 1/2°N.

Sun-synchronous orbit satellite orbit in which the orbital plane remains fixed in position with respect to the Sun.

supercontinent single world continent, formed when *plate tectonic* motions move continents together into a single, large land mass. (See also *Pangea.*)

supercooled water water existing in the liquid state at a temperature lower than the normal *freezing* point.

surface the very thin layer of a substance that receives and radiates energy and conducts heat to and away from the substance.

surface energy balance equation equation expressing the balance among *heat* flows to and from a surface.

surface water water of the lands flowing freely (as *streams*) or impounded (as ponds, *lakes,* marshes).

surges episodes of very rapid downvalley movement within an *alpine glacier.*

suspended load that part of the *stream load* carried in *turbulent suspension.*

suspension (See *turbulent suspension.*)

suture (See *continental suture.*)

swash surge of water up the *beach* slope (landward) following collapse of a *breaker.*

symbiosis form of positive interaction between *species* that is beneficial to one of the species and does not harm the other.

sympatric speciation type of *speciation* in which speciation occurs within a larger population.

synclinal mountain steep-sided ridge or elongate mountain developed by erosion of a *syncline.*

synclinal valley valley eroded on weak *strata* along the central trough or axis of a *syncline.*

syncline downfold of *strata* (or other layered *rock*) in a trough-like structure; a class of *folds.* (See also *anticline.*)

system (1) a collection of things that are somehow related or organized; (2) a scheme for naming, as in a classification system; (3) a flow system of *matter* and *energy.*

systematic geography the study of the physical, economic, and social processes that differentiate the Earth's surface into *places.*

systems approach the study of the interconnections among natural processes by focusing on how, where, and when natural systems are linked and interconnected.

systems theory body of knowledge explaining how systems work.

taiga plant *formation class* consisting of *woodland* with low, widely spaced *trees* and a ground cover of lichens and mosses, found along the northern fringes of the region of *boreal forest climate* ⑪; also called cold woodland.

tailings   (See *spoil*.)

tall-grass prairie   a *formation class* of the *grassland biome* that consists of tall grasses with broad-leaved *herbs*.

talus   accumulation of loose *rock* fragments derived by fall of *rock* from a *cliff*.

talus slope   slope formed of *talus*.

tar sand   (See *bitumen*.)

tarn   small *lake* occupying a *rock* basin in a *cirque* of *glacial trough*.

tectonic activity   process of bending (*folding*) and breaking (faulting) of crustal mountains, concentrated on or near active *lithospheric plate* boundaries.

tectonic arc   long, narrow chain of islands or mountains or a narrow submarine ridge adjacent to a *subduction* boundary and its trench, formed by *tectonic processes*, such as the construction and rise of an *accretionary prism*.

tectonic crest   ridge-like summit line of a *tectonic arc* associated with an *accretionary prism*.

tectonics   branch of *geology* relating to tectonic activity and the features it produces. (See also *plate tectonics, tectonic activity*.)

temperature gradient   rate of temperature change along a selected line or direction.

temperature inversion   upward reversal of the normal *environmental temperature lapse rate*, so that the air temperature increases upward. (See also *high-level temperature inversion, low-level temperature inversion*.)

temperature regime   distinctive type of annual temperature cycle.

tephra   collective term for all size grades of solid *igneous rock* particles blown out under gas pressure from a volcanic vent.

terminal moraine   *moraine* deposited as an embankment at the terminus of an *alpine glacier* or at the leading edge of an *ice sheet*.

terrane   continental crustal *rock* unit having a distinctive set of lithologic properties, reflecting its geologic history, that distinguish it from adjacent or surrounding *continental crust*.

terrestrial ecosystems   *ecosystems* of land plants and animals found on upland surfaces of the continents.

tetraploid   having four sets of chromosomes instead of a normal two sets.

thematic map   *map* showing a single type of information.

theme   category or class of information displayed on a *map*.

thermal erosion   in regions of *permafrost*, the physical disruption of the land surface by melting of *ground ice*, brought about by removal of a protective organic layer.

thermal infrared   a portion of the *infrared radiation wavelength* band, from approximately from 3 to 20 μm, in which objects at temperatures encountered on the Earth's surface (including fires) emit *electromagnetic radiation*.

thermal pollution   form of water pollution in which heated water is discharged into a *stream* or *lake* from the cooling system of a power plant or other industrial heat source.

thermistor   electronic device that measures (air) temperature.

thermocline   water layer of a lake or the ocean in which temperature changes rapidly in the vertical direction.

thermokarst   in arctic environments, a hummocky terrain produced by thawing of *permafrost* and subsequent water erosion when the natural surface cover is disturbed.

thermometer   instrument measuring temperature.

thermometer shelter   louvered wooden cabinet of standard construction used to hold *thermometers* and other weather-monitoring equipment.

thermosphere   atmospheric layer of upwardly increasing temperature, lying above the *mesopause*.

thorntree semidesert   *formation class* within the *desert biome*, transitional from *grassland biome* and *savanna biome* and consisting of *xerophytic trees* and *shrubs*.

thorntree-tall-grass savanna   plant *formation class*, transitional between the *savanna biome* and the *grassland biome*, consisting of widely scattered *trees* in an open grassland.

thrust sheet   sheet-like mass of *rock* moving forward over a *low-angle overthrust fault*.

thunderstorm   intense, local convectional storm associated with a *cumulonimbus cloud* and yielding heavy *precipitation*, also with lightning and thunder, and sometimes the fall of *hail*.

tidal current   current set in motion by the *ocean tide*.

tidal inlet   narrow opening in a *barrier island* or baymouth *bar* through which *tidal currents* flow.

tide   (See *ocean tide*.)

tide curve   graphical presentation of the rhythmic rise and fall of ocean water because of *ocean tides*.

till   heterogeneous mixture of *rock* fragments ranging in size from *clay* to boulders, deposited beneath moving glacial ice or directly from the *melting* in place of stagnant glacial ice.

till plain   undulating, plain-like land surface underlain by glacial *till*.

time cycle   in flow systems, a regular alternation of flow rates with time.

time zones   zones or belts of given east–west (*longitudinal*) extent within which *standard time* is applied according to a uniform system.

topographic contour   *isopleth* of uniform elevation appearing on a *map*.

tornado   small, very intense wind vortex with extremely low air pressure in center, formed beneath a dense *cumulonimbus cloud* in proximity to a *cold front*.

trade winds (trades)   surface winds in low *latitudes*, representing the low-level airflow within the *tropical easterlies*.

transcurrent fault   *fault* on which the relative motion is dominantly horizontal, in the direction of the *strike* of the fault; also called a *strike-slip fault*.

transform fault   special case of a *strike-slip fault* making up the boundary of two moving *lithospheric plates*; usually found along an offset of the *mid-oceanic ridge* where seafloor spreading is in progress.

transform plate boundary   *lithospheric plate* boundary along which two plates are in contact on a *transform fault*; the relative motion is that of a *strike-slip fault*.

transform scar   linear topographic feature of the ocean floor taking the form of an irregular scarp or ridge and originating at the offset *axial rift* of the *mid-oceanic ridge*; it represents a former *transform fault* but is no longer a plate boundary.

transformation (soils)   a class of soil-forming processes that transform materials within the soil body; examples include *mineral alteration* and *humification*.

translocation a soil-forming process in which materials are moved within the soil body, usually from one horizon to another.

transpiration evaporative loss of water to the *atmosphere* from leaf pores of plants.

transportation (See *stream transportation*.)

transported regolith *regolith* formed of *mineral matter* carried by *fluid* agents from a distant source and deposited upon the *bedrock* or upon older regolith. Examples: floodplain silt, lake clay, beach sand.

transverse dunes field of wave-like *sand dunes* with crests running at right angles to the direction of the prevailing *wind*.

traveling anticyclone center of high pressure and *outspiraling* winds that travels over the Earth's surface; often associated with clear, dry weather.

traveling cyclone center of low pressure and *inspiraling* winds that travels over the Earth's surface; includes *wave cyclones*, *tropical cyclones*, and *tornadoes*.

travertine *carbonate mineral matter*, usually *calcite*, accumulating upon *limestone cavern* surfaces situated in the *unsaturated zone*.

tree large erect woody *perennial* plant typically having a single main trunk, few branches in the lower part, and a branching crown.

trellis drainage pattern *drainage pattern* characterized by a dominant parallel set of major *subsequent streams*, joined at right angles by numerous short tributaries; typical of *coastal plains* and belts of eroded *folds*.

tropic of cancer *parallel of latitude* at 23 1/2°N.

tropic of capricorn *parallel of latitude* at 23 1/2°S.

tropical cyclone intense *traveling cyclone* of tropical and subtropical *latitudes*, accompanied by high *winds* and heavy rainfall.

tropical easterlies low-latitude wind system of persistent airflow from east to west between the two *subtropical high-pressure belts*.

tropical easterly jet stream upper-air *jet stream* of seasonal occurrence, running east to west at very high altitudes over Southeast Asia.

tropical high-pressure belt a high-pressure belt occurring in tropical latitudes at a high level in the *troposphere*; extends downward and poleward to form the *subtropical high-pressure* belt, located at the surface.

tropical zones *latitude* zones centered on the *tropic of cancer* and the *tropic of capricorn*, within the latitude ranges 10° to 25° N and 10° to 25°S, respectively.

tropical-zone rainforest plant *formation class* within the *forest biome* similar to *equatorial rainforest*, but occurring farther poleward in tropical regions.

tropopause boundary between *troposphere* and *stratosphere*.

tropophyte plant that sheds its leaves and enters a dormant state during a dry or cold season when little *soil water* is available.

troposphere lowermost layer of the *atmosphere* in which air temperature falls steadily with increasing altitude.

tsunami (See *seismic sea wave*.)

tundra biome *biome* of the cold regions of *arctic tundra* and *alpine tundra*, consisting of grasses, grass-like plants, flowering *herbs*, dwarf *shrubs*, mosses, and *lichens*.

tundra climate ⑫ cold climate of the *arctic zone* with eight or more consecutive months of zero *potential evapotranspiration* (water need).

tundra soils soils of the arctic *tundra climate* ⑫ regions.

turbulence in *fluid* flow, the motion of individual water particles in complex eddies, superimposed on the average downstream flow path.

turbulent flow mode of *fluid* flow in which individual *fluid* particles (molecules) move in complex eddies, superimposed on the average downstream flow path.

turbulent suspension *stream transportation* in which particles of *sediment* are held in the body of the *stream* by turbulent eddies. (Also applies to *wind* transportation.)

typhoon *tropical cyclone* of the western North Pacific and coastal waters of Southeast Asia.

Udalfs suborder of the *soil order Alfisols*; includes Alfisols of moist regions, usually in the *midlatitude zone*, with deciduous forest as the natural vegetation.

Udolls suborder of the *soil order Mollisols*; includes Mollisols of the moist soil-water regime in the *midlatitude zone* and with no horizon of *calcium carbonate* accumulation.

Ultisols *soil order* consisting of *soils* of warm soil temperatures with an *argillic horizon* and low *base status*.

ultramafic igneous rock *igneous rock* composed almost entirely of *mafic minerals*, usually *olivine* or *pyroxene group*.

ultraviolet radiation *electromagnetic energy* in the *wavelength* range of 0.2 to 0.4 μm.

unloading process of removal of overlying *rock* load from *bedrock* by processes of *denudation*, accompanied by expansion and often leading to the development of *sheeting structure*.

unsaturated zone subsurface water zone in which pores are not fully saturated, except at times when *infiltration* is very rapid; lies above the *saturated zone*.

unstable air air with substantial content of *water vapor*; capable of breaking into spontaneous *convectional* activity leading to the development of heavy showers and *thunderstorms*.

unstable isotope elemental *isotope* that spontaneously decays to produce one or more new isotopes. (See also *daughter product*.)

upper-air westerlies system of westerly *winds* in the upper *atmosphere* over middle and high latitudes.

upwelling upward motion of cold, nutrient-rich ocean waters, often associated with cool equatorward currents occurring along *continental margins*.

Ustalfs suborder of the *soil order Alfisols*; includes Alfisols of semiarid and seasonally dry climates in which the *soil* is dry for a long period in most years.

Ustolls suborder of the *soil order Mollisols*; includes Mollisols of the semiarid climate in the *midlatitude zone*, with a horizon of *calcium carbonate* accumulation.

valley winds air movement at night down the *gradient* of valleys and the enclosing mountainsides; alternating with daytime *mountain winds*.

variability measure of the variation in a series of observations that center around a mean.

variation in the study of evolution, natural differences arising between parents and offspring as a result of *mutation* and *recombination.*

veins small, irregular, branching network of *intrusive rock* within a preexisting *rock* mass.

vernal equinox *equinox* occurring on March 20 or 21, when the *subsolar point* is at the *equator.*

Vertisols *soil order* consisting of *soils* of the *subtropical zone* and the *tropical zone* with high *clay* content, developing deep, wide cracks when dry, and showing evidence of movement between aggregates.

visible light *electromagnetic energy* in the *wavelength* range of 0.4 to 0.7 μm.

volcanic bombs boulder-sized, semisolid masses of *lava* that are ejected from an erupting *volcano.*

volcanic neck isolated, narrow steepsided peak formed by erosion of *igneous rock* previously solidified in the feeder pipe of an extinct *volcano.*

volcanism general term for *volcano* building and related forms of *extrusive igneous* activity.

volcano conical, circular structure built by accumulation of *lava* flows and *tephra.* (See also *shield volcano, strato-volcano.*)

volcano coast *coast* formed by *volcanoes* and *lava* flows built partly below and partly above sea level.

warm front moving weather *front* along which a warm *air mass* is sliding up over a cold *air mass,* leading to production of *stratiform clouds* and *precipitation.*

warm-blooded animal animal that possesses one or more adaptations to maintain a constant internal temperature despite fluctuations in the environmental temperature.

warm-core ring circular eddy of warm water, surrounded by cold water and lying adjacent to a warm, poleward moving ocean current, such as the Gulf Stream. (See also *cold-core ring.*)

washout downsweeping of atmospheric *particulates* by *precipitation.*

water gap narrow transverse *gorge* cut across a narrow ridge by a *stream,* usually in a region of eroded *folds.*

water need (See *potential evapotranspiration.*)

water resources a field of study that couples basic study of the location, distribution, and movement of water with the utilization and quality of water for human use.

water surplus water disposed of by *runoff* or percolation to the ground water zone after the *storage capacity* of the *soil* is full.

water table upper boundary surface of the *saturated zone;* the upper limit of the *ground water* body.

water use (See *actual evapotranspiration.*)

water vapor the *gaseous* state of water.

waterfall abrupt descent of a *stream* over a *bedrock* step in the *stream channel.*

waterlogging rise of a *water table* in *alluvium* to bring the zone of saturation into the root zone of plants.

watt unit of power equal to the quantity of work done at the rate of one Joule per second; symbol, W.

wave cyclone *traveling cyclone* of the midlatitudes involving interaction of cold and warm *air masses* along sharply defined *fronts.*

wave-cut notch *rock* recess at the base of a *marine cliff* where wave impact is concentrated.

wavelength distance separating one wave crest from the next in any uniform succession of traveling waves.

weak equatorial low weak, slowly moving low-pressure center *(cyclone)* accompanied by numerous convectional showers and *thunderstorms;* it forms close to the *intertropical convergence zone* in the rainy season, or *summer monsoon.*

weather physical state of the *atmosphere* at a given time and place.

weather system recurring pattern of atmospheric circulation associated with characteristic weather, such as a *cyclone* or *anticyclone.*

weathering total of all processes acting at or near the Earth's surface to cause physical disruption and chemical decomposition of *rock.* (See also *chemical weathering, physical weathering.*)

west-wind drift ocean drift current moving eastward in zone of *prevailing westerlies.*

westerlies (See *prevailing westerly winds, upper-air westerlies.*)

wet adiabatic lapse rate reduced *adiabatic lapse rate* when *condensation* is taking place in rising air; value ranges between 4 and 9°C per 1000 m (2.2 and 4.9°F per 1000 ft).

wet equatorial climate ① moist climate of the *equatorial zone* with a large annual *water surplus,* and with uniformly warm temperatures throughout the year.

wet-dry tropical climate ③ climate of the *tropical zone* characterized by a very wet season alternating with a very dry season.

wetlands land areas of poor surface drainage, such as marshes and swamps.

Wilson Cycle *plate tectonic* cycle in which continents rupture and pull apart, forming oceans and *oceanic crust,* then converge and collide with accompanying subduction of *oceanic crust.*

wilting point quantity of stored *soil water,* less than which the foliage of plants not adapted to *drought* will wilt.

wind air motion, dominantly horizontal relative to the Earth's surface.

wind abrasion mechanical wearing action of wind-driven *mineral* particles striking exposed *rock* surfaces.

wind vane *weather* instrument used to indicate *wind* direction.

winter monsoon outflow of continental air at low levels from the Siberian high, passing over Southeast Asia as a dry, cool northerly *wind.*

winter solstice solstice occurring on December 21 or 22, when the *subsolar point* is at 23 1/2°S.

Wisconsinan Glaciation last *glaciation* of the *Pleistocene Epoch.*

woodland plant *formation class,* transitional between *forest biome* and *savanna biome,* consisting of widely spaced *trees* with canopy coverage between 25 and 60 percent.

Xeralfs suborder of the *soil order Alfisols;* includes Alfisols of the *Mediterranean climate* ⑦.

xeric animals animals adapted to dry conditions typical of a *desert* climate.

Xerolls suborder of the *soil order Mollisols;* includes Mollisols of the *Mediterranean climate* ⑦.

xerophytes plants adapted to a dry environment.

PHOTO CREDITS

INDEX

Note: Page references followed by "f" indicate illustrations; while "t" indicates tables.